INTRODUCTORY
Astronomy and Astrophysics

THIRD EDITION

Requests for permission to make copies of any part of the work should be mailed to: Permissions Department, Harcourt Brace Jovanovich, Publishers, 8th Floor, Orlando, FL 32887

Text typeface: Palatino
Compositor: York Graphic Services
Acquisitions Editor: John Vondeling
Senior Developmental Editor: Lloyd W. Black
Managing Editor: Carol Field
Senior Project Manager: Marc Sherman
Manager of Art and Design: Carol Bleistine
Associate Art Director: Doris Bruey
Art Assistant: Caroline McGowan
Text Designer: Arlene Putterman
Layout Artist: B. J. Crim
Cover Designer: Lawrence R. Didona
Director of EDP: Tim Frelick
Production Manager: Bob Butler
Marketing Manager: Marjorie Waldron

Cover Credit: Imtek Imagineering/Masterfile
Endsheet Credits: Jim Riffle, Astro Works Corporation

INTRODUCTORY ASTRONOMY AND ASTROPHYSICS, third edition
0-03-031697-9

Library of Congress Catalog Number: 91-053096

Printed in the United States of America

1234   069   987654321

Preface

Astronomy can have it both ways: as the oldest observational science, born in the wonder about the starry skies; and as the youngest science, when combined with the theoretical discipline of astrophysics. Astronomy and astrophysics span the Cosmos—from the nearby planets to the distant reaches of quasars. Indeed, they cover all of time, from the beginning of the Universe to its possible end. So, astronomy encompasses both the cosmic and the human in time as well as in space.

We have written this book to convey the content and outlook of modern astronomy and astrophysics for the serious science student. By this, we mean an undergraduate willing to combine basic mathematical skills (trigonometry, algebra, and calculus) with an elementary knowledge of physics—but no formal background in astronomy. (We assume that this physics background includes electromagnetism as well as mechanics at the level that uses calculus.) Our ideal student has the goal of using these intellectual tools to broaden his or her understanding of the contents and structure of the Universe. We have aimed to provide for this person a solid coverage in modern astronomy with a solid foundation in astrophysics. This slant requires the frequent and consistent use of physical concepts that pertain to astronomical entities and situations. In this edition, we have also tried to incorporate more physical intuition and reasoning as it applies to astrophysics.

This book follows the traditional Earth-out approach, which has the advantage of starting out with the familiar and moving to the fantastic. These realms fall into the four parts of the book:

Part One, The Solar System; Part Two, The Stars; Part Three, The Milky Way Galaxy; and Part Four, The Universe. Within the context of Part One, we introduce many of the basic physical concepts, especially those of mechanics, used throughout the text. We also examine the planets as places that have evolved since the origin of the solar system. In Part Two, we introduce essential concepts about light and its remote sensing by astronomers. We then examine the light of a special star, the Sun, which serves as a model for other stars. In Part Three, we explore the local conglomerate of stars—the Milky Way Galaxy—whose structure and evolution rests so heavily on the stars that make it up. The Galaxy also serves as a model for other galaxies. Part Four then turns to the distant Universe to examine its contents and evolution on a grand scale.

Along with bringing the material up-to-date, we have enhanced the book's usefulness as a learning tool. Some of these changes are consequences of new approaches resulting from recent development in the science. In Part One, we emphasize the concept of comparative planetology as a technique to probe the evolution of the planets. Material on the origin of the Solar System has been integrated with that on the many small bodies, which give crucial clues as to how the Solar System formed. Part Two updates the chapter on telescopes and detectors and includes new material on spectroscopy. We do not dwell on the technology as such; rather, we emphasize the purposes of observations. We have also added new material on the *transfer equation*, which will be used throughout the book. Part Three updates all aspects of stellar evolution. Part Four has undergone a substantial revision that reflects the rapid growth of extragalactic astronomy and cosmology. Two new chapters (22 and 26) have been added to Part Four in the Third Edition to provide added depth of coverage.

We have tried to develop the material consistently along physical lines of argument rather than purely astronomical ones. To this end, we have expanded the number and increased the range of the chapter-end problems. We expect that they will challenge the students and deepen their understanding of the material in the chapter. Some of these problems were provided by Alan P. Marscher of Boston University; others are by Tom Balonek of Colgate University and Paul Heckert of Western Carolina University. The Instructor's Manual with Solutions for this book provides detailed solutions for all end-of-chapter problems.

A word about *units*. Generally, we use SI rather than the traditional cgs units. Why? First, SI units, despite some problems, are the international standard; we should follow it, as teachers. Second, many students will have had or will be taking a physics course, where such units are standard. We hope that following the same units will cause less confusion. Third, some equations—notably those in electromagnetism—are simpler in the SI formulation. We will confess, however, that in some areas we have used units of convenience rather than SI ones—such as atmospheres for atmospheric pressure. And we could not avoid parsecs! Appendix 6 supplies information to help with conversions between SI and cgs units.

We expect that instructors will use this book for one-year (two-semester) course sequences, perhaps with some additional materials. For a one-semester course, the selection is best left up to the instructor. We do urge, though, that Chapters 1 and 8 be included, because they contain so much of the basic physical concepts.

We received special help from the following colleagues: Jack Burns, New Mexico State University; Tom Balonek, Colgate University; Richard Teske, University of Michigan; Alma Zook, Pomona College; Alan P. Marscher, Boston University; John Cowan, University of Oklahoma; Alan Bentley, Eastern Montana College; Roger Chevalier, University of Virginia; Robert E. Eplee, Jr., University of Arizona; and Bill Tifft, University of Arizona. Detailed feedback was provided by John G. Campbell, U.S. Military Academy; Paul Bradley, University of Texas, Austin; and W. Osborn. Reviews of the final draft manuscript were kindly provided by George S. Mumford, Tufts University; Mirek J. Plavec, University of California, Los Angeles; J. Scott Shaw, University of Georgia; and John H. Simonetti, Virginia Polytechnic Institute and State University. Special thanks for comments on this edition go to David Bruning, University of Louisville; Virginia Trimble, University of Maryland and University of California, Irvine; and Lawrence Pinsky, University of Houston. Kimberly L. Zeilik assisted with the art manuscript.

We are responsible for any errors that you may find. Please send any to Zeilik at the address below—with our thanks! Small changes can be made in future printings of this edition.

Michael Zeilik and Stephen A. Gregory
Department of Physics and Astronomy
The University of New Mexico
Albuquerque, New Mexico 87131

Elske v. P. Smith
Virginia Commonwealth University
900 Park Avenue
Richmond, Virginia 23284

Contents Overview

Part One
The Solar System 1

1 CELESTIAL MECHANICS AND THE SOLAR SYSTEM 2
2 THE SOLAR SYSTEM IN PERSPECTIVE 19
3 THE DYNAMICS OF THE EARTH 34
4 THE EARTH–MOON SYSTEM 52
5 THE TERRESTRIAL PLANETS: MERCURY, VENUS, AND MARS 79
6 THE JOVIAN PLANETS 101
7 SMALL BODIES AND THE ORIGIN OF THE SOLAR SYSTEM 119

Part Two
The Stars 149

8 ELECTROMAGNETIC RADIATION AND MATTER 150
9 TELESCOPES AND DETECTORS 176
10 THE SUN: A MODEL STAR 197
11 STARS: DISTANCES AND MAGNITUDES 223
12 STARS: BINARY SYSTEMS 223
13 STARS: THE HERTZSPRUNG–RUSSELL DIAGRAM 249

Part Three
The Milky Way Galaxy 269

14 OUR GALAXY: A PREVIEW 270
15 GALACTIC ROTATION: STELLAR MOTIONS 282

16 THE EVOLUTION OF STARS 295
17 STAR DEATHS 317
18 VARIABLE AND VIOLENT STARS 337
19 THE INTERSTELLAR MEDIUM AND STAR BIRTH 366
20 THE EVOLUTION OF OUR GALAXY 393

Part Four
The Universe 411

21 GALAXIES BEYOND THE MILKY WAY 412
22 HUBBLE'S LAW AND THE DISTANCE SCALE 429
23 LARGE-SCALE STRUCTURE IN THE UNIVERSE 439
24 ACTIVE GALAXIES AND QUASARS 455
25 COSMOLOGY: THE BIG BANG AND BEYOND 480
26 THE NEW COSMOLOGY 493

Appendices A-1

1 THE MESSIER CATALOG A-1
2 CONSTELLATIONS A-4
3 SOLAR SYSTEM DATA A-7
4 STELLAR DATA A-10
5 ATOMIC ELEMENTS A-13
6 CONVERSION OF UNITS A-15
7 CONSTANTS AND UNITS A-17
8 THE GREEK ALPHABET A-18
9 MATHEMATICAL OPERATIONS A-19
10 THE CELESTIAL SPHERE A-30

Contents

PREFACE   v

INTRODUCTION   xiii

Part One
The Solar System   1

1
CELESTIAL MECHANICS AND THE SOLAR SYSTEM   2

1–1 The Historical Basis of Solar System Models, 2
1–2 Planetary Orbits, 7
1–3 Newton's Mechanics, 10
1–4 Newton's Law of Universal Gravitation, 12
1–5 Physical Interpretations of Kepler's Laws, 14
Problems, 18

2
THE SOLAR SYSTEM IN PERSPECTIVE   19

2–1 Planets, 19
2–2 Moons, Rings, and Debris, 26
2–3 Mechanics Applied to the Solar System, 29
2–4 Essential Questions, 31
Problems, 32

3
THE DYNAMICS OF THE EARTH   34

3–1 Time and the Seasons, 34
3–2 Evidence of the Earth's Rotation, 38
3–3 Evidence of the Earth's Revolution About the Sun, 41
3–4 Differential Gravitational Forces, 44
Problems, 49

4
THE EARTH–MOON SYSTEM   52

4–1 Dimensions, 52
4–2 Dynamics, 54
4–3 Interiors, 56
4–4 Surface Features, 59
4–5 Atmospheres, 67
4–6 Magnetic Fields, 71
4–7 The Evolution of the Earth–Moon System, 75
Problems, 77

5
THE TERRESTRIAL PLANETS: MERCURY, VENUS, AND MARS   79

5–1 Modern Planetology, 79
5–2 Mercury, 79
5–3 Venus, 85
5–4 Mars, 92
5–5 Comparative Evolution of the Terrestrial Planets, 98
Problems, 100

6
THE JOVIAN PLANETS   101

6–1 Jupiter, 101
6–2 Saturn, 108
6–3 Uranus, 111
6–4 Neptune, 113
6–5 Pluto and Charon, 114
Problems, 117

7

SMALL BODIES AND THE ORIGIN OF THE SOLAR SYSTEM   119

7–1 Moons and Rings, 119
7–2 Asteroids, 132
7–3 Comets, 134
7–4 Meteoroids and Meteorites, 137
7–5 Interplanetary Gas and Dust, 141
7–6 The Formation of the Solar System, 141
Problems, 147

Part Two
The Stars   149

8

ELECTROMAGNETIC RADIATION AND MATTER   150

8–1 Electromagnetic Radiation, 150
8–2 Atomic Structure, 156
8–3 The Spectra of Atoms, Ions, and Molecules, 162
8–4 Spectral-Line Intensities, 163
8–5 Spectral-Line Broadening, 167
8–6 Blackbody Radiation, 168
8–7 The Transfer Equation, 171
Problems, 174

9

TELESCOPES AND DETECTORS   176

9–1 Optical Telescopes, 176
9–2 Invisible Astronomy, 180
9–3 Detectors and Image Processing, 187
9–4 Spectroscopy, 190
9–5 A New Generation of Telescopes, 193
Problems, 195

10

THE SUN: A MODEL STAR   197

10–1 The Structure of the Sun, 197
10–2 The Photosphere, 199
10–3 The Chromosphere, 204
10–4 The Corona, 207
10–5 The Solar Wind, 210
10–6 Solar Activity, 211
Problems, 221

11

STARS: DISTANCES AND MAGNITUDES   223

11–1 The Distances to Stars, 223
11–2 The Stellar Magnitude Scale, 225

11–3 Absolute Magnitude and Distance Modulus, 226
11–4 Magnitudes at Different Wavelengths, 226
Problems, 231

12

STARS: BINARY SYSTEMS   233

12–1 Classification of Binary Systems, 233
12–2 Visual Binaries, 234
12–3 Spectroscopic Binaries, 237
12–4 Eclipsing Binaries, 240
12–5 Interferometric Stellar Diameters and Effective Temperatures, 245
Problems, 247

13

STARS: THE HERTZSPRUNG–RUSSELL DIAGRAM   249

13–1 Stellar Atmospheres, 249
13–2 Classifying Stellar Spectra, 253
13–3 Hertzsprung–Russell Diagrams, 256
Problems, 267

Part Three
The Milky Way Galaxy   269

14

OUR GALAXY: A PREVIEW   270

14–1 The Shape of the Galaxy, 270
14–2 The Distribution of Stars, 274
14–3 Stellar Populations, 276
14–4 Galactic Dynamics: Spiral Features, 277
14–5 A Model of the Galaxy, 278
Problems, 280

15

GALACTIC ROTATION: STELLAR MOTIONS   282

15–1 Components of Stellar Motions, 282
15–2 The Local Standard of Rest, 285
15–3 Moving Clusters, 286
15–4 Galactic Rotation, 287
Problems, 293

16

THE EVOLUTION OF STARS   295

16–1 The Physical Laws of Stellar Structure, 295
16–2 Theoretical Stellar Models, 301
16–3 Stellar Evolution, 303
16–4 Interpreting the H–R Diagrams of Clusters, 313

16–5 The Synthesis of Elements in Stars, 313
Problems, 316

17
STAR DEATHS   317

17–1 White Dwarfs and Brown Dwarfs, 317
17–2 Neutron Stars, 323
17–3 Black Holes, 331
Problems, 335

18
VARIABLE AND VIOLENT STARS   337

18–1 Naming Variable Stars, 337
18–2 Pulsating Stars, 338
18–3 Nonpulsating Variables, 341
18–4 Extended Stellar Atmospheres: Mass Loss, 344
18–5 Cataclysmic and Eruptive Variables, 348
18–6 X-Ray Sources: Binary and Variable, 358
Problems, 364

19
THE INTERSTELLAR MEDIUM AND STAR BIRTH   366

19–1 Interstellar Dust, 366
19–2 Interstellar Gas, 372
19–3 Star Formation, 383
Problems, 391

20
THE EVOLUTION OF OUR GALAXY   393

20–1 The Structure of Our Galaxy from Radio Studies, 393
20–2 The Distribution of Stars and Gas in Our Galaxy, 397
20–3 Evolution of the Galaxy's Structure, 405
20–4 Cosmic Rays and Galactic Magnetic Fields, 405
Problems, 407

Part Four
The Universe   411

21
THE GALAXIES BEYOND THE MILKY WAY   412

21–1 Galaxies as Seen in Visible Light, 412
21–2 Galaxies at Radio Wavelengths, 422
21–3 Infrared Observations of Galaxies, 423
21–4 X-Ray Emission from Normal Galaxies, 423
21–5 Some Basic Theoretical Considerations, 424
Problems, 428

22
HUBBLE'S LAW AND THE DISTANCE SCALE   429

22–1 The Period/Luminosity Relationship for Cepheids, 429
22–2 Hubble's Law, 430
22–3 Distances to Galaxies—The Distance Scale, 433
Problems, 438

23
LARGE-SCALE STRUCTURE IN THE UNIVERSE   439

23–1 Clusters of Galaxies, 439
23–2 Superclusters, 446
23–3 Intergalactic Matter, 451
23–4 Masses—Round 3: The Missing Mass, 453
23–5 Summary, 453
Problems, 454

24
ACTIVE GALAXIES AND QUASARS   455

24–1 Radiation Mechanisms, 455
24–2 Moderately Active Galaxies, 456
24–3 AGNs, 457
24–4 Quasars: Discovery and Description, 467
24–5 Problems with Quasars, 470
24–6 Astrophysical Jets, 476
Problems, 477

25
COSMOLOGY: THE BIG BANG AND BEYOND   480

25–1 Steps Toward General Relativity, 480
25–2 Einstein's Theory of General Relativity, 482
25–3 The Primeval Fireball, 486
25–4 The Standard Big Bang Model, 490
Problems, 491

26
THE NEW COSMOLOGY   493

26–1 Problems with the Existing Model, 493
26–2 Cosmic Nucleosynthesis, 495
26–3 Particle Physics, 496
26–4 Inflation Theory, 499
26–5 Galaxy Formation, 500
26–6 History of the Universe, 502
26–7 Summary, 503
Problems, 504

Appendices   A-1

1
THE MESSIER CATALOG   **A-1**

2
CONSTELLATIONS   **A-4**

3
SOLAR SYSTEM DATA   **A-7**

4
STELLAR DATA   **A-10**

5
ATOMIC ELEMENTS   **A-13**

6
CONVERSION OF UNITS   **A-15**

7
CONSTANTS AND UNITS   **A-17**

8
THE GREEK ALPHABET   **A-18**

9
MATHEMATICAL OPERATIONS   **A-19**

10
THE CELESTIAL SPHERE   **A-30**

GLOSSARY   G-1
INDEX   I-1

Introduction: Astronomy, Astrophysics, and Physics

Most colleges and universities with extensive programs in astronomy offer one course entitled "Astrophysics." What does this title mean? Are the contents of this course widely agreed upon? How does astrophysics relate to physics?

We target this textbook for such an astrophysics course. The book will illustrate many instances in which well-known principles of physics, such as gravitation and electromagnetism, can be used to investigate important objects that one encounters when exploring the Universe with telescopes or space probes. The book will also discuss areas in which "new physics" will result from astronomical observations.

The applications that are discussed in this book reflect the current interests of the astrophysics community. Therefore, the contents conform to the "standards" as much as possible. Albeit, inevitably, there will be a few topics either covered or not covered by the authors that our colleagues disagree with. First, we would like to set our subject matter into the larger confines of the sciences.

ASTRONOMY AND ASTROPHYSICS

The distinction between **astronomy** and **astrophysics** can sometimes cause controversy. Both study various celestial bodies, the physical processes that govern them, or perhaps the whole Universe. For those who distinguish between astronomy and astrophysics, the difference usually lies in believing that *astronomy* refers to activities or observations of a purely phenomenological nature. In these, very little new science can be obtained even though the activity might be impressive visually.

One example is the public's interest in lunar eclipses. They are hauntingly beautiful, with the moon's color having a reddish-orange cast, but they add very little to existing scientific knowledge. A second example is the visual observation of variable stars. There is a worldwide network of amateur variable-star observers that is largely unknown to professional astrophysicists. These volunteers monitor well-known variable systems and particularly enjoy observing them at minimum or maximum light.

For a variety of reasons, we choose not to distinguish between the two words; in this book both *astronomy* and *astrophysics* refer to a common study. Partly, we do this because there is value in all astronomical observation; the amateur variable-star observers are often of great value in helping the professional to cover important, time-critical events. Also, although we do not aim this as a textbook in phenomenological astronomy, we stress the need for professionals to understand the underlying physics behind astronomical phenomena. For example, astrophysics students should have a good grasp of why eclipses occur, about how often each year they happen, and how far from the nodes of its orbit the moon must be in order for an eclipse to occur. We are most interested in advancing the body of scientific knowledge, but we also hope that serious astrophysics students will appreciate phenomenological astronomy.

ASTROPHYSICS AND PHYSICS

How then does astronomy or astrophysics relate to **physics?** Although astronomy is often described as having been the first science, and only an all-embracing **natural philosophy** existed just a century or two ago, physics is presently regarded as the broadest of the sciences. Astronomy, chemistry, engineering, geology, and so on, all study concepts such as force, energy, atoms, and nuclei that a student in our educational system first meets comprehensively in a physics course.

A glance at the course offerings in a physics department illustrates the conceptual divisions within the body of physics as a whole. A typical list of course titles and subject matter includes the following:

classical mechanics—motion and forces

electricity and magnetism—the static and dynamic properties of individual or assembled charged particles

optics—both macroscopic systems such as lenses and mirrors and quantum systems or coherent waves

solid state—condensed matter

statistical mechanics and thermodynamics—interactions of particle assemblies

plasma physics—the state of matter in which most particles are ionized

relativity—the special theory (high-speed motion) and the general theory (an alternative to Newtonian gravitation)

quantum mechanics—the microscopic world of atoms, nuclei, and particles

This is only a partial list of course topics, but it illustrates the breadth of physics. One of the most fascinating aspects of physics is that all these different subjects result from just four known forces in nature. The forces will be discussed more fully in Chapter 26, but we can list them here as the strong nuclear, electromagnetic, weak nuclear, and gravitational forces. (The order given is that of their relative strengths at a distance of about 10^{-15} meter; in this situation the strong nuclear force is fully 10^{40} times stronger than the force of gravity.)

Perhaps even more fascinating is that most of our everyday experiences deal with interactions from only two of these—the *electromagnetic* and *gravitational* forces. Although the two nuclear forces are certainly crucial to the very structure of atoms and therefore to the world we know, we do not often experience phenomena in which they

dominate. This also means that most physics courses discuss phenomena related only to the electromagnetic and gravitational forces. A student encounters significant details of the nuclear forces only in courses dealing specifically with nuclear physics. Undergraduates may hardly encounter the nuclear forces at all.

We note that research in astrophysics involves *all* of the physics disciplines listed earlier. Some astronomers study such subjects as the motions of stars in clusters or the plasma interactions of the solar wind with the earth's magnetosphere. Still others design telescopes, examine theories of the crusts of neutron stars or contemplate the thermodynamics of the Universe itself. The variety of astrophysical inquiry is astounding! Astrophysics is a means of using physical laws learned here on earth to study the cosmos and then checking whether the cosmos adds more. Also, just as most everyday experiences come from only two of the forces, most astrophysical studies involve the electromagnetic and gravitational forces.

By its very nature, astronomy is a passive science; we cannot make many experiments because our subjects are too far away. We therefore must rely on information that we receive, and almost all of this information comes from photons, whether they be radio waves, visible, light, or X-rays, for example. As we will see in several chapters in this book, photons are created and destroyed by electromagnetic interactions. Indeed, photons are the particles that mediate the electromagnetic force. This force dominates most useful interactions. For example, the differences between the common states of matter (solid, liquid, gas, and plasma) come from differing electrical binding properties, and most of astrophysics coursework involves learning how the different states of matter behave in various astronomical environments.

The second most important force in astrophysics is gravity. Although it is by far the weakest of the four forces, it dominates the macroscopic dynamics of almost all systems. This happens, of course, because most large objects have a net electrical charge of zero and also because the two nuclear forces have a very limited effective distance range (roughly the size of an atomic nucleus). Clearly, the orbits of moons, planets, stars, and galaxies are caused by the gravitational interaction. We will also see that the dynamics of the largest known galaxy superclusters and even the Universe as a whole are governed by gravity.

Although astrophysicists study and use these two forces more than the others, there are a few important physical processes where the nuclear forces come to the forefront. The most common of these is the examination of the energy production mechanism of stars. Here nuclear fusion processes turn low-mass nuclei into higher-mass nuclei with a release of energy. This is primarily a strong-nuclear-force phenomenon, but the weak force is also part of the process. In the extreme violence of some supernova explosions, the major component of released energy comes out in the form of neutrinos, which interact primarily via the weak nuclear force.

ASTROPHYSICS AND NEW PHYSICS

In some ways astrophysics sounds as if it might be a subdiscipline of physics, and if we only used traditional physical principles and applied them to celestial systems, that would indeed be the case. However, the history of science and the leading edge of today's research clearly show that *astrophysics often leads the way and introduces new ideas to physics*. A well-known historical example is Newton's realization that the same force (gravity) keeps the moon in orbit around the earth as makes objects fall if released near the earth's surface. Another, less familiar, example is the analysis by Sir Arthur Eddington that the sun could not burn by means of chemical interactions for geological times (and therefore could not account for fossils with ages of hundreds of millions of years). Although his work did not specify details, it showed the need for some new kind of physical process decades before the discovery of nuclear energy.

At present, a new understanding of the intimate relationship between physics and astrophysics comes from the exciting world of elementary particle physics. As we penetrate atoms and nuclei to study their internal structures and then further to the internal makeup of particles themselves, we find increasingly profound physical insight. This is a world of theories that invoke strong symmetry principles and experiments with large and highly energetic accelerators. A recent triumph has been the prediction and subsequent demonstration that the electromagnetic and weak nuclear forces are unified; *i.e.*, there is a more fundamental electroweak force that exists under certain conditions. Logical steps beyond this unification are Grand Unified Theories (GUTs) and the ultimate Theory of Everything (TOE). We do not yet have a successful GUT, which would unify the strong nuclear and electroweak forces, but considerable progress is slowly accumulating with time. Much less success has so far appeared for TOEs—theories that would unify gravity with the GUT force.

The astrophysical implication of this discussion is that along with theoretical progress in GUTs and TOEs must come experimental verification, and this verification demands bigger, more energetic, and ever more expensive particle accelerators. Since higher energy scales with the size (and cost) of the accelerator, there is a physical (and economic) limit to the physics that can be examined here on earth. Yet, in many places within the study of astrophysics, natural accelerators may be working. These include collapsed objects such as neutron stars and black holes as well as Active Galaxy Nuclei, which may likely be supermassive black holes with masses on the order of a million times that of our Sun. These conditions offer temperatures, densities, and energies that are found nowhere else. Finally, the Universe itself expanded from a hot, condensed state that surpasses even black holes in one's list of exotica. Within the first ticks of the Universe's clock appeared particles that have not interacted with matter since that moment. If we can directly detect or at least infer the presence of these particles, then we are experimenting with the ultimate particle accelerator. Astrophysics has an important, leading role in the future of all physical thought!

PART ONE
The Solar System

Chapter 1

Celestial Mechanics and the Solar System

Viewed from the Earth, the Sun slides eastward on its annual journey relative to the stars of the Zodiac while the wandering planets perform periodic gyrations within the Zodiac. (The **Zodiac** contains the traditional twelve constellations through which the sun travels on its annual journey with respect to the stars as seen from the earth; see Figure 1–1A.) The planets fall into two groups, based on their observed motions. The first, Mercury and Venus, can move eastward of the Sun, each appearing higher and higher in the western sky at sunset as an **evening star.** At greatest eastern elongation, each planet attains maximum angular distance east of the Sun; it then moves closer to the western horizon at sunset as it sweeps toward the Sun. Moving westward of the Sun, it rises as a predawn **morning star** in the eastern sky.

The second group visible to the naked eye is Mars, Jupiter, and Saturn. They progress steadily eastward across the sky relative to the stars until they approach **opposition**—180° away from the Sun in the sky. Then they slow to a halt at a stationary point, trace a **retrograde** loop toward the

west, slow to another stationary point, and finally resume their eastward march (Figure 1–1B).

These planetary phenomena fascinated and frustrated the ancient astronomers. This chapter briefly discusses the historical efforts to explain these planetary motions—efforts that gave birth to physics and astrophysics and culminated in the creation of **celestial mechanics** by Isaac Newton in the seventeenth century.

1–1 ◗
THE HISTORICAL BASIS OF SOLAR SYSTEM MODELS

(A) THE HELIOCENTRIC MODEL OF COPERNICUS

In the sixteenth century, the Polish astronomer Nicolaus Copernicus grew dissatisfied with the traditional **geocentric** (Earth-centered) model of the Solar System and introduced a new **heliocentric** (Sun-centered) one. This model forms the foundation of the one we use today. Copernicus placed the Sun at the center of the Solar System

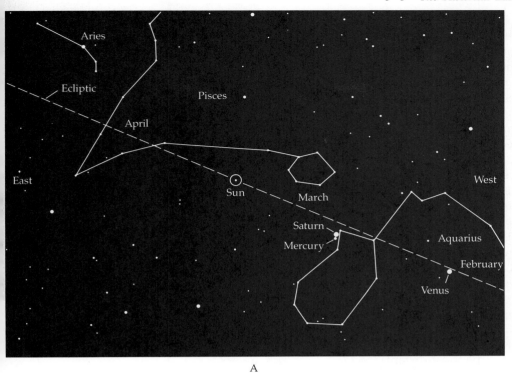

A

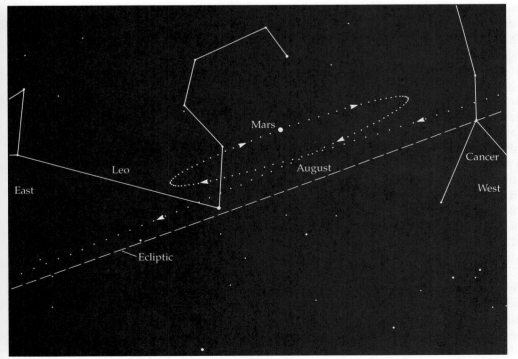

B

FIGURE 1–1 (A) The Sun on the ecliptic with Zodiacal constellations in the background. Note that Mercury, Saturn, and Venus lie close to the ecliptic (but would be invisible, as this view shows daytime). The field of view is 57° by 83° and displays approximately three months worth of the ecliptic. *(Image generated by Voyager, the Interactive Desktop Planetarium™)*

(B) Retrograde motion of Mars in 1994–95. The planet's path is indicated relative to the stars of the constellation Leo. This view, which spans 28° by 39°, shows Mars about in the middle of its retrograde loop in February, 1995, when it shines the brightest. The retrograde loop begins in December, 1993 at the first stationary point, and ends in March, 1995, at the second stationary point. *(Image generated by Voyager, the Interactive Desktop Planetarium™)*

and had the planets (including the Earth) orbit it along circles. This heliocentric model was slow to be accepted because its predictions were not better than those of the geocentric model, but eventually it caught on, primarily for aesthetic reasons of simplicity and harmony.

To understand the explanatory purposes of the Copernican model, you need a little background on naked-eye observations of the Sun, Moon, and planets, which appear geocentric. Daily, the sky turns *westward* with respect to the horizon. The Sun appears to move *eastward* with respect to the stars and circles the sky in a year. The imaginary path traced out by the Sun is called the **ecliptic;** behind it lies the special set of 12 constellations of the Zodiac. The planets (and Sun and Moon) move with respect to the stars within the band of the Zodiac.

In distance from the Sun in the heliocentric model, the planets range from Mercury (the closest) through Venus, Earth, Mars, Jupiter, Saturn, Uranus, and Neptune to Pluto (the most distant). The planets closer to the Sun than the Earth are termed **inferior planets;** these are Mercury and Venus. The planets orbiting farther from the Sun than the Earth are called **superior planets;** these are Mars through Pluto. The motions of the inferior planets as seen in our night sky differ markedly from the motions of the superior planets.

We define **elongation** (Figure 1–2) as the angle seen at the Earth between the direction to the Sun's center and the direction to a planet. We speak of **eastern** or **western elongation** according to whether the planet lies east or west of the Sun, as seen from the Earth. Elongations of particular geocentric significance are given special names: an elongation of 0° is termed **conjunction** (**inferior conjunction** when the planet lies between the Earth and the Sun and **superior conjunction** when the planet lies on the opposite side of the Sun from the Earth). An elongation of 180° is called **opposition,** and one of 90° is **quadrature.** When an inferior planet attains its maximum elongation, we refer to **greatest elongation.** Only inferior planets may be at inferior conjunction or at greatest elongation (28° for Mercury and 48° for Venus), but they may never be at either quadrature or opposition. Inferior conjunction can never occur for superior planets, and their greatest elongation is 180° (when they are in opposition). Note that our Moon

passes through inferior conjunction, quadrature, and opposition (its greatest elongation) because it is a satellite of the Earth.

Copernicus correctly stated that the farther a planet lies from the Sun, the slower it moves around the Sun. When the Earth and another planet pass each other on the same side of the Sun, the apparent retrograde loop occurs (Figure 1–3) from the relative motions of the other planet and the Earth. As we view the planet from the moving Earth, our line of sight reverses its angular motion twice, and the three-dimensional aspect of the loop comes about because the orbits of the two planets are not coplanar. This passing situation is the same for inferior or superior planets.

Copernicus derived a relationship between the synodic and sidereal periods of a planet in the heliocentric model. The **synodic period** *S* is the time it takes the planet to return to the same position in the sky relative to the Sun, as seen from the Earth.

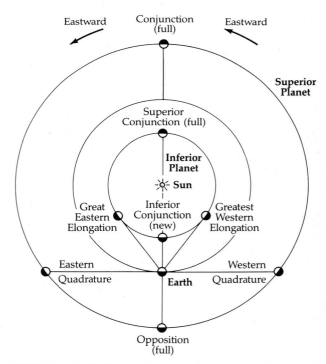

FIGURE 1–2 Heliocentric planetary configurations. Arrows indicate the direction of orbital motion as well as the rotational direction of the Earth. The phases of the planets' illumination are also shown.

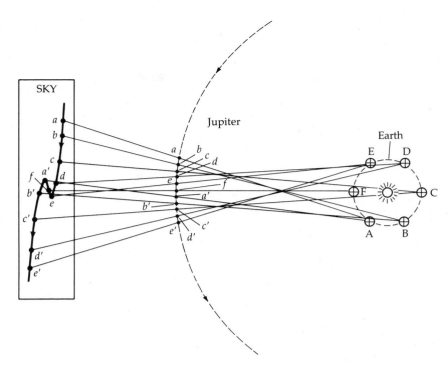

FIGURE 1–3 Retrograde motion in a heliocentric model. As the Earth passes a superior planet, that planet appears to move opposite its normal eastward direction with respect to the stars. Here the Earth passes Jupiter at point *Ff*, which marks the middle of the retrograde motion.

For the inferior planets, Mercury and Venus, this time is the interval between successive inferior conjunctions; for the superior planets, it is the interval between successive oppositions. The **sidereal period** P is the time it takes the planet to complete one orbit of the Sun with respect to the stars (Figure 1–4). The Earth's sidereal period E is 365.26 days. The Earth moves at the rate of $360°/E$ degrees per day in its orbit, while a planet's rate of angular motion is $360°/P$ as viewed from the Earth. For a superior planet, the Earth completes one orbit and must then traverse the additional angle $S \times (360°/P)$ in the time $S - E$ to catch up to the superior planet at opposition again. Hence,

$$(S - E)(360°/E) = S(360°/P)$$

or

$$1/S = 1/E - 1/P$$

For an inferior planet, the Earth is a superior planet, and so we interchange E and P to arrive at Copernicus' result.

$$1/S = 1/P - 1/E \quad \text{(inferior)}$$
$$1/S = 1/E - 1/P \quad \text{(superior)}$$

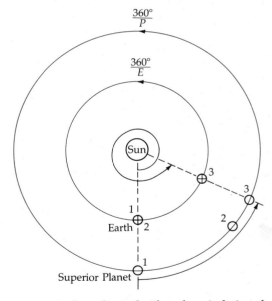

FIGURE 1–4 Synodic and sidereal periods in a heliocentric model. As the Earth orbits the Sun at an angular speed of $360°/E$ degrees per day, a superior planet moves at $360°/P$ degrees per day (as seen from the Sun). The Earth moves from position 1 to position 2 after one orbit and has $S - E$ days to reach the next opposition (at position 3). During this time, the superior planet has moved from position 1 to position 3.

As an example, consider Venus, an inferior planet with an observed synodic period of $S = 583.92$ days. The appropriate relationship is

$$1/583.92 = 1/P - 1/365.26$$
$$1/P = 0.00171 + 0.00274 = 0.00445$$
$$P = 224.7 \text{ days}$$

The telescopic observations reported by Galileo Galilei in his *Siderius Nuncius* in 1610 strongly supported the heliocentric model of the Solar System. His drawings of the wrinkled lunar surface and the moving sunspots on the Sun weakened the ancient belief in perfect and immutable heavens. Galileo also discovered the four largest moons of Jupiter and showed that they orbited Jupiter, not the Earth. This crack in the wall of geocentricism became an irreparable breach with his discovery of the phases of Venus.

Here's why. The planets and the Moon shine from the sunlight they reflect. Half of a planet is always sunlit while the other half is dark. The fraction of the sunlit hemisphere seen from the Earth, however, varies with the configuration. So the phase termed **new** occurs when we see only the dark hemisphere (at inferior conjunction for the Moon, Mercury, and Venus), and **full** phase takes place at opposition when the entire sunlit hemisphere faces us. The superior planets can never be in **crescent** phase (when less than half the observable hemisphere is sunlit) and are almost always in **gibbous** phase (when more than half the planet appears sunlit). Galileo observed that Venus shows all phases—hence it must orbit the Sun. This observation confirmed Copernicus' model.

(B) THE METHODS OF KEPLER

Using the heliocentric model of Copernicus and the positional observations of Mars that Tycho Brahe had painstakingly accumulated over 20 years, Johannes Kepler discovered the need for elliptical planetary orbits (Section 1–2). In 1609 and 1619, he published his three empirical laws of planetary motion. These set the stage for Newton's great scheme of gravitation.

Let's examine Kepler's methods for determining distance to the planets. We use the mean Sun–Earth distance—called the **astronomical unit** (AU)—as the unit of distance.

The Sun–planet distance r (in AUs) may be found when an inferior planet reaches greatest elongation (Figure 1–5A). The angle SEP is observed (call it α) and the angle EPS is $90°$; hence, trigonometry yields $r = \sin \alpha$. (This method was first worked out by Copernicus.)

Kepler's method for finding the distance to a superior planet is more complicated than the preceding procedure (Figure 1–5B). The planet is at P at the beginning and end of one sidereal period, and the Earth is at E and E' at these two times. Note that point P is on the planet's orbit but is otherwise arbitrary. Because we know the planet's sidereal period, we also know the angle ESE'; we must observe the angles PES and $PE'S$. We can solve the triangle ESE' using the law of cosines and trigonometry (see Appendix 9, "Mathematical Operations") to obtain EE' and the angles SEE' and $SE'E$. By subtraction, the angles PEE' and $PE'E$ are known, and we may solve the triangle EPE'. Enough information is now available to solve for r, using either triangle SEP or triangle

FIGURE 1–5 Distance determinations in a heliocentric model. (A) When an inferior planet reaches greatest elongation (P), we know angle SEP and can find r because angle SPE is a right angle. (B) A superior planet is at P at the start and end of one sidereal period; at these times, the Earth is at E and E' Angles PES and $PE'S$ are observed; they are the elongations of the planet from the Sun.

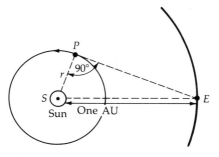
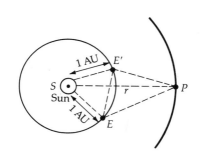

SE'P. This process was the method by which Kepler first traced out the orbit of Mars to find that it is elliptical—a significant break with the astronomical tradition of circular orbits.

1-2 ◗
PLANETARY ORBITS

(A) KEPLER'S THREE EMPIRICAL LAWS

Kepler tested many models of orbital shapes— even ovals—but discarded them all. He finally showed that the orbital planes of the planets pass through the Sun and discovered that the orbital shape was an **ellipse** [Section 1–2(b)]. These find-

ings were announced in 1609 as Kepler's first law— the law of ellipses: *the orbit of each planet is an ellipse with the Sun at one focus* (Figure 1–6A).

Kepler also investigated the speeds of the planets and found that the closer in its orbit a planet is to the Sun, the faster it moves. Drawing a straight line connecting the Sun and the planet (the **radius vector**), he discovered that he could express this fact in Kepler's second law—the law of areas: *the radius vector to a planet sweeps out equal areas in equal intervals of time* (Figure 1–6B).

Ever seeking a greater harmony in the motions of the planets, Kepler toiled for another decade and in 1619 put forth his third law—the harmonic law: *the squares of the sidereal periods of the planets are*

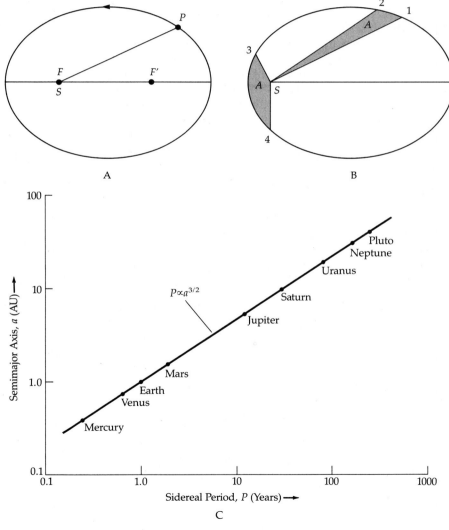

FIGURE 1–6 Kepler's laws of planetary motion. (A) Each planet (*P*) traces an elliptical orbit around the Sun (*S*), which is at one focus (*F*) of the ellipse. (B) Consider two equal time intervals, that from 1 to 2 and that from 3 to 4. The radius vector to the planet (*SP*) sweeps out the same area (*A*) during these times. (C) For all major planets, this log-log plot of semimajor axes (*a*) versus sidereal periods (*P*) falls very close to a straight line of slope 3/2, confirming Kepler's third law.

proportional to the cubes of the semimajor axes (mean radii) of their orbits (Figure 1–6C).

The third law may be written algebraically as

$$P^2 = ka^3$$

where P is a planet's sidereal period and a is its average distance from the Sun (the semimajor axis of an elliptical orbit; see below); the constant k has the same value for every body orbiting the Sun. By 1621, Kepler had shown that the four moons of Jupiter discovered by Galileo obeyed the third law (with a different value of k), confirming its wide applicability.

(B) GEOMETRIC PROPERTIES OF ELLIPTICAL ORBITS

An ellipse is defined mathematically as the locus of all points such that the sum of the distances from two foci to any point on the ellipse is a constant (Figure 1–7); hence

$$r + r' = 2a = \text{constant} \qquad \textbf{(1–1)}$$

The line joining the two foci F and F' intersects the ellipse at the two vertices A and A'. Note that a is half the distance between the vertices; a defines the **semimajor axis** of the ellipse. The shape of the ellipse is determined by its **eccentricity** e, such that the distance from each focus to the center of the ellipse is ae. When $e = 0$, we have a circle; for $e = 1$, a straight line. One-half of the perpendicular bisector of the major axis is the **semiminor axis** b. Using the dashed lines ($r = r' = a$) in Figure 1–7 and the Pythagorean theorem, we find

$$b^2 = a^2 - a^2e^2 = a^2(1 - e^2) \qquad \textbf{(1–2)}$$

Kepler's first law places the Sun at one focus, F. Then vertex A is termed the **perihelion** of the orbit (point nearest the Sun), and vertex A' is called the **aphelion** (point farthest from the Sun). The perihelion distance AF is $a(1 - e)$, and the aphelion distance $A'F$ is $a(1 + e)$. The mean (average) distance from the Sun to a planet in elliptical orbit is just the semimajor axis a. We prove this fact by noting that for each point P on the ellipse at a distance r from focus F, there is a symmetrical point P' a distance r' from F; the average of these distances is $(r + r')/2 = a$. This result holds for any arbitrary but symmetrical pair of points.

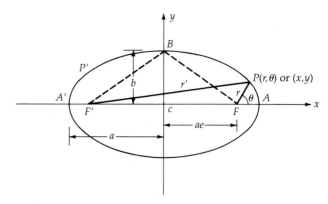

FIGURE 1–7 An ellipse. Important properties labeled here are AF, perihelion distance; $A'F$, aphelion distance; a, semimajor axis; b, semiminor axis; and c, center.

We need to know the distance from one focus to a point on the ellipse (such as the Sun–planet or planet–satellite distance) as a function of the position of that point. Center a polar coordinate system (r, θ) at F and let the line FA correspond to $\theta = 0$. Now r measures the distance FP, and then θ—the **true anomaly**—measures the counterclockwise angle AFP. Using

$$\cos(\pi - \theta) = -\cos\theta$$

and the law of cosines, we have

$$r^2 = r^2 + (2ae)^2 + 2r(2ae)\cos\theta$$

From Equation 1–1, however, $r' = 2a - r$, and so

$$r = a(1 - e^2)/(1 + e\cos\theta) \qquad \textbf{(1–3)}$$

Equation 1–3 is the equation for an ellipse in polar coordinates for $0 \le e < 1$.

To derive the area of an ellipse, we find the analog of Equation 1–3 in Cartesian coordinates (x, y) positioned at the center of the ellipse. Then Figure 1–7 and the Pythagorean theorem give

$$r'^2 = (x + ae)^2 + y^2$$
$$r^2 = (x - ae)^2 + y^2$$

Subtracting these two equations and using Equation 1–1, we find $r' = a + ex$. Substituting back into the first of the above two equations and employing Equation 1–2, we obtain

$$(x/a)^2 + (y/b)^2 = 1 \qquad \textbf{(1–4)}$$

which is the equation for an ellipse in Cartesian coordinates. The area of the ellipse is given by the double integral

$$A = 4 \int_0^b dy \int_0^x dx$$

where, from Equation 1–4,

$$x = a[1 - (y/b)^2]^{1/2}$$

The integration is easy if we use the substitution $y = b \sin z$ (so that $dy = b \cos z \, dz$) and the relationship $\sin^2 z + \cos^2 z = 1$; the final answer is

$$A = \pi ab \qquad (1\text{–}5)$$

The ellipse is one example of a class of curves called **conic sections.** This family of curves, all of which result from slicing a cone at different angles with a plane, includes the circle, ellipse, parabola, and hyperbola (Figure 1–8). From Equation 1–3, note that the ellipse degenerates to a **circle** of radius $r = a$ when $e = 0$. If we increase e, the foci move apart. When $e = 1$, one of the foci is at infinity and we have the **parabola** specified by

$$r = 2p/(1 + \cos \theta) \qquad (1\text{–}6)$$

where p is the distance of closest approach (at $\theta = 0$) to the remaining focus. When the eccentricity is greater than unity, the open **hyperbola** results:

$$r = a(e^2 - 1)/(1 + e \cos \theta) \qquad (1\text{–}7)$$

Its distance of nearest approach to the sole focus is $a(e - 1)$.

When one body moves under the gravitational influence of another, the relative orbit of the moving body must be a conic section. (The relative orbit is that seen from an observer on the more massive body.) Planets, satellites, and asteroids have elliptical orbits; many comets have eccentricities so close to unity that they follow essentially parabolic orbits. A few comets have hyperbolic orbits; after one perihelion passage, such comets leave the Solar System forever. Space probes have been launched into hyperbolic orbits with respect to the Earth, but they are nearly always captured into elliptical orbits about the Sun. Pioneer 10 was

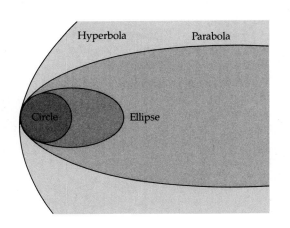

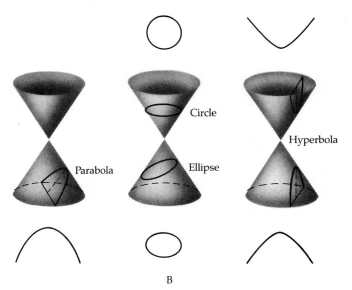

A   B

FIGURE 1–8 Conic sections. (A) The family of conic-section curves includes the circle ($e = 0$), the ellipse ($0 < e < 1$), the parabola ($e = 1$), and the hyperbola ($e > 1$). (B) Conic sections are formed when a cone is cut with a plane. When the plane is perpendicular to the cone's axis, the result is a circle; when it is parallel to one side, the result is a parabola; intermediate angles result in ellipses. A hyperbola results when the angle the plane makes with the cone's side is greater than the opening angle of the cone.

the first spacecraft that, when perturbed by Jupiter, escaped from the Solar System.

1–3 ◑
NEWTON'S MECHANICS

Using Kepler's empirical deductions about planetary orbits, Sir Isaac Newton created his unified scheme of dynamics and gravitation, which he published in his *Principia* in 1687. Newton's brilliant insight and elegant formulation laid the foundations for the Newtonian physics that we know today. This section presents the first half of his unified structure—mechanics.

Newton assumed that the arena within which motions take place is three-dimensional, Euclidean space. These motions occur in time, which passes steadily and is unaffected by any phenomenon in the Universe. The basic entity in the scheme is the point particle, which has mass but no extent (Figure 1–9A). The position of the particle at time t, relative to some origin, is indicated by a vector $\mathbf{x}(t)$; the length of this vector is measured in units such as meters. At a slightly later time $t + \Delta t$, the particle has moved to $\mathbf{x} + \Delta\mathbf{x}$ at approximately the velocity

$$\mathbf{v} \approx [(\mathbf{x} + \Delta\mathbf{x}) - \mathbf{x}]/[(t + \Delta t) - t] = \Delta\mathbf{x}/\Delta t$$

As we let $\Delta t \to 0$, the velocity vector becomes tangent to the trajectory at point $\mathbf{x}$, where it is defined by the derivative

$$\mathbf{v} \equiv d\mathbf{x}/dt \qquad (1\text{–}8)$$

The magnitude of the velocity vector is called the *speed*, and its units are distance/time, such as meters per second (m/s) or kilometers per second (km/s). Noting that the velocity of the particle at t

is $\mathbf{v}$, while at $t + \Delta t$ it is $\mathbf{v} + \Delta\mathbf{v}$, we may express the change in velocity by the acceleration vector (Figure 1–9B):

$$\mathbf{a} \approx [(\mathbf{v} + \Delta\mathbf{v}) - \mathbf{v}]/[(t + \Delta t) - t] = \Delta\mathbf{v}/\Delta t$$

$$\mathbf{a} = d\mathbf{v}/dt \qquad (\text{as } t \to 0) \qquad (1\text{–}9)$$

The units of acceleration are speed/time, which is distance/time^2, such as meters/second^2 or kilometers/second^2. Newton also considered the **linear momentum** vector (units of mass times speed—for example, kg $\cdot$ m/s) of the particle, defined by the product

$$\mathbf{p} = m\mathbf{v} \qquad (1\text{–}10)$$

where m is the particle's mass and $\mathbf{v}$ is its instantaneous velocity. With these kinematic fundamentals in mind, let's turn to Newton's laws of motion.

(A) THE LAW OF INERTIA

In his *Physics*, Aristotle attempted to show that the natural state of a body is one of rest. Our daily experience seems to verify his observation that all moving objects eventually slow to a halt. Indeed, to explain the flight of an arrow, Aristotle said that the diverted air rushing in upon the tail of the arrow "pushed" it along.

Galileo came to a quite different conclusion. He released balls so that they rolled down smooth inclined planes and observed that they rolled up adjacent inclined planes to approximately the same height as that from which he had released them. As he made the planes smoother and inclined the second plane less to the horizontal, he found that the balls rolled farther. Galileo attributed any slowing down of the balls to friction and conjectured

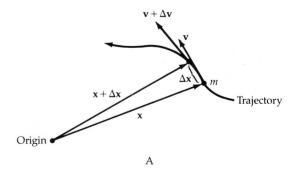

A

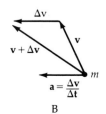

B

FIGURE 1–9 Motion of a particle. (A) A mass m at time t is at position x in its trajectory. At time $t + \Delta t$, it is at position $x + \Delta x$. The instantaneous velocity at $x + \Delta x$ is $v + \Delta v$. (B) The change in velocity between t and Δt is Δv, from which instantaneous acceleration a is defined.

that a smooth ball on a horizontal plane would roll forever at a constant speed.

René Descartes later formulated this principle in the form Newton adopted as his first law of motion, the law of inertia: *the velocity of a body remains constant (in both magnitude and direction) unless a force acts upon the body.* For a freely moving body, the first law may be written **v** = constant; when the constant is zero, a body initially at rest will remain at rest unless acted upon by a force.

The modern form of Newton's first law is known as the **law of conservation of linear momentum.** For a body of mass m, we may write **p** = m**v** = constant, which is equivalent to

$$d\mathbf{p}/dt = 0 \qquad \text{(force-free)} \qquad (1–11)$$

This relation is true even when a body's mass changes, as in the case of a rocket ship. In the case of a constant mass, then,

$$m \, d\mathbf{v}/dt = 0$$

and

$$m\mathbf{a} = 0 \qquad \text{(force-free, constant mass)}$$

(B) THE DEFINITION OF FORCE

Newton's second law is already implied by the proviso "unless a force acts upon the body" in the first law, for a force causes a change in velocity. Such a change in velocity (in speed or direction or both) is indicated by the acceleration vector (Equation 1–9). An important special case of accelerated motion is circular motion, in which the speed remains constant while the direction of motion changes.

The concept of **force** was defined in Newton's second law of motion (the force law): *the acceleration imparted to a body is proportional to and in the direction of the force applied and inversely proportional to the mass of the body.* So we may write **a** = **F**/m or, more commonly,

$$\mathbf{F} = m\mathbf{a} \qquad (1–12)$$

Note that force is a vector, with the units mass times acceleration, such as kg · m/s². If several forces act upon a single body, the resultant acceleration is determined by Equation 1–12, using the force that is the vector sum of the individual forces—this is the principle of superposition. Two

forces $\mathbf{F}_1$ and $\mathbf{F}_2$ add to give the resultant force **F** (Figure 1–10); we may reverse this procedure to decompose the force **F** into any two or more component forces; two are indicated here by F_x and F_y.

In Equation 1–12, the body's mass m must remain constant. This restriction vanishes in the modern statement of the second law, which is formulated using the body's linear momentum:

$$\mathbf{F} = d\mathbf{p}/dt \qquad (1–13)$$

We may recover Newton's form of the second law by using Equations 1–9, 1–10, and 1–13 when m is constant. Note that the first law of motion (Equation 1–11) now results from the second.

What is the meaning here of **mass**? In dynamics, mass represents the **inertia** of a body, that is, that body's resistance to any change in its state of motion. If we apply the same force to two bodies, the more massive body will change velocity at a lower rate than the less massive body. In everyday terms, think of mass as the amount of material comprising a body. **Mass** is a scalar quantity characterizing a body, and it does not depend on the body's location or state of motion [Section 1–4(b)].

(C) ACTION AND REACTION

To complete his dynamic theory, Newton developed his third law of motion, the law of action–reaction: *for every force acting on a body (in a closed system), there is an equal and opposite force exerted by*

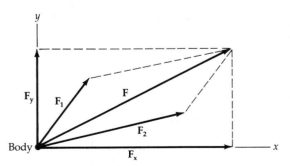

FIGURE 1–10 Superposition of forces. Two forces $\mathbf{F}_1$ and $\mathbf{F}_2$ act on a body. The resulting motion depends on **F**, the vector sum of $\mathbf{F}_1$ and $\mathbf{F}_2$. The orthogonal components $\mathbf{F}_x$ and $\mathbf{F}_y$ yield the same motion because their sum results in **F**.

that body. A simple example: The weight (a force) of a book lying on a table must be exactly balanced by the force that the table exerts on the book; otherwise, according to the second law, the book would accelerate off or through the table. The third law describes the static situation of balanced forces.

The modern version of the third law is the law of *conservation of total linear momentum.* Though we may treat any number of bodies, consider only two. The total linear momentum of the system is given by $\mathbf{P} = \mathbf{p}_1 + \mathbf{p}_2 = $ constant, when no external force acts upon the system (the first two laws). Considering two instants of time (the latter instant indicated by primes), we have

$$\mathbf{p}_1 + \mathbf{p}_2 = \mathbf{p}_1' + \mathbf{p}_2'$$

If the time interval is Δt and if we call $\Delta \mathbf{p}_1 = \mathbf{p}_1' - \mathbf{p}_1$ and $\Delta \mathbf{p}_2 = \mathbf{p}_2' - \mathbf{p}_2$, we can arrange the above equation and divide by Δt to get

$$\Delta \mathbf{p}_1 / \Delta t = -\Delta \mathbf{p}_2 / \Delta t$$

For arbitrarily small Δt, the deltas become differentials ($\Delta t \rightarrow dt$) and Equation 1–13 yields the third law:

$$\mathbf{F}_1 = -\mathbf{F}_2$$

(D) SUMMARY: NEWTON'S LAWS OF MOTION

We present here the modern version of Newton's three laws of mechanics. Recall that the Newtonian context is absolute space and time, wherein a point particle of mass m describes a trajectory $\mathbf{x}(t)$ with instantaneous velocity $\mathbf{v}(t)$, linear momentum $\mathbf{p} = m\mathbf{v}$, and acceleration $\mathbf{a}(t)$.

First law (inertia): $\mathbf{v}$ and $\mathbf{p} = $ constant	The velocity (if the mass is constant) and linear momentum of a body remain constant (in both magnitude and direction) unless a force acts upon the body.
Second law (force): $\mathbf{F} = d\mathbf{p}/dt$	The time rate of change of the linear momentum of a body (or system of bodies) equals the force acting upon the body (or system).

Third law (action–reaction) $\mathbf{F}_{\text{exerted}} = -\mathbf{F}_{\text{acting}}$ $\mathbf{P} = $ constant	In a closed system, the force exerted by a body is equal and opposite to the force acting upon the body, or, the total linear momentum of a closed system of bodies is constant in time.

1–4 ◉ NEWTON'S LAW OF UNIVERSAL GRAVITATION

Before Newton, Kepler suspected that some force acted to keep the planets in orbit about the Sun; he attributed the elliptical orbits to the force of magnetic attraction. Following a train of thought similar to the simplified derivation that follows, Newton discovered his law of universal gravitation, tested it on the motion of the Moon, and then explained the motions of the planets in detail.

(A) CENTRIPETAL FORCE AND GRAVITATION

Consider a body moving in a circular orbit of radius r about a center of force (Figure 1–11A). From symmetry, the speed v of the body must be constant, but the direction of the velocity vector is constantly changing. Such a changing velocity represents an acceleration—the **centripetal acceleration** that maintains the circular orbit; from the geometry of the figure, we deduce the acceleration. The time is t at point A, where the body's velocity is $\mathbf{v}$. At an infinitesimal time interval Δt later, the body has traversed the angle $\Delta \theta$ to B, where the velocity is $\mathbf{v}'$. In Figure 1–11B, we show the change in velocity $\Delta \mathbf{v} = \mathbf{v}' - \mathbf{v}$ by joining the tails of the two velocity vectors; the angle between $\mathbf{v}$ and $\mathbf{v}'$ is $\Delta \theta$. Recalling that the magnitude of both $\mathbf{v}$ and $\mathbf{v}'$ is the speed v, we use trigonometry to deduce, for small values of $\Delta \theta$ (Figure 1–11A),

$$\Delta \theta = s/r = v \, \Delta t/r$$

and (Figure 1–11B)

$$\Delta \theta = \Delta v/v$$

where the arc length s approximates the chord joining points A and B. So the centripetal acceleration has the magnitude

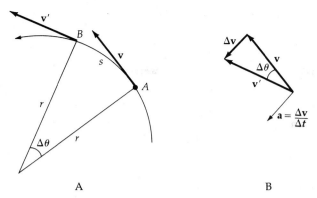

FIGURE 1-11 Centripetal acceleration. (A) An object moves in a circular orbit of radius *r* at speed *v* along a short arc *S* during the time interval Δ*t*. (B) The centripetal acceleration *a* results from the change in the direction of the velocity during time Δ*t*.

$$a = \Delta v/\Delta t = v^2/r$$

and it points directly toward the center of the circle as Δ*t* vanishes. If *m* is the mass of the orbiting body, Newton's second law gives the magnitude of the centripetal force as

$$F_{\text{cent}} = ma = mv^2/r \qquad \textbf{(1-14)}$$

If *P* is the orbital period of the body, then its speed is

$$v = 2\pi r/P$$

but Kepler's third law relates the period and the orbital radius by

$$P^2 = kr^3$$

where *k* is the proportionality constant. Substituting these two results into Equation 1-14, we find

$$F = 4\pi^2 m/kr^2$$

that is, the force maintaining the orbit is *inversely proportional to the square of the radius*. According to Newton's third law, the body (of mass *M*) at the center of the orbit feels an equal but opposite force. Because the centripetal force acting on the central body must be proportional to *M*, the mutual gravitational force is proportional to the product of the two masses. Redefining the constant of proportionality, we have

$$\mathbf{F}_{\text{grav}} = GMm/r^2 \qquad \textbf{(1-15)}$$

which is Newton's law of universal gravitation; the direction of the gravitational force is along the line joining the two bodies (from the third law of motion). The gravitation constant *G* has the measured value 6.67×10^{-11} m³/kg · s² in SI units.

Equation 1-15 expresses the attractive gravitational force between two *point* masses. To find the gravitational attraction from an extended body, we must sum the vectorial contributions from each small piece of the body. In general, this process is rather difficult, but for a spherically symmetric body acting on a point mass, symmetry arguments alone tell us that the gravitational force must act along the line joining the centers of the bodies. By integrating the effects of all parts of the spherical body, we find that *it behaves gravitationally as though its entire mass were concentrated at its center*. This important result will be used many times in this book.

(B) WEIGHT AND GRAVITATIONAL ACCELERATION

Near the Earth's surface, bodies have a constant downward gravitational acceleration of magnitude (from Equation 1-15)

$$g = GM_{\oplus}/R_{\oplus}^2 \approx 9.8 \text{ m/s}^3$$

where $M_{\oplus}$ is the mass of the Earth and $R_{\oplus}$ is its radius. The Earth's oblate surface and rapid rotation cause variations in the measured value of *g* from 9.781 m/s² at the equator to 9.832 m/s² at the poles.

The **weight** of a body is the force on it in a gravitational field. At the Earth's surface, we weigh a body on a scale, and if the mass of the body is *m*, we find

$$\text{Weight} = mg$$

In contrast to its mass, the weight of a body depends upon its *location*. An astronaut on the Moon's surface will weigh approximately one-sixth her normal Earth weight; in a satellite orbit, her weight will be zero since she is falling freely in the gravitational field. In both cases, her mass is the same. Weight and force have the same units, and the conventional SI unit is the **newton** (Appendix 6), where

$$1 \text{ newton} = 1 \text{ kg} \cdot \text{m/s}^2$$

Keep in mind that *weight is a force.*

(C) DETERMINATION OF *G* AND $M_\oplus$

Let's describe how Henry Cavendish measured the gravitational constant *G* in 1798 and how Phillip von Jolly established the mass of the Earth $M_\oplus$ in 1881. These two methods represent early procedures to determine these important constants.

To find *G*, Cavendish used an apparatus consisting of two small balls of equal mass *m* hung from a torsion beam and two larger balls of equal mass *M* attached to an independently suspended but coaxially aligned beam (Figure 1–12A). Each adjacent *M* − *m* pair is initially placed a distance *D* apart, but the gravitational force between the balls twists the torsion bars to a static equilibrium distance *d* between each *M* − *m* pair. From symmetry, the gravitational force causing the deflection is

$F_{\text{tot}} = 2GMm/d^2$; by directly measuring F_{tot}, *M*, *m*, and *d*, Cavendish found *G*.

Von Jolly's apparatus (Figure 1–12B) consisted of a balance bearing two small masses *m*, with the rotation axis of the balance beam aligned horizontally. When he placed a large mass *M* below one of these small masses, the balance tipped; to restore the original balance, he placed a small mass *n* on the pan alongside the other mass *m*. If the equilibrium *M* − *m* distance is *d*, the forces acting on each side of the torsion beam are

$$F_1 = GMm/d^2 + GM_\oplus m/R_\oplus^2$$
$$F_2 = GM_\oplus n/R_\oplus^2 + GM_\oplus m/R_\oplus^2$$

The beam is horizontal again when $F_1 = F_2$, and so we find

$$M_\oplus = (Mm/n)(R_\oplus/d)^2 = 5.976 \times 10^{24} \text{ kg}$$

1–5 ◐
PHYSICAL INTERPRETATIONS OF KEPLER'S LAWS

Newton combined his laws of motion and gravitation to derive all three of Kepler's empirical laws.

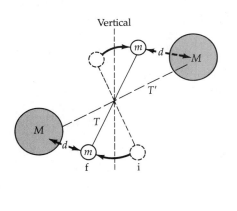

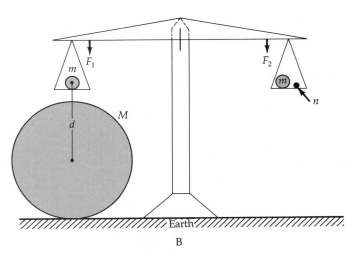

A   B

FIGURE 1–12 Measuring Newton's constant of gravitation, *G*. (A) The Cavendish experiment uses two torsion bars *T* and *T'*, which are free to rotate about a vertical axis. After the gravitational force between adjacent *M* − *m* spheres swings the bars to their equilibrium positions (to position *f* from position *i*), the separations between *M* and *m* are *d*. (B) Von Jolly's experiment has a horizontal balance that is in an equilibrium horizontal position when masses *M* and *n* are absent. When mass *M* is placed below the left-hand mass *m*, mass *n* must be added to the right-hand mass *m* to restore equilibrium.

We could derive the elliptical orbits of Equation 1–3, using Equations 1–13 and 1–15; this requires a knowledge of vector differential equations, so we won't do so. Newton's form of Kepler's laws conceptually views orbits as conic sections (Figure 1–8) with the center of mass of the system at one focus. We will accept Kepler's first law and follow Newton's footsteps in deducing Kepler's second and third laws.

(A) THE LAW OF AREAS AND ANGULAR MOMENTUM

Let's illustrate Kepler's law of areas for an elliptical orbit (Figure 1–13). A body orbits the focus F at the position r and velocity $\mathbf{v}$. During an infinitesimal time interval Δt, the body moves from P to Q and the radius vector sweeps through the angle $\Delta\theta$. This small angle is $\Delta\theta \approx v_t \Delta t/r$, where v_t is the component of $\mathbf{v}$ perpendicular to r. During this time, the radius vector has swept out the triangle FPQ, the area of which is $\Delta A \approx rv_t \Delta t/2$. Therefore, as $t \to 0$,

$$dA/dt = rv_t/2 = r^2(d\theta/dt)/2 = H/2 \quad \textbf{(1–16)}$$

where the constant H (the angular momentum per unit mass) appears because Kepler's second law states that the rate of change of area with time is a constant.

Note that $A/P = H/2$, where $A = \pi ab$ is the total area of the ellipse and P is the orbital period; this comes by integrating Equation 1–16. By combining this result with Equation 1–16 and noting that v_t is the total speed at perihelion or aphelion, we deduce the perihelion and aphelion speeds of a planet orbiting the Sun. For example, at perihelion we have

$$v = H/r = 2A/Pr = 2\pi ab/Pa(1 - e)$$

where Equation 1–3 with $\theta = 0$ is used in the last equality. Carrying through the derivation at aphelion and using Equation 1–2, we get

$$v_{\text{per}} = (2\pi a/P)[(1 + e)/(1 - e)]^{1/2}$$
$$\text{(perihelion)} \quad \textbf{(1–17a)}$$
$$v_{\text{ap}} = (2\pi a/P)[(1 - e)/(1 + e)]^{1/2}$$
$$\text{(aphelion)} \quad \textbf{(1–17b)}$$

For the Earth, a is 1 AU (1.496×10^8 km), P is one year (3.156×10^7 s), and the orbital eccentricity is $e = 0.0167$; hence, the orbital speed varies from 30.3 km/s at perihelion to 29.3 km/s at aphelion.

A modern Newtonian derivation of Kepler's second law requires the concept of the orbiting body's *angular momentum*:

$$\mathbf{L} = \mathbf{r} \times \mathbf{p} = m(\mathbf{r} \times \mathbf{v}) \quad \textbf{(1–18)}$$

where m is the body's mass, $\mathbf{r}$ its position vector, and $\mathbf{p}$ its linear momentum (Equation 1–10). The vector cross product (denoted by $\times$) in Equation 1–18 is an operation that yields the product of the perpendicular components of two vectors (see Appendix 9, "Mathematical Operations"); hence, if $\mathbf{r}$ and $\mathbf{p}$ are parallel, then $\mathbf{r} \times \mathbf{p} = 0$. Angular momentum is a vector quantity $\mathbf{L}$ with the units kg · m²/s. Differentiating Equation 1–18, we have

$$d\mathbf{L}/dt = \mathbf{v} \times \mathbf{p} + \mathbf{r} \times (d\mathbf{p}/dt) = \mathbf{r} \times \mathbf{F} \quad \textbf{(1–19)}$$

since $\mathbf{v}$ is parallel to $\mathbf{p}$ and $d\mathbf{p}/dt$ defines force. We call $d\mathbf{L}/dt$ the **torque** (with units kg · m²/s²) and see that when $\mathbf{F}$ is collinear with $\mathbf{r}$—a central force, such as gravitation—the torque vanishes. Hence, $\mathbf{L}$ is constant in time so that angular momentum is conserved for all central forces. Applying Equation 1–18 to the situation in Figure 1–13, we find

$$\mathbf{L}/m = \mathbf{r} \times \mathbf{v_t} = H = \text{constant}$$

which is Kepler's second law.

(B) NEWTON'S FORM OF KEPLER'S THIRD LAW

The external forces that act upon the Solar System are essentially negligible; hence, the total linear momentum of the Solar System is constant. So the Sun must move about the center of mass of the Solar System. We apply this idea to an isolated sys-

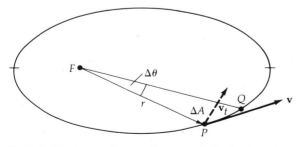

FIGURE 1–13 Law of areas for an elliptical orbit. A body at r moves at velocity $\mathbf{v}$ in an elliptical orbit about focus F and moves from P to Q in time Δt. The component of $\mathbf{v}$ perpendicular to r is v_t.

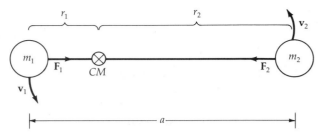

FIGURE 1–14   Center of mass. Two masses, m_1 and m_2, orbit a common center of mass, *CM*, with a total separation *a*.

tem of two bodies moving in circular orbits from their mutual gravitational attraction; our final result, Newton's form of Kepler's third law, is also applicable to elliptical orbits.

Consider two bodies of masses m_1 and m_2, orbiting their stationary center of mass at distances r_1 and r_2 (Figure 1–14). Because the gravitational force acts only along the line joining the centers of the bodies, both bodies must complete one orbit in the same period *P* (though they move at different speeds v_1 and v_2). The centripetal forces of the orbits are therefore

$$F_1 = m_1 v_1^2 / r_1 = 4\pi^2 m_1 r_1 / P^2 \quad \textbf{(1–20a)}$$
$$F_2 = m_2 v_2^2 / r_2 = 4\pi^2 m_2 r_2 / P^2 \quad \textbf{(1–20b)}$$

Newton's third law requires $F_1 = F_2$, and so we obtain

$$r_1 / r_2 = m_2 / m_1 \quad \textbf{(1–21)}$$

The more massive body orbits closer to the center of mass than does the less massive one; Equation 1–21 defines the position of the center of mass.

The total separation of the two bodies $a = r_1 + r_2$ is also the radius of their relative orbits. Now Equation 1–21 may be expressed in the form

$$r_1 = m_2 a / (m_1 + m_2) \quad \textbf{(1–22)}$$

However, the mutual gravitational force $F_{grav} = F_1 = F_2$ is

$$F_{grav} = G m_1 m_2 / a^2 \quad \textbf{(1–23)}$$

so that combining Equations 1–20a, 1–22, and 1–23 gives **Newton's form of Kepler's third law:**

$$P^2 = 4\pi^2 a^3 / G(m_1 + m_2) \quad \textbf{(1–24)}$$

If body 1 is the Sun and body 2 any planet, then $m_1 \gg m_2$; therefore $k = 4\pi^2 / GM_\odot$ is the proportionality "constant" (to a good approximation) in Kepler's third law. Note that the third law presents a way to describe gravitational effects without specifically discussing forces.

(C) ORBITAL VELOCITY

To better understand elliptical orbits, consider the **orbital velocity v.** We may decompose this velocity into two perpendicular components (Figure 1–15): v_r, the radial speed, and v_θ, the angular speed. Now, from Equation 1–16 and the discussion following it, we have

$$d\theta / dt = (2\pi / P)(a/r)^2 (1 - e^2)^{1/2} \quad \textbf{(1–25)}$$

Using Equation 1–3, which is the polar equation of an ellipse, and Equation 1–25, we compute the time derivatives below to find:

$$v_r \equiv dr/dt$$
$$= (2\pi a / P)(e \sin \theta)(1 - e^2)^{-1/2} \quad \textbf{(1–26a)}$$
$$v_\theta \equiv r(d\theta/dt)$$
$$= (2\pi a / P)(1 + e \cos \theta)(1 - e^2)^{-1/2} \quad \textbf{(1–26b)}$$

Note that Equation 1–26b reduces to Equation 1–17 at perihelion and aphelion. The total orbital speed now follows from Equations 1–26 as

$$v^2 = v_1^2 + v_\theta^2$$
$$= (2\pi a / P)^2 (1 + 2e \cos \theta + e^2)/(1 - e^2) \quad \textbf{(1–27)}$$

Rearranging the polar equation for an ellipse, we get

$$e \cos \theta = [a(1 - e^2) - r]/r$$

and substituting this result into Equation 1–27, we finally obtain (with the help of Equation 1–24)

$$v^2 = G(m_1 + m_2)[(2/r) - (1/a)] \quad \textbf{(1–28)}$$

Therefore, for given masses, *the total orbital speed depends only upon the separation and the orbit's semimajor axis.* This useful result is commonly called the **vis viva equation.**

(D) THE CONSERVATION OF TOTAL ENERGY

The concept of energy provides an alternative approach to that of forces and Newtonian mechanics

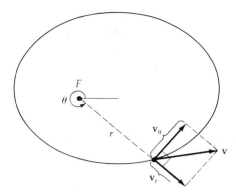

FIGURE 1–15 Components of orbital velocity. At any point in an elliptical orbit, the orbital velocity **v** may be decomposed into two perpendicular components, the radial speed *parallel* to the radius vector and the angular speed *perpendicular* to it.

for problems relating to orbits. Energy often proves a more powerful way to gain insight into orbital problems.

Energy is a quantity assigned to one body that indicates that body's ability to change the state of another body. **Heat** is a form of energy, for a hot body will warm a cold body if the two are brought into contact. **Electrical energy** causes the filament of a light bulb to glow (become hot), showing that one type of energy may be converted to another. **Kinetic energy (KE)** is a body's energy of *motion*, but if we decide to move along with the body, it has no relative motion; it has no kinetic energy in this reference frame. **Potential energy (PE)** is due to the *position* of the body; if the body is free to move, this energy may be converted to kinetic energy. In celestial mechanics, we sum the kinetic and potential energies to obtain the **total energy:** $TE = KE + PE$.

Assume that a force **F** acts upon a body of mass m that is moving in the trajectory $\mathbf{x}(t)$ about a central force. In the infinitesimal time dt, the body moves through the vector distance $d\mathbf{x}$. As the body moves from position A to position B, we define the work W (SI units = kg · m²/s² = joules; see Appendix 6) done on the body by the force as

$$W = \int_A^B \mathbf{F} \cdot d\mathbf{x} \qquad (1\text{–}29)$$

The **vector dot product** operation in Equation 1–29 yields the product of the *parallel components* of **F** and $d\mathbf{x}$ (Appendix 9, "Mathematical Operations"); when **F** and $d\mathbf{x}$ are mutually perpendicular, their dot product vanishes. To evaluate Equation 1–29, we note:

$$\mathbf{F} \cdot d\mathbf{x} = m(d\mathbf{v}/dt) \cdot \mathbf{v}\, dt = m(\mathbf{v} \cdot d\mathbf{v}) = d(mv^2/2)$$

where we have used Newton's second law and the definitions of velocity and speed. We integrate Equation 1–29 directly to give

$$W = (mv^2/2)_B - (mv^2/2)_A = KE_B - KE_A \qquad (1\text{–}30)$$

where the kinetic energy is specified by $KE = mv^2/2$. Therefore, the *work done by the force on the body changes the body's kinetic energy*. In a system of two bodies of masses m_1 and m_2, the kinetic energy is $KE = (m_1v_1^2/2) + (m_2v_2^2/2)$. If the force is the gravitational force between the two bodies, we have

$$\mathbf{F} \cdot d\mathbf{x} = -GM_1m_2/r^2\, dr = d(Gm_1m_2/r)$$

where r is the radial separation of the bodies. Equation 1–29 is again integrated:

$$W = (Gm_1m_2/r)_B - (Gm_1m_2/r)_A = PE_A - PE_B$$
$$(1\text{–}31)$$

where we have inserted the definition of the mutual gravitational potential energy $PE = -Gm_1m_2/r$. Thus the potential energy is the negative of the work done by the gravitational force as m_1 moves from $r = \infty$ to $r = r$ with m_2 held fixed. Note that the mutual gravitational potential energy vanishes when the two bodies are infinitely separated.

Both kinetic and potential energy have the same units (joules) as work, so we can combine Equations 1–30 and 1–31 to obtain

$$(KE + PE)_B = (KE + PE)_A$$

or, in terms of the total energy of the two-body system,

$$TE_B = TE_A = \text{constant}$$

Therefore, the total energy of our gravitating system is conserved:

$$TE = (m_1v_1^2/2) + (m_2v_2^2/2) - (Gm_1m_2/r)$$
$$= \text{constant} \qquad (1\text{–}32)$$

As the bodies move, kinetic and potential energy may be interchanged, but the total energy of the system remains constant.

Now to evaluate the constant TE in Equation 1–32 for elliptical orbits. Refer to Section 1–3(c) and recall that the total linear momentum of our isolated system is constant; we choose this constant to be zero so that $m_1v_1 = -m_2v_2$ and, in terms of the speeds of the bodies,

$$m_1v_1 = m_2v_2$$

Since $v = v_1 + v_2$ is the relative speed of either body with respect to the other, we find

$$v_1 = m_2v/(m_1 + m_2)$$
$$v_2 = m_1v/(m_1 + m_2)$$

Substituting this result into Equation 1–32 yields

$$TE = m_1m_2[v^2/2(m_1 + m_2) - (G/r)] \quad \textbf{(1–33)}$$

Using Equation 1–24, let's evaluate Equation 1–33 at the perihelion of the orbit, where $r = a(1 - e)$ and v is the perihelion speed given by Equation 1–17. The result, $TE = -Gm_1m_2/2a$, shows that the total energy is negative—the orbit is bound. Now Equation 1–33 takes its final form:

$$v^2 = G(m_1 + m_2)[(2/r) - (1/a)] \quad \textbf{(1–34)}$$

exactly the expression found in Equation 1–28. This classic result, the vis viva equation, is a statement of total energy conservation.

PROBLEMS ●

1. Assume that the orbital plane of a superior planet is inclined 10° to the ecliptic and that the planet crosses the ecliptic moving northward at opposition. Make a diagram similar to Figure 1–1B, showing the retrograde path of this superior planet.

2. Imagine you are observing the Earth from Jupiter. What would you observe the Earth's synodic orbital period to be? What would it be from Venus? (*Hint:* See Appendix 3.)

3. (a) Explicitly carry out the derivation of Equation 1–4, showing all the appropriate steps.
   (b) On graph paper, plot the following polar equations: Equation 1–3 for an ellipse, Equation 1–6 for a parabola, and Equation 1–7 for a hyperbola.

4. In terms of the gravitational acceleration g at the surface of Earth, find the surface gravitational acceleration of
   (a) the Moon ($M_m = 0.0123M_⊕$, $R_m = 1738$ km),
   (b) the Sun ($M_⊙ = 2 \times 10^{30}$ kg, $R_⊙ = 7 \times 10^8$ m), and
   (c) Jupiter ($M_J = 318M_⊕$, $R_J = 11.2 R_⊕$).

5. What are the perihelion and aphelion speeds of Mercury? What are the perihelion and aphelion distances of this planet? Compute the product vr (speed times distance) at each of these two points and interpret your result physically.

6. Find the relative position of the center of mass for (a) the Sun–Jupiter system and (b) the Earth–Moon system.

7. A television satellite is in circular orbit about the Earth, with a sidereal period of exactly 24 h. What is the distance from the Earth's surface for such a satellite? (*Hint:* Use Kepler's laws.) If the satellite appears stationary to an earthbound observer, what is the orientation of its orbital plane?

8. Using orbital data for Titan (Appendix 3), find the mass of Saturn.

9. A stone is released from rest at the Moon's orbit and falls toward Earth. What is the stone's speed when it is 192,000 km from the center of Earth?

10. An object is observed from the Earth to have a synodic period of 1.5 years. What are the two possible values for the semimajor axis of the object's orbit?

11. Using orbital data for the Earth found in Appendix 3, estimate the mass of the Sun. Does the mass of the Earth matter significantly in this calculation?

12. Compare the mutual gravitational force between you and the following two objects:
    (a) another person with mass 100 kg located 1 m from you;
    (b) Mars at opposition.
    Comment on your result.

13. (a) Venus has a maximum elongation of 47°. What is its distance from the Sun in astronomical units?
    (b) Mars has a synodic period of 779.9 days and a sidereal period of 686.98 days. On February 11, 1990, Mars had an elongation of 43° West. The elongation of Mars 687 days later, on December 30, 1991, was 15° West. What is the distance of Mars from the Sun in astronomical units?

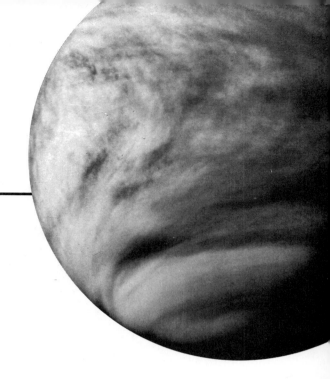

Chapter 2

The Solar System in Perspective

Our Earth is one member of the family of the Sun—the **Solar System.** Nine major planets, many natural satellites, multitudes of asteroids, a large number of comets, and an abundance of meteoroids orbit the Sun. Each class of objects has distinguishing characteristics (this chapter) and individual idiosyncrasies (Chapters 3–6), as well as systematic properties that require explanation in terms of the origin of the Solar System (Chapter 7). Here we will preview the physical aspects of the Solar System; the specifics will follow later. Celestial mechanics applied to Solar System bodies forms the physical basis of this perspective.

The Solar System comprises all objects bound in orbits by the Sun's gravity. These bodies make up a hierarchy based on mass: the Sun crowns the top, then the planets, their moons and rings, interplanetary debris (comets, asteroids, and meteoroids), down to interplanetary gas and dust. (See Appendix 3 for a summary of Solar System data.)

2–1 ◗
PLANETS

Today we place the Sun at the center of the heliocentric system of nine **planets** (in order of average distance from the Sun): Mercury, Venus, Earth, Mars, Jupiter, Saturn, Uranus, Neptune, and Pluto (Figure 2–1). (Pluto is actually a double-planet system with its moon, Charon.) Together, the planets have a total mass of but $0.0014M_\odot$, and they shine by reflected sunlight.

(A) MOTIONS

The planets obey the laws of Kepler and Newton as they move in elliptical orbits about the Sun. Their orbital semimajor axes are spaced in a fairly regular pattern. In 1766 (before Uranus, Neptune, and Pluto were discovered), Johann D. Titius (1729–1796) found an approximate empirical rule relating mean Sun–planet distances; Johann Bode (1747–1826) publicized this relationship in 1772, and it is now called **Bode's law** or the **Titius-Bode rule.** The prescription is to write down the series of numbers

$$4, \ 4 + (3 \times 2^0), \ 4 + (3 \times 2^1),$$
$$4 + (3 \times 2^2), \ 4 + (3 \times 2^3), \ \ldots$$

divide each result by 10, and arrive at the sequence

$$0.4, \ 0.7, \ 1.0, \ 1.6, \ 2.8, \ 5.2,$$
$$10.0, \ 19.6, \ 38.8, \ 77.2, \ \ldots$$

With the Sun–Earth distance (1 AU) as the unit of length, the mean Sun–planet distances (Table A3–2) are

Planet	Distance
Mercury	0.39
Venus	0.72
Earth	1.00
Mars	1.52
Jupiter	5.20
Saturn	9.54
Uranus	19.2
Neptune	30.1
Pluto	39.5

Except for the gap at 2.8 AU (where the asteroids lie), Bode's law works surprisingly well for the inner seven planets. Note the close spacing from the Sun to Mars (Figure 2–2) and the more open and regular distribution from Jupiter outward.

The planetary orbits exhibit three striking features (Table A3–1). First, all planets orbit the Sun *counterclockwise* as seen from directly above the Earth's orbital plane—the orbits are **direct.** Second, the orbital planes lie very close to the **ecliptic plane** (the plane of the Earth's orbit), so the planets (except Pluto) are always found in the 16°-wide band of the Zodiac when viewed from the Earth (the inclination of Pluto is 17°). Third, the orbital eccentricities are less than 0.1 except for the closest planet to the Sun, Mercury, and the most distant, Pluto. Because of its large eccentricity ($e = 0.249$), the orbit of Pluto ranges from 29.68 AU at perihelion (closest point to the Sun) to 49.36 AU at aphelion (farthest point from the Sun). Pluto passes closer to the Sun than Neptune on occasion (Figure 2–3). The conjecture that Pluto is an escaped satellite of Neptune rests in part on this observation.

A planet's **sidereal rotation period** refers to its rotation with respect to the stars; these periods are

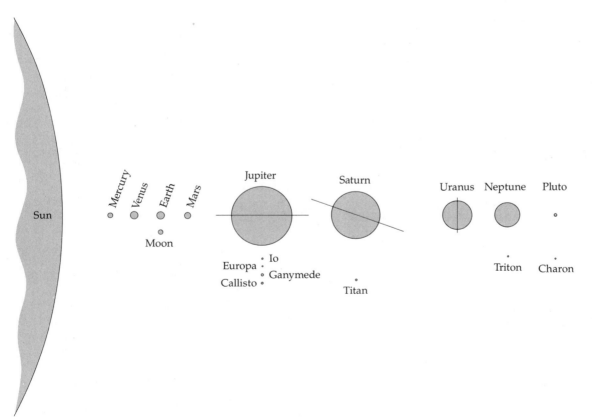

FIGURE 2–1   Relative sizes of the planets. On this scale, only a part of the limb of the Sun can be shown. Included with the planets are the eight largest moons.

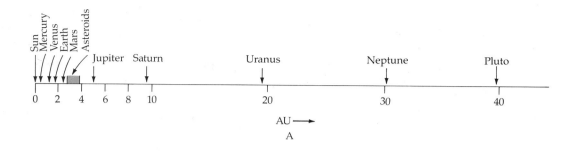

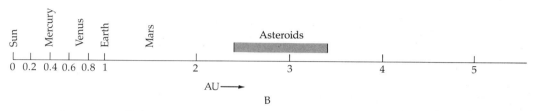

FIGURE 2–2 Relative orbital distances of the planets. (A) The distance scale in AUs from the Sun to Pluto. (B) An expanded scale shows the inner Solar System.

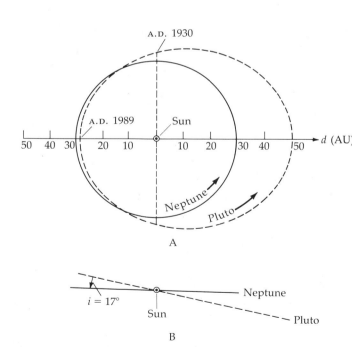

FIGURE 2–3 The orbits of Neptune and Pluto. (A) Face-on view of the orbits, with the scale in AUs. Note that at perihelion (1989), Pluto lies inside the orbit of Neptune. (B) Side view of the orbits, showing their relative inclinations.

determined in a variety of ways (Table A3–2). One main technique involves the Doppler effect, described in detail in Section 8–1(a). The essence of the Doppler shift is that when light is emitted by (or radio waves are bounced off of) a body, a radial component of velocity will be shifted in wavelength (or frequency) relative to its rest wavelength: a redshift for recession and a blueshift for approach. The amount of shift depends on the magnitude of the radial velocity (for nonrelativistic speeds):

$$\Delta\lambda/\lambda = v_r/c$$

where $\Delta\lambda$ is the wavelength change, λ the rest wavelength, v_r the radial part of the velocity, and c the speed of light.

Mercury and Venus were assumed to be in **synchronous rotation** about the Sun (that is, their sidereal rotation periods were believed to be *equal* to their sidereal orbital periods) until the mid-1960s, when Doppler radar measurements finally fixed their rotational periods. For the Moon and Mars, clear and abundant surface markings may be followed to deduce the rotation rates. The gigantic planets Jupiter, Saturn, Uranus, and Neptune show only their upper atmospheres, so that their

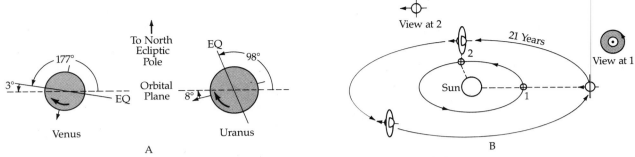

FIGURE 2–4 Equatorial inclinations. (A) Venus and Uranus rotate retrograde because the inclinations of their equators (EQ) to their orbital planes are greater than 90°. Sizes are *not* to scale. (B) The rotation axis of Uranus and its orbital plane are nearly aligned. At point 1, we see the planet pole-on; 21 years later at point 2, we see its equatorial plane edge-on. The satellites' orbits are face-on at 1 and side-on at 2.

rotation is found by following atmospheric features or by observing the Doppler-shifted radiation from their limbs. The atmospheres of these big planets rotate slightly faster at their equators than at their poles; hence, no unambiguous rotation rate exists. Sometimes that of the magnetic field is given. Because we assume that the magnetic field originates in the deep interior, this rate provides the basal value of rotation for the Jovian planets. Pluto's surface is not uniformly reflective; the amount of sunlight it reflects varies as it rotates. By observing these brightness fluctuations, we infer Pluto's rotation period.

We define a planet's **oblateness** ϵ by

$$\epsilon = (r_e - r_p)/r_e$$

where r_e is the equatorial radius and r_p the polar radius. A perfect sphere has $\epsilon = 0$. If all planets adjust to an equilibrium fluid shape, we expect that ϵ will increase as the rotation rate increases; the trend (Table A3–2) corroborates this idea. Accurate measurements of the Earth yield an oblateness of 1/298.3. For the other planets, oblateness is determined by measuring the visible disk of the planet or by analyzing perturbations on the orbits of the planet's moons. The large liquid planets are very oblate, so much so that Jupiter and Saturn appear distinctly oval in telescopes.

The equators of the planets are inclined to their orbital planes by varying amounts. The rotation axes of Mercury, our Moon, and Jupiter are nearly aligned with their revolution axes, whereas those of the Earth, Mars, Saturn, and Neptune are tilted

about 25°. The **equatorial inclination** is less than 90° for these bodies, so that they all rotate from west to east. West-to-east rotation is **direct** (in the same sense that all the planets orbit the Sun). The rotation of Venus and Uranus is from east to west, or **retrograde** (Figure 2–4A). The rotation axis of Uranus lies essentially in its orbital plane, so that if Uranus presents its pole toward us now, in 21

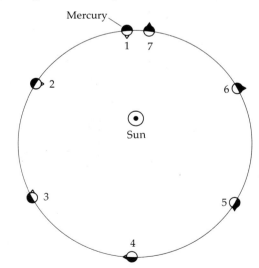

FIGURE 2–5 Mercury's orbit. Because this planet's rotation rate is two-thirds its orbital period, it completes three sidereal rotations in two revolutions. The marker on the planet goes from noon to midnight from position 1 to 7; it will return to noon after one more revolution.

years we will lie in its equatorial plane (Figure 2–4B).

Mercury and Venus have low rotation rates, which may be explained by a **spin–orbit coupling** and **resonance.** The synodic rotation period of Venus (its rotation period with respect to the Earth) is 146 days, which is one-quarter of its synodic orbital period. The sidereal rotation period of Mercury is exactly two-thirds its sidereal orbital period (Table A3–2). Hence, Venus is in a roughly 4:1 synodic resonance with the Earth, and Mercury exhibits a 3:2 sidereal lock with the Sun. Both planets are slightly nonspherical. Solar tidal forces interact with these deformations, slowing each planet's direct rotation until a resonance is achieved. Mercury's highly eccentric orbit resulted in the Sun–Mercury resonance (Figure 2–5): Mercury points alternate longitudes toward the Sun at each perihelion, rotating three times in two sidereal orbital periods.

(B) INTERIORS

The planets naturally divide into two groups (Table A3–3): (1) the small, solid **terrestrial** planets (Mercury, Venus, Earth, Moon, and Mars), with masses no greater than the Earth's, and (2) the large, liquid **Jovian** planets (Jupiter, Saturn, Uranus, and Neptune), which range in mass from $15M_\oplus$ to $318M_\oplus$. (Pluto, a binary planet, does not fall cleanly into either class.) The split between terrestrial and Jovian reflects fundamental differences in composition between the two groups.

Planetary masses are determined (1) by applying Kepler's third law to satellite orbits and, when there are no natural satellites, (2) by observing a planet's gravitational perturbations on the orbits of other planets, asteroids, comets, and artificial space probes. Planetary radii are deduced (1) by measuring the apparent optical size of the planetary disk, (2) by accurately timing the **occultations** (when the planet passes in front of an object) of stars, the planet's moons, and space probes, and, for the closer planets, (3) by precisely timing radar pulses reflected from various points on the planet's surface.

By dividing a planet's mass M by its total volume $4\pi R^3/3$, we may define its **average density** $\langle \rho \rangle$ (SI units = kg/m³):

$$\langle \rho \rangle = M/(4\pi R^3/3) \qquad \text{(2–1)}$$

The density of pure water is 1000 kg/m³. The high densities of the terrestrial planets, in the range of 3400 to 5500 kg/m³, reflect the fact that they are composed of heavy, nonvolatile elements such as iron, silicon, and magnesium. The very low densities of the Jovian planets (Saturn could *float* on water!) imply a composition similar to that of our Sun, with hydrogen and helium dominating.

The internal structure of a planet depends upon the distribution of its chemical composition, density, temperature, and pressure. In general, the pressure increases closer to the center of a planet, and the temperature also rises as a result of the greater pressure. Different materials are stable under different conditions, so that the chemical composition is radially layered—the interiors are **differentiated,** rather than homogeneous. The Earth has a metallic core above which lie lighter silicates; we expect Venus to exhibit a similar interior because its mass and composition are similar to the Earth's. The Jovian planets have thick interiors of mostly hydrogen and helium, probably with rocky cores.

Using the observed average density, composition, and oblateness of a planet, we may construct physically consistent models of its interior. Usually a range of models accommodate our data, so that only improved theory and observation can lead us to a unique picture.

(C) SURFACES

Whereas the Jovian planets and Venus show only their upper cloudy atmospheres, the other terrestrial planets and larger moons reveal surface markings. The most important general data on planetary surfaces include color, albedo, and temperature.

A planet's **color** relates to the composition of its surface and atmosphere. The oceans and continents give the Earth a blue color mottled with green, brown, and orange; large areas of cloud or snow cover appear white. The basaltic surface of the Moon looks dark gray with some tan, whereas the deserts of Mars give its characteristic brown-orange color. The surface of Io (a large moon of Jupiter) has a yellowish cast because of outflows from sulfur volcanoes.

The **albedo** of an object is the fraction of the incident sunlight reflected by it. Astronomers usually write this reflectivity as

A = amount reflected/amount incident

For planets with little or no atmosphere (Mercury, our Moon, and Mars), the albedo is very low because basaltic rocks are poor reflectors. Icy surfaces, as on most of Saturn's moons, have moderate albedos. The high reflectivity of clouds leads to the high albedos of the Jovian planets and Venus. The Earth's albedo is variable since it depends upon the season and upon snow and cloud cover; it averages about 0.35. Note that albedos are a function of wavelength; a rocky surface has different reflectivities at optical and infrared wavelengths, for example.

An important characteristic of planetary surfaces is **surface temperature.** Let's see how surface temperatures are observed and computed for **blackbody radiators.** Chapter 8 contains a complete discussion about blackbody radiators. Here we will introduce you to a few basics of these hypothetical objects. You can think of blackbodies in two ways: by their absorption properties and by their emissive ones. As the name implies, a blackbody completely absorbs all forms of electromagnetic radiation striking it; none is reflected, and so its albedo is zero. When a blackbody heats up to some temperature, its spectrum of emitted light (Figure 2–6) has a characteristic shape, called a **Planck curve,** with one maximum. A blackbody's emission peaks at the wavelength

$$\lambda_{max} = (0.002898 \text{ m})/T = (2898 \text{ }\mu m)/T \quad \textbf{(2–2)}$$

where T is the absolute temperature (kelvins) of the blackbody; one micrometer (μm) equals 10^{-6} m. This relation is **Wien's law,** which tells us that solar radiation ($T \approx 6000$ K) peaks at 0.5 μm; at normal room temperature ($T \approx 290$ K), the peak is at the longer infrared wavelength $\lambda_{max} = 10$ μm (Figure 2–6). Hence, we can determine a planet's surface temperature by observing its peak of emission.

Radiation of different wavelengths comes from different parts of a planet's surface. Therefore, we can probe the temperature and composition of varying levels of a planet's surface by observing the amount of radiation at different wavelengths. Because the Moon and Mercury have no atmo-

sphere, infrared comes from the top few millimeters of the visible surfaces, and longer-wavelength radiation originates several centimeters beneath the surface. We can use these thermal emissions to probe subsurface properties and conditions.

A planet that radiates more energy per second than it receives from the Sun must have an *internal heat source.* Delicate measurements at the Earth's surface reveal heat flowing from the hot interior; the same occurs with the Moon. The other planets known (from infrared observations) to produce excess heat are Jupiter, Saturn, and Neptune.

Now we will show how planetary blackbody temperatures are calculated (Table A3–3). The basic idea is to find the temperature at which a small blackbody (a planet) must radiate to balance the energy input from the Sun. We use **Stefan's law** of blackbody radiation (Chapter 8). This law relates the **energy flux** E (energy radiated per unit area per unit time, or W/m^2) to the temperature T of a blackbody:

$$E = \sigma T^4 \text{ W/m}^2 \quad \textbf{(2–3)}$$

where the constant is $\sigma = 5.67 \times 10^{-8}$ W $\cdot$ m$^{-2} \cdot$ K^{-4}. Because the area of a spherical surface of radius R is $4\pi R^2$, our Sun radiates energy like a blackbody at the rate ($T_\odot = 5800$ K)

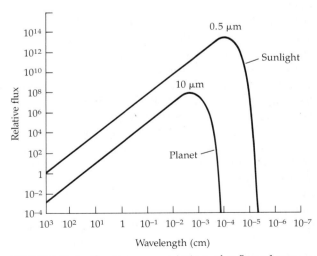

FIGURE 2–6 Continuous emission of reflected sunlight and a planet. Radiating like a blackbody, the Sun's emission peaks near 0.5 μm. For a planet at 290 K, the peak occurs at 10 μm.

$$4\pi R_\odot^2 E_\odot = 4\pi R_\odot^2 \sigma T_\odot^4 = 3.8 \times 10^{26} \text{ W}$$

which is the Sun's luminosity ($L_\odot$)—its total radiative power. This same energy flows through a sphere of area $4\pi r_p^2$ at the Sun–planet distance r_p, so that the energy flux there is

$$E_p = 4\pi R_\odot^2 E_\odot / 4\pi r_p^2 = (R_\odot/r_p)^2 E_\odot$$

If 1 m^2 of blackbody material intercepts this energy head-on with the Sun overhead, it will be raised to the **subsolar temperature** (in kelvins):

$$T_{ss} = (R_\odot/r_p)^{1/2} T_\odot \approx 394 (r_p)^{-1/2} \quad \textbf{(2–4)}$$

since $E_p = \sigma T_{ss}^4$. In the last equality, we have inserted the appropriate solar values and expressed r_p in AU. The subsolar temperatures are essentially the equilibrium **noontime** temperatures.

Although subsolar temperatures apply to the local noon on the surfaces of very slowly rotating planets (Mercury, the Moon, and Pluto), they are not appropriate for planets with atmospheres or for planets in rapid rotation. For these, assume that the effective absorbing area is the cross section πR_p^2 and that the effective radiating area is the total surface area $4\pi R_p^2$, where R_p is the radius of the planet. The albedo A is the fraction of incident solar radiation that is reflected, so that only the fraction $1 - A$ is absorbed. Therefore, energy absorbed per second is

$$(1 - A)\pi R_p^2 E_p = (1 - A)\pi R_p^2 (R_\odot/r_p)^2 \sigma T_\odot^4$$

while the planet radiates the power $4\pi R_p^2 \sigma T_p^4$. We again use the concept of equilibrium: that the rate of absorption of energy equals the rate at which it is radiated away fixes an equilibrium temperature. (Otherwise the temperature would either rise or fall.) Equating these and solving for the temperature, we have (in kelvins)

$$T_p = (1 - A)^{1/4}(R_\odot/2r_p)^{1/2} T_\odot$$
$$\approx 279(1 - A)^{1/4}(r_p)^{-1/2} \quad \textbf{(2–5a)}$$

with r_p expressed in AU. The equilibrium blackbody temperatures (Table A3–3) follow when $A = 0$. For the subsolar temperature, we get

$$T_{ss} \approx 394(1 - A)^{1/4}(r_p)^{-1/2} \quad \textbf{(2–5b)}$$

By including the albedo effect, we may more closely approximate the observed planetary temperatures, but remember that these estimates neglect important complications, such as the circulation of planetary atmospheres, their heat retention, internal heat sources, and the variation of A with wavelength.

(D) ATMOSPHERES

Of the terrestrial planets, Mercury and the Moon have essentially no atmosphere, Venus and Mars possess a carbon dioxide (CO_2) atmosphere, and the Earth's atmosphere is primarily molecular nitrogen (N_2) and oxygen (O_2). The principal constituents of the atmospheres of the Jovian planets are molecular hydrogen (H_2) and helium (He). In general, planetary atmospheres are densest near the planet's surface and thin rapidly with increasing altitude. The composition of an atmosphere may be stratified, with the densest gases residing closest to the surface of the planet, but turbulent mixing and winds can lead to regions of homogeneous composition. Far from the planet's surface, incoming solar ultraviolet rays and X-rays usually ionize atmospheric atoms or dissociate molecules to form the layered **ionosphere** (Chapter 4).

To gain some understanding of planetary atmospheres, consider a simple model for their retention. To a first approximation, an atmosphere behaves like a **perfect gas,** that is, as particles that interact only through elastic collisions. Such a gas obeys a special relationship between pressure, temperature, and density:

$$P = nkT$$

where P is the pressure (the rate of change of the particles' momenta from collisions) in units of force per unit area (N/m^2), n is the number density of particles ($\#/m^3$), T is the absolute temperature (K), and k is Boltzmann's constant, equal to 1.38×10^{-23} J/K. Now, from the continuous collisions, the particles of the gas achieve, at a given temperature, an equilibrium distribution of velocities so that

$$F(v)\,dv \propto \exp\left(-\tfrac{1}{2}mv^2/kT\right)v^2\,dv$$

known as the **Maxwellian distribution** of velocities in a gas (Figure 2–7). Note that, because of the exponential decrease, the distribution has a long tail at large velocities—a few of the particles have been boosted to high speeds by the collisions. The peak of this distribution defines a **most probable speed:**

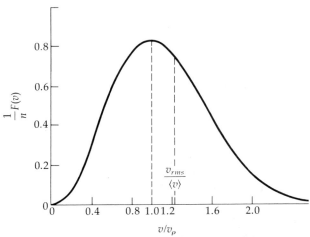

FIGURE 2–7 Maxwellian distribution of particles in a gas. The velocities of the particles spread around the most probable velocity v_p, which marks the peak of the distribution. The numbers on the vertical axis have been normalized to the total, n, in the gas.

$$v_p = (2kT/m)^{1/2}$$

where m is the mass of a gas particle. While the speeds of these particles (atoms or molecules) are distributed over a large range and change violently in each collision, the *average* kinetic energy per particle is

$$\langle KE \rangle = (m/2)\langle v^2 \rangle = 3kT/2 \qquad (2\text{–}6)$$

Here m is the particle's mass and T is the kinetic absolute temperature of the gas. From Equation 2–6 we obtain the **root mean square speed** v_{rms}:

$$v_{rms} = \langle v^2 \rangle^{1/2} = (3kT/m)^{1/2} \qquad (2\text{–}7)$$

which tells us that the mean speed of the particles increases with temperature and decreases with mass. In the very thin upper regions of atmospheres, a particle that moves outward with the *escape speed* v_e has an excellent chance of leaving the atmosphere, with v_e being

$$v_e = (2GM/R)^{1/2} \qquad (2\text{–}8)$$

where M is the planet's mass and R is its radius. If $v_{rms} = v_e$ for a given particle species, that gas will leave the atmosphere in only a few days. To retain an atmosphere for several billion years (approximately the age of the Solar System), a planet must have $v_e \geq 10 v_{rms}$. (The factor of 10 takes into account the high-speed tail of the Maxwellian distribution of speeds.) Therefore a given type of molecule is retained indefinitely when (Equations 2–7 and 2–8)

$$T \leq GMm/150kR \qquad (2\text{–}9)$$

Figure 2–8 shows points corresponding to the equilibrium blackbody temperature (Table A3–3) and v_e for the planets and some moons; the dashed lines represent $10v_{rms}$ for various molecular species. In terms of this crude model, a planet retains all gases with lines passing below its point, and the other gases escape. This model reasonably explains that the Jovian planets have retained all gases and the Earth, Venus, and Mars have lost their hydrogen and helium but retained nitrogen and carbon dioxide; Mercury and the Moon have essentially no atmosphere; and the largest moons have thin atmospheres. (Titan, in fact, has a dense atmosphere composed largely of nitrogen but made smoggy because of photochemical reactions of methane and other compounds.)

2–2 ◑
MOONS, RINGS, AND DEBRIS

Natural planetary satellites, or **moons,** are many in the Solar System, but together they comprise a total mass of only $0.1 M_{\oplus}$. Eight satellites are about the size of our Moon, while the others are much smaller and resemble large asteroids. Our Moon, the two satellites of Mars, the five inner satellites of Jupiter, the eight innermost of Saturn, and five of Uranus' moons have nearly circular orbits lying essentially in their planet's equatorial plane. Observations show that these 21 moons exhibit synchronous rotation from tidal friction (as does our Moon).

(A) MOONS

Only three moons circle the terrestrial planets, whereas the Jovian planets possess at least 51. The larger the mass of a planet, the greater the range of its gravitational influence. This coupled with the Jovian planet's proximity to the asteroid belt means that they can capture asteroids gravitationally. The Jovian satellites with small masses,

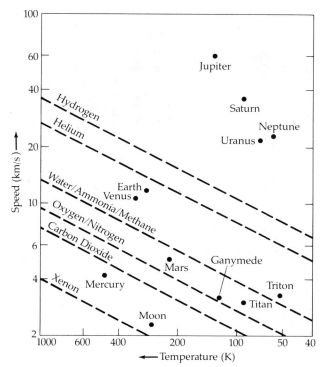

FIGURE 2–8 Retention of atmospheric gases. Mean molecular speeds are given as a function of temperature, along with the escape speeds for the indicated bodies. The dashed lines show ten times the mean molecular speeds, which defines an essentially infinite lifetime for that component in the atmosphere.

highly eccentric and inclined orbits, and retrograde motion are probably captured asteroids, as are the Martian moons.

(B) RINGS

Four of the Jovian planets are known to have rings. Of these ring systems, Saturn's is by far the most spectacular. In each case, the ring system is within the Roche limit [Section 3–4(d)]. Some rings appear to be kept in line by small satellites just outside and inside their boundaries.

(C) ASTEROIDS

Bode's law "predicted" the existence of a planet at 2.8 AU, between Mars and Jupiter, but it was not

until 1801 that Giuseppe Piazzi (1746–1826) discovered the **minor planet** Ceres in this region. By 1980, more than 300 **asteroids** had been found with semimajor orbital axes between 2.3 and 3.3 AU—they made up the **asteroid belt.** Today, the orbital elements of more than 3000 asteroids are known, and each such body is numbered in order of orbit determination and given a name, such as asteroid *1000 Piazza.*

Asteroids are too small to retain atmospheres, for their observed diameters range from relatively large (Ceres, 1020 km; Pallas, 538 km; Vesta, 549 km) to the more abundant smaller ones (about 1 km); certainly a multitude of small rocks also swarm in the asteroid belt. The total mass in asteroids is probably about a few percent of our Moon's mass. The largest asteroids tend to reside farthest from the Sun, and the smallest closest; they are all in direct orbit about the Sun. Asteroids exhibit orbital eccentricities up to 0.83, with most in the 0.1 to 0.3 range, and orbital inclinations as large as 68° but more typically less than 30°.

The asteroid belt shows distinctly depleted regions, called **Kirkwood gaps,** at semimajor axes where the orbital period is a simple fraction (such as 1/2, 1/3, 1/4, 2/5, 3/7) of Jupiter's orbital period. Periodic gravitational perturbations from Jupiter have removed all asteroids from these gaps. Where the ratio of periods is 2/3 and 1/1, asteroids accumulate in **groups,** or **families;** the 15-member Trojan groups are situated at Jupiter's orbit on the vertices of equilateral triangles with the Sun and Jupiter at the other two vertices. The Trojans are bound here where they are stable against perturbations about their equilibrium positions. Some asteroids, which are known as the Apollo group and include Daedalus, Icarus, and Geographos, have perihelia within the Earth's orbit and may pass near the Earth at times. On June 15, 1968, Icarus careened past the Earth at a distance of a mere 6.4×10^6 km. Some people feared a collision and predicted the world would end. It did not.

Just as with Pluto, the rotation periods of asteroids are determined from fluctuations in their reflected sunlight. Some observed light curves exhibit two maxima and two minima per cycle, corresponding to oblong bodies that tumble every 3 to 20 h (the average rotation period for asteroids is 7 h). So we picture irregular fragments of rock, several kilometers in diameter, spinning in a few hours and colliding with one another once every million years or so.

(D) COMETS

Comets are small and infrequent interlopers in the inner Solar System (Figure 2–9). Bright visual comets have tails that can stretch 90° across the sky. A comet is named after its discoverer (Comet Oterma) or co-discoverers (Comet Ikeya-Seki) and also by year in the order of its discovery (1971a was the first discovered in 1971, 1971b the next, and so on). After an orbit has been computed, the numbering is in the order of perihelion passage (Comet 1971 I was the first to pass perihelion in 1971).

All comets move in elliptical orbits (some retrograde) about the Sun and are members of the Solar System. They fall into two distinct classes: (1) the **long-period** comets, which are the great majority and which have orbital eccentricities very close to unity (almost parabolic), and (2) the much smaller group of **short-period** comets, which are periodic in their returns. Most observed comets reach perihelion around 1 or 2 AU from the Sun, and the inclinations of their orbits range through all values; hence, comets move in a spherical volume centered on the Sun while the planets move near the ecliptic plane. The aphelia of long-period comets may extend as far as 50,000 AU, with orbital periods as long as a million years or even greater; these comets are seen only during one perihelion passage. In some cases, planetary gravitational perturbations make comet orbits hyperbolic ($e > 1$) so that they escape the Solar System; while in other cases the orbits are changed into small ellipses. In this way, Jupiter has captured a family of about 45 comets that now orbit the Sun with aphelia near Jupiter's orbit. Short-period comets move around within the planetary system: Comet Encke has the shortest period, 3.3 years ($a \approx 2.2$ AU). The most famous short-period comet is **Halley's Comet,** which returns to perihelion about every 76 years.

With each perihelion passage, a comet loses material as a result of the intense solar heating and tidal forces; eventually the comet vanishes. Halley's Comet, for example, sheds about 0.001 of its mass at each perihelion passage. Some short-period comets have been observed to split into several pieces or even disintegrate.

FIGURE 2–9   Comet Arend-Roland. Note the thin countertail pointing toward the Sun, while the main tail points away. *(Lick Observatory)*

Because comets waste away, the supply is replenished if we are to see them today. We presume that cometary nuclei were formed with the Solar System about five billion years ago. To account for their continued existence, Jan Oort hypothesized a spherical **comet cloud** out 50,000 AU from the Sun with about 10^{11} cometary bodies with a total mass of $\approx 1 M_{\oplus}$. At its outer edge, this reservoir loses nuclei as a result of perturbations from passing stars; at its inner edge, planetary perturbations deflect nuclei toward the Sun. Further perturbations may cause a nucleus to either escape from the Sun altogether or enter a periodic orbit, where it slowly disintegrates. Recent estimates suggest as many as 10^{14} cometary bodies, including large numbers in more or less permanent orbits at 10,000 to 20,000 AU.

(E) METEOROIDS

Roaming throughout the Solar System with orbits of all inclinations are the **meteoroids.** These range in size from small asteroids (10 km) down to micrometeoroids (<1 mm) and interplanetary dust (≈ 1 μm). Pieces from asteroid collisions form the larger rocks, while the smaller pieces come from disintegrated comets. When a meteoroid enters the Earth's atmosphere, friction heats it to incandescence and visibility at an altitude of about 120 km—a **meteor**. From the brightness of meteors, we deduce that their average density is 200 to 1000 kg/m^3 (similar to a comet's nuclear material!). A meteoroid loses mass by vaporization, melting, and fragmentation, and it ionizes the air through which it passes. Some meteors are bright enough to cast shadows—these are called **fireballs.** A typical meteor particle is about the size of a grain of sand; a fireball, the size of a pebble.

Meteoroids that are not totally consumed in the atmosphere strike the ground as meteorites. Large meteorites produce craters, such as those preserved on the Moon and the kilometer-wide Barringer Meteor Crater in Arizona (Figure 2–10). Smaller meteorites do not annihilate themselves upon reaching the ground, and the micrometeorites (with diameters of 0.5 to 200 mm) simply drift through the atmosphere to deposit a total mass of about 10^6 kg per day on the Earth's surface.

The Earth occasionally passes through a group of meteoroids—pieces of solid material from a comet. When this intercept occurs, we see a **meteor shower** in the sky (Figure 2–11). During such a shower, the meteors appear to come from a particular point in the sky, called the **radiant.** Showers are usually named after the constellation in which their radiant lies. For instance, the Perseid meteor

FIGURE 2–11 The 1966 Leonid meteor shower. Tracing the trails back gives an approximate position of the radiant in Leo. *(D. Milon)*

shower, appearing in August, seems to come from the constellation Perseus. The debris from a meteor shower comes from a comet. For instance, that for the Perseids is associated with a comet named 1862 III.

(F) INTERPLANETARY DUST

Finally we turn to **interplanetary dust** (size ≈1 to 100 μm), which probably comes from the dust tails of comets. Mariner space probes revealed dust clouds around the Earth and Mars. All planets have probably captured dust clouds by gravitational attraction. The faint band of **zodiacal light,** best seen before dawn, results from the sunlight reflected by a marked concentration of dust in the ecliptic plane; the intensity of this scattered light diminishes with distance from the Sun (Figure 2–12).

2–3 ◗
MECHANICS APPLIED TO THE SOLAR SYSTEM

Now that we've given a brief overview of the solar system, let's apply some of the concepts that were introduced in Section 1–5 to specific situations.

FIGURE 2–10 The Barringer Meteor Crater. Located in Arizona near Flagstaff, this crater was formed by an impact some 40,000 years ago. *(U.S. Air Force)*

FIGURE 2–12   The zodiacal light. This short time exposure (note the stars are trailed) shows the light above Mt. Haleakala, Hawaii. *(A. Peterson and L. Kieffaber)*

(A) USING KEPLER'S THIRD LAW

The modern form of the harmonic law is Equation 1–24:

$$P^2 = 4\pi^2 a^3 / G(m_1 + m_2) \qquad \textbf{(2–10)}$$

In applying this relation to the Sun ($M_\odot$) and planets (m_p), considerable simplification occurs when we measure the sidereal periods P in years and the semimajor axes a in AU, for then

$$a^3 / P^2 = 1 + (m_p / M_\odot)$$

We have simplified by using Earth units; always use the units most appropriate to the system under consideration! Do this by forming ratios of Equation 2–10 as

$$[(m_1 + m_2)/(m_1' + m_2')](P/P')^2 = (a/a')^3$$

where the system m_1 and m_2 (with period P and semimajor axis a) is compared with the standard system m_1' and m_2' (with P' and a'). For objects orbiting the Sun or for binary stars (Chapter 12), the standard system is the Sun–Earth system: P is expressed in years, a in AU, and all masses m in solar masses ($M_\odot$). In these units, $P^2 = a^3$; that is, $k = 1$, so $G = 4\pi^2$. For planetary satellites (either natural moons or artificial satellites), we use the Earth–Moon system: set $P' = 27.3$ days, $a' = 3.84 \times$

10^5 km, and $(m_1' + m_2') = M_\oplus$ (or 5.976×10^{24} kg); we then obtain P in days, a in kilometers, and the masses m in Earth masses (or kilograms).

An example: consider a comet with an elliptical orbital period of seven years; we want to find the semimajor axis of its orbit. Since $M_\odot + m_{\text{comet}} \approx M_\odot$, Equation 2–10 gives $a = 7^{2/3} = 3.7$ AU. Another example: find the mass of Uranus (M_U) in terms of the Earth's mass. Miranda (whose mass is m_m), a moon of Uranus, orbits the planet in about 1.4 days at a mean distance of 128,000 km. Using the Earth–Moon standard system we have

$$[(M_U + m_m)/(M_\oplus + M_m)](P_m/P)^2$$
$$\approx (M_U/M_\oplus)(P_m/P)^2 = (a_m/a)^3$$

where we have neglected the masses of Miranda and the Moon with respect to the much larger masses of their primary planets, Uranus and Earth. Substituting the relevant data, we obtain

$$M_U \approx (27.3/1.4)^2 (128{,}000/384{,}000)^3 M_\oplus \approx 14 M_\oplus$$

(B) LAUNCHING ROCKETS

Consider projectiles launched vertically upward from the Earth's surface. Neglecting air friction, which continuously decreases the projectile's speed, we can use the conservation of total energy to find the height to which the projectile finally rises.

Near the Earth's surface, the strength of the downward gravitational force is $F = mg$, where m is the mass of the projectile. Hence, the potential energy at altitude h is $PE = mgh$. If the speed of the body is v at ground level, its kinetic energy there is $KE = mv^2/2$. Conservation of total energy requires

$$TE = (KE + PE)_{\text{ground}} = \text{constant} = (KE + PE)_h$$

so that by evaluating TE at $h = 0$ and at maximum height h, we find $mv^2/2 = mgh$, or,

$$h = v^2/2g \qquad \textbf{(2–11)}$$

With $g = 9.8$ m/s^2, a rock thrown upward with speed $v = 14$ m/s will rise to a height of 10 m before falling back to the ground.

When we consider altitudes greater than about one Earth radius ($h \gtrsim R_\oplus$), our previous approximation breaks down and we must use Equation 1–32. Evaluating this expression at the ground ($R_\oplus$) and at maximum height ($R_\oplus + h$), we find

$$(mv^2/2) - (GM_\oplus m/R_\oplus) = -GM_\oplus m/(R_\oplus + h)$$

or

$$
\begin{aligned}
h &= R_\oplus\{(v^2 R_\oplus/2GM_\oplus)/[1 - (v^2 R_\oplus/2GM_\oplus)]\}\\
&= (v^2/2g)\{R_\oplus/[R_\oplus - (v^2/2g)]\} \qquad \textbf{(2–12)}
\end{aligned}
$$

Equation 2–11 follows from Equation 2–12 in the limit $v^2/2g \ll R_\oplus$, that is, $h \ll R_\oplus$. Note that when $v^2/2g = R_\oplus$, at the speed $v = (2gR_\oplus)^{1/2} = 11.2$ km s^{-1}, the projectile escapes to $h = \infty$. This critical speed is called the **escape speed.**

(C) ORBITS OF ARTIFICIAL SATELLITES AND SPACE PROBES

Most major space vehicles are placed into Earth orbit—a *parking orbit* from which deep-space probes are accelerated toward the Moon and the planets. Multistage rockets lift the vehicle beyond the Earth's atmosphere, where the final stages accelerate the payload horizontally to the desired orbital speed. The Earth's rotational speed (0.46 km/s at the equator) aids launchings toward the east. For a circular orbit at distance r from the center of the Earth ($r = R_\oplus + h$, if h is the altitude of the orbit), the circular speed v_c may be found by equating the centripetal and gravitational forces:

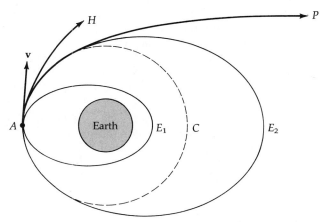

FIGURE 2–13 Orbits around the Earth. At A, a spacecraft has burnout velocity **v**. If the velocity equals that for a circular orbit at that distance from the Earth's center, a circular orbit C results. When **v** = 1.44 v_c, a parabolic orbit P ensues; for a greater velocity, the escape orbit H is hyperbolic. When **v** falls between these two limits, A is the perigee of elliptical orbit E_2; when $0 < v < v_c$, A is the apogee of elliptical orbit E_1.

$$v_c = (GM_\oplus/r)^{1/2} = v_0(1 + h/R_\oplus)^{-1/2} \quad \textbf{(2–13)}$$

where $v_0 = (GM_\oplus/R_\oplus)^{1/2} = 7.86$ km/s is the circular speed at the Earth's surface (neglecting atmospheric friction). Rearranging Equation 2–13, we find

$$v = 2^{1/2} v_0[(R_\oplus/h + 1)]^{-1/2}$$

so that $h \to \infty$ at the escape speed $v_{\text{escape}} = 2^{1/2} v_0$; note that this is the escape speed from the Earth's surface, not from a satellite orbit.

The orbit of a space probe depends critically upon the velocity at burnout. Consider the simple case of a projectile moving *parallel* to the Earth's surface at this point (point A in Figure 2–13). The energy equation (Equation 1–28) gives the semimajor axis a of the orbit if we know the burnout speed v:

$$a = r/[2 - (v/v_c)^2] \qquad \textbf{(2–14a)}$$

where r is the distance from the Earth's center to point A. We can now invert Equation 2–14a to find the *injection speed* needed to attain an orbit of semimajor axis a:

$$v = v_c(2 - r/a)^{1/2} \qquad \textbf{(2–14b)}$$

A circular orbit of radius $r = a$ results when $v = v_c$, as expected. When we have $v = 2^{1/2} v_c = 1.44 v_c$, the semimajor axis becomes infinite and the projectile escapes along a parabolic orbit; hence, this is the escape speed appropriate to radius r. When $v > 1.44 v_c$, the space probe escapes on a hyperbolic trajectory. If $v < 1.44 v_c$, the projectile enters an elliptical orbit and point A is either the perigee (point closest to Earth) or apogee (point farthest from Earth) of the orbit, since the velocity is perpendicular to the radius vector at point A. For $v_c < v < 1.44 v_c$, Equation 2–14a tells us that $a > r$ at A; hence, we insert at perigee, and the orbit never approaches the Earth closer than A. For $0 < v < v_c$, we have $a < r$, so that A is the apogee of the orbit. The satellite will collide with the Earth if the perigee distance is less than $R_\oplus$.

2–4 ◐ ESSENTIAL QUESTIONS

Our brief survey has revealed the structure and content of the Solar System, but the story is not closed. We can successfully interpret some observed features, such as the sources of meteorites. However, many questions remain unanswered concerning the origin of these features. These unresolved questions prompt tentative answers (Chapter 7).

The fundamental question is how the Solar System originated and evolved to its current state? This global question then raises many specific ones: Why are there five tiny terrestrial planets and four huge Jovian planets? What produced the distribution of planetary orbits; is Bode's law fundamental in terms of the physical processes?

The **angular momentum distribution** of the Solar System is also a puzzle. The rotational angular momentum of the Sun is $\mathbf{L} = \mathbf{I}\omega$ or

$$L = 2M_\odot\, R_\odot^2 \omega_\odot/5 \approx 10^{42} \text{ kg} \cdot \text{m}^2 \cdot \text{s}^{-1}$$

while each planet has an orbital angular momentum $2\pi ma^2/P$, where m is the planet's mass, a is its orbital semimajor axis, and P is its sidereal orbital period; the planetary total is about 33×10^{42} kg $\cdot$ m$^2 \cdot$ s^{-1}. Hence, the planets contribute 97% of the angular momentum of the system. Why doesn't the much more massive Sun provide the dominant contribution? How did it lose its original angular momentum?

How did it come about that the terrestrial planets rotate much more slowly than the Jovian planets? What caused the rotation axis of Uranus to lie in its orbital plane? Are any planetary atmospheres primordial? Why do Venus, Titan, and the Jovian planets have such dense atmospheres? What sculpted the surfaces of planets and satellites? How did planetary satellites originate, with their observed number, masses, and orbital distributions? Why do the rings of Saturn, Jupiter, Neptune, and Uranus differ so? What created the asteroid belt? How were comets formed, and how many are out there? What produced the meteoritic compositions we observe?

We think the most basic questions must be answered by any model pretending to account for our Solar System. We will address many of these questions in terms of current ideas in an attempt to decipher the riddle before us.

PROBLEMS ☽

1. (a) Describe the apparent path of the Sun across the sky of Mercury during one solar "day," as seen by an observer at that planet's equator.
   (b) Describe the seasons of Uranus, giving their durations where appropriate.

2. Consider the planets Uranus, Neptune, and Pluto.
   (a) On a graph, compare their orbital semimajor axes with the distances predicted by Bode's law.
   (b) Show that the ratios of their orbital periods are approximately *commensurable*, that is, nearly fractions such as $\frac{3}{2}$.

3. Show that the two satellites of Mars, Phobos and Deimos, obey Kepler's third law; deduce the mass of Mars from the orbits of these moons.

4. (a) How much does a spaceship having a mass of 10^3 kg weigh on the Earth's equator? At its poles? (*Hint*: consider centripetal force.)
   (b) Remembering Jupiter's rapid rotation, how much will this craft weigh at the equator of Jupiter's surface? At the poles?

5. Derive the formula that gives the synodic rotation period (solar day) of Venus with respect to the Earth, and verify the number quoted in the text.

6. In the seventeenth century, O. Roemer concluded that the speed of light is finite by observing that the satellite Jupiter I (Io) was occulted by the planet approximately 16.7 min *earlier* when Jupiter was in opposition than when it was near superior conjunction. Use this information to draw an appropriate diagram and to calculate an approximate value for the speed of light (which Roemer did *not* do!).

7. How far from the star would we have to be (in AU) to find a substellar temperature comparable to that of the Earth for
   (a) Rigel (surface temperature $T = 12,000$ K, radius $R = 35R_\odot$)?
   (b) Barnard's star ($T = 3000$ K, $R = 0.5R_\odot$)?

8. Formaldehyde (H_2CO) has been discovered in interstellar space.
   (a) Calculate its "mean" molecular speed for $T = 280$ K. Would our Moon retain this gas for billions of years?
   (b) Would Saturn's satellite Titan retain formaldehyde? (For Titan, radius ≈ 2600 km, mass $\approx 10^{23}$ kg.)

9. Consider a comet with an aphelion distance of 5×10^4 AU and an orbital eccentricity of 0.995.

(a) What are the perihelion distance and orbital period?

(b) What is the comet's speed at perihelion and at aphelion?

(c) What is the escape speed from the Solar System at the comet's aphelion, and what do you conclude from this result?

10. Draw a diagram to explain why some meteor showers are consistent from year to year whereas others are spectacular on occasion and feeble at other times (such as the Perseids in 1985).

11. The albedo of Venus is about 0.77 because of the cloudy atmosphere. What would the noontime temperature be? (The measured temperature is 750 K.)

12. (a) At noon, Mercury's surface temperature is roughly 700 K; at midnight, 125 K. Calculate the peak wavelength at which the surface emits at noon and at midnight.

(b) Calculate the energy output per square meter of the surface at midnight and noon.

13. The Earth, Venus, and Mars all have carbon dioxide in their atmospheres. Find the ratio of the root mean square speed to escape velocity for each and make a statement about the retention of carbon dioxide for these planets.

14. What is the approximate orbital period for a cometary nucleus in the Oort cloud?

15. Calculate the blackbody equilibrium temperature of a fast-rotating asteroid with a radius of 100 km and an albedo of 0.5 in an orbit between Mars and Jupiter of semimajor axis 2.8 AU.

16. Roughly speaking, the spin angular momentum of a sphere is MVR, where V is the equatorial velocity. Make an approximate calculation of the spin and orbital angular momenta of Jupiter and compare them with the spin angular momentum of the Sun.

17. (a) What is the semimajor axis of the least-energy elliptical orbit of a space probe from Earth to Venus?

(b) Relative to the Earth, what is the velocity of such a probe at the Earth's orbit?

(c) When the probe reaches Venus, what is its velocity relative to that planet?

18. Suppose that a projectile has a burnout speed $1.44v$ (where $v_c/1.44 < v < v_c$) at a distance r from the Earth's center ($r > R_\oplus$). If the velocity vector points 45° above the local parallel to the Earth's surface, find the semimajor axis a, the sidereal period P, and the eccentricity e of the resulting elliptical orbit in terms of r, v, and constants. Can you also find, the angle from burnout to the orbit's perigee? [*Hint*: Section 2–3(c) is very helpful.]

19. Compare the escape velocity of a rocket launched from the Earth with the escape velocity of one at a distance of 1 AU from the Sun (that is, the escape velocity from the Solar System at the Earth's distance from the Sun).

20. The inner (terrestrial) planets have relatively small amounts of hydrogen in their atmospheres, yet the outer (Jovian) planets are predominantly hydrogen. Taking the Earth and Jupiter as typical examples of each class, calculate the ratio of the root mean square speed to the escape velocity of hydrogen for both the Earth and Jupiter. How do these values affect the relative observed amounts of hydrogen in the atmosphere of each of these planets?

21. From information given in the chapter, what is the approximate lifetime of Halley's Comet assuming a constant mass loss rate? What does this lifetime tell us about the existence of the Oort Comet Cloud?

22. Comet Halley has an orbital period of 76 years and an orbital eccentricity of 0.967.

(a) What is the comet's perihelion distance? Aphelion distance?

(b) What is the subsolar temperature on Comet Halley at perihelion? At aphelion?

(c) The albedo of Comet Halley is 3%. What is the equilibrium blackbody temperature at perihelion? At aphelion?

Chapter 3

The Dynamics of the Earth

Before investigating the Solar System in detail, we present the dynamics of our Earth. Chapter 4 deals with the physical properties of Earth and the Moon. Both the dynamics and physics of the Earth lay the base for an investigation of the other terrestrial planets, because we know the Earth better than any other planet.

3–1 ◗
TIME AND THE SEASONS

You are familiar with the words *second, minute, hour, day, week, month,* and *year,* but what exactly do they mean? Measuring time is arbitrary and conventional. Astronomers historically defined the second, minute, hour, and day in terms of the Earth's rotation, the week and month in terms of the Moon's orbital motion, and the year in terms of the Earth's revolution about the Sun. (We suggest you read Appendix 10 before proceeding.)

Today, the second is the fundamental unit of time in the International System of Units (Appendix 6). Atomic clocks define the second; specifically, as 9,192,631,770 periods of the transition be-

tween two energy levels in the cesium-133 atom. Such clocks form the basis of Coordinated Universal Time (UTC). Hydrogen maser clocks, which astronomers use sometimes as local standards, keep time with a precision of one second in 30 million years!

(A) TERRESTRIAL TIME SYSTEMS

A **day** is historically defined as that time interval between two successive upper transits of a given celestial reference point. An **upper transit** occurs when a celestial reference point or body crosses the celestial meridian moving westward; the **celestial meridian** is the imaginary line drawn through the north point of the horizon, the zenith, and the south point of the horizon. The **vernal equinox** is a point in space used as the zero point for **sidereal time,** and the sidereal day is arbitrarily divided into 24 sidereal hours of equal length, each of which consists of 60 sidereal minutes with 60 sidereal seconds per minute. Now 1^h corresponds to 15° of the Earth's rotation with respect to the stars, so that the **local sidereal time** is $0^h\ 0^m\ 0^s$ when the vernal equinox lies on our celestial meridian and 2^h when the vernal equinox is 30° (2 h) west of the celestial meridian. We define **local sidereal time** as the *hour angle of the vernal equinox* (Figure 3–1). The **hour angle** is how far west (positive) or east (negative) of the meridian a celestial object is. The vernal equinox precesses westward along the celestial equator at the rate of about 50″ per year [Section 3–4(c)], and so the actual period of the Earth's rotation (measured with respect to the stars) is 0.008^s longer than our definition for one sidereal day.

Civil timekeeping on the Earth is based in a complicated way upon the position of the Sun in the sky. **Apparent solar time** is the hour angle of the real Sun plus 12^h, so local apparent noon always occurs at 12^h and marks the start of the apparent solar day. The length of the apparent solar day is *not* constant during the year, however, even to a given observer. These variations are caused by the eccentricity of the Earth's orbit and the inclination of the Earth's equatorial plane to the ecliptic, that is, its orbital plane (Figure 3–2). The Earth's orbital speed reaches a maximum at perihelion (about January 2) and a minimum at aphelion (about July 3), in accordance with Kepler's second law. The Sun reflects this variable motion by moving eastward along the ecliptic faster at perihelion

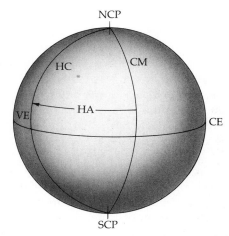

FIGURE 3–1   Reference circles on the sky. Shown on the celestial sphere are the observer's celestial meridian (CM), the celestial equator (CE), the north and south celestial poles (NCP and SCP), and the hour angle (HA) of the vernal equinox (VE) along its hour circle (HC).

than at aphelion. To return the Sun to noon, the Earth must turn through a greater angle, and so the apparent solar day is longer at perihelion than at aphelion.

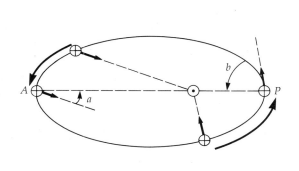

A

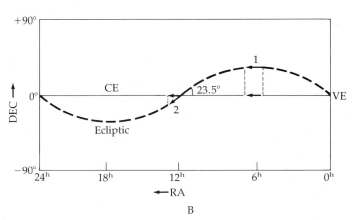

B

FIGURE 3–2   Variations in apparent solar time. (A) A schematic view of the Earth's orbital motion in one day. Near aphelion (*A*), the Earth rotates through 360° plus angle *a* to complete one apparent solar day. Near perihelion (*P*), the Earth completes an apparent solar day by turning through 360° plus angle *b*. Hence, the solar day is longer at perihelion than at aphelion because $b > a$. (B) A rectangular map of the sky shows the ecliptic inclined 23.5° to the celestial equator (CE). The Sun's eastward motion is shown at two points (1 and 2). Only the components of this motion parallel to the CE cause the apparent solar time to vary.

Another factor is that the Sun moves on the ecliptic but apparent solar time is measured along the celestial equator. So only that component of the Sun's eastward motion *parallel* to the celestial equator affects apparent solar time. To avoid the inconveniences of this variable solar time, we define **mean solar time** as the hour angle of a fictitious point (the **mean Sun**) that moves eastward along the **celestial equator** at the average angular rate of the true Sun. The **mean solar day** begins at midnight (with respect to this point), and its length is 1/365.2564 a sidereal year. The difference between apparent solar time and mean solar time is called the **equation of time;** the mean Sun may lead or lag the true Sun by as much as 16 min.

Now to compare sidereal time with mean solar time (Figure 3–3). The Earth's rotation returns the vernal equinox to upper transit as the Earth moves the distance A in its orbit—one sidereal day has passed. Since the Earth has moved about $360/365 \approx 1°$ around its orbit, it must rotate through this angle before the Sun returns to the local meridian and a mean solar day has passed (B). However, 1° corresponds to 4^m of sidereal time, and so the mean solar day is about 4 min *longer* than the sidereal day. The length of the sidereal day is $23^h 56^m 4.09^s$ in mean solar time units, and so stars appear to rise about 4 min *earlier* each night in terms of the mean solar time. A star that is in upper transit at midnight tonight will reach the meridian at 10 P.M. one month from now ($30 \times 4^m = 2^h$).

Mean solar time differs at every longitude on the Earth's surface because the hour angle of the fictitious Sun depends on the observer's location. The practical difficulties of such a timekeeping system have been avoided by the establishment of 24 **time zones** around the world. Within each longitudinal zone, which are approximately 15° (or 1^h) wide, all locations have the same **standard time;** the boundaries of each zone are adjusted for maximum convenience (for example, a city is usually placed wholly within one time zone). The reference zone is centered on Greenwich, England, at 0° longitude. Standard time at Greenwich is referred to as **Greenwich mean time** (GMT) or, equivalently, **universal time** (UT); GMT and UTC are close but not the same. New York City lies 5 h west of Greenwich in the eastern standard time (EST) zone, and so we subtract 5^h from universal time to find the local New York time for such events. To take advantage of the extra hours of daylight during the summer [Section 3–1(b)], an hour is added to local standard time from mid-spring to mid-fall in many parts of the world. Hence, 11 P.M. **Pacific daylight savings time** (PDT) in San Francisco corresponds to 10 P.M. Pacific standard time (PST), and we add 8^h to Pacific standard time to find universal time since San Francisco is eight time zones west of Greenwich (universal time equals Pacific daylight savings time plus 7^h).

The Earth's rotation rate is subject to small, unpredictable variations (Section 3–4), especially when compared to atomic clocks. To predict precisely the positions of bodies in the Solar System, we require a steady time standard, and so UT is replaced by **ephemeris time** (ET) in celestial mechanics. At the beginning of A.D. 1900, an **ephemeris second** was defined as 1/31,556,925.97474 the length of the tropical year 1900; universal and ephemeris time were in agreement. Today these times differ by about 40^s.

The **month** is of ancient origin and derives from the Moon's synodic orbital period of 29.53 days. To fit within the seasonal year, months in western calendars have been given conventional lengths of 28, 30, and 31 days. The week of seven days (each named after a planet in the astrological tradition) may be based upon the quarter phases of the Moon ($29.53/4 = 7.38 \approx 7^d$), but the exact origin is unclear. In most Native American traditions, the

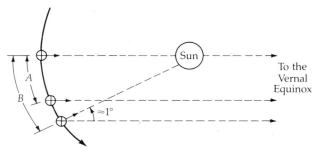

FIGURE 3–3   Sidereal and solar days. As the Earth rotates once with respect to the vernal equinox, it moves through distance A in its orbit. To complete a mean solar day, it rotates about 1° more to bring the Sun back to the meridian. The Earth has then moved through distance B.

days of the month are counted from the first to the last visible crescent, and so a month averages about 28 days (the days of invisibility are not part of the count).

The **year** is the time it takes the Earth to orbit the Sun, but different definitions give three types of years. With respect to the stars, the Earth's revolution takes one **sidereal year** of 365.2564 mean solar days ($365^d\ 6^h\ 9^m\ 10^s$), whereas the **tropical year** (the year of the seasons) of 365.2422 mean solar days ($365^d\ 5^h\ 48^m\ 46^s$) is the period with respect to the vernal equinox that precesses about 50″ westward along the ecliptic each year. Finally, because planetary perturbations cause the Earth's perihelion to precess in the direction of orbital motion compared to the direction of the vernal equinox, we call the time between successive perihelion passages the **anomalistic year** of 365.2596 mean solar days ($365^d\ 6^h\ 13^m\ 53^s$).

The **Gregorian calendar,** which attempts to approximate the year of seasons (the tropical year), contains 365 days per common year and 366 days in years divisible by four (leap years). To achieve an accuracy of one day in 20,000 years, only those century years divisible by 400 are leap years (A.D. 2000); century years divisible by 4000 remain common (A.D. 8000).

(B) THE SEASONS

The Earth's seasons—spring, summer, autumn, and winter—arise because the Earth's equatorial plane is inclined 23.5° to the ecliptic plane (Figure 3–4). The eccentricity of the Earth's orbit is too small ($e = 0.017$) to affect the seasons greatly; note that perihelion now occurs during the northern winter (January 2). The number of daylight hours and the noon altitude of the Sun lead to the characteristic temperatures of the seasons; both depend upon the latitude of the observer and the Sun's position on the ecliptic. Let's consider these two causes independently.

When the Sun is at the **vernal equinox** (about March 21) or the **autumnal equinox** (about September 23) along the ecliptic, its declination is 0° and there are 12 h of day and 12 h of night at all points on the Earth's surface. The noon altitude of the Sun is 90° (the zenith) at the equator and diminishes to 0° at the poles. At the **summer solstice** (about June 22), the Sun attains its greatest declination, +23.5°, and passes directly overhead at noon for all observers at latitude 23.5°N—the **Tropic of Cancer.** (About 3000 years ago, the Sun was in the constellation Cancer at the summer solstice.) On this date, the days are longest in the Northern Hemisphere, and summer starts there; at the same time, winter begins in the Southern Hemisphere and the days are shortest there.

The situation is reversed at the **winter solstice** (about December 22), when the Sun's declination is −23.5°. The Sun then passes directly overhead for all observers at latitude 23.5°S—the **Tropic of Capricorn** (the winter solstice 3000 years ago was in the constellation Capricornus); it is winter in the

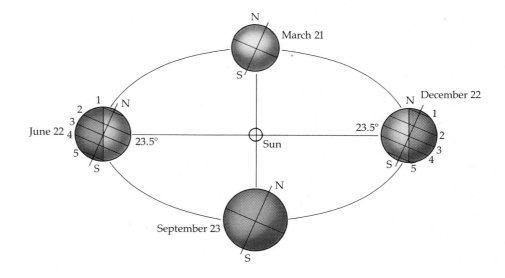

FIGURE 3–4 The Earth's equatorial inclination and the seasons. The Earth's equator inclines 23.5° to the ecliptic, the plane of its orbit. Day and night are shown for the solstices and the equinoxes. The latitudes marked are (1) Arctic Circle (66.5° N), (2) Tropic of Cancer (23.5° N), (3) Equator (0°), (4) Tropic of Capricorn (23.5° S), and (5) Antarctic Circle (63.5° S).

Northern Hemisphere and summer in the Southern.

Latitude 66.5°N is called the **Arctic Circle.** As spring becomes summer there, the days lengthen until the summer solstice, at which time the Sun doesn't set for 24 h; from fall to winter, the days shorten until the 24-h night at the winter solstice. North of the Arctic Circle, the Sun does not set for many days (the **midnight Sun**) in the summer and does not rise for many days in the winter. At the north pole, a six-month "day" begins at the vernal equinox and a six-month "night" begins at the autumnal equinox. South of the **Antarctic Circle** (latitude 66.5°S), exactly the same events occur six months after they have taken place in the north.

The overall amount of solar energy received by the Earth varies annually with distance from the sun. From day to day, the local heating effectiveness of this energy—the **solar insolation**—depends upon both latitude and time. The altitude of the Sun determines over what area a given amount of radiation is spread (Figure 3–5). Suppose that a unit of energy falls upon the area A when the Sun is at the zenith. When the Sun's altitude is θ, this same amount of energy is spread over the area $A/\sin\theta$, so the heating efficiency decreases as θ decreases. Because of this projection effect, the summer is warmer than the winter because the Sun is higher in the sky for more hours in the summer. The Earth's equatorial regions are always warm because the noon Sun never passes far from the zenith. The poles have **polar icecaps** because the Sun spends many months below the horizon and stays at a low altitude when it does rise.

The Earth's surface (especially the oceans because of the high heat capacity of water) and atmosphere are good thermal reservoirs and respond slowly to solar heating. As a result, temperature variations are moderated by daily and seasonal time lags between the extremes of solar insolation and the extremes of temperature. For example, the early afternoon is usually the warmest part of the day, even though solar insolation peaks at noon. February is the coldest month of northern winter, but December is the month of least insolation.

3–2 ◐
EVIDENCE OF THE EARTH'S ROTATION

How to prove the Earth rotates? The westward circling of the celestial sphere could be a reflection of the daily eastward turning of the Earth, but this is no proof because we could equally justify the concept of a rotating celestial sphere centered upon a stationary Earth. We can prove the Earth's rotation only by basing our arguments upon the well-verified dynamic laws of Newton.

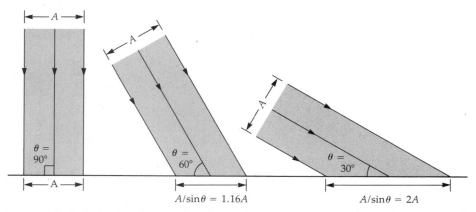

FIGURE 3–5 Solar insolation. A unit of solar energy strikes the Earth over area A when the Sun is at its zenith. For other altitudes θ, the sunlight is spread over a larger area, $A/\sin\theta$.

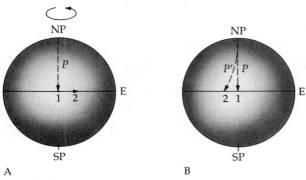

FIGURE 3–6 Projectile trajectories on the Earth. (A) On a rotating Earth, a rocket launched from the north pole (NP) is aimed at a target located at point 1 along path *P*. During the flight time, the Earth (and the target) rotates from 1 to 2. (B) The view from the Earth's surface shows the rocket's trajectory if the Earth did not rotate (*P*) and the actual one (*P'*). Note that the path curves to the right as seen by an observer at the north pole.

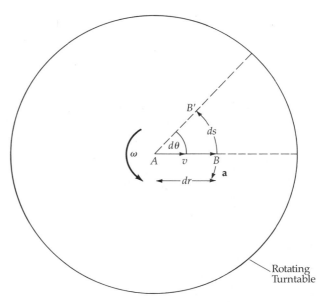

FIGURE 3–7 The Coriolis effect. A rotating turntable moves through angle $d\theta$ while a body on it moves from *A* to *B* (distance dr) in time dt. At the same time, *B* rotates to *B'* through distance ds. The body appears to be deflected to the right at acceleration **a** relative to the turntable.

(A) THE CORIOLIS EFFECT

The apparent trajectories of rockets and Earth satellites can only be understood if the Earth rotates. Consider a projectile launched from the north pole to land at the equator (Figure 3–6). On a nonrotating Earth, the projectile would clearly follow a single meridian of longitude during its entire flight. On a rotating Earth, however, the target on the equator moves eastward at 0.46 km/s and the projectile lands west of the target. Though the projectile's motion is due south, it appears deflected to the right with respect to the Earth's surface. The fictitious acceleration that produces this effect— the **Coriolis effect**—was deduced by Gaspard Gustave de Coriolis (1792–1843) in 1835.

Moving bodies always appear to be deflected to the right in the Northern Hemisphere and to the left in the Southern Hemisphere as seen from the point of origin of the motion. If the projectile's velocity is **v** and the Earth's vector angular velocity is ω (its direction is toward the north celestial pole, and its magnitude measures the Earth's spin in units of radians/s), then these observations are summarized in terms of the **Coriolis acceleration:**

$$\mathbf{a}_{\text{Coriolis}} = 2(\mathbf{v} \times \omega) \qquad (3\text{–}1)$$

The cross product yields the product of the perpendicular components of **v** and ω, and the direction of $\mathbf{a}_{\text{Coriolis}}$ is the direction in which your thumb points when you align the fingers of your right hand along **v** and rotate them through the smallest angle to ω (the **right-hand rule**).

Let's derive Equation 3–1. A body moves with a constant radial velocity **v** above a turntable rotating with angular speed ω (Figure 3–7). At time *t*, the body leaves the origin (*A*) and moves a distance dr to point *B* in an infinitesimal time dt. Meanwhile, point *B* has revolved through the angle

$$d\theta = \omega \, dt$$

to *B'*. From the geometry, we have

$$dr = v \, dt \qquad \text{and} \qquad ds = dr \, d\theta$$

Therefore

$$ds = (v \, dt)(\omega \, dt) = v\omega(dt)^2$$

But Newton's second law implies that a body moves the distance

$$ds = \mathbf{a}(dt)^2/2$$

in time dt when it experiences the *constant acceleration* $\mathbf{a}$, so that

$$\mathbf{a}_{\text{Coriolis}} = 2\,\mathbf{v} \times \boldsymbol{\omega}$$

In addition, the direction of the apparent deflection is to the right, as with Equation 3–1.

The Coriolis effect controls the characteristics of large-scale wind patterns in the Earth's atmosphere (as well as ocean currents). A **cyclone** is a local counterclockwise circulation of air in the Northern Hemisphere (clockwise in the Southern) produced by the rightward Coriolis deflection of air flowing toward the center of a low-pressure region. An **anticyclone** arises when air flowing away from the center of a high-pressure region is deflected into a local clockwise circulation in the Northern Hemisphere (counterclockwise in the Southern). Solar heating produces large-scale vertical cells of wind motion called **Hadley cells** (Figure 3–8); note the triple structure in each hemisphere. At the Earth's surface, the Coriolis effect

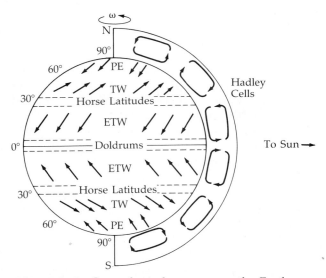

FIGURE 3–8 General wind patterns on the Earth. Solar heating produces the vertical cells of convecting air, and the Coriolis effect deflects this motion to the right (northern hemisphere) or left (southern hemisphere) to establish the easterly trade winds (ETW), the temperate westerlies (TW), and the polar easterlies (PE). Note that the pattern has three Hadley cells in each hemisphere.

causes these winds to flow in the known directions of the easterly trade winds (5 to 30°N and S), the temperate westerlies (35 to 50°N and S), and the polar easterlies (60 to 90°N and S). At low latitudes, the bands of relatively calm air are known as the doldrums (0 to 5°N and S) and the horse latitudes (30 to 35°N and S).

Note that the Coriolis effect combined with convection creates the overall pattern of atmospheric flow. **Convection** is one way to transport thermal energy from one location to another. (The other two processes are radiation and conduction.) Convection takes place in the Earth's atmosphere when the Sun's radiation heats the ground and the air in contact with it. The air expands and its density decreases. Cooler, denser air descends to displace the hotter air, which cools as it rises. The falling air heats upon contact with the ground, and so eventually is displaced upward. This circuit of rising and falling air transfers heat from the ground into the atmosphere and also establishes the up-down flow.

(B) FOUCAULT'S PENDULUM

In 1851, Bernard Léon Foucault (1819–1868) hung a pendulum from the ceiling of the Pantheon in Paris and proved the Earth's rotation by noting that the pendulum's plane of oscillation rotated during the day. If the Earth did not rotate, this rotation of the oscillation plane would not occur because all forces acting on the ball (the Earth's gravity and the tension in the wire) would lie in the plane of oscillation.

Step off the Earth and observe the pendulum swinging as the Earth turns. At the north pole, the pendulum oscillates in a fixed plane while the Earth rotates below it every 24 sidereal hours; the pendulum appears to rotate **westward** with the period $P = 24^{\text{h}}$. A pendulum swinging in the equatorial plane at the Earth's equator feels no forces perpendicular to the plane of oscillation and so doesn't rotate at all ($P = \infty$). At an intermediate latitude ϕ (Figure 3–9), the vertical component ω of the Earth's angular speed is $\omega \sin \phi$. The angular speed is inversely proportional to the period of rotation ($\omega = 2\pi/P$), however, so that the pendulum appears to rotate westward with the period $P = 24^{\text{h}}/\sin \phi$. Precise measurements of Foucault pen-

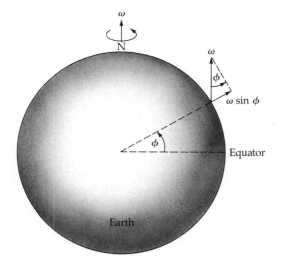

FIGURE 3–9 Foucault's pendulum. At latitude ϕ, the vertical component of the Earth's angular speed is $\omega \sin \phi$, and so the Earth rotates below the pendulum in a period proportional to $1/\sin \phi$.

dula verify this relationship in detail and permit a purely dynamical determination of the Earth's period of rotation. (They also supply a means to measure the Earth's mass.)

(C) THE OBLATE EARTH

The shape of the Earth's surface is that of an **oblate spheroid**; the polar radius ($r_p = 6356.8$ km) is 21.4 km less than the equatorial radius ($r_e = 6378.2$ km), so that the Earth's **oblateness** is $\epsilon = (r_e - r_p)/r_e = 21.4/5378.2 = 1/298.3$. This oblateness indicates that the Earth rotates. If the Earth were a fluid body, its shape would prove its rotation since a fluid must adjust its shape to all external forces—a nonrotating fluid body is spherical.

Today we know that the Earth's materials have an average strength close to that of steel, but they are plastic and so maintain the equilibrium shape of a rotating fluid body. A mass m at latitude ϕ on a spherical, rotating Earth (Figure 3–10) experiences two accelerations [Section 1–4(a)]: (1) the gravitation acceleration $GM_\oplus/r^2$ directed toward the Earth's center and (2) the centripetal acceleration $\omega^2 r \cos \phi$ in the circular orbit of radius $r \cos \phi$. If a

mass is free to move (a fluid mass), the centripetal acceleration affects it. The vertical component of this centripetal acceleration affects it. The vertical component of this centripetal acceleration, $a = \omega^2 r \cos^2 \phi$, reduces the weight of a mass while the horizontal component, $b = \omega^2 r \sin \phi \cos \phi$, causes a mass to migrate to the Earth's equator. Considering many such fluid masses (in fact, the whole Earth), we see that a bulge grows at the Earth's equator until fluid masses can no longer climb this equatorial "hill"—establishing the equilibrium oblate shape.

3–3 ◗
EVIDENCE OF THE EARTH'S REVOLUTION ABOUT THE SUN

As with the Earth's rotation, we must be careful to prove the Earth's revolution about the Sun. The following three proofs were unavailable at the time of Copernicus and Kepler, and so the ideas of these two scientists were considered suspect and were slow to be accepted. Today these proofs are

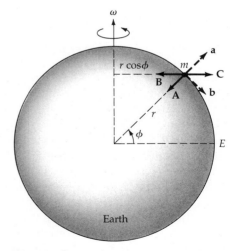

FIGURE 3–10 The Earth's oblateness. A mass m at latitude ϕ has a gravitational acceleration **A** and centripetal acceleration **B** (from its rotation around the Earth's axis). The vector sum of these forces results in a vertical component, **a**, that decreases the mass's weight and a horizontal component, **b**, that causes the mass to slide toward the equator (E).

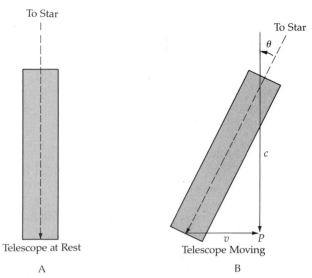

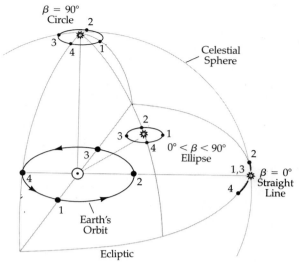

FIGURE 3–11 The aberration of starlight. (A) A telescope is at rest and so pointed upward to observe the light from a star. (B) When a telescope moves at speed v (as a result of the Earth's revolution), it is tilted through an angle θ, so that the starlight reaches P at the same time as the bottom of the telescope does.

FIGURE 3–12 Apparent stellar orbits resulting from the aberration of starlight. The Earth is shown at four positions in orbit around the Sun (1 to 4). The apparent paths traced out by stars because of the change in the direction of aberration during a year are shown for stars directly above the orbital plane (ecliptic pole), in the plane (the ecliptic), and at an intermediate position on the celestial sphere. The sizes of the apparent orbits are independent of the distances to the stars.

indisputable evidence for the heliocentric model of the Solar System.

(A) THE ABERRATION OF STARLIGHT

In 1729, the English astronomer James Bradley (1693–1762) discovered the **aberration of starlight,** and using the finite speed of light (now known to be $c \approx 3 \times 10^5$ km/s; Bradley knew it was finite, but he did not have this value), he explained this phenomenon as caused by the orbital motion of the Earth.

Suppose you walk in a vertically falling rain with an umbrella over your head. The faster you walk, the farther you must lower the umbrella in front of you to prevent the rain from striking your face. When starlight enters a telescope, an analogous phenomenon occurs (Figure 3–11). If the Earth were at rest, we would point our telescope toward the zenith to see a star situated there. If the Earth is in motion at speed v, however, we must tilt the telescope in the direction of motion by an angle (where θ is small and in radians)

$$\theta \approx \tan \theta = v/c \qquad \textbf{(3–2)}$$

so that the bottom of the telescope can meet a light ray which has entered the top of the telescope. Bradley observed this very small angle of tilt, $\theta = 20.49''$. Using Equation 3–2, we then deduce the Earth's orbital speed as $v = \theta \times c = (9.934 \times 10^{-5}$ rad$)(3.0 \times 10^5$ km/s$) = 29.80$ km/s.

The direction in which we tilt our telescope constantly changes as the Earth moves around the Sun. Because the Earth's orbit is essentially circular, stars appear to trace out annual **aberration orbits** on the celestial sphere (Figure 3–12). A star at the ecliptic pole is seen to move around a circle of angular radius $20.49''$ once a year. Stars on the ecliptic oscillate to and fro along lines of angular half-length $20.49''$. At an intermediate celestial latitude β (angle from the ecliptic), the aberration orbit is an ellipse with semimajor axis $20.49''$ and semiminor axis $(20.49'') \sin \beta$. The Earth revolving around the Sun explains this observed behavior of the aberration orbits.

(B) STELLAR PARALLAX

As you drive along a highway in your car, notice that nearby objects seem to be moving backward with respect to more distant objects (Figure 3–13A). This perspective effect of our line of sight is termed **parallax.** According to the heliocentric model of the Solar System, nearby stars should exhibit parallax effects on the celestial sphere since the Earth is in motion about the Sun (Figure 3–13B). If the observed parallax angle of a star is π'' (in seconds of arc), then, from Figure 3–13B, that star's distance is

$$d = (206,265/\pi'')\ \text{AU} \qquad (3\text{–}3)$$

Note that 206,265 is the number of arcseconds in 1 radian. (See Chapter 11 for a complete discussion of stellar parallaxes and distances.)

The heliocentric model of Copernicus remained on shaky ground until the first stellar parallax was

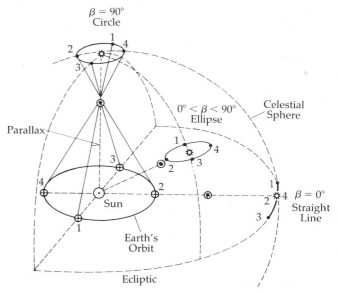

FIGURE 3–14 Parallactic orbits. The Earth's orbital change results in a periodic motion of stellar positions. Note that these are 90° out of phase with respect to the aberration orbits (Figure 3–12) and their angular size depends on the distance to the star. Their shapes depend on their celestial latitude, β.

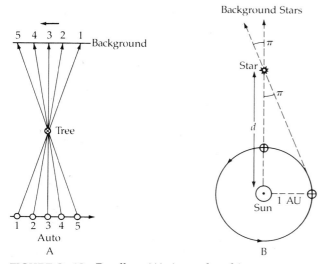

FIGURE 3–13 Parallax. (A) A nearby object appears to move with respect to the background as our baseline position changes from 1 to 5. (B) A stellar parallax occurs (angle π) as a result of the motion of the Earth around the Sun. The annual change in baseline position, from one side of the Earth's orbit to the other, results in an apparent displacement of the star relative to background stars.

observed in the nineteenth century. In 1838, Friedrich Wilhelm Bessel (1784–1846) published the first observed stellar parallax: 0.294" for the star 61 Cygni; at about the same time, F.G.W. Struve found the parallax of Vega (Alpha Lyrae) and T. Henderson found that of Alpha Centauri. Today we know that the nearest star is Proxima Centauri with a parallax of 0.764" and a distance of 270,000 AU (about 4×10^{13} km or 4 lightyears). All stellar parallaxes are less than 1.0."

As the Earth moves around the Sun in its orbit, each star traces out a yearly **parallactic orbit** on the celestial sphere (Figure 3–14). Stars at the ecliptic poles move in circles with radii dependent upon their distances from the Sun, and stars on the ecliptic oscillate along lines. For the general parallactic ellipse, the ratio of semiminor to semimajor axis is sin β, where β is the celestial latitude of the star, just as for aberration orbits. Note, however, that for a given direction β, the aberration orbit is always the same but the parallactic orbit depends upon the star's distance. Also, the aberration orbit of a star is 90° out of phase with the parallactic orbit.

(C) THE DOPPLER EFFECT

Our final proof of the Earth's revolution uses the Doppler effect. Chapter 8 derives the fact that the wavelength of electromagnetic radiation (light) is shifted in proportion to the relative line-of-sight speed of the object observed. If a star emits radiation at the wavelength λ_0 and we observe this radiation at the wavelength λ, then the Doppler formula for velocities much less than that of light is

$$\Delta\lambda/\lambda_0 = (\lambda - \lambda_0)/\lambda_0 = v_r/c \qquad (3-4)$$

where v_r is the relative line-of-sight speed (positive for recession; negative for approach) between the observer and the observed and c is the speed of light.

For stars at the ecliptic pole, no Doppler shift occurs because we have no radial component of the velocity. At an intermediate celestial latitude, β, the shift is sinusoidal with a period of one year and the amplitude $\Delta\lambda$ varies as $\cos \beta$. The maximum amplitude of this Doppler shift occurs for stars on the ecliptic, where the full magnitude of the Earth's orbital velocity comes into play; a standard method of determining the Earth's speed of revolution is to measure this maximum shift and to use Equation 3–4 to deduce $v_\oplus = 29.80$ km/s. The spatial motion of a star with respect to the Sun superimposes a constant Doppler shift on the time-varying shift caused by the Earth's revolution; the two effects are easily untangled. The Earth's rotation causes a diurnal Doppler shift with an amplitude proportional to $\cos \phi \cos \delta$, where ϕ is the terrestrial latitude of the observer and δ is the declination of the observed source.

3–4 ◗
DIFFERENTIAL GRAVITATIONAL FORCES

Two spherical bodies behave gravitationally as point masses. If the bodies are elastic or nonspherical, or if several more bodies are on the scene, **differential gravitational forces** may become important. This effect arises because gravitation depends upon the distance between bodies, and different parts of an extended body (or system) will therefore experience different gravitational accelerations.

(A) TIDES

If you have spent time near a large body of water, you know that the level of the water rises and falls twice daily in the **tides** and that a given tide occurs about an hour later each day. (Local conditions determine whether or not two tides occur every day at specific places.) The Moon is the principal cause of the tides because it returns to upper transit 53 min later each day—at about the same time relative to high tide. The solid body of the Earth, which has much greater cohesion than water, also responds quickly to the Moon's tidal forces. The Earth has a rigidity about that of steel, so that it would respond to the Moon's tidal forces by raising body tides several centimeters in height.

Let's use Newton's laws to explain the tides in a general way. Assume that the Earth is essentially spherical. Cover this sphere with a uniform depth of water and ask what effect the Moon's gravitational force has on this water (Figure 3–15A). By subtracting the vector acceleration at the Earth's center (C) from each of the surface vector accelerations, we obtain the **differential tidal accelerations** (Figure 3–15B). These tidal forces raise water tides about 1 m high at points A and B on the line of centers, and the Earth rotates below this configuration once each day. Actual tides arise from forced oscillations in the Earth's ocean basins, so that the height and timing of tides may differ markedly from the theoretical case. In some bays and estuaries, the tidal waters may accumulate to heights greater than 10 m.

Now to be more quantitative in our derivation of tidal accelerations. As a first approximation, we will neglect the small centripetal acceleration at the Earth's surface due to the orbital motion of the Earth and Moon about their barycenter every sidereal month (27.32^d). The centers of Earth and the Moon are separated by the distance d in Figure 3–16, and a small particle on the Earth's surface is at an angle ϕ to the line of centers. The gravitational acceleration of the Earth's center due to the Moon has the magnitude $A = GM_m/d^2$, and the particle's acceleration is $B = GM_m/r^2$. Subtract A

FIGURE 3–15 Tidal forces on the Earth. (A) The gravitational attraction of the Moon on the Earth is shown by selected force vectors. (B) The vector acceleration of the Earth's center (C) is subtracted from the surface accelerations shown in A. The resulting vectors show the tidal differential forces.

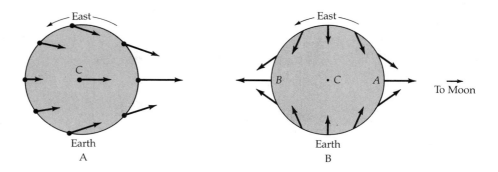

from *B* vectorially. The component of **B** perpendicular to the line of centers is unaffected by this subtraction and has the magnitude

$$a = B \sin \theta = GM_m R_\oplus \sin \phi / r^3 \quad \text{(3–5)}$$

The component of **B** parallel to the line of centers is

$$b = B \cos \theta = GM_m(d - R_\oplus \cos \phi)/r^3$$

and the remainder after subtracting **A** is the differential of the force:

$$b' = (GM_m/r^3)[d - R_\oplus \cos \phi - (r^3/d^2)] \quad \text{(3–6)}$$

Using the law of cosines and the fact that $R_\oplus/d \ll 1$, we have

$$r^3 = d^3[1 - 2(R_\oplus/d) \cos \phi + R_\oplus^2/d^2]^{3/2} \approx d^3$$

Therefore, Equation 3–5 finally becomes (to lowest order)

$$a \approx GM_m R_\oplus \sin \phi / d^3 \quad \text{(3–7a)}$$

Equation 3–6 may be written as

$$b' \approx (GM_m/d^2)[1 - (R_\oplus/d) \cos \phi - (r/d)^3]$$

and, expanding our result for r^3 using the **binomial theorem** (Appendix 9), we obtain

$$b' \approx 2GM_m R_\oplus \cos \phi / d^3 \quad \text{(3–7b)}$$

Equations 3–7 verify the qualitative picture (Figure 3–15B) and indicate that tidal gravitational forces go as MR/d^3, where M is the mass of the source of tidal force, R is the size of the body being tidally influenced, and d is the separation of the two bodies.

We can show this relationship for tidal forces generally by differentiating Newton's law of gravitation with respect to R:

$$dF/dR = -2GM_m/R^3$$

and moving the dR to the right-hand side of the equation so that

$$dF = -(2GM_m/R^3)dR \quad \text{(3–8)}$$

where dF is the differential gravitational force directed along R. Then dR is the diameter of a single solid body or the separation between two bodies close together that are being acted upon by the tidal forces.

The Sun also produces tidal effects on the Earth. Since the differential accelerations go as MR/d^3 and since $R = R_\oplus$ in both cases, the tide-raising force of

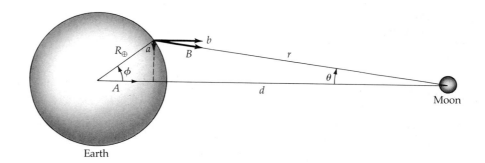

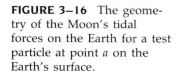

FIGURE 3–16 The geometry of the Moon's tidal forces on the Earth for a test particle at point a on the Earth's surface.

the Sun relative to that of the Moon is

$$(M_\odot/M_m)(r_m/r_\odot)^3 =$$
$$(1.99 \times 10^{30} \text{ kg})/(7.36 \times 10^{22} \text{ kg})$$
$$\times (3.84 \times 10^5 \text{ km})^3/(1.50 \times 10^8 \text{ km})^3$$
$$\approx 5/11$$

The tidal effects of the Sun and Moon combine vectorially, so that the resultant tides depend upon the elongation of the Moon. When the Moon is at conjunction or opposition, the two forces add to produce the very high **spring tides;** when the Moon is at quadratures, the two forces partially cancel to give the unusually low **neap tides.**

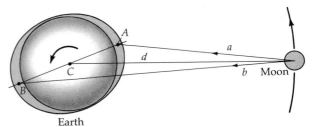

FIGURE 3–17 Tidal evolution. The Earth's tidal bulges (*A* and *B*) are driven by frictional coupling to the Earth's rotation to be ahead of the Moon's orbital position. The friction slows the Earth's rotation, and the bulges accelerate the Moon in its orbit.

(B) CONSEQUENCES OF TIDAL FRICTION

When the Earth and the oceans yield to the tide-raising forces, energy is dissipated (in the form of heat) as a result of friction; most of this energy is lost in shallow seas and at shorelines, where ocean tides abut against the continents. This **tidal friction** reduces the energy of the Earth's rotation, so that the length of the day increases at a current rate of about 0.002^s per century. (This rate varies with time.) Tidal friction causes two fascinating phenomena in the Earth–Moon system: the Moon's **synchronous rotation** and **tidal evolution.**

The Earth raises body tides on the Moon that are about $M_\oplus R_m/M_m R_\oplus \approx 20$ times higher than the Earth's body tides. The enormous energy dissipation that results slowed the Moon's rotation until the Moon was forced into **synchronous rotation,** where its sidereal rotation period is exactly the same as its sidereal period of revolution about the Earth.

The average external torques exerted on the Earth–Moon system are negligible, so that the total angular momentum (Section 1.5) of the system must remain constant. The angular momentum of the Earth decreases as tidal friction slows its rotation; hence, the Moon increases its angular momentum by moving away from the Earth. Kepler's third law then implies that the month must be lengthening. In the distant future, the "day" and "month" will become the same and equal to about 50 present days. The tide-raising forces accelerate the water, but the oceans do not respond instantly to the Moon's tidal forces (Figure 3–17). As the

Earth rotates beneath the Moon, the tidal bulges rise slightly eastward to the Earth–Moon center line. Bulge *A* is nearer the Moon than bulge *B*, and so *A* exerts a slightly greater force on the Moon. The resulting noncentral force accelerates the Moon, causing it to spiral away from the Earth. At the same time, the Moon's force on the bulges decelerates the Earth's rotation (Newton's third law). The Earth must have been spinning faster in the past and the month must have been shorter than today. Indeed, paleontological studies of fossilized corals which lived about 10^8 years ago show that there were 400 "days" in each year and that ocean tides were more vigorous than today—the Earth was rotating faster and the Moon was closer in the past.

Let's show more explicitly that the moon must increase its orbital radius to gain orbital angular momentum. Kepler's laws require that the larger an orbit, the longer its period. So, if a body moves to a larger orbit, does its angular momentum (= mass × orbital radius × orbital speed) increase or decrease? It increases, proportional to the square root of the distance.

Assume a circular orbit of radius R. Then

$$L \propto RV$$

where

$$V = 2\pi R/P$$

so

$$V^2 = 4\pi^2 R^2/P^2$$

But Kepler's third law requires that

$$P^2 \propto R^3$$

$$1/V^2 \propto R$$

or

$$V \propto 1/r^{1/2}$$

Go back to

$$L \propto RV$$
$$\propto R/R^{1/2}$$
$$\propto R^{1/2}$$

as stated.

(C) PRECESSION AND NUTATION

You are probably familiar with the behavior of a spinning top or a gyroscope. When the top's spin axis is not aligned with the vertical, we expect the top to fall on its side, but instead the spin axis maintains the same angle to the vertical and simply rotates slowly about the vertical. This **precession** of the top is explained by Newton's laws of motion.

The Earth's gravitational attraction **F** on the top produces the *horizontal* torque $\mathbf{N} = \mathbf{r} \times \mathbf{F}$ [Section 1–5(a)]. Since torque is just the time rate of change of the top's vector angular momentum **L** ($\mathbf{N} = d\mathbf{L}/dt$), there is no vertical component to topple the top and so it can only rotate, or **precess,** about the vertical. Differential gravitational forces

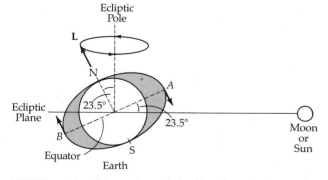

FIGURE 3–18 Precession of the Earth's axis. The differential tidal forces on the Earth's equatorial bulge (*A* and *B*) result in a torque (pointing into the page) that causes the Earth's angular momentum vector (**L**) to precess westward around the ecliptic pole.

acting upon the oblate Earth's (rotational) equatorial bulge produce torques that lead to a similar phenomenon—the precession of the Earth.

The Moon and Sun cause the Earth's lunisolar precession, with the Moon's effect dominating. The Moon's orbit is inclined about 5° to the ecliptic, but its average force is centered on the ecliptic. The Earth's equatorial plane is inclined 23.5° to the ecliptic (Figure 3–18), with the prominent equatorial bulges at *A* and *B*. Bulge *A* is more strongly attracted to the Moon than bulge *B*, and the differential forces are indicated. The torque that results points into the page, so that the Earth's angular momentum vector **L** precesses toward the west. The Sun induces a similar but slightly weaker effect, and the perturbing torques from the other plants produce a **planetary precession** that amounts to about 2% of the total. As a result of these torques, the celestial poles remain inclined 23.5° to the ecliptic pole but trace a circular path around the ecliptic pole once every 26,000 years. At present, the modestly bright star Polaris (Alpha Ursae Minoris) roughly marks the position of the north celestial pole, but in about A.D. 14,000, the pole star will be Vega (Alpha Lyrae).

As the celestial poles precess, the intersection of the celestial equator and the ecliptic (vernal and autumnal equinoxes) progresses westward at the rate of 360°/26,000 years ≈ 50″ per year along the ecliptic (about 50″ cos 23.5° ≈ 46″ per year along the celestial equator). This phenomenon of the **precession of the equinoxes** has several important consequences: it affects terrestrial timekeeping systems through the definition of the day and the year; it changes those stars that a given observer would consider circumpolar; and it significantly affects the celestial coordinate positions (Appendix 10) of all celestial objects. Because the precession effect is readily observable, the positions of objects in the celestial equatorial coordinate system (Appendix 10) must be constantly updated to the current epoch.

Because the Moon and Sun move above and below the Earth's equatorial plane, periodic variations occur in the torques acting on the Earth's equatorial bulge. These variations lead to a **nutation,** or wobbling, of the Earth's rotation axis. Small contributions to the nutation have monthly and yearly periods, but the principal contribution (discovered by James Bradley of aberration fame) of amplitude 9″ and period 18.6 years is from the regression of the nodes of the Moon's orbit (Chapter 4).

(D) THE ROCHE AND INSTABILITY LIMITS

We have discussed the relative orbits of two gravitating (rigid) spherical masses and obtained Kepler's laws. Here we consider the effects of differential gravitational forces upon nonrigid bodies and upon systems of more than two bodies. In general, a satellite cannot approach its primary planet too closely (the **Roche limit**) or stray too far (the **instability limit**) without dire dynamical consequences.

Consider a spherical satellite of mass m and radius r orbiting at a distance d from its very massive ($M \gg m$) primary planet of mass M and radius R (Figure 3–19A). If the satellite is large enough ($r \geq$ 500 km), its self-gravitation dominates all cohesive forces and determines its shape and strength. In 1850, Edouard Roche (1820–1883) demonstrated that such a satellite would be torn asunder by tidal forces if it approached the primary closer than

$$d = 2.44(\rho_M/\rho_m)^{1/3}R \qquad (3\text{–}9)$$

where ρ_M is the average density (SI units = kg/m³) of the primary and ρ_m is the average density of the satellite. This distance is the **Roche limit.** For example, if our Moon were to come closer than $d = 2.44(5.5/3.3)^{1/3} \approx 2.9R_\oplus \approx$ 18,500 km from the center of the Earth, it would tidally disrupt into small fragments.

Equation 3–9 was derived for a fluid satellite with the shape of a prolate (football-shaped)

spheriod in response to the primary's tidal forces. Now consider a *rigid* spherical satellite to derive roughly this result (Figure 3–19A). Since the satellite's orbital centripetal acceleration $\omega^2 d$ is produced by the gravitational attraction from the primary GM/d^2, the angular speed of the satellite about the massive primary is (Kepler's third law)

$$\omega = (GM/d^3)^{1/2}$$

The differential gravitational acceleration between the center of the satellite (point 1) and the outer edge (point 2) due to the primary is

$$A = (GM/d^2) - GM/(d + r)^2 \approx 2GMr/d^3$$

and the differential centripetal acceleration between these two points is

$$B = \omega^2(d + r) - \omega^2 d = \omega^2 r = GMr/d^3$$

The combination $A + B = 3GMr/d^3$ must be balanced by the satellite's self-gravitational acceleration, Gm/r^2, if the satellite is not to be pulled apart. That disruption occurs at

$$d = r(3M/m)^{1/3} \qquad (3\text{–}10)$$

Because the average density is defined as total mass divided by volume, we have $\rho_M = 3M/4\pi R^3$ and $\rho_m = 3m/4\pi r^3$. Substituting into Equation 3–10 yields

$$d = R(3\rho_M/\rho_m)^{1/3} \approx 1.44(\rho_M/\rho_m)^{1/3}R \quad (3\text{–}11)$$

The numerical difference between Equations 3–9 and 3–11 arises from our assumption of a rigid (nonfluid) satellite.

We can see this result more elegantly by considering the differential form of tidal forces

A

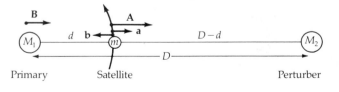

B

FIGURE 3–19 Roche and instability limits. (A) Geometry for the Roche limit, when the differential gravitational force **A** and differential centripetal acceleration **B** between points 1 and 2 exceed a body's self-gravitation. (B) A satellite m escapes from its primary planet of mass M_1 as a result of differential perturbations, $\mathbf{a} = \mathbf{A} - \mathbf{B}$, arising from the perturber of mass M_2, when d exceeds the instability limit.

(Equation 3–8) for two particles of equal size and mass that are just touching. Their attractive force is

$$F = Gmm/(dr)^2$$

and we equate this to the tidal forces to find Roche's limit d from mass M:

$$GMm/(dr)^2 = 2GMm\,dr/d^3$$

so that

$$d = (2M/m)^{1/3}\,dr$$

which becomes, for bodies with uniform densities,

$$d \cong 2.5(\rho_M/\rho_m)^{1/3}R \qquad \textbf{(3–12)}$$

This result demonstrates that various cases differ only in the proportionality constant, which occurs in part for the internal density distribution and the cohesive strength of the materials. For rocky or icy bodies greater than 40 km in diameter, the numerical coefficient is 1.38, and for a body falling right into a planet, the coefficient is 1.19. In any case, the functional relationship is

$$d \propto (\rho_M/\rho_m)^{1/3}R$$

The larger natural satellites in the Solar System orbit beyond the Roche limit of their primaries, though some of Saturn's tiny moonlets actually orbit within it. Saturn's beautiful system of rings lies about 80,000 to 136,000 km from the center of the planet and so is entirely within the Roche limit of about 150,000 km. The Jupiter, Neptune, and Uranus ring systems are also within the Roche limits for those planets. No rocky or icy satellite with a diameter in excess of about 40 km can exist or could have been formed within these limits. That these rings are composed of multitudes of small solid particles implies that their cohesive strength

is greater than the disrupting tidal forces. Artifical satellites certainly do orbit within the Earth's Roche limit, but they are held together by the tensile strength of their materials. A solid sphere of steel 1 m in diameter can approach within 100 m of a point mass of $1M_\oplus$ before the tidal forces overwhelm its internal tensile strength!

As a body orbits farther and farther from its primary, **differential perturbations** from other bodies become more important (Figure 3–19B). Beyond the **instability limit,** the body escapes from its primary. The perturber M_2 produces the differential acceleration

$$a = A - B = [GM_2/(D - d)^2] - GM_2/D^2$$
$$\approx 2GM_2d/D^3$$

between the orbiting body and its primary (when $d \ll D$). When this acceleration equals the gravitational acceleration from the primary $b = GM_1/d^2$, the orbiting body is at the instability limit:

$$d = (M_1/2M_2)^{1/3}D \qquad \textbf{(3–13)}$$

Note that Equation 3–13 is valid only when $M_1 \ll M_2$; when $M_1 \geq M_2$, we must use the exact relation

$$d^3(2D - d) = (M_1/M_2)D^2(D - d)^2 \quad \textbf{(3–14)}$$

to solve for d in terms of D. From Equation 3–13, the instability limit for our Moon—with the Sun as perturber—is 1.7×10^6 km; this is four times the Moon's present distance from the Earth, and so our Moon is now stable against escape. Comets may escape from the Solar system as a result of perturbations from other stars ($M_1 \approx M_2$) when the comets' aphelia lie at distances greater than about 10^5 AU (since Equation 3–14 implies $d \approx D/2$ and $D \approx 2 \times 10^5$ AU).

PROBLEMS ●

1. (a) How much does a sidereal clock gain (or lose) on a mean solar clock in five mean solar hours?
   (b) What is the approximate sidereal time when it is noon apparent solar time on the following days: (i) the first day of spring, (ii) the first day of summer, (iii) April 21, (iv) January 2?

2. In terms of azimuth and altitude (Appendix 10), describe the Sun's daily path across the sky during every season of the year at

   (a) the equator
   (b) latitude 35°N
   (c) the north pole
   Use such descriptive terms as noon altitude, sunrise azimuth, sunset azimuth, and angle at which Sun meets horizon.

3. Cape Canaveral is at longitude 80°23′W and latitude 28°30′N. A rocket is launched from there due south,

and it lands on the equator 10 min later. What is the longitude of impact?

4. In discussing the Coriolis effect, we mentioned that a body which undergoes a constant acceleration a will travel a distance $s = at^2/2$ in a time t. Show that the speed of the body is proportional to time and that the body's acceleration is indeed a.

5. The Earth's rotation also produces an aberration of starlight.
   (a) What is the maximum value of this daily aberration?
   (b) Where on the Earth is this effect maximum?
   (c) For what stars (location on celestial sphere) is the effect maximum?

6. (a) If the centripetal acceleration from the Earth's rotation is $\omega^2 R_\oplus$ at the equator and if $\omega = 2\pi/P$, where P is one day, then by what percentage is a person's weight reduced as he or she walks from the north pole to the equator? Ignore the Earth's oblateness.
   (b) Now ignore the Earth's rotation and find by what percentage a person's weight increases as he or she walks from equator to pole on the oblate Earth.
   (c) Compare your results from (a) and (b) and combine them to deduce how the *effective g* varies from the Earth's poles to the equator.

7. The Earth's orbital speed is approximately 30 km/s. A star emits a spectral line at wavelength $\lambda_e = 517.3$ nm (1 nm $= 10^{-9}$ m). Over what amplitude does this wavelength oscillate as the Earth orbits the Sun when the star is located at the ecliptic (celestial latitude $\beta = 0°$)?

8. In Section 3–4(a), we computed the Moon's differential tidal forces at the Earth's surface, ignoring the motion of the Earth–Moon system about its center of mass every month. Choose a coordinate system centered upon this center of mass and rotating eastward with the angular speed ω (due to this sidereal monthly motion) and include the centripetal acceleration at the Earth's surface to deduce the correct dependence of the total tidal acceleration at the Earth's surface. Ignore the daily rotation of the Earth. (*Hint:* The Earth–Moon center of mass is located within the Earth.)

9. On a large piece of graph paper, plot *to scale* the distances of the Roche limit and orbits of the innermost planetary satellites and rings for (i) the Earth, (ii) Mars, (iii) Jupiter, (iv) Saturn, and (v) Uranus.

Assume that $\rho_M = \rho_m$ in every case. Write a brief statement summarizing your results.

10. Compare the tidal forces the Moon exerts on the Earth (at perigee) with those the Sun exerts on the Earth (at perihelion) and those Venus exerts on the Earth (at closest approach).

11. Assume that the Earth and Moon are spherical and that the Moon orbits the Earth in a circle. Calculate the *spin* angular momenta of the Earth and Moon and compare these with the *orbital* angular momentum of the Moon. A spherical mass of uniform density has a spin angular momentum of $(2/5)MVR$, where V is the equatorial velocity and R the radius. The sum of these momenta must be a constant for the Earth–Moon system (if we ignore external torques). From the rate of loss of the Earth's spin angular momentum from tidal friction, estimate the rate at which the Moon moves radially away from the Earth.

12. The Moon will move away from the Earth until it no longer lags the tidal bulges, and the angular momentum transfer will stop. Computer calculations indicate that this will occur at an Earth–Moon distance of 6.45×10^5 km. Calculate the Moon's orbital period then.

13. Compare the solar insolation at noon in Albuquerque, New Mexico, (latitude about 35°N) for the day of the summer solstice and the day of the winter solstice.

14. An astronomer from Cullowhee, North Carolina (latitude 35° N, longitude 83° W) wants to visit a colleague in Hamilton, New York (latitude 43° N, longitude 75.5° W). He hops into a plane, but finds on taking off that the weather is overcast and it is impossible to navigate on the basis of landmarks. The compass is also broken, so he is able to navigate only by flying above the clouds in the direction of the North Star. He relies on the Coriolis effect to allow for east-west motion while aiming his plane due north.
   (a) Will the Coriolis effect operate in the correct direction?
   (b) If so, at what average speed must he fly in order to be at the longitude of Hamilton when he has reached the proper latitude?

15. (a) Because of the eccentricity of the Earth's orbit, the magnitude of the aberration of starlight due to the Earth's orbital motion is not constant. Determine the aberration at perihelion and aphelion.

(b) Determine the aberration of starlight as seen from Mars at perihelion and aphelion.

16. In the early 1980s the planets were all located on the same side of the Sun, with a maximum angular separation of roughly 90° as seen from the Sun. This rough "alignment" was sufficient to make possible the Voyager spacecraft grand tour. Some people claimed that this planetary alignment would produce destructive earthquakes, triggered by the cumulative tidal effects of all the planets acting together. Very few scientists took this prediction seriously! To understand why, compute the maximum tidal effects on the Earth produced by Jupiter (the most massive planet) and Venus (the closest planet). Compare these tidal effects to those caused by the Moon each month.

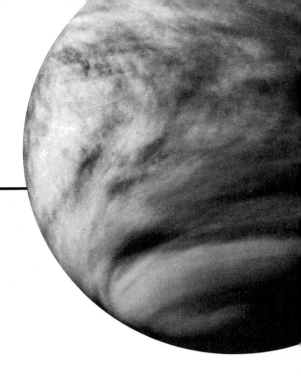

Chapter 4

The Earth–Moon System

Our detailed examination of the Solar System begins with our home planet, the Earth, and its satellite, the Moon. Centuries of probing the Earth have led to a good understanding of our planet, and the study of the Moon has been revolutionized by the space age since 1957, with manned and unmanned landings. This chapter outlines our knowledge of our terrestrial neighborhood to find the Earth as the most evolved terrestrial planet and the Moon as a fossil world.

4–1 ◐
DIMENSIONS

The size of the Earth was first determined by the Greek astronomer Eratosthenes (276–195 B.C.), who noted that, at the summer solstice, the noon altitude of the Sun in the city of Syene differed by 7°12′ from the altitude in Alexandria (Figure 4–1). Assuming that the Earth was spherical and noting that Alexandria was 5000 stadia (and if 1 stadium ≈0.16 km) due north of Syene, he found the Earth's circumference to be (360°/7.2°) × 5000 = 250,000 stadia (≈40,000 km). This value gives a

radius within 1% of the Earth's equatorial radius $R_\oplus = 6378.2$ km. But we don't really know the exact value of the stadium, and so the agreement with the modern value may be an accident. Eratosthenes certainly had the right idea in this method, for he got the correct ratio of the distance to the circumference. Modern measurements include techniques such as satellite geodesy and radar.

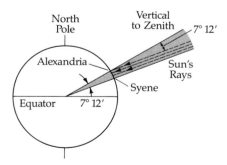

FIGURE 4–1 Geometry for Eratosthenes' method for determining the size of the Earth. Note that the Sun's rays come in parallel to each other.

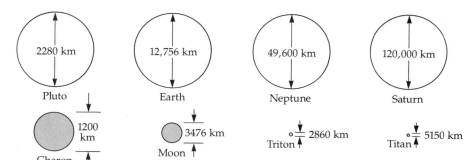

FIGURE 4–2 Satellite sizes. The largest moons, relative to the sizes of their parent planets, are shown to scale. Note the large size of Charon compared to Pluto.

The Earth is the major partner in the **Earth–Moon system.** The Moon is one of the largest and most massive satellites in the Solar System, relative to its primary planet (Figure 4–2). Pluto's satellite, Charon, is first in this regard with Charon's radius about 0.5 Pluto's and mass 1/7 Pluto's (Chapter 6). Our Moon's radius is 1738 km ($0.272R_{\oplus}$), and its mass is 7.35×10^{22} kg ($0.0123M_{\oplus}$); the closest runners-up are Neptune's Triton ($0.056R_N$, $0.0013M_N$) and Saturn's Titan ($0.043R_S$, $0.00024M_S$). (See Appendix 3 for planetary data.) The mean distance between the centers of the Earth and the Moon is 384,405 km, or about $60.3R_{\oplus}$ (Figure 4–3A). The **barycenter** (center of mass) of the system is located $M_m a_m/(M_{\oplus} + M_m) = (0.0123) \times (384,400)/(1.0123) = 4671$ km from the Earth's center; the Earth and Moon monthly orbit this point, which is buried 1707 km *below* the Earth's surface. (Figure 4–3B).

We can estimate the mass of the Earth from artificial satellite orbits if we notice that their mass is small relative to that of the Earth and use Kepler's third law: $M_{\oplus} + m_s \approx M_{\oplus} = 4\pi^2 a_s^3/GP_s^2$. The result is $M_{\oplus} = 5.98 \times 10^{24}$ kg. The Moon's mass is determined by observing the Earth's motion about the barycenter. By knowing the Earth's mass and the location of the barycenter, we compute $M_m = (d_{\oplus}/d_m)M_{\oplus} = 7.35 \times 10^{22}$ kg $= (1/81.3)M_{\oplus}$, where $d_{\oplus}$ is the distance from the barycenter to the Earth's center (=4671 km) and d_m is the distance from the barycenter to the Moon's center. Today the orbits of lunar spacecraft are used to measure M_m accurately as well as to deduce the internal mass distribution of the Moon. The oblateness of the Moon is 0.006; the Moon is slightly oblong, with its long axis (tidal bulge!) pointed toward the Earth.

The Earth–Moon distance is now determined by reflecting radar pulses from the lunar surface (accuracy of a few meters) or laser light pulses from mirrors set up by the Apollo astronauts (accuracy of about 1 cm). Using the distance to the Moon, we may find the size of the Moon by (1) timing occultations of stars and planets or of solar eclipses and (2) measuring the apparent angular size of the Moon in the sky. From the Moon's apparent diameter, which averages 31' of arc (about 1/2°), and the average Earth–Moon distance, we calculate its diameter to be 3476 km. Because of its eccentricity ($e = 0.055$), the Moon's orbital distance ranges from 363,263 km at perigee to 405,547 km at apogee. The angular diameter of the Moon varies from 32.9' to 29.5'.

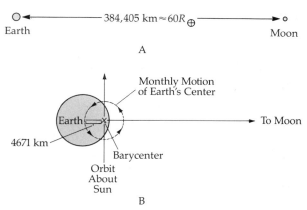

FIGURE 4–3 (A) The Earth–Moon system drawn to scale. (B) The location of the barycenter (*center of mass*) of the Earth–Moon system; it lies within the mantle.

4–2 ◗
DYNAMICS

(A) MOTIONS

Let's summarize the principal motions of the Earth (Chapter 3). The barycenter of the Earth–Moon system orbits the Sun at about 1 AU in one sidereal year of 365.2564 mean solar days. The Earth rotates once every 24 sidereal hours, and its center orbits the barycenter in one sidereal month of 27.322 days. The Earth's rotation axis precesses about the ecliptic pole in roughly 26,000 years, and its body axis wobbles slightly about the rotation axis. Tidal friction slows the Earth's rotation by roughly 0.002 s per century.

The Moon's motions are much more complicated. With respect to the stars, the Moon orbits the barycenter in one **sidereal month** (27.322^d); with respect to the Sun, this orbit takes one **synodic month** (the month of phases, 29.531^d). With respect to the Moon's line of nodes (see below), the orbital period is one **nodical** or **draconic month** (27.212^d), and with respect to its perigee, the period is one **anomalistic month** (27.555^d). The Moon's **synchronous** rotation period is one sidereal month, and so we see only one hemisphere of the Moon. The far side was first photographed by the Russian spacecraft Luna 3 in October 1959.

Actually, we can see about 59% of the total lunar surface as a result of the Moon's **librations** (first known and interpreted by Galileo). Lunar librations are geometric effects caused by the inclination of the Moon's orbit and equator. The Moon rotates at a fairly steady rate, but it moves at different speeds in its eccentric orbit, this variation leads to an apparent east–west "rocking" of the Moon by about 6°17'—the **libration in longitude** (Figure 4–4A). Although the Moon's equator lies essentially parallel to the ecliptic plane, the lunar orbit is inclined 5°9' to the ecliptic; hence, we see an apparent north–south "nodding" motion of about 6°41'—the **libration in latitude** (Figure 4–4B).

Differential solar gravitational forces on the Earth–Moon system produce three significant effects: (1) they tend to elongate the Moon's orbit at quadrature, (2) they cause the perigee of the Moon's orbit to precess eastward (direct) with a period of 8.85 years, and (3) they produce a torque on the inclined orbit, which causes the **line of**

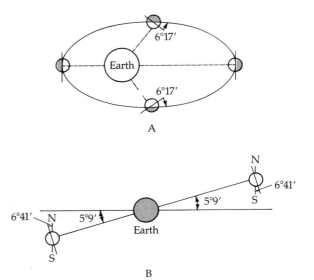

FIGURE 4–4 Lunar librations. (A) A libration in longitude occurs as a result of the Moon's constant rotation in an elliptical orbit, where the orbital speed varies. (B) A libration in latitude results from a combination of the inclination of the Moon's orbit to the ecliptic and the inclination of its equator to its orbital plane. Both effects allow us to see more than half of the lunar surface.

nodes (intersection of the Moon's orbital plane and the ecliptic plane) to regress *westward* along the ecliptic with a period of 18.6 years.

(B) PHASES

The Sun always illuminates one hemisphere of the Moon, but from the Earth we see varying fractions of the sunlit hemisphere depending upon the Moon's elongation. The cycle of **geocentric phases** of the Moon lasts one synodic month (Figure 4–5); these phases occur in the following sequence: **new** (inferior conjunction), **waxing crescent, first quarter** (quadrature), **waxing gibbous, full** (opposition), **waning gibbous, third quarter** (quadrature), **waning crescent,** and back to new. The Moon's position in the sky correlates with its phase. The first-quarter Moon rises in the east at noon, transits the local meridian at sunset, and sets in the west at midnight. As seen from the Moon, the Earth exhibits similar phases in a reverse cycle (new Moon occurs at the time of full Earth), and

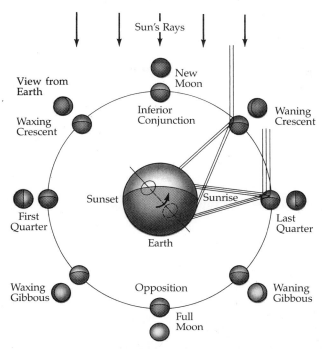

FIGURE 4–5 Lunar phases. The Moon is shown here at eight positions in its synodic phase cycle as it orbits the Earth. The shaded portion is the dark, unilluminated half of the Moon. The lunar images outside of those on the orbital path show the Moon's appearance from the Earth.

relative to the Moon's horizon, the Earth remains roughly at the same place in the sky.

(C) ECLIPSES

In general, an **eclipse** occurs when the shadow of one celestial body falls on another body. Eclipses in the Earth–Moon system depend upon the Moon's phase and the orientation of the line of nodes of the Moon's inclined orbit, since the Moon must be near the ecliptic plane to pass directly between the Sun and the Earth in a solar eclipse and to pass through the Earth's shadow in a lunar eclipse (Figure 4–6).

Solar Eclipses

The Sun's apparent angular diameter is 32′, almost the same as that of the Moon! The line of nodes of the Moon's orbit points toward the Sun twice each year, so that new Moon can occur close enough to the ecliptic for the Moon to cover the Sun—a **solar eclipse.** When the Moon passes to one side of the center of the Sun, we see a **partial solar eclipse.** If the Moon is not near perigee in its orbit, its angular size is slightly smaller than that of the Sun and an **annular solar eclipse** can take place with a ring of the Sun's disk visible. Near its perigee, when the Moon's center crosses the Sun's center, we have a spectacular **total solar eclipse.**

FIGURE 4–6 Lunar nodes and eclipses. The alignments of the Sun, Earth, and Moon are shown for one year. The shadows cast by the Earth and Moon are the shaded regions. The lines of nodes (dashed line) must point to the Sun for an eclipse to occur. Lunar (L) and solar (S) eclipses are indicated.

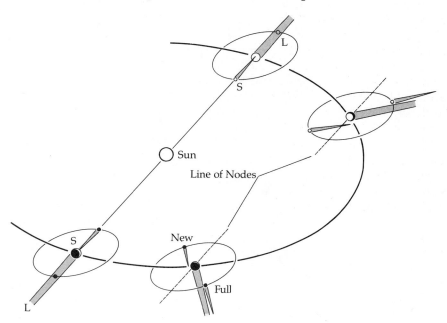

Lunar Eclipses

At the Moon's orbit, the Earth's shadow (where the Sun is totally hidden) is 9212 km across, which is equivalent to an angular diameter of 1°22.4' as viewed from the Moon. When the line of nodes points down this shadow, a **lunar eclipse** can take place only at full Moon (Figure 4–7). Both **partial** and **total lunar eclipses** are possible, with the Moon remaining dark for up to 1^{h}40^m in a central total lunar eclipse.

The Greek astronomer Meton (c. 400 B.C.) noted that the Moon exhibits the same phase on the same day of the month at intervals of 18.6 years—this is known as the **Metonic cycle.** A similar phenomenon, also due to the **regression of the nodes** of the Moon's orbit, is the **Saros cycle.** Similar solar and lunar eclipses take place at intervals of 223 synodic months (18 years, 10 days); since the Saros interval is 6585.32 days, we must wait three Saros cycles to see an eclipse repeat at the same place on the Earth. The Saros cycle comes about from the near equality of 223 synodic months, 242 nodical months, and 239 anomalistic months.

4–3 ◗
INTERIORS

(A) THE EARTH

The **average** or **bulk density** of the Earth is

$$\langle \rho \rangle = 3M_{\oplus}/4\pi R_{\oplus}^3 = 5520 \text{ kg/m}^3$$

(the density of pure water is 1000 kg/m³). Because the density of rocks at the Earth's surface is about

FIGURE 4–7 A time sequence of the later stages of a total lunar eclipse. From left to right, the series of exposures on the same negative shows the moon emerging from the Earth's shadow. *(B. Walski)*

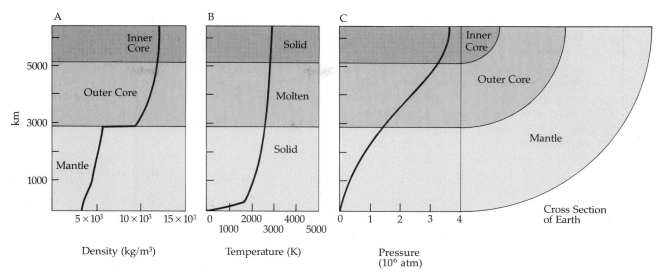

FIGURE 4–8 The Earth's interior. This model is inferred from seismic wave observations. (A) The run of density from the surface to the center; note the discontinuity between the core and mantle. (B) The variation in temperature, which rises quickly then levels off. (C) The trend in pressure, which has its greatest value at the center of the core.

2800 kg/m^3, the interior of the Earth must be *very* dense. The Earth's interior is stratified: a 35-km-thick surface layer, the crust, of density 3300 kg/m^3, then the **solid mantle** of silicates such as olivine [(Mg, Fe)$_2$SiO$_4$] with densities ranging from 3400 to 5500 kg/m^3 to a depth of 2900 km, next the 2200-km-thick **liquid outer core** with densities from about 9900 to 12,000 kg/m^3, and finally the solid inner core of radius 1300 km with densities around 13,000 kg/m^3 (Figure 4–8).

This picture of the Earth's interior has been deduced from the study of the propagation of earthquake waves **(seismology).** Earthquakes generate both **longitudinal compression (P) waves** and **transverse distortion (S) waves,** which travel on paths determined by the density and composition of the interior materials (Figure 4–9). Both types of waves are refracted by changes in the density of the medium through which they are traveling and reflected by density discontinuities; only the P waves, however, can propagate through a liquid region. By measuring whether the wave has compressional (P) or up–down (S) characteristics and its time of travel from the center of an earthquake, we infer the wave's path through the Earth. No S

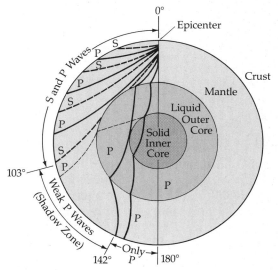

FIGURE 4–9 Seismic waves in the Earth. Both S (dashed line) and P (solid line) waves are shown. The S waves absorbed by the liquid outer core, and the P waves are reflected by the inner core (light dashed line).

waves are seen through the core (they are absorbed by the liquid outer core); P waves are refracted at the mantle–core interface, and weak P waves reflected by the *solid* inner core are observed.

We do know that part of the Earth's core is liquid, but its exact composition is not really known. Most geophysicists attribute the density jump to a change from silicates in the mantle to nickel–iron or some iron alloy in the core. Whatever its exact composition, the core is metallic in contrast to the rocky mantle; so the interior of the Earth is differentiated.

We can estimate some properties of the interior as follows. The Earth is neither expanding nor contracting. At each layer in its interior, the downward force of gravity is just balanced by the upward pressure of the materials—a condition called **hydrostatic equilibrium.** Consider a small slab of the interior of thickness Δr and area A. The gravitational force pulling the slab down is

$$F_g = ma = \rho(r)A\,\Delta r\,(GM/r^2)$$

where $\rho(r)$ is the density of the material in the slab; note we calculate the mass of this slab from its density times $A\,\Delta r$, the volume of the slab. The *net* upward pressure on the slab is $A\,\Delta P$, where ΔP is the pressure *difference* between the top and bottom of the slab. In equilibrium, these forces balance so that

$$A\,\Delta P = -\rho(r)A\,\Delta r\,(GM/r^2)$$

and

$$\Delta P/\Delta r = -\rho(r)(GM/r^2)$$

Take the limit as $\Delta r \to 0$; then

$$dP/dr = -\rho(r)(GM/r^2) \qquad \textbf{(4–1)}$$

is the **equation of hydrostatic equilibrium.** It applies to any situation of a pressure–gravity balance. For the interior of a stable planet, this condition then requires that the internal pressure increase as one goes deeper into a planet's interior (because a greater weight is supported).

Now we use the equation of hydrostatic equilibrium to estimate roughly the central pressure of a planet. Take M as the mass within radius r:

$$M = (4/3)\pi r^3 \langle \rho \rangle$$

and $\langle \rho \rangle$ as the average density within r. To simplify the calculation, assume that the density is constant throughout the planet and equals the average density, $\langle \rho \rangle$, and that the surface pressure is essentially zero compared to the central pressure, $P_S \approx 0$. Then

$$dP = -\langle \rho \rangle G(4/3)\pi\langle \rho \rangle r\,dr$$

which we then integrate from the center to the surface:

$$\int_{P_c}^{0} dP = -\langle \rho \rangle^2\,(4/3)\pi G \int_{0}^{R} r\,dr$$

so

$$P_c = (2/3)\pi G\langle \rho \rangle^2 R^2 = (1.4 \times 10^{-10})\langle \rho \rangle^2 R^2$$

in SI units (Pa for pressure). Then for the Earth, we have a radius of about 6400 km and an average density of roughly 5500 kg/m^3, so

$$\begin{aligned} P_c &\cong (1.4 \times 10^{-10})(5500\ \text{kg/m}^3)^2(6.4 \times 10^6\ \text{m})^2 \\ &= (1.4 \times 10^{-10})(3.0 \times 10^7)(4.1 \times 10^{13}) \\ &= 1.7 \times 10^{11}\ \text{Pa} = 1.7 \times 10^6\ \text{atm} \end{aligned}$$

Detailed models give a central pressure of some 3.7×10^6 atm, so our estimate from hydrostatic equilibrium is pretty reasonable.

(B) THE MOON

From its mass and radius, we find the Moon's average density is 3370 kg/m^3, similar to the density of the Earth's crust. The Apollo missions have returned lunar surface samples with densities near 3000 kg/m^3. The composition of these samples is like that of basaltic silicates, so the density cannot increase much toward the Moon's center because the bulk density is not much higher than the surface density.

Seismic stations set up by Apollo astronauts have revealed that the Moon is seismically quiet; moonquakes measured less than 2 on the Richter scale, compared with 7 or 8 for major earthquakes. Some lunar quakes are attributed to meteorite impacts. A few feeble vibrations originate at lunar perigee, and the occurrence of moonquake groups—a series of quakes lasting several days—has been observed. The origin of these quake groups is unknown, however, and apparently unrelated to tidal effects. Vibrations caused by the deliberate impacts of Apollo components on the

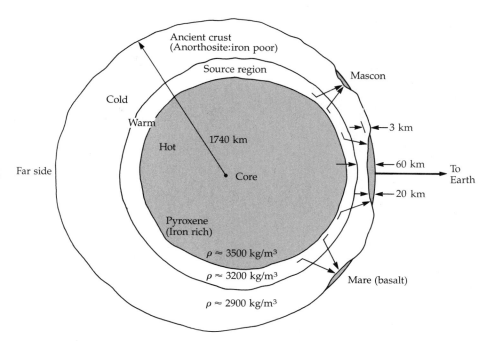

FIGURE 4–10 The Moon's interior. This model is based on Apollo measurements of seismic waves and heat flow; it shows typical values for sizes and densities for each region. The "source region" marks the location of Moonquakes. Note the asymmetry of the interior along the line to the Earth. Size variations are exaggerated for emphasis.

lunar surface damp out very slowly (the Moon resonates for almost an hour!).

One current model of the Moon's interior based on this lunar seismology includes a solid, inactive core, a mantle in which partial melting has occurred (but is not now liquid), and a crust at least 60 to 70 km deep in some places (Figure 4–10). The region of the origin of moonquakes appears to be a warm layer between the mantle and core. So, the outer regions of the Moon's interior appear to be stratified and differentiated. The core is *not* metallic, as expected from the low bulk density.

4–4 ◗
SURFACE FEATURES

The surfaces of the Earth and Moon differ radically from each other. The Earth's surface undergoes rapid and extensive evolution as a result of the hot interior and the thick, erosive atmosphere. Because the Moon's interior is cold and because the Moon has no atmosphere, its surface has preserved its early development. Let's briefly outline the characteristics of the Earth's surface and then examine the lunar surface in contrast.

(A) EARTH'S SURFACE AND AGE

The surface of the Earth is the interface between the low-density **crust** and the **atmosphere.** The crust is composed of (1) the solid **lithosphere** of average density 3300 kg/m³, with light granitic continental blocks reaching to a depth of 35 km and the denser basaltic ocean floors extending down only 5 km, and (2) the liquid-water **hydrosphere,** which covers approximately 70% of the Earth's surface area to an average depth of 3.5 km (Figure 4–11A). The crust floats on the mantle, in buoyant equilibrium. Slow motions of the upper mantle (caused by the heat released by radioactive decay of long-lived isotopes) drive continental drift, mountain building, and earthquakes. The Earth's surface consists of a number of separate continental and oceanic plates. Along ridges in the oceanic plates, lava from the lower crust wells up to push the plates apart and move the continents around the surface (Figure 4–11B); this evolutionary process for the crust is called **plate tectonics.**

Plate tectonics profoundly changes the face of the Earth over time scales that are astronomically short. The present continents, for instance, have moved considerably since their breakup from a

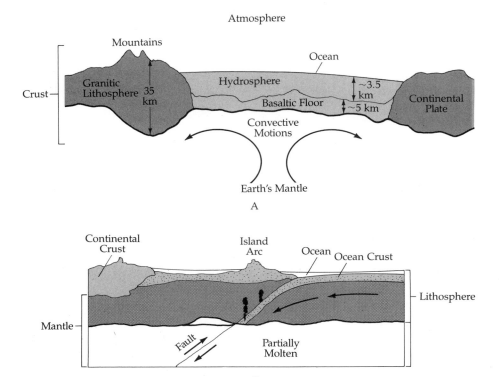

FIGURE 4–11 The Earth's crust and plate tectonics. (A) The upper mantle, crust (lithosphere and hydrosphere), and plates, with convective motions in the mantle moving the plates. (B) Collisions between crustal and oceanic plates drive the evolution of the surface structure.

single landmass 200 million years ago. Geologic evidence indicates that the rate of continental drift now amounts to only a few centimeters per year. That estimate has been confirmed by very-long-baseline interferometry (Chapter 9), which has detected a drift of 1 to 2 cm/year now. Plate tectonics ensures that the ocean basins are the youngest part of the Earth's crust because that is where new lava flows up from the mantle.

The surface of the lithosphere, with an observed temperature range of 200 to 340 K, is the home of humankind. Seen from space, the Earth's cloud cover and polar icecaps appear bright, and the continents show up light green and tan. The oceans are very dark blue. Our home is a beautiful but complex ship in space.

How old is the Earth? Core parts of the continental masses are the oldest parts of the crust. We can date the ages of these rocks by **radioactive decay** and so infer the Earth's age. The **radiometric dating** technique works because of the natural instability of the nuclei of radioactive elements. When these nuclei decay, they break apart into simpler nuclei. The original isotope is called the **parent;** the resulting products are **daughter** atoms.

With a large number of radioactive atoms, you can determine a gross rate of disintegration. Half a piece of uranium-238 (^{238}U) decays to lead (^{206}Pb) in 4.5 billion years, half again in the next 4.5 billion years, and so on (Figure 4–12). You can calculate the amount of uranium left at any time, even though the decay time for any one uranium atom cannot be specified. Conversely, given a rock sample containing ^{238}U and ^{206}Pb and knowing the half-life of uranium, you can calculate the age of the sample. This technique works because the products are rare isotopes in nature.

Elements in addition to uranium that can serve as radioactive clocks include rubidium (^{87}Rb), which decays to strontium (^{87}Sr) with a half-life of 47 billion years, and potassium (^{40}K), which decays to the inert gas argon (^{40}Ar) with a half-life of 1.3 billion years. The derived data is the time elapsed since the rocks were last *solidified* and the parent and daughter atoms trapped. Remelting and resolidification of rocks *after* their origin would reset their dating clocks.

Let's examine the functional form of radioactive decay. Given *n*, a large number of radioactive atoms, the average number decaying in a time interval dt will be dn, which is proportional to the number remaining at time t. So

$$-dn = \lambda n \, dt$$
$$dn/n = -\lambda \, dt$$

where λ is the **decay constant.** We integrate this equation to get

$$\ln n = -\lambda t + \text{constant}$$

At $t = 0$, $n = n_0$, and so

$$\ln n = -\lambda t + \ln n_0$$
$$\ln n - \ln n_0 = -\lambda t$$
$$\ln (n/n_0) = -\lambda t$$

and finally,

$$n/n_0 = \exp(-\lambda t) \qquad (4-2)$$

The **half-life** occurs when $n = n_0/2$ at time τ, so that

$$\tau = (\ln 2)/\lambda = 0.693/\lambda$$

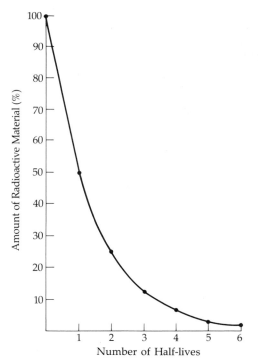

FIGURE 4–12 Radioactive decay. During each half-life, one half of the parent isotope decays. The form of this curve is that of an exponential decline.

Then

$$n/n_0 = \exp(-0.693t/\tau) \qquad (4-3)$$

Suppose, for example, that a rock has an ^{40}Ar to ^{40}K ratio of 1/4. How old is it (assuming that the initial fraction of ^{40}Ar was negligible and no argon has escaped)? If the ratio Ar/K now is 1/4, then $K_{original} = K_{now} + K_{decayed} = K_{now} + 1/4 \, K_{now} = 5/4 \, K_{now}$; so $n(K_{now})/n(K_{original}) = 4/5$. Then

$$4/5 = \exp(-0.693t/\tau)$$
$$\ln 0.80 = -0.693t/\tau$$
$$-0.223 = -0.693t/\tau$$
$$0.322 = t/\tau$$
$$t = 2\tau = (0.322)(1.3 \times 10^9 \text{ years})$$
$$= 4.2 \times 10^8 \text{ years}$$

(B) LUNAR SURFACE

Through remote sensing and lunar landings, the surface and the first few meters of the Moon's soil have been carefully studied. The surface is made up of gray-tan materials of basaltic composition with a very low average albedo ($A = 0.07$). As the Moon's **terminator** (the line separating the dark and sunlit hemispheres) moves along the lunar equator at 15.4 km/h, the surface temperature drops rapidly from a maximum of 380 K to a minimum of 110 K. We can calculate the subsolar (noon) temperature by Equation 2–5B with the albedo factor

$$T = 394[(1 - A)/D^2]^{1/4}$$

where A is the Moon's average albedo and D its distance from the Sun (1 AU). So

$$T = 394[(1 - 0.07)/1^2]^{1/4} = 387 \, K$$

a bit higher than what is observed. (Note we have assumed an equilibrium situation for a blackbody directly illuminated by the Sun.)

Visible Structures

In A.D. 1609, Galileo first discerned (Figure 4–13) the dark lunar **maria,** or "seas" (which were given Latin names by Riccioli in 1651), the light lunar **highlands** with their numerous **craters,** and lunar **mountains**—all through his crude telescope. Spacecraft, manned and unmanned, have revealed a great deal about the lunar surface.

FIGURE 4–13 Moon at last quarter. The terminator between the sunlit and dark regions separates the western (day) and eastern (night) sides of the Moon. The crater Copernicus with its extensive ray system is near the center. *(Lick Observatory)*

Craters The entire lunar surface is pockmarked with **craters:** concave depressions with rims elevated above the exterior level, sometimes with a peak in the center of the floor. Craters range in size from microscopic pits to the largest on the nearside, Bailly, with a diameter of 295 km. The large crater Copernicus (Figure 4–14) dominates in the Ocean of Storms. The juxtaposition of different features permits us to determine relative ages, since young craters will clearly overlap older craters. The lunar maria have far fewer craters than do the lunar highlands, and so the maria are younger than the highlands. Many craters, especially Tycho and Copernicus (Figure 4–14), exhibit patterns of **rays,** which include numerous **secondary craters** from the blown-out debris, which is called **ejecta.**

Craters evolve and disappear slowly as material slides down their walls and as the walls themselves slump (Figure 4–15); meteorite bombardment produces new craters, which fill, obliterate, and degrade the older craters. The lifetime against such erosion has been estimated at several million years for craters 1 cm in diameter, and longer than the age of the Moon for large craters (tens of kilometers in diameter). Although large craters do not disappear, their walls do slump and their rims and rays are eroded by meteoritic impact, temperature changes, and moonquakes.

FIGURE 4–14 Copernicus crater. Note the ray system emerging from the crater, the rugged terrain outside the rim, and the central mountains. *(Lick Observatory)*

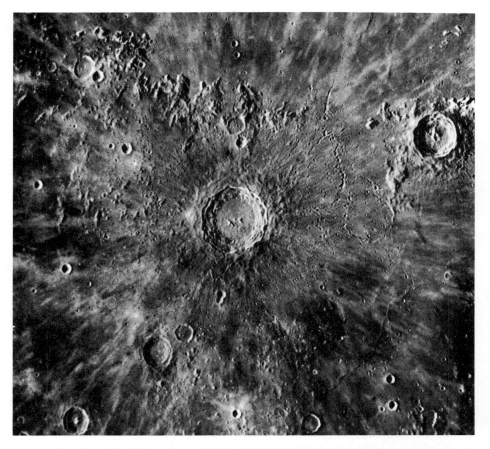

FIGURE 4–15 Slumping and ejecta blanket of an impact crater. This close-up view of the north wall of Copernicus shows several ledges formed by slumping and the downward flow of the rim's material. Note the ejecta outside the rim and the shocked appearance of the surface there. *(NASA)*

The accepted model of crater origin is **meteorite impact,** with the largest craters arising from small asteroids that hit the Moon. A few lunar features, including some craters, such as Alphonsus, betray evidence of lava flows or volcanic origin, but very few are not impact craters.

An impact crater is produced when an incoming meteoroid hits the surface and converts much of its kinetic energy to explosive thermal energy; the remainder of the kinetic energy fractures the surface materials and creates seismic waves. The meteoroid is vaporized upon impact. Lunar material scatters in analogy to chemical (TNT) or nuclear explosions. A shock wave is formed as the meteorite strikes the surface, compressing the surface rock so that some of it is pulverized and behaves like a fluid. This rocky material is ejected explosively, excavating a hole to form a crater, and is piled up to form an elevated rim. Some is flung out as an **ejecta blanket** of debris beyond the rim (Figure 4–15). Sprays of high-velocity particles land farther out to produce rays of secondary craters, and the rebound from the initial shock frequently raises a central peak (similar to the rebound when an object drops into a liquid). The meteoroid is destroyed in the process.

An asteroid 10 km in radius contains about 2×10^{15} kg of rock, and when it hits the Moon at a speed of 45 km/s, the kinetic energy of $mv^2/2 \cong 10^{24}$ J vaporizes the rock to create a crater about the size of Copernicus. (Note that 45 km/s is the escape velocity from the Moon and that from the Sun 1 AU away from it; this sum is the minimum speed with which an object falling in from "infinity" would hit the Moon's surface.) Large craters such as Tycho, Calvius, and Mare Orientale required energies in the range 10^{21} to 10^{26} J for their excavation.

Highlands   Most of the lunar surface consists of the light-colored, heavily cratered **lunar highlands,** especially the lunar farside (Figure 4–16A). The highlands have been radiometrically dated at 4.0×10^9 years; this represents the oldest surviving lunar surface. In general, the highlands have a greater elevation than the maria, with an average difference of 3 km.

Lowlands (Maria)   The large dark areas of the Moon's surface are called **maria** or "seas" (the sin-

gular is **mare**), though they are really flat plains of dark basaltic lava. Of the 30 maria known, only four are found on the farside of the Moon. Several are roughly circular and range in diameter from about 300 to 1000 km. These circular **mare basins** are impact features that subsequently filled with molten lava (Figure 4–16B). The presence of old, flooded craters implies that some cratering occurred between the time of formation of the mare basin and the extrusion of the lava. The impact model of mare basins is supported by the curved shape of some of the mountain chains (such as the Apennines) that border and rise above the maria and by the observation of ejecta sculpting on the nearby highlands. The Apennine Mountains resulted from uplifting of the crust at the time of the impact that produced Mare Imbrium.

The maria have fewer craters than the lunar highlands (Figure 4–16C); this lower crater density implies a younger age—an idea confirmed by the radiometric dating of lunar samples, which are about 3.2×10^9 years old. Fresh, young craters (ages $\approx 10^8$ years), such as Copernicus (Figures 4–13 and 4–14), are seen superimposed on the maria.

Gravitational perturbations of the lunar orbiter spacecraft imply that the mare basins on the nearside are filled with dense material—these gravitational anomalies are called **mascons** (**mas**s **con**centrations). They occur because the lava filling the basins is of higher density and different composition than the rest of the lunar surface. Their existence indicates that the Moon's crust is so rigid and deep that the mascons do not sink into it. Also, the nearside of the moon contains the majority of mascons, which implies a tidal locking with the Earth. Along the line to the Earth, the Moon's center of mass is actually offset from its center (Figure 4–10). The asymmetrical distribution of mare basins may have resulted from a series of impacts from a preferential direction.

Composition of Surface Materials

American and Russian landers have probed and analyzed the lunar surface; the manned Apollo missions also returned hundreds of kilograms of surface soil and rocks. The bedrock of the lunar surface is covered by a thin, well-mixed layer of loose soil and rocks—the **regolith.** The regolith was produced by meteorite impacts that frag-

A

B

C

FIGURE 4–16 The lunar surface. (A) This photograph of the lunar farside shows the cratered highland regions. The dark-floored crater on the east (right) is Tsiolkovsky. *(NASA)* (B) A lunar mare basin; note the overall circular shape. *(Lick Observatory)* (C) A closer view of the surface of a mare as the Apollo lunar module descends for a landing. *(NASA)*

mented and scattered the bedrock and pulverized the soil over the eons. This soil ranges in depth from 2 to 10 m in the maria and is perhaps 10 m deep in the highlands because of extensive cratering; large chunks of bedrock are frequently seen as boulders at the bottoms of and near deep impact craters. The lunar soil has the texture and strength of cohesive sand, and its color is dark gray with a shade of tan. Churning by meteorite impacts mixes the regolith thoroughly in about 50 million years.

Materials brought back from the Moon (Figure 4–17) include (1) samples of the cohesive, fine-grained **soil** containing particles of average size 10 μm, with an admixture of small (50 μm) glass or obsidian spheres; (2) small **igneous** rocks from maria that show evidence of melting, recrystallization, and shock fracture; (3) small chunks of coarse-grained composite **breccias,** which resemble a compressed conglomerate of many materials; and (4) light-colored igneous rocks from the highlands, called **anorthosites.** Note that both highland and lowland rocks are igneous and so formed by the solidification of lava. They differ strikingly in composition, however. The anorthosites are low-density aluminum and calcium silicates; the mare basalts are high-density magnesium and iron silicates. The anorthosites show large crystals, which indicate that they cooled more slowly than the mare basalts. The igneous rocks are fragments of the lunar bedrock brought to the surface by cratering, and the breccias were produced when the regolith was compressed into agglomerates by meteorite impact. Most samples returned from the maria have been dated to 3.2 billion years, while some fragments in the highlands have ages as high as 4.6 billion years. Generally, highland rocks have ages of 3.8 to 4.0 billion years.

One important conclusion derived from these ages is that the major part of the evolution of the Moon's surface must have occurred within a very short time span—within the first billion years after the Moon's origin. The mare basins formed well after the highlands, and their flooding by lava came later still. The radiometric dating of maria material gives it an age of 3.2 billion years.

The chemical composition of the lunar surface is basically **basaltic,** with the following key features: (1) the maria are lavas of igneous (molten) origin; (2) the content of free iron is low, with the maria containing more iron, cobalt, and nickel than the

A

B

C

FIGURE 4–17 Lunar surface samples. (A) A mare basalt, showing the holes from gas bubbles in the lava. The pit in the center resulted from the impact of a small meteorite. (B) A highlands rock containing anorthosite (white areas). This sample from Apollo 15 is known as the Genesis Rock. (C) A breccia, in which large pieces of other rocks are embedded in the darker material that had melted from an impact. *(NASA)*

highlands; (3) most of the volatile elements are underabundant, and some chemical differentiation has apparently taken place; (4) the lunar lavas have a higher content of iron and titanium bound into minerals than do terrestrial basalts; (5) the lunar materials are *anhydrous*; that is, they are not formed in the presence of water or water vapor. The areas sampled indicate that the maria are chemically homogeneous and differ somewhat from the highlands in that they have a considerably lower abundance of aluminum.

4–5 ◖ ATMOSPHERES

(A) THE MOON

Only a very small lunar atmosphere has ever been measured. Indeed, the Moon appears to have too little mass to undergo extensive internal heating, though rare instances of outgassing (like volcanic vapors) have possibly been observed. Any atmosphere generated by the lava flows of the maria has long since escaped as a result of the Moon's low surface gravity and high surface temperatures. In addition, the solar wind is very efficient in sweeping away any trace gases that might emerge at the surface. Even the exhaust gases from the Apollo landers did not linger for long, and they briefly contributed a major addition upon landing and takeoff. The major constituents to the atmosphere come from material outgassed from rocks: about 40% Ne, 40% Ar, and 20% He, with a surface pressure of only some 10^{-14} atm.

The following example demonstrates this point. Consider water dumped on the lunar surface by the Apollo astronauts. At lunar noon ($T \approx 400$ K), the root mean square speed of water molecules (Equation 2–7) is

$$V_{rms} = (3kT/m)^{1/2}$$
$$= [3(1.4 \times 10^{-23})(400)/(18) \times (1.7 \times 10^{-27})]^{1/2}$$
$$\cong 740 \text{ m/s}$$

so that $V_{esc}/V_{rms} \approx 3$, which means that the lifetime is only a few years as the molecules escape from the high-speed tail of the Maxwellian velocity distribution. (Recall that in Chapter 2 we used the criterion that $V_{esc}/V_{rms} \geq 10$.)

TABLE 4–1 *Chemical Composition of the Earth's Atmosphere Near Ground Level*

Gas	Percentage (by Volume)
Nitrogen (N_2)	78.08
Oxygen (O_2)	20.95
Argon (Ar)	0.934
Carbon dioxide (CO_2)	0.033
Neon (Ne)	0.0018
Helium (He)	0.00052
Methane (CH_4)	0.00015
Krypton (Kr)	0.00011
Hydrogen (H_2), carbon monoxide (CO), xenon (Xe), ozone (O_3), radon (Rn), *etc.*	<0.0001
Water vapor (H_2O)	Variable (0 to 4)

(B) THE EARTH

The Earth lost its original atmosphere (of H, He, and H compounds) but produced a **secondary** atmosphere as a result of vulcanism and outgassing. The major components of outgassed material include H_2O, CO_2, N_2, and Ar. The large mass and gravitational attraction of the Earth permit it to retain this atmosphere, which was modified into the atmosphere we breathe today, mostly by the development of life on our planet (Table 4–1).

Atmospheric Structure
The Earth's atmosphere is layered (Figure 4–18). Nearest the surface is the dense, well-mixed **troposphere**, where most weather takes place. The temperature in the troposphere decreases steadily with height until we reach the **tropopause** at 15 km. Then the temperature rises slightly in the thin, stable **stratosphere**, which extends to 40 km into the **mesophere**. A second temperature minimum occurs near 90 km (at about 190 K), and then the temperature increases steadily through the **thermosphere** (90 to 250 km) until it levels off near 1500 to 2000 K at the base of the **exosphere**. The exosphere marks the region where the atmosphere can escape into space.

Atmospheric density decreases rapidly in the troposphere and then more gradually at higher altitudes, the decline typical of an exponential function. The Earth's atmosphere is in equilibrium, so it obeys the hydrostatic equilibrium relation, Equa-

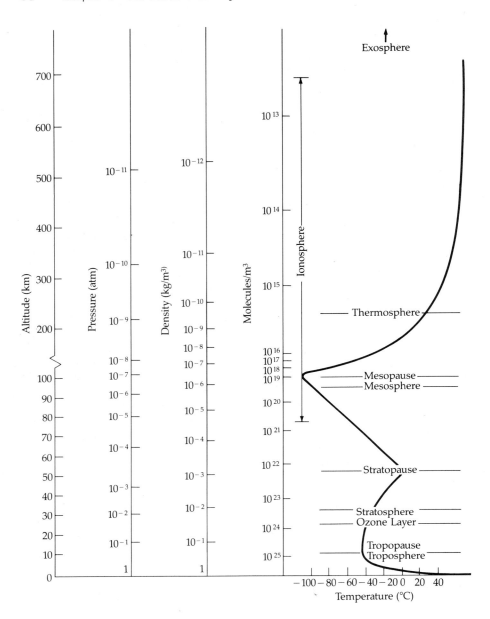

FIGURE 4–18 Pressure, density, and temperature change with altitude through each of the layers of the Earth's atmosphere. Note the change in scale between 100 and 200 km.

tion 2–4. Assume that the atmosphere is an ideal gas, so that it has the equation of state

$$P = nkT$$

where $n = \rho/m$. Here, n is the number density (#/m³), ρ the mass density (kg/m³), and m the **mean molecular weight** (in units of hydrogen masses) of the atmosphere gas. So

$$P = \rho kT/m$$

or

$$\rho = mP/kT$$

We substitute this for ρ in Equation 4–3:

$$dP/dr = -P(m/kT)(GM/r^2)$$

or

$$dP/P = -(m/kT)(GM/r^2)\,dr$$

Now recall that

$$g(r) = GM/r^2$$

where g is the acceleration from gravitation at distance r from the Earth's center and M is the Earth's mass within r. Then

$$dP/P = -g(r)(m/kT)\, dr$$

and, if we integrate this equation from some r_0 to r:

$$P(r)/P(r_0) = \exp[-g(m/kT)(r - r_0)]$$

where we have assumed that g, T, and m are roughly constant over the range $r - r_0$. Now define $r - r_0 = h$, the height above the surface (r_0), and

$$H = kT/gm \cong \text{constant}$$

as the **scale height.** Then

$$P(h) = P(h_0) \exp(-h/H)$$

where h is any height above a reference level h_0. This relation is called the **barometric equation;** it applies to regions in planetary atmospheres where the temperature and mean molecular weight do *not* change rapidly. Note that H, the scale height, has the units of length; it is the distance to move up in the atmosphere for the pressure to decrease by $1/e$. At the Earth's surface $H \approx 8$ km, and so the pressure (and density) at an 8-km elevation is roughly $1/e$, 2.7 times less than at the surface. The pressure at the Earth's surface is 1 **atmosphere** (atm) and equals 1.01×10^5 Pa.

Whereas atmospheric density decreases steadily with height, the temperature profile exhibits three maxima from absorption of sunlight. Solar ultraviolet rays and X-rays interact with the compositionally stratified atmosphere to dissociate oxygen and nitrogen molecules and to ionize some atoms to produce the stratified density of free electrons in the **ionosphere** (about 50 to 300 km). The ionization balance of the ionosphere is self-regulating since free electrons recombine faster with ions as more electrons and ions become available; hence, a steady-state equilibrium is set up. Concentrations of atomic oxygen (O), atomic nitrogen (N), ozone (O_3), and nitric oxide (NO, NO^+) are found at different levels of the ionosphere, and the entire system acts as a shield that absorbs most of the solar radiation dangerous to life.

Human Activity and the Atmosphere

In the modern era, people have disturbed the natural CO_2 balance by extracting fossil fuels from the earth, burning them for energy, and so adding to the CO_2 in the atmosphere. In addition, our destruction of forests has eliminated a substantial part of the green plants that take in atmospheric CO_2 and has added some of their carbon to the atmosphere. The ocean can absorb only a part of the excess.

Our activities have a net result of increasing the percentage of CO_2 in the earth's atmosphere (Figure 4–19). Observations indicate an increase of 4.3% per year since 1860 to a level now of about 350 μL/L. (During the last ice age, the level was about 200 μL/L.) That amounts to few *gigatons* of carbon added to the atmosphere in a year. Although there is still much controversy over what the impact of this increase will be, it may result in a temperature increase of another 2°C in 40 years, if the overall water vapor and cloud cover do not change. From 1880 to 1988, the average global temperature increased at a rate of 0.0055°C/y. This sounds like a small change, but it could have serious effects on climate and atmospheric circulation.

Human activity has also resulted in a sharp drop in the O_3 content of the atmosphere. As mea-

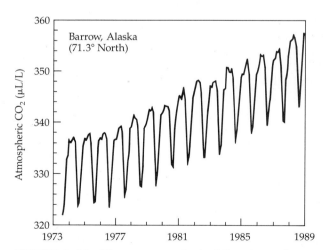

FIGURE 4–19 The ozone content of the atmosphere measure at Barrow, Alaska, since the 1970s. Note the seasonal changes, which reflect photosynthesis in the summer and plant respiration in the winter. The long-term trend is a gradual increase.

sured from Antarctica from the 1950s to the 1980s, the amount of total O_3 decreased by about a half, resulting in the so-called "ozone hole" above that region. Mario Molina and Sherwood Rowland proposed in 1975 that chlorofluorocarbons released in aerosol propellants, refrigerants, and other industrial applications were the possible culprits to deplete atmospheric O_3—a concept now strongly confirmed by observations. The final thieves are chlorine atoms. In a process that takes a year or two from first release of the gases, each chlorine atoms destroys some 10^5 molecules of O_3. Each year some 10^6 tons of chlorofluorocarbons are emitted by human actions; these remain in the atmosphere for about 100 years. The emission so far could result in concentrations about 10 times that right now, and so decrease the O_3 amount even more. The United Nations has developed a protocol to severely limit the use of chlorofluorocarbons by 1999; yet, because of the longevity of the carbons, much of the damage is already in the air.

Observed Effects of the Atmosphere

Electromagnetic radiation is absorbed, scattered, and refracted by the Earth's atmosphere, and all these phenomena depend upon the wavelength of the incoming radiation. The atmosphere is opaque to most wavelengths, transmitting (1) radio waves with wavelengths in the range 1 mm to 20 m (which are detected by radio telescopes) and (2) visible light from 290 nm (near ultraviolet) to about 1 μm (near infrared)—the "window" through which we see the Universe with our eyes and telescopes. Short-wavelength radiation (less than 290 nm) dissociates and ionizes the atmosphere and is absorbed in the process. For example, the ozone layer (10 to 40 km) absorbs radiation from 300 to 210 nm. In the infrared region (about 1 μm to 1 mm), radiation is absorbed when it excites molecules such as N_2, O_2, CO_2, and especially H_2O. At wavelengths longer than about 20 m, the ionosphere acts as a conducting shield that absorbs and reflects incoming radiation.

How the atmosphere affects visible light, from violet (410 nm) to red (650 nm), directly influences our view of the Universe. The most important effects are scattering, extinction, refraction (and see-

ing), and dispersion. Light is **scattered** when it interacts with a particle, and how it scatters depends upon its wavelength and on the size L of the scatterer. When light is scattered by atmospheric molecules ($L \ll \lambda$), the intensity of the scattered radiation obeys the **Rayleigh scattering law:**

$$I_{\mathrm{scat}} \propto 1/\lambda^4$$

Hence, more blue light is scattered out of incident sunlight than red, and we see a blue sky. This preferential scattering causes the Sun (or any other star) to appear *reddened* when viewed through a considerable thickness of atmosphere, such as at sunset. The scattering depletes the blue light along the line of sight. Some radiation is scattered at every wavelength, even the red, so that a star's brightness is dimmed by the atmosphere—this we term **extinction** of the light. Astronomical observations must be corrected to account for reddening and extinction. When $L \approx \lambda$, such as for 1-μm dust particles scattering red light, the scattering law becomes

$$I_{\mathrm{scat}} \propto 1/\lambda$$

Finally, when $L \gg \lambda$, as for water droplets in clouds, the scattering is essentially *independent of wavelength*—clouds appear white since they scatter sunlight this way.

When light passes from one medium to another, it is *refracted*, or bent. We characterize a medium by its **index of refraction.** The index of refraction of air increases with its density. Hence, starlight coming from a star at a true altitude θ is refracted downward toward the Earth, so that it appears to reach the surface at the altitude angle $\theta' > \theta$ (Figure 4–20). This effect, which vanishes at the zenith, becomes more important as we approach the horizon. At the horizon, a celestial object appears to be elevated about 35' above its true position, so that the setting Sun is actually *below* the horizon when we see its bottom resting on the horizon.

Density inhomogeneities in the atmosphere cause randomly fluctuating amounts of refraction for incoming light, so that stars appear to dance about *(twinkle)* in the sky. This effect limits the **angular resolution** of telescopes to about 1" (0.25" in exceptional cases), and we term this **astronomical**

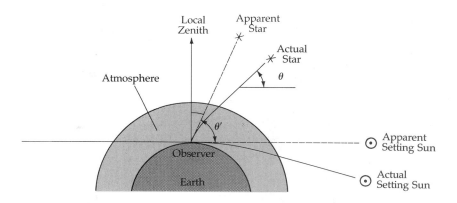

FIGURE 4–20 Atmospheric refraction. Light rays from celestial objects, such as the Sun or any other star, are bent by the atmosphere so that they appear higher than they actually are. The effect is larger closer to the horizon because the path length is longer.

seeing. The seeing is good when the atmosphere is stable and the twinkling is small. Planets and satellites with angular diameters exceeding the seeing limit shine steadily since only their edges twinkle.

With all these effects interfering with light from space (especially with the opaqueness of the Earth's atmosphere to most wavelengths of electromagnetic radiation), you should not be surprised that observers locate their instruments on mountaintops or strive to view the Universe from spacecraft. The ideal observatory is clearly one located above the Earth's obscuring atmosphere, such as on the Moon! There, the lack of atmosphere means no seeing problems, no refraction or extinction, and no wind and weather—a remote observing site of the highest quality in the inner part of the Solar System.

4–6 ◑
MAGNETIC FIELDS

(A) LUNAR MAGNETISM

The Earth's magnetic field, with a strength of about 4×10^{-5} T at the surface, has been studied for several hundred years. In contrast, lunar magnetism could be investigated only with space probes to the Moon's surface. Magnetometers placed there by the Apollo astronauts reveal that the intrinsic lunar magnetic field is less than 10^{-9} T, but localized sources of magnetism exist on the Moon with strengths around 10^{-8} T. In any case, the Moon's *global* magnetism is negligibly small now. If the dynamo model [Section 4–6(b)] for the origin of a planet's magnetic field is correct, then the Moon's core cannot be molten or composed mainly of iron and nickel. (Also, the Moon's density is too low for it to have a metallic core.) On the other hand, some surface rock samples are magnetized much more than you would expect from such a weak magnetic field. Iron minerals in an igneous rock preserve the magnetic field present at the time of solidification. So in the past the Moon's magnetic field was stronger than it is now; how so remains a puzzle.

(B) THE EARTH'S MAGNETOSPHERE

The Earth has a **dipolar** magnetic field whose axis is inclined 12° to the Earth's rotation axis, with field lines that emerge at the north magnetic pole and re-enter at the south magnetic pole. The Earth's rotation helps produce the magnetism. Fluid motions in the metallic (electrically conducting) outer core, made into stable convective zones by the Coriolis effect, are believed responsible for the magnetic field, through some type of **dynamo** (moving electric charges produce a magnetic field). Paleomagnetic investigations have shown that the field reverses its direction in a random way with an average period of about 10^4 to 10^5 years, and today we observe that the strength of the field is slowly decreasing.

Because the strength of a dipolar magnetic field drops roughly as $1/r^3$ with distance r from the Earth whereas the density of the atmosphere falls exponentially, the magnetic field is important far beyond the atmosphere. The domain of the Earth's magnetic field is the **magnetosphere,** which ends at the **magnetopause** boundary, where it meets

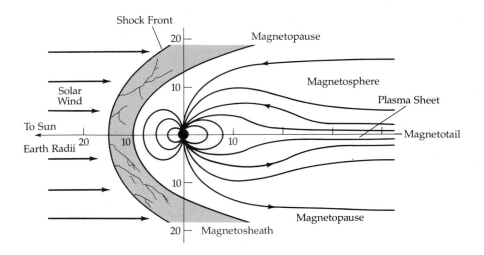

FIGURE 4–21 The Earth's magnetosphere. The solar wind encounters the Earth's dipolar magnetic field to create a shock wave behind which lies the magnetosphere. The wind creates a long tail, within which lies a sheet of plasma. Units are in Earth radii.

and interacts with the solar wind (Figure 4–21). The magnetosphere has a cometlike structure caused by the solar wind's pushing the Earth's magnetic field away from the Sun. Within $10R_\oplus$ of the Earth, the dipolar field is still evident; beyond there lies an abrupt shock-and-transition region on the sunward side and the magnetotail longer than $1000R_\oplus$ pointing away from the Sun.

The Earth's magnetosphere contains complex physical processes involving plasmas and electromagnetic fields, which we are just beginning to understand. Luckily, this magnetosphere is the closest one and has been closely studied since the beginning of space exploration. It serves as the model for other magnetospheres in the Solar Sys-

tem, which are found around other planets and with comets. The key point is this: the interaction of the solar wind and the magnetosphere creates a planetary electromagnetic generator that converts the kinetic energy of the solar wind into electricity at a rate of some 10^6 megawatts!

(C) VAN ALLEN RADIATION BELTS

The magnetopause deflects the solar wind away from the Earth, but many protons and electrons leak into the magnetosphere. There they are trapped by the Earth's dipolar magnetic field in toroidal radiation belts, concentric to the magnetic axis (Figure 4–22). The discovery of these belts

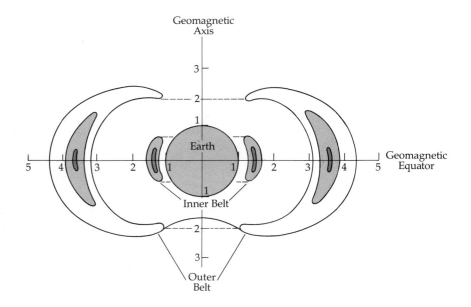

FIGURE 4–22 The Van Allen radiation belts. Dimensions are in Earth radii. The shaded contours represent the relative densities of charged particles.

FIGURE 4–23 Particle motion in the Van Allen belts. Charged particles from the Sun spiral along the magnetic field lines and bounce between the northern and southern *mirror points*. Protons and electrons drift in opposite directions because of their opposite charges.

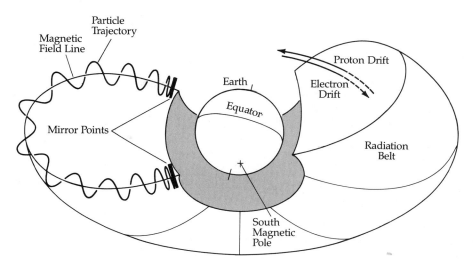

(Van Allen radiation belts) by a group headed by James A. Van Allen came from satellite observations in 1958.

Though both protons and electrons are found throughout the magnetosphere, there are two belts of particular concentration: (1) the small *inner belt* between $1R_\oplus$ and $2R_\oplus$, where protons of energy 50 MeV and electrons with energies greater than 30 MeV tend to reside, then a distinct gap, and (2) the large **outer belt** from $3R_\oplus$ to $4R_\oplus$, where less-energetic protons and electrons are concentrated. (The **electron volt**, eV, is a convenient energy unit in atomic and particle physics; 1 eV = 1.6021×10^{-19} J, 1 keV = 10^3 eV, and 1 MeV = 10^6 eV. It is the energy gained by an electron accelerating between two plates separated by a distance of 1 m and a potential difference of 1 V.) The inner belt is relatively stable, but the outer belt varies in its number of particles by as much as a factor of 100. The particles trapped in the belts come mostly from the solar wind.

The charged particles trapped in the radiation belts spiral along lines of magnetic force while bouncing (with periods from 0.1 to 3 s) between the northern and southern **mirror points** (Figure 4–23). Particles in the inner belt may interact with the thin upper atmosphere at these mirror points to produce the aurorae [Section 4–6(d)]; such particles are lost from the belts. In addition to spiraling and oscillating north–south, the particles drift in longitude owing to the decreasing strengths of both magnetic and gravitational fields with increasing distance from the Earth. High-energy protons drift westward around the Earth in about

0.1 s, and low-energy electrons drift eastward in about 1 to 10 h. This drift leads to the longitudinal uniformity of the radiation belts.

Let's investigate these charged particle motions quantitatively. The **Lorentz force law**

$$\mathbf{F} = q(\mathbf{v} \times \mathbf{B}) \qquad (4\text{–}4)$$

tells us the force $\mathbf{F}$ (in newtons) experienced by a particle of **charge** q (in coulombs, where an electron's charge is 1.60×10^{-19} C) moving with velocity $\mathbf{v}$ (in m/s) through a magnetic field of strength $\mathbf{B}$ (in units of teslas). Note that $\mathbf{F}$ is perpendicular to both $\mathbf{v}$ and $\mathbf{B}$, in accordance with the right-hand rule (Figure 4–24A). Although the mass of the proton (1.673×10^{-27} kg) is much larger than the mass of the electron (9.109×10^{-31} kg), the charge of the proton is equal to the charge of the electron but of opposite sign. No work can be done on a charged particle by the magnetic field since $\mathbf{F} \cdot d\mathbf{r} = (\mathbf{F} \cdot \mathbf{v}) \, dt = 0$ ($\mathbf{F}$ is perpendicular to $\mathbf{v}$!); hence, the kinetic energy of a particle is *not* changed by a magnetic field (the total speed v remains constant). The particle moves in a circular orbit in a *uniform* magnetic field (Figure 4–24B). The Lorentz force (Equation 4–4) provides the centripetal force for the orbit of radius r, so that

$$mv^2/r = qvB$$
$$r = mv/qB \qquad (4\text{–}5)$$

For example, a proton with speed $v = 10^8$ m/s circles the magnetic field lines with a radius $r = 10$ km when $B = 10^{-4}$ T. This radius is often called the **gyroradius**.

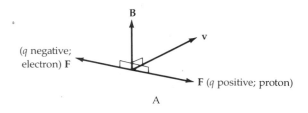

A

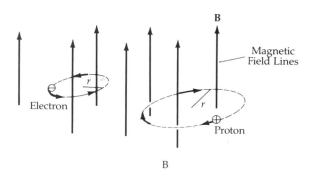

B

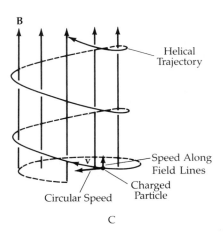

C

FIGURE 4–24 Motions of charged particles in a magnetic field. (A) Vector diagram for the direction of the Lorentz force. (B) Circular orbits of electrons and protons in a uniform magnetic field. (C) Helical orbit of a charged particle with some component of its velocity along uniform field lines.

If the particle has some speed along a magnetic field line, it will follow a helical trajectory (Figure 4–24C). When the particle moves into a region of higher field strength (B increases), the circular orbit shrinks while the circular speed increases. Since the particle's total kinetic energy cannot change, its motion along the field line must slow down; eventually the forward motion reverses if one particle collides with another. Note that charged particles moving into a region of narrowing magnetic field lines are effectively accelerated to very high speeds. This process occurs in a number of astrophysical situations, such as aurorae, at the mirror point.

(D) AURORAE

The inner radiation belt interacts with the Earth's upper atmosphere to produce the colorful **aurorae: aurora borealis,** or northern lights, in the high northern latitudes and **aurora australis** in the high southern latitudes. Diffuse luminous forms over large areas of the sky have been observed since ancient times in northern regions between 15 and 30° from the magnetic pole. Auroral displays may be flickering streaks or curtains of light, or steady lacework draperies that change intensity and color (pale pink, blue, and green) in the course of hours (Figure 4–25). Triangulation measurements of the heights of aurorae show that most occur between 80 and 160 km, with a few as high as 1000 km.

Auroral emissions result when low-energy electrons drop out of the inner radiation zone and collisionally excite and ionize atmospheric gases. As these excited gases, principally oxygen and nitrogen, return to their stable forms, visible light is emitted with characteristic colors. The solar wind plays an important role in aurorae, by supplying the necessary electrons and by perturbing the magnetosphere so that the particles trapped in the radiation belts are dumped into the atmosphere near the mirror points. That solar activity induces aurorae is inferred from the fact that it is so strongly correlated with the appearance of auroral displays.

Recent models picture aurorae as a result of the release of energy from the magnetotail by a process called **magnetic reconnection.** This process takes place when regions of opposite magnetic fields come together and the magnetic field lines can break and reconnect in new combinations. One site of such reconnections is the plasma sheet in the magnetotail; the point of reconnection usu-

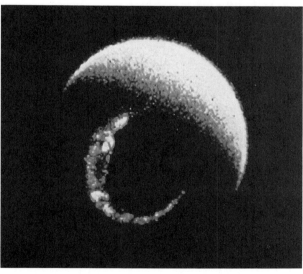

FIGURE 4–25 The aurorae borealis. (A) View from the Earth's surface. Note the drape-like structure in this time exposure. *(National Research Council of Canada)* (B) View in November 1981 from a satellite at a distance of 23,000 km. Note the ring-like structure. Atomic oxygen lines at 130.4 and 135.6 nm produce most of the emission. *(L. Frank, University of Iowa)*

ally lies about $100R_\oplus$ downstream from the Earth. The net magnetic field at this point is zero. Occasionally, the solar wind adds sufficient magnetic energy to the magnetosphere so that the field lines there overstretch and a new reconnection takes place only $15R_\oplus$ from the Earth. The field on the Earthside collapses, injecting electrons into the atmosphere to drive the auroral emissions. Meanwhile, a blob of plasma containing looping field lines is free of magnetic attachment from the Earth and moves downstream. The equilibrium magnetosphere is then re-established. The process of magnetic reconnection in a plasma applies to the environment of comets (Section 7–3) and other astrophysical situations.

4–7 ◗
THE EVOLUTION OF THE EARTH–MOON SYSTEM

We are now in a position to rely upon the physical properties of the Earth and Moon—especially radioactive dating of rocks—to infer their evolution. This comparison will highlight the Earth as the most evolved—and still evolving—terrestrial planet; the Moon, as a fossil world, preserves evidence of the early stages in terrestrial planetary evolution. [See Section 7–6(f) for a presentation about the origin of the Earth and Moon.]

(A) LUNAR HISTORY

About 4.6 billion years ago, the Moon probably formed by the accretion of chunks of material. These pieces continued to plunge into the Moon after most of its mass had gathered. During the first 200 million years after formation of the Moon, these projectiles from space bombarded the surface and, along with radioactive decay of short-lived isotopes (such as ^{26}Al), heated it enough to melt it. Less-dense materials floated to the surface of the melted shell; volatile materials were lost to space. The crust began to solidify from this melted shell about 4.5 billion years ago. From 4.5 to 4.1 billion years ago, the crust slowly cooled. The bombardment from space continued but began to taper off. This falling debris made many of the craters now found in the highland areas.

Just below the surface, the Moon's material remained molten. About 4.1 to 3.9 billion years ago, a few huge chunks smashed the crust to produce basins which later became maria. For example, the Mare Orientale basin formed 4 billion years ago when an object about 25 km across smashed into the Moon. Only later did the basins fill with lava. Though the crust lost its original heat of accretion, short-lived radioactive elements that decay rapidly reheated sections of it. From 3.9 to 3.0 billion years ago, lava formed by the radioactive reheating punctured the thin crust beneath the basins, flowing into them to make the maria.

For the past 3 billion years, the crust has been inactive. However, small particles from space have incessantly plowed into the surface since it solidified. These sand-sized grains scoured the surface, smoothed it down, and pulverized it. Continued bombardment by larger bodies churned the fragmented surface. Impacts melted the soil, which swiftly cooled to form breccias and glass spheres.

Note that in this evolutionary history (Table 4–2) the highlands make up the oldest sections of the Moon's crust and the lowlands (maria) make up the youngest. Once the maria formed, the

TABLE 4–2   *Evolution of the Moon: A Contemporary Model*

Event	Time (Gy)	Processes
Formation	4.6	Accretion of small chunks of material
Melted shell	4.6–4.5	Melting of outer layer by heat from infall of material and/or radioactive decay; volatile elements lost
Cratered highlands	4.5–4.1	Solidification of crust while debris still falls to form craters
Large basins	4.1–3.9	Reduced infall, but formation of basins by impact of a few large pieces; outflow of basalts from lava below solid crust
Maria flooding	3.9–3.0	Flooding of basins by lava produced by radioactive decay
Quiet crust	3.0–now	Bombardment by small particles to pulverize and erode surface

Moon's interior and crustal evolution halted because the interior lost much of its primordial heat in 1 billion years. This internal heat, from the original accretion and radioactive decay, drives the evolution of a terrestrial planet. In general the rate of heat loss is proportional to a planet's surface area; the reservoir of heat is proportional to its mass, or to $\langle\rho\rangle R^3$. Hence the lifetime of a terrestrial planet's evolution will be roughly proportional to the total amount of energy available divided by the loss rate, or $t \propto R^3/R^2 \propto R$. So larger terrestrial planets should be the more evolved in the sense that the processes driven by internal heat will last a longer time. Also, the more massive planet will retain an atmosphere longer, resulting in wind and rain erosion of the surface. Let's examine the Earth with this point in mind.

(B) THE EARTH'S HISTORY

Driven by its internal heat, evolutionary processes have moved the Earth through six stages of evolution. The first stage began 4.6 billion years ago, when the Earth accreted from small, rocky bodies in only a few million years. That left a pockmarked planet of more or less uniform composition, since each piece was made of about the same combination of materials. The atmosphere at the time of accretion was rich in hydrogen and inert gases. Accretion and radioactive decay produced rapid internal heating. The low-mass gases escaped into space.

In the second stage, beginning about 4.5 billion years ago, radioactive heating (and perhaps gravitational contraction) melted the interior. It differentiated to form a core of dense materials and a crust of light ones. In the third and fourth stages, volcanic activity caused by interior heating created the second atmosphere, containing outgassed water, methane, ammonia, sulfur dioxide, and carbon dioxide. An infall of large objects continued, fracturing the crust. Ocean basins were formed, and the Earth's surface cooled enough for rain to fall and begin filling the basins.

In the fifth stage, about 3.7 billion years ago, the first continents appeared and plate tectonics began. Mountains grew, only to succumb to weathering by wind and rain. Slowly the atmosphere evolved, affected in part by the presence of life. Roughly 2.2 billion years ago, crustal cooling thickened the crust enough to allow plate activity as we see it today. By 600 million years ago, the planet had entered its sixth state of evolution. The processes that had begun in the fifth stage continued at a lower rate until the Earth came to look much as it does today.

The evolution of terrestrial planets involves change to their interiors, surfaces, and atmospheres. These changes are driven mostly by the planets' internal heat: the more massive a planet, the more internal heat it generates (from radioactive decay), the longer it retains this heat, and the greater the degree to which it evolves.

Surfaces are modified by several processes: impact of interplanetary debris, outflow of internal heat, volcanism, erosion by wind and water, and crustal movements (if the mantle is hot). The atmospheres change as a result of interaction with sunlight, escape into space, degassing (from a hot interior), and life (if it exists). In particular, life on Earth has transformed its second atmosphere (which came from outgassing) into the oxygen-rich atmosphere that we enjoy now.

PROBLEMS ◑

1. (a) An Apollo spaceship heads for the Moon. At what point between the Earth and the Moon will the ship experience no net gravitational acceleration?
   (b) How long will it take the ship to circumnavigate the Moon in a circular orbit 50 km above the lunar surface?

2. Using the length of the sidereal month (27.322^d) and the periods of the Earth's revolution (365.26^d), the regression of the nodes (18.6 years), and the precession of the Moon's perigee (8.85 years), compute the lengths of
   (a) the synodic month
   (b) the nodical month
   (c) the anomalistic month

3. Assume that the Moon's orbit is circular and lies in the ecliptic plane. Find the difference in the solar attraction on the Moon's orbit at opposition and inferior conjunction; compare this with the gravitational attraction from the Earth. Can you now understand why the lunar orbit is not a simple ellipse?

4. (a) If the Moon had a core of radius $R_m/10$ and the remainder of its interior had a uniform density $\rho = 3000 \text{ kg/m}^3$, what must be the uniform density of this core?
   (b) Compare the Earth's tidal forces on the Moon at the perigee and apogee of the Moon's orbit; comment on your results.

5. The mean kinetic energy per molecule in a gas at temperature T is $mv^2/2 = 3kT/2$, where $k = 1.380 \times 10^{-23}$ J/K. A particle with a vertical speed v at the Earth's surface will rise to a height $h = v^2/2g$ before it falls back to the Earth.
   (a) Show that the characteristic height for a molecule of mass m at temperature T is $h = 3kT/2mg$.
   (b) At $T = 250$ K, compute the characteristic heights for nitrogen (N_2), oxygen (O_2), carbon dioxide (CO_2), and hydrogen (H_2). What does this tell you about the *compositional* structure of the Earth's atmosphere?
   (c) How does this calculation differ from that of the scale height done in the chapter?

6. (a) If a star emits the same intensity of radiation at all visible wavelengths, what will be its apparent color at the Earth's surface?
   (b) Explain why the Sun appears flattened (like an ellipse) at sunset.

7. Show that the circular period P (in seconds) for a charge in a uniform magnetic field B does *not* depend upon the radius of the orbit. Evaluate this period for a proton moving at speed $v = 10^7$ m/s in a magnetic field of 10^{-4} T.

8. Draw a diagram and use the Lorentz force law to explain why electrons in the radiation belts drift *eastward* owing to the decrease in the magnetic field strength with distance from the Earth.

9. Find the radius of the Earth's core relative to the total radius if the core density is 10,000 kg/m³, the mantle density is 4500 kg/m³, and the average density is 5500 kg/m³.

10. In the Earth's exosphere, the temperature can reach 2000 K. Estimate the lifetime of water vapor here by comparing its mean velocity with the Earth's escape velocity.

11. Use the equation of hydrostatic equilibrium to esti-

mate the central pressure of the Moon and compare it to the Earth's central pressure.

12. (a) Use the strength of the magnetic field at the Earth's surface to estimate the strength in the Van Allen belts.
    (b) Use this estimate to calculate the radius of curvature of a 50-MeV proton in the belts.

13. Demonstrate that a dipolar field decreasing as $1/r^3$ beats out an atmospheric density falloff that is exponential.

14. Assume that a rock sample contains ^{40}Ar and ^{40}K in the ratio of ^{40}Ar/^{40}K = 3. How old is the rock? What assumptions do you have to make in order to calculate an age?

15. Assuming that the average Sun rises at 6 a.m., crosses the meridian at noon, and sets at 6 p.m., what are the average times for the Moon's rising, crossing the meridian, and setting for each of the lunar phases; new, waxing crescent, first quarter, waxing gibbous, full, waning gibbous, third quarter, and waning crescent. Assume each of the phases is 45° apart in elongation, and ignore complicating factors.

16. You and a friend decide to determine the radius of the Earth. You synchronize watches; then your friend drives 50 km due west, at latitude 40°. Each of you determines the time when the Sun lies due south—on the meridian. Your friend observes the Sun to be on her meridian 140 seconds after you observe the Sun on your meridian. What is your estimate of the Earth's radius?

17. The summit of Mauna Kea, Hawaii (4200 m above sea level) is considered one of the world's premier observing sites. It is especially good as an infrared site because of the low water vapor and carbon dioxide content in the atmosphere above this altitude. (Water and carbon dioxide have lots of infrared wavelength absorption bands.) However, working at these altitudes can be hard on astronomers who are not acclimated to the lower abundance of molecular oxygen.
    (a) To understand both of the above effects, calculate the scale height for H_2O, CO_2, and O_2 in the Earth's atmosphere.
    (b) What percentage of each of these gases in the atmosphere is below 4200 m? (Assume the content decreases uniformly for each gas.)

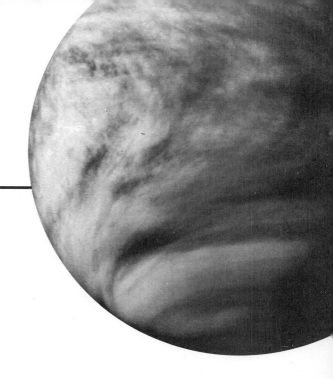

Chapter 5

The Terrestrial Planets: Mercury, Venus, and Mars

We have so far presented the general properties of the Solar System (Chapters 1 and 2) and the basics of two special planets, the Earth and Moon (Chapters 3 and 4). We will use these two familiar worlds as the basis for the **comparative planetology** of the other terrestrial planets: Mercury, Venus, and Mars. We will try to infer the evolutionary histories of these bodies by drawing appropriate analogies to the Earth and Moon.

5–1 ◑
MODERN PLANETOLOGY

Because the human eye alone cannot discern details smaller than 1 arcmin, it took the invention of the telescope in the seventeenth century to spark the investigation of the planets—**planetology.** In 1610, Galileo described telescopic observations of the four largest (Galilean) moons of Jupiter, the phases of Venus, the cratered surface of our Moon, the dark spots on the Sun, and the strange non-spherical shape of Saturn (its rings). The Earth's atmospheric seeing limits the visible resolution of telescopes to about 1 arcsec, however, so that the surface markings and atmospheres of only those

objects of much greater angular size can be studied in any detail. The maximum angular diameter of Mercury seen from the Earth is only 12″; in such cases, optical viewing is marginal at best (only 10 or so resolution elements across the disk).

New technologies and techniques characterize modern planetology. The optical window has been expanded to the ultraviolet (photography) and infrared (solid-state detectors). Radio telescopes and radar now peer through the radio window. High-altitude balloons, rockets, artificial satellites, and space probes rise above the Earth's atmosphere to give a broader and clearer view of the Solar System. Spacecraft sent to the Moon and planets have dramatically improved our picture of these bodies. The main point here is that a flyby spacecraft improves our resolving power enormously—by a factor of 10^4 for the Voyager missions to Jupiter and Saturn.

5–2 ◑
MERCURY

Mercury is the planet closest to the Sun and the second smallest of all the known planets. Each year, it appears about three times as a bright evening star near the sunset horizon and as a predawn morning star. Because of the swiftness of its celestial motion, it is named after the mythological god of flight. At times, Mercury rivals Jupiter in brightness, but it is usually obscured by the brilliance of the nearby Sun.

(A) MOTIONS

Mercury revolves about the Sun in a highly eccentric ($e = 0.2056$) and inclined (7.00° to the ecliptic plane) orbit of semimajor axis 0.3871 AU, with a sidereal orbital period of 87.96^{d}. The greatest elongation of this planet viewed from the Earth ranges from 18° (perihelion) to 28° (aphelion); it averages 23°.

The sidereal rotation period of Mercury was thought to be either 24^{h} (Earthlike) or the *synchronous* value of 88^{d}. However, in the early 1960s, radar pulses reflected from the surface of Mercury were first detected, and by 1965 G. H. Pettengill

and R. B. Dyce had shown the rotation period to be about 59^{d} direct, using Doppler radar techniques. A re-analysis of the visual data verified the radar period and resulted in a measured period of $58.65 \pm 0.01^{\mathrm{d}}$, with the planet's equator essentially parallel to its orbital plane. Mercury is in a spin-orbit resonance with the Sun [Section 2–1(a); Figure 2–5]; its sidereal rotation period is 58.64^{d}, which is two-thirds of its sidereal orbital period. The earlier periods were in error because surface features on Mercury can be observed visually only near greatest elongation, which occurs at the synodic orbital period of 115.88^{d}—essentially twice the sidereal rotation period.

Radar astronomy permits us to deduce a nearby planet's distance, size, rotation, and large-scale surface features by analyzing radar pulses (wavelengths $\approx$1 to 10 cm) reflected from the planetary surface. Whereas lunar surface features, such as craters and mountains, are easily resolved, a radar pulse directed at a terrestrial planet is usually wider in angle than the planet itself. Hence, we

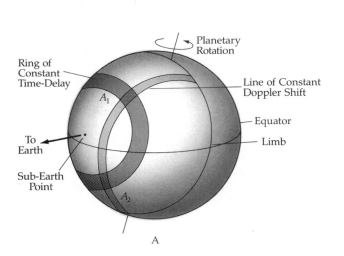

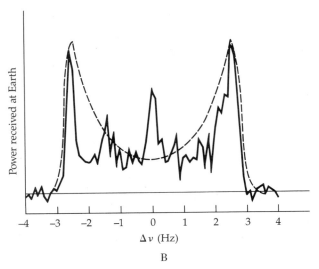

FIGURE 5–1 (A) Radar mapping. Imagine a radar pulse reflecting from a planet's hemisphere facing the Earth. Different time delays form a series of concentric rings. They intersect lines of constant Doppler shift at two points (A_1 and A_2), so that a feature at either A_1 or A_2 has an ambiguity in position. (B) Radar signal reflected from Venus. The solid curve shows the actual power received relative to the rest frequency. The dashed curve shows the expected power received from a uniformly rough surface. *(G. H. Pettengill and I. I. Shapiro)*

must resolve the return pulse into its *time-delayed* and *Doppler-shifted* components (Figure 5–1A). The shortest time delay occurs at the sub-Earth point on the planet and tells us the planet's distance from the Earth. Successively longer time delays originate from rings concentric about the sun-Earth point at constant distances, with the longest delay defining the planet's **limb** (edge) and so its radius. As the planet rotates, the reflected radar waves are Doppler-shifted in frequency (Figure 5–1B), so that higher frequencies than the original return from the approaching limb and lower frequencies from the receding limb; these Doppler shifts reveal the rotation speed of the planet. The shifts are maximum at the limbs and zero along the line going through the poles and the sub-Earth point. Radar mapping of the planetary surface is limited by uncertainties in signal strength, time delay, and frequency shift; in addition, a twofold positional ambiguity (Figure 5–1A) exists, but this may be resolved as the Earth–planet direction changes.

We can see the relationship between Doppler shift and rotation period as follows. The shift in wavelength, the difference between that sent out and that received, is

$$\Delta\lambda/\lambda_0 = v/c$$

where λ_0 is the emitted wavelength, v the radial velocity, and c the speed of light. For a spherical planet, the maximum linear rotation rate occurs at the equator:

$$v_{eq} = 2\pi R/P$$

Because each limb contributes to the total shift,

$$\Delta\lambda/\lambda_0 = 2v_{eq}/c = 4\pi R/Pc$$

from which we calculate the rotation rate from the planet's radius and the measured shift. Note we have assumed that we are viewing the planet directly on the equator (it is the sub-Earth point).

How long is the solar day on Mercury? We derive a relationship between the solar day and the sidereal day in a manner very similar to the derivation of the relationship between the synodic and sidereal orbital periods of the planets. Suppose you were on Mercury with the sun directly overhead (Figure 5–2). Let T be the period of revolution of Mercury about the sun (88 days), P the sidereal period of rotation (58.7 days), and S the synodic

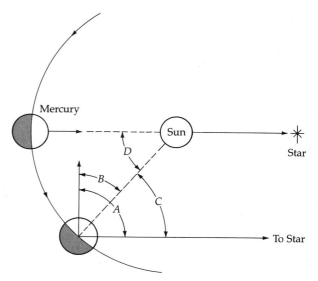

FIGURE 5–2 Geometry for the solar day on Mercury.

period of rotation (the solar day). After one Earth-day, Mercury will have rotated 360°/P (angle A) with respect to the stars but only 360°/S (angle B) with respect to the Sun. The difference between these two angles (angle C) is equal to the angular distance Mercury has moved in its orbit, 360°/T (angle D). So

$$360°/P - 360°/S = 360°/T$$

or

$$1/S = 1/P - 1/T$$

Putting in the numbers,

$$1/S = 1/58.7 - 1/88.0 = 0.00567$$
$$S = 176 \text{ days}$$

Mercury's solar day is just twice the length of its sidereal year.

(B) PHYSICAL CHARACTERISTICS

Mercury's radius is 2440 km. Its mass of $0.055M_\oplus$ (3.3×10^{23} kg) is deduced from gravitational perturbations upon spacecraft (Mercury has no natural satellites). The average density is 5420 kg/m³, typical of a terrestrial planet but high for small one. Because Mercury's overall gravity is less than the

Earth's (compressing it less) but its bulk density is about the same, it must contain a greater proportion of metals. We infer that Mercury's interior (Figure 5–3) has a rocky mantle and a large metallic (probably nickel–iron) core, which encompasses about 75% of the radius.

We can use hydrostatic equilibrium to estimate Mercury's central pressure from

$$P_c = (2/3)\pi G\langle\rho\rangle^2 R^2 = (1.4 \times 10^{-10})\langle\rho\rangle^2 R^2$$

in Pa. For a radius of about 2400 km and an average density of roughly 5400 kg/m³, we have

$$\begin{aligned}
P_c &\cong (1.4 \times 10^{-10})(5400 \text{ kg/m}^3)^2(2.4 \times 10^6 \text{ m})^2 \\
&= (1.4 \times 10^{-10})(2.9 \times 10^7)(5.8 \times 10^{12}) \\
&= 2.3 \times 10^{10} \text{ Pa}
\end{aligned}$$

or about 0.1 that of the Earth's.

The surface albedo is very low (0.056 at visual wavelengths), indicative of rocky material even darker than the Moon's surface. Surface temperatures vary from about 700 K at the perihelion subsolar point to about 100 K on the dark side. The long, hot solar day and low escape velocity (4.2 km/s) make it unlikely that Mercury has much of an atmosphere. Gas molecules, even the more massive ones, would easily reach escape velocity,

and so any atmosphere could not be expected to last long. The Mariner 10 space probe had on board an ultraviolet spectrometer to search for an atmosphere. This device detected a thin atmosphere of helium and hydrogen, but the surface atmospheric pressure was very small, about 10^{-12} atm. Here is the atmospheric escape calculation. For helium at noon,

$$\begin{aligned}
V_{rms} &= (3kT/m)^{1/2} \\
&= [3(1.4 \times 10^{-23})(700)/(4)(1.7 \times 10^{-27})]^{1/2} \\
&= 2.1 \text{ km/s}
\end{aligned}$$

so that

$$V_{rms}/V_{esc} = 2.1/4.2 = 0.5$$

and the lifetime is but a few hours.

New ground-based spectroscopic observations have revealed that sodium and potassium exist in

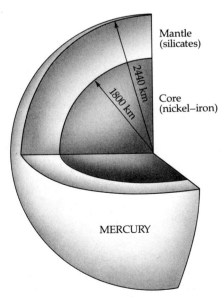

FIGURE 5–3   Model of Mercury's interior. Note the large fraction of the interior taken up by the metallic core.

Mantle (silicates)

Core (nickel–iron)

2440 km

1800 km

MERCURY

FIGURE 5–4   Mercury's cratered highlands. Note the bright, rayed crater (lower left) and the double-ringed crater above it. The outer ring is about 170 km across. *(NASA)*

the atmosphere on the dayside. These gases are likely released by the surface rocks when they absorb solar ultraviolet. The sodium gas dominates the atmosphere during the day.

No significant atmosphere means no insulation from space. That's why the range on noon-to-midnight temperatures on Mercury is so severe. Let's contrast them to the Moon's. Night on the Moon and Mercury differ in duration but their slow rotation allows infrared radiation on the nightside to escape directly into space during the long night. Thus both bodies have about the same midnight surface temperature: 100 K. At noon, the surface temperature depends on the surface albedo and distance from the Sun. The Moon and Mercury have about the same albedo, but Mercury averages 2.5 times closer to the Sun than is the Moon, and so is much hotter at noon because the Sun's flux is 6.25 times greater. From Equation 2–5B, if we say that the albedoes are the same, T_{ss} (Mercury) $\approx 1.6\ T_{ss}$ (Moon) ≈ 630 K.

With the day so hot and the escape velocity so low, why does Mercury have any atmosphere at all? One answer is the solar wind. The influx of material from the solar wind, which contains some 10% helium, could possibly replenish the loss. Another possible source, at least for helium, is decay of radioactive elements in Mercury's subsurface layers.

(C) SURFACE FEATURES

Television cameras on board Mariner 10 scanned 50% of Mercury to reveal a surface like our Moon's. Key differences are fewer mid-sized craters; no mountain ranges; many shallow, scalloped cliffs, called **scarps;** fewer basins and large lava flows; and more relatively uncratered plains amid the heavily cratered regions.

These differences are important, and yet Mercury's surface clearly resembles the farside of the Moon. Mercury's highlands (Figure 5–4) are riddled with craters like the Moon's bleak highlands. Light-colored rays spring from some of the craters, an indication that these were formed by violent impacts more recently than others. Some craters are more than 200 km in diameter, comparable to the largest lunar craters. Mariner 10 found but one mare basin: the Caloris ("hot") Basin on the north-

FIGURE 5–5 Caloris Basin. One half of this 1300-km-diameter, multiringed basin is visible at the terminator, the line between night and day on the left. Wrinkled ridges and flooded lowlands are visible to the right. *(NASA)*

west part of the planet (Figure 5–5). It probably has an overall diameter of 1300 km. The basin is bounded by rings of mountains about 2 km high. In size and structure, the Caloris Basin strongly resembles the Moon's Orientale Basin.

All these similarities do not mean that the Moon and Mercury are identical. Their surfaces differ in at least three ways: (1) Mercury's surface has scarps hundreds of kilometers long, (2) even the most heavily cratered regions of Mercury are not completely saturated with craters but are interspersed with intercrater plains, and (3) Mercury has three times the number of craters 10 km or larger in diameter than does the Moon. From these last differences, we can infer that the cratering material had a different range of sizes for Mercury than for the Moon.

Mercury's scarps (Figure 5–6) vary in length from 20 to 500 km and have heights from a few hundred meters up to 2 km. Individual scarps often traverse different types of terrain. If the regions photographed by Mariner 10 represent Mercury's overall surface, the characteristics of the

FIGURE 5–6 A scarp in Mercury. This ridge (arrow) extends for hundreds of kilometers. *(NASA)*

scarps imply that Mercury's radius has shrunk by 1 to 3 km, from cooling of either its core or its crust, or both, thereby forming the *thrust faults* that created the scarps.

The moon and Mercury may have an affinity in their origins. Mercury has a *much* higher percentage of metals than is predicted by models for the origin of the solar system (Section 7–6). As a differentiated terrestrial planet, Mercury has too small a mantle. A giant impact early in Mercury's history could have stripped off a rocky mantle, leaving mostly a metallic core behind. Hence, the formation of the Moon and Mercury may have involved violent impacts in the early days of the Solar System.

(D) MAGNETIC FIELD

Mariner 10 detected a weak planetary magnetic field with a strength of about 300 nT (1 nT = 10^{-9} T) at the equator. Although small, this is sufficient to carve out a magnetosphere in the solar wind. Here the magnetic field deflects the charged particles (mostly protons) of the solar wind around the planet (Figure 5–7). The overall shape of the magnetosphere resembles that of the Earth; we expect that reconnection events, driven by the accumulation of the solar wind plasma, can trigger magnetic substorms that would inject plasma toward Mercury. Since Mercury lacks an atmosphere, however, the energetic particles would strike the surface directly, perhaps heating the surface regions preferentially in "auroral" zones (Figure 5–7).

Mercury's field appears to be a dipole, more or less aligned with the planet's spin axis. In general, then, Mercury's magnetic field is similar to the Earth's, only weaker. The presence of a magnetic field as well as the planet's high density indicates that Mercury, like the Earth, has a metallic core. This core is presumed to be relatively cold and solid now because a small planet loses heat quickly. Recall that the Earth's magnetic field supposedly arises from an internal dynamo whose churning motions are thought to be driven by the Coriolis effect from the Earth's spinning. Because Mercury rotates much more slowly than the Earth, a planet-wide magnetic field was not really expected; its source is not clear. Perhaps Mercury's core is still partially molten in a shell.

(E) EVOLUTION OF THE SURFACE

Because both the Moon and Mercury lack substantial atmospheres, weathering does not erode their surfaces. Both are tiny worlds with interiors cooler than Earth's. Neither has much (if any) volcanic activity now, and neither has undergone the continual surface evolution the Earth experiences from the shifting of crustal plates.

The lack of atmospheres and the short time of crustal evolution are both related to the small masses of Mercury and the Moon. Their surface gravities are so low that most gases achieve escape velocities, and atmospheres are not retained for long. The small masses also imply that internal heating from radioactive decay would be less than that for the Earth and the flow of heat outward would be so fast that both bodies would cool off quickly. The Earth is hot in its interior, and out-

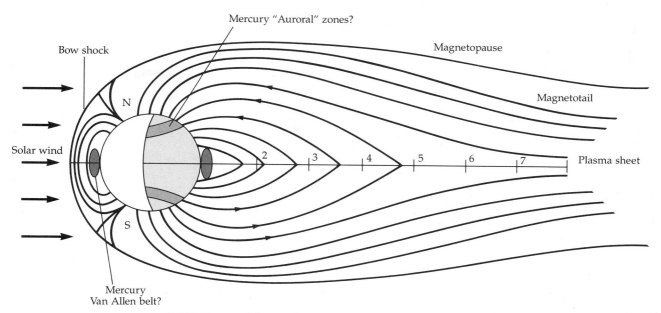

FIGURE 5–7 Mercury's magnetosphere, inferred from Mariner 10 data. Units are in Mercury radii. *(D. Baker, J. Borovsky, G. Gisler, J. Burns, and M. Zeilik)*

ward flow of heat sets up currents in the plastic mantle; these power the evolution of the Earth's crust. Both Mercury and the Moon lack this combination of hot interior and plastic mantle.

Using lunar analogies, we can set up a working model for the evolution of Mercury, with the following general stages: (1) heating of the surface (by impacts and/or radioactive decay) and formation of a solid crust, (2) heavy cratering, (3) formation of impact basins, (4) filling in of basins by lava flows, and (5) low-intensity cratering.

A lunar comparison suggests that the intense sculpting of Mercury's surface took place about 4 billion years ago, not long after the planet formed (stage 1). The earliest phase of cratering (stage 2) was wiped out by later volcanism. We come to this conclusion because the intercrater plains seem to have covered up any scars of the earliest accretion. At about the same time, the scarps developed from global shrinking of the crust, the interior, or both.

Basin formation (stage 3) must have come at the end of the heavy bombardment of the surface. A few large pieces crashed into the surface; one made the Caloris Basin. Not long after this, widespread lava flows (similar to those which made the lunar maria) created the broad, smooth plains,

such as those adjacent to the Caloris Basin (stage 4). Since then, a few impacts have punched out the fresh rayed craters (stage 5).

5–3 ◑ VENUS

Venus is the second terrestrial planet from the Sun and at times the closest planet to the Earth. As a morning and evening star, Venus rises to a greatest elongation of 48°. Its maximum brilliance is exceeded only by those of the Sun and Moon. Named for the goddess of love, Venus closely resembles the Earth in size and mass but otherwise differs enormously.

(A) MOTIONS

The nearly circular ($e = 0.0068$) orbit of Venus is inclined 3.39° to the ecliptic, with a semimajor axis of 0.7233 AU and a sidereal orbital period of 224.70 days.

Doppler-shift radar studies show that the planet rotates *retrograde* with a sidereal period of 243.01 days and with its equator inclined only 3° to its orbital plane (conventionally this inclination is written as $+177°$ or $-87°$ to indicate the retrograde behavior).

(B) PHYSICAL CHARACTERISTICS

We can measure the Earth–Venus distance directly by radar; then from the planet's angular diameter, we can calculate its physical radius. Venus turns out to have a radius of 6052 km, only 5% smaller than the Earth's.

Like Mercury, Venus has no known natural moon, and so we can accurately determine its mass only when a spacecraft passes or orbits it. If the craft is in orbit, we simply use Kepler's third law. During a flyby, we measure the acceleration of the spacecraft (from the Doppler shift of its radio signals) and use Newton's law of gravitation. Venus' mass comes out to be 4.86×10^{24} kg (82% of the Earth's mass). The resulting bulk density is 5200 kg/m³, almost the same as the Earth's. A similar mass and bulk density imply that the interior of Venus closely resembles the Earth's interior (Figure 5–8): a rocky crust (which Venus landers confirmed), a mantle, and a metallic core. Because Venus has a somewhat lower density than the Earth, we imagine that it has a somewhat smaller core (some 0.4 of the radius); its central pressure should be roughly the same.

Interplanetary probes launched by both the United States and the Soviet Union indicate that the atmosphere of Venus contains about 96% carbon dioxide, 3% nitrogen; variable traces of water vapor (0.1–0.4%); and small amounts of oxygen, hydrogen chloride, argon, hydrogen fluoride, hydrogen sulfide, sulfur dioxide, helium, and carbon monoxide. Venus has a much greater abundance of nonradiogenic, inert gases than does the Earth; in particular, the atmosphere contains about 80 times as much ^{36}Ar and ^{38}Ar. (In contrast, it has only 1/3 as much radiogenic ^{40}Ar.)

The Russian Venera landers found the surface atmospheric pressure to be 95 atm and the sunlit surface temperature about 730 K. This high temperature probably results from the effective trapping of surface heat because carbon dioxide absorbs infrared radiation well (the so-called **greenhouse effect**). Atmospheric winds on Venus blow from the day to the night side and from the equator to the polar regions; the wind flow carries heat. Along with the very effective greenhouse effect, this upper atmospheric transport of heat keeps the surface temperatures fairly constant over

FIGURE 5–9 Features in the upper atmosphere of Venus. The "Y" feature, which is a combination of recurring clouds, has changed in these four photographs, taken one day apart. Note the cap of clouds at each pole. *(NASA)*

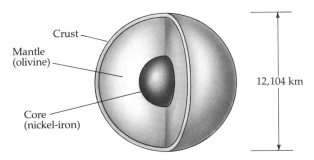

FIGURE 5–8 Interior of Venus. This model is based on the planet's bulk density and the assumption that its interior is similar to that of the Earth, with a large, metallic core.

Crust
Mantle (olivine)
Core (nickel-iron)
12,104 km

Venus' surface; they vary 10 K or less. Hence, Venus has little or no surface winds, which result from differential heating.

The visual brightness of Venus results from the high albedo (0.76) of its clouds and from its thick atmosphere. The structure and motion of the clouds can be seen in ultraviolet photography (Figure 5–9). The Pioneer Venus probes found that clouds float in two broad layers (Figure 5–10). The upper cloud deck tops are at roughly 65 km; this layer has a thickness of some 5 km. The liquid droplets in the clouds have diameters of about 1 μm. Below the upper deck floats a thin haze layer. Below that, at 50 km, is the lower cloud deck, by far the densest layer. The cloud particles here are both liquid and solid, some 10 μm in diameter. Below an altitude of 50 km, the clouds gradually thin out; below 33 km, the atmosphere is clear of any particles down to the surface. Models of the clouds designed to match their observed in-

frared spectra imply that they contain a solution of 90% sulfuric acid mixed with water. Although the atmosphere and clouds of Venus do contain some water vapor, it would amount to a layer only 30 cm thick if it all condensed out uniformly on the surface.

Ultraviolet photography by Mariner 10 revealed that these cloud tops flow with the upper atmosphere (Figure 5-9) in patterns similar to the jet streams of the Earth. Ringing the equator, the clouds whiz around at roughly 300 km/h—fast enough to circle the planet in only four days. In addition to this planet-wide circulation, winds also blow from the equator to the poles in large cyclones 100 to 500 km in diameter. They culminate in two gigantic cloud vortices capping the polar regions.

The circulation pattern of Venus' atmosphere is in large part explained in terms of Hadley cells. Air rises over warm regions—the equator—and flows on top of cooler air at higher latitudes. The cooler air then replaces the rising warm air at the surface. On Earth, instabilities caused by the rapid rotation break up the Hadley circulation into three cells in each hemisphere. On Venus a single-cell model describes the flow in the lower atmosphere. The rising warm air can be understood in terms of the dependence of scale height H on temperature [Section 4–5(b)].

(C) SURFACE FEATURES

We probe the surface of Venus by sending landers to photograph it or by bouncing off radar signals to map the terrain. This recent work has revealed high plateaus, gigantic volcanoes, impact craters, and long rift valleys. Overall, Venus looks fairly flat. Elevation differences are small, only 2 to 3 km, with the exception of three highland regions. Here the land rises up to 12 km, compared with 4-km highland–lowland differences on the Moon and Mercury, 25-km differences on Mars, and 9- to 20-km differences on the Earth (Mt. Everest is 8.85 km high, and the Marianas trench is 11.03 km deep).

The southern and northern halves of the mapped face of Venus differ remarkably from each other. The northern region is mountainous, with uncratered **upland plateaus.** In contrast, the southern part consists of relatively flat **cratered ter-**

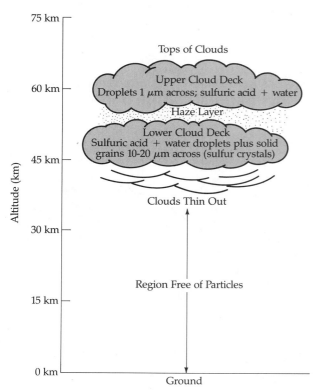

FIGURE 5–10 Structure of Venus' clouds, which come in two high layers with haze between them. Below the clouds, the atmosphere is clear. The clouds are made of H_2O and H_2SO_4. *(NASA)*

rain. The great (1000 by 1500 km) northern plateau is called Ishtar Terra. It is larger than the biggest upland plateau on the Earth, the Himalayan Plateau. Three mountain ranges border Ishtar on the west, north, and east. The eastern range, called Maxwell Montes (Figure 5–11), contains the highest elevations on Venus seen to date: 12 km. Radar shows a rough terrain, as expected from a lava flow, and a volcanic cone near Maxwell's center. The northern mountain range rises about 3 km above Ishtar; the western one reaches only 2 km above the plateau. These three mountain ranges may have folded and risen from moving plates in Venus' crust and so may be similar to mountains built from plate tectonics on the Earth.

The southern half of Venus' face consists of low, rolling plains punctuated by craters, both large craters (up to 800 km in diameter) and smaller ones (less than 1 km across). Craters smaller than this do not form on Venus because the dense atmosphere completely burns up incoming objects before they reach the ground. The craters of Venus resemble those of the Moon, Mercury, and Mars, and so they most likely have an impact origin. Venera 8 landed in the midst of the southern terrain. Its instruments indicated that the rock was more granitic than basaltic—one clue that at least some of the surface here is old. Later lava flows

could have poured out basaltic rock. Also, the craters here imply the surface must be old; otherwise volcanoes and mountain building would have obliterated the craters.

Beta Regio, which contains at least two separate volcanoes, is an enormous volcanic complex that formed from a great north–south fracture zone (Figure 5–12). The volcanoes here have gentle slopes; they are called **shield volcanoes.** (Mauna Loa and Mauna Kea in Hawaii are shield volcanoes, which are relatively flat. Often shield volcanoes have a collapsed central crater—a caldera—at their summits.) One volcano in Beta Regio has a diameter of 820 km, a height of 5 km, and a summit crater 60 by 90 km. In contrast, the island of Hawaii (a volcanic island) is 200 km across and 9 km high. (The largest volcano on Mars, Olympus Mons, is 550 km in diameter and 25 km high.) Overall, about 75% of Venus' surface appears to be volcanic in character.

There are hints that some of the volcanoes on Venus may be active now. The evidence, however, is very indirect. First, the levels of sulfur dioxide (SO_2) detected by the Pioneer Venus ultraviolet spectrometer have shown a steady decline with occasional bursts to higher amounts. (Sulfur dioxide has a limited lifetime in the atmosphere unless it is replenished.) Terrestrial volcanoes spew out

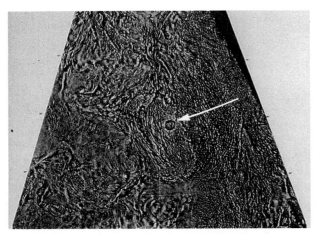

FIGURE 5–11 Radar map of Maxwell Montes from Venera 15 and 16. This high mountain appears to be volcanic; note the volcanic crater at the top (arrow) and the irregular surface of the sides. *(U.S. Geological Survey)*

FIGURE 5–12 The Beta Regio plateau (longitude 280°, latitude 30°). This computer-enhanced radar map shows two large shield volcanoes with flows radiating from them. Theia Mons is at the bottom; Rhea Mons at the top. The vertical relief has been exaggerated. *(USGS)*

sulfur dioxide (as well as hydrogen chloride and hydrogen fluoride, which are minor constituents of Venus' atmosphere). One interpretation of these results is the episodic injection of sulfur dioxide by volcanic eruptions. The haze above the clouds may also be supplied by volcanic injection, at levels equivalent to the largest such eruptions on the Earth (Krakatoa in 1883, for instance). Second, Pioneer Venus also recorded low-frequency radio bursts believed to be emitted by lightning strokes. On Earth, lightning discharges often flicker through the plumes of erupting volcanoes. The Venus bursts were clustered over three regions thought to be volcanic, including Beta Regio.

Venus has huge canyons, extending for more than 1000 km. These canyons appear to be rift valleys, which form along fault zones rather than by water erosion. Rift valleys appear on the Earth where the crust is spreading apart. About 25% of Venus' surface appears to be tectonic in character, where the original lava flows are distorted by surface movement. This limited tectonic activity, compared to the Earth, may indicate that the mantle is much stiffer than the Earth's.

Overall, the surface of Venus is much flatter than the Earth's: only 18% of the mapped surface extends above 7 km, 11% above 10 km. In contrast, about 30% of the Earth's surface reaches above 10 km. Venus does not appear to have lunar-type basins, lowlands filled by lava flows. As indicated by the presence of craters, the lowlands of Venus must be older than the rest of the crust. The highlands are more evolved. That's just the opposite of what we find on Earth, where the lowlands (the ocean basins) make up the youngest part of the crust.

The Soviet Union has sent four spacecraft (Veneras 9, 10, 13, and 14) to the planet's surface. The pictures from Veneras 9 and 10 show slabby rocks about 60 cm long and 20 cm wide. A few rocks are vesicular, which implies a volcanic origin. Some rocks show jagged edges, indicating little erosion, and others show blunted, rounded edges, indicating much erosion. The rocks rest on loose, coarse-grained dirt. Lander measurements of gamma rays emitted by the radioactive potassium, uranium, and thorium in the rocks indicate that at one lander site they are basaltic, like those lining the Earth's ocean basins, but at the other site they are granitic, like the Earth's mountains. Both are igneous rocks formed from lava.

The photographs from Veneras 13 and 14 show quite different views (Figure 5–13). Those from Venera 13 reveal a rocky plain with clustered outcroppings; the rocks are basaltic, similar to those on Earth around continents. In contrast, those from Venera 14 show even layers of broken, rocky plates that are basically volcanic, like the basalts in the Earth's ocean floor near mid-oceanic ridges. So in both places (separated by 900 km), the rocks appear to be relatively young.

(D) MAGELLAN AT VENUS

In August, 1990, the Magellan spacecraft entered into an orbit around Venus, and one month later began its mission of high-resolution surface mapping—100 m at best, or about 10 times better than previously! This mission will last almost a year, and it will take many more years to analyze the data. Already, though, we can see the surface with a clarity never before achieved.

The most remarkable features are the striking impact craters (Figure 5–14A). They generally appear like lunar craters, with central peaks, terraced walls, shocked surfaces, and flooded floors (Figure 5–14B). But they lack extensive ray systems because of the dense atmosphere. They also tend to have a dark shock ring made by a shock front in the atmosphere during the impact.

Overall, Venus appears a fresher, more violent world than was visible in earlier images. The degree of volcanism seems more widespread, with many little features along with large ones (Figure 5–14C), and more surface fracturing (Figure 5–14D) is revealed by Magellan images. As the mission progresses, we will finally be able to compare the Earth and Venus in geologic detail.

(E) MAGNETIC FIELD

A metallic core, liquid in part, implies by comparison with the Earth that Venus should have a planetary magnetic field. Because Venus rotates 243 times more slowly than the Earth, we expect its internal dynamo to be weaker and so the magnetic field to be less intense but still present. No probe to date has detected any magnetic field. If one exists, measurements imply an upper limit of $0.5 \times$

FIGURE 5–13 Close-ups of Venus' surface. The upper two frames were taken by the Venera 13 lander. Note the slabs of rock, patches of dark soil, and pebbles close to the edge of the lander. The view is distorted by a wide-angle lens. The lower two frames were taken by Venera 14. Here the flat expanses of rock are unbroken by soil or pebbles.

10^{-8} T! That's much weaker than expected from a simple dynamo model. One possible explanation: we know that the Earth's magnetic field reverses its polarity; in the middle of a reversal, the magnetic field is essentially zero. That may be the situation with Venus now. (Recent reversals on the Earth have occurred roughly every 10^5 years or so.)

Because Venus lacks a global magnetic field, it does not have a magnetosphere that stands off the solar wind. Without this buffer, the solar wind strikes the upper atmosphere; the magnetic field carried by the solar wind picks up some of the charged particles of the ionosphere and carries them away downstream of the planet (Figure 5–15). The wind creates a long tail a few Venus radii in length, a magnetotail that resembles that made by the solar wind interactions with comets.

(F) EVOLUTION OF THE SURFACE

The crust of Venus has experienced some crustal plate movement, as evidenced by rift valleys, where a few plates have pulled apart, and by mountain plateaus, where plates have collided. The widespread cratered terrain indicates that plate movement has not been a planet-wide process, however, as it has been on the Earth; its upper mantle is more rigid.

What do these observations imply about the geologic history of Venus? Its early history (earlier than 4 billion years ago) followed the Earth's early history because the two planets have similar masses, densities, and sizes. We infer that Venus formed about 4.6 billion years ago with other terrestrial planets. As happened to the Earth, Venus' interior differentiated from internal heating. Dur-

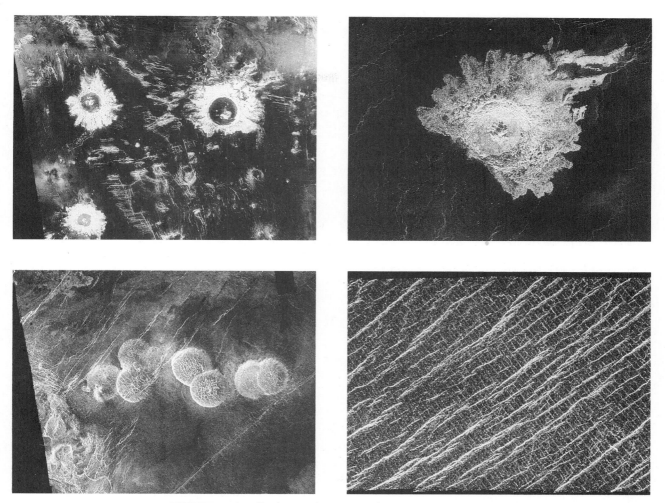

FIGURE 5–14 Surface radar maps of Venus by Magellan. (A) Three large impact craters and fractured plains in the Lavinia region. Their diameters range from 37 to 50 km. Note the rough, bright ejecta, terraced walls, and central peaks. (B) Close-up view of the impact crater Aurelia. Its diameter is 31.9 km; note the bright ejecta is lacking on one side. The impacting body was probably moving at a shallow angle in this direction. (C) Pancake volcanic domes in the southern hemisphere; they average 25 km in diameter. The walls appear bright because they are rough and reflect radar well. (D) Unusual pattern of lines, which are fractures or fault features. Regions like these, called gridded plains, are associated with volcanic features on Venus. *(NASA-JPL)*

ing the first 500 million years, a crust—part basalt and part granite—formed and solidified. About 4 to 5 billion years ago, large masses bombarded the surface and fractured the crust. Volcanoes erupted. Bombardment by smaller bodies from space cratered the surface. About 4 billion years ago, this intensive bombardment ended, and erosion has somewhat altered the ancient surface. Since then, plate movement helped to push up some highland regions. Rift valleys appeared. Huge volcanoes vented through cracks in the surface; their cones formed the shield volcanoes of

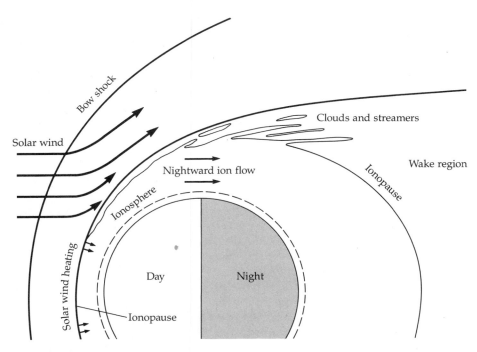

FIGURE 5–15 The solar wind interacting with Venus. Because of the lack of a global magnetic field to hold off the solar wind, it flows directly into the ionosphere and rips off pieces of it in clouds and streamers.

today. Smaller features mark the widespread nature of this activity.

Venus seems to have evolved in a sequence similar to that of the Earth but more slowly and not as much in the later stages, when plate tectonics has dramatically altered the Earth's surface.

5–4 ◐
MARS

Mars shows a ruddy red-orange color, which historically links it to the Greek god of war. This fourth planet from the Sun runs the Earth a close race in its orbit, so that it exhibits the most pronounced retrograde motion at opposition. Opposition occurs with a synodic period of 779.9^d, or about 26 months, and at this time the planet has an apparent angular size of 18″ and may be viewed at full phase throughout the entire night.

(A) MOTIONS

Mars' orbit of semimajor axis 1.5237 AU is inclined only 1.85° to the ecliptic. The orbital eccentricity of $e = 0.0934$ implies that at opposition the Earth–Mars distance may range from about 10^8 km (at which time Mars shows an angular diameter of 14″) to 5.5×10^7 km (25″). The thin atmosphere of Mars permits well-defined surface markings to be

followed clearly; the Martian sidereal rotation period is $24^h 37^m 22.6^s$, and the rotation axis is inclined 25°12′ to the orbital plane. This near coincidence with the Earth's rotational properties ($23^h 56^m 4.09^s$ and 23°27′) implies that you might feel very much at home on Mars, except that the Martian seasons last about twice as long as the Earth's (and, overall, it's much colder!).

(B) PHYSICAL CHARACTERISTICS

Mars has a radius of 3394 km, which is only 53% of the Earth's radius. Mars has two moons, and so we can use Newton's form of Kepler's third law and

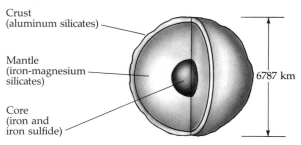

FIGURE 5–16 Interior model of Mars. Note the relatively small size of the core compared to the mantle.

the orbits of these moons to infer Mars' mass. It is only 6.4×10^{23} kg, about 11% of the Earth's mass. The density is 3900 kg/m^3, only a bit higher than the Moon's (3300 kg/m^3) and much lower than the Earth's (5500 kg/m^3).

This comparatively low density implies that Mars' interior (Figure 5–16) must be different from the Earth's. In particular, its core must be smaller and probably consists of a mixture of iron and iron sulfide, which has a lower density than the materials in the Earth's core (4600 to 5000 kg/m^3 compared to 7000 kg/m^3). The Martian mantle probably has the same density as the Earth's. The exact composition of the mantle is not known; one model has a mantle with olivine (an iron–magnesium silicate), iron oxide, and some water (0.3%).

Astronomers have known for a long time that Mars has an atmosphere. The Viking landers found average surface pressures of roughly 0.02 atm. In the Earth's atmosphere, the pressure falls this low at an altitude of 40 km. This thin atmosphere consists of 95% carbon dioxide, 0.1 to 0.4% molecular oxygen, 2 to 3% molecular nitrogen, and about 1 to 2% argon—a composition very similar to the atmosphere of Venus.

The Viking orbiters measured the amount of water vapor in the atmosphere and found the greatest amounts in the high northern latitudes. Peak concentrations were about 0.01 mm of precipitable water (the thickness of the water layer if all the water were taken out of the atmosphere and spread over the surface). On Earth, the atmospheric water vapor is typically several centimeters of precipitable water, and of course the oceans are

several kilometers thick. Mars is a much drier planet than the Earth. Liquid water cannot exist on Mars today because of the low surface pressure. Only in the deepest canyons, where the atmospheric pressure is higher, could water be liquid on the surface. However, it is common to have water ice on Mars, either on the surface or in the clouds. Some evidence suggests water in a permanent frost layer beneath the surface.

Although the atmosphere contains mostly carbon dioxide, its low density does not provide much of a greenhouse effect against temperature extremes. (The overall greenhouse increase is only about 5 K.) At the Martian equator, the difference between noon and midnight surface temperature amounts to almost 100 K when Mars is closest to the Sun. The summer tropical high temperature of 310 K is exceptional. For a period of two Martian months, the surface temperature remains below the freezing point of water both day and night over the entire surface. At the Viking 1 site, 23°N latitude, the air temperature near the ground ranges from 190 to 245 K. At the Viking 2 site, which is farther north, the temperature falls even lower and water condenses as frost on the surface. The layer of water ice that coated the rocks and soil is less than 1 mm thick.

(C) SURFACE FEATURES FROM AFAR

The visible surface features on Mars are dark, apparently greenish gray areas, in contrast to the reddish orange color of the rest of the surface (Figure 5–17). This greenness turns out to be an illusion

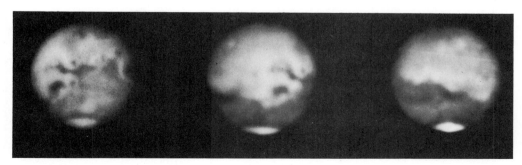

FIGURE 5–17 Mars in 1971. These photographs were made during the 1971 opposition, when the Earth and Mars were very close to each other. Note the polar cap and dark regions. *(R. Minton and S. Larson, Lunar and Planetary Laboratory, University of Arizona)*

caused by the contrast of the light and dark areas. The dark regions are not really green; they are actually red, with greatest intensity in the red region of the spectrum.

The light orange and yellow-brown regions make up almost 70% of the Martian surface. They give Mars its striking reddish appearance. In 1934, the American astronomer Rupert Wildt (1905–1976) suggested that these areas contain ferric oxide (Fe_2O_3). Iron oxides come in many forms on the Earth; all are characteristically brown, yellow, and orange. The Viking landers' measurements indicated a surface composition of about 19% ferric oxide. In addition, the landers measured about 44% silica, which leads to the conclusion that silicate minerals make up a major part of the surface.

The rusty sand, some of which is much finer than that on Earth's beaches, is blown up by fierce winds (speeds greater than 100 km/h) to create planet-wide dust storms. These storms occur most violently when Mars is closest to the Sun. Then the dust clouds, whipped up to heights of 50 km, shroud the entire planet, covering it in a yellow haze for about one month. It takes many months for the fine dust to completely settle to the surface. These global storms sandblast the surface and mix it up so much that the surface composition over the planet becomes essentially uniform. We now

know that this wind-driven dust caused most of the changes in Martian surface features seen in the past.

In 1877, Schiaparelli recorded Martian surface features in great detail. He charted a number of dark, almost straight features, which he called **canali,** Italian for "channels." This word was translated into English as "canals," which implied to some people that they were artificial structures. These so-called canals ignited the curiosity of the American astronomer Percival Lowell (1855–1915). To pursue his interest in Mars, Lowell in 1894 founded an observatory near Flagstaff, Arizona, to take advantage of the excellent observing conditions there. Shortly afterward, he published Martian maps showing a mosaic of more than 500 canals. In a series of popular books, he argued that the canals were artificial waterways constructed by Martians to carry water from the polar caps to irrigate arid regions for farming.

We know now that the north polar cap does indeed contain mostly water ice, especially the residual cap left in the summer, which ranges in thickness from 1 m to 1 km (Figure 5–18). The outer reaches of the cap, prominent in winter, consist of carbon dioxide ice, which condenses at a lower temperature than water ice. (At Martian surface pressures, water ice condenses at about 190 K, car-

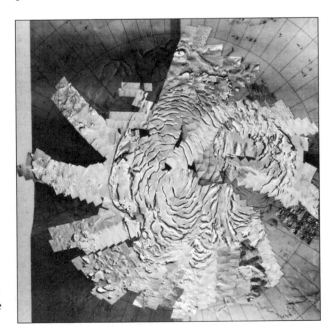

FIGURE 5–18 A summer polar cap. The residual cap of water ice in the summer shows the layered structure of the polar region. *(NASA)*

bon dioxide ice at 150 K.) In contrast, the south polar cap consists of CO_2 ice. The total amount of water in the polar caps is only some 10 m of precipitable liquid.

Lowell was wrong about the canals, however. Planetary astronomers now believe that wind-blown dust deposits might have created temporary features that were seen as the largest and fuzziest of the canals. A comparison of Lowell's canal maps with orbiter photographs indicates that only one real feature (part of Valles Marineris) corresponds to any of the so-called canals.

(D) SURFACE FEATURES CLOSE UP

In 1969, data from Mariners 6 and 7 reinforced the view (first developed from Mariner 4 pictures of 1965) that parts of the surface of Mars resemble that of the Moon. These craft photographed abundant Martian craters visible even under the thinner regions of the polar caps. The Martian craters do look like impact craters but tend to be shallower than lunar ones, in part because the Martian surface gravity is twice that of the Moon. Their flat floors and low rims indicate that they have been strongly eroded, and important clue to conditions in the past.

In contrast, the photographs taken by Mariner 9 (1971) also showed spectacular features of a geologically once-active planet. The two Martian hemispheres have different topological characteristics: the southern is relatively flat, older, and heavily cratered; the northern is younger, with extensive lava flows, collapsed depressions and huge volcanoes. Near the equator separating the two hemispheres lies a huge canyon, called Valles Marineris (Figure 5-19). This chasm is 5000 km long (about the length of the United States) and 500 km wide in places.

The Viking landers touched down on Mars in 1976. They photographed a bleak, dry surface with large rock boulders strewn about amid gravel, sand, and silt. The boulders are basaltic. Some contain small holes (Figure 5-20) from which gas has apparently escaped. On Earth, such basalts originate in frothy, gas-filled lava; the Martian rocks probably had a similar origin.

Both landers uncovered indirect evidence of once-flowing Martian surface water. The region

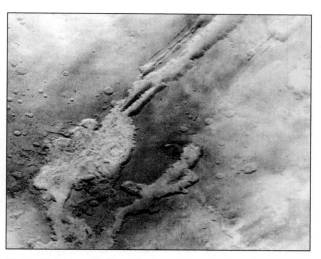

FIGURE 5-19 Valles Marineris. This view shows an area 1500 by 2000 km; the total length is 5000 km. *(NASA)*

around Viking 1 seems to be a flood plain where water sorted the smaller rocks into gravel, sand, and silt. The ground there also resembles the hardened soil of Earth's deserts. Such soil forms when underground water percolates upward and evaporates at the surface. Minerals left behind when this water evaporates harden the soil.

The Mariner 9 mission discovered and the Viking orbiters confirmed a number of sinuous channels that appear to have been cut in the surface by running water. The largest ones have lengths up to 1500 km and widths up to 100 km. (These channels are not the canals seen by Lowell and others; they are crooked and too small to be visible from the Earth.) Some resemble the arroyos commonly found in the southwestern United States. An **arroyo** is a channel in which water flows only occasionally; it shows a downhill flow with meandering patterns and tributary structures where several flows merged to form a larger one. The presence of Martian arroyos requires extensive running water for short periods of time. Since Mars does not have liquid surface water now, conditions for its presence must have occurred in the past and would have required a denser, warmer atmosphere.

The largest of the Martian channels are dry river beds with many branching tributaries (Figure

FIGURE 5–20 View of the Viking 1 site. The rocks are a few tens of centimeters in diameter. Note the dark rock with holes to the left of the scoop; it is a volcanic rock. The scoop has sampled the soil just to the left of this rock. *(NASA)*

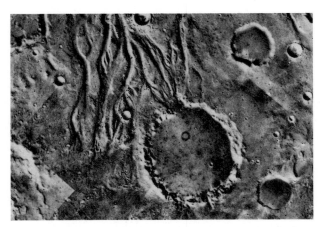

FIGURE 5–21 Outflows channels on Mars, made by the catastrophic release of groundwater. Here the terrain has a slope of a few kilometers. Note how some of the channels cut through craters, an indication that they formed after the cratering. *(NASA)*

5–21). The Mangala Vallis region in the northern hemisphere was a major river junction in wetter times. It covers hundreds of square kilometers. Craters in the channels, some of which show islands from the old flows, give an estimated date of 3 to 3.8 billion years ago as the time of activity.

By far the most striking Martian surface features are the shield volcanoes clustered on and near the Tharsis ridge. The largest is Olympus Mons, 550 to 600 km across at its base (Figure 5–22). The cone's surface shows a wavy texture that is the result of lava flows. The cone reaches 25 km above the surrounding plain, and its base would span the bases of the islands of Hawaii, which are made of several volcanoes. Olympus Mons soars more than 2.5 times the height of Mr. Everest above sea level. The huge mass of Olympus Mons requires that the Martian crust beneath it be 120 to 130 km thick,

FIGURE 5–22 Olympus Mons. Reaching about 25 km high, this shield volcano is wreathed in clouds along its flank. *(NASA)*

about twice the thickness of the Earth's crust, despite the lower Martian surface gravity.

The Tharsis ridge is a hallmark of Mars' northern hemisphere, which differs dramatically from the southern one. The ridge rises about 10 km above the average surface height of the planet and contains numerous volcanic structures. Very few impact craters are visible. In contrast, the southern hemisphere is basically a desert pockmarked by old, eroded craters. The geologic inference from this difference is that, about 3 billion years ago, a huge mass of lava oozed out from under the surface in the northern hemisphere, creating the volcanic plains and the volcanoes over a long period of time. This flow wiped out the older deserts and craters. Other flows have since taken place in this region.

The southern hemisphere of Mars has a cratered terrain (Figure 5–23) that resembles the ancient highlands of the Moon or the intercrater plains of Mercury. The landscape contains impact craters that range in size from huge, lava-filled basins down to some only a few meters across. The Martian craters come in the same variations as lunar

FIGURE 5–23 Argyre Planitia in the Martian southern hemisphere. Argyre is the large impact basin to the left of center. Note the many craters in the surrounding region. *(NASA)*

and Mercurian ones, some with central peaks that mark their impact origin. In general, the Martian craters are shallower than the ones on the Moon and Mercury because of wind erosion and greater surface gravity.

(E) MAGNETIC FIELD

Mars has an extremely weak planet-wide magnetic field, the magnitude of which is only 60 nT at the surface. (Venus, though, has a weaker field!) That small a value presents a puzzle if the dynamo model correctly describes planetary magnetic fields. Mars rotates as fast as the Earth. Though the Martian core is smaller, it should contain a substantial amount of metals. We have no direct evidence that the core is liquid now, but the evidence for past volcanic activity implies a hot mantle and therefore a hot core. So Mars should have a moderately strong field, but it does not. Perhaps, as for Venus, we are viewing Mars in the middle of a magnetic field reversal, but it is unlikely that we would catch both planets in the process of changing polarity.

(F) EVOLUTION OF THE SURFACE

The evolution of Mars has reached a level between that of the Moon and that of the Earth. After the formation of Mars by accretion, impact craters covered the surface. Shortly thereafter, the planet differentiated to form a crust, mantle, and core. Regions of thicker crust rose to higher elevations. In the second phase, thin regions of the crust fractured and the Tharsis ridge uplifted, cracking the surface around it. During this time, a primitive atmosphere, denser and warmer than at present, held large amounts of water vapor from the volcanic outgassing. Rainfall may have eroded the surface in furrows and then percolated to a depth of a few kilometers. Decreasing temperatures formed ice at shallow depths. When heated (perhaps by volcanic activity), this ice melted, leading to the formation of collapse and flow features. Planet-wide water erosion carved the surface.

In the next phase, extensive volcanic activity occurred, especially in the northern hemisphere. The Tharsis region continued to uplift, generating

more faults. Valles Marineris formed at this time. Finally, recent volcanism—most of it concentrated on the Tharsis ridge—broke the surface and spewed out great flows of lava. Since those great eruptions, it is mainly wind erosion that has sculpted the Martian surface. A few small impact craters have probably formed from time to time.

5–5 ◑
COMPARATIVE EVOLUTION OF THE TERRESTRIAL PLANETS

Internal heat, generated by radioactive decay now or left over from the planet's formation, drives the evolutionary processes of Earth-like planets. As we argued in Chapter 4, the lifetime of these processes is roughly proportional to the planet's radius, so that we expect—and find—the Earth the most evolved (and still evolving!) of the terrestrial planets.

We can divide the evolutionary sequence of terrestrial planets into five main stages:

I. Formation, heating of crust and interior
II. Crust solidification, intense cratering
III. Basin formation and flooding
IV. Low-intensity cratering, atmosphere by outgassing
V. Volcanoes, continents, crustal movement

We can compare and contrast the overall evolutionary status of the terrestrial planets in this scheme. Both Mercury and the Moon evolved to the end of stage III; they provide fossil records of the early history of all the terrestrial planets. In particular, they show that an intense bombardment heavily cratered the surfaces 4 billion years ago. On Mars, some of those impact craters are still visible, though highly modified. On the Earth and Venus, subsequent evolution has wiped out evidence of that era. Mars appears, in fact, to have reached an evolutionary status between that for the Moon and Mercury and that for Venus and the Earth; it reached the beginning of stage V. Venus has entered stage V, and the Earth is far into this stage.

Note that the differentiated interior structure of the terrestrial planets implies that all of them became hot enough in their interiors to melt the materials there or at least make them plastic. The original mixture then separated into layers on the basis of different densities.

Why do the Earth and Venus differ so dramatically now even though their masses are almost the same? In particular, why does Venus have an atmosphere that is almost all carbon dioxide, and why did this planet never develop oceans on its surface? The answer probably relates to the fact that Venus is closer to the Sun than the Earth is. Consider the atmospheric carbon dioxide. If the carbon dioxide in the Earth's crust and oceans is added to that in its atmosphere, the total is about the same as that in the atmosphere of Venus. The carbon dioxide on Venus did not end up in the crust because of the lack of water. (Carbon dioxide in solution in the oceans reacts with silicates to form carbonates, and shells of sea life also contain carbonates.) We infer that Venus is so close to the Sun that the greenhouse effect has always kept the surface temperature too high for liquid water. Also, any water vapor in Venus' atmosphere has probably escaped into space. Ultraviolet light from the Sun has enough energy to dissociate water into hydrogen and oxygen; the hydrogen is then lost as it has high enough speeds to escape rapidly to space.

Finally, a comment about cratering, which clearly contributed to the early evolution of all the terrestrial planets. As you will see in the next two chapters, nearly every solid surface in the solar system is cratered. The number of craters of various sizes on a surface allows us, with some assumptions, to infer the range of sizes and infall rates of the impacting masses and the ages of the cratered surfaces.

Basically, the diameter of a crater relates to the kinetic energy of the impacting object, which is a stronger function of infall velocity than mass. Hence, the fact that on most planetary surfaces smaller craters outnumber larger ones implies that smaller objects were in greater abundance. The largest craters and basins were formed by objects more than 100 km in size. Note that, once formed, craters are modified by later impacts, lava flows, and erosion (for a planet with a substantial atmosphere).

The Moon serves as the baseline for estimating cratering rates as a function of time because we can

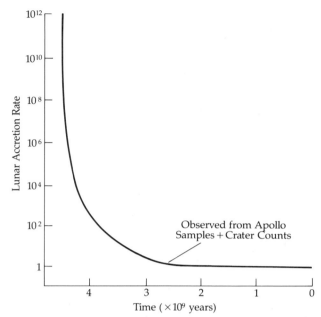

FIGURE 5–24 Schematic reconstruction of the relative rate of cratering of the Moon. Present rates are observed for the Moon; the rates from 4.0 to 4.5 billion years are very uncertain and may be a factor of 100 lower than shown here. *(Adapted from a figure by W. K. Hartmann)*

date the surface areas. One interpretation of the data (Figure 5–24) envisions a sharp but smooth decline in the rate over the past 4.5 billion years. The rate now may be as much as a million times less than that during the era the planets formed—that is what is meant by the early intense bombardment of the surfaces. Impact rates varied from planet to planet, but should be roughly the same (within a factor of 2) for at least the terrestrial planets. It is probable that the evolution of the cratering rate (Figure 5–24) was more or less the same (in a relative sense) for all the planets and satellites.

For the terrestrial planets, we can use the spectrum of crater sizes and surface densities to estimate different degrees of surface evolution. Our Moon ranks as the least modified (along with Deimos and Phobos, the moons of Mars; see Section 7–1), followed by Mercury, Mars, Venus, and the Earth.

PROBLEMS

1. **(a)** Calculate and compare Mercury's surface temperature at noon at perihelion and at aphelion.
   (b) At what wavelength would Mercury's thermal emission peak on the dayside? On the nightside?

2. At a frequency of 10 GHz, calculate the difference (from rotation) between the Doppler shift of a radio signal bounced off one side of Mercury and that of a signal bounced off the other side.

3. Assume Mercury has a satellite whose composition is the same as Mercury's and whose mass is 1% of the mass of Mercury. How closely must it orbit Mercury to remain bound against the tidal force of the Sun?

4. Calculate the length of the solar day on Venus. Watch out for the retrograde rotation!

5. **(a)** Calculate the maximum radar Doppler shift (at 10 GHz) due to rotation from Venus and Mars.
   (b) What accuracy in timing of radio signals must be obtained to detect the minimum height differences on Venus?

6. At what distance from the center of Venus would you expect a magnetic field strength equivalent to that of the Earth's Van Allen belts? Do the same calculations for Mars. What do you conclude?

7. Compare the appearance of the Earth–Moon system viewed from Mars with that viewed from Mercury.

8. Calculate the Roche limit for Mars (with the planetary and satellite densities equal) and compare your results with the orbits of Deimos and Phobos, the moons of Mars.

9. Estimate the lifetime of CO_2 in the atmospheres of Mars and Venus.

10. Compare the scale heights of CO_2 in the atmospheres of Mars and Venus. At what altitude in Venus' atmosphere does the pressure drop to 1 atm?

11. For a magnetic dipole, the field strength far from the dipole varies as $1/R^3$. At what distance from Mercury's center does that planet's magnetic field have the same strength as the Earth's at the Van Allen belts?

12. Use hydrostatic equilibrium to compare the central pressures of Mercury, Venus, and Mars. (*Hint:* Call the surface pressure zero and use the average density.)

13. The mean albedo of Mars is 0.16. What is the noontime temperature of Mars at perihelion? At aphelion? Compare your results to the observed temperature range of 210–300 K.

14. In the late nineteenth century, Percival Lowell claimed that he was able to observe extensive canal structure on Mars. However, other astronomers at the time were unable to see the canals. Evaluate these claims by considering the size of features visible on Mars and the angular resolution of Earth-based telescopes. For a good telescope, angular resolution is limited by the atmosphere "seeing" and not the telescope optics. On a night that is considered excellent seeing, it is possible to resolve structures to an angular size of 1 arcsecond. [On the best sites on the best nights 0.3 arcseconds is possible.]
    (a) What is the angular size of Mars at opposition?
    (b) Valles Marineris is one of the larger features on the Martian surface, with dimensions 5000 km by 500 km. At opposition what are the angular dimensions of Valles Marineris? Should it be visible from the Earth?
    (c) At opposition what would be the angular resolution required to detect something roughly the size of a 1-km-wide canal? Could Lowell have seen this size canal?
    (d) What is the minimum size feature visible with 1 arcsecond seeing?

15. At what distance from the Sun would Mercury be pulled apart by tidal forces? How does this compare with Mercury's actual distance from the Sun?

16. **(a)** To understand the magnitude of the greenhouse effect on Venus, calculate the equilibrium blackbody and subsolar blackbody temperatures of Venus. Compare these temperatures to the observed temperature of 750 K. Make a statement about the importance of the greenhouse effect.
    (b) Do the same for the Earth. Comment on the relative importance of the greenhouse affect for the Earth compared to Venus.

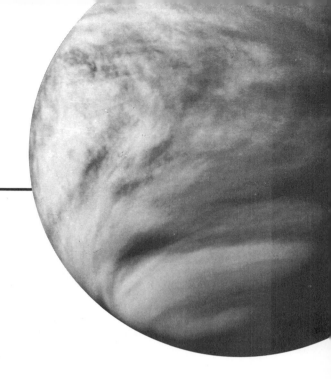

Chapter 6

The Jovian Planets

Based on physical properties, the planets of the Solar System fall into two categories: **terrestrial** and **Jovian.** The previous chapter highlighted the rocky, earthlike worlds. This chapter turns to the gigantic, liquid worlds that make up the Jovian planets—places quite different from the terrestrial planets. The main point of the comparison is this: the Jovian planets are primitive worlds that have evolved less since their time of formation than have the active terrestrial planets (Earth, Venus, and Mars).

6–1 ◗
JUPITER

Beyond Mars, we pass the **asteroid belt** (Section 7–2) near 3 AU and finally come to the lord of the Jovian planets, Jupiter, named after the king of the Olympian gods. Because of its enormous size and high albedo (0.51), Jupiter is a bright planet in the Earth's night sky, especially at opposition.

(A) MOTIONS

Jupiter's orbit about the Sun has a small eccentricity (0.0484) and is inclined only 1.30° to the plane

of the ecliptic; at a semimajor axis of 5.2028 AU, the planet completes one sidereal orbit in 11.862 Earth years. The synodic orbital period of 398.88^d implies that Jupiter returns to opposition (at full phase) about one month later each year.

Because we can see only the upper atmosphere of Jupiter (Figure 6–1), the planet's rotation period is determined by following the rotation period of atmospheric features, by measuring the Doppler shifts of light from the approaching and receding limbs, and by studying the rotation of the magnetic field structure. (The magnetic field period is

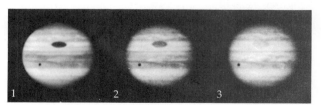

FIGURE 6–1 Jupiter. Taken in 1969, these photographs show Jupiter through blue (1), green (2), and red filters (3). The dark spot at lower left is the shadow of Io. (*New Mexico State University Observatory*)

the most common base rotation rate, called the *body rotation period*.) We find that Jupiter's rotation axis is inclined 3°7' to its orbital axis, but the sidereal rotation period varies from 9^h 50^m near the equator to 9^h 55^m at higher latitudes. The body rate is 9^h 55.5^m. Hence, the gaseous Jovian atmosphere exhibits **differential rotation:** fastest at the equator and slowest at the poles. (The Sun also rotates differentially because it is a fluid.) The rotation structure of Jupiter's atmosphere is reminiscent of the Hadley cells and characteristic trade winds in the Earth's atmosphere [Section 3–2(a)]. Jupiter's extremely rapid rotation results in its large oblateness (0.062).

(B) PHYSICAL CHARACTERISTICS

The equatorial radius ($11.19R_\oplus$) and mass ($318M_\oplus$) of Jupiter have been accurately determined by observing the orbits and occultations of its moons, by noting its gravitational perturbations upon the orbits of comets and asteroids, by measuring the angular diameter of its visible disk (47″ at opposition), and by Voyager flyby measurements. This gigantic prototype of the Jovian planets has a mean density of only 1330 kg/m^3. This low density implies that the composition of Jupiter is similar to the solar abundances of about 75% hydrogen, 24% helium, and 1% all heavier elements (by mass).

The disk of Jupiter shows bands of white, blue, red, and yellow clouds (Figure 6–2). The colors result from the various chemical compounds formed there. These banks change their structure with time, but the relatively stable Great Red Spot (first reported in 1664) has been constantly observed since the nineteenth century. This enormous atmospheric feature measures about 20,000 km by 50,000 km (larger than the Earth!), and it changes shape, position, and intensity. Jupiter's cloudy atmosphere reflects light well and results in the planet's high albedo.

The alternating strips of light and dark regions that run parallel to the equator are called **zones** and **belts.** Infrared observations show that the zones have lower temperatures than the belts. Be-

FIGURE 6–2 Jupiter, Voyager 1 and 2 views. The two top photographs, taken by Voyager 2, show changes in the atmosphere that occurred after the small center photograph had been taken by Voyager 1 four months earlier. *(NASA)*

cause the atmospheric temperature falls at greater altitudes, the zones are higher up than the belts. These differences in temperature imply that the zones mark the tops of rising regions of high pressure and the belts descending regions of low pressure. (The Coriolis effect stretches these convective regions out parallel to the equator.) The convective atmospheric flow transports heat out to space from the planet's interior. The Voyager missions zoomed in on these complex streams and swirls of Jupiter's upper cloud layer. The stunning photographs show the turbulent atmospheric flow (Figure 6–2). Earth-based telescopes have revealed complex changes in the belts and zones. Occasionally, dark blue, red, brown, and white ovals appear against the banded background. These small oval spots last as long as one or two years.

From a variety of observations, we know that the Red Spot is a few degrees cooler than, and extends about 8 km above, the surrounding zone; it is a rising region of high pressure. Also, it rotates counterclockwise like a vortex, just as expected from a high-pressure region in Jupiter's southern hemisphere. A model emerges in which the Red Spot turns like a huge wheel pushed by the surrounding atmospheric flow and in turn deflects nearby clouds to force them around it. So the Red Spot is a huge, long-lived atmospheric eddy with a turbulent region flowing past it (Figure 6–3). As in terrestrial storms, the lifetime scales roughly as the size of the eddy.

Infrared spectroscopy reveals the atmospheric composition above the clouds. In 1934, methane (CH_4) was discovered, the first molecule definitely identified. Later, ammonia (NH_3), molecular hydrogen (H_2), and atmospheric helium (He) were found. Voyager also observed acetylene (C_2H_2), ethane (C_2H_6), phosphine (PH_3), water (H_2O), and germane (G_2H_4). Some of these molecules had been previously detected in spectra taken from the Earth, as had CO and HCN. A few molecules containing deuterium are also known to be present. An analysis of Voyager spectroscopic data implies that Jupiter's upper atmosphere contains (by mass) about 78% hydrogen, 20% helium, and 2% all other elements, a composition essentially the same as the Sun's. Most of this material exists as molecules.

The visible clouds at the tops of the zones are most likely ammonia ice crystals because the temperature here is about 130 K. Below them, according to one model, lies a layer of ammonia hydrosulfide (NH_4HS) clouds. Below these float ammonia vapor and water ice clouds. This model

FIGURE 6–3 The Red Spot. This computer-enhanced view emphasizes the turbulent flows around the Red Spot. The smallest details are 30 km across. *(NASA)*

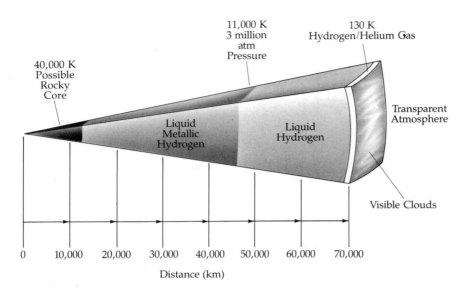

40,000 K
Possible
Rocky
Core

11,000 K
3 million
atm
Pressure

130 K
Hydrogen/Helium Gas

Liquid
Metallic
Hydrogen

Liquid
Hydrogen

Transparent
Atmosphere

Visible Clouds

0   10,000   20,000   30,000   40,000   50,000   60,000   70,000

Distance (km)

FIGURE 6–4 A model of Jupiter's interior. Note the large zone of liquid metallic hydrogen. *(NASA)*

describes three separate cloud layers making up the upper atmosphere. The top layer is ammonia, below that are ammonia and hydrogen sulfide, and the bottom layer is water ice. The colors of the Red Spot and other atmospheric features probably result from chemical reactions of major and trace molecules, with photoionization and dissociation driving the reactions to make hydrocarbons that provide the coloration.

We infer Jupiter's internal structure from physical models that include a solar mix of material throughout and the fact that Jupiter radiates into space more energy than it receives from the Sun (about twice as much) because its temperature is higher than that of a blackbody at the same distance from the Sun and having the same albedo. The internal heat is probably left over from Jupiter's formation.

Models of Jupiter show that the density, temperature, and pressure increase inward (as expected from hydrostatic equilibrium), and so the hydrogen exists in a liquid state. At a pressure of about 3 million atm, the hydrogen is squeezed so tightly that the molecules are separated into protons and electrons that freely move around and can conduct electricity. This state is called **metallic hydrogen,** which has recently been observed in a laboratory on the Earth; its existence is predicted from quantum physics. This strange state persists to within about 14,000 km of the planet's center.

Here, if Jupiter does have a solar composition, lies a solid core of heavy elements (Figure 6–4).

Note that the equation of hydrostatic equilibrium (Equation 4–3) can be applied to the interior of a stable planet, which is neither expanding nor contracting. This condition then requires that the internal pressure increase as one goes deeper into a planet's interior (because a greater weight has to be supported). We can use the equation of hydrostatic equilibrium to estimate roughly the central pressure of Jupiter, as we did for the terrestrial planets. Take

$$P_c = (2/3)\pi G \langle \rho \rangle^2 R^2 = (1.4 \times 10^{-10}) \langle \rho^2 \rangle R^2$$

in SI units (Pa for pressure). Then for Jupiter, we have a radius of about 70,000 km and an average density of roughly 1300 kg/m³, so

$$P_c \cong (1.4 \times 10^{-10})(1300)^2(7 \times 10)^2$$
$$= (1.4 \times 10^{-10})(8.3 \times 10^{21})$$
$$= 1.2 \times 10^{12} \text{ Pa} = 1.2 \times 10^7 \text{ atm}$$

As mentioned earlier, Jupiter (along with Saturn and Neptune, Table 6–1) emits more infrared energy than the amount absorbed from incoming sunlight. The total internal excess power amounts of about 4×10^{17} W (roughly equal to the solar energy input). How can we conclude that the source of this excess is energy from the time of

TABLE 6–1   *Internal Heat of the Jovian Planets*

Planet	T_B (K) (blackbody temperature)	Ratio Radiated to Absorbed Energy	Internal Power (W)
Jupiter	124.4	$(1.67 \pm 0.08) \times$ solar input	4×10^{17}
Saturn	95.0	$(1.79 \pm 0.10) \times$ solar input	2×10^{17}
Uranus	59.1	$\leq 1.4 \times$ solar input	$\leq 10^{15}$
Neptune	59.3	$(2.7 \pm 0.3) \times$ solar input	3×10^{15}

Jupiter's formation? Consider an alternative: the heat may come from a very slow gravitational contraction as the central metallic layer grows at the expense of less dense material. How much shrinkage is needed? Roughly, the gravitational potential energy of a spherical mass is

$$PE \approx GM^2/R$$

so that

$$\text{Energy loss} = d(PE)/dt \approx (-GM^2/R^2)dR/dt$$

and

$$(4 \times 10^{17} \text{ W})(3 \times 10^7 \text{ s/year}) \approx 10^{25} \text{ J/year}$$
$$\approx [(7 \times 10^{-11})(1.9 \times 10^{27})^2/(7.1 \times 10^7)^2](dR/dt)$$

which implies that

$$dR/dt \approx 10^{-4} \text{ m/year}$$

This amount of shrinkage would be undetectable. However, the problem with this idea is that Jupiter's liquid interior is basically incompressible, and so this shrinkage, though small, could not happen for long. In contrast, if the thermal conductivity of metallic hydrogen is not high, then Jupiter could easily retain its primordial internal energy for billions of years. The period of gravitational contraction then could have occurred early in the planet's history until it could no longer contract.

Most of Jupiter is hydrogen, and most of that hydrogen is liquid—quite a contrast to the Earth's interior and those of the other terrestrial planets. The core temperature may be as high as 40,000 K. The flow of heat outward from the core drives the convective circulation of the atmosphere. The rapid rotation of the planet creates a large Coriolis acceleration that produces the beautiful banded atmosphere.

(C) MAGNETIC FIELD

Jupiter exhibits radio emissions that have been linked to a *Jovian magnetic field* of about 4×10^4 T at the surface. (See Table 6–2 for a comparison of the magnetic fields.) This strong magnetic field arises from a dynamo mechanism in a rapidly rotating liquid core of metallic hydrogen. At wavelengths from 3 to 75 cm, the planet radiates nonthermally; this decimeter radiation (1 decimeter = 10^{-1} m) is

TABLE 6–2   *Comparison of Planetary Magnetic Fields*

Property	Earth	Jupiter	Saturn	Uranus	Neptune
Body rotation period (h)	23.9345	9.9249	10.6562	17.24	16.11
Tilt of equator to orbit (°)	23.45	3.08	26.73	97.92	18.8
Mean surface field (T)	0.31×10^{-4}	4.28×10^{-4}	0.22×10^{-4}	0.23×10^{-4}	0.13×10^{-4}
Magnetic dipole tilt (°)	11.4	9.6	0.0	58.6	46.8
Magnetic dipole offset (radii)	0.0725	0.0	0.04	0.3	0.55

synchrotron radiation from relativistic electrons spiraling at speeds very near the speed of light in Jovian **radiation belts** trapped by Jupiter's magnetic field. Radio interferometer measurements and the Voyager missions reveal radiation belts similar to the Earth's Van Allen belts (Figure 6–5), extending beyond three Jovian radii at the magnetic equator; the magnetic axis is inclined 9.6° to Jupiter's rotation axis. This intense field creates around Jupiter a huge magnetosphere that deflects the solar wind.

In 1955, scientists at the Carnegie Institute observed sporadic radio bursts at decameter wavelengths (1 decameter = 10 m); these bursts correlated with the transit of Jupiter. In 1964, some radiation was found associated with the position of the Jovian satellite Io relative to Jupiter's magnetic axis. The probability of radio bursts is greatly enhanced when Io is in one of two positions relative to Jupiter and the Earth: either 90° or 240° from the direction of superior conjunction. This Io-related emission likely is generated in and above Jupiter's ionosphere on magnetic field lines that thread to Io [Section 7–1(b)] along a magnetic flux tube. Other radiation occurs as sharp, intermittent bursts that last about 1 s. A strong burst generates approximately 10^{11} J; in comparison, an intense lightning bolt on Earth discharges about 10^5 J. These radio bursts are likely generated by superbolts of lightning. Some Voyager pictures show bright regions on the planet's nightside that may be lightning. Violent updrafts and turbulence in Jupiter's atmosphere probably generate this lightning in much the same way as thunderstorms do on the Earth.

Earth-based radio observations and satellite flybys reveal that Jupiter's magnetosphere has a complex structure. If it were visible to us optically, the magnetosphere seen head-on would subtend an angle of 2°—four times larger than the angular size of the Sun or Moon (Figure 6–6)! Jupiter's magnetosphere falls into three distinct zones. An *inner magnetosphere* marks the region where the magnetic field generated by currents within the planet dominates; it extends out to about $6R_J$.

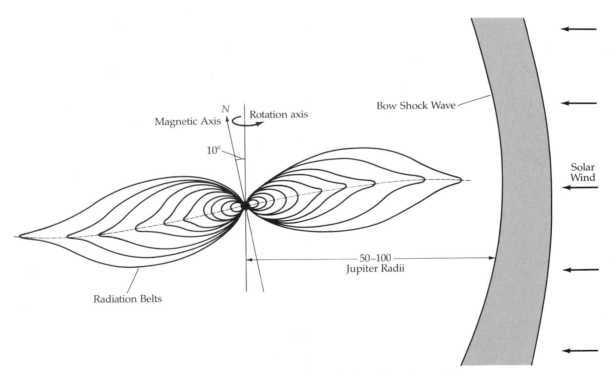

FIGURE 6–5 Jupiter's magnetosphere, based on spacecraft data. The size and shape vary with time depending on the gustiness of the solar wind.

FIGURE 6–6 Radio map of part of Jupiter's magnetosphere. This radio map at 1.5 GHz shows the synchrotron emission from the relativistic electrons in the radiation belts. *(VLA, NRAO/AUI)*

From there to $30R_J$ to $50R_J$ extends the *middle magnetosphere*, where equatorial azimuthal currents control the field configuration. Beyond $50R_J$, the geometry of the *outer magnetosphere* depends on the sunward or nightside orientation. The sunside fields act as a buffer zone that expands and contracts with variations in the solar wind's intensity. The nightside shows a long magnetic tail, about $400R_J$ in diameter and a few AU in length.

The planet's dipolar field controls the inner magnetosphere. It is most simply modeled by a dipole tilted with respect to the spin axis. The magnetic field strength is 4.3×10^{-4} T at $R_J = 1$. As this field spins, it accelerates electrons to relativistic speeds—to tens of MeV. Those in the region from $1.3R_J$ to $3R_J$ generate the synchrotron emission.

From our previous discussion [Section 4–6(c)] of the motion of charged particles in the Earth's magnetic field, you know that the particles will gyrate around the magnetic field lines with a radius

$$r = mv/qB$$

where m is the particle's mass, v its velocity, q its charge (in coulombs), and B the magnetic field strength (in teslas). The frequency of its motion is

$$f = v/2\pi r$$
$$= v/2\pi(mv/qB)$$
$$= qB/2\pi m$$

This frequency is often called the **cyclotron frequency,** for it is the frequency with which charged particles travel around a cyclotron, which uses a uniform magnetic field to contain them. For electrons, the cyclotron frequency is

$$f_c = Be/2\pi m_e \approx (2.8 \times 10^4)B \qquad \text{(MHz)}$$

where B is in teslas. These accelerated particles emit electromagnetic radiation. Such radiation moving at nonrelativistic speeds is emitted in all directions. For electrons moving at relativistic speeds, the emission is concentrated in the forward direction in a beam of opening angle

$$\theta = (1 - v^2/c^2)^{1/2} \qquad \text{(radians)}$$

or

$$\theta \approx 56/E \qquad \text{(degrees)}$$

when E is in MeV. An observer located in the electron's plane of gyration would see one pulse per revolution at frequency f_c, and the emission has linear polarization. This polarized radiation is called **synchrotron radiation.**

The actual synchrotron emission of a group of electrons at various energies consists of a broad envelope characterized by the spread in f_c. If their velocities are not perpendicular to the magnetic field, the particle paths are helices with a pitch angle of α (α is 90° when an electron moves perpendicular to the magnetic field). Then the peak in the synchrotron emission spectrum occurs at approximately

$$f_{max} \approx (4.8 \times 10^{-4})E^2B \sin \alpha \qquad \text{(MHz)}$$

where E is in MeV and B in teslas The flat peak in the emission from Jupiter occurs at roughly 1 GHz ($=1000$ MHz). For a 90° pitch angle, this implies that in a field of 10^{-4} T, the electron energies are

$$E^2 \approx f_{max}/(4.8 \times 10^{-4})B \sin \alpha$$
$$\approx (1000)/(10^{-4})(1)(4.8 \times 10^{-4})$$
$$E \approx 140 \text{ MeV}$$

and the emission is beamed within a cone angle of 0.4°.

Voyager pictures of Jupiter's nightside show polar *aurorae* for the first time. The photographs indicate that the aurorae occur in at least three layers: at 700, 1400, and 2300 km above the cloud tops. We presume that these aurorae happen for the same reason as on the Earth: excitation of the upper atmosphere by energetic charged particles pouring in near the north and south magnetic poles. Some of these particles flow from Io and are trapped by Jupiter's magnetic field; others are solar wind particles.

6–2 ◑
SATURN

Beyond Jupiter orbits the last of the seven planets known to the ancients: Saturn, named after the father of Jupiter. Saturn is the second largest Jovian planet in the Solar System, and it is girdled by a most magnificent system of rings (Figure 6–7). The splendor of the Saturnian sky is marked by the bright bands of its rings and its many moons.

(A) MOTIONS

At the orbital semimajor axis of 9.539 AU, Saturn's sidereal revolution period is 29.458 years in a moderately eccentric (0.0556) orbit inclined 2.49° to the ecliptic. From the Earth, Saturn's angular diameter at opposition is about 20″.

Like Jupiter, Saturn has a thick, cloud-filled atmosphere that rotates differentially. By observing the Doppler shifts across the planet and accurately timing atmospheric markings, we find the sidereal rotation period to be $10^h 14^m$ near the equator and $10^h 38^m$ at high latitudes. The body rotation period is $10^h 39^m$. Again, we have a differential rotation similar to Jupiter's. Saturn's equator is inclined 26°44′ to its orbital plane, so that alternate poles of the planet are tilted toward the Earth at intervals of about 15 years. The rotation causes the large oblateness (0.096) of Saturn; the polar and equatorial radii are in the ratio of about 9:10.

(B) PHYSICAL CHARACTERISTICS

Saturn bears considerable resemblance to Jupiter. Saturn is slightly smaller ($9.0R_\oplus$) and less massive ($95M_\oplus$) than Jupiter. It has the lowest bulk density of any of the planets—only 680 kg/m^3, less than that of water!

The atmospheric structure of Saturn also resembles that of Jupiter with its belts running parallel to the equator (Figure 6–8A). Disturbances in Saturn's belts are much rarer (only ten spots have been observed to date from the Earth) than on Jupiter; occasionally large storms do occur (Figure 6–8B). Voyager 1 discovered a reddish spot, but it is much smaller than the Great Red Spot on Jupiter; clouds only a few hundred kilometers across were detected at high latitudes.

In September 1990, amateur astronomers discovered a rare outburst in Saturn's atmosphere: a giant "white spot." HST photographed (Figure 6–8B) this turbulent storm in the northern hemi-

FIGURE 6–7 Saturn viewed from the Earth. Note the banded structure of the atmosphere. *(Palomar Observatory, California Institute of Technology)*

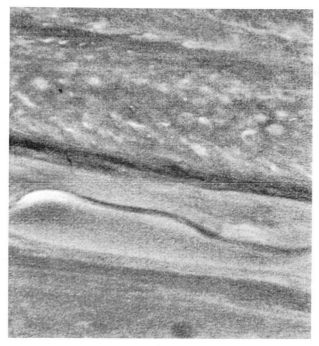

A

B

FIGURE 6–8 Saturn's atmosphere. (A) This Voyager photograph shows jet streams and turbulence in the upper atmosphere. (B) A large "white spot" disturbance in Saturn's atmosphere in 1990. The initial cloud grew rapidly, spread along the equator, and climbed higher. *(NASA)*

sphere, which has thrust gas upward from lower layers into the higher, colder regions of the atmosphere. Such episodes have taken place at a regular interval in 1876, 1903, 1933, and 1960 at midsummer, when the atmosphere is heated by solar radiation. The winds in Saturn's upper atmosphere blow as fast as 1800 km/s, so that cloud is stretched out in a long band near the equator.

The atmosphere of Saturn probably has much the same composition as that of Jupiter. So far, methane (CH_4), ammonia (NH_3), ethane (C_2H_6), phosphine (PH_3), acetylene (C_2H_2), methylacetylene (C_3H_4), propane (C_3H_8), and molecular hydrogen (H_2) have been detected. The percentage of ammonia is less than that found on Jupiter; probably just as much exists, but at the lower temperature of Saturn (100 K) it has frozen and fallen out of the upper atmosphere as ammonia snow. Infrared spectroscopy has detected abundant molecular hydrogen and a substantial percentage of helium. However, Voyager spectrometers measured only 11% helium by mass, compared with 20% for Jupiter. Some of the helium may have condensed and settled out toward the interior.

Saturn's clouds appear far less colorful than those of Jupiter—the former being mostly a faint yellow and orange. Because temperatures on Saturn are lower than on Jupiter, the Saturnian clouds lie lower in the atmosphere. The Voyager photographs show much of the same complexity of cloud patterns seen on Jupiter, with wind speeds much higher (up to 500 m/s near Saturn's equator, compared to 100 m/s on Jupiter). Voyager 2 photographs show that the weather on Saturn can change enormously in a time equivalent to 1 week on the Earth. Large storm systems changed shape, but still remained visible—a hint that Saturn's storms, like Jupiter's, are longer-lived than Earth's storms.

The Voyager photographs also show that the upper atmospheric wind flow of Saturn is different from Jupiter's. Near the equator, the winds all blow eastward at speeds four times those found on Jupiter. At the higher latitudes, the pattern follows an east–west flow alternation, as found on Jupiter. The wind velocities for both planets fall off rapidly away from the equator, but Saturn's atmospheric bands do not mark jet stream flows as they do on Jupiter.

Saturn resembles Jupiter in another important

respect: infrared observations show that Saturn emits more energy, as infrared radiation, than it receives from the Sun. The excess is about twice the energy Saturn receives from the Sun; the infrared blackbody temperature is about 95 K, but we expect it to be only 82 K from the amount of incoming sunlight and Saturn's albedo. The excess heat from Saturn is somewhat of a puzzle. For Jupiter, the emission can be accounted for as that left over from a period of gravitational contraction during its formation; a similar model for Saturn fails to account for the infrared excess.

What is the source? One speculative idea uses helium rain to release gravitational energy. On earth, water condenses and raindrops fall when the air cools enough. The condensation releases the heat that previously vaporized the water. Saturn's rain is made of helium droplets; as the helium raindrops fall through liquid hydrogen in the interior, their friction produces heat. In essence, gravitational energy is converted to heat by this process, which must have started some 2 billion years ago to account for the excess seen now. About half the original helium would have fallen inward during this time, a consequence supported by the Voyager observations that the atmosphere of Saturn contains about half as much helium (in percentage) as is found in the Jovian and solar atmospheres.

Saturn's interior (Figure 6–9) probably reflects Jupiter's composition. Theoretical estimates are about 74% hydrogen, 24% helium, and 2% heavier elements. This composition is also roughly the same as that of the Sun. Saturn may have a small, rock and ice core 14,000 km in diameter and having a mass of $20M_\oplus$ (25% of the total mass). Other models have the metallic hydrogen region extending right to the center. The level at which hydrogen becomes metallic (pressure of 3 million atm) is much deeper in Saturn than in Jupiter because Saturn has a smaller mass and density and so the internal pressure does not rise as fast as it does in Jupiter.

(C) MAGNETIC FIELD

Saturn's magnetic field has a total moment only 1/34 that of Jupiter's, but that is still strong enough to generate a Jovian-type magnetosphere with

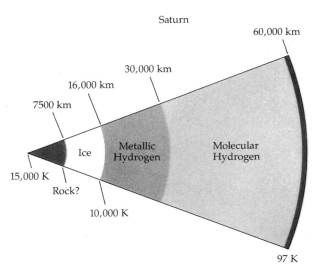

FIGURE 6–9 Interior of Saturn. Note that the liquid metallic hydrogen zone is smaller than Jupiter's.

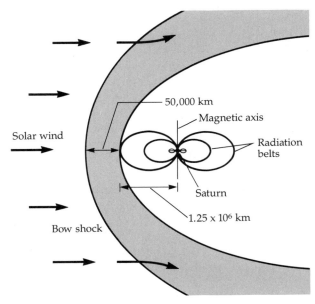

FIGURE 6–10 Saturn's magnetosphere, based on the observations of Voyager 1 as it passed through the magnetosphere. The rotation and magnetic axis are lined up. Note the general shape resembles that of Jupiter. *(NASA)*

Earth-type radiation belts (Figure 6–10). The magnetic dipole moment aligns with Saturn's spin axis, in contrast to the clear tilt (about 10°) of Earth's and Jupiter's magnetic axes. The cloudtop field strength is 2.1×10^{-5} T.

Saturn's magnetosphere stores far fewer particles than Jupiter's. Two reasons for this difference are (1) the lack of a local source of charged particles, as provided by Io's eruptions for Jupiter and (2) Saturn's rings, which effectively absorb charged particles and so sweep clean the inner magnetosphere. Outside the edge of the rings, the density of charged particles rises quickly; it hits a peak at about $5R_S$ to $10R_S$. Here, the charged particles are tightly coupled to the rapidly spinning magnetic field; this interaction generates a plasma sheet about $2R_S$ thick that extends out to $15R_S$. Beyond this, the magnetosphere lacks structure; its size varies with the solar wind. At high solar wind pressures, the magnetosphere can shrink to a radius of $20R_S$; at low pressures, it balloons to $30R_S$ and larger.

6–3 ◑
URANUS

Uranus (Figure 6–11) is named after the progenitor of the Titans and father of Saturn; it is the seventh planet from the Sun and the third Jovian planet. William Herschel discovered it in 1781; at first he thought it was a comet, but his observations implied a low-eccentricity elliptical—hence, planetary—orbit about the Sun. Uranus is just at the limit of naked-eye visibility from the Earth, with an angular diameter at opposition of only 3.6″.

(A) MOTIONS

The orbit of Uranus has a semimajor axis of 19.182 AU, an eccentricity of 0.0472, and an inclination of only 0.77° to the ecliptic; the sidereal orbital period is 84.013 years. Chapter 2 discussed the bizarre rotational behavior of Uranus; with its equatorial plane inclined 98° to its orbital plane, Uranus rotates *retrograde* in 17ʰ 14ᵐ. Since the rotation axis lies essentially in the ecliptic plane, we observe the following phenomena: if we see one pole of the planet now, the equatorial plane will be seen edge-on in 21 years, and in 42 years the opposite pole will point toward the Earth.

FIGURE 6–11 Uranus and its five major moons, as viewed from the Earth. *(W. Liller, NOAO)*

Measuring the rate of Uranus' rotation has been a frustration to astronomers for years, in part because the axial tilt limits the usefulness of Doppler studies. Earth-based observations reported rotation rates of 10.8, 15, 15.6, 23, and 24 hours; infrared observations implied a period close to either 15 or 17 h. The Voyager mission finally provided the body rotation period from measurements of the magnetic field.

(B) PHYSICAL CHARACTERISTICS

Because it is so far from the Sun, Uranus' atmosphere must be very cold. Infrared observations put the blackbody temperature at 59 K. In such a deep freeze, all the ammonia has frozen out of the atmosphere and cannot be detected spectroscopically. Methane and hydrogen do appear in the spectrum. Helium may also have been detected, but this result has not been confirmed.

Viewed through a telescope, Uranus has a distinctive bluish-green color, which comes from sunlight that penetrates deep into the planet's atmosphere; some red light is absorbed in the atmosphere, and much of the green is reflected

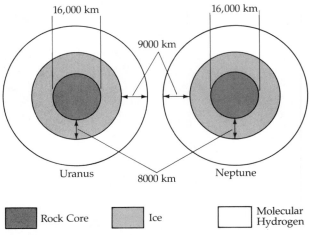

16,000 km 16,000 km

9000 km

Uranus 8000 km Neptune

Rock Core | Ice | Molecular Hydrogen

FIGURE 6–12 Interiors of Uranus and Neptune. Note that the internal structures and compositions are essentially the same, as would be expected from their similar sizes and bulk densities.

back into space. This selective absorption and reflection occur because the atmosphere's spectrum is dominated by absorption banks of methane.

The low bulk density of Uranus, 1600 kg/m³, implies that it contains mostly lightweight elements. Uranus (Figure 6–12) may consist of roughly 15% hydrogen and helium, 60% icy materials (H_2O, CH_4, and NH_3), and 25% earthy materials (silicates) by mass.

(C) VOYAGER RESULTS

The Voyager 2 flyby of Uranus in January 1986 confirmed some of our ideas about this planet and provided some new ones. The planet's rotation rate is now known to be to $17^h 14^m$ from observations of a magnetic field. Curiously, this field is tilted 58.6° to the spin axis, and the north magnetic pole is closest to the south geographic pole (Figure 6–13). (This offset amounts to about 0.3 radius from the center.) The offset of the dipole may result from a dynamo in the mantle rather than the core of the planet.

The spacecraft measured high-altitude winds at speeds of 300 to 400 m/s and an atmospheric content of helium of 10 to 15% (by number)—consistent with the values for Jupiter and Saturn. Voyager also noted emissions from the daylight side of the atmosphere, which are very puzzling because they are not auroral in nature. They seem to occur from molecules excited by low-energy electrons, but the source of the electrons is unclear. Overall, these observations support a model

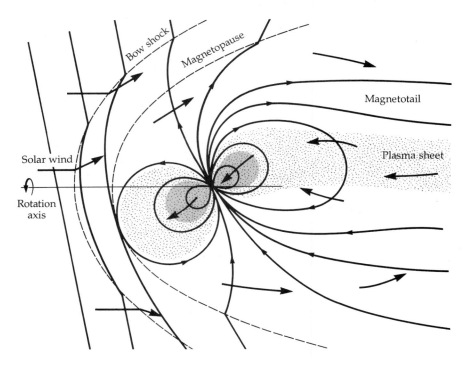

Bow shock

Magnetopause

Magnetotail

Plasma sheet

Solar wind

Rotation axis

FIGURE 6–13 The magnetosphere of Uranus. Note the large tilt (almost 60°) of the magnetic axis with respect to the rotation axis. Plasma of the solar wind is picked up by the magnetosphere to form a long tail and also Van Allen-type belts.

of a planet with a rock and liquid core surrounded by an ocean of water and dissolved ammonia.

Computer-enhanced photographs showed that the ammonia clouds lie low in Uranus' atmosphere below deep layers of haze (Figure 6–14). Images taken through special filters revealed a banded structure to the clouds; those near the equator rotated once in 17^h, near the pole in 15^h—a resolution to the ground-based rotation rate measurements of both periods. (Note that the differential rotation here is the reverse of that for Jupiter and Saturn.) Atmospheric winds generally blow the clouds in the same direction that the planet rotates. Strong plumes appeared in the upper atmosphere, probably generated by violent convection lower down.

Charged particle detectors onboard the spacecraft confirmed that a substantial magnetosphere surrounds the planet. Its presence was first suspected from synchrotron emission measured on the incoming flight path. Overall, the total magnetic field moment is about 1/400 that of Jupiter, but the surface field strength is about the same as Saturn's. The magnetosphere contains energetic particles, many with energies greater than 1 MeV. As expected from a planet with a strong magnetosphere, aurorae actually were observed on the dark side of Uranus.

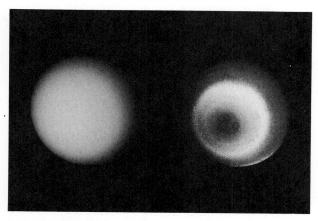

FIGURE 6–14 Voyager 2 image of the south polar region of Uranus. The image on the right has been computer-processed to bring out the structure (such as the dark polar hood) in the upper atmosphere. The image on the left lacks the enhancement. *(NASA)*

6–4 ◑
NEPTUNE

The last of the Jovian planets, and the eighth planet from the Sun, is Neptune. This near-twin of Uranus is named for the god of the sea. Between 1790 and 1840, the orbit of Uranus exhibited perturbations from an unknown source, and the existence of a more distant planet was suspected. J. C. Adams (in 1843) and U. J. Leverrier (in 1846) independently used Newtonian celestial mechanics to deduce the mass and orbit of this eighth planet from the observed perturbations on Uranus. In 1846, Johann G. Galle at the Berlin Observatory found Neptune within 1° of the predicted position! Neptune may have been first seen by none other than Galileo, however, 234 years earlier. Calculations of Neptune's orbit show that it should have been very close to Jupiter in January 1613. Galileo's journals have entries showing that he observed an object in the vicinity of Jupiter near Neptune's predicted position on both December 27, 1612, and January 28, 1613, when he detected a small motion of Neptune with respect to a nearby star. Inexplicably, Galileo never followed up on this discovery and so failed to recognize the object as a new planet.

(A) MOTIONS

Neptune moves around its low-eccentricity (0.0086), low-inclination (1.77°) orbit with a semi-major axis of 30.06 AU in a sidereal period of 164.79 years. Since its discovery, Neptune has traversed only three-quarters of its orbit. Its average distance from the Sun is less than Pluto's, but Pluto has such an eccentric orbit that it can—and has—come within Neptune's orbit. That occurred in January 1979, and until March 1999, Neptune will be the outermost planet in the Solar System!

Neptune's rotation period has been hard to pin down. Ground-based images in the near infrared showed atmospheric features that rotate in $17^h 50^m$, in good agreement with infrared photometry of rotation variations with a period of $17^h 43^m$. Voyager 2 found the body rotation rate of 16.1 hours.

(B) PHYSICAL CHARACTERISTICS

In many ways, Neptune is the twin of Uranus. Like Uranus, Neptune has a light green color from selective methane absorption. The upper atmosphere displays faint cloud bands. This cold atmosphere (≈74 K) probably contains water ice and ammonia ice mixed with gaseous methane, hydrogen, and helium. One difference: ethane (C_2H_6) has been detected in Neptune's atmosphere but not in Uranus'. The internal structure of Neptune (Figure 6–12) probably resembles closely that of Uranus because the bulk densities and masses are similar. Infrared observations show that Neptune's temperature is about 59 K; the expected blackbody value is 44 K if Neptune were heated only by the Sun. So Neptune, unlike Uranus, has internal heat.

(C) VOYAGER AT NEPTUNE

In August 1989, Voyager 2 skirted a mere 5000 km above Neptune's clouds—a fitting flourish to end its 12-year mission of discovery. Voyager revealed a Neptune that enlarged the strange realm of the Jovian worlds, in which planets, rings, and moons make up a complex part of the solar system.

Months before the closest encounter, Voyager began imaging conspicuous markings in the upper atmosphere of Neptune. The most striking became known as the Great Dark Spot (Figure 6–15), which is a storm some 30,000 km across rotating counterclockwise in a few days. A region of high pressure, the Dark Spot lacks the typical atmospheric methane; here, we are looking into Neptune's atmosphere. Bright, cirrus-like clouds accompany the Dark Spot (which is actually bluish in color) and also appear in some other latitude bands. Most of these clouds change size or shape from one rotation to the next. Believed to be condensed methane, the clouds lie about 50 km above the general cloud layer (Figure 6–15), which consist of hydrogen sulfide. Compared to that on bland Uranus, the atmospheric activity seen on Neptune came as a surprise. It is likely driven by the outflow of Neptune's internal heat. A few other dark and bright spots are present, but the complex swirls and banded structure seen on both Jupiter and Saturn seem to be absent on Neptune. In general, the atmosphere contains molecular hydrogen and helium.

Six days before reaching closest approach, much later than expected, Voyager finally picked up radio signals from Neptune's magnetosphere.

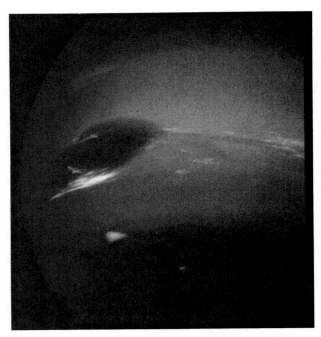

FIGURE 6–15 Voyager 2 image of Neptune's Great Dark Spot and smaller dark spot. Note the white, high-altitude clouds between them. *(NASA)*

Tracking these signals gave an accurate measure of the rotation period: $16^h 3^m$. A surprise was the discovery that the magnetic axis is tilted about 46.8° from Neptune's axis of rotation, almost as much as the 58.6° tilt of Uranus' magnetic axis. The reason for these large tilts is not yet known. The magnetic field strength is about one-fifth that of the earth's. The dipole is offset about 0.55 Neptune's radius. The magnetosphere has a very low density of trapped, charged particles.

6–5 ◐
PLUTO AND CHARON

Pluto, the ninth planet from the Sun (sometimes!), is named after the god of the underworld (Hades). From the Earth, Pluto presents only a faint stellar image at the telescope; from Pluto, the rest of the Solar System is distant and close to the Sun, which appears only as a very bright star in the sky.

After the discovery of Neptune, small unexplained perturbations appeared in the orbit of Uranus. Because the predicted position of a ninth planet that would cause these perturbations was very uncertain, initial attempts to find the planet were unsuccessful. It was not until March 1930 that Clyde W. Tombaugh found Pluto near a position predicted by Percival Lowell. Today we know that this discovery was a fluke because Pluto's low mass could not have caused the apparent perturbations of Uranus. (Pluto is in this chapter mostly because of its proximity to the Jovian planets. Pluto is much smaller than the Jovian planets, but it has a somewhat similar density.)

Pluto's average distance from the Sun is 39.44 AU. Since it has a highly eccentric orbit ($e = 0.25$), it ranges from 29.7 to 49.3 AU from the Sun and so is never closer to the Earth than 28.7 AU at opposition. Because of its great distance and small diameter, Pluto presents a difficult object to observe well, and attempts to measure its diameter have been frustrating. Infrared spectral observations show that methane ice coats come of Pluto's surface. The methane ice there means that the surface temperature is no more than 40 K. In addition, Pluto has a darker equatorial region compared to its polar caps. Observations of Pluto's brightness have revealed cyclic variation every 6.4 days. It is the only evidence of rotation, and 6.4 days is generally accepted as Pluto's rotation period. Pluto's axis of rotation lies close to the ecliptic, like that of Uranus.

On 5 September 1989, Pluto reached perihelion in its orbit. A little more than a year earlier, on 9 June 1988, Pluto passed in front of a star as observed in Australia. Coordinated observations from this region confirmed that Pluto has an atmosphere, which stretches over 600 km from the planet's surface. This atmosphere probably consists of methane gas (with a surface pressure of a mere 10^{-8} atm or so) that has been released from the ice on the surface as the planet is heated by its closest approach to the sun in 248 years.

In June 1978, James Christy of the U.S. Naval Observatory in Flagstaff, Arizona, noticed what appeared to be a bump on Pluto's image in a photograph (Figure 6–16). Checking older photographs, Christy found seven showing the same bump, always oriented approximately north–south. He proposed that the bump was the faint image of Pluto's moon partially merged with the image of the planet. Christy named this moon

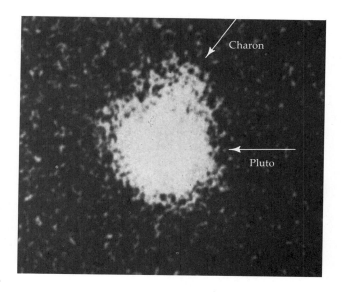

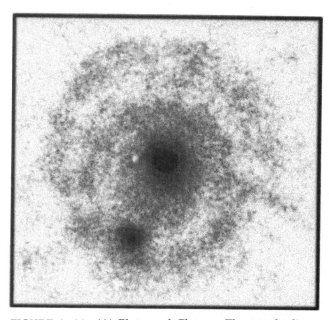

FIGURE 6–16 (A) Pluto and Charon. The two bodies are so close together that their images are merged; Charon is a bump (arrow) off Pluto. *(J. Christy, U.S. Naval Observatory)* (B) Pluto and Charon imaged by HST. At the time of the observation, Charon, at the lower left, was about 0.9 arcsec from Pluto. *(NASA/ESA)*

Charon, after the mythological boatman who ferried souls across the river Styx to the underworld god Pluto.

A few years after this discovery, Pluto and Charon came into an orientation so that they eclipsed each other as seen from earth. Such alignments occur only twice each 248 years. These eclipses (Figure 6–17) indicated that Pluto has a diameter of 2240 km; Charon has a diameter of roughly 1120 km, about half the size of Pluto. The observations of Charon show a revolution period of 6.4 days (the same as Pluto's rotation period, so it is in synchronous rotation) at a distance of 19,600 km from Pluto. That the revolution period of Charon is the same as Pluto's rotation indicates that the two bodies are tidally locked.

Knowing the orbital properties of the Pluto–Charon system, we can find Pluto's mass by Kepler's third law. The orbital period is 6.4 days. The separation is 19,600 km. Let's compare the Earth–Moon system with the Pluto–Charon system. For the Earth and Moon,

$$M_E + M_M = (4\pi^2/G)(a_{EM}{}^3/P_{EM}{}^2)$$

and for Pluto and Charon,

$$M_P + M_C = (4\pi^2/G)(a_{PC}{}^3/P_{PC}{}^2)$$

Divide the second equation by the first to get

$$(M_P + M_C)/(M_E + M_M) = (a_{PC}/a_{EM})^3(P_{EM}/P_{PC})^2$$

Assume that the masses of the moons are much smaller than those of their parent planets to approximate $M_E + M_M$ by M_E and $M_P + M_C$ by M_P. Then

$$
\begin{aligned}
M_P/M_E &= [(1.96 \times 10^4 \text{ km})/(3.8 \times 10^5 \text{ km})]^3 \\
&\quad \times [(27.3 \text{ days})/(6.4 \text{ days})]^2 \\
&= (1.4 \times 10^{-4})(18.2) = 2.5 \times 10^{-3}
\end{aligned}
$$

The Earth has a mass of 6×10^{24} kg, and so

$$
\begin{aligned}
M_P &= (6.0 \times 10^{24} \text{ kg})(2.5 \times 10^{-3}) \\
&= 1.5 \times 10^{22} \text{ kg}
\end{aligned}
$$

We note that using Kepler's third law with the usual approximation does *not* give an *accurate* value for Pluto's mass because Charon has a mass about 10% that of Pluto. We find this mass ratio from observations of the center of mass of the

FIGURE 6–17 Pluto and Charon eclipses. Charon's orbit around Pluto as viewed from the Earth created mutual eclipses every 3.2 days from 1985 until 1991. These events allow us to measure the sizes and also see variations in surface brightness on both bodies. These computer-generated images show an eclipse sequence and surface features. *(K. Horne, M. Buie, and D. Tholen)*

Pluto–Charon system; Kepler's third law provides only the sum of the masses.

Together, Pluto and Charon have an average density of some 2100 kg/m³. Pluto may contain as much as 75% rocky material. Its internal structure (Figure 6–18) may have a mantle a few hundreds of kilometers thick of water; below that, a core of

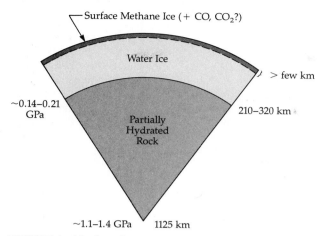

Surface Methane Ice (+ CO, CO_2?)

Water Ice

Partially Hydrated Rock

~0.14–0.21 GPa

> few km

210–320 km

~1.1–1.4 GPa   1125 km

FIGURE 6–18 Interior structure of Pluto today, as inferred from its bulk density and size and assumed equations of state of the materials. *(Based on a model by W. B. McKinnon and S. Mueller)*

partially hydrated rock. This contrasts to the mostly water compositions of the Jovian moons (Chapter 7). One scenario is that a collision with a large body could have stripped Pluto of some of its icy materials and changed the rock-to-ice ratio from that of a lower-density Jovian moon. (The analogy here is to a giant impact stripping the rocky mantle of Mercury, Section 5–2).

The search at Lowell Observatory for other planets beyond Neptune ended in 1945 without yielding further positive results. A planet with the same characteristics as Pluto but placed at a greater distance from the Sun has little change of discovery. Some astronomers have postulated a tenth planet orbiting retrograde beyond Pluto (from perturbations observed for comets), but nothing has been observed to date. Tombaugh's search, which covered 13 years, would have uncovered a planet like Neptune as far as 100 AU from the Sun.

PROBLEMS

1. Determine the orbital periods of particles at the inner and outer edges of Saturn's rings. At what distance from the center of Saturn will a particle orbit the planet in 10^h 14^m? Show that the inner particles of the rings rise in the west and set in the east of Saturn's sky and the outer particles rise in the east and set in the west. Is this result paradoxical? Explain.

2. Show how the orbits of Uranus' moons appear from the Earth over a period of 100 years.

3. Show that the moons of Neptune obey Kepler's third (harmonic) law and deduce the mass of Neptune. (*Hint*: Use appropriate units or ratios.)

4. If Pluto's radius is 1120 km, what must its mass be to give the planet the same density as an icy moon of Saturn?

5. Jupiter has a strong magnetic field, about 10^{-5} T at a distance of 25×10^3 km from the surface (Pioneer 11). Estimate the size of Jupiter's magnetosphere and compare it with that of the Earth. Assume that the field is a dipole and that the solar wind pressure falls off as $1/R^2$ with distance R from the Sun. (*Hint*: Magnetic field pressure is proportional to the *square* of field intensity.)

6. Infrared observations indicate that Saturn gives off 2.8 times the energy it receives from the Sun for a total internal power loss of 2×10^{17} W. Assume that gravitational contraction releases this thermal energy. How much must Saturn shrink per year to account for this output?

7. Assume that Saturn's internal heat is left over from primordial contraction. Calculate the *maximum* bulk thermal conductivity the planet would need to retain enough of its internal energy to account for its present luminosity. Theoretical calculations indicate that Saturn's maximum luminosity was about 10^{20} W 4.5 billion years ago. The *thermal conductivity*, κ, is the flow of heat energy per unit time per unit area per unit temperature gradient (units: J/s · m · K), so

$$\kappa = -H/A(\Delta T/\Delta x)$$

where H is the flow of heat energy (J/s), $\Delta T/\Delta x$ is the temperature gradient (K/m), and A is the surface area (m^2).

8. Spectroscopic observations suggest that Pluto is covered with icy frost and thus has a high albedo (0.5). The brightness of Pluto at opposition (38 AU from the Earth) is 2×10^{-17} as bright as the Sun (1 AU

from the Earth). From these two observations, calculate a radius of Pluto.

9. Imagine that you are viewing an eclipse of Charon by Pluto.
   (a) How could you use your observations to infer a diameter for Pluto?
   (b) By what percentage would the total brightness of the system dim at mid-eclipse?

10. Estimate the lifetime of methane in the atmospheres of Jupiter and Uranus.

11. Use the equation of hydrostatic equilibrium to compare the central pressures of Saturn and Uranus.

12. Calculate the blackbody equilibrium temperatures of Uranus and Neptune and compare your values with the measured temperatures given in the text.

13. Calculate the relative Doppler shift from the approaching and receding edges of Jupiter due to planetary rotation for a spectral line at rest wavelength $\lambda_0 = 500$ nm.

14. (a) At what distance from Jupiter's surface is its magnetic field strength equal to the Earth's surface field strength? Assume both planets have dipole magnetic fields and consider field strengths at the equator.
    (b) Assuming ideal dipole behavior (even in Jupiter's interior) what would be the magnetic field strength at one Earth radius from Jupiter's center? [*Note:* The ideal dipole assumption would not really be valid at this point because it is inside the metallic hydrogen layer that generates Jupiter's magnetic field. But use it anyway.]

15. Qualitatively compare the characteristics (size, temperature, rotation, meteorological characteristics) of Jupiter's Great Red Spot with Neptune's Great Dark Spot, and with the smaller white spots and brown ovals on Jupiter.

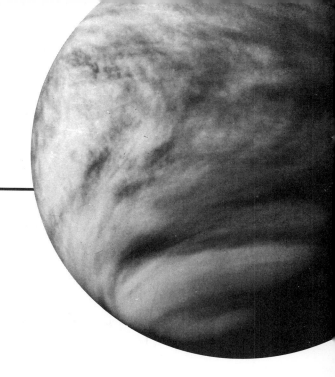

Chapter 7

Small Bodies and the Origin of the Solar System

The previous four chapters put forth our current understanding of the planets. We have focused on the evolution of these large bodies of the Solar System but so far have only hinted at their formation. Key clues to the origin of the Solar System are locked in its small bodies: moons, rings, asteroids, meteoroids, comets, and interplanetary dust. This chapter treats the properties of this interplanetary debris and connects them to a contemporary model of the Solar System's formation, which took place 4.6 billion years ago.

7–1 ◗
MOONS AND RINGS

Only three moons orbit all the terrestrial planets (our Moon and Deimos and Phobos of Mars). In contrast, the Jovian planets carry at least 50 moons (not counting Charon) as well as many rings, which contain a multitude of tiny moons. This section examines these objects, which range from small rocks to planetary-sized bodies.

(A) THE MOONS OF MARS

Two moons encircle the planet Mars; they are named Phobos and Deimos ("fear" and "panic") after the mythological companions of the god Mars. Asaph Hall (1829–1907) at the U.S. Naval Observatory discovered the two moons in 1877, both of which lie close to Mars and orbit the planet rapidly (Figure 7–1). Deimos, the outer moon, circles Mars in 30.3^h; Phobos, the inner moon, takes a mere 7.67^h. In fact, Phobos is one of few moons (including Jupiter's innermost satellites, Metis and Andrastea, discovered by Voyager) that orbit their parent planets faster than the planets spin. So Phobos rises in the west and sets in the east as seen from the Martian surface! Like the Earth's Moon, the Martian moons keep the same face to the planet in synchronous rotation.

Deimos and Phobos have ellipsoidal shapes with three axes. Phobos, the larger, has axes about 27, 21, and 19 km long; Deimos' axes are only 15, 12, and 11 km. Phobos (Figure 7–2) and Deimos have heavily cratered surfaces. The sizes and num-

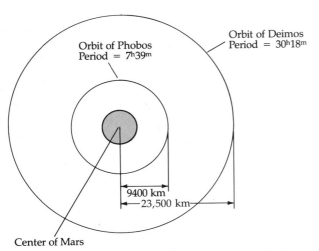

FIGURE 7–1 Orbits of Deimos and Phobos relative to Mars.

bers of these craters indicate that the surfaces of these satellites are at least 4 billion years old and little modified since the cratering.

What is the origin of these miniature moons? One hint comes from the overall albedos: 0.022 for Deimos and 0.018 for Phobos in the visual. These dark surfaces resemble a certain class of meteorite (carbonaceous chondrites, Section 7–4) and asteroids (such as Ceres, Section 7–2). Also, recall that Mars is the closest of the terrestrial planets to the asteroid belt (the average distance between Mars and the asteroids is 1.3 AU). Hence, Mars may have captured both moons from high-eccentricity asteroids that passed close by. Theoretical calculations indicate that such captures are possible. (And you will see that Jupiter has a set of asteroidal moons.)

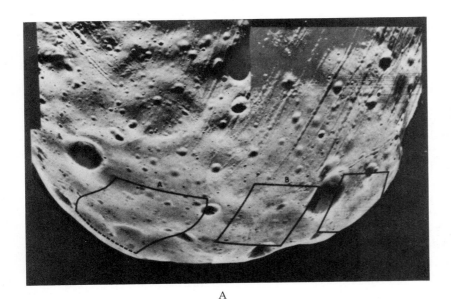

A

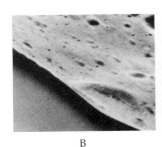

B

C

D

FIGURE 7–2 Phobos. The areas outlined in the upper photo are shown in more detail in the lower ones. These close-up photographs show craters as small as 10 m in diameter. Their flat bottoms indicate a surface soil hundreds of meters deep. *(NASA)*

TABLE 7–1   *Properties of the Galilean Satellites*

Name	Diameter (km)	Distance (km)	Orbital Period (days)	Bulk Density (kg/m^3)	Mass (Moon = 1)
Io	3632	4.22×10^5	1.77	3530	1.21
Europa	3138	6.71×10^5	3.55	3030	0.66
Ganymede	5262	1.07×10^6	7.16	1930	2.03
Callisto	4800	1.883×10^6	16.69	1790	1.45

(B) THE MOONS AND RINGS OF JUPITER

Jupiter possesses an entourage of at least 16 moons, the brightest and largest of which were first discovered with a telescope by Galileo. These *Galilean moons* (Table 7–1) have orbits that lie within 3° of Jupiter's equatorial plane, close to our line of sight, in the following order: Io, Europa, Ganymede, and Callisto (Figure 7–3). Each rotates synchronously.

Their bulk densities are: Io, 3500 kg/m^3; Europa, 3000 kg/m^3; Ganymede, 1900 kg/m^3; and Callisto, 1800 kg/m^3. Note the pattern; density decreases with increasing distance from Jupiter. Such density differences show that the compositions of Io and Europa are mostly rock, with perhaps a little icy material. In contrast, Ganymede and Callisto must contain substantial amounts of water ice or other low-density icy materials and proportionally much less rocky material.

Io

Io has a thin atmosphere, with a surface pressure of about 10^{-10} atm. (Only two other satellites, Saturn's Titan and Neptune's Triton, are known to have atmospheres.) Io's atmosphere gives off a yellow glow from emission by sodium atoms, which surround Io out to a distance of about 30,000 km. The cloud extends about 200,000 km along Io's orbit, forming a partial ring of gas around Jupiter.

Volcanic eruptions, at least in part, produce Io's sodium cloud. Io has at least 11 active volcanoes; in fact, it is the most volcanically active body in the Solar System, which implies that the interior is now hot. The volcanoes eject plumes of gas and dust to heights of 250 km at velocities of up to 1000 m/s. (In contrast, the Earth's large volcanoes spit out material at about 50 m/s.) On a nearly airless body like Io, the volcanic gas and dust crest like a fountain plume in several minutes and then spread and fall in a dome shape (Figure 7–4).

Io's volcanoes are not shields or cones like those commonly found on the terrestrial planets. Instead, they resemble collapsed volcanic craters from which lava simply pours from a crater vent and spreads outward for hundreds of kilometers. Multicolored lava lakes surround many of Io's volcanoes; the temperatures in these lava lakes are about 330 K. The red, black, yellow, orange, and white coloration comes from sulfur and sulfur compounds. No impact craters appear on Io; volcanic flows have covered them up. Its surface is

FIGURE 7–3   The Galilean moons of Jupiter in their correct relative sizes: Io (top left), Europa (top right), Ganymede (bottom left), and Callisto (bottom right). *(NASA)*

A

FIGURE 7–4 Volcanic eruptions on Io. (A) An eruption plume (arrow), darker than the surface, is visible at the top right. (B) An eruption of Pele, in a view looking down on the black fissure in a complex of hills that is the source of the outburst. *(NASA)*

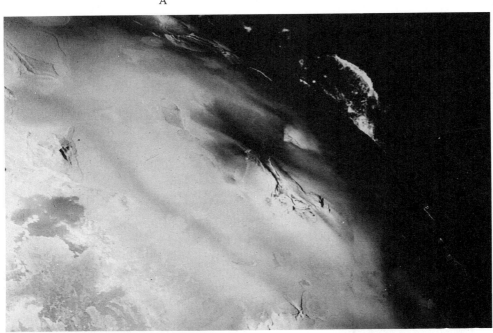

B

the youngest in the Solar System, probably less than 1 million years old (Figure 7–5).

Why is Io's interior hot? The other Galilean moons force Io by gravitation into an eccentric orbit, and so its distance from Jupiter changes significantly. These distance variations cause large and variable tidal forces from Jupiter to act on Io, whose interior heats up from the continual internal stress from the tidal forces.

Europa

The surface features of Europa consist of bright areas of water ice among darker orange-brown areas. Europa's surface is criss-crossed by stripes and bands that are filled fractures in the icy crust, making it look like a cracked eggshell (Figure 7–6). Some of these cracks extend for thousands of kilometers, splitting to widths of 50 to 200 km but reaching depths of only 100 m or so. Europa's sur-

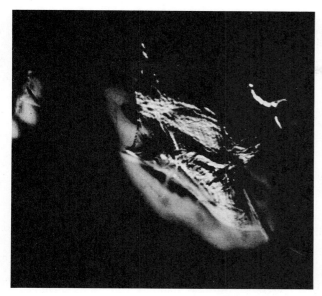

FIGURE 7–5 Haemus Mons on Io, an unusual, isolated high mountain, visible here under oblique illumination. *(NASA)*

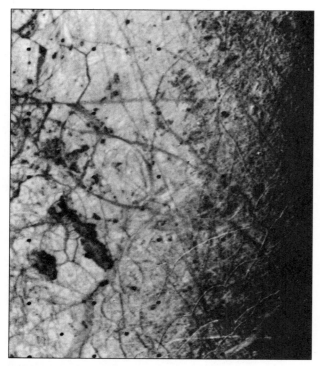

FIGURE 7–6 Europa's surface. This close-up view shows an area 600 by 800 km. The bright ridges are 5 to 10 km wide, the dark bands 20 to 40 km wide. *(NASA)*

face is almost devoid of impact craters and so cannot be a primitive one. The crust must have been warm and soft sometime after formation to wipe out evidence of the early, intense bombardment.

Europa's cracked surface indicates that its solid, icy crust is thin and its interior hot and primarily molten. One model proposes that its crust long ago may have been a slush kept partially melted by a hot interior. As Europa cooled, its crust turned to smooth, glassy ice that later cracked.

Ganymede

The largest moon of Jupiter (radius 2630 km), Ganymede ranks overall as the largest moon in the Solar System. (Titan of Saturn is second.) Its surface looks vaguely like our Moon's, with dark, maria-like regions. It also has huge fault lines along its surface, as Europa does.

Ganymede has two basic types of terrain (Figure 7–7): cratered and grooved. Craters up to 150 km in diameter densely mark the cratered terrain, which is some 4 billion years old. Compared with those on the Moon and Mercury, the craters are shallow for their size, and some have convex rather than concave floors. Many craters on Ganymede have very bright rays extending from them (Figure 7–7), attesting to their formation by impacts on an icy surface.

The grooved terrain separates the cratered terrain into polygon-shaped segments. The grooved terrain consists of a mosaic of light ridges and darker grooves where the ground has slid, sheared, and torn apart. Long cracks, where the surface has moved sideways for hundreds of kilometers, also abound. No large mountainous regions or large basins exist on Ganymede; nowhere does any relief amount to more than about 1 km. This suggests that the crust is somewhat plastic, probably from large fraction of water ice. Ganymede's bulk density implies that its interior contains about half water and half rock. Occasional stresses on the water-rock crust have created the fracture pattern. Some ridges and grooves overlie others, an indication that there have been many episodes of crustal deformation.

Callisto

Farthest out of the Galilean moons, Callisto (Figure 7–8) has a surface riddled with craters of a wide range of sizes. Some have bright ice rays; others

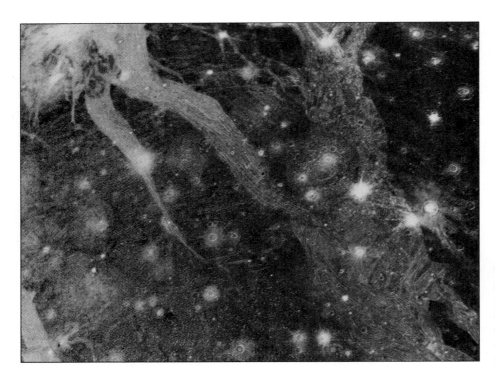

FIGURE 7–7 Ganymede's surface. This photograph shows features as small as 5 km. Note the fresh impact craters with bright rays. The dark regions are the oldest part of the surface. (*NASA*)

are filled with ice. Callisto's craters are shallow, less than several hundred meters deep, because the surface is a mixture of ice and rock. The surface slowly flows, flattening out any relief.

Callisto has eight multiringed basins; the largest is *Valhalla* (Figure 7–8). Its central floor is 600 km in diameter, and it is surrounded by 20 to 30 mountainous rings having diameters up to 4000 km. The rings look like a series of frozen waves formed in a stupendous collision that melted subsurface ice, then caused the water to spread in waves that quickly froze in the 100-K surface temperature. The ripple marks are preserved as rings: frozen blast waves. The central floor of this ringed feature has fewer craters than the rest of the terrain. This difference indicates that the impact forming the rings occurred after much of the surface cratering, probably 4 billion years ago.

Asteroidal Moons

Jupiter's other moons are asteroid-like bodies, and we expect that they are indeed captured asteroids. (Recall that Jupiter lies just outside the asteroid belt.) There are two groups of four moons each,

one group at a distance of about 12×10^6 km that orbits counterclockwise (direct) and another at about 23×10^6 km that orbits clockwise (retrograde).

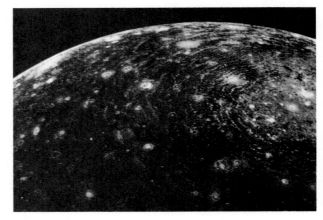

FIGURE 7–8 Callisto's surface. The outer ring of the large impact basin, called Valhalla, is about 2600 km across. Note the fresh impact craters on the dark, icy surface. (*NASA*)

We have observed closely one other asteroidal moon—Amalthea. Only 181,000 km out of Jupiter, it orbits once every 12^h. It is elongated, 270 km by 155 km; the surface is cratered and has a dark red color. This moon's irregular shape, small size, and dark, cratered surface imply its asteroid-like character (more about asteroids in Section 7–2). It physically resembles Deimos and Phobos (except that it is larger).

Rings

Voyager 1 discovered Jupiter's ring system. The rings are so thin (less than 30 km thick) that they are essentially transparent. They are most visible when viewed edge on; then the particles scatter light well, which implies that the particles must be small, about 3 μm in diameter. We do not yet know what they are made of, but on the basis of their infrared properties, they are likely a rocky material.

Dramatic pictures of the backlit rings (Figure 7–9) show them to have a definite structure. The outermost, brightest part is 800 km wide and lies about 128,500 km from Jupiter's center. Within it is a broader ring 6000 km wide, and within that ring lies a faint sheet of material that extends from 119,000 km out from Jupiter's center down to the cloud tops. The main ring extends from 1.72R_J to

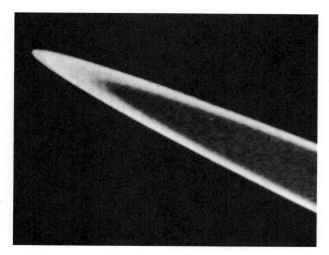

FIGURE 7–9 Jupiter's rings. This backlit view shows that the outer edge of the rings is thicker than the inner edge and that there is a sheet of particles between the two. *(NASA)*

1.81R_J, which places the entire ring system well inside the Roche limit (Equation 3–9) for a fluid moon.

Computer processing of Voyager's wide-angle photographs of the rings shows that a faint, outer ring surrounds the system. Called the *gossamer ring*, it lies mostly within the orbit of Amalthea, though some of its material extends out to 210,000 km. The ring's thickness is less than 200 km, and it is made of micrometer-size particles, which collide with the plasma in the magnetosphere and are swept out by these collisions in only about 1000 years. Hence, unless the gossamer ring is very young, its particles must be replenished from moons that lie close by.

(C) THE MOONS AND RINGS OF SATURN

Saturn's band of moons totals at least 19. With two exceptions (Phoebe and Iapetus), all the moons stick close to Saturn's equatorial plane. Masses for some of the moons were determined from their gravitational attraction on spacecraft. The densities range from 1200 kg/m^3 for Tethys to 1400 kg/m^3 for Dione, similar to the densities of the outer Galilean moons of Jupiter.

The moons of Saturn fall into three groups: Titan by itself, the six large, icy moons (Mimas, Enceladus, Tethys, Dione, Rhea, and Iapetus, in order outward from Saturn), and the ten small moons (Phoebe, Hyperion, and the rest). Overall, their densities are less than 2000 kg/m^3, which implies that they are mostly ice (60 to 70%) with some rock (30 to 40%). In contrast to the Galilean moons, no trend of densities follows with the distance from Saturn. Like Jupiter's moons, all of Saturn's moons except one (Phoebe) orbit in synchronous rotation.

Most of the moons are cratered. Some cratered terrain has been modified on the larger moons, which implies internal heating to melt parts of the icy surfaces. In contrast, the small moons, which are also cratered, show no changes—they still have their original surfaces. Note that the craters we see here imply that intense bombardment 4 billion years ago took place *throughout* the Solar System.

Titan

Titan, the largest moon, has a mass of 1.37 × 10^{23} kg and a radius of 2575 km. Its density is

1900 kg/m³, which implies a composition of half ice and half rock. Titan was the first moon found to have an atmosphere, which consists mostly of nitrogen (99%) with about 1% methane. Several hydrocarbons have also been detected, including ethane, acetylene, ethylene, and hydrogen cyanide. The atmosphere's surface pressure is about 1.5 atm; the surface temperature is roughly 94 K.

Color photographs show a stratospheric layer of orange smog that varies to a blue color along Titan's edge. This variation indicates that the atmosphere varies in composition. No surface features

were seen (Figure 7–10A). Voyager's pressure and temperature data, along with the spectroscopic detections of nitrogen and hydrocarbons, have led to models of a surface covered with a frigid ocean of nitrogen, methane, and ethane up to 1 km deep.

Other Moons

After Titan, Saturn's four largest moons are Iapetus, Rhea, Dione, and Tethys, with diameters ranging from 1020 to 1530 km. They appear heavily cratered (Figure 7–10B). In a few cases, wispy white streaks form rayed patterns around impact craters. These streaks are probably deposits of frozen ice, but whether from material emanating from the interior or from debris deposited by colliding bodies is unknown. Iapetus (Figure 7–10C) has the most extremes of surface cover. The hemisphere leading in its orbit is only 1/15 as bright as that following. The leading surface seems covered with dark debris picked up during its journey around the planet. Only Enceladus does *not* have a surface thick with craters, a sure sign that this satellite has suffered large-scale modification of the surface.

The rest of the moons are all small bodies, a few hundred kilometers or less in diameter. The largest is Hyperion, 400 km in diameter. It has a strange

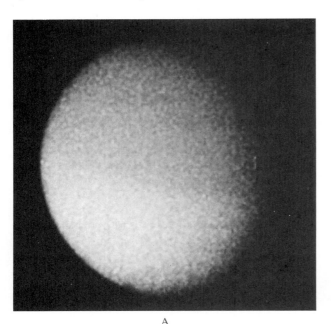

A

B

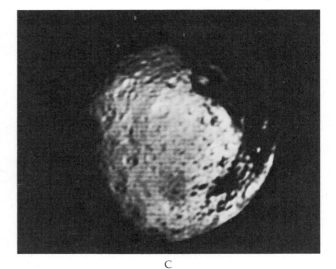

C

FIGURE 7–10 Some of the larger moons of Saturn. (A) Titan, showing its hazy atmosphere. (B) Dione, showing its icy, cratered surface. (C) Iapetus, showing a black layer on one side covering its icy crust. *(NASA)*

shape, like that of a thick hamburger, and a cratered surface. The other moons are also cratered but much smaller, less than 30 km in diameter (Figure 7–11). We presume that all these bodies are basically ice, as are the larger moons.

Ring System

Saturn's rings lie in the planet's equatorial plane and therefore are tipped about 26° to the orbital plane; because of their tilt, they change their appearance as viewed from the Earth during the course of Saturn's revolution about the Sun. The near disappearances of the edge-on rings indicate that they are very thin, no more than a few kilometers thick. Although thin, the rings are wide; the three main ones visible from the Earth reach from 71,000 to 140,000 km from Saturn's center (Figure 7–12). The largest gap is known as **Cassini's division.**

FIGURE 7–11 The smaller moons of Saturn. This composite photograph shows the heavily cratered surfaces of these irregularly shaped moons, whose sizes are in the correct relative scale. Actual sizes range from 25 km to 220 by 160 km. (*NASA*)

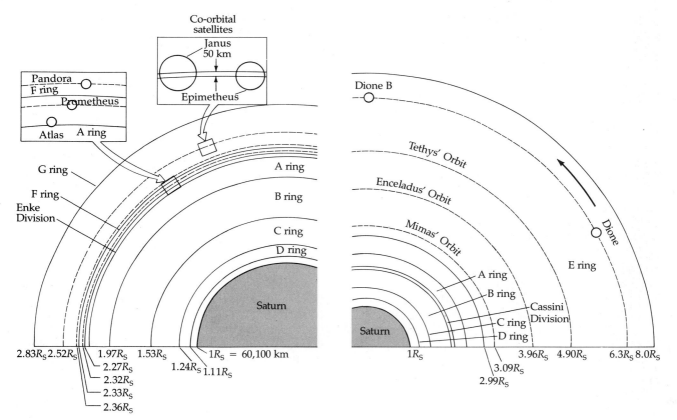

FIGURE 7–12 Saturn's rings. These diagrams show the ring system and some orbits of selected satellites as viewed from above Saturn's north pole. Size units are Saturn radii (R_S). Note the change in scale from the left to the right figure.

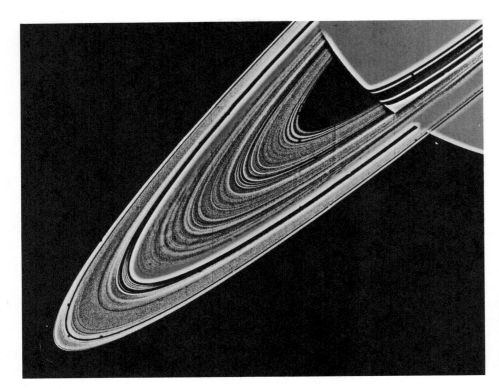

FIGURE 7–13 Ringlets. Saturn's main rings contain many smaller rings within them; this image was computer-enhanced to bring out the ringlets. *(NASA)*

Although the A ring is relatively smooth, the B and C rings break up into numerous small ringlets (Figure 7–13). Many hundreds, perhaps a thousand, light and dark ringlets surround the planet, with widths as small as 2 km, the best resolution of the Voyager cameras. Some (in the C ring) appear elliptical rather than circular. Even the Cassini division, apparently empty as seen from Earth, was found to be filled with at least 20 ringlets.

Dark, spoke-like features occur in the B ring (Figure 7–14). Typically, the spokes are about 10,000 km long and 1000 km wide. They consist of very small particles, much smaller than the average particle in the rings. Because the inner particles orbit faster than the outer ones, the spokes last only a few hours. They may be small, darker particles with electric charges lifted out of the main ring plane by Saturn's magnetic field. Note that these spokes do *not* obey Kepler's laws, because different orbital periods would break up the spokes in one or two revolutions. Hence, the particles producing the spokes cannot be attached to the ring particles.

Pioneer 11 discovered a new ring out beyond the previously known ones. Called the F ring, it lies 3500 km outside the edge of the rings visible from the Earth. The F ring appears to be 320 km wide and a mere 3 to 4 km thick. Voyager 1 photographs resolve this ring into a complex system of knots and a braided structure of at least three strands; Voyager 2 photographs taken nine months later show that the braiding had disappeared. Apparently, the braiding was a dynamically unstable situation. Another such very narrow ring (the G ring) is 10,000 km farther out. Two other extremely faint rings are known. The E ring extends out beyond the F ring to at least 6.5 Saturn radii (400,000 km). A ring inside the C ring, called the D ring, extends at least halfway to the surface of Saturn.

The gravitational effects of the satellite just outside the A ring (Atlas) and the two that straddle the F ring (Prometheus and Pandora) play important roles in the dynamics of the rings. The F-ring moons, in particular, are called the **shepherd satellites** because they keep the ring particles in a narrow range of orbits. The inner moon accelerates the inner ring particles as it passes them (as expected from Kepler's third law, it has a shorter or-

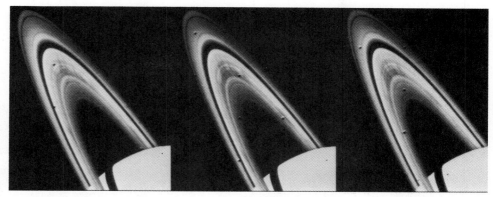

FIGURE 7–14 Spoke-like features in the rings. This sequence was taken at intervals of 15 min. Note the motion of the spokes. *(NASA)*

bital period). They spiral outward to larger orbits just as the tidal force of the Earth on the Moon forces the Moon into a larger orbit [Section 3–4(b)]. In a physically similar way, the slower-moving outer moon decelerates outer-ring particles as they pass by, so that they spiral inward. The balance of these interactions constrains the particles' motions and preserves the narrowness of the F ring. Likewise, the A-ring shepherd causes the sharp outer edge of the A ring. Because tidal forces tend to spread rings out, shepherd satellites work to preserve the sharp edges.

The rotation rate of the rings varies according to Doppler-shift data. The velocities range from 16 km/s at the outer boundary of the A ring to 20 km/s at the inner boundary of the B ring. The measured velocities agree with those expected from Kepler's third law for individual masses placed at the ring distances from Saturn; this agreement indicates that separate particles make up the rings.

Infrared observations of Saturn's rings show that they are particles of water ice or rocky particles coated with water ice. The ice does not evaporate because the surface temperature of the particles is only about 70 K. At this equilibrium temperature, the rings' icy material has a very low vapor pressure, and so the ice stays in solid form. Radio signals from Voyager reflected by the rings indicate that the particles are about 1 m in diameter, but a range of sizes, from centimeters to tens of meters, probably exists. Although covering a large area of space, the rings have a total mass estimated to be only 10^{16} kg, about 10^{-10} the mass of Saturn.

(D) THE MOONS AND RINGS OF URANUS

The five major moons of Uranus—Miranda, Ariel, Umbriel, Titania, and Oberon—move in the planet's equatorial plane and revolve in the same direction as the planet rotates. Because the moons lie in the same plane as Uranus' equator, their orbits as seen from the Earth are alternately edge-on and fully open every 21 years; in 1987 they appeared as circles, but in 2008 they will appear edge-on.

Miranda is the smallest (less than 320 km in diameter) and closest to Uranus. The others range in diameter from 1190 km (Umbriel) to 1550 km (Oberon). Their surfaces appear to be made of a dirty ice, very much like that of Saturn's Hyperion. The bulk densities range from 1300 to 2700 kg/m^3, which implies that these are bodies made of rock and ice.

Voyager 2 transformed our view of Uranus' moons much as it did our view of the moons of Jupiter and Saturn. First, the spacecraft discovered ten more moons, for a total of at least 15. The largest is Puck; it has a diameter of only 170 km and orbits within 86,000 km of Uranus. The other nine moons orbit between Uranus and Puck. These innermost moons have diameters between 40 and 80 km. Second, it provided close-up views of the largest moons (Figure 7–15A through E). Miranda (Figure 7–15A) has the most complex surface, with

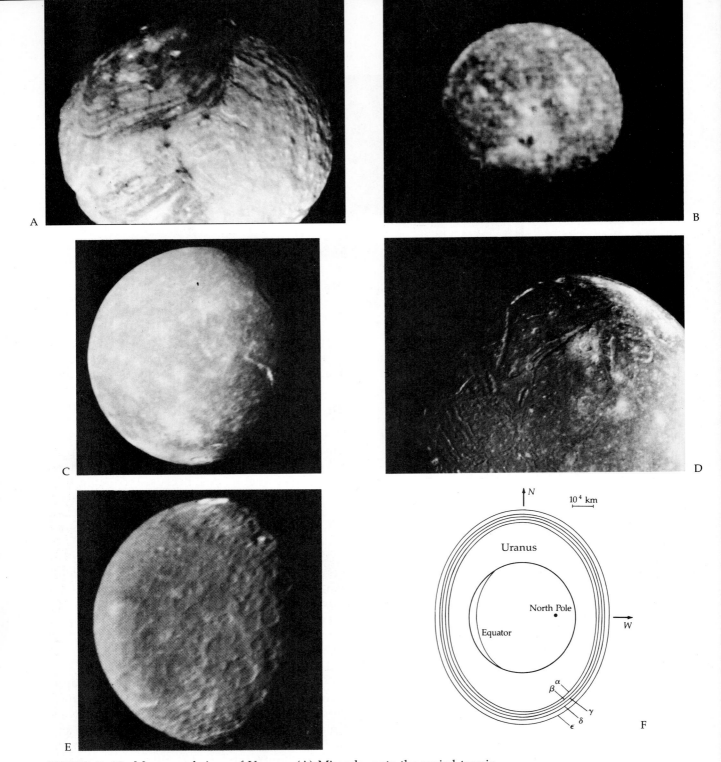

FIGURE 7–15 Moons and rings of Uranus. (A) Miranda; note the varied terrain. Resolution is 2.7 km. (B) Oberon; the visible features are probably impact craters. (C) Titania; note the abundance of impact craters. (D) Ariel, showing complex terrain with faults and valleys. The largest craters are about 30 km in diameter. (E) Umbriel, with its heavily cratered terrain. *(NASA)* (F) Rings of Uranus. The five major rings of the nine total are shown here in this pole-on view.

many types of terrain that appear tectonically shuffled (perhaps by tidal forces of Uranus). Oberon (Figure 7–15B) is densely covered with impact craters and has at least one mountain peak—probably volcanic—about 5 km high. Titania (Figure 7–15C) has a surface plastered with impact craters and strewn with valleys 50 to 100 km wide and hundreds of kilometers long (one of them cuts across the entire surface). Ariel also has impact craters and large fractures and valleys (Figure 7–15D). Finally, Umbriel shows the least dramatic surface, with overlapping impact caters but no special dramatic features (Figure 7–15E). We again find evidence here for the era of torrential impacts early in the history of the Solar System.

Uranus has rings discovered accidentally in March 1977 while Uranus occulted a faint star. Voyager 2 found more rings for a total of at least 11. These rings circle the planet in roughly three groups (Figure 7–15F): rings 6, 5, and 4 at about 42,500 km; α and β at 45,000 km; η, γ, and δ at 48,000 km; and the ϵ ring and two others at 51,000 km from the center of Uranus. The narrowest rings have widths of only about 5 km; the ϵ ring is 100 km wide. The outermost ϵ ring is flanked by two shepherd satellites (called Cordelia and Ophelia), which serve to keep the ring stable and intact. The rings are dark (visual albedo less than 0.03). Voyager 2 radar observations showed that the particles in the rings of Uranus are black ice with a size of about 1 m.

(E) THE MOONS AND RINGS OF NEPTUNE

Before the Voyager 2 mission, Neptune had only two known moons, Triton and Nereid. Triton has a diameter of about 2720 km, which makes it one of the largest satellites in the Solar System. Nereid has a diameter of 355 km. Triton revolves with a period of about 5 days in a retrograde (east-to-west) orbit that is inclined 20° to the plane of Neptune's equator. No other planet has a close moon moving retrograde and so steeply inclined. Nereid, the outer moon, has an orbital eccentricity of 0.75—two times larger than that of any other solar system satellite; its distance from Neptune ranges from 1 million to 10 million km. Nereid's orbital velocity of 3 km/s at closest approach to Neptune is only 0.2 km/s shy of escape velocity.

The Voyager images of Triton displayed a fascinating face (Figure 7–16). The cratering here is not too heavy, which means that the surface must be relatively young and recently modified—subject to meltings and refreezings. The overall temperature is low, a mere 37 K. On parts of the surface lie frozen ice lakes, which resemble lunar maria in shape. Some are stepped, which suggests a series of meltings and freezings. But in general the surface relief is quite low—less than 200 m. Triton's atmosphere is about 800 km thick; it contains mostly nitrogen with a trace of methane.

Near Triton's south pole (which is now in a summer season), the surface ice (consisting of methane and nitrogen) appears evaporated in spots (Figure 7–16). In other regions, small flows have filled valleys and fissures—slow moving glaciers of methane and nitrogen. In other sections, the icy surface appears to have melted and collapsed. Dark, elongated streaks tens of kilometers across seem to be the trails of ice volcanoes. Just 30

FIGURE 7–16 Voyager 2 image of the southern hemisphere of Triton. The southern pole cap has rugged terrain, which smooths out to the north. *(NASA)*

meters below the surface, the pressure is high enough to liquify nitrogen ice. When the surface cracks, the liquid can burst out and turn to gas, which shoots up several kilometers above the surface. The vapor condenses and falls back down, creating a thin, dark layer on the surface. One such geyser was seen actively erupting during the Voyager encounter.

Triton's bulk density turns out to be 2020 kg/m³ (about the same as Pluto and Charon). That implies a rocky core surrounded by a mantle of methane and water ice. The icy surface, which is primarily molecular nitrogen (N_2) ice, results in an average albedo of roughly 70%. Triton's orbit results in a seasonal cycle of 165 years; summer in the southern hemisphere has already lasted some 30 years. Still, so far from the sun, the surface remains so cold that regions of nitrogen ice are still visible here.

Voyager also captured images of six small moons. They range in size from 60 to 415 km and exhibit the usual rugged surfaces of Jovian moons (Figure 7–17). Their albedoes are all low, about 5% or so. Two of them lie close to the outer and middle rings, and so may well act as shepherding moons. Except for the smallest and innermost moon (called Naiad), all lie within the plane of Neptune's

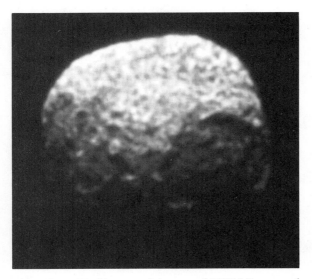

FIGURE 7–17 Voyager 2 image of 1989 N1, one of the moons of Neptune discovered during the flyby. It has a diameter of 400 km; note its cratered surface. *(NASA)*

equator. Proteus has an ellipsoid shape some 400 by 435 km across. It now displaces Nereid as Neptune's second largest moon.

Voyager showed clearly for the first time Neptune's somewhat mysterious ring system, which contains five individual rings (Figure 7–18). The two brightest, outer rings have radii of 53,000 and 62,000 km with a tenuous ring spread between them. Closer in lies a wide ring about 2000 km across. The outer ring is clearly clumpy (Figure 7–18), with three brighter segments strung along a fainter but complete ring like sausages on a string. This structure explains the earlier ground-based observations that were interpreted as a series of arcs around the planet.

7–2
ASTEROIDS

Chapter 2 presented the general orbital properties of asteroids. Here we will focus on their physical characteristics because these inform us about the conditions of solid matter early in the Solar System's history.

Basically, an asteroid is an irregular, rocky body both smaller and less massive than a planet. (Only about 200 asteroids have diameters greater than 100 km; some 10^6 are thought to be in the asteroid belt.) Ceres, the largest known asteroid, has a diameter of only 940 km. Asteroid sizes can be measured directly when an asteroid occults a star. We can also measure an asteroid's reflectivity of visible light and its infrared emission (typically at 10 μm) when it is at a known distance from the Sun. These two measures in an equilibrium state allow an indirect estimate of the asteroid's size.

The flux of solar energy falling on an asteroid is

$$(L_\odot/4\pi D^2)\pi R^2$$

where $L_\odot$ is the luminosity of the sun, D the distance from the Sun, and R the radius of the asteroid. A fraction A (the albedo) is reflected back into space. If the Earth is a distance d from the asteroid, the flux of reflected light at the Earth is

$$F_{vis} = (L_\odot/4\pi D^2)\pi R^2(A/4\pi d^2)$$

We can measure this flux and so determine the quantity R^2A. The fraction of energy absorbed, $1 - A$, heats up the asteroid and is re-emitted into

FIGURE 7–18 Backlit image of the main rings of Neptune, with the disk of the planet partially blocked out. Note the segmentation of the outer ring and the fuzziness of the inner one. *(NASA)*

space as infrared radiation. We can observe this infrared flux at the Earth. The ratio of visible to infrared flux is

$$F_{\text{vis}}/F_{IR} = A/(1 - A)$$

From this measurement, we can determine A and then use this value with the previous determination of R^2A to calculate R.

Observations of the surface reflectivity of the larger asteroids give some hints about their compositions. Asteroids have a wide range of reflectivities, from Nysa (diameter 82 km), whose albedo is 35%, to Cybele (diameter 280 km), which reflects only 2%. Nysa's surface reflects sunlight about as well as the icy satellites of Jupiter and Saturn; Cybele has an albedo similar to that of the rings of Uranus.

The reflectivities indicate that most asteroids fall into two major compositional classes. Some are relatively bright with albedos of about 15%, and oth-

ers are much darker, with albedos of 2 to 5%, indicating that they contain a substantial percentage of dark compounds, such as carbon or the black mineral magnetite (Fe_3O_4). These dark asteroids resemble a class of meteorites (Section 7–4), the carbonaceous chondrites, which are dark because they contain carbon compounds (roughly 1 to 5% carbon). The lighter class is dubbed **S-type** asteroids, the darker ones **C-type** asteroids. The S type, in addition to having higher albedos, also show spectral absorption bands indicative of silicate materials. A third class, called **M-type** asteroids, has characteristics suggestive of metallic substances. They have albedos of about 10%. Only 5% of known asteroids belong to this last class.

Based on albedos, compositions in the asteroid belt vary with distance from the Sun. Near the orbit of Mars, almost all asteroids have S-type characteristics. Farther out are fewer high-albedo ones and more dark ones. At the outer edge of the belt, 3 AU from the sun, 80% of the asteroids are C type.

Do asteroids have moons? That depends on the stability of a binary (double) asteroid system to tidal forces. Asteroids have very small masses; Ceres, for example, if it were all rock, would have a mass of only about 10^{21} kg. A typical asteroid is about 10 times smaller and so has a mass of only 10^{-3} that of Ceres. This small mass means that the gravitational force between the bodies would be so weak that the tidal force of the Sun or Jupiter could easily disrupt the binary system. Any existing system is likely to have a very small separation between asteroid and moon, and so it would be hard to see the two separate objects telescopically. The two may be visible during an occultation. A few such moon-occultation observations have been reported. None has yet been repeated.

In 1989, the Arecibo telescope made radar images of the asteroid 1989 PB, one of 140 or so near-Earth asteroids. These asteroids are thought to be extinct cometary nuclei as well as fragments of main-belt asteroids, with sizes ranging from 50 m to 50 km. At closest approach, 1989 PB was only 0.027 AU from the Earth. The images showed a double-lobed shape, which has been hypothesized to be two separate pieces that suffered a low-velocity collision to make a contact-binary configuration.

7–3 ◑
COMETS

When first sighted telescopically, a comet typically appears as a small, hazy dot. This bright head of the comet is called the **coma** (Figure 7–19A). Sometimes the coma contains a small, starlike point called the **nucleus** (Figure 7–19B). Cometary nuclei are very small, perhaps no larger than a few kilometers across; Halley's nucleus is only about 10 km in diameter. As a comet moves toward perihelion, it grows brighter and sprouts a **tail** (Figure 7–18A). A comet's tail may stretch for millions of kilometers and always points away from the Sun. Surrounding the entire comet when it is in the inner Solar System is a gigantic **halo** of hydrogen gas, spanning millions of kilometers. This halo, detectable from spacecraft in the ultraviolet, results from the photodissociation by sunlight of the hydroxyl radical (OH^+) in the coma.

A

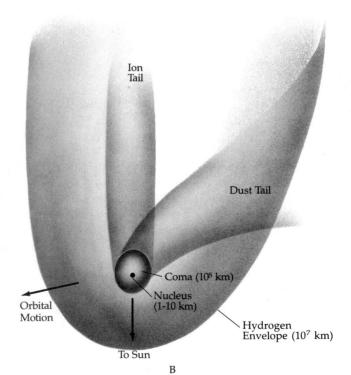

B

FIGURE 7–19 Cometary structure. (A) Head and tail of Comet Kohoutek in January 1974. *(Hale Observatory, California Institute of Technology)* (B) Main parts of a comet. Note that the comet is much larger than visible optically, but it originates from a very small nucleus.

Tails

Comets may have two types of tails: ionized gas (plasma) and dust (Figure 7–18B). The physical difference between the two shows up in their spectra. The spectrum of the plasma tail has emission lines. The dust tail's spectrum is that of sunlight reflected from dust expelled from the coma. The radiation pressure from sunlight detaches dust from the coma and pushes it out to form a tail; these particles follow Keplerian orbits around the Sun.

In the spectrum of the plasma tail, the most conspicuous spectral lines are those produced by carbon monoxide (CO), carbon dioxide (CO_2), nitrogen (N_2), and radicals of ammonia (NH_3) and methane (CH_4). Puffs of gas sometimes shoot through the tail. The German astronomer Ludwig Biermann (1907–1986) suggested in 1951 that the solar wind has a major effect on the ionized tails. Measurements of the solar wind confirmed that the magnetic fields carried by the wind's particles indeed drag ions from the comet's coma so the tail points away from the Sun.

The ICE (International Cometary Explorer) intercept of Comet Giacobini-Zinner in September 1985 revealed the true complexity of the interaction between comets and the solar wind. ICE's passage through the comet's tail verified that the interaction produces many energetic ions around the comet. These ions are picked up by the wind; the additional mass slows down the flow behind the comet and produces the long, filamentary ion tail. Giacobini-Zinner's tail was 25,000 km thick where it was crossed by ICE (7800 km downstream from the nucleus) and contained an induced bipolar magnetotail split by a neutral sheet between the opposite current flows (Figure 7–20). Electron densities in the heart of the tail exceed $10^9/m^3$. The magnetic field here comes from the field carried by the solar wind. The magnetic field–plasma interactions fix the form of a visible comet.

Cometary Nuclei

At great distances from the Sun, the coma shows a reflected solar spectrum, and so the cometary head must also contain solid particles that reflect sunlight. At about 1 AU from the Sun, the head exhibits molecular emission bands of carbon (C_2), cyanogen (CN), oxygen (O_2), hydroxyl (OH), and

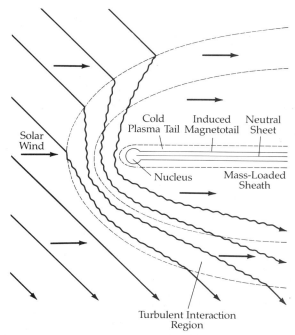

FIGURE 7–20 Schematic diagram of the structure of the interaction of Comet Giacobini-Zinner with the solar wind. *(NASA)*

hydrides of nitrogen (NH and NH_2). As the comet speeds nearer to the Sun, emission lines of silicon, calcium, sodium, potassium, and nickel appear. Table 7–2 summarizes the materials that have been observed to date in the heads and tails of comets.

The observed brightness B of a comet depends upon its distance from the Sun R (which determines its fluorescence and the amount of sunlight reflected) and its distance from the Earth r (which

TABLE 7–2 *Observed Composition of Comets*

Coma	Tail
H, C, C_2, C_3, CH, CN, HCN, CH_3, CN, NH, NH_2, O, OH, H_2O, Na, K, Ca, V, Cr, Mn, Fe, Co, Ni, Cu plus dust particles with silicates	CH^+, CO^+, CO_2^+, N_2^+, OH^+, H_2O^+, Ca^+ plus dust particles with silicates

determines the flux of radiation we receive):

$$B \propto R^{-n}r^{-2} \qquad (7–1)$$

Far from the Sun, no fluorescence occurs and so $n = 2$. Near the Sun, however, $n \approx 4$ and may range from 2 to 6, depending upon the particular comet; for Halley's Comet, $n \approx 5$ for distances closer than 6 AU. How much a comet glows (primarily in its coma) depends on the composition and amount of gases released by the nucleus.

For all their stunning length against the sky, comets have very small masses. Halley's Comet, one of the largest, has an estimated mass of only about 10^{14} kg, and it loses about 10^{11} kg during each perihelion passage. In 1910, the tail of Halley's Comet stretched about 100°, reaching from horizon to horizon; the lack of any noticeable effects as the Earth traveled through the tail indicates that the gas was quite rarefied, less than about 10^{-7} kg/m^3. The density of the coma is estimated at 10^{-17} kg/m^3, much higher than the tail's density but still a good vacuum.

The mass expelled from a comet, mostly as gas, comes from the nucleus. Fred L. Whipple has developed a **dirty-snowball cometary model** in which the comet nuclei are compact, solid bodies made of frozen gases (ices) of water, carbon dioxide, ammonia, and methane embedded with rocky material. Beyond Jupiter, low temperatures allow the ice–rock conglomerate to persist unchanged for long periods of time. As the comet nears the Sun, the icy material vaporizes. This released material enlarges the coma and creates the tail. As the ice evaporates, a thin coating of rocky material remains to form a solid but fragile crust on the nucleus. The heating of the subsurface material creates gas jets that blow off puffs of gas and act as small rockets that slightly change the comet's orbit.

The Comet Cloud

Periodic comets lose a little material each time they pass the Sun and eventually are completely outgassed. How, then, can we explain their abundance? Comets are gravitationally attached to the Sun, most in long orbital periods. From the observed orbits, we find that the average value of their semimajor axes is about 50,000 AU and the corresponding orbital period is about 10^7 years. The orbits are highly elliptical; in accord with Kep-

ler's second law, the comets travel very slowly at aphelion, only a few kilometers per day, and so such comets spend most of their time coasting far from the Sun.

A cometary cloud, proposed by the Dutch astronomer Jan Oort in 1950, is sometimes known as **Oort's cloud.** It makes up the Solar System's cometary reservoir. According to Oort's model, most comets never come very near the Sun and we never see them. Occasionally, however, the gravitational action of passing stars pushes (or pulls) a comet into an orbit that does bring it closer. Comets are eventually lost either by vaporization of the nucleus or by Jupiter's perturbation of their orbits. So the supply of Sun-approaching comets must be replenished with new comets from the cloud. To ensure sufficient input to make up the losses, Oort's picture requires at least 10^{11} (and perhaps as many as 10^{14}) comets to be clustered in the cloud.

Halley's Comet

Halley's Comet returned to the earth's neighborhood in 1985–1986, when it reached perihelion on February 9, 1986, at a distance of 0.59 AU. After rounding the sun, the comet passed closest to the earth (0.42 AU) on April 11, 1986. Unfortunately, this return of Halley's Comet was the worst for viewing in the past 2000 years! But for astronomers, this encounter marked the first time that they could study a comet in detail with modern equipment and discover its true nature (Figure 7–21). On the ground, they organized an interna-

FIGURE 7–21 Comet Halley, showing fine details in the plasma tail in March, 1986. *(Royal Observatory, Edinburgh)*

tional network of telescopes. In space, a small armada of flybys intercepted Halley's Comet for close-up views. The ground-based network was able to record a time-series of observations, which caught physical processes such as magnetic reconnections and plasma tail separations.

The most spectacular spacecraft mission was Giotto. Launched by the European Space Agency (ESA), Giotto swooped within 600 km of the core of Halley's Comet on March 14, 1986. Well before that closest approach, Giotto sampled the environment around the comet's head. In a zone thousands of kilometers thick, it found gas ions moving at speeds greater than the escape speed from the nucleus and coma but trapped by magnetic fields. Most of the ions were related to water; in fact, most of the comet's gas was water. What was not expected was that the gas contained a fair amount of sulfur and also of carbon compounds. An infrared spectrometer found absorption bands due to water, carbon dioxide, CO^+, and formaldehyde (H_2CO). Remarkably, the inner coma was found to contain short polymer chains of formaldehyde, $(H_2CO)_5$, the first polymer to be found in space.

Giotto and the two Soviet Vega spacecraft measured magnetic fields and ion densities as they traversed the coma. As expected, they encountered a bow shock far from the comet, where the solar wind is first deflected. Farther in was another boundary surface, within which were found cometary ions, not those of the solar wind. Finally, a region near the nucleus was encountered which, surprisingly, contained *no* magnetic field.

Giotto took almost 3000 pictures during its approach to the nucleus. The last, taken before the camera was destroyed by collision with particles in the coma, came from a distance of about 1700 km. It revealed (Figure 7–22A) the nucleus was peanut-shaped and about 16 by 8 by 8 km in ellipsoidal dimensions. If we assume a density of 500 kg/m^3, appropriate for icy material, these dimensions give a mass of 3×10^{14} kg.

The comet's surface was dark—black as velvet, reflecting only 4% of the sunlight that hits it. The surface also appeared rough, with hills and craters. With a resolution of a few hundred meters, the best images showed the most prominent crater-like structure to have a width of 2 km and a depth of 150 m. They also show a chain of hills, each about 0.5 km in size, a mountain some 1 km

high, and a central depression in the middle of the elongated nucleus. Observations of these structures over time showed that the nucleus rotated once every 54^h.

Most dramatic were the gas and dust jets (Figure 7–22B). Before the Halley missions, most astronomers had imagined that a comet's nucleus was a dirty snowball of ices and dust. They were basically right, but the amount and nature of the dust surprised everyone. As expected, the sunlit side of the comet blows off heated materials—ice that vaporizes to gas, and the dust with it. These blow off in jets only a few kilometers wide—at least nine were visible during the Giotto encounter. The surface sources of the jets were less than a kilometer in diameter. Comparison of pictures of the 1986 Halley encounter with photographs from 1910 indicates that the jets emanated from the same areas at both apparitions; the fractures producing the jets seem to be relatively permanent. The outspray from the jets contained about 80% water vapor and 20% dust, which was almost all carbon with a small amount of sand-like material mixed in.

The comet's carbon showed up in another way— in a long molecule, the polymer polyoxymethylene, which contains a chain of formaldehyde (H_2CO) units. The molecule, called *POM* for short, was detected as Giotto plunged through the inner coma. Because Halley's Comet still expels the POM molecules after many passes of the sun, we infer that they were incorporated in the nucleus at the time of its formation. Hence we have a strong, albeit indirect, clue that POM might exist in interstellar clouds as well.

7–4 ◗
METEOROIDS AND METEORITES

When a meteoroid—interplanetary debris—enters the Earth's atmosphere, it generates the streak of light in the sky called a **meteor.** If any material survives the plunge through air, it strikes the ground as a **meteorite.** Meteoroids hit the atmosphere with speeds ranging from 12 to 72 km/s, and more meteors are seen after midnight than before (Figure 7–23). Meteors belong to the Solar System, so that their speed at the Earth's orbit cannot exceed the solar escape speed there of 42 km/s. Before

A

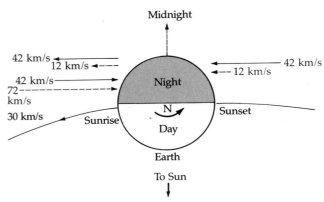

FIGURE 7–23 Meteor speeds. The solar escape velocity at the Earth's orbit is 42 km/s; meteoroids cannot move faster than this speed. In early evening, meteors catch up to the Earth. After midnight, the Earth (orbital speed 30 km/s) catches up to all but the fastest meteoroids moving along its orbit.

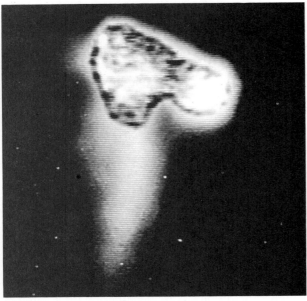

B

FIGURE 7–22 Nucleus of Comet Halley. (A) A composite, detailed image taken with the Halley Multicolour Camera during the ESA Giotto encounter on March 13, 1986. (*Max-Planck-Institute für Aeronomie, Lindau/Harz, Germany*) (B) VEGA-2 image of the jets from the nucleus in early March, 1986. The strongest jets are emitted sunward and originate from a long, linear feature at the waist of the nucleus. (*Research Institute for Particle and Nuclear Physics, Hungarian Academy of Sciences*)

midnight, only those meteoroids moving faster than the Earth (30 km/s) can catch up with it from behind; the relative speed of the fastest such meteor is 12 km/s. After midnight, all meteoroids except those moving faster than the Earth along its orbit will be seen; now velocities add to give a maximum relative speed of 72 km/s.

Because a meteor's tail provides a brief record of the body's disintegration, astronomers have been able to determine the orbit and general physical characteristics of meteoroids. Most meteoroids are fragile particles that crumble quickly when they come in contact with air (Figure 7–24). A block of meteoroid material having a volume of 1 m³ would crumble under its own weight, for it is no stronger than cigarette ash.

What is the source of this low-density meteoroid material? Comets. During a comet's successive passages by the Sun, solar heating causes a continual loss of icy material from the cometary nucleus. The dust and solid particles interspersed in the ice flake off and scatter in an array around the comet. This solid debris is very fragile and has a low density. About 99% of all meteors are of cometary origin.

The remainder are probably associated with asteroids. Only those meteors over a certain size can survive vaporization in the atmosphere and hit the

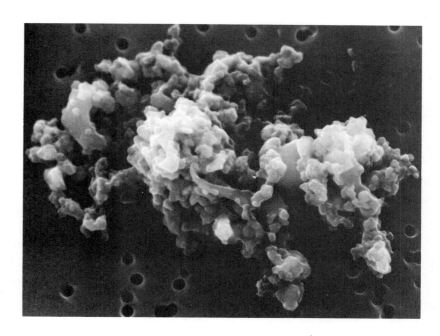

FIGURE 7–24 A meteoroid, obtained in the Earth's upper atmosphere. This particle of interplanetary dust is about 10 μm across. Note its flaky structure. *(D. Brownlee, University of Washington)*

ground as meteorites. In terms of physical and chemical composition, astronomers divide meteorites into three broad classifications: irons, stones, and stony irons. The **irons,** which are generally about 90% iron and 9% nickel with a trace of other elements, are the most common finds. The **stones** are composed of light silicate materials similar to the Earth's crustal rocks. Though actually the most common kind of meteorite seen to fall, they are difficult to distinguish from an ordinary terrestrial stone and so make up a small fraction of the finds (Figure 7–25A). When examined under a microscope, many stones are seen to contain silicate spheres called **chondrules** embedded in a smooth matrix. These stones are known as **chondrites.** The **stony irons** represent a cross between the irons and the stones and commonly exhibit small stone pieces set in iron. Irons are the densest meteorites, with densities ranging from 7500 to 8000 kg/m³. Stones are the least dense, averaging from 3000 to 3500 kg/m³. Stony irons, because they are a mixture of stones and irons, have an intermediate density of 5500 to 6000 kg/m³.

One curious kind of chondrite is the **carbonaceous chondrite.** The chondrules in these meteorites are embedded in material that contains much more carbon than other stony chondrites, typically from 1 to 4% carbon by mass, which gives these

meteorites a dark appearance. Carbonaceous chondrites also contain water (ranging from 3 to 20%) and volatile materials. In addition, the relative abundances of condensable elements in carbonaceous chondrites are closer to those found in the Sun's photosphere than to those found in the crust of the Earth. This similarity suggests that carbonaceous chondrites formed out of the same primordial material as the Sun and have suffered no major bulk heating since that time.

An important clue about the origin of iron meteorites comes from etching then with acid on a polished surface. Large crystalline patterns called **Widmanstätten figures** become visible (Figure 7–25B). Widmanstätten figures give a clue to the history of the meteoritic material. A nickel–iron mixture, when cooled slowly under low pressures from a melting temperature of about 1600 K, forms large crystals. The cooling must be very gradual (about 1 K every 10^6 years). Metals conduct heat well, however, and in space, a molten mass of nickel and iron would cool rapidly. Nickel–iron meteorites could grow Widmanstätten figures only if they had protection from the cold, and so it's likely that nickel–iron meteorite material solidified inside small bodies, termed **parent meteorite bodies.** To allow slow cooling, these bodies were at least 10 km in diameter.

A

B

FIGURE 7–25 Different types of meteorites. (A) A stony meteorite. A piece from the Allende, Mexico, fall. Note the light-colored inclusions. (B) An iron meteorite. The crystalline structure is visible in the flat, polished section of this meteorite, which fell in Dora, New Mexico. *(M. Zeilik, Institute of Meteoritics, University of New Mexico)*

Such bodies probably formed at the origin of the Solar System. Parent meteorite bodies are envisioned as having been only a few hundred kilometers across. Once formed, they could be heated by the radioactive decay of such short-lived isotopes as ^{26}Al. When heated to melting, a parent meteorite body differentiates; the densest material falls to the center and the least dense comes to the sur-

face. So the object ends up with a core of metals and a cover of rocky material, which cools to form a crust. This crust insulates the molten metals and allows them to cool slowly and form large crystals. Later, the parent meteorite bodies collide and fragment. Pieces from the outer crust make stony meteorites, pieces from farther down become stony-iron meteorites, and the core produces iron meteorites.

This scenario requires that the parent meteor bodies were among the first solid objects to form in the young Solar System. Radiometric dating of meteorites backs up this idea: the ages of all fall close to 4.6 billion years. So meteorites provide us with the most direct evidence of primitive solids—chemically, isotopically, and texturally.

Not all meteorites originate from asteroids. Some (very few) come from comets—perhaps selected carbonaceous chondrites. However, we have a few meteorites that very likely came from the Moon! Since 1981, scientists from the United States and Japan have recovered thousands of meteorites buried in old ice in Antarctica. A few of these—each only a few centimeters across—have a texture similar to that of lunar breccias; their chemical composition is similar to that of lunar rocks (especially in noble gases, potassium, magnesium, and iron), and their isotopic abundances of oxygen also appear similar to the Moon's. Hence, some lunar impacts provided enough energy to boost pieces above lunar escape velocity (2.4 km/s), and these pieces then traveled to the Earth. Those trapped in the antarctic ice have been well preserved for examination.

Other meteorites may have come from Mars. The suspects are a group of eight rocks gathered from around the world. They resemble some basaltic rocks, which suggests a volcanic origin. But they also have an oxygen isotopic composition quite different from that of the Earth or any other meteorite. Two properties point to Mars as the source: the radiometric ages are young, 1.3 billion years or newer; and the noble gas ratios in them match that detected at Mars by the Viking landers. How could these meteorites get from Mars to the Earth? A basin-forming impact on Mars could have blasted material into space at speeds greater than the Martian escape velocity. These pieces orbited the sun until they chanced to intersect the Earth.

7–5 ◗
INTERPLANETARY GAS AND DUST

Interplanetary gas comes from various sources. Some has escaped from planetary atmospheres, and some has been released in the demise of comets. Most, however, comes from the Sun. The solar wind, essentially an expanding extension of the atmosphere (Chapter 10), rushes past the Earth at about 500 km/s. The plasma does not stop there; it sweeps onward beyond Pluto's orbit until it slows down and dissipates in interstellar space.

The solar wind has only a minor effect on the heavier dust particles in the Solar System. These particles, which orbit approximately in the plane of the ecliptic, produce a phenomenon known as the **zodiacal light,** a faint cloud of light that extends in a roughly triangular shape above the horizon before sunrise and after sunset. The zodiacal light shines so faintly that the lights of even a small town can completely obscure it. Satellite observations show that the zodiacal dust particles in the Earth's neighborhood have a concentration of about 10^{-8} particles/m^3 and that they are composed mostly of silicates, iron, and nickel. When this cosmic dust falls to the Earth, it is found as micrometeorites. It adds perhaps a few million tons to the Earth's mass each year. The total mass of the zodiacal cloud is about 10^{16} kg.

Although dust is constantly being fed into the Solar System by disintegrating comets (and perhaps asteroid collisions), two physical processes efficiently remove it: radiation pressure and the Poynting–Robertson effect. **Radiation pressure** occurs because electromagnetic radiation carries momentum at the speed of light. The radiant energy flux E (J/m^2 · s) corresponds to a **momentum flux** $P = E/c$ (J/m^3), where c is the speed of light (Chapter 8). The momentum flux is the light pressure on the particle. When solar radiation interacts with a dust particle of effective area or cross section A, the rate of change of the particle's momentum is the **radiation force:**

$$F_R = PA = AE/c$$

From the discussion of planetary temperatures, $E = (R_\odot/d)^2 \sigma T_\odot^4$, where d is the distance from the Sun. If we assume that $A = \pi r^2$, where r is the particle's radius, then the radiation force pushing

the particle *away* from the Sun is

$$F_R = (\pi \sigma r^2 R_\odot^2 T_\odot^4/c)/d^2 \qquad (7\text{–}2)$$

but the Sun's gravitational attractive force on the particle is

$$F_G = GM_\odot(4\pi r^3 \rho/3)d^2 \qquad (7\text{–}3)$$

where ρ is the density of the particle. We form the *ratio* of Equations 7–2 and 7–3 and substitute the appropriate numbers to find

$$F_R/F_G = 3\sigma R_\odot^2 T_\odot^4/4cGM_\odot \rho r \qquad (7\text{–}4)$$
$$= 5.78 \times 10^{-5}/\rho r$$

where the ratio is dimensionless if we express r in meters and ρ in kilograms per cubic meter. Note that the particle's distance d from the Sun has disappeared. We see that $F_R = F_G$ for reasonable densities ($\rho \approx 1000$ to 6000 kg/m^3) when $r \approx 0.1 \ \mu m$, so that dust particles smaller than 1 μm are blown out of the Solar System.

For larger particles, $F_G \gg F_R$, the **Poynting-Robertson effect** becomes important. Just as the Earth's orbital motion leads to the aberration of starlight, so also will a particle's Keplerian orbit cause solar radiation to appear to be coming from slightly in front of the particle. If v is the speed of the particle in circular solar orbit, the angle between the incoming radiation and the radius vector to the Sun is clearly $\theta \approx v/c$ so that the component $(v/c)F_R$ of the radiation force is impeding the particle's motion. So a particle will spiral into the Sun. A particle originally orbiting at distance d (AU) will fall into the Sun in a time

$$t = (7 \times 10^5)\rho r d^2 \text{ years}$$

For example, a particle of size $r = 1 \ \mu m$ and density $\rho = 4300$ kg/m^3 takes only 3×10^3 years to spiral into the Sun from a distance of 1 AU and about 5×10^6 years from 40 AU. In this fashion, the Solar System is purged of its small dust, but that material is replenished by the breakup of larger bodies, such as comets.

7–6 ◗
THE FORMATION OF THE
SOLAR SYSTEM

We end this chapter with a composite sketch of the development of our Solar System based on the best models we have today. These models are basically **nebular models,** where a cloud of interstellar gas and dust contracts to form the Sun and planets. The overall features of this picture are probably correct even though some details are still vague and uncertain. We will focus on two major aspects of these models: dynamics and chemistry.

(A) DYNAMICS

The Solar System displays a regular structure in terms of its dynamic properties. Viewed from above the Sun's north pole, the Solar System shows the following regularities:

1. The planets revolve counterclockwise around the Sun; the Sun rotates in the same direction.
2. With the exceptions of Mercury and Pluto, the major planets have orbital planes that are only slightly inclined to the plane of the ecliptic; the orbits are nearly **coplanar.**
3. With the exceptions of Mercury and Pluto, the planets move in orbits that are very nearly circular.
4. With the exceptions of Venus and Uranus, the planets rotate counterclockwise, in the same direction as their orbital motions.
5. The planets' orbital distances from the Sun follow a regular spacing; roughly, each planet lies twice as far out as the previous one.
6. Most satellites revolve in the same direction as their parent planets rotate and lie close to their planets' equatorial planes.
7. The planets together contain much more angular momentum than does the Sun.
8. Long-period comets have orbits that come in from all directions and angles, in contrast to the coplanar orbits of the planets, satellites, asteroids, and short-period comets.

The essential feature of nebular models is that the Sun and then the planets form from a cloud of interstellar material. The Sun's formation takes place in the center of a flattened cloud. The planets grow from the disk of the cloud. So the problem has two basic parts: (1) how to make a flat solar system and (2) how to get the planets to grow out of the cloud.

To tackle the first part, consider the conservation of angular momentum. The basic point is this:

once a body starts spinning, it will keep on spinning as long as no torque affects it. The amount of spin angular momentum depends on how much mass the body has and how much is spread out. If by itself the body changes size—for instance, if it contracts gravitationally—it will naturally spin faster to keep its spin angular momentum the same. And it will flatten as it falls in along the rotation axis, where the spin angular momentum per unit mass is the least. As a natural result of contraction with spin, we get the planets' orbits aligned in a thin disk and the Sun rotating in the same direction as the planets revolve.

We then have to tackle the present distribution of angular momentum. Although the sun holds 99% of the system's mass, it contains less than 1% of the angular momentum. The Jovian planets have the most, 99% of the total. A good nebular model requires a process to redistribute the angular momentum. One idea that has been worked on in some detail involves the interaction of magnetic fields and charged particles to rearrange the distribution of angular momentum, so that the spin of the central part of the nebula is deceased and transferred to the outer regions.

Plasmas and magnetic fields interact in such a way that the charged particles spiral along the magnetic lines of force. As the Sun forms, it heats up the interior regions of the nebula, and the gas is ionized. As the Sun rotates, it carries its magnetic field lines with it; these drag along the charged particles, which in turn interact with and drag along the rest of the gas and dust. So the magnetic field spins up the material near the Sun. At the same time, the inertia of the nebula resists the rotation. This drag on the magnetic field lines stretches them into a spiral shape (Figure 7–26). So the nebular material gains rotation (and angular momentum) and causes a torque on the Sun's rotation, which slows it down.

(B) CHEMISTRY

Planet formation is a multistep process. First, solid grains condense out of the solar nebula's gas. Second, these particles accrete into large bodies called **planetesimals,** which then collide and accrete to make **protoplanets,** which evolved into the planets of today. The condensation of grains determined

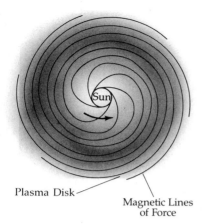

FIGURE 7–26 Magnetic fields in the solar nebula. A possible configuration of the twisting of magnetic field lines from the Sun trapped by ionized gas in the solar nebula.

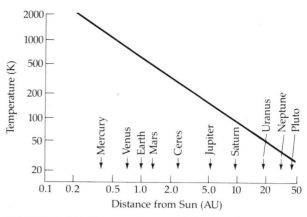

FIGURE 7–27 Temperature and the condensation sequence. To result in the differences in composition from one planet to another, the temperature in the solar nebula followed the trend shown here.

the chemical composition of planets by a process called the **condensation sequence.**

The basic idea of the condensation sequence is this: the nebula's center had a temperature of a few thousand kelvins. Here solid grains, even iron compounds and silicates, could not condense. Elsewhere, which materials would condense as new grains depended on the temperature. Just below 2000 K, grains made of terrestrial materials would condense; below 273 K, grains of both terrestrial and icy materials could form. At different temperatures, the gases available and the solids present react chemically to produce a variety of compounds (Table 7–3). The densities and compositions of the planets can be well explained with

the condensation sequence *if* the temperature of the nebula dropped *rapidly* from the center outward. Then, at different distances from the Sun, different temperatures allowed different chemical compounds to condense and form grains that eventually made up the protoplanets (Figure 7–27).

In general, the condensation sequence requires a certain *minimum* temperature to be reached to account for the known chemical composition of the planets. Roughly, these temperatures are 1400 K or Mercury, 900 K for Venus, 600 K for the Earth, 400 K for Mars, and 200 K for Jupiter. Note that, at a given distance from the Sun, the temperature varies with time, and so these values are the *minimums* reached during the interval of planetary formation.

(C) ACCRETION

Once the grains condense, they accrete into larger masses. Accretion processes fall into two distinct physical categories: (1) growth by collision and sticking due to the *geometric* cross section, and (2) growth by collision due to gravitational attraction, with a *gravitational* cross section.

A grain's geometric cross section is simply πR^2 for a spherical grain of radius R. We can define a gravitational cross section (or impact parameter) as follows. Consider a test particle at velocity V_0 ap-

TABLE 7–3 *Planetary Chemical Condensation Sequence*

Compound/Mineral	Temperature (K)
Al_2O_3	1743
Fe-Si	1458
Mg_2SiO_4	1433
Al_2SiO_5	1068
FeS (trolite)	703
Fe_3O_4	403
Carbonaceous compounds	373–473
Hydrated Mg-silicates	273–373
Ices	<273

Note: For a pressure of 10^{-3} atm.

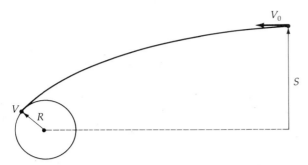

FIGURE 7–28 The geometry for particle impacts with gravity increasing the effective impact cross section.

proaching a grain of radius R. When the test particle hits the grain, it has velocity V. Let S be the maximum transverse distance from the center of the grain that a test particle can have and still strike the grain (Figure 7–28). The test particle starts out at position $S \gg R$ so that its potential energy at S is zero. Conservation of angular momentum says that

$$VR = V_0 S$$

Conservation of energy requires that

$$mV_0^2/2 = mV^2/2 - GmM/R$$

where m is the test particle's mass and M and the grain's mass. Then

$$V^2/2 = V_0^2/2 + GM/R$$

and

$$V = (V_0^2 + 2GM/R)^{1/2}$$

Substitute this expression for V back into the angular momentum conservation equation:

$$V_0 S = (V_0^2 + 2GM/R)^{1/2} R$$

so that

$$S = (R/V_0)(V_0^2 + 2GM/R)^{1/2}$$
$$= (R^2 + 2GMR/V_0^2)^{1/2}$$

This is the equation for the *gravitational impact parameter*; note that it is a sum of a geometric term and a gravitational term.

Suppose particles grow by adding material of the *same* density so that the particle's density is constant. Then

$$dM/dt = d/dt[(4/3)\pi\rho R^3]$$
$$= 4\pi\rho R^2 (dR/dt)$$

The particles grow in size to increase their mass. Assume that gravity dominates the growth. Now consider how a particle grows as it moves through a group of other particles. The collision rate depends on the velocity (greater velocity, more collisions in a given time), cross section (the larger the cross section, the larger the number of collisions in a given time), and the number density of the other particles (greater density means more collisions in a given time). So

$$dM/dt = \text{velocity} \times \text{particle density} \times \text{cross section}$$
$$= V_0 \rho_0 \pi S^2$$
$$= V_0 \rho_0 \pi (R^2 + 2GMR/V_0^2)$$
$$= V_0 \rho_0 \pi R^2 + 2\pi\rho_0 GMR/V_0$$

Ignore the first (geometric) term for the case where the gravitational cross section dominates. Then

$$dM/dt \approx (2\pi G\rho_0/V_0)MR$$
$$= (2\pi G\rho_0^2/V_0)[(4/3)\pi R^3]R$$

or

$$dM/dt \propto R^4$$

As the particles grow, they add material at a very fast and accelerating rate.

At what particle size does the gravitational cross section dominate? Let's say that happens when $S^2 = 2R^2$, *twice* the geometric one. Then

$$S = 2^{1/2}R = (R^2 + 2GMR/V_0^2)^{1/2}$$
$$2R^2 = R^2 + 2GMR/V_0^2$$

so

$$R^2 = 2GMR/V_0^2$$

and

$$R = 2GM/V_0^2$$
$$= (2G/V_0^2)(4/3)(\pi\rho R^3)$$

or

$$R = (3V_0^2/8\pi G\rho)^{1/2}$$

Note that V_0 is the *relative* velocity of particles and ρ the particles' density. For $V_0 = 1$ km/s and $\rho \approx 3000$ kg/m³, we have

$$R \approx 1000 \text{ km}$$

as the transition size. An object larger than this is probably a planetesimal.

(D) THE FORMATION OF JUPITER AND SATURN

To make matters somewhat confusing, Jupiter and Saturn may have formed in a way other than that described by the planetesimal accretion model. In analogy with star birth, Jupiter and Saturn may have condensed gravitationally from single, large blobs of material in the nebula rather than by accretion of planetesimals. The heat they gain comes from the conversion of gravitational potential energy to heat during gravitational contraction.

A proto-Jupiter with a solar mixture of material 16 times Jupiter's present size has a central temperature of 16,000 K, a surface temperature of 1000 K, and a luminosity of almost 10^{-2} the Sun's present luminosity. Gravity quickly collapses the proto-Jupiter at first. The shrinking slows down when the planet's interior is liquid, as liquids are difficult to compress. In the next 4.5 billion years, Jupiter contracts to its present size and its central temperature drops as it loses some of its heat of formation to space at a rate of $1.8 \times 10^{-9} L_\odot$. (Similar models describe the early evolution of Saturn.) Soon after its formation, the proto-Jupiter went through a brief phase of high luminosity, which may explain, in the context of a condensation sequence, why the Galilean satellites decrease in density going outward from Jupiter. So the Galilean moons may mimic the condensation and accretion of the terrestrial planets.

(E) ASTEROIDS, METEORITES, AND COMETS

Asteroids are probably planetesimals that did not accrete to make a planet. Recall the compositional variation across the asteroid belt. On the inner edge, it contains mostly S-type asteroids; at the outer edge, mostly C types. These albedo differences fit in nicely with the condensation sequence if the C types contain more carbon than the S types. At the inside of the belt, the temperatures were low enough for silicates to condense but too high for carbon-bearing materials to do so. Farther out, both types of materials condensed to end up in planetesimals. Why didn't they form a planet? Probably because of the tidal influence of the proto-Jupiter. Solar tidal forces also helped to change the orbits of the planetesimals from circular to elliptical. Some crashed into others, shattering them into smaller pieces. Some of these pieces caromed into the inner part of the Solar System and eventually rained onto the surfaces of Mercury, Venus, the Moon, the Earth, and Mars, forming craters—the epoch of bombardment 4 billion years ago.

The characteristics of chondritic meteorites support the condensation picture. Their chemical composition and unmixed structure suggest that they are the original condensed material of the nebula. The turbulence in the nebula may have created shock waves that swiftly melted the grains. After its passage, the drops cooled and solidified to form chondrules. The glassy spheres that resulted accumulated in planetesimals. Chondrules suggest that the bulk of their condensation took place at temperatures around 600 K. About 1 million years after formation, radioactive decay reheated some planetesimals, melting them to some extent and allowing them to differentiate into iron cores and stony mantles. The planetesimals that were not gathered into a protoplanet possibly became the parent meteorite bodies. These bodies later collided and fragmented.

Near the Jovian planets, planetesimals would be mostly icy materials; these may have formed the nuclei of comets. These icy bodies may have then been gravitationally directed into the Oort cloud. We expect from the condensation sequence that bodies formed near the Jovian planets would have an icy composition. Estimates for the actual compositions of Uranus and Neptune are 60 to 70% icy materials. Comets are believed to have almost the same relative percentage. The inference: these two planets grew by the accretion of icy planetesimals. The remaining planetesimals have their orbits perturbed by the Jovian planets and ejected into orbits (20,000 to 50,000 AU) typical for the Oort cloud. Uranus and Neptune have the most efficient gravitational effects and account for some 75% of the objects in the comet cloud. Hence, the Oort cloud is a by-product of planetary accretion in the outer Solar System.

(F) FORMATION OF THE EARTH AND MOON

Because the Moon is a simpler, more primitive body than the Earth, we have a better idea of its history—but not of its formation! Attempts to explain the Moon's formation have usually fallen into one of three categories: capture, fission, or binary accretion. In the capture model, the Moon forms by accretion some distance from the Earth and later is captured gravitationally by it. In the simple fission model, a rapidly rotating, plastic Earth throws off a large part of its mantle (right after its core is formed); this piece cools to become the Moon. The binary accretion model views the formation of the Earth and Moon as taking place more or less simultaneously from the same part of the solar nebula. Each model has different strengths and weaknesses, and planetary scientists now feel that the best synthesis to date is a fourth model, the giant-impact model.

The capture model has largely fallen out of favor. It focuses on the differences in the chemical composition of the Earth and Moon (the Moon's lack of water and iron and its higher abundance of uranium and rare-earth elements). But both bodies have the same isotopic abundances of oxygen (as do some types of meteorites) so that they probably formed nearby in the solar nebular. Also, the dynamics of capture require very special conditions that are hard to arrange—in particular, a massive third body is needed nearby during the capture to ensure the conservation of momentum.

In contrast, the fission model relies on similarities between the Earth and the Moon (specifically, the chemical composition of the Earth's mantle). But substantial differences exist for both volatiles and major elements. In particular, the Mg/Si ratio for the Moon is outside the range of possible values for the Earth's mantle; refractory elements (such as aluminum and calcium) are enriched by a factor of two in the Moon; volatile elements are depleted; and the Ni/Fe ratio is too low to be derived from the Earth's mantle.

Binary accretion models handle many of these problems. The Earth and Moon grow by accretion in the same region of the solar nebula from nearby material. Some matter is ejected off the accreting body by extraordinary impacts, adding debris to

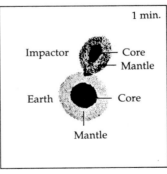
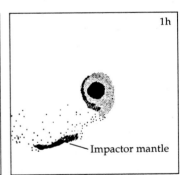
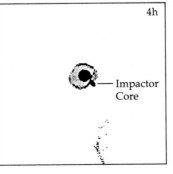
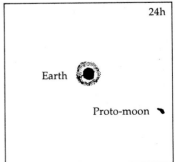

FIGURE 7–29 Computer simulation of a giant impact formation of the Moon. (A) The impactor, of 0.14 Earth mass, strikes the Earth with a glancing blow at a speed of 5 km/s. (B) The impactor spreads out into space; its mantle eventually falls onto the Earth after 1 hour. (C) The impactor core accretes to the Earth about 4 h after the collision. (D) A silicate lump, mostly from the mantle of the impactor, remains in orbit to form the proto-Moon 24 hours after impact. *(A. G. W. Cameron and W. Benz)*

the area. The Earth accretes most of the iron, which falls into the core that forms early in the sequence. Large impacts then blast off mantle-like material. Once the Moon forms, radioactive decay (from ^{26}Al) heats it enough to deplete the volatiles. The end result is that the Earth and Moon share many chemical similarities with some notable differences.

The **giant-impact model** envisions the impact of a Mars-sized body into the young Earth. The energy of this tremendous impact injects a large fraction of the mass of the Earth and the impacting body into a disk of material around the Earth from which the Moon accreted. In order to account for the angular momentum of the Earth–Moon system, the incoming body would have to strike the Earth tangentially at about 10 km/s. This idea has not encountered any serious objections, and computer simulations support it (Figure 7–29).

We now turn to the Earth. After the era of accretion (which may have lasted but a few million years), the Earth was a pockmarked planet (a surface cratered like the lunar highlands) of roughly uniform composition; its atmosphere was mainly hydrogen. Radioactive heating then melted the interior and the core formed. This heating promoted degassing from the interior to create a second atmosphere rich in water, carbon dioxide, methane, and ammonia. When the Earth's surface had cooled enough, intense rains fell to begin the formation of the oceans. Perhaps a billion years after formation, plate tectonics began to modify the crust, which cooled and thickened. The ocean basins were mostly filled. Life developed and modified the atmosphere into the one we have today. Some 200 million years ago, the continents broke up and were driven by plate tectonics into the positions we see now.

PROBLEMS ◑

1. What effects do radiation pressure and the Poynting-Robertson effect have upon an artificial space probe of mean density 1000 kg/m^3 and radius 1 m? Justify your answer quantitatively and state your assumptions.

2. At the origin of the Solar System, the Sun's tidal forces and Roche limit might have played an important role. To avoid tidal disruption, what is the minimum density a protoplanet must have at the distance d (in AU) from the Sun? Comment upon your result.

3. According to theoretical models, 4 billion years ago Jupiter may have had a luminosity as high as 10^{-2} $L_\odot$ and a surface temperature of 1000 K. What was Jupiter's radius then if it radiated as a blackbody?

4. (a) Estimate the present rate at which the Earth is accreting interplanetary dust.
   (b) Calculate how long it will take the Earth at this rate to increase its mass by 50%.

5. Calculate how fast the sun would rotate if all the angular momentum of the planets were added to it.

6. Compare the gravitational impact parameters for the Earth (surface) and for Jupiter (top of atmosphere) for bodies falling in at their respective escape velocities.

7. Compare the escape velocities from Titan, Dione, and Hyperion. Estimate the lifetime of methane on each.

8. Ceres has an infrared flux (at 10 μm) that is 4 times its visible flux. What is the albedo of Ceres? What is its radius?

9. Assume the asteroid Herculina has a moon of radius 100 km, orbiting 1000 km from Herculina. Make both bodies rocky.
   (a) Calculate the moon's orbital period.
   (b) Compare the tidal force of the Sun pulling the bodies apart with the mutual gravitational forces holding them together. (*Hint:* assume that both bodies are of the same density.)

10. On a clear night, if you are patient and away from city lights, you can see about 10 meteors per hour. The meteoroid that creates a meteor has a mass of about 1 g and burns about 100 km above the Earth's surface. *Estimate* the amount of mass of meteoroids that enters the Earth's atmosphere per year. If this influx has been constant, how much mass has been added to the Earth since it was formed?

11. Comet Ikeya-Seki (the great sun-grazing comet of 1965) had an elliptical orbit with a period of about 700 years. As it rounded the Sun, the comet had a perihelion distance of 0.008 AU.

(a) How far away from the Sun (in AU) will the comet be at aphelion?

(b) If its perihelion velocity was 500 km/s, what will its orbital velocity be at aphelion?

(c) Make some reasonable assumption about the average density of the comet's nucleus. Did the comet pass close enough to the Sun to be ripped apart by tidal gravitational forces? (*Hint*: Work out the Roche limit for the Sun.)

12. The crushing strength (pressure for material deformation) for iron meteorites is about 4×10^8 Pa. At what size would an iron asteroid have this central pressure? What would happen to it if it were this size or larger? (*Hint*: Use the equation of hydrostatic equilibrium.)

13. Assume that the particles in Saturn's rings have circular orbits ranging from 87×10^3 to 137×10^3 km from the center of Saturn.

(a) Calculate the orbital velocities at the inner and outer edges of the rings, assuming that they are made of individual particles obeying Kepler's laws.

(b) What resolution (what accuracy in wavelength measurement) would a spectrograph have to have in order to distinguish between the inner and outer edge speeds of the rings?

14. Compare the expected brightness of Halley's Comet at 3 AU from the Earth to that at 1 AU.

15. Assume that a typical semimajor axis is 50,000 AU for a cometary nuclei in the Oort Cloud. What is the typical orbital period?

16. Compare the characteristics for the ring systems of the four Jovian planets.

17. (a) The Voyager mission found that Saturn's moon Titan has a substantial nitrogen atmosphere. Calculate the ratio of the root mean square speed to escape velocity of nitrogen in Titan's atmosphere. Make a statement about the implications of your result.

(b) Do the same for Neptune's moon Triton, which was found by Voyager to have a less substantial nitrogen atmosphere.

18. The Orionid meteor shower is believed to be produced by dust in the orbit of Halley's Comet. What is the orbital speed of meteors seen during the Orionid meteor shower?

19. A newly discovered faint asteroid and Mars are observed at opposition. Mars is observed to be 100 million times brighter than the asteroid. Assume Mars and the asteroid are 1.5 and 3.0 AU from the Sun, respectively.

(a) Estimate the radius of the asteroid if it has an albedo of 16%, similar to some S-type asteroids.

(b) Estimate the radius of the asteroid if it has an albedo of 4%, typical of C-type asteroids.

PART TWO
The Stars

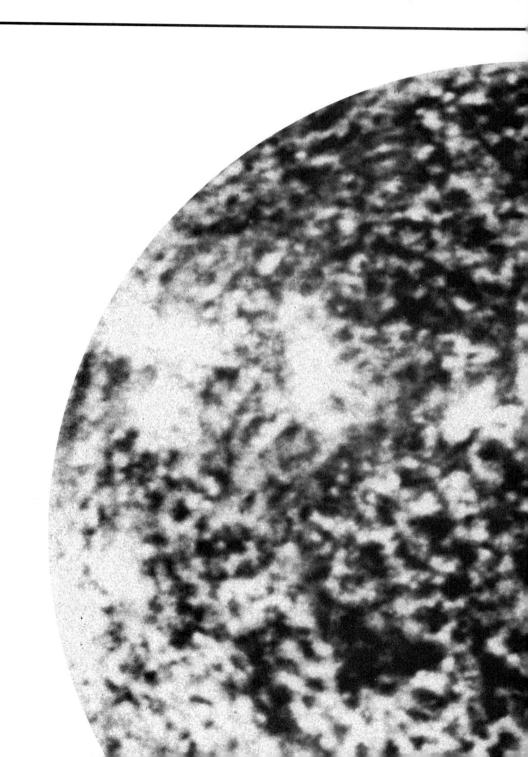

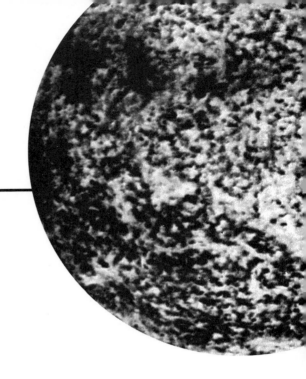

Chapter 8

Electromagnetic Radiation and Matter

We turn now from planets to stars. The nearest star is our own Sun, which shows a visible disk. The other stars are mere points of light scattered across the night sky. How can we study such objects? Stars betray their presence by the light they emit, and by detecting and deciphering these radiations can we infer the properties of stars.

This chapter deals with the characteristics of **electromagnetic radiation** (most familiar in the form of **visible light**), the atomic structure of matter, and the all-important interactions between matter and radiation. Astronomers handle light in two ways: by measuring the total intensity (over selected ranges of wavelengths) and by dispersing the light into a **spectrum** and examining it in detail. Both techniques resolve the astronomer's problem of how to infer the physical properties of the faraway stars because they permit an investigation of the full range of the electromagnetic spectrum.

8–1 ◑
ELECTROMAGNETIC RADIATION

Keeping in mind our ultimate goal of understanding stars, let's begin by investigating the nature of light as electromagnetic waves; later, we will turn to its particle aspect.

(A) THE WAVE NATURE OF LIGHT

What is a **wave?** You are familiar with water waves, which are undulatory disturbances traveling along the surface of the liquid. Place corks along the line of the wave's motion, and you may characterize such a wave (a transverse, or up–down, wave) by the height h of the corks above the average surface level (Figure 8–1) in the mathematical form

$$h = h_0 \sin\left[(2\pi/\lambda)(x - vt)\right] \qquad (8\text{–}1)$$

Note that in a **transverse wave** the disturbance of the medium through which the wave travels is perpendicular to the direction of motion of the wave. In contrast, a **longitudinal wave** involves a disturbance that lies along the direction of motion, such as the compression of a spring.

Equation 8–1 represents a sinusoidal wave of **amplitude** h_0 progressing with time t along the positive x axis at the speed v; the distance between

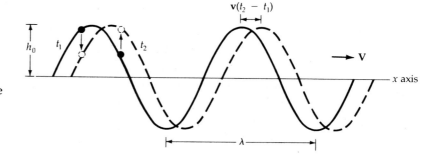

FIGURE 8–1 A traveling wave. A water wave moving with speed v to the right is shown at two times, t_1 and t_2. Floating corks mark the oscillating surface.

successive wave crests is called the **wavelength** λ. At a given time (say, $t = 0$), the corks betray the oscillatory pattern

$$h = h_0 \sin (2\pi x/\lambda) \qquad (8\text{--}2)$$

If you observe the cork at $x = 0$, it rises and falls periodically. Note that a given cork completes one oscillation in the time λ/v second; in 1 s, v/λ wave crests will have passed a given point, and so the **frequency,** ν, of oscillation is just $\nu = v/\lambda$ per second. Therefore, we completely characterize such a wave by the fundamental relationship

$$\lambda\nu = v \qquad (8\text{--}3)$$

For example, water waves of wavelength $\lambda = 5$ cm moving at a speed $v = 10$ cm/s will pass a given point at a frequency $\nu = 2/s$ (twice per second). The SI unit for frequency is the **hertz** (Hz), with 1 Hz equaling one cycle per second.

From Equation 8–3, we see that

$$\lambda\nu = c \qquad (8\text{--}4)$$

is the fundamental relationship between the wavelength and frequency of an electromagnetic wave (in vacuum), where c is 299,792 km/s—the speed of light.

Most waves require a material medium in which to be transmitted: water waves travel along the surface of water, sound waves move through air, and earthquake waves, both compressional and transverse, propagate through the solid Earth. Electromagnetic waves, however, may propagate through a pure vacuum at speed c. What is it that is propagating? The space around an electric charge may be characterized by an **electric field vector, E,** which manifests itself as a force on a test charge placed nearby. If an electromagnetic wave encounters such a test charge, that charge will os-

cillate, so that we may ascribe the sinusoidal electric field

$$\mathbf{E} = \mathbf{E}_0 \sin [(2\pi/\lambda)(x - ct)] \qquad (8\text{--}5)$$

to an electromagnetic wave traveling in the positive x direction along the x axis. Maxwell's equations say that a time-varying electric field produces a perpendicular time-varying **magnetic field, B,** however, and so an electromagnetic wave is a self-propagating disturbance of electric and magnetic fields in a vacuum (Figure 8–2). If the electric field always oscillates in a single plane, the wave is **plane-polarized;** otherwise, it is **elliptically polarized** (a special case of which is **circular polarization**). That is, if we write the components of the **E** vector as

$$E_x = E_{x1} \sin (kz - \omega t - \delta_1)$$

and

$$E_y = E_{y1} \sin (kz - \omega t - \delta_2)$$

then the waves are polarized if $\delta_1 - \delta_2 =$ constant.

Polarization plays an important role in astronomical observations, and so we deal with it in some detail here. Because light is a transverse wave, the orientations of the wave amplitude perpendicular to the direction of motion are the possible directions of polarization. Most light sources are unpolarized, which means that the electric field vector vibrates in a random orientation in the plane perpendicular to the propagation direction.

If we consider that the electric field has two components in this plane (E_x and E_y), then the phase differences between them determine the kind of polarization the wave has. If they are in phase, the wave is plane-polarized; if they are not in phase, the wave is elliptically polarized; if they

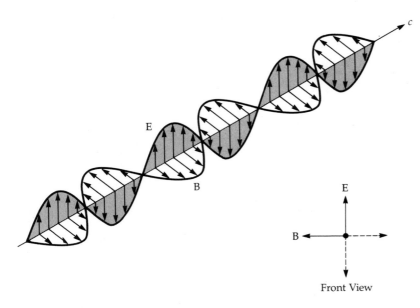

FIGURE 8–2 An electromagnetic wave. The electric field **E** and magnetic field **B** propagate at speed c through a vacuum. The two fields oscillate perpendicular to each other. The front view shows this orientation for a linearly polarized wave.

are 90° out of phase, the wave has a circular polarization.

The Electromagnetic Spectrum

Light is that class of electromagnetic radiation occupying the wavelength range $\lambda \approx 390$ nm (violet) to 720 nm (far red)—the **optical spectrum.** A wavelength λ is a length with units of centimeters or meters (or kilometers or miles); three convenient units of length in astronomy are the micrometer (sometimes called the micron), $1\ \mu m = 10^{-6}$ m; the nanometer, $1 nm = 10^{-9}$ m; and the angstrom, $\mathring{A} = 10^{-10}$ m $= 10^{-4}\ \mu m$ (Appendix 7). Electromagnetic waves of different wavelengths are detected in different ways (Table 8–1), so that we give characteristic names to various parts of the electromag-

TABLE 8–1   *The Electromagnetic Spectrum*

Wavelength	Energy or Frequency	Type of Radiation	Detectors
10^{-6} nm	1240 MeV	Gamma rays	Geiger counters
10^{-5} nm	124.0 MeV	Gamma rays	Scintillators
10^{-4} nm	12.4 MeV	Gamma rays	Proportional counters
10^{-3} nm	1.24 MeV	Gamma rays	Cerenkov light detectors
10^{-2} nm	124 KeV	X-rays	Photo emulsions
10^{-1} nm	12.4 KeV	X-rays	Photodetectors
1.0 nm $= 10^{-9}$ m	1.24 KeV	Ultraviolet	Photo emulsions
10 nm	124 eV	Ultraviolet	TV camera (vidicon)
100 nm	12.4 eV	Ultraviolet	Microchannel plates
1000 nm $= 1\ \mu m$	1.24 eV	Optical	Photoconductors, CCDs
10 μm	0.124 eV	Infrared	Photovoltaic detectors
100 μm	0.0124 eV	Infrared	Thermal detectors
1000 $\mu m = 1mm$	0.0012 eV	Infrared	Bolometers
10 mm $= 1$ cm	30,000 MHz	Microwave (radar)	Hetrodyne radio receivers
10 cm	3,000 MHz	UHF	Radio receivers
100 cm $= 1$ m	300 MHz	FM	Radio receivers
10 m	30 MHz	Short-wave radio	Radio receivers
100 m	3 MHz	Short-wave radio	Radio receivers
1000 m $= 1$ km	300 kHz	Long-wave radio	Radio receivers
10 km	30 kHz	Long-wave radio	Radio receivers
100 km	3 kHz	Long-wave radio	Radio receivers

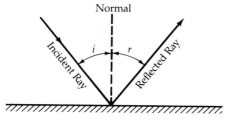

FIGURE 8–3   Reflection of light. The angle of incidence i equals the angle of reflection r.

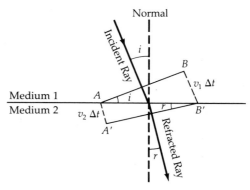

FIGURE 8–4   Refraction of light. The line AB represents a plane wavefront incident on the interface between medium 1 and medium 2 with angle of incidence i. Point A travels to A' in time Δt, a distance of $v_2 \Delta t$. Point B travels $v_1 \Delta t$. The ray of light is bent across $A'B'$.

netic spectrum: gamma rays ($\lambda \leq 10^{-3}$ nm), X-rays (10^{-3} to 10 nm), ultraviolet (10 to 300 nm), visible light (400 to 800 nm), infrared (1 to 10^3 μm), microwaves (1 mm to 10 cm), and radio waves ($\lambda \geq$ 1 cm). Visible light occupies less than one decade of the 21 decades of wavelength shown in Table 8–1, and yet visible light still dominates observational astronomy because it easily penetrates the Earth's atmosphere and it is readily perceived by the human eye—and so we have a natural bias!

Reflection and Refraction

Now to consider some of the properties of light that make it possible for us to manipulate it—the goal of **optics**. A ray (or beam or pencil) of light falling upon a mirror is *reflected* in accordance with the rule (Figure 8–3) that the angle of reflection, r, equals the angle of incidence, i. Note that these angles are defined with respect to the **normal** (perpendicular) to the reflecting surface. Reflecting telescopes utilize the law of reflection to collect, direct, and focus light.

When light is reflected, both the incident and reflected waves travel in the same medium at the same speed, but what happens when light passes from one medium into another? The speed of light v in a medium is generally different from the speed of light c in a vacuum. We may characterize a given medium by its **index of refraction:** $n = c/v$. The index of refraction of air is $n = 1.0003$, which is practically the same as that of a vacuum ($n = 1.0000$). Crown glass, however, has $n = 1.5$, and so in this medium $v = 2c/3 \approx 2 \times 10^5$ km/s. Light passing from a medium with index of refraction n_1 into a medium with index n_2 is bent, or *refracted*, in accordance with **Snell's law** (Figure 8–4):

$$n_1 \sin i = n_2 \sin r \qquad \text{(8–6)}$$

where i is the angle of incidence in medium 1 and r is the angle of refraction in medium 2 (both with respect to the normal to the interface between the media). A refracting telescope uses refraction guided by lenses to make images.

In general, the index of refraction of a medium depends upon wavelength: $n = n(\lambda)$. That is, light of different wavelengths (or *colors*) is refracted through different angles $r(\lambda)$ when the incident beams all have the same angle of incidence i. So an incident mixed (white) beam is dispersed into separate beams of pure colors, as Newton demonstrated. The phenomenon of **dispersion** enables us to decompose light into its component colors—in the form of a spectrum—and is the basis of spectroscopy.

Diffraction and Interference

When water waves encounter an island, they *diffract* around the sides of the island to converge, and when these converging waves meet one another, they *interfere*. In the case of sound waves, we are also familiar with these phenomena of **diffraction** (sound bends around sharp corners) and **interference** (recall the silent spots in a large auditorium). Light exhibits these same phenomena, which are understandable in terms of a wave model.

Consider the situations where light waves meet an opaque wall having one or two apertures, with a viewing screen placed beyond (Figure 8–5). The sharp images of the apertures, which we expect to see on the viewing screen, are distorted (a) by the diffraction and spreading out of the light waves and (b) by the constructive and destructive interference of different light waves. We may understand both processes by considering how the intensity patterns at the viewing screen are formed. The intensity I of light is proportional to the square of the electric field:

$$I \propto |\mathbf{E}|^2 \qquad (8\text{–}7)$$

At an arbitrary point on the viewing screen (Figure 8–5B), the electric field of the wave is given by Equation 8–5:

$$E = E_0 [\sin a + \sin (a + b)] \qquad (8\text{–}8)$$

where a refers to the wave from one aperture and $a + b$ to the wave from the other aperture. The **phase shift,** b, corresponds to the difference in path lengths from the apertures to the point on the viewing screen. From Equation 8–8, the intensity is

$$I \propto E_0^2 [\sin^2 a + \sin^2 (a + b) \\ + 2 \sin a \sin (a + b)] \qquad (8\text{–}9)$$

The first two terms on the right are just the intensities attributable to the separate apertures; the third term is responsible for the oscillatory interference pattern on the screen (as b changes with position).

Every point within the single aperture (Figure 8–5A) contributes to the diffraction pattern at the viewing screen. Application of Equations 8–7 and 8–8 shows that the **angular width,** θ, of the principal diffraction image is

$$\theta \approx \lambda / d \qquad (8\text{–}10)$$

where θ is in radians, λ is the wavelength of the light, and d is the size of the aperture. No optical image can be smaller than the **diffraction limit** of Equation 8–10, and so we say that this is the optimum **angular resolution** of the system. For example, the approximate resolution of a telescope of aperture $d = 1$ m viewing light of wavelength 500 nm is

$$\theta \approx (500 \text{ nm})/(10^9 \text{ nm}) \\ = 5 \times 10^{-7} \text{ rad} = 0.1 \text{ arcsecond}$$

This theoretical resolving power is usually not achieved by a ground-based telescope because of turbulence in the Earth's atmosphere.

The Doppler Effect

So far we have considered waves where both the source and the observer are at rest. When either is in motion along the line connecting them, both the wavelength and the frequency of the wave are altered by the **Doppler effect,** named for the Austrian physicist C. J. Doppler (1803–1853). You are familiar with the Doppler phenomenon in water and air waves; for example, as a police car with its siren blaring approaches, the pitch (frequency) of the sound is high, but the frequency drops noticeably to a lower tone as the car passes you and then recedes.

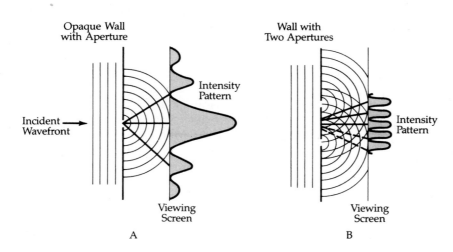

FIGURE 8–5 Diffraction and interference. (A) Diffraction of light waves with a single slit. (B) Interference of light waves with two slits. The intensity peaks where the wavefronts constructively reinforce.

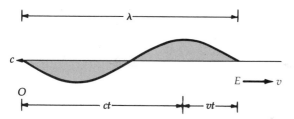

FIGURE 8–6 The Doppler effect. An observer at O sees a wave emitted by a source E moving at speed v to the right.

Most familiar wave motions are associated with a material medium, which may complicate the picture by itself being in motion. Let's consider electromagnetic waves propagating in a vacuum. Armand Fizeau (1819–1896) correctly explained the classical Doppler effect of light in 1848, and Albert Einstein gave the relativistic explanation in 1905.

Imagine a light source E (Figure 8–6) receding at speed v from an observer O while emitting radiation of wavelength λ_0 and frequency ν_0. In the time $t = 1/\nu_0$, one wavelength (λ_0) emerges from the source but, as seen by the observer, that wave has the length

$$\lambda = (c + v)t$$
$$= c[1 + (v/c)]/\nu_0 = \lambda_0[1 + (v/c)] \quad \textbf{(8–11)}$$

since the source has traveled the distance vt to the right. Note that the fundamental Equation (8–4) has been used in the last equality. The observed frequency is

$$\nu = c/\lambda = \nu_0/[1 + (v/c)] \quad \textbf{(8–12)}$$

so that $\lambda > \lambda_0$ and $\nu < \nu_0$ and the light is **redshifted** (that is, shifted to longer wavelengths, or toward the red). When the source *approaches* the observer, v changes sign to give $\lambda < \lambda_0$ and $\nu > \nu_0$, and the radiation is **blueshifted.**

From Equation 8–11, we extract the wavelength shift, $\Delta\lambda/\lambda_0$:

$$\Delta\lambda/\lambda_0 = (\lambda - \lambda_0)/\lambda_0 = v/c \quad \textbf{(8–13)}$$

Remember that the sign of v is positive for recession and negative for approach.

When v approaches c, Einstein's theory of special relativity must be used. The two basic postulates of this theory are (1) the speed of light is independent of the motion of either source or observer

and (2) only relative motions are observable. So v is the *relative* speed of the source and the observer, and no relative speed greater than the speed of light is possible ($v \leq c$). Equations 8–11 and 8–12 then necessarily take the symmetrical forms

$$\lambda = \lambda_0[(1 + v/c)/(1 - v/c)]^{1/2} \quad \textbf{(8–14a)}$$
$$\nu = \nu_0[(1 - v/c)/(1 + v/c)]^{1/2} \quad \textbf{(8–14b)}$$

The nonrelativistic shift in wavelength (Equation 8–13) follows from Equation 8–14a in the limit when $v \ll c$.

(B) THE QUANTUM NATURE OF LIGHT: PHOTONS

At the beginning of the eighteenth century, Newton proposed a particle theory of light, but it was not until the end of the nineteenth century that the particle-like manifestations of light began to appear in experiments.

Light is neither a particle nor a wave, but it can manifest itself as either one or the other! This apparently paradoxical behavior explains such phenomena is the photoelectric effect, Compton scattering, and blackbody radiation [Section 8–6(a)]. In addition, the interaction of light with atoms and molecules is understandable only if electromagnetic energy propagates in the form of discrete bundles, which we call **photons** or **quanta**. The energy of a light quantum (E) is proportional to the frequency characterizing the light wave:

$$E = h\nu \quad \textbf{(8–15)}$$

where $h = 6.626 \times 10^{-34}$ J · s is known as **Planck's constant.** We may crudely picture a classical light wave of wavelength λ and frequency ν as composed of many quanta, each with the energy given in Equation 8–15.

(C) INTENSITY VERSUS FLUX

In detecting the energy (or counting the photons) coming from a distant light source, we must be careful to distinguish between intensity and flux. **Intensity** depends upon direction, in the sense that the intensity I of a source is the amount of energy emitted per unit time Δt, per unit area of the source ΔA, per unit frequency interval $\Delta\nu$, per unit solid angle $\Delta\Omega$ in a given direction. The **solid**

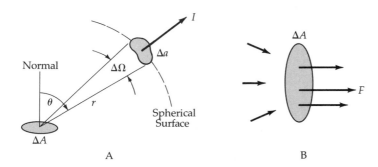

FIGURE 8–7 Intensity and flux. (A) The intensity of light depends upon direction θ and solid angle $\Delta\Omega$. (B) Flux depends only upon the energy passing through an area ΔA per unit time.

angle (Figure 8–7A) of the beam is related to the area Δa intercepted by the beam at the spherical surface of radius r by

$$\Delta\Omega = \Delta a/r^2 \qquad (8\text{–}16)$$

The dimensionless unit of solid angle is the **steradian** (sr), with the entire spherical surface subtending 4π sr—since $\Delta a = 4\pi r^2$ for the surface area of a sphere. (A steradian is essentially 1 rad²; since 1 rad = 57.3°, 1 sr = 3283 square degrees of arc.) Common units of intensity are W/m² · Hz · sr. For example, the total energy flow, E, from a spherical star of surface area A, which astronomers call luminosity, L, is just

$$E = L = 4\pi \int_0^\infty I(\nu)d\nu$$

where $I(\nu)$ is called the **monochromatic intensity,** the intensity at a specific frequency per unit area per unit time per unit solid angle in a specific direction; E or L is in watts.

Flux, F, relates directly to what we measure with a telescope. The flux of energy through a surface (or into a detector) is the amount of energy per unit time passing through a unit area of the surface per unit frequency interval: $F(\nu) = \text{energy}/\Delta A \cdot \Delta t \cdot \Delta\nu$ (Figure 8–7B). Hence, the unit of monochromatic flux $F(\nu)$ is W/s · Hz. For example, if a star emits energy at the rate L (in watts), then the flux of energy (in W/m²) through a concentric spherical surface of radius R is

$$F = L/4\pi R^2 \qquad (8\text{–}17)$$

Equation 8–17 is a form of the well-known inverse-square law for the diminution of radiant flux with distance.

We may relate intensity and flux by noting that the energy passing through a certain area is composed of beams entering at different angles to the normal to the surface; to a beam entering at the angle θ, the surface area appears reduced by a factor cos θ, so that

$$F(\nu) = \int_0^\infty \int_0^{4\pi} I(\nu, \theta) \cos \theta \, d\Omega$$

Note that the intensity is independent of distance, while flux depends on the inverse-square of the distance. For instance, if we measure the Sun's flux with a telescope from the Earth and then from Mars, we find it decreases by $(1/1.5)^2$, or 44%. But if we are looking at a specific area of the Sun's disk, the measured intensity is the same from both planets.

8–2 ⦾
ATOMIC STRUCTURE

If we divide matter into smaller and smaller pieces, we find that at scales near 10^{-10} m, all forms of matter are composed of **atoms** (and molecules, which are bound aggregates of atoms). An elemental substance, such as gold, consists solely of atoms of the **element** gold; there are but 92 natural elements (hydrogen to uranium; Appendix 5), but this list has been extended to 105 with the addition of the synthetic transuranic elements.

(A) ATOMIC BUILDING BLOCKS

What is an atom, and how may we characterize the elements? An atom is composed of a small **nucleus,** which contains most of the atom's mass in a size of the order of 10^{-14} m, surrounded by a dif-

TABLE 8-2 *Properties of Atomic Particles*

Name	Mass (kg)	Mass (electron mass)	Electric Charge
Proton	1.6725×10^{-27}	1836	$+e$
Neutron	1.6748×10^{-27}	1838	0
Electron	9.1091×10^{-31}	1	$-e$

fuse cloud of **electrons** extending out to about 10^{-10} m. The atomic nucleus consists of **protons** and **neutrons** bound together by the strong interaction. These three elementary particles—the proton, the neutron, and the electron—are the fundamental building blocks of an atom. Table 8-2 lists the important parameters of each. Both the proton and the neutron are about 2000 times more massive than the electron. The neutron is uncharged, and the proton (positive) and the electron (negative) have electric charges of opposite sign and equal magnitude $e = 1.602 \times 10^{-19}$ coulomb (C). A normal neutral atom consists of an equal number of protons and electrons, with approximately the same number of neutrons as protons (for the lighter elements); so an element is determined by the number of protons Z (called the **atomic number**) in the atomic nucleus. Heavier elements have a higher proportion of neutrons. For example, hydrogen ($Z = 1$) is one electron orbiting one proton; uranium ($Z = 92$) has 92 protons and 92 electrons, with about 146 neutrons in its nucleus.

A given element may exist in several different forms, called **isotopes.** All isotopes of an element have the same atomic number but differing numbers of neutrons (N). We term the different isotopic nuclei **nuclides.** To characterize a nuclide, the notation ^{Z+N}X or ^{A}X is used, where X is the symbol of the element having Z protons and $A = Z + N$ is the **atomic mass** (number of protons plus neutrons). For example, three isotopes of hydrogen are known: ordinary hydrogen, ^1H; deuterium, ^2H; and tritium, ^3H. The mass of an atom is conveniently given in terms of atomic mass units (amu); the standard is ^{12}C, which has a mass of exactly 12 amu. Since 1 amu is essentially the mass of a proton, an atom's mass is A amu. When atomic masses are tabulated (Appendix 5), they frequently are not integers because it is the average mass of the naturally occurring isotopes that is being listed.

(B) THE BOHR ATOM

What dynamical configuration of electrons about a nucleus leads to a stable atom? Because the attractive electric force between a proton and an electron (Coulomb's law) obeys an inverse-square law, scientists envisaged electrons orbiting the nucleus. Electrons are charged particles, however, and charged particles radiate energy when they are accelerated (in circular orbit). So the very property that permits a bound atom should cause it to collapse almost at once because the electron's orbit would decay from the loss of energy! In 1913, Niels Bohr (1885–1962) advanced a simple theory that offered a way out of this problem; it resulted in the modern theory of **quantum mechanics** in the 1920s. Let's discuss Bohr's atomic theory and apply it to the hydrogen atom (which it accounts for well).

Quantized Orbits

In 1911, Sir Ernest Rutherford (1871–1937) proposed the nuclear model of the atom. Bohr spent a year with Rutherford and then stated two astounding postulates that breathed life into atomic theory. Of the infinite number of possible electron orbits in Rutherford's model, Bohr first postulated that *only a discrete number of orbits are allowed to the electron and when in those orbits, the electron cannot radiate.* The permitted orbits are those in which the orbital angular momentum of the electron is an *integral* multiple of $h/2\pi$, where h is Planck's constant.

Let's apply Bohr's postulate of quantized orbits to an electron in a circular orbit of radius r about a nucleus of charge Ze. The centripetal force maintaining the orbit (where m is the electron's mass),

$$mv^2/r$$

is provided by the coulombic attraction between the electron and the nucleus:

$$(kZe)e/r^2$$

where $k = 1/4\pi\epsilon_0 = 8.99 \times 10^9$ N · m²/C² in SI units. Equating these gives

$$mv^2/r = (kZe)e/r^2$$

But Bohr's postulate implies the additional constraint

$$mvr = n(h/2\pi) \qquad n = 1, 2, 3, \ldots \quad \textbf{(8-18)}$$

Combining these equations, we find

$$r = kZe^2/mv^2 = nh/2\pi mv$$

or

$$r = n^2(h^2/4\pi^2me^2kZ) \qquad \textbf{(8–19)}$$

Therefore, the permitted discrete orbits occur at geometrically increasing (n^2) distances, with the smallest Bohr orbit occurring when the **principal quantum number,** n, equals unity.

We may now find the **total energy,** E, of these orbits. If E is negative, the system is bound and we have an atom. We know that E = kinetic energy + potential energy, where the potential energy for an atom is

$$PE = \int_\infty^r \frac{kZe^2 \, dr}{r^2}$$

so that

$$E = (mv^2/2) - (kZe^2/r) \qquad \textbf{(8–20)}$$

By using Equations 8–18 and 8–19 to evaluate Equation 8–20, we find

$$E(n) = -(2\pi 2me^4k^2Z^2)/n^2h^2 \qquad \textbf{(8–21)}$$

(That these energies are negative indicates that the orbits are **bound.**) The smallest Bohr orbit ($n = 1$) is the most strongly bound, and all orbits are bound until $n \to \infty$ where $E \to 0$. For $E > 0$, a *continuum* of unbound orbits is available to the electron.

Quantized Radiation

Bohr's second postulate concerns the absorption and emission of radiation by an atom—the fundamental interaction between matter and radiation. According to the first postulate, the electron cannot radiate while in one of the allowed discrete orbits, and so the second postulate states that (a) *radiation in the form of a single discrete quantum is emitted or absorbed as the electron jumps from one orbit to another* and (b) *the energy of this radiation equals the energy difference between the orbits.*

When an electron makes a **transition** from a higher orbit (n_a) to a lower orbit (n_b), a photon is emitted. The energies of this process may be symbolized by

$$E(n_a) = E(n_b) + h\nu \qquad \text{(emission)} \quad \textbf{(8–22a)}$$

where $n_a > n_b$. For the electron to make a transition from the lower orbit to the upper orbit, the atom must absorb a photon of exactly the correct energy:

$$E(n_b) + h\nu = E(n_a) \qquad \text{(absorption)} \quad \textbf{(8–22b)}$$

In both cases, the frequency of the photon involved is (Equations 8–15 and 8–21)

$$\begin{aligned} v_{ab} &= [E(n_a) - E(n_b)]/h \\ &= (2\pi^2me^4k^2/h^3)Z^2[(1/n_b^2) - (1/n_a^2)] \end{aligned} \quad \textbf{(8–23)}$$

Note that only one quantum is emitted or absorbed, even if $n_a > n_b + 1$; the electron may jump over several intermediate orbits. On the other hand, an electron may **cascade** to the lowest orbit, emitting several photons of different energy as it jumps through a series of adjacent orbits. The lowest orbit, called the **ground state,** corresponds to $n = 1$.

(C) THE BOHR MODEL OF THE HYDROGEN ATOM

Now to apply Bohr's picture to the simplest atom, hydrogen ($Z = 1$), with its single electron. The electron's permitted orbital energies are, from Equation 8–21,

$$E(n) = -(2\pi^2me^4k^2/h^2)(1/n^2) \equiv -R'(1/n^2) \quad \textbf{(8–24)}$$

where $R' = 2.18 \times 10^{-18}$ J incorporates all the other constants.

It is convenient to express Equation 8–23 in terms of the **wave number** (reciprocal wavelength):

$$\begin{aligned} 1/\lambda_{ab} &= v_{ab}/c = (R'/ch)(1/n_b^2 - 1/n_a^2) \\ &= R(1/n_b^2 - 1/n_a^2) \end{aligned} \quad \textbf{(8–25)}$$

The **Rydberg constant** $R(= R'/ch)$ has the value 10.96776 μm^{-1} for hydrogen. In the case of a difference of one level, where $n_a = n_b - 1$, then

$$\begin{aligned} 1/\lambda_{ab} &= R[1/n_b^2 - 1/(n_b - 1)^2] \\ &= R\{(n_b^2 - 2n_b + 1 - n_b^2)/[n_b^2 - 1)]\} \\ &\approx 2R/n_b^3 \end{aligned}$$

Equation 8–25 says that there is a series of wavelengths of *each* value of n_b when we consider a sequence of increasing values of n_a beginning with

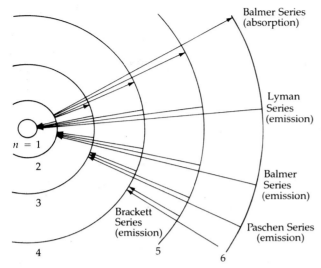

Balmer Series
(absorption)

Lyman
Series
(emission)

Balmer
Series
(emission)

Paschen Series
(emission)

Brackett
Series
(emission)

$n = 1$
2
3
4
5
6

FIGURE 8–8 The Bohr model for hydrogen. The orbitals are the stable energy levels of the electron, from $n = 1$ (ground state) and up. Shown are the sequences of electronic transitions for the Balmer (absorption and emission), Lyman, and Paschen series.

$n_a = n_b + 1$ (Figure 8–8). Transitions of the **Lyman series** (in the ultraviolet) all have the ground state ($n_b = 1$) as their lowest orbit, with $n_a > 2$. The visible spectral lines of the **Balmer series** have $n_b = 2$ and $n_a \geq 3$. Other important series are the **Paschen, Brackett,** and **Pfund** series, with $n_b = 3$, 4, and 5, respectively. All of these series are named for the physicists who first observed the spectral lines corresponding to the indicated transitions.

The first hydrogen lines discovered were the Balmer series, and they are designated H_α for $n_a = 3$, H_β for $n_a = 4$, H_λ for $n_a = 5$, and so forth. Use Equation 8–25 to compute the wavelength of the H_α line, which corresponds to $n_b = 2$ and $n_a = 3$:

$$1/\lambda_{H_\alpha} = 10.96776(1/4 - 1/9) \; \mu\text{m}^{-1} = 1.52330 \; \mu\text{m}^{-1}$$
$$\lambda_{H_\alpha} = 656.3 \text{ nm}$$

Similarly, the Lyman-α line ($n_b = 1$, $n_a = 2$) has the wavelength $\lambda_{L_\alpha} = 121.6$ nm (in the ultraviolet).

(D) ENERGY-LEVEL DIAGRAMS

We must warn you that Bohr's simple theory is an approximation of the actual dynamics of atomic phenomena. The theory encounters insurmount-

able difficulties when extended to atoms more complicated than hydrogen, and the full mathematical machinery of *quantum mechanics* is necessary to understand atoms in detail.

Atomic particles (such as electrons) also exhibit a wave nature, just as photons do; hence, an inherent uncertainty exists in both the position and the velocity of an electron (the **Heisenberg uncertainty principle**). The electron in the hydrogen atom may be portrayed as a cloud surrounding the proton, with the most probable position of the electron being one of the Bohr orbits. A multielectron atom has several such clouds, with the electrons tending to occupy fuzzy **shells** about the nucleus. The simplest shells are spherical, but in general more complicated shapes occur.

We will sidestep the sophistications of quantum mechanics in this book by abandoning spatial models of atoms. Instead, we represent atoms abstractly by means of an **energy-level diagram** (Figure 8–9). Such a diagram is directly related to atomic transitions, and so it can be constructed even for complicated atoms. As an example, consider the energy-level diagram for hydrogen. The permitted energies of a bound electron in the hydrogen atom (Equation 8–24) are negative. Because we observe the positive-energy photons corresponding to electronic transitions, let's normalize to a positive energy scale by subtracting the ground-state energy $E(1)$ from all energies $E(n)$ to obtain

$$E(n) = R'[1 - (1/n^2)]$$

Note that $E(1)$ now equals zero. Change our energy units to **electron volts,** where 1 eV = 1.602 $\times$ 10^{-19} J. The electron volt is the energy acquired by an electron (or any particle of charge e) when accelerated through a voltage difference of 1 V; this unit is convenient in atomic and particle physics. Then

$$E(\infty) = R' = 2.18 \times 10^{-18} \text{ J} = 13.6 \text{ eV}$$

The energy levels for several values of n are shown in Figure 8–9 [$E(1) = 0$, $E(2) = 10.2$ eV, $E(3) = 12.1$ eV, and so on]. When the atom is in any level above the ground state, it is in an **excited state,** and the energy of such a level is called its **excitation potential.** To reach a higher level, the atom must be **excited**—an excessive energy of excitation leads to **ionization**—and in dropping back toward the ground state, the atom is then **de-excited.** Note

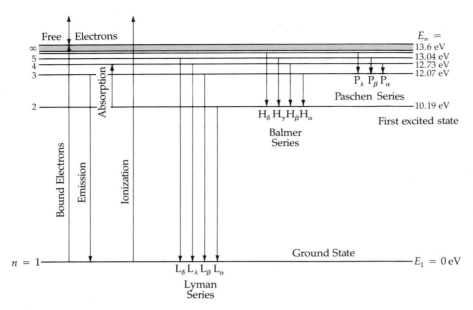

FIGURE 8–9 Hydrogen energy-level diagram. Transitions for the first three hydrogen series are shown in emission. On the left are shown the general types of transitions for absorption, emission, and ionization. The levels here are drawn to scale.

that the energy difference between two levels, m and n, is

$$E(n) - E(m) = R'(1/n^2 - 1/m^2)$$

Excitation

An atom may be excited to a higher energy level in two ways: radiatively or collisionally. **Radiative excitation** occurs when a photon is absorbed by the atom; the photon's energy must correspond exactly to the energy difference between two energy levels of the atom. This process produces **ab-sorption lines** superimposed on a background continuous spectrum (Section 8–6).

An atom generally remains in an excited state for only an extremely short time (about 10^{-8} s) before re-emitting a photon. How then can an absorption line be produced? Recall that the electron may cascade through several energy levels on its way to the ground state, and so several lower-energy photons may be emitted for each photon absorbed. The input wavelength is converted to longer wavelengths, depleting the spectrum at the input wavelength. Also (Figure 8–10), the ab-

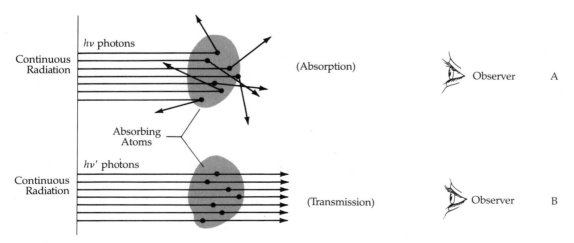

FIGURE 8–10 Absorption of light. (A) If E is the excitation energy of an atom, then only photons with this energy ($=h\nu$) can be absorbed. (B) Photons with energy E' ($=h\nu$) cannot be absorbed and reach an observer through the cloud.

sorbed photons come predominantly from one direction, the direction of their source, whereas the emitted photons can travel away in *any* direction. Hence, fewer photons at the absorption wavelength reach an observer than photons at other wavelengths. An absorption line is darker than an unabsorbed continuum but not completely black because some photons of the critical wavelength still reach the observer.

Collisional excitation takes place when a free particle (an electron or another atom) collides with an atom, giving part of its kinetic energy to the atom. Such an inelastic collision does not involve any photons. A particle approaching the atom with speed v_i and leaving with speed v_f has deposited the energy $E = m(v_i^2 - v_f^2)/2$ with the atom; if E corresponds to the energy of an electronic transition, the atom is collisionally excited to a higher state. Such an excited atom returns to its ground state by emitting photons, producing an **emission line spectrum** in the process.

De-excitation

Atoms are always interacting with the electromagnetic field. This interaction causes an excited atom to jump spontaneously to a lower energy level (de-excite) in a characteristic time of order 10^{-8} s. Because a photon is emitted, we call the process **radiative de-excitation.** Another form of de-excitation, where the phenomenon is not announced by a photon, is **collisional de-excitation;** this is the exact inverse of collisional excitation, for the colliding particle *gains* kinetic energy in the exchange (a superelastic collision). In astrophysical situations, these two modes of de-excitation compete with one another.

Although most spontaneous downward transitions occur on short time scales, certain transitions—because of quantum mechanical rules—take place much more slowly. These transitions are called **forbidden transitions** and result in **forbidden lines.** They usually involve jumps from excited metastable states to the ground state of an atom. Under most conditions, collisions de-excite the metastable state before it loses energy by a radiative process. Hence, forbidden lines usually are produced when gas densities are low, so that the chances of collision are small during the time interval between excitation and radiative de-excitation.

Ionization

With sufficient energy (either radiative or collisoinal) to liberate an electron from a neutral atom, the atom is **ionized.** Schematically, this reaction is

$$X + energy \rightarrow X^+ + e^-$$

where X represents the atom. To denote neutral hydrogen (with *no* electrons removed), we write H or H I, where the Roman numeral I signifies the neutral state; similarly, for neutral helium, He or He I. The singly ionized atom (with *one* electron removed) is in its first ionization state, such as $H^+ = H\ II$ or $He^+ = He\ II$. Hydrogen possesses but one electron and so may not be ionized beyond H II, but we can have $O^{++} = O\ III$ (doubly ionized oxygen) and $O^{+++} = O\ IV$ (oxygen with three electrons removed). For high stages of ionization (such as Fe XIV), the Roman numeral system is more common in astronomy than the multiple-sign convention.

The energy required to ionize an atom depends upon the ionization state of the atom, the particular electron to be liberated, and the excitation levels of that electron. For simplicity, consider the hydrogen atom with its sole electron. An electron in the ground state ($n = 1$; Figure 8–9) is removed from the atom when we supply it the energy $E \geq 13.6$ eV (the **ionization potential**); note that a **continuum** of energy states is available to the free electron above $E(n = \infty)$. If the electron is in the first excited state ($n = 2$), it will be freed when given energy $E \geq E(\infty) - E(2) = 13.6 - 10.2 = 3.4$ eV. In general for hydrogen, the ionization potential for an electron in excitation level n is

$$IP(n) = E(\infty) - E(n) = 13.6/n^2 \text{ eV} \quad \textbf{(8–26)}$$

The kinetic energy available to the departing electron is the difference between the energy provided and the ionization potential ($E - IP$).

Radiative ionization (by photons) leads to spectral absorption continua because an infinite number of states are available above the ionization level. For example, hydrogen atoms in the ground state absorb discrete-wavelength photons to produce the Lyman absorption series, but this series ends at the **series limit** $\lambda = 91.2$ nm. For wavelengths shorter than the series limit, we observe the Lyman **absorption continuum,** corresponding to photons that can ionize hydrogen from its ground state. Similarly, the Balmer, Paschen, Brackett, and Pfund series limits and their associated absorption continua for ionizations arise from levels $n = 2, 3, 4$, and so forth.

Free electrons can **recombine** with ions by emitting a photon of the appropriate energy. Because this process is just the reverse of ionization, the various hydrogen emission series may end in an **emission continuum** if the conditions are right.

8–3 ◗
THE SPECTRA OF ATOMS, IONS, AND MOLECULES

(A) ATOMIC SPECTRA

In multielectron atoms, quantum mechanics and the **Pauli exclusion principle** dictate that only two electrons may occupy the innermost shell, eight the next, 18 the third, and so on. When a shell contains the maximum allowed number of electrons, it is filled; in this case, the atom is extremely stable and hard to excite (helium, neon, argon, . . .). Innermost shells tend to be the first to be filled with electrons, and any excess electrons **(valence electrons)** are available for chemical interactions. For example, calcium (with 20 electrons) acts as if there are only two electrons in the outermost shell since the two inner shells $(2 + 8)$ are closed and eight of the electrons in the third shell form two **subshells** $(2 + 6)$.

Filled shells and subshells are tightly bound and shield the nucleus from the outer electrons, and so these electrons are only lightly bound (easily excited and ionized). The spectra of atoms with one outer electron, such as lithium and sodium, are similar to that of hydrogen but do show the effects of the inner closed shells. Atoms with more than one outer electron have increasingly complex spectra. To illustrate the various possibilities in the **periodic table,** Table 8–3 lists, for several representative atoms, (1) the excitation potential of the first excited state, (2) the ionization potential from the ground state, and (3) the wavelengths corresponding to these transitions.

(B) IONIC SPECTRA

Ions possessing at least one bound electron behave spectrally just like neutral atoms: they may be excited, de-excited, and ionized further. Except for overall wavelength modifications imposed by the greater charge of the nucleus, the spectrum of an ion closely resembles that of a neutral atom with the same number of outer electrons.

Consider ions with one electron remaining, such as He II, Li III, O VIII, and even Fe XXVI, which is called an **isoelectronic series.** The Bohr wave number relationship for these cases is

$$1/\lambda_{ab} = RZ^2(1/n_b^2 - 1/n_a^2) \qquad \textbf{(8–27)}$$

TABLE 8–3 *Excitation and Ionization Potentials for Some Atoms*

Atom/Ion	Excitation Potential* (eV)	λ (nm)	Ionization Potential† (eV)	$\lambda_{\text{series limit}}$ (nm)
Hydrogen (1 e)	10.2	121.6	13.6	91.2
Helium (1 closed shell)	20.9	58.4	24.5	48.8
Lithium (1 filled shell, 1 outer e)	1.8	670.8	5.4	225.0
Neon (2 filled shells)	16.6	73.5	21.5	57.6
Sodium (2 filled shells, 1 outer e)	2.1	589.0	5.1	243.0
Magnesium (2 filled shells, 2 outer e)	2.7	457.1	7.6	163.0
Calcium (2 filled shells, 2 filled subshells, 2 outer e)	1.9	657.3	6.1	203.0

*From ground state to first excited state.
†From ground state of neutral atom.

Here Z equals the value of the ionization state. By analogy, each such ion exhibits Lyman, Balmer, and the other series, but the wavelengths differ from those of the hydrogen spectral lines by a factor of Z^{-2}. So the He II Lyman-α line lies at 30.4 nm instead of 121.6 nm (since $Z = 2$). The mass of the nucleus also affects the sequence of levels, so that

$$R = R'[1/(1 + m/M)] = \mu R'/M$$

where μ is the *reduced mass.*

(C) MOLECULAR SPECTRA

Molecules are formed when atoms bind together (by the electric chemical bond). Quantum mechanics applies to such a union, and three types of discrete energy levels are exhibited by molecules. (1) Electronic energy states are in the combined electron cloud surrounding the nuclei. **Electronic transitions** similar to those in an atom can take place between these states, leading to excitation, de-excitation, and ionization of the molecule (such as $H_2 \rightarrow H_2^+ + e^-$). (2) The internuclear distances are quantized in discrete **vibrational energy states,** with consequent vibrational transitions. When the separation becomes so great that the atoms are no longer bound together, we say that the molecule has **dissociated.** (3) A molecule may rotate about various axes in space, resulting in discrete **rotational energy states.**

We will illustrate the concept of rotational molecular spectra with the example of the carbon monoxide (CO) molecule—one of the most common in interstellar space. Consider a two-mass rotator with two masses M and m separated by a distance r, spinning at angular velocity ω with

moment of inertia I. The rotational energy is

$$E = I\omega^2/2 = \mu r^2 \omega^2/2$$

where μ is the reduced mass of the system:

$$\mu = mM/(m + M)$$

Now, the rotational states of molecules are quantized so that the angular momentum can have only discrete values given by

$$I\omega = (h/2\pi)J$$

where J is the total angular momentum quantum number. Then the energy states corresponding to the possible states of J are

$$E = (h/2\pi)^2 J(J + 1)/2\mu r^2$$

Hence, for a diatomic molecule, the energy levels are equally spaced since J must vary by integral increments. For CO, the important transitions are from $J = 1$ to 0 and from 2 to 1. The former corresponds to a frequency of 115.2712 GHz, the latter to 230.5424 GHz.

The three classes of molecular transitions lead to numerous spectral lines superimposed upon one another. Since the vibrational and rotational transitions involve small differences in energy (usually much less than 1 eV), their spectral lines are closely spaced in wavelength and appear as **bands.** Molecular spectra are much more complex than atomic spectra, but they are easily recognized by their banded structure.

8–4 ◗ SPECTRAL-LINE INTENSITIES

Let's now discuss the strengths of emission and absorption spectral lines (Figure 8–11). The inten-

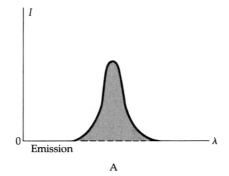
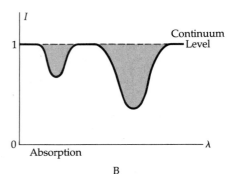

FIGURE 8–11 Generic spectral-line profiles. (A) An emission line rising above in flux level zero. (B) Two absorption lines. Note that the absorption is measured relative to the continuum level.

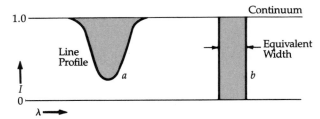

FIGURE 8–12 Equivalent width. The area of rectangle *b* is identical to that of line profile *a*. The width (in wavelength units) of the rectangle from 0 to 1 defines the equivalent width of the line. The continuum has been arbitrarily set to a level of 1.0.

sity of an emission line is proportional to the number of photons emitted in that particular transition. Similarly, the strength of an absorption line relative to the adjacent continuum depends upon the number of photons absorbed. Because the majority of astronomical spectra are absorption spectra, we will consider the case for absorption; similar arguments follow for emission spectra.

An absorption line is never infinitely sharp. It exhibits a **profile,** the **intensity** (residual radiation in terms of the continuum as unity) of which varies with wavelength (Figure 8–11). If we represent the distribution of energy with wavelength on a two-dimensional diagram (like that of Figure 8–11), we may measure an area in the λ, I plane. The total strength of a line is proportional to its area, which may be represented by the line's **equivalent width** (Figure 8–12). We replace the area of the line profile by a rectangle of equal area, with one dimension of the rectangle being the height of the continuum and the other dimension being the equivalent width (usually in milliangstroms, mÅ, in astronomy). Note that the equivalent width increases with the strength of the line. When the center of the line profile reaches zero intensity, we say that the line has become **saturated;** any further increase in the equivalent width comes only from the wings of the line.

(A) EXCITATION EQUILIBRIUM: BOLTZMANN'S EQUATION

The equivalent width of the spectral line depends directly upon the number of atoms in the energy state from which the transition occurs. So we would like to know the fraction of all the atoms of a given element that are excited to that energy state. Recall that excitation and de-excitation may occur collisionally and/or radiatively (a radiative de-excitation is also termed a **spontaneous transition**). Both processes depend upon temperature since the mean kinetic energy of a gas particle goes as

$$\langle mv^2/2 \rangle = 3kT/2 \qquad \text{(8–28)}$$

where m is the particle's mass, v its speed, $k = 1.380 \times 10^{-23}$ J/K is **Boltzmann's constant,** and T is the gas temperature in kelvins. The number of photons of a given energy increases rapidly with increasing temperature. So absorption lines originating from excited levels tend to be stronger in hot gases than in cool gases.

For simplicity, consider the situation where **thermal equilibrium** prevails and the average number of atoms in a given state remains unchanged in time (a steady-state situation). Each excitation is, on the average, balanced by a de-excitation. In this case, statistical mechanics shows that the **number density** (number per unit volume) of atoms in state B is related to the number density in state A $(B > A)$ by **Boltzmann's equation** [named after the Austrian physicist Ludwig Boltzmann (1844–1906) who discovered the relationship]:

$$N_B/N_A = (g_B/g_A) \exp\left[(E_A - E_B)/kT\right] \qquad \text{(8–29)}$$

where N is the number density in the level, g is the multiplicity of the level (often called the *statistical weight function*), and E is the energy of the level. For $E_B > E_A$, the bracketed quantity in Equation 8–29 is always negative, so that the ratio N_B/N_A increases as the temperature increases $(N_B/N_A \rightarrow g_B/g_A$ as $T \rightarrow \infty)$. For a given temperature, the **excitation ratio** N_B/N_A *increases* as the excitation potential $E_B - E_A$ *decreases* between the two energy levels.

Because exp $(-\infty) = 0$, exp $(-1) = 0.368$, and exp $(0) = 1$, a significant population of the upper level occurs when $T \approx (E_B - E_A)/k$; an excitation potential of 1 eV corresponds to a temperature of 11,600 K. As an example, consider a volume of gas that contains the same number of hydrogen and helium atoms at a temperature such that the number of hydrogen atoms in the first excited state (N_2) equals one-tenth the number in the ground state

(N_1), or $N_2/N_1 = 0.1$. On the other hand, the ratio N_2/N_1 for helium will be very low. So at a given temperature, the fraction of atoms in the second level differs widely from element to element, depending on the excitation potential. In this case, the absorption lines arising from transitions from $n = 2$ to $n = 3$ will be strong for hydrogen and very weak for helium. The equivalent width of the line is a function of both the abundance of the particular element and the temperature. In this example, we have unrealistically ignored the ionization of the atoms; for a complete picture, we must include both ionization and excitation to other levels.

(B) IONIZATION EQUILIBRIUM: SAHA'S EQUATION

As the temperature of a gas is increased, more and more energy (either radiative or collisional) becomes available to ionize the atoms. In general, the hot gas consists of neutral atoms, ions, and free electrons. The greater the **electron density** ($N_e =$ number of electrons per unit volume), the greater the probability that an ion will capture an electron and become a neutral atom. These two competing processes, ionization ($\rightarrow$) and recombination ($\leftarrow$), are written as

$$X \leftrightarrow X^+ + e^-$$

A steady-state condition of **ionization equilibrium** is achieved in the gas when the rate of ionization equals the rate of recombination. A quantitative expression of this ionization equilibrium is given by **Saha's equation** [named after the Indian physicist Meghnad N. Saha (1893–1956)]:

$$N_+/N_0 = [A(kT)^{3/2}/N_e] \exp(-\chi_0/kT) \quad \textbf{(8–30)}$$

where N_+ is the number density of ions, N_0 is the number density of neutral atoms in the ground state, the constant A includes several atomic constants and incorporates the probability of different states of ionization, T is the absolute temperature, N_e is the electron density, and χ_0 is the **ionization potential** (in electron volts) from the ground state of the neutral atom. Equation 8–30 is very similar to Boltzmann's Equation 8–29 except for the dependence upon N_e and the additional factor of $T^{3/2}$, which arises because a continuum of energies above χ_0 will ionize the atom and the liberated

electron is more likely to escape the ion as the electron's kinetic energy increases. Note that for pure hydrogen gas, $N_e = N_+$.

The Boltzmann excitation equation (8–29) applies to *any two levels of excitation*, those of an ion as well as those of a neutral atom. Similarly, the Saha ionization equation (8–30) can be generalized to give the ratio N_{i+1}/N_i for any stage of ionization $i + 1$ and the next lower stage i. The appropriate form of Saha's equation is

$$N_{i+1}/N_i = [A(kT)^{3/2}/N_e] \exp(-\chi_i/kT) \quad \textbf{(8–31)}$$

where χ_i is the ionization potential of the lower stage (the energy needed to ionize the particle in the i state to the $i + 1$ state. For example, Equation 8–31 applies to the ionization balance between Ca III ($i + 1 = 3$) and Ca II ($i = 2$). The relative population of the upper ionization stage increases rapidly with increasing temperature or smaller values of χ_i.

(C) THE BOLTZMANN AND SAHA EQUATIONS COMBINED

The Boltzmann equation gives the number of atoms in an excited state relative to the number in any other state; this applies to both neutral and ionized atoms. The Saha equation tells us the relative populations of two adjacent stages of ionization. We combine these two equations to calculate the number of atoms available to make a certain transition and so produce a given spectral line.

Consider the Balmer absorption lines of neutral hydrogen. Their strength is proportional to the number of atoms in the first excited state (N_2) of the neutral atom relative to the total number of hydrogen atoms in all stages of ionization (N). Hydrogen has only two stages of ionization, however: neutral (N_0) and singly ionized (N_+); hence, we know that $N = N_0 + N_+$. The proportion of N_2/N is

$$N_2/N = N_2/(N_0 + N_+)$$
$$\approx (N_2/N_1)/[1 + (N_+/N_0)] \quad \textbf{(8–32)}$$

where we have used the reasonable approximation $N_0 \approx N_1$ in the last equality. The Boltzmann equation yields N_2/N_1, the ratio of neutral atoms in the first excited state to those in the ground state; the Saha equation gives N_+/N_0, the ratio of ionized

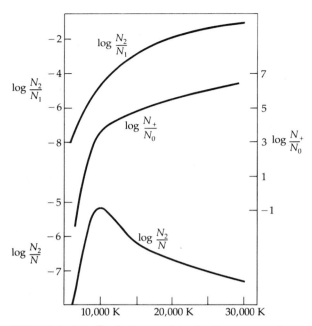

FIGURE 8–13 Excitation and ionization curves for hydrogen Balmer lines. The relative populations of the energy levels (N_2/N_1) from the Boltzmann equation and the ionization stages (N_+/N_0) from the Saha equation are calculated for equilibrium at the indicated temperatures. The lower curve shows the combination of the upper two with $N = N_0 + N_+$.

to neutral atoms. The approximate form of Equation 8–32 is good enough for our present purposes.

A plot of N_+/N_0 as a function of temperature (Figure 8–13) shows that most of the hydrogen is neutral at temperatures below 7000 K, but at higher temperatures, ionization increases to the point where the number of neutral atoms becomes negligible. The exponential increase of N_2/N_1 with increasing temperature is therefore countered by the lack of neutral atoms at high temperatures. As a result, the N_2/N curve has a maximum around 10,000 K. The strength of the Balmer absorption lines of hydrogen is greatest near $\approx$10,000 K, decreasing at both higher and lower temperatures. At 6000 K—the approximate surface temperature of the Sun—the ratio N_2/N_1 is about 10^{-8}, but reasonably strong Balmer absorption lines are seen as a result of the great abundance of hydrogen in the Sun. At 20,000 K—the temperature of very hot

stars—the ratio is about 10^{-2}, but the Balmer lines are similar in strength to those of the Sun because most of the hydrogen atoms are now ionized!

The Boltzmann and Saha equations have wide application in astrophysics. Through them, we may interpret stellar absorption (and emission) spectra in order to deduce the surface temperatures and pressures of stars. For example, at temperatures in the range 5000 to 7000 K, calcium should be predominantly in the form Ca II (singly ionized). Stars with strong Ca II lines but weak Ca I lines (our Sun, for example) must then have temperatures of this order. On the other hand, a star that has much lower density than our Sun but produces equally strong Ca II lines must actually have a lower temperature to compensate for the smaller electron density N_e (Equation 8–31), assuming that the chemical composition is the same.

Now to generalize the previous formulas to obtain $N_{i,s}$, the relative number of atoms in any state of excitation, s, of a stage of ionization i. The ratio of interest is $N_{i,s}/N$, where N is summed over all stages of ionization:

$$N = N_0 + N_1 + N_2 + \cdots + N_n$$
$$= \sum_{i=0}^{n} N_i$$

In general, n is the number of electrons in the neutral atom, but in practice, only two or three stages of ionization need be considered, for the number of ions in other stages will be negligible at a given temperature. To a reasonable first approximation

$$
\begin{aligned}
N_{i,s}/N &\approx N_{i,s}/(N_{i-1} + N_i + N_{i+1}) \\
&= (N_{i,s}/N_i)/[(N_{i-1}/N_i) \\
&\quad + 1 + (N_{i+1}/N_i)] \quad \textbf{(8–33)}
\end{aligned}
$$

The numerator of the last expression is given by the Boltzmann equation:

$$N_{i,s}/N_i \propto \exp(-\chi_s/kT)$$

and the denominator can be found from the Saha equation:

$$N_{j+1}/N_j \propto [(kT)^{3/2}/N_e] \exp(-\chi_j/kT)$$

Although Equation 8–33 is useful for the strongest spectral lines of a gas, in the general case extensive numerical computations are necessary to accurately reproduce spectral-line strengths. Our approximation holds only when *i* is the predominant stage of ionization for the prevailing temperature.

8–5 ◐
SPECTRAL-LINE BROADENING

Spectral lines are never perfectly sharp; their profiles always have finite width. The fundamental tenets of quantum mechanics account for the minimum width (natural broadening) of a spectral feature, and various physical processes further broaden the line profile. By interpreting the observed profile of a spectral line in terms of these broadening mechanisms, we can deduce some characteristics of the radiation from a star.

(A) NATURAL BROADENING

Quantum mechanics ascribes a wave nature to all atomic particles; an electron in an atomic energy level is such a particle. The Heisenberg uncertainty principle implies that the energy of a given state may not be specified more accurately than

$$\Delta E = (1/2\pi)(h/\Delta t) \tag{8–34}$$

where *h* is Planck's constant and Δt is the lifetime of the state. So an assemblage of atoms will produce an absorption or emission line with a minimum spread in photon frequencies—the natural width—of order $\Delta \nu = \Delta E/h \approx 1/\Delta t$. Typical excited states live about 10^{-8} s before decaying (in contrast, a ground state may last forever), so that a normal natural width is near 5×10^{-5} nm (0.05 mÅ) for visible light. Much smaller natural widths occur for the so-called metastable states, some of which last more than one second ($\Delta t \geq$ 1 s).

(B) THERMAL DOPPLER BROADENING

Whereas natural broadening depends only upon the intrinsic lifetime of an energy level, thermal Doppler broadening depends upon the temperature and composition of a gas. When a gas is at a certain temperature, *T*, gas particles (each of mass *m*) move about with a Maxwellian distribution of velocities characterized by the mean kinetic energy of Equation 8–28:

$$\langle mv^2/2 \rangle = 3kT/2$$

Atomic motions along our line of sight imply Doppler shifts in the radiation absorbed or emitted in atomic transitions. At a given temperature, the spectral lines of heavy elements are narrower than those of light elements since, on the average, the heavy particles move more slowly than the light particles. For example, neutral hydrogen at 6000 K moves at the mean speed $v \approx 12$ km/s, corresponding to a fractional Doppler broadening of $\Delta \lambda / \lambda \approx v/c \approx 4 \times 10^{-5}$; hence, the thermal Doppler width of the Balmer-α line (656.3 nm) is approximately 0.025 nm.

(C) COLLISIONAL BROADENING

The energy levels of an atom are shifted by neighboring particles, especially charged particles, such as ions and electrons (called the *Stark effect*). In a gas, these perturbations are random and result in a broadening of spectral lines. Because the perturbations are larger the nearer the perturbing particle, there is a direct dependence of this collisional (or pressure) broadening upon particle density. The greater the density (and hence pressure) of the gas, the greater the width of the spectral lines.

(D) THE ZEEMAN EFFECT

When an atom is placed in a magnetic field, the atomic energy levels each separate into three or more sublevels—this is the **Zeeman effect,** after the Dutchman Pieter Zeeman (1865–1943). Instead of a single atomic transition and a single spectral feature, we now have three or more closely spaced lines (the spacing is proportional to the magnetic field strength). If the Zeeman components are not resolved, we see only a broadened spectral line. In those cases in which the magnetic field is very strong and is highly ordered and uniform (such as sunspots and magnetic stars), we may resolve the Zeeman splitting and deduce the magnetic field strength and orientation of the source.

(E) OTHER BROADENING MECHANISMS

Finally, we mention three macroscopic broadening mechanisms based upon the Doppler effect. Consider a typical star whose image cannot be resolved. Large-scale random motions at the surface of such a star imply Doppler shifts that appear as turbulence broadening of spectral lines. If the atmosphere of the star is expanding, we simultaneously see gas moving in all directions; the integrated effect of all the Doppler shifts is expansion broadening of the observed spectral lines. A rapidly rotating star (not seen pole-on) will have rotationally broadened lines since one limb of the star is approaching us while the other limb is receding. Because all spectral features exhibit the same rotational broadening, we may identify the stellar rotation and determine its rate (or period).

8–6 ◗
BLACKBODY RADIATION

So far we have dealt with individual atomic transitions and the spectral lines they produce. Where then does a *thermal* continuous spectrum come from? (Recall that spectral absorption lines result when photons are selectively absorbed from such a continuum.) We noted that various emission and absorption continua can originate from individual atoms and that spectral features become more broadened as atoms interact more strongly with one another. When an aggregate of atoms interact so strongly (such as in a solid, a liquid, or an opaque gas) that all detailed spectral features are washed out, a thermal continuum results.

Such a continuous spectrum comes from a blackbody, whose spectrum depends only upon the absolute temperature. A **blackbody** is so named because it absorbs all electromagnetic energy incident upon it—it is completely black. To be in perfect thermal equilibrium, however, such a body must radiate energy at exactly the same rate that it absorbs energy; otherwise, the body will heat up or cool down (its temperature will change). Ideally, a blackbody is a perfectly insulated enclosure within which radiation has come into thermal equilibrium with the walls of the enclosure. Practically, blackbody radiation may be sampled by observing the enclosure through a tiny pinhole in one of the walls. The gases in the interior of a star are opaque (highly absorbent) to all radiation (otherwise, we would see the stellar interior at some wavelength!); hence, the radiation there is blackbody in character. We sample this radiation as it slowly leaks from the surface of the

star—to a rough approximation, the continuum radiation from some stars in blackbody in nature.

(A) PLANCK'S RADIATION LAW

After Maxwell's theory of electromagnetism appeared in 1864, many attempts were made to understand blackbody radiation theoretically. None succeeded until, in 1900, Max K. E. L. Planck (1858–1947) postulated that electromagnetic energy can propagate only in discrete quanta, or photons, each of energy $E = h\nu$. He then derived the spectral intensity relationship, or Planck blackbody radiation law:

$$I(\nu)\Delta\nu = (2h\nu^3/c^2)[1/(e^{h\nu/kT} - 1)] \quad \textbf{(8–35a)}$$

where $I(\nu)\Delta\nu$ is the intensity (J/m^2 · s · sr) of radiation from a blackbody at temperature T in the frequency range between ν and $\nu + \Delta\nu$, h is Planck's constant, c is the speed of light, and k is Boltzmann's constant. Note the exponential in the denominator.

Because the frequency ν and wavelength λ of electromagnetic radiation are related by $\lambda\nu = c$, we may also express Planck's formula (Equation 8–35a) in terms of the intensity emitted per unit wavelength interval:

$$I(\lambda)\Delta\lambda = (2hc^2/\lambda^5)[1/(e^{hc/\lambda kT} - 1)] \quad \textbf{(8–35b)}$$

Equation 8–35b follows because the intensity $I(\lambda)\Delta\lambda$ equals the intensity $I(\nu)\Delta\nu$ in the corresponding wavelength interval, where

$$\nu = c/\lambda \rightarrow |\Delta\nu| = |c\,\Delta\lambda/\lambda^2|$$

This expression follows by differentiation or by noting that, in the limit as $\Delta\lambda \rightarrow 0$ and $\Delta\nu \rightarrow 0$,

$$\lambda\nu = (\lambda + \Delta\lambda)(\nu + \Delta\nu) = c$$

where $\Delta\lambda\,\Delta\nu$ is negligible.

Equation 8–35b is illustrated in Figure 8–14 for several values of T. Note that both $I(\lambda)$ and $I(\nu)$ increase as the blackbody temperature increases—the blackbody becomes brighter. This effect is easily interpreted in terms of Equation 8–35a, when we note that $I(\nu)\Delta\nu$ is directly proportional to the number of photons emitted per second near the energy $h\nu$. The Planck function is special enough so that its given its own symbol, $B(\lambda)$ or $B(\nu)$, for intensity.

FIGURE 8–14 Blackbody radiation. (A) Linear plot of B(λ) versus λ for three temperatures typical of stars; the wavelength range spans from UV to near IR. (B) A log–log plot of the Planck curves for a wide range of temperatures. Note that the wavelengths run from longer to shorter in units of centimeters.

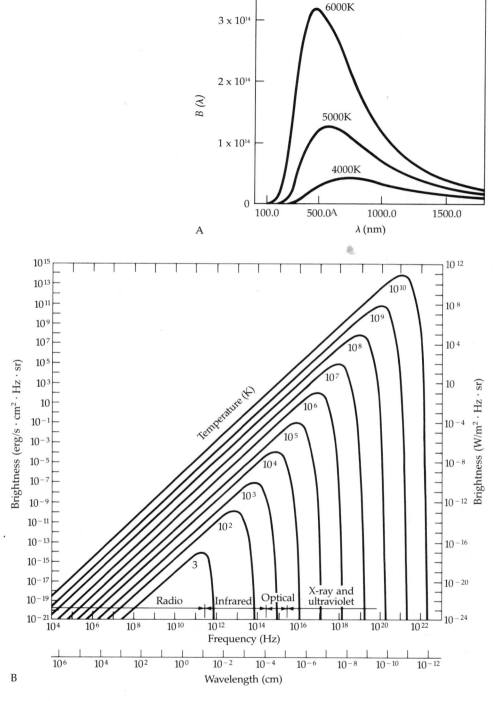

A

B

To better understand the Planck blackbody formulas (Equations 8–35), let's outline a derivation by Albert Einstein. At a given frequency, ν, the radiation will be in thermal equilibrium with the atoms that compose the walls of the blackbody cavity. Because of the strong atomic interactions in the walls, all possible energy states are available there. Consider the upper (U) and lower (L) energy states corresponding to the quantum of energy $E_U - E_L = h\nu$. The Boltzmann equation relates the number of atoms in each state by

$$N_U = (g_U/g_L)N_L \exp(-h\nu/kT)$$

If the probability per atom for a downward transition (U → L) is a_{UL} and that for an upward transition (L → U) is a_{LU}, then a steady-state equilibrium is attained when

$$N_L a_{LU} = N_U a_{UL}$$

But a_{LU}—radiative excitation—depends upon the energy density $\rho(\nu)$ of radiation at the frequency ν and upon the atomic parameter B_{LU} (the **Einstein absorption coefficient**): $a_{LU} = B_{LU}\rho(\nu)$. De-excitation (a_{UL}) can occur in two ways: spontaneously (A_{UL}) or through the stimulating influence of the radiation field [$B_{UL}\rho(\nu)$]; hence, we have $a_{UL} = A_{UL} + B_{UL}\rho(\nu)$. Combining these various results and solving for $\rho(\nu)$, we find

$$\rho(\nu) = (A_{UL}/B_{UL})/[(B_{LU}g_L/B_{UL}g_U)e^{h\nu/kT} - 1]$$

The result is independent of the number of atoms involved. Einstein then substituted suitable expressions for the atomic coefficients A_{UL}, B_{UL}, and B_{LU} to obtain the **Planck blackbody radiation density law:**

$$\rho(\nu)\Delta\nu = (8\pi h\nu^3/c^3)[1/(e^{h\nu/kT} - 1)] \quad \textbf{(8–36)}$$

where $\rho(\nu)\Delta\nu$ is an energy density (J/m^3). In converting to intensity radiated from the blackbody, $I(\nu)$, a factor of $c/4\pi$ is introduced (Equation 8–35a).

Before we leave the topic of blackbody radiation, which is so useful in astrophysics, we put forth some useful approximations. We have so far considered formulas for the monochromatic intensities (Equations 8–35). If we evaluate these in the case where the exponentials become very large (much greater than unity), then we find

$$B_\nu(T) = I_\nu(T) = (2h\nu^3/c^2) \exp(-h\nu/kT) \quad \textbf{(8–37a)}$$
$$B_\lambda(T) = I_\lambda(T) = (2hc^2/\lambda^5) \exp(-hc/\lambda kT) \quad \textbf{(8–37b)}$$

which is an appropriate approximation when the temperature is low and the wavelengths are short. It is sometimes called the **Wien distribution.** In the opposite case, when the value of the exponential is very small (much less than unity), we use the

power series expansion for the exponential to note that $e^x - 1 \approx x$, and

$$B_\nu(T) = I_\nu(T) = 2\nu^2 kT/c^2 \quad \textbf{(8–38a)}$$
$$B_\lambda(T) = I_\lambda(T) = 2ckT/\lambda^4 \quad \textbf{(8–38b)}$$

which is known as the **Rayleigh-Jeans distribution** and works at high temperatures and long wavelengths (low frequencies).

(B) WIEN'S LAW

A blackbody emits at a peak intensity that shifts to shorter wavelengths as its temperature increases (Figure 8–14). Wilhelm Wien (1864–1928) expressed the wavelength at which the maximum intensity of blackbody radiation is emitted—the peak (that wavelength for which $dI(\lambda)/d\lambda = 0$) of the Planck curve (found from taking the first derivative of Planck's law)—by **Wien's displacement law:**

$$\lambda_{max} = 2.898 \times 10^{-3}/T \quad \textbf{(8–39)}$$

where λ_{max} is in meters when T is in kelvins. For example, the continuum spectrum from our Sun is approximately blackbody, peaking at $\lambda_{max} \approx 500$ nm; therefore, the surface temperature is near 5800 K. Note that because $\lambda_{max}T = $ constant, increasing one proportionally decreases the other.

(C) THE LAW OF STEFAN AND BOLTZMANN

The area under the Planck curve (integrating the Planck function) represents the **total energy flux,** F (W/m^2), emitted by a blackbody when we sum over all wavelengths and solid angles:

$$F(T) = \sigma T^4 = \pi B(T) \quad \textbf{(8–40)}$$

where $\sigma = 5.669 \times 10^{-8}$ W/m$^2 \cdot$ K^4. The strong temperature dependence of Equation 8–40 was first deduced from thermodynamics in 1879 by Josef Stefan (1835–1893) and was derived from statistical mechanics in 1884 by Boltzmann; therefore we call the expression the Stefan-Boltzmann law. The brightness of a blackbody increases as the fourth power of its temperature. If we approximate a star by a blackbody, the total energy output per unit time of the star (its **power** or **luminosity** in watts) is just $L = 4\pi R^2 \sigma T^4$ since the surface area of a sphere of radius R is $4\pi R^2$.

TABLE 8–4 *Types of Temperature*

Temperature	Basic Law/Equation	Observations
Brightness	Planck curve	Flux at one wavelength
Color	Planck curve	Flux at two or more wavelengths
Effective	Stefan-Boltzmann law	Luminosity (power) and radius
Excitation	Boltzmann equation	Relative equivalent widths of spectral lines of the same element
Ionization	Saha equation	Relative equivalent widths of spectral lines of adjacent stages of ionization
Kinetic	Thermal Doppler broadening	Widths and profiles of spectral lines

To summarize: A blackbody radiator has a number of special characteristics. One, a blackbody emits *some* energy at *all* wavelengths. Two, a hotter blackbody emits *more* energy per unit area and time at *all* wavelengths than does a cooler one. Three, a hotter blackbody emits a *greater proportion* of its radiation at *shorter* wavelengths than does a cooler one. Four, the amount of radiation emitted per second by a unit surface area of a blackbody depends on the *fourth power* of its temperature.

(D) TEMPERATURE

We end with a word of caution concerning temperature. In general, you cannot define temperature uniquely—it depends on the process under consideration. Also, a temperature can be assigned to a radiation field and to matter. For radiation, if we are dealing with a true blackbody emitter, we may establish the temperature using (1) the shape of the Planck curve, or at least two points on the curve, or one point on the curve; (2) Wien's law and the wavelength of peak emission; or (3) the Stefan-Boltzmann law and the total power. Because no astrophysical object is a perfect blackbody, we will obtain slightly different temperatures from each of these three methods.

Temperature can also be assigned to matter, based on the velocity distribution (a kinetic temperature), the degree of excitation, and the degree of ionization. In general, temperatures based on continuum spectra may differ from the temperatures derived from the relative intensities of spectral lines by using the Boltzmann and Saha equations. It is therefore a good practice to stipulate

what kind of temperature one is using (see the examples in Table 8–4). All these temperatures will be the same only if *thermodynamic equilibrium* holds true, so that all these proceses and their inverses go on at the same rate. This condition will not usually be the case; then one attempts to look at the local detailed balancing to see if local thermodynamic equilibrium (LTE) holds; in many astrophysical circumstances, LTE will apply.

8–7 ◗
THE TRANSFER EQUATION

One of the most common problems in astrophysics is to calculate the expected intensity of radiation after it passes through a cloud of gas or the atmosphere of a planet or star. The differential equation that describes this process is called the **transfer equation.** Some of its solutions will be discussed since they are commonly used to describe interesting sources.

First, we need to introduce the following quantities. Note that the subscript, ν, emphasizes the frequency dependence of the quantity.

Intensity
$$dI_\nu = (dV \, d\Omega \, d\nu \, dt)^{-1} \, dE$$

I_ν is the amount of energy per unit volume emitted in a given solid angle, $d\Omega$, in a given frequency range, $d\nu$, per unit time.

Extinction coefficient
$$\chi_\nu$$

The dimensions are cm^{-1}. The inverse of χ_ν is the mean free path. This measures the average distance over which a photon travels before being absorbed.

Emission coefficient
η_ν

This quantity gives the energy added to the radiation field. Dimensions are joules cm^{-3} sr^{-1} Hz^{-1} s^{-1}.

Source function
$S_\nu = \eta_\nu/\chi_\nu$

The source function is the ratio of total emissivity to total opacity. In cases of thermodynamic equilibrium, S_ν is given by the Planck distribution.

Optical depth
$\tau_\nu = \int \chi_\nu \, dl$

τ_ν gives the number of mean free path lengths along the line of sight. It is a dimensionless quantity.

Figure 8–15 illustrates the geometry of the problem. In this drawing the linear depth dl and the optical depth increase from the right-hand side of the gas. We have indicated that the problem could be of either finite or infinite optical depth by drawing the left-hand side of the diagram in an indefinite manner. We first write the differential equation in terms of the physical length element dl.

$$\cos\theta[dI_\nu/dl] = \eta_\nu - \chi_\nu I_\nu$$

This expression satisfies physical intuition. The change in intensity over the incremental length dl is the contribution of the emission coefficient,

which is independent of the intensity, minus the product of the intensity and the extinction coefficient, which is just the fraction of the intensity that was extinguished in dl.

Now if we define $\mu = \cos\theta$ and divide through by χ_ν, we have the standard form of the transfer equation.

$$\mu[dI_\nu/dl] = I_\nu - S_\nu \qquad \textbf{(8–41)}$$

A negative sign was also introduced in order to represent the case of an observer looking *into* the source. An integrating factor for this equation is $e^{-\tau/\mu}$. So if we multiply both sides by this quantity and integrate, we arrive at the *general solution*:

$$I_\nu = I_\nu(0)e^{-(\tau 2 - \tau 1)/\mu} + \mu^{-1}\int_{\tau 1}^{\tau 2} S_\nu e^{-(\tau - \tau 1)/\mu} \, d\tau \quad \textbf{(8–42)}$$

(A) APPLICATION TO STELLAR ATMOSPHERES

One of the most important applications of the transfer equation is that of no incident radiation and infinite optical depth—a stellar atmosphere. In this situation the formal solution is

$$I_\nu = \mu^{-1}\int_0^\infty S_\nu e^{-\tau/\mu} \, d\tau \qquad \textbf{(8–43)}$$

As a first approximation, let's assume that the source function is a linear function of the optical depth

$$S_\nu = S0_\nu + S1_\nu\tau_\nu \qquad \textbf{(8–44)}$$

If we introduce this into Equation 8–43, then the solution is

$I_\nu = S0_\nu + S1_\nu\mu$; the same as
$$I_\nu = S_\nu(\tau = \mu) \quad \textbf{(8–45)}$$

These last equations are known as the **Eddington-Barbier relation.** As a first approximation they provide exceedingly illuminating insight into the observable radiation field. Consider the Sun, which is the only star for which we can easily see variations across the disk. If we look at the center of the disk, then $\mu = \cos\theta = 1$. From this we see that *the intensity is approximately given by the source function at an optical depth of 1.*

This result is quite general for stellar atmospheres. As we move our observations from the

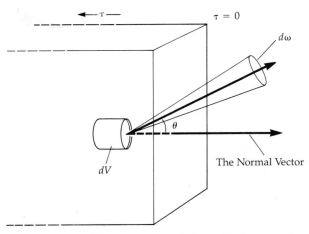

FIGURE 8–15   The geometry of the radiative transfer equation. The intensity emergent from the volume element dV is given by Equation 8–42.

center to the edge (limb) of the Sun's disk, the geometry departs from normal emergence. Therefore, the linear depth into the atmosphere which corresponds to an optical depth of 1 becomes less. From simple physical arguments, we expect the temperature at these lesser depths to be lower and the source function (approximately given by the Planck function) to provide a less intense radiation field. Therefore, we should see **limb darkening.** This is, in fact, observed (details in Section 10–1) and can be used to estimate the run of temperature with depth in the solar atmosphere.

Similar reasoning provides an excellent first description of stellar absorption lines. These can be thought of as small regions of the spectrum over which the optical depth rapidly changes. At the line center the opacity is largest, and an optical depth of 1 lies relatively high in the cooler regions of the atmosphere. So the gas is emitting photons at that wavelength, they are just *less intense* than in the line wings.

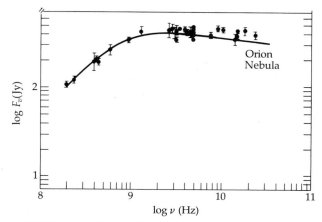

FIGURE 8–16 The radio wavelength spectrum of the Orion Nebula. In the low-frequency regime, the source is optically thick and the intensity goes as ν^2. In the high-frequency regime, the source is optically thin and is nearly constant.

(B) GASEOUS NEBULAE

A second useful example is a slab of material in which S_ν is constant with τ. If we consider only normally emergent radiation, then $\mu = 1$, and the integrated solution to Equation 8–43 is

$$I_\nu = S_\nu(1 + e^{-\tau}) \qquad (8–46)$$

where τ is now the total optical depth. Two cases are of interest:

Case I: $\tau \gg 1$ (optically thick)
Then $e^{-\tau}$ vanishes and

$$I_\nu = S_\nu$$

The emergent intensity is identical to the source function. In the particular instance of thermal equilibrium, the source function—and therefore the emergent intensity—is the same as the Planck blackbody function.

Case II: $\tau \ll 1$ (optically thin)

$$I_\nu = S_\nu\tau$$

Now the emergent intensity is the product of the source function and the optical depth. Since τ will

generally have a frequency dependence, the emergent intensity will *not* be identical to the source function.

The distinction between an optically thin or optically thick gas is one of the critical analyses that an astrophysicist makes. Since all the important quantities generally have a frequency dependence, a given nebula will typically be optically thin at some spectral regions but optically thick at others.

For example, an H II region (a cloud of ionized hydrogen gas) is a thermal source of radio waves. At long wavelengths the blackbody spectrum is approximated by the Rayleigh-Jeans distribution which has a ν^2 dependence. In this same spectral region, the optical depth is proportional to ν^{-2}. Figure 8–16 presents the radio continuum spectrum of one particular H II region, the Orion Nebula. At lower frequencies τ is large, and the optically thick regime holds. The observed spectrum rises with a slope of 2 in this log–log plot. At a frequency of about 1000 MHz, τ passes from the optically thick to optically thin regime. Since the source function and optical depth have inversely related frequency dependence, the emergent intensity (their product) is nearly constant with frequency as seen in the high ν region of the spectrum. (More about H II regions comes in Section 19–2.)

PROBLEMS ◗

1. **(a)** Show that a beam of light obliquely incident upon and passing through a plane-parallel piece of glass is simply displaced without changing direction when it emerges from the glass.
   (b) If the glass has thickness d and index of refraction n, what is the linear displacement of the beam as a function of n and θ?

2. What aperture is required to give 1 arcsecond resolution for a wavelength of
   (a) 500 nm (visible)
   (b) 21 cm (radio)
   Can you make a generalization from your results?

3. **(a)** At what wavelengths will the following spectral lines be observed:
   (i) a line emitted at 500 nm by a star moving toward us at 100 km/s
   (ii) the Ca II line (undisplaced wavelength of 397.0 nm) emitted by a galaxy receding at 60,000 km/s
   (b) A cloud of neutral hydrogen (H I) emits the 21-cm radio line (at rest frequency 1420.4 MHz) while moving away at 200 km/s. At what frequency will we observe this line?

4. A simple form of the binomial theorem states that
$$(1 + x)^n = 1 + nx + [n(n - 1)x^2/2]$$
$$+ n(n - 1)(n - 2)x^3/6 + \cdots$$
   when $x^2 < 1$. Starting with the relativistic expression, use this theorem to derive the classical equation for the Doppler shift in wavelength for the case in which $v \ll c$.

5. **(a)** What is the energy of one photon of wavelength $\lambda = 300$ nm? Express your answer both in joules and in electron volts.
   (b) An atom in the second excited state ($n = 3$) of hydrogen is just barely ionized when a photon strikes the atom. What is the wavelength of the photon if all of its energy is transferred to the atom?

6. The emission line of He II at 468.6 nm corresponds to what electronic transition?

7. By applying the Boltzmann equation to the neutral hydrogen atom (neglect ionization), derive an expression for the population of the nth energy level relative to that of the ground state, at temperature T. Now, assuming that the multiplicity of each level is unity ($g_n = 1$), construct an appropriate graph showing your results for $T = 6000$ K. Do the same for $T = 15,000$ K. What are the major differences between them?

8. To understand the relative importance of the different parameters in the Saha equation, perform the following experiment. Assume that $T = 5000$ K, $N_e = 10^{15}/cm^3$, and $\chi = 12$ eV. By what factor does the ionization ratio (N_+/N_0) change when we separately
   (a) double the temperature
   (b) double the electron density
   (c) double the ionization potential
   Which is more important during the temperature change, the exponential term or the $T^{3/2}$ term?

9. Let N_2 be the number of second-level (first excited state) hydrogen atoms and N_1 be the number in the ground state. Using Figure 8–13, find the excitation ratio (N_2/N_1) and the excited fraction (N_2/N) for each of the following stars:
   (a) Sirius, $T = 10,000$ K
   (b) Rigel, $T = 15,000$ K
   (c) the Sun, $T = 5800$ K
   Which star will exhibit the strongest Balmer absorption lines? Explain your reasoning in arriving at this answer.

10. **(a)** What is the speed of an electron with just sufficient energy to ionize by collision a sodium atom in the ground state?
    (b) What is the speed of a proton ionizing this atom?
    (c) What is the corresponding gas temperature?
    (d) At this temperature, what is the fractional thermal Doppler broadening ($\Delta\lambda/\lambda_0$) of a sodium spectral line?

11. **(a)** How much more energy is emitted by a star at 20,000 K than one at 5000 K?
    (b) What is the predominant color of each star in part **(a)**? Use the Wien displacement law, and express your answers in wavelengths.

12. Deduce an approximate expression for Planck's radiation law (Equation 8–35a)
    (a) at high frequencies ($h\nu/kT \gg 1$), the Wien distribution.
    (b) at low frequencies ($h\nu/kT \ll 1$), the Rayleigh-Jeans approximation.

    You may use the approximation $\exp(h\nu/kT) \approx 1 + h\nu/kT$ when $h/kT \ll 1$. To what wavelength does $h\nu/kT = 1$ correspond?

13. Estimate the line width from thermal Doppler broadening for the Ca II K line at 393.3 nm for a star at 3000 K; a star at 6000 K; and one at 12,000 K. Comment on the degree of importance of temperature on the Ca II K line broadening.

14. A star behind a nebula is 25% as bright at 500 nm as it would be if it were not behind the nebula. What is the optical depth of the nebula at 500 nm? (Assume that the nebula does not contribute to the light observed at 500 nm.)

15. Given that the $J = 1$ to 0 rotational transition of carbon monoxide (CO) corresponds to a photon frequency of 115.27 GHz, estimate the distance between the carbon and oxygen atoms in a carbon monoxide molecule.

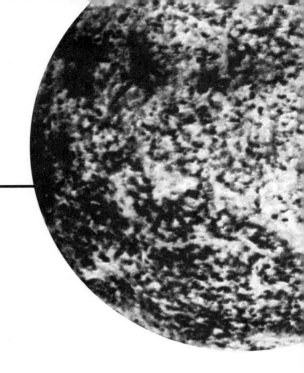

Chapter 9

Telescopes and Detectors

We have discussed the nature and properties of light and how to use an understanding of these facts to infer the physical properties of astronomical objects. To underscore that process, we describe briefly in this chapter how astronomers gather and detect light—the basis of observational astronomy. In many respects, astronomy is much more an observational science than an experimental one (as physics and chemistry are).

9–1 ◐
OPTICAL TELESCOPES

As extensions of the human eye, optical telescopes amplified the power of detection without extending the spectral range of our vision. Today we can sense much more than the visible part of the electromagnetic spectrum, but the discussion in this section restricts itself to optical telescopes—those that manipulate light detectable by the eye. Be warned, however, that the light that reaches the Earth's surface has passed through many filters before it reaches us: intergalactic and interstellar media, the Earth's atmosphere, and always the telescope and detection system.

Optical components, such as lenses and mirrors, are used to control the paths of light rays. For a telescope, such optical components bring light to a **focus,** usually to make an **image.** Light can be concentrated in a focus by either a curved mirror (using reflection) or a lens (by refraction). A lens brings rays from a point source to a point image at the focus of the lens. The lens makes an image of an object of finite size by focusing rays from each point on the object onto a separate point in the image (Figure 9–1). This image is generally smaller than the object and upside down. For objects at large distances, the distance from lens to image is approximately the same for all objects. This distance is termed the **focal length.** A smoothly curved mirror—one whose surface, for instance, follows the curve of a parabola—brings all the light to a focus. The distance from the mirror's surface to its focus is its focal length.

For either a lens or a mirror, the ratio of focal length to diameter is called the f **ratio:**

$$f \text{ ratio} = f/d$$

where f is the focal length and d the diameter. The f ratio essentially gives the brightness of the image.

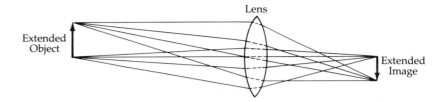

FIGURE 9–1 Images. A convex lens will make an image of an extended object that is smaller than the object and inverted.

A small f ratio, such as $f/3$, results in a brighter image than a large f ratio, such as $f/15$. That is, an $f/15$ lens or mirror will spread the image out more than an $f/3$ one will and so decrease the brightness.

If we call the distance to the object (for which we want an image) as d_o, and the distance from the objective to the image as d_i, then the focal length relates to these two quantities (for a simple lens) as

$$1/d_o + 1/d_i = 1/f \qquad (9–1)$$

This is often called the *lens-maker's formula*; it also applies to spherically curved mirrors.

Another property related to focal length is the linear size of the image of an extended object. For example, the Moon subtends a $1/2°$ arc in the sky; if its image covers 10 cm at the focal surface of the lens, then the scale is $0.05°/cm$, or 200 cm per degree. For the size s of an image at the focus that corresponds to $1°$ in the sky.

$$s = 0.01745f$$

where f is the focal length; s then comes out in units of f per degree; it is usually called the **plate scale**. An example: Capilla Peak Observatory of the University of New Mexico has an $f/15$ reflecting telescope with a 60-cm-diameter mirror. Its focal length is

$$f/d = 15$$
$$f = 15 \times 60 \text{ cm} = 900 \text{ cm}$$

The plate scale is

$$s = 0.01745 \times 900 \text{ cm} = 15.7 \text{ cm}/°$$

A telescope is essentially an instrument that gathers light and makes an image at a focus. A lens or mirror, called the **objective,** brings the light to a focus. A lens at the focus, called the **eyepiece,** allows visual examination of the image. There are two basic types of telescopes, distinguished by their objectives: **refracting telescopes** (or **refractors**), which use a lens, and **reflecting telescopes** (or **reflectors**), which use a mirror.

A telescope is basically a light bucket for collecting photons. That's the main reason astronomers want large telescopes—for their greater light-gathering capacity. A telescope's **light-gathering power** is directly proportional to the square of its diameter. The amount of light a lens or mirror catches depends on its surface area, and its area with diameter d is $(\pi/4)d^2$. Light-gathering power is a relative measure, not an absolute one. It states how two instruments compare with one another, not how much light is gathered. Because the factor $\pi/4$ is a constant, we need use only the diameters of the instruments to arrive at a comparative figure. For example, compared with your eye, which has a diameter of about 0.5 cm, a telescope with a 50-cm objective has a light-gathering power of

$$LGP = (50/0.5)^2 = 100^2 = 10,000$$

Similarly, the 5-m Hale telescope at Mount Palomar outdoes the 0.6-m Capilla Peak telescope by

$$LGP = (5/0.6)^2 = 8.3^2 = 69$$

A second important function of a telescope is to produce an image in which objects that are close together in the sky can be seen as clearly separate. This ability is called **resolving power** and is sometimes expressed as the inverse of the minimum angle there must be between two points in order for them to be easily separated:

$$RP = 1/\theta_{min}$$

The resolving power and the minimum angle depend on the diameter of the objective and also on the wavelength of the light. For the same wavelength, the resolving power depends directly on the objective's diameter. The minimum resolvable angle depends on both the diameter of the telescope's objective and the wavelength of light being observed. In general,

$$\theta_{min} = 206,265\lambda/d$$

where θ_{min} is the **minimum resolvable angle,** or

resolution in seconds of arc (206,265 is the number of arcseconds in 1 rad), λ is the wavelength, and d is the diameter of the objective in the same length units. As a slight complication, the circular aperture of a telescope produces a diffraction pattern that spreads out the images. To take this diffraction into account, at least at visible wavelengths, requires multiplying θ_{min} by 1.22. For example, a 10-cm (0.1-m) telescope working at a wavelength of 500 nm has a resolution angle of

$$1.22\theta_{min} = (1.22)(206,265)(5 \times 10^{-7}/0.1)$$
$$= 1.25 \text{ arcseconds}$$

This result means that if this telescope is pointed at two stars that are more than 1.25" of arc apart, you will see two distinguishable stellar images.

The example above gives the **theoretical** resolving power of a 10-cm telescope, but such performance rarely, if ever, is attained by ground-based telescopes. The resolving power of a large telescope is limited not by its optics but by the Earth's atmosphere. Stars twinkle because turbulence in the air makes the atmosphere act like an imperfect, distorted lens. The motion of blobs of air distorts and blurs images seen through a telescope; this is an effect referred to as **seeing.** (It results from cells in the atmosphere on the order of tens of centimeters across.) Even on the best of nights, the 5-m Hale telescope does not resolve better than a 10-cm telescope; it is a rare night when star images are smaller than 1" of arc. The limit the Earth's atmosphere sets on the resolving power of big telescopes makes a strong case for placing a large telescope in space. Here a telescope's resolving power would be limited by its optics, not the atmosphere.

The third function of a visual telescope is to magnify the image. **Magnifying power,** the apparent increase in the size of an object compared with unaided visual observation, depends on the ratio of the focal length of the objective to the focal length of the eyepiece:

$$MP = F/f$$

where F is the focal length of the objective and f is the focal length of the eyepiece in the same length units. For example, an eyepiece having a focal length of 5 cm used with an objective having a focal length of 780 cm gives a magnifying power of

$$MP = 780/5 = 156$$

If you insert an eyepiece with *half* the focal length, you *double* the magnifying power. You could choose eyepieces to produce as high a magnifying power as you like, but there is little point in using a magnifying power any greater than that necessary to see clearly the smallest detail in the image. The degree of detail is determined by the resolving power and the seeing. Extremely high magnification of an image produced by a telescope with a low resolving power merely makes the fuzziness worse. And higher magnification makes extended objects dimmer.

The Keck Telescope is the largest ground-based optical telescope (Table 9–1). Located on Mauna

TABLE 9–1 *Large Ground-Based Optical Telescopes*

Aperture (cm)	Focal Length (m)	Primary f-ratio	Elevation (m)	Name
1000	17.5	1.75	4145	W. M. Keck Telescope, Mauna Kea, U.S.A.
600	24.0	4.0	2070	Bolshoi Teleskop, Mt. Pastukov, U.S.S.R.
508	16.8	3.3	1706	George E. Hale Telescope, Palomar Mt., U.S.A.
450	4.9	2.7	2600	Multiple Mirror Telescope, Mt. Hopkins, U.S.A.
400	11.2	2.8	2399	CTIO, Cerro Tololo, Chile
389	12.7	3.3	1164	Anglo-Australian Telescope, Australia
381	10.7	2.8	2064	N. U. Mayall Reflector, KPNO, U.S.A.

FIGURE 9–2 The Keck telescope. The dome of the Keck telescope (foreground) on Mauna Kea.

Kea at an elevation of 4145 m (Figure 9–2), the Keck has innovative designs throughout—especially its 10-m mirror, which consists of 36 hexagonal segments (Figure 9–3) with a total weight of 14.4 tons. Each segment has a diameter of 1.8 m and a thickness of 75 mm, and is individually controlled by an active system to maintain the mirror's shape under a variety of load conditions. The dome and the telescope's structure minimizes local seeing to take advantage of the site's natural potential—occasionally as good as 0.3″. Early images show that the telescope should achieve its potential (Figure 9–4). The Keck can come close to diffraction-limited seeing in the infrared, for which

FIGURE 9–3 Nine of the hexagonal structures of the Keck mirror installed. (*California Association for Research in Astronomy*)

FIGURE 9–4 An early photograph of a spiral galaxy (NGC1232) taken by the Keck telescope in 1990 with nine mirror segments in place. The image has been computer enhanced. (*California Association for Research in Astronomy*)

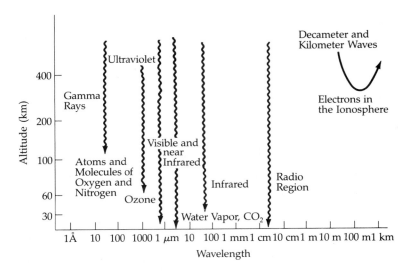

FIGURE 9–5 Atmospheric transmission. The Earth's atmosphere is transparent to limited regions of the electromagnetic spectrum. The incoming wavy lines show the altitude at which various parts of the spectrum penetrate the atmosphere.

Mauna Kea is a superior site in atmospheric transparency. The altitude is a drawback for people, who can easily become ill at the summit. Hence, most of the observing will be done remotely rather than at the site.

Astronomers are developing various techniques of active optics to try to beat the seeing limitations of the Earth's atmosphere. The New Technology Telescope (NNT) at the European Southern Observatory (ESO) in La Silla, Chile, was dedicated in 1990 and is currently the most functional. Its 3.6-m mirror can flex under servo control based on monitoring the image of a bright reference star. So far, the NNT can achieve resolutions of 0.3″ regularly and occasionally as good at 0.18″.

9–2 ◖
INVISIBLE ASTRONOMY

The human eye senses only the visible range of the electromagnetic spectrum. To cover the complete range, from radio to gamma rays, requires a variety of detectors sensitive to various wavelengths; current detector technology drives new developments in observational astronomy. Invisible astronomy involves techniques that enable us to go beyond the wavelength limits of optical astronomy.

The practice of invisible astronomy also relates to the transparency of the Earth's atmosphere, which effectively absorbs large segments of the electromagnetic spectrum, especially ultraviolet light, X-rays, some infrared wavelengths, and short-wavelength (millimeter) radio waves (Figure 9–5). Infrared radiation is primarily absorbed by water vapor, which is found concentrated in the lower portions of the atmosphere, below 20 km. (Carbon dioxide also absorbs a lesser amount.) The ultraviolet and X-ray radiations are primarily absorbed in the ionosphere, at an altitude of 100 km, well above the levels that can be reached by balloons and airplanes. The obvious way to get around atmospheric absorption is to go above the atmosphere. This is space astronomy, which makes use of rockets, balloons, and airplanes as well as satellites and spacecraft. So invisible astronomy has two natural divisions—that which can be done from the ground and that which must be accomplished in space.

(A) GROUND-BASED RADIO ASTRONOMY

Radio astronomy was born in 1930 when Karl Jansky (1905–1950) undertook a study for the Bell Telephone Company to identify sources of static affecting transoceanic radiotelephone communications. Jansky identified one source of noise as a celestial object: the Milky Way in Sagittarius. Jansky's discovery was published in 1932 but had little impact on the astronomers of the day. However, an American engineer, Grote Reber, read Jansky's work and decided to search for cosmic radio static in his spare time. By the 1940s, Reber had made detailed maps of the radio sky and had detected the Sun. He sensed that a new astronomy was in the making and so took an astrophysics course at the University of Chicago to learn more about astronomy and to discuss his discoveries with astronomers—only a few of whom were impressed. World War II forced technical developments in radio and radar work. John S. Hey in Britain accidentally discovered that the Sun strongly emits radio waves. After the war, he continued his astronomical pursuits at radio wavelengths. So did other groups in Britain, the Netherlands, and Australia.

A common type of radio telescope, the *radio dish* (Figure 9–6), functions like a reflecting telescope. Essentially, it's a radio-wave bucket with a detector (a radio receiver) at its focus. It reflects and con-

centrates radio waves the same way a mirror does in a reflecting telescope. The radio receiver translates incoming radio waves into a voltage that can be measured and recorded on magnetic tape. Finally, the measurements are usually made into a contour map (Figure 9–7) or a false-color intensity map.

Radio telescopes naturally have low resolving power. Remember, $\theta_{min} \propto \lambda/d$; radio waves are much longer than the wavelengths of visible light, typically by 10^5. So if an optical and a radio telescope have the same diameter, the radio one will have 10^5 times *less* resolving power. For example, for a radio telescope to have the same theoretical resolving power as the 5-m Hale optical telescope, it would be 10^5 times the diameter, about 500 km! Obviously, a single dish of this size cannot be built on Earth.

Radio astronomers use *interferometry* to make small radio telescopes function as if they were large ones—and so gain in resolving power. Imagine two radio telescopes placed, say, 10 km apart. If the signals received by the two are synchronized, the separate units can be made to act like a single dish with a diameter of 10 km, but only for a

FIGURE 9–6 A radio dish. A view from the back of the NRAO millimeter-wave radio telescope on Kitt Peak. (*M. Zeilik*)

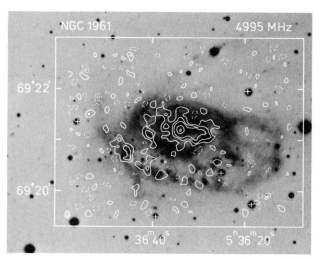

FIGURE 9–7 Radio contour map. The emission is 21 cm from hydrogen atoms, made with the VLA. The radio contours have been superimposed on a negative optical print. The coordinates (Appendix 10) give the position on the sky. (*E. Hummel, University of New Mexico and National Radio Astronomy Observatory*)

narrow strip across the sky. (That's because they act like two small pieces at the opposite ends of a large dish.) The essence of interferometry is to consider the phase of the wavefronts reaching the two dishes of a simple interferometer. If a source is directly overhead, the waves arrive at both telescopes in phase. When the signals are combined (electronically in a mixer), they constructively reinforce to give a strong signal. Now assume that the source has moved somewhat west of overhead so that the path lengths to the two antennas differ by exactly $\lambda/2$. They arrive 180° out of phase and destructively interfere. So the two telescopes "see" the sky as a series of bands of constructive and destructive interference—an interference pattern exactly like that for light passing through two slits. The angular separation of the peaks in the intensity pattern establishes the resolving power of the interferometer.

Consider two antennas (Figure 9–8) separated by a distance L, which is an integral multiple of the observing wavelength λ, so that

$$L = n\lambda$$

where n is any integer. Now imagine a radiation source at O, a large distance from the antennas. The waves from this source travel along the different path lengths P_1 and P_2, where $P_1 > P_2$ and P' is the extra amount of travel along $P_1 = P + P'$. Note that

$$P' = L \sin \theta$$

and for constructive interference, P' must be integral multiples of the wavelength, so that

$$L \sin \theta = m\lambda$$

or

$$\sin \theta = m\lambda/L$$

where m is any integer. As the Earth turns, θ varies. The fringes are evenly spaced by the angle for which $m = 1$:

$$\sin \theta_f = \lambda/L = \lambda/n\lambda = 1/n$$

and because θ_f will be small, $\sin \theta_f \approx \theta_f$ (in radians) and

$$\theta_f = 1/n \text{ rad}$$

where n is just the number of wavelengths (λ) in the baseline (L). For $L = 21$ km at an observing wavelength of 21 cm,

$$\theta_f \approx (21/21) \times 10^5 \approx 10^{-5} \text{ rad} \approx 2''$$

comparable to an optical telescope. Note that the fringe spacing essentially sets the resolving power of the interferometer.

A two-antenna interferometer gives only one part of the separation between points in the source (in the same direction as the baseline). To completely map a section of the sky with uniform resolving power requires a little trick: twisting the object with respect to the interferometer's pattern, a technique known as **aperture synthesis.** This twisting can be accomplished naturally by observing the source move across the sky; its motion will twist its orientation with respect to the interferometer's baseline. To help this process, the baseline of the interferometer can also be twisted or the antennas can be arranged in a **Y** pattern [as in the Very Large Array (VLA) in New Mexico, Figure 9–9]. The Earth's rotation effectively synthesizes an aperture with a resolution comparable to a single antenna with a diameter equal to the largest baseline between the dishes. Eventually, sets of interference maps are combined by a computer to give a composite picture of the source with high resolution. The VLA and other interferometer arrays consist essentially of many two-antenna arrays at different angles to each other, operating on the same source at the same time. A complete radio picture can be constructed in about 1 day or less, depending on the resolution required, the area of the sky covered, and how long the object is above the horizon.

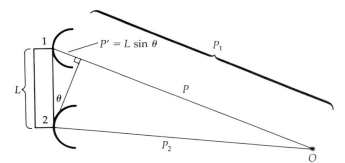

FIGURE 9–8 Geometry for an interferometer. The path length differences to the two antennas from the source (at O) generate the interference pattern.

FIGURE 9–9 A radio interferometer. A view down one arm of the VLA. Each antenna is 25 m in diameter. The spacings are varied by moving the antennas between various stations. *(M. Zeilik)*

Even higher resolution can be obtained by Very-Long-Baseline Interferometry (VLBI). The VLBI signals received by very distant antennas (even located on different continents) are recorded on magnetic tape and combined later in a computer. The maximum baseline for such VLBI is the diameter of the Earth, $L = 12,000$ km, and so (for the example above) a resolving power of $\theta_f = 3 \times 10^{-3}$ arcsecond is possible at 21 cm. In the future, radio telescopes in space separated by larger baselines will give even better resolution; some proposals even envision one element of an interferometer in Earth orbit or on the Moon—a separation 30 times greater than is possible on the Earth. Resolution could then approach about 100 microarcseconds and would be high enough to see the disks of nearby solar-type stars! As a compromise for this grand scheme, the National Science Foundation has begun construction of a VLBA (Very-Large-Baseline Array) that will operate like the VLA but will have greater antenna separations (up to a few thousand kilometers). Present plans call for about ten 25-m antennas distributed from the Virgin Islands to Hawaii and controlled by telephone links

from an operation center. The receivers will be synchronized by atomic clocks, and the data tapes will be shipped to a computer in New Mexico for image processing.

(B) GROUND-BASED INFRARED ASTRONOMY

Carbon dioxide and water vapor in the Earth's atmosphere limit what ground-based infrared astronomers can observe to only a few wavelength ranges: 2 to 25, 30 to 40, and 350 to 450 μm. Such observations are best made from high sites in dry climates, where there is a minimum of water vapor in the atmosphere above the telescope.

An infrared telescope differs from an optical one in the detector at the telescope's focus. Special infrared detectors suitable for astronomical work have been around for about 20 years. A common infrared detector is a thermal detector called a **bolometer**, a tiny chip of germanium (about the size of the head of a very small nail) cooled to about 2 K. When infrared radiation strikes it, a bolometer heats up and its resistance to an electric current changes. Such changes can be measured electronically, and the amount of variation indicates how much infrared energy the bolometer is absorbing.

Infrared observing has at least two distinct advantages over optical observing. First, infrared radiation is hindered less by interstellar dust. Second, cool celestial objects (3000 K and cooler) give off most of their radiation in the infrared (as expected from Wien's law for blackbody radiation). Another practical advantage is that much infrared observing can be done during the day, when telescopes are not used by optical astronomers. This advantage occurs because very little sunlight at infrared wavelengths is scattered by air molecules, leaving the infrared sky dark both day and night (except looking directly at the Sun). However, infrared astronomy has the problem that a telescope emits like a blackbody at 300 K, so that, as expected from Wien's law, it generates an intense infrared flux at 10 μm. The Earth's atmosphere also contributes a strong background flux at this wavelength. Special techniques have been developed to compensate for these strong local fluxes in order to measure the much smaller infrared flux from celestial objects.

(C) SPACE ASTRONOMY

Most infrared, ultraviolet, and X-ray emission can be detected only above the Earth's atmosphere from airplanes, balloons, rockets, satellites, or spacecraft. For example, most of the far-infrared radiation (wavelengths longer than about 40 μm) does not penetrate to the ground. At altitudes of 15 to 20 km or so, however, very little of the Earth's atmosphere remains above the observer. Far-infrared observations can be made at these altitudes from airplanes or balloons. The equipment is simply a reflecting telescope equipped with a bolometer.

Low-energy infrared astronomy had a space champion: the Infrared Astronomical Satellite or IRAS. An international collaboration of the United States, United Kingdom, and the Netherlands, IRAS mapped the sky at infrared wavelengths, in bands at 12, 25, 60, and 100 μm, with a 57-cm telescope and an array of 62 detectors. The satellite functioned for about nine months until it ran out of the liquid helium that kept the infrared detectors cooled down to 2 K (and the whole telescope below 5 K). A stunning success, IRAS scanned almost all the sky close to three times. It detected over 200,000 infrared sources, which are kept in a computer archive. Many of these related to the process of starbirth in the Milky Way (more in Chapter 19).

For ultraviolet astronomy, methods of light gathering and detection are similar to those of optical astronomy. Special electronic devices respond well to ultraviolet light, and so detection presents no serious problem. Because glass absorbs ultraviolet light, refracting telescopes cannot be used, but reflectors work perfectly well. Good ultraviolet astronomy must be done from space, for the absorbing layer of the atmosphere (the ionosphere) is higher than balloons or aircraft can reach. Probably the most successful ultraviolet telescope has been the International Ultraviolet Explorer (IUE) launched in 1978 and still (sufficiently) operational. It has a 0.45-m telescope with detectors that operate from 115 to 320 nm; these are secondary electron conducting (SEC) vidicon tubes.

For the high-energy realm of X-rays, special gathering and focusing techniques are required. X-ray telescopes use the fact that X-rays can be reflected from certain surfaces if they strike at very small angles, almost parallel to the reflecting surface. This design is called a **grazing incidence telescope;** it provides reflectivities of about 50% from silvered surfaces. Such reflections by a complex series of concentric parabolic and hyperbolic reflecting surfaces produce images, which are sensed by electronic detectors. HEAO-B, which is called the Einstein Observatory and was launched in 1978 and used until 1981, was probably the most productive of recent X-ray telescopes. The 58-cm telescope (which contained four nested pairs of parabolic/hyperbolic mirrors with a total effective area of 400 cm^2) in the satellite produced high-resolution images (with grazing angles of 40 to 70 arcmins) of X-ray sources in the wavelength range from 0.3 to 5 nm. The data are available in a computer archive for examination and analysis by astronomers.

The current functional X-ray telescope is called ROSAT (the Roentgen Satellite), launched in 1990; it is a joint project of the United States, Germany, and the United Kingdom. It can provide high-resolution X-ray images covering areas about 5 arcmin square. ROSAT's first priority is a complete survey of X-ray sources in the sky; it will then observe selected target sources.

Because of their higher energies, gamma rays present an even more difficult imaging problem. Basically, because of these problems, gamma-ray telescopes do not yet exist —no images are made. Crystals that fluoresce when they absorb gamma rays are used. The flash of visible light is then detected by phototubes. Various techniques are used to limit the field of view, which typically covers a few degrees of the sky.

(D) THE HUBBLE SPACE TELESCOPE

On April 24, 1990, the space shuttle Discovery transported Hubble Space Telescope (HST) into space. The astronauts deployed it in the next day (Figure 9–10). It was a long time coming. The godfather of the concept in the United States was Princeton astrophysicist Lyman Spitzer, Jr., who in a 1946 report envisioned the use of powerful rockets for astronomy. In 1962, NASA commissioned study groups for future astronomy payloads; and in 1965, the Space Science Board of the U.S. National Academy of Sciences recommended that NASA pursue a large space telescope. In 1969, a committee chaired by Spitzer proposed scientific

FIGURE 9–10 The Hubble Space Telescope deployed by the Space Shuttle Discovery in April 1990. *(NASA)*

objectives for a 3-meter telescope. The U.S. Congress gave its approval in 1977. The launch was originally planned for 1982; this date slipped to 1986; the Challenger explosion halted efforts until the spring of 1990—almost three decades after conception. At a developmental cost of $1.5 billion, the HST is the most expensive astronomical project ever. (For comparison, the VLA cost a bit under $80 million when completed in 1981.)

After all that time and money, astronomers were shocked to find out during tests that the main mirror is flawed. Simply put, the overall shape of the 2.4-meter primary mirror is too shallow, from edge to center, by about 2 μm and so suffers from spherical aberration. As small as this error seems, it is large compared to the wavelengths of ultraviolet and visible light. The spherical aberration means that light rays come to a focus at different distances (along the focal path) that depend on the radius at which the light rays strike the primary mirror. Light from the edge of the mirror comes to a focus about 38 mm beyond the point where the innermost rays converge. So the images of stars cannot be brought into good focus. In particular, the guide star images are out of focus, so that the telescope often cannot lock onto these for long exposures, nor can it use faint stars for pointing.

The original specifications mandated that most of the starlight should fall into an image with a radius of 0.1 arcsec. The actual performance is a radius of 0.7 arcsec, about that of a ground-based telescope at a good site. At certain focal positions,

stars show a sharp central core of about 0.07 arcsec, but most of the flux is scattered in a large halo 1.5 arcsec in diameter (Figure 9–11).

Spherical aberration degrades the science capacity of HST. In the short term, two instruments have suffered the most: the Wide Field and Plane-

FIGURE 9–11 A stellar image taken by the Planetary Camera of HST at 487 nm. Note the bright core (about 0.1″ in radius) and the extended halo (diameter about 4.0″). Most of the flux falls into the halo. *(C. Burrows, STScI)*

tary Camera (WFPC) and the Faint Object Camera (FOC). Both instruments were to have turned out high-resolution optical images that excited astronomers. That won't come about, though the FOC can still do imaging in the ultraviolet that is not possible from the ground. Meanwhile, the instruments designed for optical and ultraviolet spectroscopy will largely be able to achieve their goals, but not for faint objects.

But the long term will present solutions to the imaging problems. The HST was designed for a 15-year mission, with new and upgraded instruments ferried up by future shuttles. A general optical solution is a pair of mirrors for the field of view of each scientific instrument to correct the spherical aberration. The first mirror forms an image of the primary mirror onto the second one, which has a spherical aberration of precisely the same amount as the primary, but with the opposite sign. Hence, it cancels the effect of the aberration. A second WFPC will be installed in a few years; its internal optics are designed to compensate for the mirror's flaws. Before that, image reconstruction by

new computer software is possible. The analogy is the VLA; new computer programs have made its radio images much better than even its original designers imagined.

Unrelated to the optical problem is a guiding one. During day to night transitions (or vice versa), thermal shocks in the solar arrays create oscillations in the vehicle. These excited modes are large enough to result in a loss of lock on faint guide stars. Recovery usually takes place in five minutes, but the performance is degraded. A fix to the loss of lock will have a high priority in the first servicing mission.

Although working below anticipated limits, HST still will be able to discover the unexpected, especially for nearby, bright objects, such as planets or regions of bright stars. A good example is a WFPC image of the star cluster in the 30 Doradus Nebula (Figure 9–12), which is located in the Large Magellanic cloud. Earth-based photos did not clearly reveal the concentration of stars in the core of the cluster; HST did (after a lot of image processing work!).

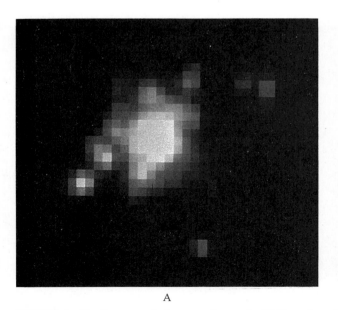

A

B

FIGURE 9–12 Images of the star cluster in 30 Doradus. (A) A close-up view of the central region, showing the compact cluster called R136, taken from a ground-based telescope. The stellar images are 0.6″ in diameter. (B) A computer-enhanced HST image of the region of R136; the cores of the images are 0.1″ in diameter. At least 60 stars are visible in a region only about a parsec across. *(NASA)*

9–3 ◑
DETECTORS AND IMAGE PROCESSING

You may have the romantic image of an astronomer glued to the eyepiece of a telescope through a cloudless night working to produce spectacular photographs. Today, most astronomers observe through the telescope on a television screen from a warm room! And most optical observations (which encompass a narrow region of the electromagnetic spectrum) no longer involve direct photography. Here we will give you a taste of modern technology's influence on astronomy.

(A) PHOTOGRAPHY

This old standby still cannot be beat if the goal is to gather a large amount of storable information in a short time. Special photographic emulsions used by astronomers are usually coated on a glass plate, whose size depends on the characteristics of the telescope employed. (The glass plate does not flex and can be measured accurately.) The emulsion can integrate the light striking it to build up an image of very faint objects over a long period of time (Figure 9–13)—occasionally many hours spread over several nights. The plates are often specially treated to increase their insensitivity to light, and yet they rarely have quantum efficiencies greater than a few percent. We use **quantum efficiency (QE)** to mean the percentage of photons striking a detector that activate it relative to the incoming total. The QE of any detector is a function of wavelength:

$$QE(\lambda) = \text{\# photons detected}/\text{\# photons incident}$$

A perfect detector would have $QE(\lambda) = 1.0$ or 100%; in practice, detectors differ considerably in how QE varies with wavelength. The human eye, for example, has a peak QE ~ 1% at 550 nm (Figure 9–14). A photographic plate achieves a typical QE of 1% in the range from 400 to 650 nm (Figure 9–14).

Although a photograph's quantum efficiency is relatively low, its large area collects information from a large part of the entire field of view of the telescope. The resolution of the plate is limited by the grains in the emulsion; these are best 20 μm

FIGURE 9–13 Time exposures and information. Longer times for exposing a photographic plate reveal fainter details. The times here are 1, 5, 30, and 45 min for the Andromeda galaxy. The first exposure is similar to the visual view through a small telescope. *(National Optical Astronomy Observatories)*

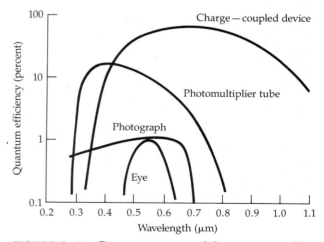

FIGURE 9–14 Response curves of the quantum efficiency (QE) of a charge-coupled device (CCD), photomultiplier tube, photographic plate, and the human eye. *(J. Kristian)*

FIGURE 9–15 Image processing of an archival photograph. A computer-processed image of a 1910 photograph (taken at Lowell Observatory) of Halley's Comet. The computer enhances the differences of brightness in the tail, which are not obvious in the original photograph. *(National Optical Astronomy Observatories)*

across. Also, the image results from the density of the grains, so that it is a function of the logarithm of the density rather than directly as the number of photons. This means that a photographic plate is *not* a linear detector (in this sense, it mimics the eye).

The image can later be converted to a digital form that can be manipulated by computer to enhance specific aspects of the original data (Figure 9–15). This process, called **image processing,** plays a central role in all types of astronomy today, especially invisible astronomy, where the computer-processed image quickly conveys significant information to the human brain.

(B) PHOTOTUBES

The quantum nature of light shows up dramatically in the **photoelectric effect.** Light striking the surfaces of certain materials can be absorbed and an electron dislodged. The photons must have a certain minimum energy to knock out the electrons. The displaced electrons can flow in a current, or they can be individually counted. A device that does this counting is called a **phototube.** A phototube that can transform *one* electron into many ($\approx 10^5$) for ease of detection is called a **photomultiplier.** Phototubes do not produce images; they simply measure the flux of photons through the focal plane of a telescope, usually through a small aperture that restricts the field of view to, say, a single star or galaxy. Filters can be placed in the light path in front of the phototube to restrict the detection to certain wavelength bands. The active materials in phototubes have quantum efficiencies that sometimes reach 20%, though 10% is more typical (Figure 9–14). The response of a photomultiplier is generally linear—twice the flux gives twice the current. Hence, light intensity can be measured with great accuracy by measuring an electric current or by counting the rate of production of individual photons.

A television camera includes a phototube designed to produce images by a magnetic focusing technique. However, commercial television cameras are not very sensitive, cannot accumulate images over long times, and are not linear in response. (Linearity is important because you want the output to be directly proportional to the number of photons captured.) Astronomers have modified television systems for certain applications—for a telescope's acquisition and guiding, for example.

(C) CHARGE-COUPLED DEVICES

Astronomers would love an ideal detector for their demanding low-light applications. It would have a high quantum efficiency (to make good use of those few photons), the ability to integrate, a linear response over a wide range of photon fluxes, and the ability to cover a large angular field of view (as does a photographic plate). Recent advances in solid-state microelectronics have resulted in the development of **charge-coupled devices** (CCDs), which may be the astronomers' dream detector. A CCD is a small chip, a few to 10 mm on a side, made of a thin silicon wafer. (It uses the same microelectronic technology employed in integrated circuits.) It consists of a large number of small regions each of which makes up a picture element called a **pixel.** A typical chip may contain 800 by 800 pixels (640,000 total) arranged in rows and columns that can be electronically controlled. Each pixel independently can accumulate a charge as photons are absorbed and dislodge electrons. In a sense, each pixel acts like a very small phototube, but a CCD can integrate the charge in the pixels for a long time and then move the charges out in a regimented way to preserve digitally the spatial pattern of light intensity falling on the chip. Computers then process the image from each exposure of the chip.

CCDs offer the following advantages over other detectors. One, they are highly quantum efficient—close to 100% in the red region of the spectrum (Figure 9–14). This great quantum efficiency means that a small telescope can act like a large one. Two, CCDs are very linear, and so they can easily measure light intensity accurately. Three, though small so far, CCDs are area detectors. They cover a good fraction of a telescope's field of view (as if many small phototubes were operating at once, a multichannel detector that makes use of most of the light coming through a telescope at a given time). Fourth, the image emerges in a naturally digital form, ready for computer processing.

Along with the inexpensive microcomputers that are a key part in the operation, CCDs are the best current example of technology expanding the capabilities of older, smaller telescopes. The sky subtends a large solid angle, but any one telescope sees only a small piece of that solid angle at any one time. And the number of telescopes is extremely limited; many are in great demand. CCDs enhance the power of small telescopes and may well re-elevate the status of the small observatory (Figure 9–16). For example, in the ability to detect faint objects, a 1-h exposure on a 61-cm telescope with a CCD equals a 1-h exposure on a 500-cm telescope with a photographic plate.

FIGURE 9–16 CCD image, taken with a 0.6-m telescope, of a cluster of galaxies in the constellation of Corona Borealis. Because of the CCD's high quantum efficiency, these faint galaxies covering a region of the sky of a few arcminutes were recorded in a short exposure with a small telescope. *(Jay Moody, Capilla Peak Observatory, University of New Mexico)*

(D) SIGNAL TO NOISE

Just detecting photons is not sufficient to make a high-quality astronomical observation. The real goal is that observations should contain the most information possible, which is usually discussed in terms of the **signal-to-noise ratio.** All observations contain noise, even if the instruments are perfect, because of the statistical fluctuations in a beam of photons acting as particles. Consider σ_m as the standard deviation from the mean in a series of measurements, say that of counting photons. Then if $\langle N \rangle$ is the mean number of photons counted, the signal-to-noise ratio is

$$S/N = \langle N \rangle / \sigma_m$$

The larger this number, the higher the quality of the observation, and the more information it contains.

Now, photons obey a Poisson probability distribution, so that if these fluctuations alone are the source of noise, then

$$S/N = \langle N \rangle^{1/2}$$

Hence, if we count 10^4 photons, the S/N = 100.

Consider a detector on which falls a photon flux, f_p, photons per second. Then the total number of photons detected is the product of the detector's QE, the photon flux, and the length of time of the observation (usually called the integration time, t):

$$\langle N \rangle = QE \times f_p \times t$$

and so

$$S/N = (QE \times f_p \times t)^{1/2}$$

This fundamental equation tells you of the essential limits of observational astronomy and the critical role of a detector's QE. For instance, if we double QE, the S/N increases by 1.414 for the same integration time. Or we can double the integration time to achieve the same increase in S/N. Although it appears that increasing t will result in any S/N we want, the reality is that we have a limited amount of telescope time; the higher the QE, the shorter the integration time and the more efficient the use of the telescope. So the best strategy would be to know what information content we must achieve to make sense of the observations, and aim for that level of S/N.

9–4 ◗
SPECTROSCOPY

Broad band and narrow band photometry are very useful tools for analyzing astronomical sources, but there are many instances when we need to examine the wavelength distribution of intensity with much higher resolution. When we look at the intensity versus wavelength, we are observing the **spectrum** of the object. The two principal ways we use to separate the "white light" of an object into its spectrum are by passing the light through a prism or by using a diffraction grating.

(A) PRISM SPECTROSCOPY

Figure 9–17 illustrates the path of light through a prism. The refraction of light at the air/glass inter-

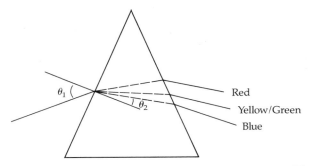

FIGURE 9–17 The path of light through a prism. The wavelength dependence of the index of refraction causes each color to pass through the prism at a slightly different angle—thereby producing a spectrum.

faces is governed by Snell's law [Section 8–1(A)].

$$n_a \sin \theta_1 = n_g \sin \theta_2$$

The index of refraction in air, n_a, is very close to 1. The ability of a prism to disperse white light into a spectrum comes from the fact that the index of refraction in glass, n_g, is a function of wavelength. Therefore, the refracted angle, θ_2, is also a function of wavelength.

Most prism spectrographs are designed with both the incident angle and emergent angle at approximately 60°. This represents a compromise between high **dispersion** (that is, greater separation of the colors) and transmission efficiency.

Prism spectrographs are not very common in astronomy. The reason for this is two-fold. First, refractive optics are difficult to construct because the glass must be perfect and must be finished on more than one side. The second reason is more important. Since the index of refraction in glass is a very complicated function of λ, the spectrum is highly nonlinear. This makes it very difficult to calibrate (to relate the position in the observed spectrum to λ).

(B) GRATING SPECTROGRAPHS

Figure 9–18 illustrates a magnified view of the surface of a reflective **diffraction grating**. The grooves act very much like the small holes in a double-slit

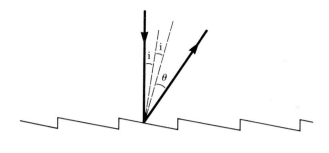

FIGURE 9–18 A magnified cross section of the surface of a reflective diffraction grating. Because of interference effects for light paths reflected from different grooves, the angle of reflectance is greater for longer wavelengths—thereby producing a spectrum. The pitch angle of the grooves is called the *blaze angle* and is adjusted to put a maximum amount of light into the desired order at which the grating will be used.

interference experiment in that they cause the many reflected light paths to interfere with each other. In some directions the paths interfere constructively and in others destructively. The angles for constructive interference are given by

$$\sin \theta = (n\lambda)/d,$$

where λ is the wavelength, d the groove spacing, and n is the **order** of the spectrum. Since these are the angles at which the interference adds, and since the angle is dependent upon λ, the result is a series (determined by n) of bright spectra. Note that the zeroth order has no λ-dependence and is therefore just a white light image. A second important point is that the dispersion of the spectra increases with increasing order. Thirdly, since $\sin \theta \approx \theta$ for small angles (in radians), the dispersion is approximately linear with λ.

Figure 9–19 shows the light path of a grating spectrograph. The observer usually has several options for the gratings, which range from 300 to 1800 grooves mm^{-1}. One problem with gratings is that the presence of many orders means that not all of the light goes into the order chosen for observation, which decreases the efficiency of the observations. This can be somewhat mitigated by carefully choosing not only the number of grooves but also their angle. The proper **blaze angle** can concentrate as much as 80% of the light into one order.

In a normal grating spectrograph at higher orders, hence higher dispersions, the orders overlap and are not useful. An **echelle** grating is an alternative that is designed for high dispersion. In this kind of spectrograph the grating is illuminated at a very oblique angle. (The grooves are more like steps in this case.) Typical useful orders are in the range of 50–100. By adding a low dispersion grating acting perpendicular to the echelle's axis of dispersion, the overlapping orders are placed side by side. Therefore, a two-dimensional detector is essential.

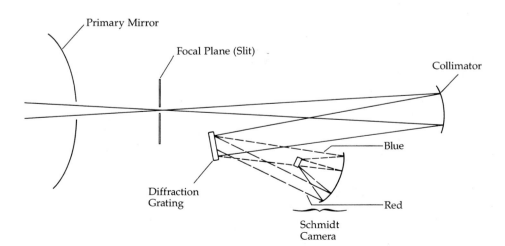

FIGURE 9–19 The optical path for a typical grating spectrograph.

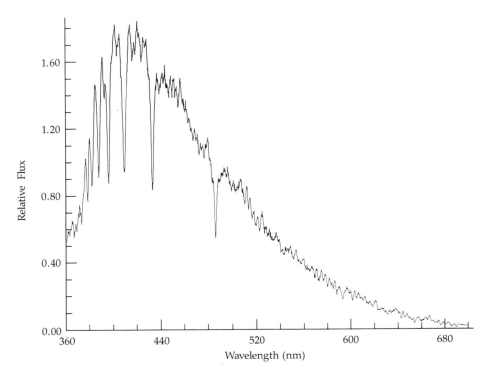

FIGURE 9–20 An example of the spectrum of an A spectral type star observed with the Boller and Chivens grating spectrograph at the Steward Observatory 90″ telescope. Note the strong, regularly spaced absorption line of the hydrogen Balmer series.

(C) OBSERVING WITH SPECTROGRAPHS

Most modern spectroscopy is done with a normal grating spectrograph, a CCD detector, and a long slit. The spectrum can be thought of as an image of the slit at each wavelength. For extended objects such as galaxies or gaseous nebulae, the image *along* the slit is resolved. So the *distribution* of spectra at various parts of the object is observed.

The astronomer may be looking for the **radial velocities** of the observed objects. In this case one would optimize the resolution of the spectrograph. Higher resolution is obtained by employing a grating with a higher density of grooves and by narrowing the slit. For faint objects, a compromise must be reached and resolution might have to be sacrificed for a reasonable S/N ratio. Observations of radial velocities require concerted efforts at calibrating the spectra by frequent observations of a gas discharge tube (often using iron, helium, neon, and argon) that provides a bright line comparison spectrum at zero velocity.

A second type of observation focuses on the spectral intensities more than the line positions. In this case, **spectrophotometry,** one must be more concerned with calibrating the effects of non-uniform detector response and the effects of the Earth's atmosphere. This demands frequent observations of standard flux objects such as well-observed stars. Figures 9–20 and 9–21 show examples of stellar spectra observed with a grating spectrograph. In Chapter 21 and 22 you will find sample galaxy spectra.

For various special projects, alternatives to the standard observing technique may be chosen. One of the major problems with long-slit spectroscopy is that only one object is observed at a time. One relatively well proven means of obtaining many spectra simultaneously is to use an objective prism in front of a Schmidt telescope. The prism covers the whole diameter of the telescope tube, and the design of Schmidt telescopes emphasizes their ability to make wide-field observations. Therefore, the photographic plate records the spectra of all objects in a very wide field. Although the spectra

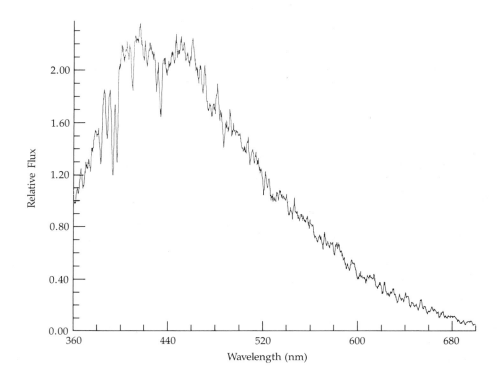

FIGURE 9–21 An example of the spectrum of a G spectral type star observed with the same instrument as in Figure 9–20.

are of low dispersion since the prism must have a quite narrow apex angle, they are very useful for classifications. Hundreds of thousands of stars have had their spectral types determined in this manner. In the extragalactic field, objective prism surveys have been employed to observe unusual objects such as quasars and emission line galaxies.

Another, more modern, means of obtaining many simultaneous spectra is multi-slit spectroscopy. This type of spectroscopy comes in two forms. In one, a mask is made for each field of view planned for an observing run. The mask is often photographically generated and is opaque in all regions except for tiny slits at the positions of the objects. A CCD image of the focal plane then provides a number of spectra centered at the position of each object. The second form of multi-slit spectroscope really uses only one long slit and brings the light from each desired object to the slit by means of fiber optics. One end of each fiber has a small lens glued onto it and can be positioned anywhere in the focal plane by means of computer-controlled arms. A CCD image of the one long slit then has several spectra lying side by side along the length of the slit.

9–5 ◗
A NEW GENERATION OF TELESCOPES

For the last several decades, the largest telescopes were in the 4- to 5-meter-diameter class. New progress in the detection of faint objects during this period was made by improved detector technology. The most important of the new detectors is the charge coupled device (CCD). Its quantum efficiency is about 2 orders of magnitude better than that of photographic plates, which it replaced. Since the newest examples have quantum efficiencies that approach 100% and also have nearly negligible readout noise, there is little room left for improvement in detectors.

Astronomers have not, however, reached the end of their need for improvements. The two main goals are for better resolution and for the ability to detect fainter objects. Large space telescopes, such as the Hubble Space Telescope, only partially alleviate the resolution problem. Although HST is above the blurring atmosphere, its expense and the limited number of objects it can observe each day prevent it from doing some very important types of observations. Similarly, the largest ground-based telescopes can gather more photons than HST; their light gathering power from larger mirrors more than makes up for the distortions of the atmosphere.

A new generation of telescopes with mirror diameters of 7 to 10 meters is currently planned or actually under construction. Some of these new observatories place more than one mirror into the same telescope structure or couple the output from more than one telescope. These telescopes can clearly improve the detection of faint objects, but some of them can also achieve resolutions that compete with those obtainable from orbit. In addition to the Keck 10-meter telescope (Section 9–1), other new observatories are the Columbus twin 8.4-meter telescope (Italy, the University of Arizona, and Ohio State University), the Magellan 8-meter telescope (Carnegie, Johns Hopkins, and the University of Arizona), the 7.5-meter Japanese National Large Telescope, the National Optical Astronomy Observatory 8-meter telescopes for both the Northern and Southern hemispheres, and the four 8.2-meter telescopes of the European Southern Observatory.

One way to improve resolution in the near infrared comes simply through the larger apertures. In the visible window, the dominant image degradation results from atmospheric turbulence which arises from convective motions of "cells" with typical diameters of 15 to 30 centimeters. Since these atmospheric cells have different temperatures and hence different indices of refraction, the image moves as it passes through different cells. A somewhat different regime holds in the near infrared, where the image resolution is limited by diffraction caused by the telescope aperture itself. Larger diameters will narrow the diffraction limitation down to the point where the seeing becomes the principal problem in the infrared as well as in the visible.

A second way to improve resolution is to account for the motion of the atmospheric cells in real time. Since the turbulence can be thought of as deforming the initially planar form of the wavefront from the astronomical source, we could, in principle, deform the telescope optics to compensate for the waveform—**adaptive optics.** The engi-

neering difficulties of this procedure are substantial. One has to sample the incoming waveform and move a large number of actuators at the base of the primary mirror to alter its shape. All of this has to happen with a frequency of about 100 hertz!

A third method is to perform **aperture synthesis** interferometry in a manner similar to that used by radio interferometers like the VLA. This is also an extremely challenging engineering problem. Since the wavelength of visible light is many orders of magnitude less than that of radio waves, the necessity of combining the different light paths while keeping their relative phase differences intact is quite difficult. The most ambitious effort in this direction is the Very Large Telescope project of the European Southern Observatory. The four 8-meter diameter telescopes are planned with separations on the order of tens of meters. The telescopes can be pointed toward different objects, but there will also be a facility for combining their light as an interferometer.

All of the new telescopes demand new technologies in mirror construction. In the past, mirror blanks were cast as monolithic disks having diameter-to-thickness ratios of 6:1 to 8:1. At the diameters of the new generation of telescopes, this thickness ratio would make the new mirrors far too heavy to support with the needed accuracies. The amount of mass of large, thick mirrors would also make them unsuitable because of their high **thermal inertias.** If the mirrors are not at the ambient temperature of the air in the telescope dome (tolerances of less than 0.1 K for some applications), air currents can worsen the convective turbulence. Since the modern mirrors are also to be quite fast, having f-ratios of 1.2 to 2.0, the deep cuts that would have to be made in the initial flat surface would be very expensive; up to \$1,000,000 of glass would have to be ground away!

Some of the new methods for mirror construction include **segmented mirrors** like those of the Keck telescope, **thin meniscus mirrors,** and **honeycomb mirrors.** The latter two methods demand that the mirror and its oven be rotated during the casting and annealing stages in order to approximate the final figure—hence much less grinding. With these techniques the diameter-width ratios of the new mirrors can be as large as 40:1.

In addition to improved methods of casting, new materials are also important for large mirrors. These new materials include *fused silica, glass-ceramics* and *borosilicates;* they replace such older materials as pyrex and quartz.

The ultimate goal of the coming generation of telescopes is for high resolution to go with the large apertures. Hence, these telescopes will be located only at sites with excellent seeing potential. The best sites are located in the Southwestern United States, Hawaii, and the Chilean Andes on relatively high mountains where the seeing disk can often be as sharp as 0.3 arcseconds. In addition to a good site, astronomers have found that it is necessary to make sure that the telescope dome does not contribute to poor seeing. This necessitates removal of all heat sources within the dome and maximizing the ventilation. The whole struc-

ture, including dome, telescope support structure, and mirror needs to be brought to the ambient temperature, and some of the new domes can be opened for 360° around the telescope by collapsing flexible walls or opening louvers.

Astronomers have found that each large telescope needs to be supported by smaller telescopes. Typically, the smaller instruments make surveys to find the relatively small number of critically interesting objects that only the giant telescopes can properly examine. With the new generation of large telescopes, a wise addition to the astronomy community would be an increased number of smaller ground-based telescopes. Several analyses have shown that the 4-meter class offers the best compromise between modest cost and good scientific rewards, and we hope that the number of these instruments will soon increase.

PROMBLEMS

1. Compare the resolving power and the light-gathering power of the human eye with those of a
   (a) 10-cm telescope
   (b) 4-m telescope

2. For what astronomical purposes would one use a telescope having a long focal length and a large f ratio?

3. The maximum possible antenna separation of the VLA is 40 km. What is its resolving power when operating at 1.5 cm? Why is it that the Earth's atmosphere does not limit the resolving power?

4. The transmission of the Earth's atmosphere at 3 μm is about 10%. Astronomers define the optical depth τ at some wavelength λ such that

$$I_\tau = I_o \exp (-\tau)$$

   where I_o is the original intensity in a light beam and I_τ the intensity after passing through a material with optical depth τ. What is the optical depth through the Earth's atmosphere at 3 μm?

5. Calculate the scale height for water vapor in the Earth's atmosphere. Use this information to state whether telescopes atop Mauna Kea would be better for infrared astronomy than ones at sea level.

6. The shortest wavelength at which VLBI is routinely done is 1.3 cm.
   (a) What is the minimum resolvable angle of an interferometer operating at 1.3 cm if the baseline is about one Earth diameter?
   (b) VLBI experiments are now being attempted at a wavelength of 3 mm. What is the resolving power at this wavelength for an Earth-diameter baseline?

7. One of the disadvantages of an interferometer is that it is insensitive to objects with angular sizes much larger than the minimum resolvable angle of the shortest baseline (essentially because different portions of the source destructively interfere with each other). The flux from a source of angular diameter θ'' measured by an interferometer of baseline L (in kilometers) is approximately

$$F_{meas} \approx F_{true} \exp [-0.3(L\theta/\lambda)^2]$$

   where λ is measured in centimeters. For a wavelength of 6 cm, what is the largest angular size of a source that can be observed with an interferometer of minimum baseline 1000 km if the lowest detectable flux is $0.1F_{true}$? What would be the properties of sources that one would *not* want to observe using an interferometer?

8. **(a)** What is the theoretical resolution of the Hubble Space Telescope (objective diameter = 2.4 m) at
    (i) λ = 500 nm (visible)
    (ii) λ = 200 nm (ultraviolet)
    (iii) λ = 2000 nm (infrared)
   (b) Considering your answers to **(a)**, why would it ever be advantageous to observe in the infrared with the space telescope?

9. What are the observational characteristics of an object that is better studied by the Hubble Space Telescope than by a very large ground-based optical telescope? What are the characteristics best suited for a very large ground-based optical telescope?

10. One of the main advantages of photographic plates or CCDs is their ability to integrate, that is, to collect light for a time much longer than the human eye can. The human eye effectively integrates for about 0.2 s (which is why we don't notice the individual frames of movie films). The flux of the faintest object that can be detected is proportional to $1/t^{1/2}$, where t is the integration time. (The square root arises because the flux of the background sky against which the object must be contrasted also increases with integration time.) How much fainter an object can be detected using a 1-h integration on the Palomar 5-m telescope than can be seen with the naked eye?

11. One of the major difficulties with gamma-ray astronomy is the poor resolving power, with angles less than about 2° being unresolved even for the best instruments. Nevertheless, gamma-ray images of our Galaxy have been produced using satellite data. Discuss the difficulty in interpreting these images in light of the poor resolving power.

12. Compare the theoretical resolution of a 5-m optical telescope at 500 nm with a 300-m radio telescope at 21 cm. Comment on why radio astronomers regu-larly use the technique of interferometry in order to map small angular scale structures.

13. You are observing a faint star and would like 1% accuracy in your photometric measurement (e.g., signal to noise = 100:1). After 5 minutes of integration time, you have a signal-to-noise ratio of 10:1 (10% accuracy).
   (a) What is the total integration time required to achieve the desired signal to noise ratio of 100:1? Do you think it is worth striving for this accuracy of measurement?
   (b) If you could increase the quantum efficiency of your detector by a factor of 10, what would be the total integration time required to achieve the desired signal to noise ratio of 100:1? Do you think it would now be worth striving for this accuracy of measurement?

14. A charge-coupled device (CCD) detector is mounted at the focus of an $f/14$ reflecting telescope with a 40-cm-diameter mirror. The CCD chip contains 500 by 500 pixels, with each pixel being 20 μm square.
   (a) What is the angular size (in arcseconds) of the sky which is imaged on each pixel?
   (b) What is the angular field (in arcminutes) of the CCD chip?

15. To appreciate how advances in telescope design and instrumentation have allowed astronomers to study previously undetectable faint objects, compare a present day imaging system with that of two decades ago. Given identical exposure times, how much fainter an object can the Keck 10-m telescope equipped with a charge-coupled device detector of quantum efficiency 70% detect than the Mt. Palomar 5-m telescope equipped with a photographic plate (quantum efficiency 1%)?

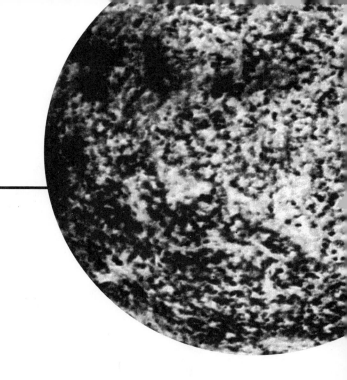

Chapter 10

The Sun:
A Model Star

Our Sun is the nearest star. The fascinating properties and phenomena of the solar surface layers are easily observed and have been studied intensely. Unfortunately, models for understanding solar phenomena have not kept pace with such detailed data. Because the Sun is a fairly typical star and because it is the only star that spans a large angular diameter as seen from the Earth, the discussion here serves as the physical basis to investigate the other stars (Table 10–1).

TABLE 10–1 Key Properties of The Sun

Property	Value
Mass	1.99×10^{30} kg
Radius	6.96×10^{8} m
Mean density	1410 kg/m³
Surface gravity	2.74×10^{2} m/s²
Escape speed	6.18×10^{5} m/s
Effective Temperature	5770 K
Luminosity	3.86×10^{26} W
Magnetic fields:	
Sunspots	0.3 T
Global	0.0001 T
Network	0.002 T
Plages	0.02 T

10–1 ◗
THE STRUCTURE OF THE SUN

At 1 AU from the Earth, the Sun provides the energy necessary for life. This gigantic sphere of gas, with a radius of 6.96×10^5 km ($\approx 109 R_\oplus$) and a mass of 1.99×10^{30} kg ($\approx 333,000 M_\oplus$) has a **luminosity,** or rate of total radiative energy output, of 3.86×10^{26} W. The average density of the Sun is only 1400 kg/m³, a value consistent with a composition of mostly gaseous hydrogen and helium.

The pressure, and also the temperature, in the solar interior must be extremely high just to support the weight of the Sun. The body of the Sun is gaseous and hot; in fact, the gases (primarily hydrogen and helium) are almost completely ionized (a **plasma**). Temperature, pressure, and density increase from the Sun's surface inward to the center, where energy is liberated by thermonuclear reactions. As hydrogen is transformed to helium in the Sun's **core,** vast quantities of energy are released in the form of photons and thermal motions (Figure 10–1). The photons diffuse outward through the large **radiative zone** until they reach the outer **convective zone,** where most of the transport of energy takes place by boiling motions of the gas. The visible surface of the Sun (the **pho-**

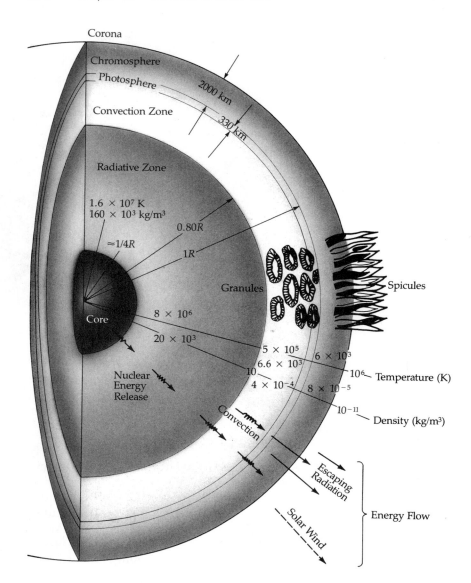

FIGURE 10–1 The structure of the Sun. This model shows the main regions of the Sun and values of important physical parameters. The granules and spicules are drawn out of proportion to show them clearly.

tosphere) occurs at the top of the convective layer, where the complex and extended solar atmosphere begins.

The base of the solar atmosphere is the photosphere (Figure 10–1), a thin layer of gas that represents the greatest depth to which we observe and from which emanates the bulk of the visible radiation. Sunspots appear on the photosphere. The next layer outward is the **chromosphere;** from the top of the chromosphere emerge the sharp spicules and the graceful prominences. Beyond this region lies the tenuous extended **corona** with its ghostly light merging into the outward-flowing **solar wind** and the interplanetary medium.

Although we cannot see optically into the interior of the Sun, there are two classes of observations that allow us to make some educated guesses about the solar depths. One is the observation of the neutrino flux from the Sun's core. You will see in Section 16–1(d) that the observed flux is only about one-third that predicted by theoretical models. This deficiency has challenged astrophysicists and physicists into proposing a number of explanations, including modifications of the solar models and new theories of the characteristics of the neutrino.

The other probe is the analysis of the radial motions of gases at the solar surface. Oscillations are observed as periodic changes in the Doppler shifts of photospheric and chromospheric spectral lines. These oscillations have periods ranging from less than 5 min to 2 h 40 min. The first to be discovered and in many ways the most important are the 5-min oscillations, which are observed as vertical motions of areas of the Sun, much like small boats on ocean waves. At peak amplitude, the velocities are 0.4 km/s. These oscillations result from sound waves from the interior; reflection and refraction below the Sun's surface confine these waves within acoustic cavities. Such waves and their harmonics can tell us some of the properties of the interior. Thus it may be said that the Sun rings like a bell and that its interior can be probed in a manner similar to the way terrestrial seismic waves are used to study the interior of the Earth. This field of research has been given the name **helioseismology.** Acoustic waves within the Sun are visible as oscillations on its surface, which can be measured by Doppler shifts in photospheric lines. Their pattern and period hold clues to the Sun's interior (Figure 10–2).

From helioseismologic studies have come inferences concerning the details of the convection-layer structure. Some vibrations are excited by the turbulence in the convective zone. They involve millions of tones turned on by the noise of convection, and allow a probe of the physical properties of the convection zone. To date, astronomers have measured thousands of frequencies to accuracies close to 0.01 percent. They indicate that the convective zone may go somewhat deeper than previously believed—perhaps about 30% of the radius rather than 20%. The remainder of this chapter presents a detailed examination of the more directly observable phenomena in the outer layers of our Sun.

10–2 ◖ THE PHOTOSPHERE

(A) GRANULATION

We cannot peer optically deeper into the Sun than the base of the photosphere, which defines the layer of the Sun's atmosphere where the gases become opaque to visible light. Here (Figure 10–3) we see a patchwork pattern of small (average diameter about 700 km), transient (average lifetime from five to tens of minutes) **granules:** bright irregular formations surrounded by darker lanes.

FIGURE 10–2   Solar vibrations. A computer model of the surface vibrations for one possible set of modes of oscillations. Solid lines represent zones of expansion; dotted lines those of contraction.

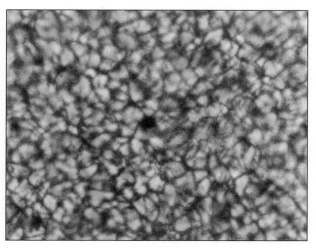

FIGURE 10–3   Photospheric granulation. The bright areas are about 1500 km across. The dark circle is a small sunspot. (*National Optical Astronomy Observatories*)

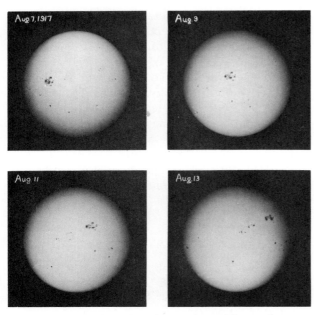

FIGURE 10–4 The solar photosphere. These photographs show limb darkening at the edges of the disk and also the Sun's rotation over a seven-day period, as tracked by the position of the sunspots. *(Yerkes Observatory)*

This solar granulation marks the top layer of the Sun's **convective zone,** a gaseous layer about 0.2 to $0.3R_\odot$ thick located just below the base of the photosphere. In this zone, heat energy is transported by **convection;** hot masses of gas (**convection cells**) rise, appear as bright granules, and dissipate their energy at the photosphere; the cooler gases sink back down. The resulting transfer of energy imparts motions of the order of tenths of a kilometer per second to the lower layers of the photosphere.

(B) PHOTOSPHERIC TEMPERATURES

The continuum spectrum of the entire solar disk defines a Stefan-Boltzmann effective temperature (Table 8–4) of 5800 K for the photosphere, but how does the temperature vary in the photosphere? A clue is evident in a white-light photograph of the Sun (Figure 10–4), for we see that the brightness of the solar disk decreases from the center to the limb—this effect is termed **limb darkening.** (Look back at Section 8–7 for the basic physics.)

Limb darkening arises because we see deeper, hotter gas layers when we look directly at the center of the disk and higher, cooler layers when we look near the limb (Figure 10–5A). Assume that we can see only a fixed distance *d* through the solar atmosphere. The limb appears darkened as the temperature *decreases* from the lower to the upper photosphere because, according to the Stefan-Boltzmann law (Section 8–6), a cool gas radiates

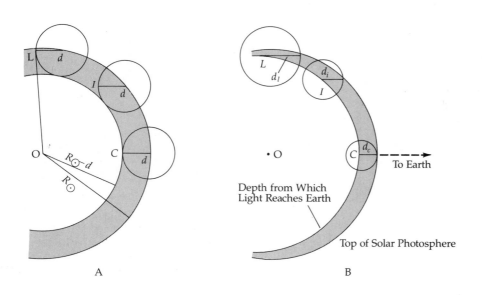

FIGURE 10–5 Geometry for limb darkening. The center of the disk is denoted by *C,* the limb by *L,* and an intermediate position by *I.* (A) All circles have the same radius *d,* but the line of sight ends higher in the atmosphere than at the center. (B) High-density gas low in the atmosphere is more opaque than the thinner gas higher up; here each circle corresponds to the same optical depth.

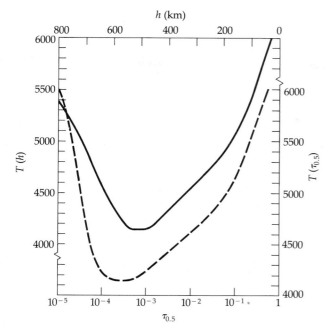

FIGURE 10–6 Temperature structure of the photosphere and chromosphere. The solid curve gives the temperature as a function of height, h, above the photosphere (use left and top scales), and the dashed curve for the temperature as a function of optical depth at a wavelength of 500 nm, $\tau_{0.5}$ (use bottom and right scales). *(J. Vernazza, E. Avrett, and R. Loeser)*

less energy per unit area than does a hot gas. The top of the photosphere, or bottom of the chromosphere, is defined as height = 0 km. Outward through the photosphere (Figure 10–6), the temperature drops rapidly then again starts to rise at about 500 km into the chromosphere, reaching very high temperatures in the corona.

At this point, you may have discerned an apparent paradox: how can the solar limb appear darkened when the temperature rises rapidly through the chromosphere? Answering this question requires an understanding of the concepts of **opacity** and **optical depth.** Simply put, the chromosphere is almost optically transparent relative to the photosphere. Hence, the Sun appears to end sharply at its photospheric surface—within the outer 300 km of its 700,000-km radius.

Our line of sight penetrates the solar atmosphere only to the depth from which radiation can escape unhindered (where the optical depth is

small). Interior to this point, solar radiation is constantly absorbed and re-emitted (and so scattered) by atoms and ions. To characterize this absorption by the gas at a given wavelength, we use the **opacity** k_λ (in m²/kg) of the gas. The simplest way to understand opacity is to consider what happens when radiation of flux F_λ strikes a slab of gas of thickness dx. Let the mass density ρ (in kg/m³) describe this gas. Part of this flux is absorbed by the slab, and so

$$dF_\lambda = -k_\lambda \rho F_\lambda \, dx \qquad \textbf{(10–1)}$$

In a uniform medium, Equation 10–1 is integrated to yield

$$F_\lambda(x) = F_\lambda(0) \exp\,(-k_\lambda \rho x) \qquad \textbf{(10–2)}$$

so that the flux diminishes exponentially with depth of penetration. For convenience, astronomers define another measure of this absorption by the **optical depth,** τ_λ, where

$$d\tau_\lambda = k_\lambda \rho \, dx \qquad \textbf{(10–3)}$$

Note that τ_λ is dimensionless and Equation 10–2 becomes

$$F_\lambda(\tau_\lambda) = F_\lambda(0) \exp\,(-\tau_\lambda) \qquad \textbf{(10–4)}$$

For $k_\lambda < 1$, the flux is constant, so that the gas is said to be **optically thin** (transparent) at this wavelength. In general, an **optically thick** (opaque) gas is one with $k_\lambda > 1$; the base of the photosphere corresponds to this case. Limb darkening arises from photospheric opacity (Figure 10–5B); k_λ is low in the chromosphere.

(C) H⁻ CONTINUOUS ABSORPTION

The photosphere emits a blackbody-like continuum and so must be opaque at visible wavelengths, but the densities here are far less than those needed for the gas to be opaque. How then does the photospheric continuum radiation arise?

The chief contributor to the continuous opacity is H⁻, the **negative hydrogen ion.** This ion can exist because the single electron of the neutral hydrogen atom does not completely screen the positive proton. Hence, a second electron may loosely attach itself; the ionization potential is only 0.75 eV (that of normal hydrogen is 13.54 eV). Absorption occurs by the dissociation reaction $H^- \rightarrow H + e^-$,

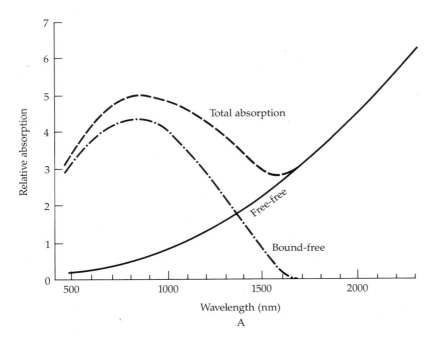

A

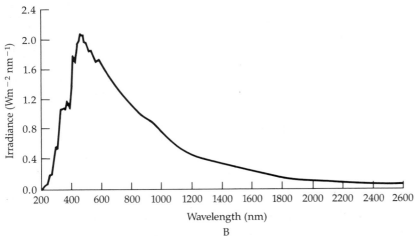

B

FIGURE 10–7 Continuum opacity in the Sun. (A) Relative absorption of H⁻ from free-free and bound-free interactions in the visible and near infrared parts of the spectrum. (B) Measured solar continuum above the Earth's atmosphere. *(U.S. Air Force Cambridge Research Laboratory)*

and emission takes place when an electron attaches itself to a neutral hydrogen atom in the reaction $H + e^- \rightarrow H^-$. The solar continuum in the infrared and optical regions is largely produced by these reactions, which are enhanced by the relatively high electron and hydrogen densities in the photosphere. The opacity comes from both free-free and bound-free interactions (Figure 10–7A).

We must note that these and other atomic processes that absorb light in the photosphere do *not* give complete opacity at all wavelengths. Hence

the solar continuum has a general blackbody trend (Figure 10–7B) but does not follow a Planck curve exactly.

(D) THE FRAUNHOFER ABSORPTION SPECTRUM

Spectral Lines

In 1814, the German physicist Joseph von Fraunhofer (1787–1826) made the first definitive map-

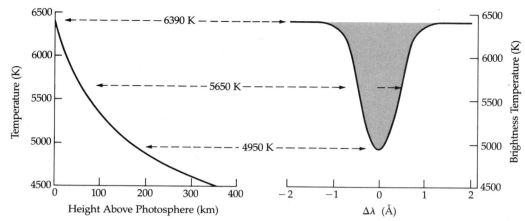

FIGURE 10–8 Photospheric depth and line profiles. Different parts of an absorption line form at different heights, the line center coming from the coolest, highest region. (*R. W. Noyes*)

ping of the vast array of absorption lines seen on the photospheric continuum—the **Fraunhofer absorption spectrum.** Not knowing the correct identifications of these features, Fraunhofer designated the strongest lines (starting from the red) with capital letters and the weaker lines with lowercase letters. (See Section 8–4 for the definition of line strength.) Today we still refer to the D lines of a resolved doublet of sodium and the H and K lines of Ca II. The earliest line identifications included the hydrogen Balmer series and absorption lines of sodium, calcium, and magnesium. The Fraunhofer absorption line spectrum extends to wavelengths as short as 165 nm. At wavelengths shorter than the near ultraviolet (<165 nm), the solar spectrum is dominated by **emission lines** that are produced in the chromosphere and corona [Sections 10–3(a) and 10–4(c)].

Because the solar continuum comes from a thin layer of H^- ions, the Fraunhofer absorption lines can and do form at the same solar atmospheric level as does the continuum. The weaker lines originate in the lower photosphere, and the stronger ones form largely in the upper photosphere; in fact, the strongest (the Balmer lines and the H and K lines of Ca II) form primarily in the lower chromosphere. The opacity is low in the chromosphere for most wavelengths. In strong lines, however, the opacity becomes large even at considerable heights in the solar atmosphere. The reasons include the abundance of the element, the number of atoms in the lower level of the transition producing the line (which in turn depends on the temperature through the Boltzmann equation; Section 8–4), and the transition probability (the intrinsic atomic parameter that determines the probability that an atom will make that particular transition). The depth of formation differs from line to line and within a given line. So the profiles of the Fraunhofer absorption lines provide a powerful tool for probing different heights of the solar atmosphere (Figure 10–8).

High-dispersion spectra have revealed fine structure wiggles in most Fraunhofer lines. These wavelength oscillations are Doppler shifts arising from vertical motions at about 0.4 km/s of small-scale (1000 km in diameter) structures in the photosphere. They are generated by gas motions induced by the granules at the bottom and continuing up into the chromosphere. In addition, horizontal motions run parallel to the solar surface with about the same velocities. These motions arise in larger structures (30,000 km across) called **supergranules.** Gas flows slowly from the center of a supergranule to the edge. Supergranules are related to higher chromospheric structures, though they actually originate below the granulation. Their flowing gases carry magnetic lines of force to their edges, and so the magnetic fields become more localized and concentrated. As you will see, solar activity is generally associated with strong localized magnetic fields.

TABLE 10–2 *Abundances of the Elements in the Photosphere Relative to Hydrogen by Number*

H	1,000,000
He	63,000
C	420
N	87
O	690
Ne	37
Na	1.7
Mg	40
Al	3
Si	45
P	0.27
S	16
K	0.11
Ca	2.1
Fe	32

Elemental Abundances

Using the ideas outlined in Chapter 8, we can analyze the Fraunhofer lines to infer properties of the photosphere. The characteristic temperatures and pressures are first found, and then the relative line strengths (Section 8–4) tell us the chemical composition. Hydrogen is by far the most abundant element (71%), helium second (27%); all of the heavier elements account for about 2% of the total mass (Table 10–2). Be warned, though, that finding photospheric abundances is a complex and difficult task. Overall, elemental abundances are known to a factor of 2 at best. For some elements (especially if they have few and blended lines in the spectrum), the values are uncertain to an order of magnitude.

10–3 ◗
THE CHROMOSPHERE

The solar **chromosphere** extends about 10,000 km above the photosphere, and its gas density is far less than that of the photosphere. This thin layer has a reddish hue—as a result of the Balmer (H_α) emission of hydrogen—visible during a total solar eclipse.

(A) THE CHROMOSPHERIC SPECTRUM

The chromospheric spectrum is completely overwhelmed by the photosphere except when the photospheric light is blocked out. Emission lines are visible in the chromosphere, notably those of helium, which requires high temperatures to become excited.

Chromospheric spectra reveal characteristic variations with altitude in the chromosphere. Atomic transitions of low excitation potential, such as some of those of neutral metals, are seen only at the base of the chromosphere. Lines of ionized iron and calcium are evident somewhat higher up. The hydrogen Balmer and neutral helium features are seen for many thousands of kilometers above the photosphere. The chromospheric lines fade with height, but the strongest line of He II (468.6 nm) fades most slowly. This decrease in line strengths has two major causes: (1) the gas density drops sharply with height and (2) the temperature increases rapidly with height above the photosphere (Figure 10–9). The high-excitation-potential lines of helium remain strong as a result of the high temperatures in the outer chromosphere. Helium (after the Greek *helios*, meaning "sun") was found in chromospheric spectra before it was discovered on Earth. Most helium lines in the optical part of the spectrum are too weak (optically thin) to be seen against the solar disk.

The hydrogen Balmer lines are formed mainly in the chromosphere, even though the chromo-

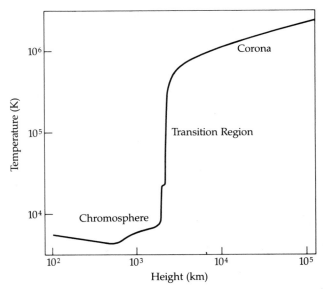

FIGURE 10–9 Overall temperature structure in the chromosphere and corona as height above the photosphere. Note that this is a log–log plot.

FIGURE 10–10 The Sun in H_α. This region near the limb shows the structure from strong local magnetic fields. *(R. B. Dunn, National Optical Astronomy Observatories)*

sphere is hotter than the photosphere. How? The photospheric continuum radiation and the temperature there are not sufficiently energetic to promote many hydrogen atoms from the ground (Lyman) state to the first excited state. (Remember that the Balmer lines must originate from the first excited state.) Only in the hotter chromosphere, in accordance with the Boltzmann equation, does the population of second-level hydrogen atoms become significant, primarily through collisional excitation. Then continuum radiation is absorbed in the chromosphere to form the Balmer absorption lines (when viewed in projection against the photosphere) and the Balmer emission lines that color the chromosphere (when seen against dark space at the Sun's limb).

(B) CHROMOSPHERIC FINE STRUCTURE

Certain absorption lines (H_α and the H and K lines of Ca II) in the chromosphere have a large optical depth. Radiation at these wavelengths cannot escape from the photosphere because the chromosphere is essentially opaque here. The chromosphere may be studied at these wavelengths because the absorption lines are not completely black; the center of each line is darker than the adjacent continuum, but some photons are still emitted in our direction from the chromosphere.

Photographs of the Sun (Figure 10–10) in H_α and Ca II K reveal large bright and dark patches. These are the **plages** and **filaments** associated with solar activity [Section 10–6(b)]. In addition, distinctive structure appears over the entire solar disk: the bright **network** (Figure 10–11) associated with the magnetic fields at the boundaries of su-

FIGURE 10–11
Chromospheric network. This photograph in the Ca II K line shows the bright network associated with the magnetic fields at the edges of supergranules. *(N. R. Sheeley and S. Y. Liu, National Optical Astronomy Observatories)*

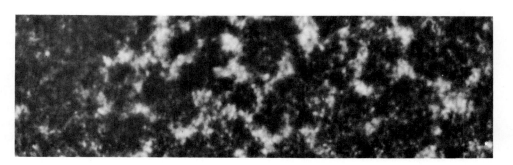

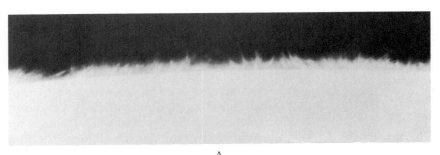

A

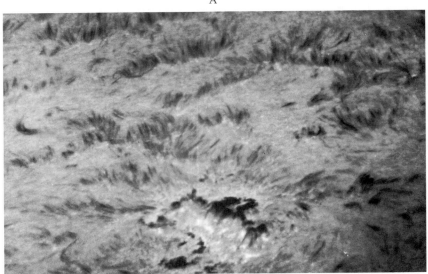

B

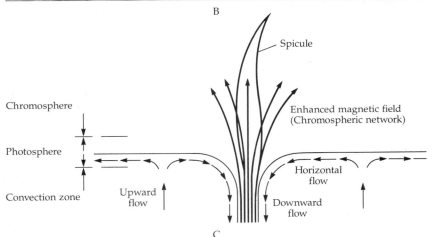

C

FIGURE 10–12 Spicules. (A) An
H_α photograph of the limb shows
the top of the chromosphere and
the spicules there. (B) In this H_α
photograph, the spicules appear
as elongated dark features against
the disk. *(R. B. Dunn, National Optical Astronomy Observatories)* (C) A
model for the formation of spicules as part of the chromospheric
network of magnetic fields.

pergranules. A brightening of the Ca II K line correlates with increased magnetic field strengths,
which range up to 3×10^{-3} T.

At the limb of the Sun, tenuous jets of glowing
gas 500 to 1500 km across extend to a distance of
10,000 km upward from the chromosphere (Figure
10–12A). In the **spicules,** which are best observed
in H_α, gas is rising at about 20 to 25 km/s. Although spicules occupy less than 1% of the Sun's

surface area and have lifetimes of 15 min or less,
they probably play a significant role in the mass
balance of the chromosphere, corona, and solar
wind. Spicules are not distributed uniformly over
the solar surface (Figure 10–12B) but form a network pattern, a part of the chromospheric network
that constitutes the boundaries of the supergranules. So spicules occur only in regions of enhanced magnetic fields (Figure 10–12C).

(C) THE TRANSITION REGION

Ultraviolet spectral features occur because of the high chromospheric temperatures; they are an excellent probe of the upper chromosphere and narrow chromosphere–corona transition region. Temperatures rise very steeply, from about 10,000 K in the chromosphere to 50,000 K within a few hundred kilometers of the transition zone to over 10^6 K in the corona (Figure 10–9). At wavelengths shorter than 150 nm, the photospheric continuum becomes undetectable (see the Planck curve for 6000 K in Figure 8–14). Because of the low flux of the photospheric continuum in the far ultraviolet, radiative excitation cannot occur. However, at the high temperatures of the Sun's upper atmosphere, atoms and ions become collisionally excited and produce emission lines as they return to their ground states. Moreover, they can be observed in projection on the face of the Sun, for they will not be overwhelmed by the photospheric background.

One of the strongest ultraviolet lines is the Lyman α line. Recall that the strong hydrogen Balmer lines imply an abundance of hydrogen atoms in the first excited state ($n = 2$). Absorption from the continuum will excite these atoms to higher levels ($n = 3, 4$, and so on), producing the Balmer absorption series [Section 10–3(a)]; however, most of the atoms will return immediately to the ground state ($n = 1$), emitting L_α photons. Hydrogen atoms can be excited to $n = 2$ only by collisions or reabsorption of L_α photons (self-absorption) because of the absence of the photospheric continuum at this wavelength. Photographs in L_α show essentially the same chromospheric network seen in Ca II.

Emission lines of C III, peaking at 70,000 K, N III at 100,000 K, and O VI at 300,000 K are used to study the structure through the transition zone. The network apparently continues through the region. It disappears, however, in images made in Mg X at 60 nm, corresponding to 1.6 million K, well into the corona.

10–4 ◑ THE CORONA

At solar eclipses, the **corona** appears as a pearly white halo extending far from the Sun's limb (Figure 10–13). A brighter inner halo hugs the solar limb, and coronal streamers extend far into space.

FIGURE 10–13 A quiet-sun corona. Note the polar plumes and the elongation at the equator. *(J. D. Bahng and K. L. Hallam)*

FIGURE 10–14 Active Sun coronal streamers, showing the magnetic field configuration and solar wind flow far from the photosphere. *(C. Keller, Los Alamos National Laboratory)*

(A) THE VISIBLE CORONA

Coronal continuum radiation at optical wavelengths is composed of two parts. The corona itself is divided into the **K corona** (dominating near the Sun) and the **F corona** (evident farther out to a few solar radii). Part of the low-intensity coronal continuum matches the wavelength dependence of the photosphere. This results from light scattered by electrons, which constitute more than half the particle number density in the K corona. No photospheric absorption lines show in this scattered component, a fact we attribute to Doppler broadening by rapidly moving electrons. Because the absorption lines are completely washed out, these electrons must have extremely high temperatures—average (quiet) coronal temperatures are 1 to 2×10^6 K.

Superimposed on the electron-scattering continuum is the F corona spectrum, which does show the Fraunhofer absorption lines. This component comes from light scattering from dust particles, identical to those grains that pervade interplanetary space. The dust is concentrated in the plane of the ecliptic, for we see the outer extension of the F corona as the *zodiacal light* [Section 2–2(f)]. We can separate the F and K coronal components because the Fraunhofer lines appear only in the former and because electrons and dust differ radically in the way they polarize the light they scatter.

Solar activity strongly affects the appearance of the K corona (Figure 10–13). At times of **sunspot maximum** [Section 10–6(a)], the corona is very bright and uniform around the solar limb and bright **coronal streamers** (Figure 10–14) and other condensations associated with active regions are much in evidence. At **sunspot minimum,** the corona extends considerably farther at the solar equator than at the poles and the coronal steamers are concentrated at the equator.

(B) THE RADIO CORONA

Later in this chapter, we will describe the radio bursts associated with solar activity; here we discuss the radio properties of the quiet Sun. In the ionized gases of the solar atmosphere, free electrons provide the emission and absorption of radio radiation. The interactions involved are *free–free transitions*: an electron sideswipes an ion or atom and absorbs or emits a low-energy photon while the electron's kinetic energy changes slightly. The electron is unbound both before and after the collision. The closer the electron passes by the scatterer, the more both the frequency and the strength of the interaction increase. Hence, the character of the free–free photons depends upon the gas density. The denser the gas, the more frequent and energetic the interaction; the resulting photons have higher energies (or shorter wavelengths). This explains why short-wave (1 to 20 cm) radiation characterizes the chromosphere and lower corona and wavelengths longer than 10 cm arise in the outer corona.

At wavelengths longer than about 20 cm, the Sun is observed to be **limb-brightened** (into the corona), confirming that the longer the wavelength observed, the higher in the corona we are looking. (Recall that, at visible wavelengths, the Sun is limb-darkened.) At wavelengths longer than 50 cm, we encounter great radio variability because the emission from the tenuous electron gas fluctuates wildly as the number density of particles varies.

Just as the H^- ion accounts for photospheric opacity, so also does the electron density lead to coronal opacity at radio wavelengths. The wavelength dependence of this electron opacity is the following. The corona is optically thin at short wavelengths (≈ 1 cm), so that such radiation can reach us from the chromosphere; at longer wavelengths, it becomes more and more optically thick, so that the corona is opaque to these wavelengths at greater heights. On the average, we can fit the radio data to the low-energy tail of a Planck blackbody distribution at temperatures of about 10^6 K.

(C) LINE EMISSION

Forbidden Lines

Superimposed on the visible coronal continuum are some emission lines that were unidentified until about 1942, when W. Grotrian of Germany and B. Edlén of Sweden interpreted them. (They were long called "coronium" lines, for they did not fit any known atomic transition.) The two strongest lines are the green line of Fe XIV (530.3 nm) and the red line of Fe X (637.4 nm); both are forbidden lines (Section 8–2).

Two significant obstacles hindered the identification of the coronal emission lines: (1) the responsible transitions are *forbidden* and (2) the temperatures of the corona are unexpectedly high. In quantum mechanics, certain energy levels of an atom are *metastable* because downward transitions from such levels are strongly prohibited. While an ordinary permitted transition takes place in about 10^{-8} s, these metastable levels may persist for seconds or even days before a forbidden transition occurs. In most laboratory and astrophysical situations, gas densities are so high that collisional de-excitation empties metastable levels very rapidly—there is just not enough time for a forbidden transition to take place. In the near vacuum of the corona, however, metastable levels populated by either photospheric radiation or collisions can decay and forbidden emission features are formed.

Very energetic collisions are required to ionize iron; say, nine and 13 times, and so the coronal gas must be very hot. To produce Fe X requires a temperature of 1.3×10^6 K, and Fe XIV requires a temperature of 2.3×10^6 K. At times of strong solar activity (such as flares), much higher temperatures occur, for the lines of Ca XV (3.6×10^6 K) are seen. The characteristic range is 10^6 K (quiet corona) to 4×10^6 K (active corona).

Extreme Ultraviolet Lines

Highly ionized atoms, such as those present in the solar corona, have lost many of the electrons that shield the atomic nucleus, and the remaining electrons are strongly attracted and tightly bound to the nucleus. Permitted transitions correspond to very high excitation potentials, and the resulting spectral photons are very energetic—at ultraviolet wavelengths. In fact, the spectral region from 5 to 50 nm (detectable only above the Earth's atmosphere) is dominated by the permitted emission lines from the coronal ions Fe VIII to XVI, Si VII to XII, Mg VIII to X, Ne VIII to IX, and S VIII to XII. These transitions permits us to deduce the relative elemental abundances in the corona; the results are consistent with the photospheric abundances.

Coronal Loops and Holes

Because the coronal gas is so hot, it emits low-energy X-rays and shows up in X-ray photographs of the Sun (Figure 10–15). These pictures show that the coronal gas has an irregular distribution above and around the Sun. The large loop structures indicate where the ionized gas flows along magnetic fields that arch high above the Sun's surface and return to it. The hot gas is trapped in

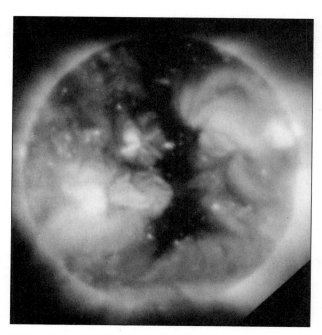

FIGURE 10–15 The X-ray sun. The hot plasma in the corona appears bright; note the loops, shaped by magnetic fields. The dark region down the middle is a coronal hole. *(G. Viana, Harvard College Observatory)*

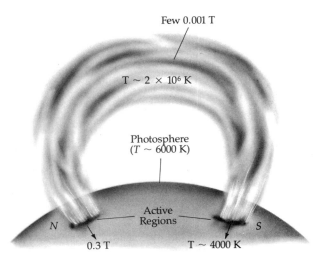

Few 0.001 T

$T \sim 2 \times 10^6$ K

Photosphere
($T \sim 6000$ K)

Active
Regions

N   S

0.3 T $T \sim 4000$ K

FIGURE 10–16 Coronal loops formed by magnetic flux tubes above active regions; these shape the plasma structure in the lower corona.

these magnetic loops. Solar physicists now view the corona as consisting primarily of such loops.

Note also that some regions of the corona appear dark (Figure 10–15), especially at the top pole and down the middle part of the Sun. Here the coronal gas must be much less dense and less hot than usual; these regions are called **coronal holes.** The coronal holes at the poles do not appear to change very much, but those above other regions seem somehow related to solar activity. X-ray bright points, lasting only a matter of hours, are scattered within these coronal holes. These are themselves apparently short-lived active regions and are also loops, albeit very small ones.

Solar astronomers believe that coronal holes mark areas where magnetic fields from the Sun continue outward into space rather than flow back to the Sun in loops. So the coronal gas, not tied down in these regions, can flow away from the sun out of the coronal holes; this flow makes the solar wind (Section 10–5).

The coronal gas does *not* follow the differential rotation of the photosphere. Rather it rotates at the same angular speed at all latitudes (as the Earth does). This fact implies that the bottoms of the magnetic loops are anchored deep below the photosphere, perhaps at the very bottom of the convective zone, where such fields might be generated by a solar dynamo.

Now back to why the corona (and chromosphere) are so hot. The magnetic loops (Figure

10–16) rising from the photosphere up to 400,000 km into the corona are believed to play the key role. The magnetic field in a loop is twisted by photospheric motions at its base. If the twisting happens slowly, it generates electric fields that then heat the coronal gas. It does not take much energy to do so because the coronal gas is so thin that it has a small heat capacity.

10–5 ◗
THE SOLAR WIND

The high coronal temperatures tend to blow the corona away from the Sun. The Sun's gravitational attraction on this gas is insufficient to retain it, and so a continuously flowing **solar wind** leaves the Sun. This flowing gas, composed of approximately equal numbers of electrons and protons, is a *plasma* (it is electrically neutral on a large scale). The thermal conductivity of the plasma is very high, so that high temperatures prevail over great distances from the Sun. Hence, the wind accelerates as it expands (large-scale speeds near 300 km/s at $30R_\odot$ and 400 km/s at 1 AU) and the particle density decreases to an average of a few million electrons and protons per cubic meter at 1 AU. These characteristics of the solar wind have been measured directly by interplanetary space probes.

A plasma couples tightly to lines of magnetic force; in fact, the magnetic field is essentially frozen into the gas. Therefore, the solar wind drags the extensions of solar magnetic fields into interplanetary space. The large-scale solar magnetic fields are directly related to the interplanetary fields by a sectored structure (Figure 10–17). The latitude band within 30° of the solar equator is frequently divided into extended longitude regions of one magnetic polarity or the other. The radially outflowing solar wind conveys these fields away from the Sun in sectors; the sector boundaries are distorted into spirals because the Sun rotates away from the receding gas and magnetic field.

The solar wind shows considerable complexity and variability. For instance, at the Earth's orbit, the proton density varies from 0.4 to $80 \times 10^6/m^3$ and the speed ranges from 300 to over 700 km/s. This variation is closely tied to coronal holes, where both the density and the temperature are less than in the normal corona; the density may be only a few percent of the usual value. Some of the energy that normally goes to heating the coronal material trapped within the magnetic loops is used

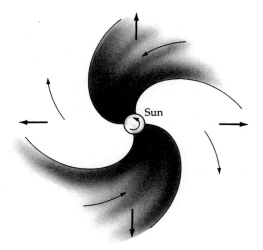

FIGURE 10–17 Magnetic fields and the solar wind. The interplanetary magnetic field extends from the Sun. The thin arrows show the direction of magnetic fields, and the thick ones show the direction of the solar wind. The shading shows two sectors of the structure.

to accelerate the gas outward through the open field lines. So bursts of solar wind may reach speeds in excess of 700 km/s.

Solar activity, especially flares, changes the magnetic field structure, sometimes dramatically. Whereas the normal solar wind is composed of low-energy protons and electrons (10^3 eV), solar flares eject clouds of high-energy protons (10^7 to 10^{10} eV). These clouds rush through the solar wind, altering its speed and density locally and distorting the magnetic field structure. These clouds are dangerous to unshielded astronauts, and they cause magnetic disturbances in the Earth's magnetosphere a few days after they leave the Sun.

10–6 ◐
SOLAR ACTIVITY

(A) THE SOLAR CYCLE

The Sun is near enough that we can easily observe transient phenomena in its atmosphere. Such phenomena are manifestations of **solar activity,** and they are linked through the solar rotation and magnetic field in the **solar cycle.** The remainder of this chapter discusses the **active Sun,** using the generic term **active region** to specify an area with sunspots, prominences, plages, and flares (see Table 10–3 for a summary of solar activity).

Sunspots
Sunspots are photospheric phenomena that appear darker than the surrounding photosphere (at

TABLE 10–3 *Summary of Solar Activity*

Photosphere	
Sunspots	Strong magnetic fields, lower temperature than photosphere
Faculae*	Denser, hotter, brighter than photosphere (in white light)
Bipolar magnetic regions	Medium to weak magnetic fields
Chromosphere	
Plages*	Brighter than chromosphere in H and Ca II lines
	Denser, hotter than chromosphere
Prominences (filaments)	Chromospheric material in corona
	Exhibit motions associated with magnetic fields
Flares†	Brief brightenings in plages (in H_α and Ca II lines)
Corona	
Condensations*	White-light features due to increased electron density, increased emission in forbidden and UV lines, associated with slowly varying radio emission
Flares	Radio bursts from fast electrons trapped in upper corona
Holes	Solar cosmic rays and enhanced solar wind
Coronal mass ejections	Sudden eruptions at speeds of hundreds of kilometers per second involving 10^9 tons of material

*Phenomena related to plages at different heights in the solar atmosphere.
†Flare-related phenomena.

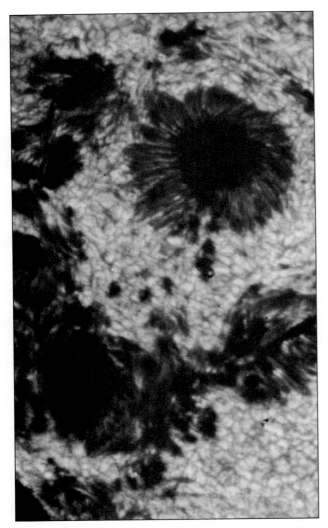

FIGURE 10–18 Sunspot group. A classic sunspot has an almost circular central umbra and a radial penumbra, determined by magnetic fields. Note the distorted spots in the lower left. Small spots without penumbras are pores. *(Project Stratoscope, Princeton University)*

portions. The largest have umbral diameters of 30,000 km and penumbral diameters more than twice as large.

The most important characteristic of a sunspot is its magnetic field. Typical field strengths are near 0.1 T, but fields as strong as 0.4 T have been measured. (The field strengths are deduced from the observed Zeeman splitting of spectral lines.) These fields by their vertical components inhibit the convective transport of energy to the photosphere, so that the sunspot is cooler than its surroundings. Related to the magnetic field is a horizontal flow of gas in the sunspot penumbra: gas moves out along the lower filaments and inward along the higher filaments (at speeds up to 6 km/s).

A given sunspot has an associated **magnetic polarity.** Lines of magnetic force diverge from a north magnetic pole and converge at a south pole; you are familiar with this characteristic of bar magnets and our Earth. A magnetic pole cannot exist in isolation, however, because magnetic lines of force must be complete (according to Maxwell's equations). So, two sunspots of complementary polarity are generally found together in a **bipolar** spot group (see the following paragraphs). Exceptions to this rule occur. Sometimes the second magnetic region is so diffuse that only one lone sunspot is seen; at other times, large, complicated groups of many sunspots appear—such a group may become the nucleus of a large active region [Section 10–6(b)] on the solar disk.

Sunspot Numbers

Counts of the number of sunspots visible at any given time have been recorded since Galileo's time (Figure 10–19). The sunspot number changes with time. A cyclic phenomenon is taking place, for successive sunspot maxima (or minima) occur every 11 years on the average (there may be a variation of as much as two or three years from cycle to cycle). A new cycle begins when the number is a minimum.

Recent investigations indicate that some historical evidence exists to show an absence of the 11-year cycle in sunspot activity in the period before 1700—a lull in activity called the **Maunder minimum.** Hardly any sunspots were seen in the 60-year period from 1645 to 1705 (Figure 10–19). This period corresponded to an unusual cold spell (sometimes called the Little Ice Age; the average temperature of the earth dipped about 0.5 K) that

about 5800 K) because they are cooler (sunspot continuum temperatures are about 3800 K, and sunspot excitation temperatures are about 3900 K). The darkest, central part (with the temperatures just mentioned) is termed the **umbra** (Figure 10–18); the umbra is usually surrounded by the lighter **penumbra** with its radial filamentary structure. Small sunspots develop from **pores,** larger-than-usual dark areas between bright granules. Although most pores and small spots soon disintegrate, some grow into true sunspots of huge pro-

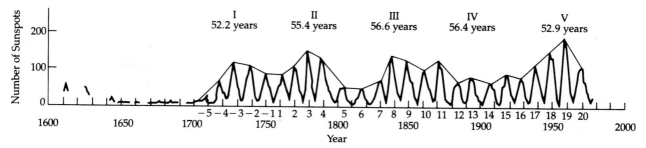

FIGURE 10–19 Sunspot cycles. Plotted here are the number of sunspots observed annually from 1600 to the 1970s. Note how few sunspots were seen prior to 1700; that was the time of the Maunder minimum.   [H. Yoshimura, Astrophysical Journal 227:1047 (1979)]

extended from the sixteenth to the eighteenth centuries. The relative consistency of the cycle in modern times may be a brief phase that recurs over longer times. Evidence in the layering of Australian rocks some 700 million years old hints at an activity cycle very much like that recorded since 1700 A.D. Overall, solar activity may have much more complex behavior than that inferred from the limited time spans investigated so far. And it may

well affect the earth's climate, albeit in complex ways.

Positional Variation

The distribution of sunspots in solar latitude varies in a characteristic way during the 11-year sunspot-number cycle. Sunspots tend to reside at high latitudes (±35°) at the start of a cycle (Figure 10–20), most spots are near ±15° at maximum, and the few

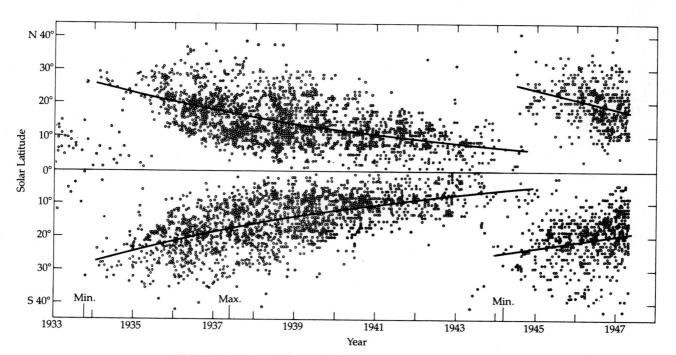

FIGURE 10–20 A diagram showing the change in latitude of sunspots as a function of phase in the solar activity cycle. Spot groups appear at high latitudes at the beginning of a cycle and drift toward the equator in active latitude bands as the cycle progresses. (G. Abetti, The Sun, Faber and Faber, 1969)

spots at the end of the cycle cluster near ±8°. Very few sunspots are ever found at latitudes greater than ±40°. The lifetime of a sunspot ranges from a few days (small spots) to months (large spots). In fact, a sunspot dies at the same latitude where it was born (a characteristic that permits us to determine the solar rotation). What takes place is this: as the cycle progresses, new spots appear at ever lower latitudes. The first high-latitude spots of a cycle appear even before the last low-latitude spots of the previous cycle have vanished.

So the sunspots follow active latitude belts during the course of a cycle. Less clear, but just as tantalizing, large active regions and spot groups seem to fall into preferred longitudes during a cycle—active longitude belts. Concentrations of magnetic fields appear to persist below the photosphere, so that new spot groups arise from about the same locations as previous ones. The active longitude belts sometimes appear about 180° apart, though not all the time; frequently they are not symmetrical across the equator. They reveal a different persistent pattern in the magnetic field structure in the convective zone.

Sunspot Polarity

Because most sunspot groups tend to be magnetically bipolar, it is useful to refer to **preceding spots** and **following spots** (in the sense of solar rotation). The Sun rotates eastward (as does the Earth), so that a preceding spot lies west of the following spot *as seen from the Earth*. After George E. Hale (1868–1938) discovered the magnetic character of sunspots in 1908, it was realized that all bipolar groups in one solar hemisphere have the same polarity and those in the other hemisphere have the opposite polarity. For example, in one solar cycle, preceding spots in the northern solar hemisphere have negative (south) polarity and preceding spots in the southern hemisphere have positive (north) polarity. Moreover, the sense of polarities reverses with each cycle, so that northern hemisphere leading spots will be positive in the next cycle. So a 22-year solar cycle is the true length of sunspot magnetic activity.

Solar Rotation

By observing sunspots with his telescope, Galileo determined that the Sun's surface rotates eastward (synodically) in about one month. Today, the same

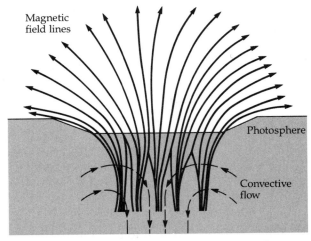

FIGURE 10–21   A model of the magnetic fields in the photosphere that generate sunspots. *(E. N. Parker)*

method is used in the sunspot zone (other methods, such as Doppler shifts, are necessary above latitude ±40°), and we know that the Sun rotates **differentially.** That is, the rotation period is shorter at the solar equator (about 25^d) than at higher latitudes (about 27^d at 40° and 30^d at 70°). The average sidereal period adopted for the sunspot zone is 25.4 days, with a corresponding synodic period (appropriate in discussions of solar–terrestrial events) of 27.3 days.

To sum up, sunspots reveal on a small scale the complexity and variability of solar magnetic phenomena. The parts of a sunspot are all transient magnetic structures—basically a sheaf of magnetic flux tubes filling the umbra and penumbra and fanning out above them (Figure 10–21). The sunspots' magnetic fields suppress the hot gas rising from the convective zone. The hot gas runs into a magnetic thicket and has trouble breaking through the surface in the sunspot's center. A convective downflow may draw away some of the hot gas, removing heat from the region. As a result, the sunspot is cooler and darker than the surrounding photosphere.

(B) ACTIVE REGIONS

As the sunspot number increases, so also does solar activity. With each sunspot group is associated a large **active region,** several hundred thousands of kilometers across. Magnetic activity on the Sun concentrates in these extended active regions.

Bipolar Magnetic Regions and Plages

The most significant property of active regions is their magnetic fields: 0.1 T in sunspots and 0.01 T overall. Even when the sunspot group is a tangle of different polarities, the enveloping region is usually bipolar in character, and so we refer to **bipolar magnetic regions** (BMRs).

A **magnetograph** maps out the magnetic field structure of BMRs (Figure 10–22). In a magnetic field, the Zeeman effect splits a spectral feature into several components, each with a characteristic optical polarization. When the splitting and intensity of these components are compared with a magnetograph based on the different polarizations, a map of the magnetic field strength and direction is produced.

In white light, photospheric **faculae** are brightenings that mark active regions. The enhanced brightness is from greater temperatures

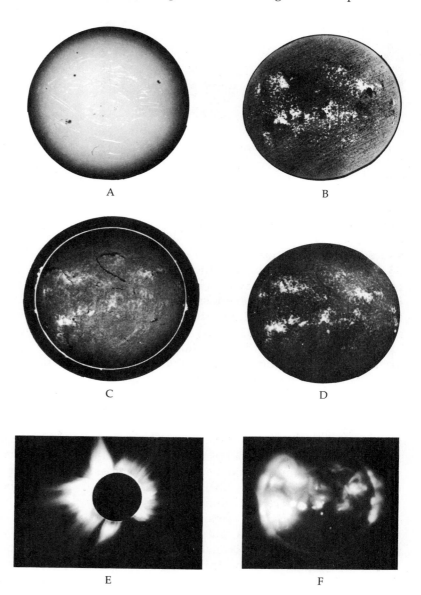

FIGURE 10–22 The Sun at various wavelengths in March 1970.
(A) White-light photograph. *(Culgoora Solar Observatory)* (B) Magnetogram of bipolar magnetic regions. *(W. Livingston, National Optical Astronomy Observatories)* (C) A photograph in H$_\alpha$. *(P. S. McIntosh, National Oceanic and Atmospheric Laboratories)* (D) A Ca II spectrogram. *(National Optical Astronomy Observatories)*
(E) The corona during an eclipse. *(G. Newkirk, High Altitude Observatory)*
(F) X-rays from the corona. *(American Science and Engineering)*

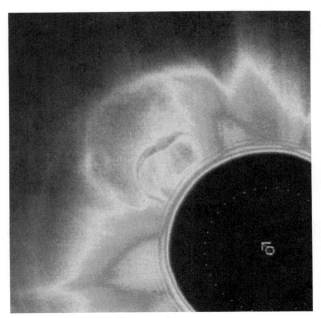

FIGURE 10–23 A coronal mass ejection. The outward speed of the bright outer loop is about 300 km/s. The coronagraph used to take this image blocks out the Sun's photosphere. *(SMM–NCAR)*

and densities than those found in nearby regions of the photosphere, similar to the chromospheric plages to which the faculae are related. Limb darkening renders faculae visible near the solar limb, though they are invisible near the center of the disk.

Above active regions in the photosphere, the bright **plages** float in the chromosphere. Plages are regions in which the density and temperature are higher than in the surrounding chromosphere and are caused by the magnetic fields of the active regions. These features appear on spectroheliograms taken in the light of the H_α line and the spectral lines of Ca II. In many respects, they look like concentrations and intensifications of the chromospheric network (Figure 10–22D).

In the corona, active regions manifest themselves again in the higher densities and temperatures of **coronal streamers** and condensations of the white-light corona. Coronal line emission is stronger over plage regions than elsewhere, and enhanced radio emission arises from increased electron densities. This radio emission characterizes the long-lived active regions.

During active times, large ejections of mass, called **coronal mass ejections,** disrupt the corona (Figure 10–23). These huge bubbles or clouds of gas, containing *billions* of tons of solar material, travel outward so forcefully that they cover millions of kilometers within a few hours. The coronal mass ejections are the most energetic events in the solar system, yet their cause is unknown. They may result from a gradual evolution of coronal magnetic fields into an unstable arrangement.

Prominences and Other Displays

Spectacular markers of active regions are the **prominences,** which appear as long, dark **filaments** when seen projected on the solar disk. Though visible in white light at a total solar eclipse, these displays are best recorded in H_α or Ca II lines.

Prominences are streams of chromospheric gas occupying coronal regions tens of thousands of kilometers above the chromosphere. Two characteristic types of prominences are **quiescent** and **active.** Quiescent prominences last for weeks and look like curtains with gas slowly descending from the corona into the chromosphere; they tend to lie along the neutral line separating the two poles of a BMR. Most active prominences survive only a few hours. Among the most active are the **loop** prominences (Figure 10–24), which are closely associated with solar flares and which survive only an hour or so, during which time gas streams down the magnetic field lines joining the BMR poles. Flares will also upon occasion disrupt quiescent prominences, causing them to erupt and be ejected through the corona at high velocities.

(C) SOLAR FLARES

Among the most puzzling, spectacular, and energetic phenomena associated with active regions are **solar flares.** Although these transient outbursts liberate tremendous quantities of energy, we still do not truly know how they originate. Flares radiate at many frequencies, from the X-rays and gamma rays to long-wavelength radio waves; in addition, they emit high-energy particles called **solar cosmic rays** (protons, electrons, and atomic nuclei; see subhead "Solar Cosmic Rays" below).

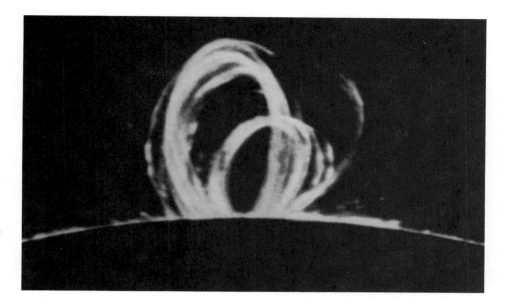

FIGURE 10–24 A loop prominence. Here the magnetic loops in the corona control the flow of the plasma. *(National Optical Astronomy Observatories)*

Flare X-rays and ultraviolet radiation disrupt terrestrial radio communication by disturbing the Earth's ionosphere. The high-energy-particle clouds, which are lethal to unprotected astronauts, reach the Earth in 30 min; clouds of low-energy particles and disturbances in the solar wind require from six to 24 h to transit from Sun to Earth.

On 6 March 1989, the most energetic solar flare in 20 years blasted ultraviolet detectors on Solar Max, which recorded hot gas in a magnetic loop some 60,000 km long. At its height, the temperature in the flare's plasma reached 10 million kelvins (Figure 10–25). A week later, storms in the earth's magnetosphere ignited auroras, disrupted radio communications, and resulted in power surges that blacked out the lights for 6 million people in Quebec, Canada. The total energy expelled by the flare is estimated at 10^{30} J. This violence was the prelude to the current activity cycle, which peaked in early 1990.

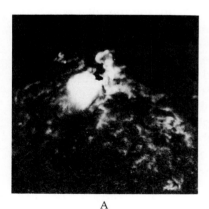
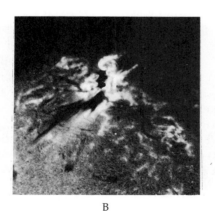
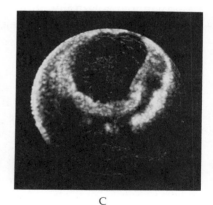

A B C

FIGURE 10–25 The great flare of March 1989. (A) This very strong flare appears to shine brightly at 15:32 UT in H_α within the active region. Dark filamentary prominences are also visible above the photosphere. (B) The ejection of charged particles from the flare, visible at the lower left as a dark, spray-like structure. *(NSO–Sacramento Peak)* (C) The aurorae borealis induced by the flare, as viewed from space. *(L. Frank and J. D. Craven, University of Iowa)*

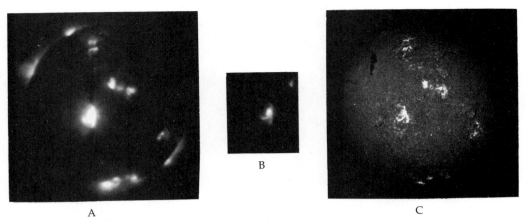

FIGURE 10–26   Active regions in X-rays. (A) Active regions in this X-ray photograph include a solar flare near the center of the disk. (B) A close-up of the flare at shorter X-ray wavelengths. *(American Science and Engineering)* (C) An H_α photograph for comparison; note that the bright active regions show up in X-rays. *(ESSA)*

Optical Manifestations

Flares usually appear in plages as brightenings in H_α (Figure 10–25). In fact, the H_α line becomes an emission feature, reaching maximum brightness within 5 min and decaying in about 20 min (3 h for the largest flares). Flares vary in size from 10,000 km to over 300,000 km; in general, the larger the flare, the more energetic it is and the longer-lived. At the peak of the solar cycle, the average occurrence of small flares is hourly and that of large flares is monthly. They are virtually nonexistent at solar minimum. Even during maximum, attempts to predict major flares are still rough.

X-Ray and Radio Bursts

In some flares, X-rays and centimeter radio waves occur together; they probably originate in the upper chromosphere or corona (Figure 10–26). Both consist of two components: (a) a *slow* component lasting about 30 min and (b) a *sudden* component (or *burst*) lasting a few minutes. The burst phenomenon is partly *nonthermal* (probably synchrotron), with X-rays more energetic than 20 keV and radio emission corresponding to 8×10^7 K, and partly thermal, causing extreme ionization. The less-energetic slow components result from a heating of the corona to 4×10^6 K; enhanced densities also stimulate emission in ultraviolet and forbidden lines. Higher in the corona, a flare produces synchrotron radio bursts at meter wavelengths. These disturbances travel through the corona at speeds up to $0.3c$. It appears that energetic electrons are trapped by coronal magnetic fields having a strength of about 10^{-4} T, leading to the radio emission.

Solar Cosmic Rays

Solar flares accelerate atomic particles leaving the Sun. A blast wave propagates through the solar wind at 1500 km/s, disturbing the solar wind flow at the Earth (Section 10–5). Protons, electrons, and atomic nuclei are accelerated to high energies in flares—we call these **solar cosmic rays.** Most of the particles observed are protons, for the electrons lose much of their energy in exciting radio bursts in the corona and the solar abundance of other nuclei is low. **Alpha particles** (helium nuclei) are the second most abundant nuclei after the protons. Solar-particle energies range from keV (10^3 eV) to about 20 GeV (1 GeV = 10^9 eV), with the bulk of the particles in the MeV (10^6 eV) range. The highest-energy particles arrive at the Earth within 30 min of the H_α flare maximum, followed by the peak number of particles 1 h later, with the low-energy cosmic rays bringing up the rear hours later. About half the flare energy (some 10^{25} J for the largest flares!) is in the H_α emission, half in the shock wave, and only 1% in solar cosmic rays.

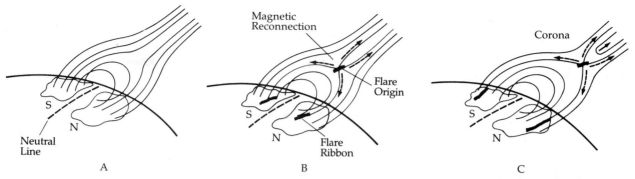

FIGURE 10–27 Model of a flare. (A) Preflare stage of a BMR. Note the unconnected magnetic field lines above the coronal loop. (B) Start of the flare as the magnetic field lines reconnect at a neutral point; a current flows down to the feet of the BMR. (C) Site of the flare moves up into the corona. *(R. Noyes)*

A Flare Model

Here's one model for the development of a flare (Figure 10–27). Magnetic loops through the corona connect the two parts of a bipolar active region; field lines from the outer parts of the region extend out indefinitely. A prominence may define a neutral line in the BMR. A stress on the magnetic field (in the convection zone?) causes an instability. Energy is released instantaneously at the top of the loops as magnetic field lines reconnect—a process similar to that in the Earth's magnetosphere (Section 4–6).

According to Lenz's law, electric currents that oppose this change in the magnetic field are established. Ohmic dissipation of the currents heats the gas. The resulting heating of that region of the corona causes the emission of X-rays, extreme ultraviolet emission, and centimeter radio bursts while electrons and protons are accelerated downward along the loops into the chromosphere, heating it to produce the H_α brightening (two-ribbon flares are quite common, corresponding to the two sides of the loop), more lower-energy X-rays, and transition-zone extreme ultraviolet radiation. At the same time, other particles are accelerated outward, with the electrons producing meter-wavelength radio bursts and the protons and other nuclei becoming solar cosmic rays. Only major flares produce all these manifestations; the most energetic may even emit gamma rays. Most flares, however, are much more modest.

(D) A MODEL OF THE SOLAR CYCLE

Since 1960, we have been struggling with a detailed magnetic model of the solar cycle that satisfies at least some of the observations. One model produces a 22-year solar cycle by coupling the Sun's magnetic fields to its differential rotation within the Solar dynamo (Figure 10–28).

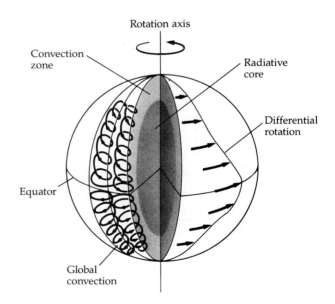

FIGURE 10–28 General model of the solar dynamo. Global convection drives the differential rotation; convection and the differential rotation together drive the magnetic dynamo. *(P. Gilman)*

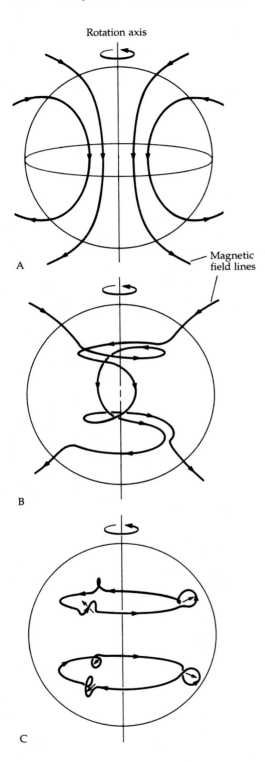

Rotation axis

Magnetic
field lines

A

B

C

FIGURE 10–29 Alternation of solar poloidal and toroidal fields. (A) The initial poloidal field is stretched by convection and differential rotation. (B) The field configuration becomes toroidal. (C) Rising convective cells produce small poloidal components; many such cells and the dominance of rising ones reforms the poloidal field.

Prior to the start of solar activity, the Sun has a weak dipolar magnetic field, with the lines of force running along meridians about $0.1R_\odot$ below the surface. Each line completes itself by emerging near the poles. This configuration is called a **poloidal field** (Figure 10–29A). The Sun's differential rotation draws each line out along the equator, wrapping it around the Sun many times. A **toroidal field** develops (Figure 10–29B). As the density of lines increases, so also does the associated magnetic pressure; inside the **flux tubes** that contain the lines, the gas pressure must decrease (to have pressure equilibrium), but then the tubes are less dense than their surroundings—they experience **magnetic buoyancy**. The field strength further increases as gas motions (convection) twist the lines. The twist of small convective cells eventually reestablishes the poloidal field and the cycle repeats itself (Figure 10–29C). This natural alternation between the poloidal and toroidal fields establishes the basic rhythm of the solar activity cycle.

The details of the coupling of surface magnetic phenomena with the interior dynamo evolution is not yet worked out in detail. The general picture goes like this (Figure 10–30): The amplification of the magnetic field is maximum at higher latitudes (about ±35°), so that a critical field strength is first attained there. At this point, a flux tube rises to the solar surface and appears as a BMR with the strongest fields in the sunspots. Supergranules facilitate the emergence of a flux tube since spots generally originate at the boundaries of supergranules. The opposite poles are a consequence of the continuity of the field lines from north to south: because the following spot is nearer to the Sun's poles both in latitude and along the flux tube, its polarity is opposite that of the nearest pole.

At lower latitudes, the critical field strength is attained later. A sunspot maximum corresponds to the time of greatest field strength over the largest latitude band; eventually, lines become mixed and recombine near the equator, so that fewer spots occur later in the cycle. Supergranulation pushes the flux tubes about, dissipating the magnetic fields. The differential rotation causes the following magnetic regions to move toward the poles while the preceding regions move toward the equator. At the equator, regions of opposite polarity from different hemispheres meet and annihilate. The following regions converge on the poles, where their opposite polarity first neutralizes and then reverses the polar fields. After 11 years, the polarity of the Sun's overall dipolar field is reversed and the stage is set for the second half of the 22-year cycle.

In summary, the analysis of the solar cycle rests on the development of solar magnetic dynamo models. To date, such models have focused on the interplay of poloidal and toroidal field configurations. The shear of differential rotation produces a toroidal field from a poloidal one. In turn, cyclonic motions in the convective zone may regenerate a poloidal field. If so, then all stars with an outer convective zone and differential rotation should have magnetic activity cycles. We now have good observations that such is the case (Section 18–3). Hence, we can conclude that we have pinned down the basic physics even though we do not know all the details.

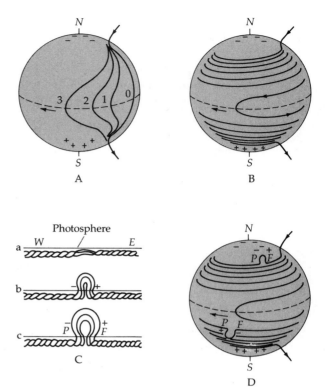

FIGURE 10–30 Model of surface phenomena in the solar cycle. (A) Differential rotation stretches the field lines out along the equator until they are tightly wound (B). Rising convective cells twist the field lines into ropes (C), where kinks develop in which the magnetic pressure increases, causing the kinks to rise to the surface and form bipolar regions (D) with preceding *(P)* and following *(F)* fields.

PROBLEMS

1. Show that the Sun is not at rest in the center of the Solar System by calculating the distance from the Sun's center to the center of mass of the Sun–Jupiter system.

2. Find the thermal Doppler width of a spectral line at 500 nm formed in the Sun's photosphere ($T \approx$ 5400 K).

3. (a) Given that the photosphere is at a temperature of 6000 K, would you expect collisional or radiative excitations to be more important in exciting hydrogen atoms to the second ($n = 2$) level?
   (b) Would you expect the Lyman α line to appear in emission or absorption?

4. Consider the following two lines of similar excitation potential: Fe I at 414.4 nm and Fe II at 417.3 nm. Explain in general terms (with reference to the Boltzmann and/or Saha equation) why the 414.4-nm line is the stronger of the two in the photospheric spectrum and the weaker of the two in the chromospheric spectrum.

5. From our description of chromospheric and coronal radio opacities, describe how you would determine the motion of a solar radio burst through the solar atmosphere.

6. How can one determine the temperature structure of a sunspot using only its continuous spectrum?

7. How would you unambiguously assign a sunspot near the sunspot minimum to the old or the new cycle?

8. Some prominences are said to have speeds greater than the escape speed from the Sun at the chromosphere. What is the critical speed?

9. Using the data on the solar wind given in this chapter, compute the average rate of mass loss of our Sun ($M_\odot$/year) from
   (a) the solar wind
   (b) energy generation

10. Calculate and compare the scale heights for hydrogen in the Sun's photosphere, chromosphere, and corona.

11. Calculate the magnetic pressure exerted by a sunspot and compare it to the kinetic gas pressure in the photosphere.

12. A sunspot's umbra has a temperature of only 4000 K. At an equal optical depth of $\tau_{0.5} = 1$, the umbra is about 2600 K cooler than the photosphere. Calculate the umbral intensity contrast, I_u/I_p, at 550 nm and 1.0 μm. Compare to the observed values of 0.1 at 550 nm and 0.23 at 1.0 μm. Comment?

13. Estimate the angular size of a granule and a supergranule as viewed from the Earth.

14. Use the blackbody radiation law to estimate the wavelength of maximum intensity for the light from the Sun's photosphere and corona.

15. Using the Stefan-Boltzmann law with reasonable values for temperature and radius, estimate the energy output of the Sun's photosphere and corona in watts. Compare your result for the photosphere and the corona to the value of the solar luminosity given in the text. Comment on the validity of using the blackbody approximation for the corona.

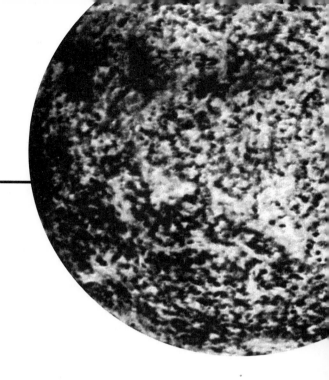

Chapter 11

Stars: Distances and Magnitudes

Having discussed the Sun, we now jump far beyond the Solar System to the stars. This chapter presents some of the methods by which the distances to the stars are determined and quantifies stellar brightness in astronomical terms. The key point is this: by comparing other stars with the Sun, we can infer their physical properties.

11–1 ◗
THE DISTANCES TO STARS

Within the Solar System, we determine absolute distances by using Newtonian celestial mechanics and radar. However, even the nearest stars are so distant, in terms of familiar measures such as the astronomical unit, that other methods must be employed to determine their distances. As you will see, some of these methods are very indirect.

(A) TRIGONOMETRIC (HELIOCENTRIC) PARALLAX

As the Earth orbits the Sun, the nearest stars appear to move relative to the more distant stars. The lack of any observation of this **heliocentric parallax** effect led astronomers such as Tycho Brahe to be skeptical of the Copernican heliocentric model of the Solar System. It was not until 1838, when F. Bessel detected the parallax of the star 61 Cygni and F. Struve detected that of Vega, that Copernicus' model was finally vindicated by direct observation. (Note that Bradley's discovery of stellar aberration [Section 3–3(a)] in 1729 also proved that the Earth is in motion.)

The parallactic displacement of a star on the sky as a result of the Earth's orbital motion permits us to determine the distance from the Sun to the star by the method of **trigonometric parallax** (Figure 11–1). We define the trigonometric parallax of the star as the angle π subtended, as seen from the star, by the Earth's orbit of radius 1 AU. If the star is at rest with respect to the Sun, the parallax is half the maximum apparent annual angular displacement of the star as seen from the Earth. Letting a denote the Sun–Earth distance and d the Sun–star distance, we have

$$\pi \ (\text{rad}) = a/d \qquad (11\text{–}1)$$

223

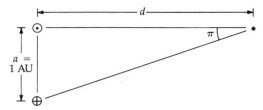

FIGURE 11–1   Geometry for trigonometric parallax. The angle subtended by the Earth's orbit of radius 1 AU is π arcseconds at a star–Sun distance of d (in parsecs).

Recall that there are 2π rad in a circle (360°), so that 1 rad equals 57°17′44.81″ (206,264.81″). Hence, if we agree to measure all angles in arcseconds and all distances in **parsecs** (abbreviation pc; 1 pc = 206,265 AU), Equation 11–1 becomes

$$\pi'' = 1/d \qquad \textbf{(11–2)}$$

Note that 1 pc = 3.086×10^{16} m = 3.26 lightyears, where 1 **lightyear** is the distance light travels in one year.

The measurement and interpretation of stellar parallaxes are a branch of **astrometry,** and the work is exacting and time-consuming. Consider that the nearest star, Alpha Centauri, at a distance of 1.3 pc, has a parallax of only 0.76″ (Table A4–1); all other stars have smaller parallaxes. Today, stellar parallaxes can be determined with a probable error of order ±0.004″, which means that a parallax 0.100″ ± 0.004″ has a 50% probability of actually being between 0.096″ and 0.104″. So present trigonometric parallaxes are believable only to distances of about 100 pc ($\pi'' = 0.01''$), which is miniscule relative to the 8.5-kpc (1 kpc = 10^3 pc) distance to the center of our Galaxy.

Technological advances (including the Hubble Space Telescope) will improve parallax accuracy to 0.001″ within a few years. Fewer than 10,000 stellar parallaxes have been measured to date (and only 500 known well), but there are about 10^{12} stars in our Galaxy! Space observations planned by the European Space Agency with the Hipparcos mission will accurately determine the parallaxes of many more stars. Though a poor orbit will limit its usefulness, Hipparcos will achieve a precision of about 0.002″. The method of trigonometric parallax

is important because it is our *only* direct distance technique for stars.

The ground-based trigonometric parallax of a star is determined by photographing a given star field from a number (about 20) of selected points in the Earth's orbit. The comparison stars selected are distant background stars of nearly the same apparent brightness as the star whose parallax is being measured. Corrections are made for atmospheric refraction and dispersion and for detectable motions of the background stars; any motion of the star relative to the Sun is then extracted. What remains is the smaller annual parallactic motion; it is recognized because it cycles annually. Because a seeing resolution of 0.25″ is considered exceptional (more typical is 1″), it may seem strange that a stellar position can be determined to ±0.01″ in one measurement; this accuracy is possible because we are determining the center of the fuzzy stellar image.

(B) OTHER GEOMETRIC METHODS

To reach distances greater than 100 pc using geometric methods, we can use (1) the Sun's motion through the nearby stars and (2) the motions of star clusters that are not too far away. Both methods depend upon stellar motions; because this topic is extensively discussed in Chapter 15, we only summarize the techniques here.

The motion of the Sun among the nearby stars—the **solar motion**—is 20 km/s (4.1 AU/year) toward the constellation Hercules. This baseline grows year by year, so that over an interval of ten years, we could measure stellar distances to about 2000 pc if the nearby stars were stationary in space. All stars do move, however, just as the Sun does, and so only average parallaxes of groups of stars are possible. By assuming that the peculiar motions (Section 15–1) of the stars in a large sample (preferably the same spectral type; Section 13–2) average to zero, we can deduce the **mean parallax** of that sample. Moreover, because the solar motion affects only that component of a star's proper motion parallel to the solar motion, we may use the other (perpendicular) component to find **statistical parallaxes** (Section 15–2). We are forced to sacrifice accuracy in order to attain greater distances, and the distances obtained relate to a *group* of stars.

The **moving cluster** method leads to precise individual stellar distances greater than 100 pc; unfortunately, there are very few clusters (the Hyades cluster in the constellation Taurus is the best studied) where the method is really applicable. A star cluster consists of many stars moving as a group through space because the stars are bound together gravitationally. If the cluster subtends an appreciable angle on the sky, the individual stellar proper motions appear to converge to (or diverge from) a single point on the sky. By measuring the average radial velocity of the cluster and using trigonometric calculations, we can determine the distance to each star in the cluster (Section 15–3).

(C) LUMINOSITY DISTANCES

Finally, we preview distance determinations based upon stellar luminosity. We must rely upon these rather indirect methods to truly probe our Galaxy (and other galaxies). Each method yields only *relative* distances until we calibrate the luminosities of the stars involved. The calibration is accomplished when we find a nearby representative star (or cluster) for which the distance and stellar luminosity can be determined by means of trigonometric parallax or the moving-cluster method. Because radiant flux decreases as the square of the distance from the source [Section 8–1(c)], the luminosity follows from the observed flux once the distance is known. Conversely, if we can estimate the luminosity and measure the flux, we can infer the distance from the inverse-square law of light. You will see in detail how these methods work in the next two chapters. Measurement of stellar distances underlies all of astronomy and astrophysics. Now to discuss stellar brightnesses quantitatively and link them to stellar distances.

11–2 ◑
THE STELLAR MAGNITUDE SCALE

The first stellar brightness scale—the **magnitude scale**—was defined by Hipparchus and refined by Ptolemy almost 2000 years ago. In this qualitative scheme, naked-eye stars fall into six categories: the *brightest* (Table A4–2) are of *first* magnitude, and the *faintest* of *sixth* magnitude. *Note that the brighter the star, the smaller the value of the magnitude.* In 1856, N. R. Pogson verified William Herschel's finding that a first-magnitude star is 100 times brighter than a sixth-magnitude star and the scale was quantified. Because an interval of five magnitudes corresponds to a factor of 100 in brightness, a one-magnitude difference corresponds to a factor of $100^{1/5} = 2.512$. (This definition reflects the operation of human vision, which converts equal *ratios* of actual intensity to equal *intervals* of perceived intensity. In other words, the eye is a logarithmic detector.) The magnitude scale has been extended to positive magnitudes larger than +6.0 to include faint stars (the 5-m telescope on Mount Palomar can reach to magnitude +23.5) and to negative magnitudes for very bright objects (the star Sirius is magnitude −1.4). The limiting magnitude of the Hubble Space Telescope is expected to be about +25.

Astronomers find it convenient to work with logarithms to base 10 (Appendix 9, "Mathematical Operations") rather than with exponents in making the conversions from brightness ratios to magnitudes and vice versa. Consider two stars of magnitude m and n with respective apparent brightnesses (fluxes) l_m and l_n. The ratio of their brightnesses l_n/l_m corresponds to the magnitude difference $m - n$. Because a one-magnitude difference means a brightness ratio of $100^{1/5}$, $m - n$ magnitudes refer to a ratio of $(100^{1/5})^{m-n} = 100^{(m-n)/5}$, or

$$l_n/l_m = 100^{(m-n)/5} \qquad (11\text{–}3)$$

Taking the logarithm to base 10 of both sides of Equation 11–3 yields, because $\log x^a = a \log x$ and $\log 10^a = a \log 10 = a$,

$$\log (l_n/l_m)$$
$$= [(m - n)/5] \log 100 = 0.4(m - n) \qquad (11\text{–}4)$$

or

$$m - n = 2.5 \log (l_n/l_m) \qquad (11\text{–}5)$$

Equation 11–5 defines **apparent magnitude;** note that $m > n$ when $l_n > l_m$, that is, brighter objects have numerically smaller magnitudes. Also note that when the brightnesses are those observed at the Earth, physically they are fluxes. Apparent magnitude is the astronomically peculiar way of talking about fluxes.

To help you use this key equation (11–5), here are a few worked examples:

(a) The apparent magnitude of the variable star RR Lyrae ranges from 7.1 to 7.8—a magnitude amplitude of 0.7. To find the relative increase in brightness from minimum to maximum, we use

$$\log (l_{max}/l_{min}) = 0.4 \times 0.7 = 0.28$$

so that

$$l_{max}/l_{min} = 10^{0.28} = 1.91$$

This star is almost twice as bright at maximum light than at minimum.

(b) A binary system consists of two stars *a* and *b*, with a brightness ratio of 2; however, we see them unresolved as a point of magnitude +5.0. We would like to find the magnitude of *each* star. The magnitude *difference* is

$$m_b - m_a = 2.5 \log (l_a/l_b) = 2.5 \log 2 = 0.75$$

Since we are dealing with brightness ratios, it is not right to put $m_a + m_b = +5.0$. The sum of the luminosities $(l_a + l_b)$ corresponds to a fifth-magnitude star. Compare this to a 100-fold brighter star, of magnitude 0.0 and luminosity l_0:

$$m_{a+b} - m_0 = 2.5 \log [l_0/(l_a + l_b)]$$

or

$$5.0 - 0.0 = 2.5 \log 100 = 5$$

but $l_a = 2l_b$, so that $l_b = (l_a + l_b)/3$; therefore

$$m_b - m_0 = 2.5 \log (l_0/l_b)$$
$$= 2.5 \log 300 = 2.5 \times 2.477 = 6.19$$

The magnitude of the fainter star is 6.19, and from our earlier result on the magnitude difference, that of the brighter star is 5.44.

11–3 ◗
ABSOLUTE MAGNITUDE AND DISTANCE MODULUS

So far we have dealt with stars as we see them, that is, their fluxes or apparent magnitudes, but we frequently want to know the luminosity of a star. A very luminous star will appear dim if it is far enough away, and a low-luminosity star may look bright if it is close enough. Our Sun is a case in point: if it were at the distance of the closest star (Alpha Centauri), the Sun would appear slightly fainter to us than Alpha Centauri does. Hence, *distance* links fluxes and luminosities.

The luminosity of a star relates to its **absolute magnitude,** which is the magnitude that would be observed if the star were placed at a distance of 10 pc from the Sun. (Note that absolute magnitude is the way of talking about luminosity peculiar to astronomy.) By convention, absolute magnitude is capitalized (M) and apparent magnitude is written lowercase (m). The inverse-square law of radiative flux links the flux l of a star at a distance d to the luminosity L it would have it if were at a distance $D = 10$ pc:

$$L/l = (d/D)^2 = (d/10)^2$$

If M corresponds to L and m corresponds to l, then Equation 11–5 becomes

$$m - M = 2.5 \log (L/l)$$
$$= 2.5 \log (d/10)^2 = 5 \log (d/10)$$

Expanding this expression, we have the useful alternative forms

$$m - M = 5 \log d - 5 \tag{11–6}$$
$$M = m + 5 - 5 \log d \tag{11–7}$$
$$M = m + 5 + 5 \log \pi'' \tag{11–8}$$

Here d is in parsecs and π'' is the parallax angle in arcseconds. The quantity $m - M$ is called the **distance modulus,** for it is directly related to the star's distance in Equation 11–6. In many applications, we refer only to the distance moduli of different objects rather than converting back to distances in parsecs or lightyears.

11–4 ◗
MAGNITUDES AT DIFFERENT WAVELENGTHS

The kind of magnitude that we measure depends on how the light is filtered anywhere along the path to the detector and on the response function of the detector itself. So that problem comes down to how to define standard magnitude systems.

(A) MAGNITUDE SYSTEMS

Detectors of electromagnetic radiation (such as the photographic plate, the photoelectric photometer, and the human eye) are sensitive only over given wavelength bands. So a given measurement samples but part of the radiation arriving from a star. Because the flux of starlight varies with wavelength, the magnitude of a star depends upon the wavelength at which we observe. Originally, photographic plates were sensitive only to blue light, and the term **photographic magnitude** (m_{pg}) still refers to magnitudes centered around 420 nm (in the blue region of the spectrum). Similarly, because the human eye is most sensitive to green and yellow, **visual magnitude** (m_v) or the photographic equivalent **photovisual magnitude** (m_{pv}) pertains to the wavelength region around 540 nm.

Today we can measure magnitudes in the infrared, as well as in the ultraviolet, by using filters in conjunction with the wide spectral sensitivity of photoelectric photometers. So systems of many different magnitudes (color combinations) are possible. In general, a photometric system requires a detector, filters, and a calibration (in energy units). The properties of the filters are typified by their effective wavelength, λ_0, and bandpass, $\Delta\lambda$, which is defined as the full width at half maximum in the transmission profile (Figure 11–2). The three main filter types are wide ($\Delta\lambda \approx 100$ nm), intermediate ($\Delta\lambda \approx 10$ nm), and narrow ($\Delta\lambda \approx 1$ nm). There is a tradeoff for the bandwidth choice: a smaller $\Delta\lambda$ provides more spectral information but admits less flux into the detector, resulting in longer integration times. For a given range of the spectrum, the design of the filters makes the greatest difference in photometric magnitude systems.

Note that the characteristics of a filter are a function of the spectrum of the measured flux. For instance, if a blackbody spectrum is observed, then

$$\lambda_0 = \int_0^\infty \frac{\lambda \; \phi_f(\lambda) \; B(\lambda) \; d\lambda}{\phi_f(\lambda) \; B(\lambda) \; d\lambda}$$

where $\phi_f(\lambda)$ is the transmission function of the fil-

TABLE 11–1 *Some Photometric Bands*

Band	Effective Wavelength (λ_0)	Bandpass ($\Delta\lambda$)
U	350 nm	100 nm
B	430 nm	100 nm
V	550 nm	100 nm
R	640 nm	150 nm
I	790 nm	150 nm
J	1.25 μm	0.12 μm
H	1.66 μm	0.16 μm
K	2.22 μm	0.22 μm
L	3.45 μm	0.35 μm
M	4.65 μm	0.46 μm
N	10.3 μm	1.0 μm

ter and $B(\lambda)$ the Planck blackbody distribution. The relation also implies that a perfectly sharp filter at a single wavelength will measure a monochromatic flux of the source.

A commonly used wide-band magnitude system (Table 11–1) is the **UBV system**: a combination of ultraviolet (U), blue (B), and visual (V) magnitudes, developed by H. L. Johnson. These three bands are centered at 365, 440, and 550 nm; each wavelength band is roughly 100 nm wide (Figure 11–3). In this system, apparent magnitudes are denoted by B or V and the corresponding absolute magnitudes are subscripted: M_B or M_V.

To be useful in measuring fluxes, the photometric system must be calibrated in energy units for

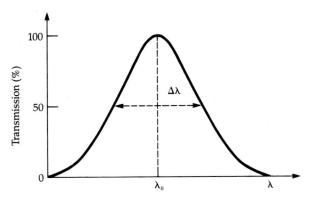

FIGURE 11–2 General properties of a transmission filter, with its effective or central wavelength, λ_0, and full width at half maximum, $\Delta\lambda$.

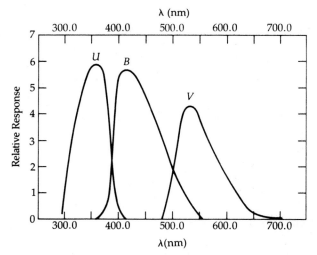

FIGURE 11–3 *UBV* filters. The transmission profile of the standard filters in the *UBV* system for a source with equal flux at all wavelengths.

each of its bandpasses. This calibration turns out to be the hardest part of the job. In general, it relies first on a set of standard stars that define the magnitudes for a particular filter set and detector; that is, these stars define the *standard magnitudes* for the photometric system to the precision with which they can be measured. For instance, the *UBV* system has about 100 standard stars measured to about ±0.01 magnitude. Then if we can calibrate the flux of just one of these stars, we have calibrated the system. The calibration is usually given for zero magnitude at each filter; all fluxes are then derived from this base level. The star usually chosen as the calibration star is Vega.

Finally, we note that the *UBV* system has been extended into the red and infrared (in part because of the development of new detectors, such as CCDs, sensitive to this region of the spectrum). The extensions are not as well standardized as that for the Johnson *UBV* system, but they tend to include *R* and *I* in the far red and *J, H, K, L,* and *M* in the infrared (Table 11–1). Other photometric systems exist for different purposes; most observations, at least of stars, have been done to date in the *UBV* system.

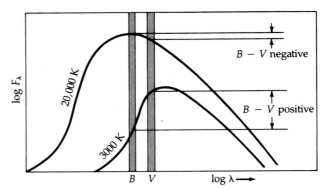

FIGURE 11–4 Color index in the *BV* system. Blackbody curves for 20,000 K and 3000 K, along with their intensities at *B* and *V* wavelengths. Note that *B − V* is negative for the hotter star, positive for the cooler one.

(B) COLOR INDEX

A quantitative measure of the color of a star is given by its **color index** (**CI**), which is defined as the difference between magnitudes at two different effective wavelengths. For example,

$$CI = m_{pg} - m_v = M_{pg} - M_v \qquad \textbf{(11–9a)}$$

or more generally

$$CI = m(\lambda_1) - m(\lambda_2) \qquad \textbf{(11–9b)}$$

where the equality in Equation 11–9a follows from Equation 11–7 because we are talking about one star. Note that the long-wavelength magnitude is subtracted from the short-wavelength magnitude and that a star's color index does not depend upon distance. Similarly, the quantities *B − V* and *U − B* are also color indices. Also, note that because the color index is a magnitude difference, it is equivalent to a flux *ratio* at the specific wavelengths involved. That is

$$CI = constant - 2.5 \log [F(\lambda_1)/F(\lambda_2)]$$

where the constant is the calibration constant for the system and $F(\lambda_1)$ and $F(\lambda_2)$ are the fluxes observed at the two different wavelengths. Again, if we are observing a blackbody emitter, then

$$CI = constant - 2.5 \log [B(\lambda_1)/B(\lambda_2)]$$

Magnitude systems and their color indices are observer-dependent because the percentage of the incident stellar energy measured depends upon both the wavelength (Figure 11–4) and the particular instruments used. So that all observations may be compared on the same basis, the systems are set so that the color index of stars (such as Vega) with surface temperatures about 10,000 K are equal to zero. By this convention, hotter stars have a *negative* color index, and cooler ones have a *positive* index.

Because stars have different temperatures, their spectral energy (Planck) curves peak at different wavelengths (Section 8–6 and Figure 11–4); therefore, hot stars are bluish and cool stars are reddish. Using the color index *B − V*, we see that a bluish star (20,000 K) has a negative color index because it is brighter in the blue (smaller *B* magnitude) than at longer wavelengths (larger *V* magnitude). A reddish star (3000 K) has a positive color index because it is brighter in *V* than in *B*. The *R − I* color index is useful for finding the temperatures of cooler stars, because more of their flux is emitted in the red and infrared.

If stars radiated as perfect blackbodies (they don't!), the relationship between their surface temperatures and $B - V$ color index would be simple and well-defined:

$$B - V = -0.71 + 7090/T \qquad \textbf{(11–10a)}$$

where the constants arise from the calibration of the system so that a star with a $T = 10{,}000$ K has a $B - V = 0.0$. We then invert this equation so that T can be found from $B - V$:

$$T = 7090/[(B - V) + 0.71] \qquad \textbf{(11–10b)}$$

As you will see in Chapter 13, the spectra of stars deviate from that of a blackbody, so that this relationship is only a rough guide. For stars like the sun, the empirical relationship for 4000 K $< T <$ 10,000 K is

$$B - V = -0.865 + 8540/T \qquad \textbf{(11–11a)}$$

and

$$T = 8540/[(B - V) + 0.865] \qquad \textbf{(11–11b)}$$

We have assumed that the observed color indices are intrinsic to the stars. However, interstellar space is pervaded by dust grains that absorb and scatter starlight, and so the observed light appears redder than when it was emitted [Section 19–1(b)]. This **interstellar reddening** is wavelength-dependent, and it therefore affects color indices. We term the difference between the observed and the intrinsic color indices the **color excess.** We mention this point here to warn you that a star's temperature is *not* uniquely determined by a *single* color index measurement between two wavelength bands—three-color (or more) photometry is necessary to separate out the effects of interstellar reddening.

It is fairly straightforward to correct for **atmospheric extinction,** that is, scattering and absorption in the Earth's atmosphere. We again use the concepts of the transfer equation of Section 8–7. Current practice is to express magnitudes after such correction, as though they were observed above the Earth's atmosphere. Let $F_0(\lambda)$ be the star's flux outside the atmosphere and the optical depth of the atmosphere τ_λ. Then the flux at the bottom of the atmosphere, $F(\lambda)$, is

$$F(\lambda) = F_0(\lambda) \exp(-\tau_\lambda) \qquad \textbf{(11–12)}$$

Assume that the Earth's atmosphere is plane-

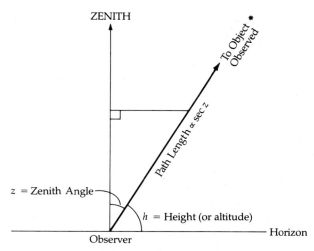

FIGURE 11–5 Geometry for atmospheric extinction based on the assumption of a plane-parallel atmosphere.

parallel (that is, ignore the Earth's curvature). Define h as the angular height of a star above the horizon and z as the angle between the direction to the zenith and the direction to the star ($z = 90° - h$); z is the **zenith angle.** The path length traversed by light through the atmosphere is proportional to sec z (Figure 11–5). So at the zenith, a star suffers minimal extinction, which increases rapidly close to ($\approx 30°$) the horizon. So the optical depth at the zenith is at a minimum; call it $\tau_0(\lambda)$. Then at other angles,

$$\tau(\lambda) = \tau_0(\lambda) \sec z \qquad \textbf{(11–13)}$$

So if we know $\tau_0(\lambda)$, we can calculate $F_0(\lambda)$ for the star, since

$$F(\lambda)/F_0(\lambda) = \exp(-\tau_\lambda) \qquad \textbf{(11–14)}$$

If possible, you want to avoid making measurements at such low elevations, where the plane-parallel approximation no longer holds true. A good rule is to keep $z < 60°$.

On a given night, selected standard stars and unknown sources suffer the same atmospheric extinction. (You hope that the extinction does *not* vary over short time periods.) To find the **extinction coefficient** for the night, you measure the flux of a standard star over a large range of zenith angles (from close to the horizon up to the zenith) and plot measured values against sec z. If things have been done correctly, you get a straight line that tells you the amount of extinction (in magnitudes) per unit sec z (which has the name of **air mass**), that is, magnitudes per air mass. You then correct your other measurements for that night with this extinction coefficient, which depends on the wavelength of observation.

Let's see how this is done to a first approximation. By the definition of magnitude:

$$m(\lambda) - m_0(\lambda) = -2.5 \log [F(\lambda)/F_0(\lambda)]$$

where $m_0(\lambda)$ is the apparent magnitude of the star above the atmosphere. Then by substituting Equation 11–14, we have

$$\begin{aligned} m(\lambda) - m_0(\lambda) &= -2.5 \log [\exp (-\tau_\lambda)] \\ &= 2.5 (\log e) \, \tau_\lambda \\ &= 1.086 \, \tau_\lambda \end{aligned}$$

and

$$m_0(\lambda) = m(\lambda) - 1.086 \, \tau_\lambda$$

We now substitute with Equation 11–13:

$$\begin{aligned} m_0(\lambda) &= m(\lambda) - 1.086 \, \tau_0(\lambda) \sec z \\ &= m(\lambda) - k_0(\lambda) \sec z \end{aligned}$$

Astronomers call $k_0(\lambda)$ the *first order extinction coefficient*; a typical range of values at V band is 0.15 to 0.2. The value varies some from night to night and depends on the local atmospheric conditions.

(C) BOLOMETRIC MAGNITUDES AND STELLAR LUMINOSITIES

The complete spectral energy distribution of a star is only sampled by the color indices we have been discussing. Of greater significance for the overall structure of the star, however, is its total rate of energy output (in watts) at all wavelengths. If we place ourselves outside the Earth's atmosphere, the radiative flux from the star per unit wavelength l_λ (W/m^2 · Å) permits us to define the total **bolometric flux** (W/m^2):

$$l_{bol} = \int_0^\infty l(\lambda) \, d\lambda \qquad \textbf{(11–15)}$$

The **apparent bolometric magnitude** of the star m_{bol} follows from Equation 11–5 as

$$m_{bol} = -2.5 \log l_{bol} + \text{constant} \qquad \textbf{(11–16)}$$

where the constant is an arbitrary zero point. The **absolute bolometric magnitude** of the star M_{bol} is the bolometric magnitude if the star were at the standard distance of 10 pc. In the notation of Equation 11–15, the **visual flux** l_v, for example, is

$$l_v = \int_0^\infty l(\lambda) \, S(\lambda) \, d\lambda \qquad \textbf{(11–17)}$$

where $S(\lambda)$ expresses the spectral sensitivity of the visual photometric system (the V bandpass of Figure 11–3). In analogy to Equation 11–16, we can then define the **visual magnitude** as

$$m_v = -2.5 \log l_v + \text{constant} \qquad \textbf{(11–18)}$$

Our Sun is the only star for which l_λ has been accurately observed. Indeed, l_{bol} is related to the **solar constant**: the total solar radiative flux received at the Earth's orbit outside our atmosphere (1370 W/m^2). The **solar luminosity** $L_\odot$ (3.86 × 10^{26} W) is calculated from the solar constant in the following manner. Using the inverse-square law, we find the radiative flux at the Sun's surface $R_\odot$. Then $L_\odot$ is just $4\pi R_\odot^2$ times this flux. The solar energy distribution curve may be approximated by a Planck blackbody curve (Section 8–6) at the **effective temperature** T_{eff}, defined as the temperature of a blackbody that would emit the same total energy as an emitting body, such as the Sun or a star. Then the Stefan-Boltzmann law (Equation 8–40) implies

$$L_\odot = 4\pi R_\odot^2 \sigma T_{eff}^4 \text{ J/s} \qquad \textbf{(11–19)}$$

where σ is the Stefan-Boltzmann constant.

If we know the absolute bolometric magnitude of a star, we can use Equation 11–5 to find that star's luminosity:

$$M_{bol}(\odot) - M_{bol}(*) = 2.5 \log (L_*/L_\odot) \qquad \textbf{(11–20a)}$$

With $M_{bol}(\odot) = +4.72$, this becomes

$$\log (L_*/L_\odot) = 1.89 - 0.4 M_{bol}(*) \qquad \textbf{(11–20b)}$$

$M_{bol}(*)$ is *not* directly observed (although this is now becoming possible with space probes and satellites), but L_* may be deduced by studying the star's spectrum (Chapter 13); then the absolute bolometric magnitude follows from Equation 11–20a.

In practice, we use the **bolometric correction** BC, which is the difference between the bolometric and visual magnitudes, to determine a star's bolometric magnitude. For example,

$$BC = m_{bol} - m_v = M_{bol} - M_v$$
$$BC = 2.5 \log (l_v/l_{bol}) \qquad \textbf{(11–21)}$$

Bolometric corrections are inferred from ground-based observations by using theoretical stellar models; these corrections have been checked and improved with the ultraviolet data from orbiting satellites. In the UBV magnitude system, the bolometric correction is a minimum for stars with $T_{eff} = 6500$ K; $BC = -0.08$ for our Sun. (Table A4–3 gives the bolometric corrections for stars of different temperatures.) For stars with surface temperatures of 6700 K, the spectral energy peaks in the *V* wavelength band, so that the greatest percentage of the star's energy is detected. For other stellar temperatures, a smaller percentage of the total radiative energy is measured in the *V* band; hence, their bolometric corrections are larger (in absolute value) than that for 6700-K stars.

Finally, a word of caution. The light we receive and measure from stars has been filtered many times before we record: by the matter in interstellar space, by the Earth's atmosphere, by the optics of our telescope, and by the detector and filters used. The goal of the observer is to correct for this filtration so that the light has the same characteristics as that just emitted by the star at its photosphere.

PROBLEMS

1. Astronomers living on Mars would define their astronomical unit in terms of the orbit of Mars. If they defined parsec in the same manner as we do, how many Martian astronomical units would such a parsec contain? How many Earth astronomical units would equal a Martian parsec? How many Earth parsecs are there in a Martian parsec?

2. A variable star changes in brightness by a factor of 4. What is the change in magnitude?

3. What is the combined apparent magnitude of a binary system consisting of two stars of apparent magnitudes 3.0 and 4.0?

4. If a star has an apparent magnitude of -0.4 and a parallax of 0.3″, what is
   (a) the distance modulus?
   (b) the absolute magnitude?

5. What is the distance (in parsecs) of a star whose absolute magnitude is $+6.0$ and whose apparent magnitude is $+16.0$?

6. What are the absolute magnitudes of the following stars:
   (a) $m = 5.0$, distance $d = 100$ pc
   (b) $m = 10.0$, $d = 1$ pc (is there such a star?)
   (c) $m = 6.5$, $d = 250$ pc
   (d) $m = -3.0$, $d = 5$ pc
   (e) $m = -1.0$, $d = 500$ pc
   (f) $m = 6.5$, parallax $\pi = 0.004''$

7. What would the expression for absolute magnitude be, in terms of apparent magnitude and distance, if absolute magnitude were defined as the magnitude a star would have at 100 pc?

8. The Sun has an apparent visual magnitude of -26.74.
   (a) Calculate its absolute visual magnitude.
   (b) Calculate its magnitude at the distance of Alpha Centauri (1.3 pc).
   (c) The Palomar Sky Survey is complete to magnitudes as faint as $+19$. How far away (in parsecs) would a star identical to the Sun have to be in order to just barely be bright enough to be visible on Sky Survey photographs?

9. Using the data from this chapter, determine the apparent magnitude difference between Sirius and the Sun, as seen from the Earth. How much more luminous is Sirius than the Sun?

10. A certain globular cluster has a total of 10^4 stars; 100 of them have $M_v = 0.0$, and the rest have $M_v = +5.0$. What is the integrated visual magnitude of the cluster?

11. The *V* magnitudes of two stars are both observed to be 7.5, but their blue magnitudes are $B_1 = 7.2$ and $B_2 = 8.7$.
    (a) What is the color index of each star?
    (b) Which star is the bluer and by what factor is it brighter at blue wavelengths than the other star?

12. What is the color index of a star at a distance of 150 pc with $m_v = 7.55$ and $M_B = 2.00$?

13. What is the absolute bolometric magnitude of a star with a luminosity of 10^{33} W?

14. Given the expressions for the luminosity of a star (Equation 11–19) and its bolometric magnitude in terms of that of the Sun (Equation 11–20a), find an

expression for the bolometric magnitude of the star as a function of its temperature and radius. The effective temperature of the Sun is 5780 K.

15. The bolometric correction for a star is -0.4, and its apparent visual magnitude is $+3.5$. Find its apparent bolometric magnitude.

16. Two astronomers, located 100 km apart along a north–south line, simultaneously observe an asteroid near the zenith. Comparison of their observations indicates that the star had a parallax of 5 arcseconds. Estimate the distance to the asteroid in kilometers. How many times more distant is it than the Moon?

17. Observational astronomers often use a rule of thumb that a 1% change of brightness roughly corresponds to a change of 0.01 magnitude. Justify this approximation and comment on the validity of this approximation.

18. What is the largest distance that a star of absolute magnitude -6 could be detected by the Palomar 5-m telescope? By the Hubble Space Telescope? [Use the limiting magnitudes given in Section 11–2.]

19. A variable star is observed to change its $B - V$ color index from 0.5 to 0.7.
    (a) Assuming the star radiated as a blackbody, what would be the temperatures corresponding to the two color indices?
    (b) Assuming the star is like the Sun, what would be the temperatures corresponding to the two color indices?

20. (a) Assuming an extinction coefficient of 0.2, how much fainter (in magnitudes) would a star appear when at an altitude (angular height above the horizon) of 30° compared to when it is at an altitude of 90°?
    (b) At what zenith angle would the star appear 1 magnitude fainter than when at the zenith?

Chapter 12

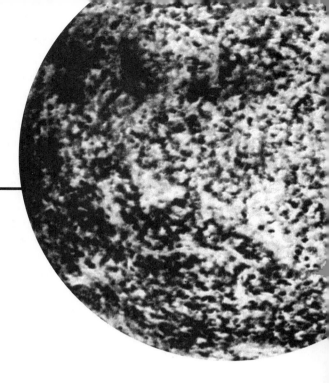

Stars: Binary Systems

Most stars clearly visible in optical telescopes turn out to be binary or multiple star systems. The stars in a multiple system are physically related; they orbit one another under the influence of their mutual gravitational attraction. That Newton's law of universal gravitation is applicable beyond our Solar System became evident when William Herschel detected the orbital motion of the Castor binary system in 1804. As you will see, known physical laws can be coupled with suitable observations of binary systems to tell us about many important stellar characteristics, especially masses and radii. In addition, they can be used to infer densities, surface temperatures and luminosities, and rotation rates.

Fortunately, most stars in the neighborhood of our Sun (well over 50%, perhaps close to 80%) belong to multiple systems, which allows us to infer the physical properties of many of them. This fact is especially important for the estimate of stellar masses, which can be done directly only in binary systems for ordinary stars. The prevalence of multiple star systems also underlines the importance of angular momentum in the process of star formation.

12–1 ◑
CLASSIFICATION OF BINARY SYSTEMS

For physical and observational reasons, astronomers classify stellar binary systems into several different types:

Apparent binary Two stars that are not physically associated but appear close together on the sky because they lie along the same line of sight. Their uncorrelated space motions soon reveal that they are not members of a physical binary system. (Sometimes these are called **optical** binaries.)

Visual binary A bound system that can be resolved into two stars at the telescope. The mutual orbital motions of these stars are observed to have periods ranging from about one year to thousands of years.

Astrometric binary Only one star is seen telescopically, but its oscillatory motion in the sky reveals that it is accompanied by an unseen companion. Both bodies are orbiting about their mutual center of mass.

Spectroscopic binary   An unresolved system whose duplicity is revealed by periodic oscillations of the lines in its spectrum. In some cases, two sets of spectral features are seen (one for each star) oscillating with opposite phases; in other cases, one of the stars is too dim to be seen, so that only one set of oscillating spectral lines is recorded. Typical orbital periods here range from hours to a few months.

Spectrum binary   An unresolved system in which spectral features do not reveal orbital motion but two clearly different spectra are superimposed. We infer that two members of a binary system are producing the observed composite spectrum.

Eclipsing binary   A binary system whose two stars periodically eclipse one another, leading to periodic changes in the apparent brightness of the system. Such systems may also be visual, astrometric, or spectroscopic binaries.

Finally, we point out that the Sun as a single star is an oddity. Among solar-type stars, the observed ratio of single:double:triple:quadruple systems is 45:46:8:1. For binary systems, orbital separations range uniformly from 3×10^9 to 3×10^{15} m (orbital periods from one day to 3×10^6 years). Roughly 10% of all stars are binaries with orbital periods of from one to ten days, another 10% with periods of from ten to 100 days, and so on.

In what follows, keep these ideas in mind. The **apparent orbit** is that traced out on the sky; most likely, it is tilted with respect to the line of sight. The **true orbit** is corrected for this tilt. When one of the stars is considered fixed while the other moves around it, the orbit is then the **relative orbit**. The **absolute orbit** is that traced out by both stars around the center of mass of the system.

12–2 ◑
VISUAL BINARIES

As a result of the Earth's turbulent atmosphere, the seeing image of a star is seldom less than 1″ in diameter. The two stars of a binary system are resolved telescopically as a **visual binary** if their centers are separated by more than 1″. The members of a visual binary must be well separated in angle at some point in their orbital motion; otherwise, the duplicity will not be resolved. In many instances, the length and orientation of an unre-

solved image provide sufficient information to identify the system as a visual binary, even if the separation is less than 1″. In any case, the observed orbital periods are necessarily long (years to hun-

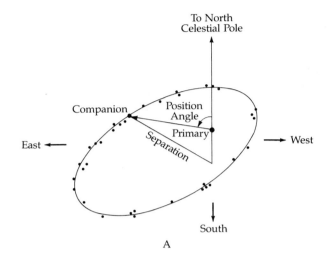

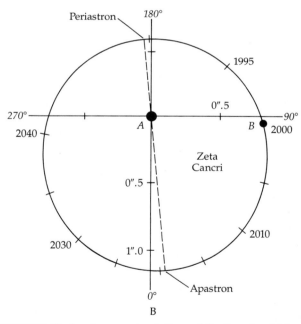

FIGURE 12–1   Apparent relative orbit of a visual binary. (A) The orbit is apparent because it is projected on the sky. It is relative because one star (the primary) is assumed to be the center of the motion of the other (the companion). (B) Actual apparent orbit of Zeta Cancri B around Zeta Cancri A; the orbital period is 60 years. Note that the orbit is almost face-on to the line of sight.

dreds or thousands of years), as expected from Kepler's third law.

(A) THE DETERMINATION OF STELLAR MASSES

A single observation of a visual binary (Figure 12–1) is specified by giving the apparent angular **separation** (in arcseconds on the sky) and the **position angle** (angle measured eastward from north, in degrees) of the fainter star (the **companion**) relative to the brighter star (the **primary**). As time passes, these points trace out the **apparent relative orbit** of the binary system on the celestial sphere.

Two gravitating bodies orbit one another, as well as their center of mass, in accordance with Kepler's laws. So the orbit is an ellipse, and the orbital motion satisfies the law of equal areas and the third law. We do not generally see the true orbit, however, for the orbital plane of a binary system may be inclined at any angle to the plane of the sky. (The **inclination** is 0° when the two planes coincide and 90° when the orbit is seen edge-on.) Fortunately, the law of equal areas holds (but with a different constant of proportionality) for the apparent (projected on the sky) orbit, and the elliptical true orbit always projects into an elliptical apparent orbit. The foci of the apparent orbit do not correspond to the true foci (in particular, the primary does not lie at one focus of the apparent ellipse). By measuring the displacement of the primary from the apparent focus, we can determine the inclination of the orbit to the local tangent plane on the celestial sphere; the true eccentricity and the true semimajor axis a″ (in arcseconds) can then be determined as well.

Having determined the true orbit of the visual binary, we may now apply Kepler's third law to deduce the masses of the member stars. The general form of the third law is

$$(M_1 + M_2)P^2 = a^3 \qquad \textbf{(12–1)}$$

where mass M is measured in solar masses ($M_\odot$), the orbital period P is measured in years, and the true orbital semimajor axis a is measured in astronomical units. Although we may observe P directly, a follows from $a″$ only when we know the distance (or parallax $\pi″$) to the visual binary. Geometrically, we have

$$a = a″/\pi″ \qquad \textbf{(12–2)}$$

so that Equation 12–1 may be written in terms of

observables as

$$(M_1 + M_2)P^2 = (a″/\pi″)^3 \qquad \textbf{(12–3)}$$

An accurate value for the sum of the stellar masses follows from Equation 12–3. To determine the individual masses, we must find the relative distance of each star from the center of mass of the system because

$$M_1 a_1 = M_2 a_2 \quad \text{(where } a_1 + a_2 = a) \qquad \textbf{(12–4)}$$

On the sky, the center of mass travels in a straight path with respect to the background stars, and the binary components weave periodically about this path (Figure 12–2). By eliminating the center-of-

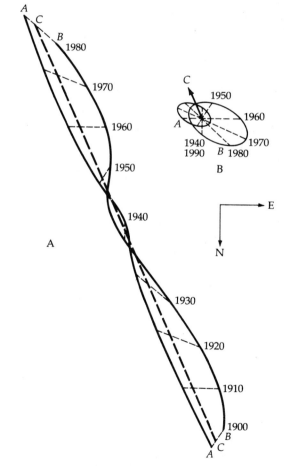

FIGURE 12–2 Motions of Sirius A and B. (A) The apparent motions relative to background stars of Sirius *(A)*, its companion *(B)*, and the center of mass of the system *(C)*. (B) Orbital motions of Sirius A and B relative to the system's center of mass.

mass motion and correcting for the orbital inclination, we obtain a_1'' and a_2'' and therefore a_1''/a_2'', which equals a_1/a_2.

A worked example. Imagine that the two stars of a visual binary are observed to have a maximum separation of 3″.0 and a trigonometric parallax of 0″.10; the apparent orbit is completed in 30 years. Both stars orbit a common center of mass. Neither star orbits at the focus of the true orbits in space, but each can lie at the focus of the relative orbits. Here the primary coincides with the focus of that orbit. Then we are seeing the true orbit, and the sum of the stellar masses is $30M_\odot$ (from Equation 12–3).

$$M_1 + M_2 = (3.0/0.1)^3/30^2 = 30$$

The companion is observed to be five times farther from the center of mass than the primary, so that $a_1/a_2 = 1/5$ and thence (Equation 12–4)

$$M_1 \text{ (primary)} = 25M_\odot$$
$$M_2 \text{ (companion)} = 5M_\odot$$

(B) THE MASS–LUMINOSITY RELATIONSHIP

Just as the determination of the period and size of the Earth's orbit (by Kepler's third law) leads to the Sun's mass, so also have we deduced binary stellar masses. Because it is necessary to know the distance to the binary system in order to establish these masses, we need only observe the radiant flux of each star to find its luminosity.

When the observed masses and luminosities for stars in binary systems are graphed, we obtain the correlation (Figure 12–3) called the **mass–luminosity relationship** (or **M–L relation** for short). In 1924, Arthur S. Eddington calculated that the mass and luminosity of normal stars like the Sun are related by

$$L/L_\odot = (M/M_\odot)^\alpha \qquad \textbf{(12–5a)}$$

His first crude theoretical models indicated that $\alpha \approx 3$. On a log–log plot, Equation 12–5 graphs as a straight line with a slope of α. So main-sequence

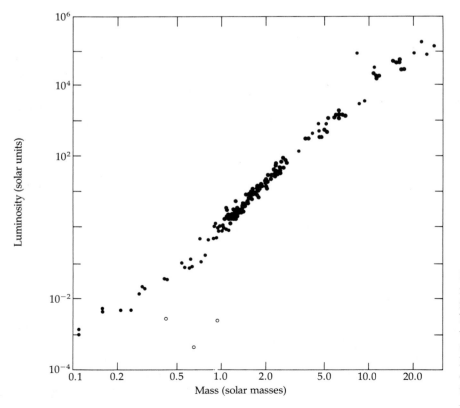

FIGURE 12–3 Mass–luminosity relationship. Masses and luminosities are shown here for stars in binary systems for which good values can be inferred. The stars (circles) below the trend line (at low luminosities) are white dwarfs made of degenerate rather than ordinary gases.

stars do seem to conform to Equation 12–5, although the exponent varies from $\alpha \approx 3$ for luminous and massive stars through $\alpha \approx 4$ for solar-type stars to $\alpha \approx 2$ for dim red stars of low mass. From a sample of 126 well-studied binary systems, we find that the break in slope occurs at a mass of $0.43 M_\odot$, and that the slope below this value is 2.26; above it, 3.99. For general use, an adequate form of the M–L relation for each regime is:

$$L/L_\odot = (M/M_\odot)^{4.0} \qquad \textbf{(12–5b)}$$
$$L/L_\odot = 0.23(M/M_\odot)^{2.3} \qquad \textbf{(12–5c)}$$

Today, astrophysical theories of stellar structure explain these results in terms of the different internal structures of stars of different mass and the opacities of stellar atmospheres at different temperatures. Note that the M–L law does not apply to highly evolved stars, such as red giants (with extended atmospheres) and white dwarfs (with degenerate matter; Figure 12–3). While most stellar masses lie in the narrow range from $0.085 M_\odot$ to $100 M_\odot$, stellar luminosities cover the vast span $10^{-4} \leq L/L_\odot \leq 10^6$!

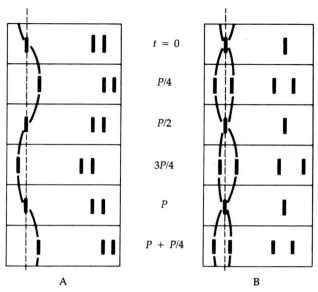

FIGURE 12–4 Spectra of spectroscopic binaries. (A) In a single-line system, only one set of lines shows an oscillation in wavelength from the Doppler shift. (B) In a double-line system, two sets of lines oscillate out of phase as the two stars revolve around their center of mass.

12–3 ◗
SPECTROSCOPIC BINARIES

If a binary system cannot be optically resolved at the telescope, its duplicity may show in its spectrum. Although orbital motion may not be detectable, we know that we are dealing with a **spectrum binary** when two different sets of line features are seen superimposed in the spectrum. A more useful, and interesting, case is the **spectroscopic binary:** here two stars orbit their center of mass closely (≤ 1 AU) and rapidly ($P \approx$ hours to a few months), and the orbital inclination is not 0°.

The spectrum of a spectroscopic binary (Figure 12–4) exhibits lines that oscillate periodically in wavelength. If the companion is so dim that its spectral features are not detected, we have a **single-line** spectroscopic binary (Figure 12–4A); two stars of more nearly equal luminosity produce two sets of spectral features that oscillate in opposite senses (in wavelengths)—we call this system a **double-line** spectroscopic binary (Figure 12–4B). About a thousand spectroscopic binaries are known, and good orbits have been determined for a few hundred.

(A) THE VELOCITY CURVE

To obtain useful information from the spectrum of a spectroscopic binary, we must interpret the behavior of the spectral lines. Because the two stars orbit in a plane inclined (at angle i) to the celestial sphere, that component of their velocity in the line of sight **Doppler-shifts** their spectral features. (Note that no Doppler shift can occur as a result of orbital motion alone when $i = 0°$; the system may then appear as a spectrum binary.) In addition, the center of mass of the system moves with respect to the Sun, so that the entire spectrum may be Doppler-shifted by some constant amount.

From Equation 8–13, the Doppler-shift formula is

$$\Delta\lambda/\lambda_0 \equiv (\lambda - \lambda_0)/\lambda_0 = v_r/c \qquad \textbf{(12–6)}$$

where λ_0 is the *emitted* wavelength (laboratory wavelength) of a spectral feature, λ is the *observed* wavelength, v_r is the radial speed (positive for recession, negative for approach) of the star, and $c = 3 \times 10^5$ km/s is the speed of light. Because of the finite width of spectral lines, we are limited at visible wavelengths to a shift resolution of $\Delta\lambda \geq$ 0.001 nm; hence, the radial speed must be $v_r \geq$ 1 km/s to be detectable. So the periods of observable spectroscopic binaries are necessarily short.

When we convert (using Equation 12–6) the Doppler shifts to radial velocities (Figure 12–5A) and plot the results as a function of time, we obtain the **velocity curve**. The simplest case is circular stellar orbits at the inclination $i = 90°$ (edge-on); the two curves (one for each star) are sinusoidal and oscillate with exactly opposite phases about the center-of-mass velocity in a period P (Figure 12–5B). In this case, we find the distances to the center of mass by noting that in one period, the primary traverses the circumference $2\pi r_1$ at constant speed V. Hence, $VP = 2\pi r_1$ and

$$r_1 = VP/2\pi \quad \text{and} \quad r_2 = vP/2\pi \quad \text{(12–7)}$$

The ratio of stellar masses is

$$M/m = r_2/r_1 = v/V$$

the relative semimajor axis a is $r_1 + r_2$, and, from Equation 12–1, the sum of the stellar masses is

$$M + m = a^3/P^2$$

The individual stellar masses follow from their sum and ratio, and the dynamical characteristics of

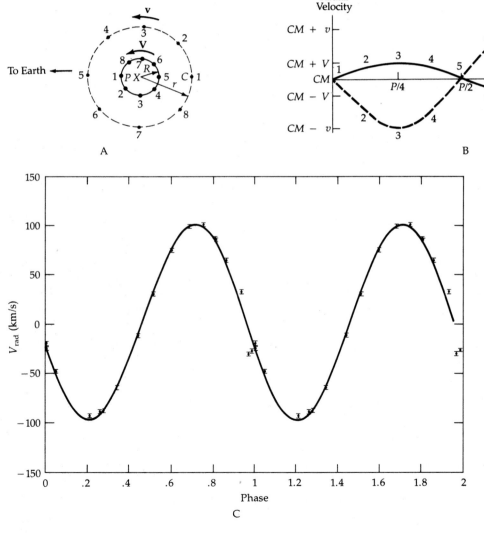

FIGURE 12–5 A binary system's radial velocity curve. (A) The primary *(P)* and companion *(C)* stars orbit the center of mass *(X)* in circular orbits of inclination 90°. (B) The center of mass is receding at a constant speed *(CM)* relative to the Sun, and the stars move at speeds V (primary) and v (companion) relative to the center of mass in one orbital period P. (C) Actual observational data (filled points) for the eclipsing binary WY Cnc with an orbital period of 0.83 d. The solid curve is the best fit to the data points, which are those for the primary star alone. *(J. Newmark)*

this spectroscopic binary are completely determined.

In general, this simple picture does not occur. For a single-line spectroscopic binary (only the primary's spectrum is seen; Figure 12–5C), we could determine only r_1 and the so-called "mass function" $m^3 \sin^3 i/(M + m)^2$ (see the next section for details); a reasonable value for M might be obtained from the primary spectral type; then the system could be approximately deciphered. A greater difficulty is that, unless the system is also an *eclipsing* binary (Section 12–4), we have no clear idea what the orbital inclination is. If the velocity curve is purely sinusoidal, we know only that we are dealing with a circular orbit whose plane is tilted at some angle i to the celestial sphere. The amplitudes of the velocity curves give (by trigonometry) the *observed* (denoted by primes) circular speeds:

$$V' = V \sin i$$
$$v' = v \sin i$$

Hence, we may determine the mass ratio exactly because

$$M/m = r_2/r_1 = v/V = v'/V'$$

but only the *lower limit*, $a \sin i$, to the relative semimajor axis is accessible.

If the orbit is not circular but has an eccentricity e, the velocity curves are distorted from pure sinusoids (Figure 12–6). The double-line curves are mirror images of one another but have different amplitudes—an orbital inclination i merely reduces all radial velocities by the same factor $\sin i$. The periodicity and characteristic shapes of these curves allow us to find P, e, and Ω (the orientation of the major axis with respect to the line of sight) immediately. When $i = 90°$, the relative semimajor axis and both stellar masses may be obtained.

(B) THE MASS FUNCTION

The strong tidal interactions between the component stars in short-period spectroscopic binaries ($P \leq 10$ days) circularize their orbits quickly (in about 5×10^9 years or less). Consider a circular (or small eccentricity, $e \ll 1$) relative orbit at inclination i. What can we say about the masses of the stars? Because we can obtain P, r_1', and r_2' (thence $a' = r_1' + r_2'$) from the velocity curve of a double-line binary, Kepler's harmonic law gives

$$(M + m) \sin^3 i = (a')^3/P^2$$

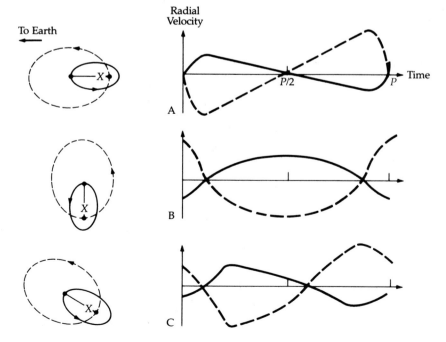

FIGURE 12–6 Velocity curves for elliptical orbits. The primary (mass M) and companion (mass m) have elliptical orbits of the same eccentricity but with semimajor axes in the ratio of M/m. The inclination is 90°, and the orbital period is P.
(A) Major axes along the line of sight. (B) Major axes at 90° to the line of sight. (C) Major axes at 45° to the line of sight.

Recall that M/m was also found in this case. If we see only the primary in a single-line binary, we can find only the **mass function** $f(M, m)$ by

$$(M + m)P^2 = a^3 = (r_1 + r_2)^3 = r_1^3(1 + r_2/r_1)^3$$
$$= r_1^3(1 + M/m)^3 = (r_1')^3(M + m)^3/m^3 \sin^3 i$$

or

$$f(M, m) \equiv m^3 \sin^3 i/(M + m)^2 = (r_1')^3/P^2 \quad \textbf{(12–8)}$$

where r_1' and P are observed and we have used the center of mass relation $Mr_1 = mr_2$.

If the orbital inclination is unknown, of what use is the mass function? We cannot evaluate individual stellar masses, but by combining many data, **statistical masses** may be obtained. If the orbital planes are randomly distributed in i, then the mean value of $\sin^3 i$ is 0.59; however, we are more likely to detect spectroscopic binaries with $i \approx 90°$ (almost edge-on), and so we correct for this observational selection effect by assigning a somewhat larger value to the mean of $\sin^3 i$, say $\approx 2/3$. In addition, we have spectral information on the visible components, which can suggest appropriate masses.

12–4 ◗
ECLIPSING BINARIES

When the inclination of a binary orbit is close to 90°, each of the stars can eclipse the other periodically—we call these **eclipsing binaries.** A few thousand such systems are known; most are also spectroscopic binaries, and very few are visual binaries. For a relative orbit of radius ρ, tilted at an angle ϕ to the line of sight ($\phi = 90° - i$), an eclipse can occur (Figure 12–7) only when $\rho \sin \phi < R$ (primary) $+ R$ (companion), where R is the stellar radius. So small orbits are favored; because these small orbits have short periods and high orbital velocities, they imply spectroscopic binaries.

(A) INTERPRETING THE LIGHT CURVE

Eclipsing binaries are most readily detectable by their periodically varying brightness. If we plot the apparent magnitude or flux of such a binary as a function of time, we obtain the **light curve,** which generally exhibits two brightness minima of different depths corresponding to the two possible eclipses per orbital period (Figure 12–8).

The deeper minimum—**primary eclipse**—occurs when the hotter star passes behind the cooler star; the other eclipse—the **secondary**—is shallower. Several types of eclipses are possible: (1) when $i = 90°$, both the **total** eclipse (smaller star behind larger star) and the **annular** eclipse (smaller star in front) are termed **central;** (2) when $\rho \cos i <$ [R (primary) $- R$ (companion)], we still have total and annular eclipses; and (3) when [R (primary) $- R$ (companion)] $< \rho \cos i <$ [R (primary) $+ R$ (companion)], only **partial** eclipses take place. Note that in every case exactly the same stellar area is covered at both primary minimum and secondary minimum, if the orbits are circular or $i = 90°$.

Consider the light curve associated with central eclipses and a circular relative stellar orbit for the situation where the larger star has a lower surface temperature than the smaller one (Figure 12–9A and B). Four points (in the time during one eclipse) exist where the limbs of the two stars are tangent; we speak of **first contact** (t_1) when the eclipse begins, **second contact** (t_2) when brightness minimum is reached, **third contact** (t_3) when the smaller star begins to leave the disk of the larger star, and **fourth contact** (t_4) when the eclipse ends. Both primary and secondary minima are flat, and they occur exactly half an orbital period apart. If we denote the stellar radii by R_l (larger star) and R_s (smaller star) and the relative orbital speed of the smaller star by v, the geometry implies

$$2R_s = v(t_2 - t_1) = v(t_4 - t_3) \quad \textbf{(12–9a)}$$
$$2(R_s + R_l) = v(t_4 - t_1) \quad \textbf{(12–9b)}$$

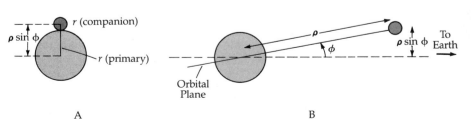

FIGURE 12–7 Eclipse geometry for binaries. (A) Front view from Earth; note that the companion must pass in front of the primary for an eclipse to occur. (B) Side view showing the permissible range of the companion for eclipses.

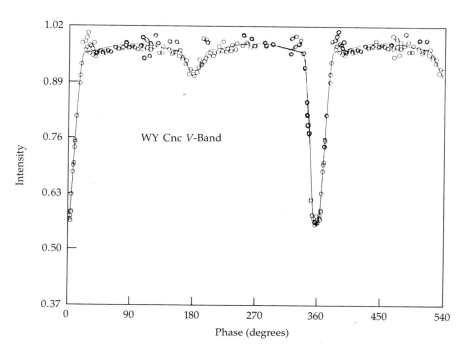

FIGURE 12–8 1989 *V*-band observations of an eclipsing binary, WY Cnc, an RS CVn system. The open circles are the actual data, the solid curve the best fit of an eclipsing binary model to the data points. The intensity units are normalized so that the sum of both stars outside of eclipses is set equal to 1.0. Note that the primary eclipse (phase 0° and 360°) is much deeper than the secondary one (phase 180°), which implies that the secondary star is much less luminous than the primary one. (*M. Zeilik; observations at KPNO*)

However, the radius *a* of the relative circular orbit is

$$a = vP/2\pi \tag{12–10}$$

where *P* is the orbital period. By combining Equations 12–9 and 12–10, we can determine the stellar radii relative to the orbital radius only:

$$R_s/a = \pi(t_2 - t_1)/P$$
$$R_1/a = \pi(t_4 - t_2)/P \tag{12–11}$$

Without consulting the stellar spectra, we may also determine the ratio of the effective surface temperatures of the two stars. Let these effective (blackbody) surface temperatures be denoted by T_1 and T_s. Surface brightness is equal to σT_{eff}^4 by the Stefan-Boltzmann law (Section 8–6); since the same stellar area (πR_s^2) is covered at each eclipse minimum, the relative depths of the two eclipse minima give us $(T_1/T_s)^4$ directly—the hotter star is eclipsed at primary minimum.

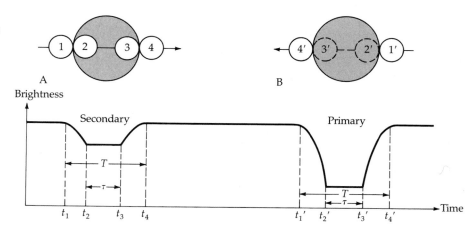

FIGURE 12–9 Central eclipses for circular orbits. The smaller star is assumed the hotter of the two. The four numbered contact points define the duration of the eclipse. These central eclipses have flat bottoms. (A) During secondary eclipse, the smaller star passes in front of the larger one. (B) During primary eclipse, one-half an orbital period later, the smaller star passes behind the larger one.

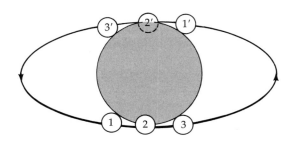

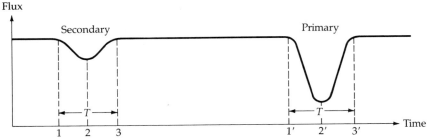

Flux

Secondary

Primary

Time

FIGURE 12–10 Partial eclipses for an inclined circular orbit. The smaller star is assumed the hotter, and the eclipses occur one-half a period apart, but note that the light curve is not flat during the eclipses.

When the eclipses are **partial** for a circular orbit (Figure 12–10), both eclipses are still of equal duration (though briefer than for the central eclipse) and the brightness minima are not flat. Because the two eclipses still occur exactly half an orbital period apart, we know that the orbit is circular. In this case, it is possible to determine (1) the orbital inclination i, (2) the relative stellar radii R_s/a and R_l/a, and (3) the relative effective surface temperatures T_l/T_s.

For elliptical orbits, different times elapse from primary to secondary eclipse and from secondary to primary eclipse. Also, in general, the eclipse durations are not equal. These features permit us to determine the eccentricity e, orientation Ω, and inclination i of the orbit.

(B) ECLIPSING-SPECTROSCOPIC BINARIES

We have seen that light curves yield only relative results. It is indeed fortunate that most eclipsing binaries are also spectroscopic binaries, for we may determine speeds in kilometers per second from their velocity curves. For Equations 12–9 and 12–10, we find absolute values (in kilometers) for a, R_s, and R_l. Because the orbital inclination follows from the light curve, we can evaluate $\sin i$ and determine the stellar masses. By plotting these results, we obtain the mass–radius correlation for these stars (Figure 12–11A). Mean stellar densities $\langle \rho \rangle$ may then be computed by

$$\langle \rho \rangle = 3M/4\pi R^3$$

Knowing the stellar radii, we may find the ratio of stellar luminosities (from the effective temperature

ratio) and the total luminosity of the system; the flux of the system then tells us the distance to the binary. Finally, we can infer the masses and luminosities of each star (Figure 12–11B; see also Figure 12–3). Table 12–1 summarizes the various data we can obtain from binary stars.

(C) OTHER USEFUL INFORMATION

Simple light curves do not tell the whole story of eclipsing binaries. Much useful information, including knowledge about stellar interiors, may be gleaned from the unusual light curves of more complicated systems (Figure 12–12).

From our experience with the Sun [Section 10–2(b)], we expect that stellar disks will exhibit **limb darkening** (Figure 12–12A). This feature rounds off the edges of the eclipses. In tight binary orbits, the hotter star will heat that part of the cooler star's atmosphere that is nearest (on the line between the stellar centers). This hotter gas is more luminous and leads to the **reflection effect:** just before and after secondary minimum, the system appears brighter than we would otherwise expect (Figure 12–12B).

In close binaries, the fluid (gaseous) bodies of the stars are distorted into prolate spheroids (like footballs) oriented along their lines of center so that they are also in synchronous rotation as a result of the strong tidal forces. This gives rise to the **ellipticity effect** (Figure 12–12C); the observed brightness varies continuously and not just during eclipses. Such systems are usually in circular orbits, also because of tidal effects. If the distortion is great enough, gas streams may be drawn from the

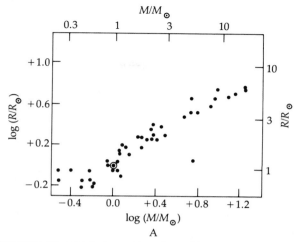

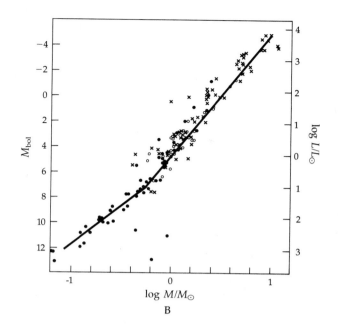

FIGURE 12–11 Information from binary stars.
(A) Main-sequence mass–radius relationship; note
that the more massive stars are larger. (B) Mass–
luminosity relationship for a variety of binaries whose
distances are known. Filled circles are visual binaries
(best data); the open circles represent visual systems
that have second-class data; crosses are eclipsing binaries. Note the break in the slope at a mass a little less
than that of the Sun. *(W. D. Heintz)*

stellar atmospheres into the space between and
around the stars—the system Beta Lyrae is one
famous example. This gas is most readily noticeable in spectra, for bright emission lines can be
seen.

Stellar rotation shows up in two ways. The velocity curve exhibits anomalous bumps (Figure 12–
12D) just before and just after each eclipse—in one
case we see the receding limb of the rotating star
(because the rest is already eclipsed) and in the

TABLE 12–1 *Stellar Data from Binary Systems*

Type of Binary	Observations Performed or Needed	Parameters Determined
Visual	(a) Apparent magnitudes and π''	Stellar luminosities
	(b) P, a'', and π''	Semimajor axis (a)
		Mass sum ($M + m$)
	(c) Motion relative to CM	M and m
Spectroscopic	(a) Single-line velocity curve	Mass function $f(M, m)$
	(b) Double-line velocity curve	Mass ratio (M/m)
		$(M + m) \sin^3 i$
		$a \sin i$
Eclipsing	(a) Shape of light curve eclipses	Orbital inclination (i)
		Relative stellar radii ($r_{l,s}/a$)
	(b) Relative times between eclipses	Orbital eccentricity (e)
	(c) Light loss at eclipse minima	Surface temperature ratio (T_l/T_s)
Eclipsing-Spectroscopic	(a) Light and velocity curves	Absolute dimensions (a, r_s, r_l)
		e and i
		M and m (also densities)
	(b) Spectroscopic parallax + apparent magnitude	Distance to binary
		Stellar luminosities
		Surface temperatures (T_l, T_s)

Note: Subscript l indicates the larger star: s, the smaller one.

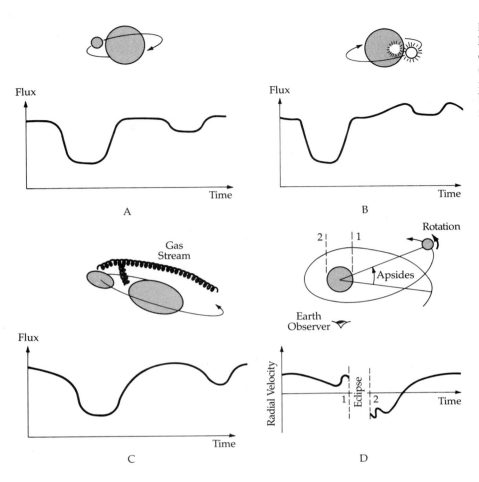

FIGURE 12–12 Variations on light curves. (A) Effect of limb darkening of the primary. (B) Reflection effect off the primary. (C) Gas streams between both stars. (D) Apsidal motion from stellar rotation.

other we see the approaching limb. In the light curve, stellar rotation appears in the rotation of the direction of the semimajor axis of elliptical orbits (Figure 12–12D). Because a rotating star is oblate (flattened), its gravitational attraction is not that of a point mass; two such stars in a binary interact to change the orientation of their semimajor axes with time.

Finally, we expand the idea of close binaries to that of **contact binaries.** Eclipsing systems with extremely short periods—only a few hours—are in physical contact. Their light curves show this interaction because their maxima are rounded and their minima have almost the same depths. In such systems, the two stars share a common envelope of material and both are severely distorted by tidal effects.

We picture the interactions of these systems by considering the effective gravity at many points

locally. The *effective gravity* results from the combination of the real gravitational attractions and the centripetal force from orbital motions. If you explore the space around the stars, you will find a certain region, shaped like a figure **8**, where the effective gravities of the two stars cancel (L_1 in Figure 12–13A). Here the effective gravity is zero. Each half of the figure **8** indicates the regions controlled by the effective gravity of each star; these are called **Roche lobes.**

We can now classify close binaries on the basis of how large each star is relative to its Roche lobe. If both stars are smaller than their Roche lobes, the system is **detached** (Figure 12–13A). If one fills its Roche lobe, the system is **semidetached;** matter can flow through the contact point L_1 to the other star (Figure 12–13B). The gas from the star that fills its Roche lobe (the donor star) is free to flow onto the other star (the recipient star). Finally, if both

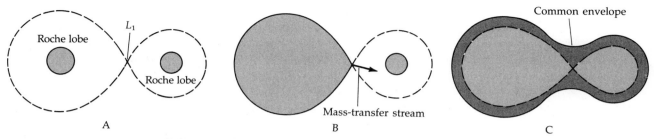

FIGURE 12–13 Contact binaries. (A) A detached system, in which both stars are smaller than their Roche lobes. (B) In a semidetached system, one star fills its Roche lobe and mass flows to the companion. (C) Both stars fill their Roche lobes in a contact system, and they are surrounded by a common envelope of uniform temperature.

stars fill their Roche lobes, they are in **contact** and a common envelope of material enshrouds them both (Figure 12–13C).

12–5 ◗ INTERFEROMETRIC STELLAR DIAMETERS AND EFFECTIVE TEMPERATURES

Finally, we briefly mention methods for determining stellar diameters that do *not* require a binary system. Such methods are basically **interferometric;** they depend upon the constructive and destructive interference of the light waves from a star (Section 8–1).

To see why such indirect procedures are necessary (except for our Sun), consider the following. At a distance of 1 pc, a star with a diameter of 1 AU (radius = $109R_\odot$) subtends an angle of 1.0″ on the sky, but this angle is just the same as the size of the seeing image (as a result of the Earth's turbulent atmosphere) seen at a telescope, so that the star's size is unresolvable. Besides that, no star is as close as 1 pc.

Within a band 10° wide centered on the ecliptic, stars may be occulted by our Moon. In such a **lunar occultation,** the star does not disappear instantaneously; instead it fades away in a few seconds (Figure 12–14). Electromagnetic wavefronts from the star are progressively screened out as time goes on, but the unobstructed portions interfere (actually, diffract) to produce a characteristic intensity-versus-time pattern at the Earth. This pattern depends directly upon the angular size of the star; if we know the stellar distance, we can deduce the

stellar diameter. Average errors are about one milliarcsec (abbreviated mas) for angular sizes that range from 2 to 20 mas.

In the 1920s, A. E. Michelson invented and used a **stellar interferometer** to measure the angular diameters of large nearby stars. In this device, widely separated (by many meters) mirrors reflect

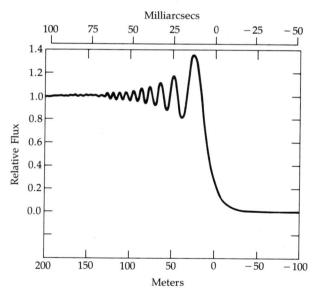

FIGURE 12–14 Occultation of a star by the Moon. As the limb of the Moon cuts in front of the star, a diffraction pattern appears before the light is completely cut out. The top scale is an angular one, the bottom a linear one. [*S. Ridgeway et al., Astronomical Journal 84:247 (1979)*]

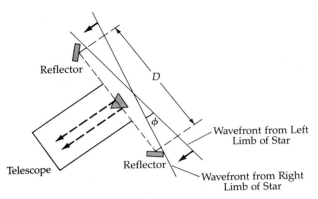

FIGURE 12–15 Schematic of a Michelson interferometer. The widely spaced mirrors send starlight into the telescope to produce the interference patterns. Light waves from opposite ends of a star arrive simultaneously, intersecting one another at angle ϕ, which is the angular diameter of the star.

the starlight to an ordinary focusing telescope, where the wavefronts from different parts of the star produce a characteristic interference pattern (Figure 12–15). This pattern depends upon the angle between the wavefronts from opposite limbs of the star; this intersection angle increases as the stellar angular diameter increases. (Modern versions of this technique often use two telescopes rather than one, but the physical principles are the same.) So to find a star's angular diameter, we use the simple small-angle approximation:

$$\alpha = D/d$$

where α is in radians, D is the star's physical diameter, and d is its distance (D and d are in the same

units). Table 12–2 gives selected observations of a few bright stars.

One star measured this way is Aldebaran (Alpha Tauri). The results average about 21 mas for the angular diameter. The distance d of Aldebaran is 20.8 pc. Then

$$D = \alpha d$$

with 21×10^{-3} arcsecond $= 1.02 \times 10^{-7}$ rad, so that

$$D = (1.02 \times 10^{-7})(20.8 \text{ pc})$$
$$= 2.12 \times 10^{-6} \text{ pc} = 6.54 \times 10^{7} \text{ km}$$

So Aldebaran's radius is 3.27×10^{7} km, about 50 times larger than the Sun's radius.

Another type of interferometry, called **speckle interferometry,** exploits the unavoidable seeing effects of the Earth's atmosphere. The image of a star observed through a telescope and the atmosphere has a diffraction pattern that splits up into many bright granules, or speckles, because of the random interference of light traveling through different atmospheric cells. The result is different phases of the incoming waves. Because the speckle effect is a function of wavelength, observations must be made at a narrow band of wavelengths. In practice, many short exposures of the speckle pattern are made rapidly. These are then averaged and the speckle pattern analyzed by Fourier techniques.

Speckle interferometry not only provides resolved images of single stars; it also offers a way to resolve the components in very close multiple star systems. For example, the star Betelgeuse (Alpha Orionis) was found to be a triple-star system by

TABLE 12–2 *Some Interferometric Stellar Diameters*

Star	Spectral Type	Limb-darkened Diameter (mas)	Distance (pc)	Radius ($R_\odot$)
Alpha Cas	K0 II–III	5.4 ± 0.6	45 ± 9	26 ± 8
Beta And	M0 III	13.2 ± 1.7	23 ± 3	33 ± 9
Gamma And	K3 II	6.3 ± 0.6	75 ± 15	50 ± 14
Alpha Cyg	A2 Ia	2.7 ± 0.3	500 ± 100	145 ± 45

Note: Based on observations by D. Bonneau, L. Koechlin, J. Oneto, and F. Vakili.

speckle observations (Figure 12–16). This technique is especially useful for spectroscopic binaries, where measurements of the radial velocity curve combined with the angular separation most directly determine the stellar masses.

We can now use the observed stellar angular sizes to find directly the effective temperatures of star. Recall that T_{eff} is a measure of the total energy integrated over all wavelengths radiated from a unit surface area. Its value is fixed by a star's radius and luminosity. Direct determination of T_{eff} is thus possible from a value of a star's angular diameter (radius) and flux (apparent magnitude). With the Sun as the standard ($T_{eff} = 5800$ K, $M_{bol} = 4.72$, and $BC = -0.08$), then for another star

$$\log T_{eff} = 4.2207 - 0.10(V_0 + BC) - 0.5 \log \theta_{ld}$$

where V_0 is the (unreddened) apparent visual magnitude and θ_{ld} the limb-darkened angular diameter (in mas). A model of the stellar disk must be used to find θ_{ld} from the observed θ; this results in about a 1% uncertainty. In contrast, the typical measurement errors are 10%. Overall, the errors in T_{eff} are about 100 to 400 K, which pins down T_{eff} pretty well so that they can be compared to those derived from stellar models.

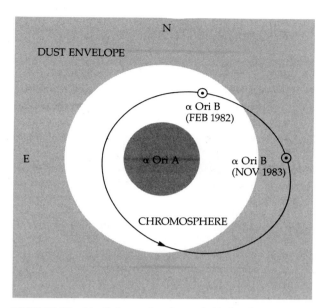

FIGURE 12–16 Schematic view of Alpha Orionis A (Betelgeuse) and its close companion, Alpha Orionis B, which orbits in a period of about 2.1 y within the chromosphere and dust envelope of Betelgeuse. The third star (component C) is about 10 times farther away than B, and so outside this view. *(M. Karovska, P. Nisenson, R. Noyes, and F. Roddier)*

PROBLEMS

1. From the information given in Section 12–2(A), show that Kepler's harmonic law holds for the apparent orbit of a visual binary.

2. Demonstrate the correctness of Equation 12–2. Use a diagram if necessary.

3. What is the sum of the stellar masses in a visual binary of period 40 years, maximum separation 5″0, and parallax 0″3? Assume an orbital inclination of zero and a circular apparent orbit.

4. Find the distance in parsecs to a visual binary that consists of stars of absolute bolometric magnitudes of +5.0 and +2.0. The mean angular separation is 0.05″, and the observed orbital period is ten years. The stars obey the mass-luminosity relation, Equation 12–5. What assumptions have you made to arrive at your answer?

5. Show that binary systems with small orbits have high orbital speeds.

6. The velocity curves of a double-line spectroscopic binary are observed to be sinusoidal, with amplitudes of 20 km/s and 60 km/s and a period of 1.5 years.
   (a) What is the orbital eccentricity?
   (b) Which star is the more massive, and what is the ratio of stellar masses?
   (c) If the orbital inclination is 90°, find the relative semimajor axis (in astronomical units) and the individual stellar masses (in solar masses).

7. An eclipsing binary has an orbital period of $2^d 22^h$, the duration of each eclipse is 18^h, and totality lasts 4^h.
   (a) Find the stellar radii in terms of the circular orbital radius a.
   (b) If spectroscopic data indicate a relative orbital speed of 200 km/s, what are the actual stellar radii (in kilometers and solar radii)?

8. The surface temperature of one component of an eclipsing binary is 15,000 K, and that of the other is 5000 K. The cooler star is a giant with a radius four times that of the hotter star.
    (a) What is the ratio of the stellar luminosities?
    (b) Which star is eclipsed at primary minimum?
    (c) Is primary minimum a total or an annular eclipse?
    (d) Primary minimum is how many times deeper than secondary minimum (in energy units).

9. The star Sirius A has a surface temperature of 10,000 K, a radius of $1.8R_\odot$, and $M_{bol} = 1.4$; the radius of its white dwarf companion, Sirius B, is $0.01R_\odot$ and $M_{bol} = 11.5$.
    (a) What is the ratio of their luminosities?
    (b) What is the ratio of their effective temperatures?
    (c) If they orbit at $i = 90°$, which star is eclipsed at primary minimum?
    (d) If your photometer can measure magnitudes to an accuracy of ≈ 0.001, would you be able to detect the hypothetical primary eclipse? (*Hint:* Use $\log_{10}(1 + x) = x/2.3$ for $x \ll 1$.)

10. Derive an expression giving the stellar angular diameter in milliarcseconds when the actual stellar diameter (in solar radii) and distance from us (in parsecs) are known.

11. Estimate the orbital separation and velocity for a binary star system composed of solar mass stars that have an orbital period of 12 hours (assume circular orbits).

12. (a) Use the mass–luminosity relation to compute the luminosity range of stars from the observed mass range of $0.085\ M_\odot$ to $100\ M_\odot$.
    (b) What is the mass of a star that is 0.1 the luminosity of the Sun? What is the mass of a star that is 1000 times the luminosity of the Sun?

13. (a) Refer to the light curve for the eclipsing binary star WY Cancri (Figure 12–8). What qualitative information can be deduced from a visual inspection of the light curve?
    (b) If WY Cancri has a period of 19.9 hours, estimate the durations of primary and secondary eclipse (R_l/a and R_s/a).

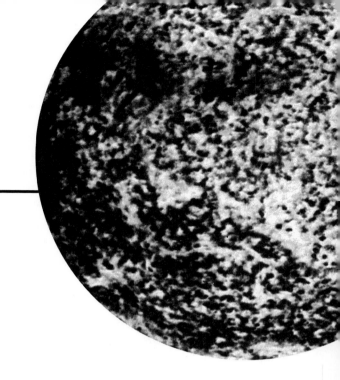

Chapter 13

Stars: The Hertzsprung-Russell Diagram

We infer the properties of stars from their light—in bulk for fluxes and spread out for spectra. This chapter deals with the wealth of information that can be discerned by studying stellar spectra. First we consider stellar atmospheres, for here the stellar spectra originate. Then we tell the story of spectral observations—how they have been made, correlated, and interpreted. Finally, we present that famous and crucial synthesis—the Hertzsprung-Russell diagram—and some of its implications. This discussion will lead to an understanding of the stars themselves (Chapter 16).

13–1 ◖
STELLAR ATMOSPHERES

The spectral energy distribution of starlight is determined in a star's **atmosphere,** the region from which radiation can freely escape. To understand stellar spectra, we first discuss a model stellar atmosphere and investigate the characteristics that determine the spectral features.

(A) PHYSICAL CHARACTERISTICS

The stellar photosphere, a thin, gaseous layer, shields the stellar interior from view. The photosphere is thin relative to the stellar radius, and so we regard it as a *uniform* shell of gas. The physical properties of this shell may be approximately specified by the average values of its pressure P, temperature T, and chemical composition μ (chemical abundances).

We make the reasonable assumption that the number density n (number/m^3) of gas particles (molecules, atoms, ions, and electrons) is high enough so that **thermodynamic equilibrium** holds. This means that particles collide so frequently that both the Boltzmann and the Saha equation apply. We also assume that the gas obeys the **perfect-gas law:**

$$P = nkT \tag{13–1}$$

where k is Boltzmann's constant. The particle number density is related to both the **mass density** ρ (kg/m^3) and the composition (or **mean molecular**

weight) μ by the following definition of μ:

$$1/\mu = m_H n/\rho \qquad (13-2)$$

where $m_H = 1.67 \times 10^{-27}$ kg is the mass of a hydrogen atom. For a star of pure atomic hydrogen, $\mu = 1$. If the hydrogen is completely ionized, $\mu = 1/2$ because electrons and protons (hydrogen nuclei) are equal in number and electrons are far less massive than protons. In general, stellar interior gases are ionized and

$$1/\mu \approx 2X + (3/4)Y + (1/2)Z \approx 1.6.$$

where X is the **mass fraction** of hydrogen, Y is that of helium, and Z is that of all heavier elements. The mass fraction is the percentage by mass of one species relative to the total.

We also assume a steady-state atmosphere: although the individual gas particles move about rapidly, nothing changes with time on the macroscopic scale (there are no mass motions). This implies **hydrostatic equilibrium,** wherein a typical volume of gas experiences no net force. We have already derived this relationship in Chapter 4 and so simply repeat the equation here:

$$dP/dr = -(GM/R^2)\rho = -g\rho \qquad (13-3)$$

where g is the gravitational acceleration (m/s^2), or gravity, at the photosphere. Note that the pressure decreases continuously outward through the atmosphere of a star.

As we did in Section 4–5(b), define

$$H = kT/gm \cong \text{constant}$$

as the **scale height.** Then, for a stellar atmosphere, we can also apply the barometric equation:

$$P(h) = P(h_0) \exp(-h/H)$$

where h is any height above a reference level h_0. It applies to regions in stellar atmospheres where the temperature and mean molecular weight do *not* change rapidly. Recall that H has the units of length, and it is the distance to move up in the atmosphere for the pressure to decrease by $1/e$. By parameterizing the pressure in terms of optical depth τ (Section 8–7) instead of radius r, where $d\tau = -\kappa\rho\, dr$ (κ is the atmospheric opacity in m^2/kg), Equation 13–3 takes the useful form

$$dP/d\tau = g/\kappa \qquad (13-4)$$

In our uniform approximation, then, we integrate Equation 13–4 to

$$P = (g/\kappa)\tau \qquad (13-5)$$

so that the atmospheric gas pressure depends upon g and κ. We take $\tau = 1$ as the atmospheric level where spectral features are formed because this opacity is the minimum needed for line formation. As for the Sun, this level defines the photosphere of a star.

(B) TEMPERATURES

From the pressure and composition of the stellar atmosphere, we now turn to the temperature. We have already noted (Section 8–6) that the continuous spectrum, or continuum, from a star may be approximated by the Planck blackbody spectral-energy distribution. For a given star, the continuum defines a temperature by fitting the appropriate Planck curve. We can also define the temperature from Wien's displacement law:

$$\lambda_{\max}T = 2.898 \times 10^{-3} \text{ m} \cdot \text{K} \qquad (13-6)$$

which states that the peak intensity of the Planck curve occurs at a wavelength $\lambda_{\max}$ that varies inversely with the Planck temperature T. The value of $\lambda_{\max}$ then defines a temperature. Also note here that the hotter a star is, the greater will be its luminous flux (in W/m^2), in accordance with the Stefan-Boltzmann law:

$$F = \sigma T^4 \qquad (13-7)$$

where $\sigma = 5.67 \times 10^{-8}$ W/m$^2 \cdot$ K^4. Then the relation

$$L = 4\pi R^2 \sigma T_{\text{eff}}^4$$

defines the effective temperature of the photosphere.

A word of caution: the effective temperature of a star is usually *not* identical to its excitation or ionization temperature because spectral-line formation redistributes radiation from the continuum. This effect is called **line blanketing** (Figure 13–1) and becomes important when the numbers and strengths of spectral lines are large. When spectral features are not numerous, we can detect the continuum between them and obtain a reasonably accurate value for the star's effective surface temperature. The line blanketing alters the atmosphere's blackbody character.

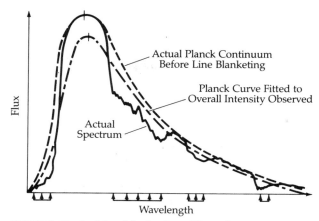

FIGURE 13–1 Line blanketing. The color temperature of a star should correspond to its Planck continuum (dashed line), but absorption lines (arrows) blanket the continuum to produce the observed spectrum (solid line). A Planck curve fitted to the observed intensity (dotted line) results in too low a color temperature.

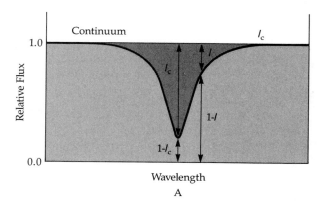

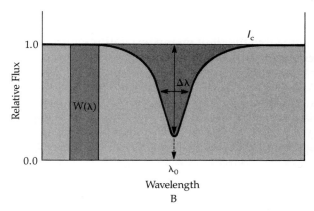

FIGURE 13–2 Spectral lines. (A) Parameters that characterize an absorption line: continuum level, $I_c = 1.0$; depth at any wavelength is $I(\lambda)$. (B) Line width, $\Delta\lambda$, equivalent width, $W(\lambda)$.

(C) SPECTRAL-LINE FORMATION

Sections 8–2 and 8–3 describe how spectral absorption features are formed when the molecules, atoms, and ions of a gas absorb continuum photons and re-emit fewer of these photons toward the observer. The composition of the gas determines which species are available to absorb the photons, and the temperature and pressure determine which spectral features are formed. For example, molecular spectral features can originate only in a cool gas, for molecules are easily dissociated by collisions with other particles; neutral atoms and their spectral lines predominate at intermediate temperatures; at high temperatures, all species are ionized so that only the spectral features arising from ions are seen.

Before we take the plunge, we will review and elaborate on the astronomical concepts relevant to spectral lines. Let's use an absorption line as an example (Figure 13–2A). The central part of the line is called the *core*, which extends into the continuum and merges with it as some intensity level, I_{con}, which is usually set equal to unity in relative units. These extended parts of the line are called the *wings*. The line has a measured intensity or depth, $I(\lambda)$, as a function of wavelength. The cen-

tral depth is l_c. Astronomers use the equivalent width, $EW(\lambda)$, to define the depth and so the strength of the line in a consistent way (Figure 13–2B):

$$EW(\lambda) = \int_{-\infty}^{\infty} \frac{I_{con} - I(\lambda)}{I_{con}} \, d\lambda$$

where the integration actually takes place over a short interval in wavelength covered by the extent of the wings of the line. So the equivalent width typically has a very small value, usually given by astronomers in milliÅngstroms (mÅ). The level to which $EW(\lambda)$ can be measured depends on the dispersion of the spectrograph; at a dispersion of 40 Å/mm, we can typically measure $EW(\lambda) \approx$ 0.1 mÅ. Note that we we write about "line strengths," we are really dealing with equivalent widths.

Now to discuss spectral line formation on a more quantitative basis. Take a cubic meter of gas in which the number of particles of each elemental species is specified by the composition μ. Now consider the particles of a particular element (such as hydrogen). Continuum photons at discrete wavelengths are absorbed when the particles (neutral atoms or ions) are excited from one atomic energy level to another—the equivalent width of each absorption feature is essentially proportional to the number of particles populating a given energy level. The relative number of atoms in energy levels B (N_B) and A (N_A), where B > A, is determined by the **Boltzmann excitation-equilibrium equation** (Section 8–4):

$$N_B/N_A \propto \exp\left[(E_A - E_B)/kT\right] \quad \textbf{(13–8)}$$

where E is the energy of the level and T is the gas temperature. At a given temperature, the higher levels are *less* populated than the lower levels, but the relative number of **ions** in a given ionization stage i (i = the number of electrons lost by an atom) is determined by both the temperature and the electron number density by the **Saha ionization-equilibrium equation** (Section 8–4):

$$N_{i+1}/N_i \propto \left[(kT)^{3/2}/N_e\right] \exp\left(-\chi_i/kT\right) \quad \textbf{(13–9)}$$

where χ_i is the ionization potential from stage i to stage $i + 1$.

To express Equations 13–8 and 13–9 in logarithmic form, we take the base-10 logarithm of both sides of each equation. Noting that $\log_{10} e = 0.4343$ and expressing T in kelvins and all energies in electron volts (eV), we have, from Equation 13–8,

$$\log (N_B/N_A) = (-5040/T)E_A - E_B + \text{constant} \quad \textbf{(13–10)}$$

and from Equation 13–9,

$$\log (N_{i+1}/N_i) = (3/2)\log T - (5040/T)\chi_i - \log N_e + \text{constant}' \quad \textbf{(13–11)}$$

In this way, the dependence upon each parameter is clear. Since the electrons also constitute a perfect gas, we may write Equation 13–11 in terms of the **electron pressure** P_e (where $P_e \propto N_e T$) as

$$\log (N_{i+1}/N_i) = (5/2)\log T - (5040/T)\chi_i - \log P_e + \text{constant}'' \quad \textbf{(13–12)}$$

The Boltzmann and Saha equations must be combined to deduce the population of every energy level and thence the spectral-line strengths. Figure 8–13 shows that the temperature behavior of the combination is the following. At low temperatures, the atoms are all neutral and in their ground states. As the temperature rises, the higher energy levels of the neutral atoms become more populated until $T \approx \chi_0/k$, when single ionization produces a significant number of ions (depleting the number of neutrals). At still higher temperatures, these ions predominate, but when $T \approx \chi_1/k$, double ionization (two electrons lost) becomes important. At the top of the ionization stages, the temperature is so high that only bare nuclei and free electrons remain—the atoms are fully ionized—and no more spectral-line absorption can take place. You should remember that this sequence depends strongly upon the exact energy-level structure of each atomic species, so that the spectral features produced at any temperature uniquely characterize the species.

The following verbal flow chart expresses the spectral-line behavior to be expected of a given stellar atmosphere. Denoting the strength of a spectral feature by its equivalent width EW, we have the functional dependence

$$EW = EW(n, T)$$

because the rate of absorption is determined by the number density n of atoms and the intensity of the continuum, which relates to the temperature T. However, the equation of state of the gas (Equation 13–1) and the composition of the gas (Equation 13–2) imply that

$$n = n(P, T, \mu)$$

Hydrostatic equilibrium (Equations 13–3 and 13–4) determines the pressure:

$$P = P(M, R, \kappa)$$

The opacity κ is clearly a function of number density, gas composition, and ionization–excitation state determined by T:

$$\kappa = \kappa(n, T, \mu)$$

while the star's luminosity L depends upon its temperature and radius through the relationship

$$L = 4\pi R^2 \sigma T^4 \quad \textbf{(13–13)}$$

If we combine the dependences of these five equations, then in general

$$EW = EW(L, T, M, \mu) \qquad \text{(13–14)}$$

For stars of a given composition, Equation 13–14 reduces to

$$EW = EW(L, T, M) \qquad \text{(13–15)}$$

and if we consider, for example, only stars with a unique mass–luminosity relation, we have

$$EW = EW(L, T) \qquad \text{(13–16)}$$

Equation 13–16 is the theoretical justification for seeking temperature sequences and two-dimensional Hertzsprung-Russell diagrams in the remainder of this chapter: Equation 13–14 covers those cases in which stellar luminosity is not related to stellar mass and those in which compositional differences are important. We note, however, that Hertzsprung-Russell diagrams were first

arrived at observationally; the theoretical understanding came later.

13–2 ◑
CLASSIFYING STELLAR SPECTRA

(A) OBSERVATIONS

A single stellar spectrum is produced when starlight is focused by a telescope onto a **spectrometer** or **spectrograph,** where it is dispersed (spread out) in wavelength and recorded photographically or electronically. If the star is bright, we may obtain a **high-dispersion** spectrum, that is, a few mÅ per millimeter on the spectrogram, because there is enough radiation to be spread broadly and thinly (the solar spectrum of Figure 13–3 is a good example). At high dispersion, a wealth of detail appears in the spectrum, but the method is slow (only one stellar spectrum at a time) and limited to fairly

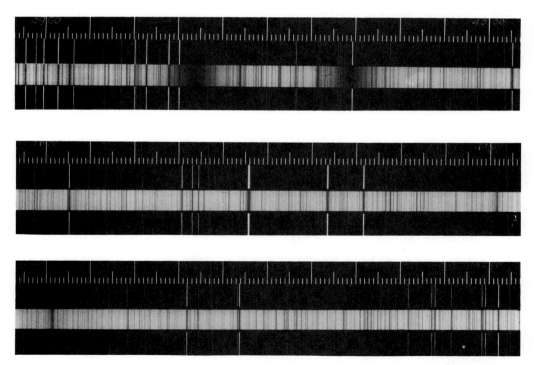

FIGURE 13–3 A solar spectrum. This high-dispersion spectrum (0.5 Å/mm) shows a wealth of detail. The bright lines above and below the absorption-line spectrum are used for wavelength calibration. *(Mount Wilson Observatory, Carnegie Institute of Technology)*

Main Sequence Spectra

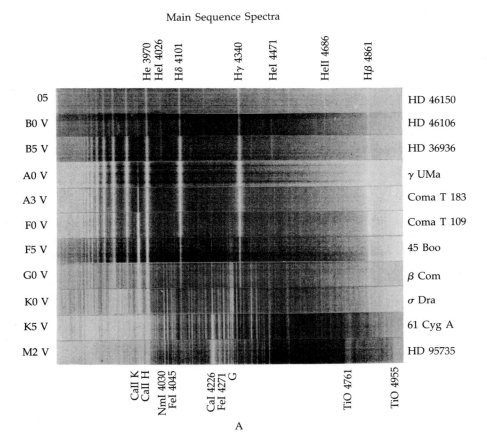

A

V

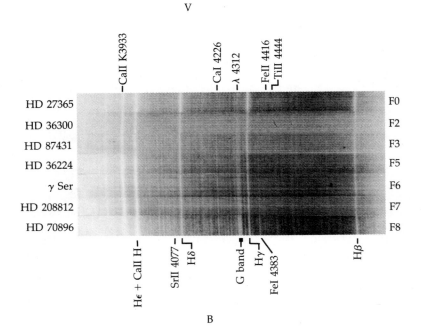

B

FIGURE 13–4 Stellar spectra; these are negative prints (absorption lines appear white; continuum dark) that show fine detail better than positive ones. The strongest lines are indicated; wavelengths are given in Ångstroms. (A) Harvard spectral sequence. Main-sequence stars show representative spectra for each spectral type. Star names are on the right. *(National Optical Astronomy Observatories)* (B) Subdivisions for the F stars. Star names are on the left. *(From An Atlas of Objective-Prism Spectra by N. Houk, N. J. Irvine, and D. Rosenbush)*

bright stars. Dispersion is the key to unlocking the information in starlight.

(B) THE SPECTRAL-LINE SEQUENCE

At first glance, the spectra of different stars seem to bear no relationship to one another. In 1863, however, Angelo Secchi found that he could crudely order the spectra and define different **spectral types.** Alternative ordering schemes appeared in the ensuing years, but the system developed at the Harvard Observatory by Annie J. Cannon and her colleagues was internationally adopted in 1910. This sequence, the **Harvard spectral classification system,** is still used today. (About 400,000 stars were classified by Cannon and published in various volumes of the *Henry Draper Catalogue,* 1910–1924, and its *Extension,* 1949. The catalog is now being reobserved for updated classifications by astronomers at the University of Michigan, headed by Nancy Houk.)

At first, the Harvard scheme was based upon the strengths of the hydrogen Balmer absorption lines in stellar spectra, and the spectral ordering was alphabetical (A through P). The A stars had the strongest Balmer lines, and the P stars had the weakest. Some letters were eventually dropped, and the ordering was rearranged to correspond to a sequence of decreasing temperatures (recall the effects of the Boltzmann and Saha equations): OBAFGKM. Stars nearer the beginning of the spectral sequence (closer to O) are sometimes called **early-type** stars, and those closer to the M end are referred to as **late-type.** Each spectral type is divided into ten parts from 0 (early) to 9 (late); for example, . . . F8 F9 G0 G1 G2 . . . G9 K0 . . . . In this scheme, our Sun is spectral type G2. In 1922, the International Astronomical Union (IAU) adopted the Harvard system (with some modifications) as the international standard.

Many mnemonics have been devised to help students retain the spectral sequence. We urge you to devise your own. A variation of the traditional one is "Oh, Be a Fine Guy, Kiss Me." Our students have come up with others, including such masterful phrases as "Only Bold Astronomers Forge Great Knowledgeable Minds" and "Optical Binary Affairs Fundamentally Generate Keplerian Marriages."

Figure 13–4A shows exemplary stellar spectra arranged in order; note how the conspicuous spectral features strengthen and diminish in a characteristic way through the spectral types. Table 13–1 summarizes those spectral characteristics that define each spectral type. Study this table and Figure 13–4B, which shows the subdivisions for the F spectral type. In general, this classification scheme relies on three basic criteria to subdivide the stellar spectra: (1) the absence of lines, (2) the strengths (equivalent widths) of lines, and (3) the ratios of line strengths (such as Ca II K-lines compared to those of the Balmer series).

(C) THE TEMPERATURE SEQUENCE

The spectral sequence is a temperature sequence, but we must carefully qualify this statement. There are many different kinds of temperatures and many ways to determine them. In Figure 13–5, the strengths of various spectral features are plotted against excitation–ionization (or Boltzmann-Saha) temperature; the spectral sequence does correlate with this temperature.

Theoretically, the temperature should correlate with spectral type and so with the star's color. From the spectra of intermediate-type stars (A to K), we find that the (continuum) color temperature does so, but difficulties occur at both ends of the sequence. For O and B stars, the continuum peaks in the far ultraviolet, where it is undetectable by ground-based observations. Through satellite observations in the far ultraviolet, we are beginning to understand the ultraviolet spectra of O and B stars. For the cool M stars, not only does the Planck curve peak in the infrared, but numerous molecular bands also blanket the spectra of these low-temperature stars.

In practice, we measure a star's **color index,** $CI = B - V$, to determine the effective stellar temperature [(Section 11–4(b)]. If the stellar continuum is Planckian and contains no spectral lines, this procedure clearly gives a unique temperature, but observational uncertainties and physical effects do lead to problems: (a) for the very hot O and B stars, CI varies slowly with T_{eff} and small uncertainties in its value lead to very large uncertainties in T; (b) for the very cool M stars, CI is large and positive, but these faint stars have not

TABLE 13–1   *The Harvard Spectral Sequence*

Spectral Type	Principal Characteristics	Spectral Criteria
O	Hottest bluish-white stars; relatively few lines; He II dominates	Strong He II lines in absorption, sometimes emission; He I lines weak but increasing in strength from O5 to O9; hydrogen Balmer lines prominent but weak relative to later types; lines of Si IV, O III, N III, and C III
B	Hot bluish-white stars; more lines; He I dominates	He I lines dominate, with maximum strength at B2; He II lines virtually absent; hydrogen lines strengthening from B0 to B9; Mg II and Si II lines
A	White stars; ionized metal lines; hydrogen Balmer lines dominate	Hydrogen lines reach maximum strength at A0; lines of ionized metals (Fe II, Si II, Mg II) at maximum strength near A5; Ca II lines strengthening; lines of neutral metals appearing weakly
F	White stars; hydrogen lines declining; neutral metal lines increasing	Hydrogen lines weakening rapidly while H and K lines of Ca II strengthen; neutral metal (Fe I and Cr I) lines gaining on ionized metal lines by late F
G	Yellowish stars; many metal lines; Ca II lines dominate	Hydrogen lines very weak; Ca II H and K lines reach maximum strength near G2; neutral metal (Fe I, Mn I, Ca I) lines strengthening while ionized metal lines diminish; molecular G band of CH becomes strong
K	Reddish stars; molecular bands appear; neutral metal lines dominate	Hydrogen lines almost gone; Ca lines strong; neutral metal lines very prominent; molecular bands of TiO begin to appear by late K
M	Coolest reddish stars; neutral metal lines strong; molecular bands dominate	Neutral metal lines very strong; molecular bands prominent, with TiO bands dominating by M5; vanadium oxide bands appear

been adequately observed and so CI is not well determined for them; (c) any instrumental deficiencies, calibration errors, or unknown blanketing in the B or V bands affect the value of CI—and thus the deduced T. Hence, it is best to define the CI versus T relation observationally (next section).

13–3 ◗
HERTZSPRUNG-RUSSELL DIAGRAMS

Examine Figure 13–6. In 1911, Ejnar Hertzsprung plotted the first such two-dimensional diagram (absolute magnitude versus spectral type) for observed stars, followed (independently) in 1913 by Henry Norris Russell; today, this plot is called a

Hertzsprung-Russell (H–R) diagram. As will soon become clear to you, this simple diagram represents one of the great observational syntheses in astrophysics. Note that any two of luminosity, magnitude, temperature, and radius could be used, but visual, magnitude, and temperature are universally obtained quantities for stars.

(A) MAGNITUDE VERSUS SPECTRAL TYPE

The first H–R diagrams considered stars in the solar neighborhood and plotted **absolute visual magnitude, M,** versus **spectral type, Sp,** which is

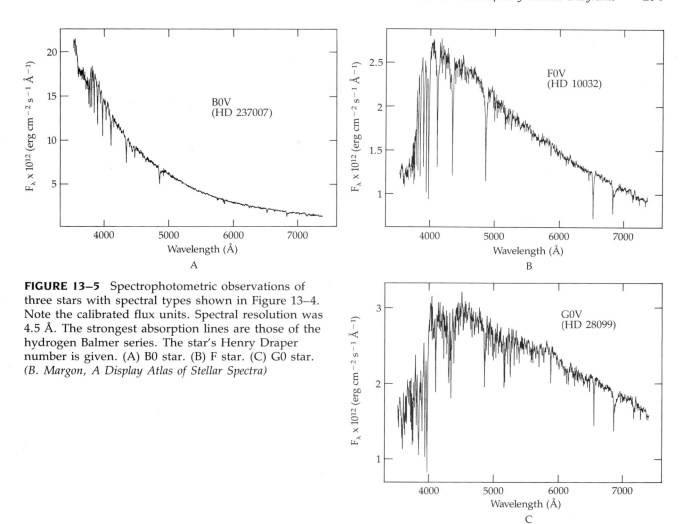

FIGURE 13–5 Spectrophotometric observations of three stars with spectral types shown in Figure 13–4. Note the calibrated flux units. Spectral resolution was 4.5 Å. The strongest absorption lines are those of the hydrogen Balmer series. The star's Henry Draper number is given. (A) B0 star. (B) F star. (C) G0 star. (*B. Margon, A Display Atlas of Stellar Spectra*)

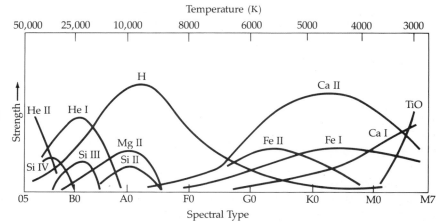

FIGURE 13–6 Absorption lines and temperature. The strengths (equivalent widths) of the absorption lines for various ionic species are shown as a function of stellar temperature. These changes result in ionization–excitation equilibria as described by the Boltzmann-Saha equation.

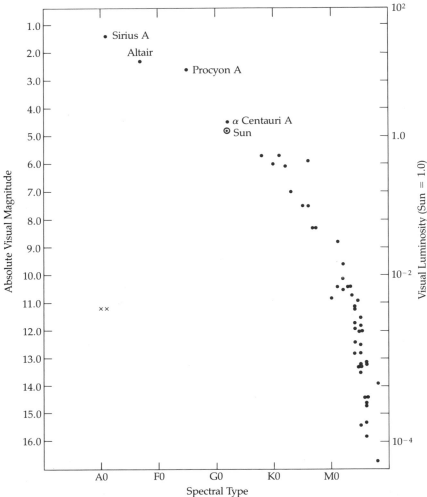

◀ **FIGURE 13–7** Hertzsprung-Russell diagrams. (A) For stars within 5 pc of the Sun. Note that most are low-luminosity, cool stars.
(B) For the brightest stars in the sky. ▶ These tend to be very luminous stars.

equivalent to luminosity versus spectral type or luminosity versus temperature. Figure 13–6 shows this type of plot for the stars with well-determined distances within about 5 pc of the Sun. Note (a) the well-defined main sequence (class V) with ever-increasing numbers of stars toward later spectral types and an absence of spectral classes earlier than A1 (Sirius), (b) the absence of giants and supergiants (classes III and I), and (c) the few white dwarfs at the lower left.

In contrast, the H–R diagram for the brightest stars includes a significant number of giants and supergiants as well as several early-type main-sequence stars (Figure 13–7). Here we have made

a selection that emphasizes very luminous stars at distances far from the Sun. Note that the H–R diagram of the nearest stars is most representative of those throughout the Galaxy: the most common stars are low-luminosity spectral type M.

Finally, Figure 13–8 presents an H–R diagrams (as well as numbers sampled at each spectral type) for selected samples of stars from the Michigan Spectral Catalog stars. Note that the low-luminosity stars bias the results for samples of nearby stars; these cannot provide the correct relative numbers in general. (White dwarfs are missing from these last two figures because of their very low luminosities.)

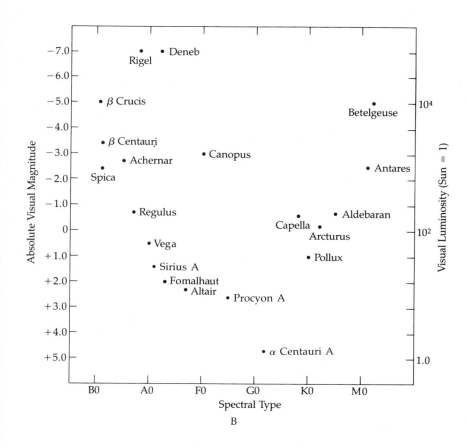

B

(B) MAGNITUDE VERSUS COLOR

Because stellar colors and spectral types are roughly correlated, we may construct a plot of absolute magnitude versus color—called a **color-magnitude diagram.** The relative ease and convenience with which color indices (such as $B - V$) may be determined for vast numbers of stars dictates the popularity of color-magnitude plots. The resulting diagrams are very similar to the magnitude-spectral type H–R diagrams considered above. Let's see what information we can glean from them.

(C) METAL ABUNDANCES AND STELLAR POPULATIONS

There are two extreme stellar populations: the young, metal-rich Population I and the old, metal-poor Population II. How has this been determined? Consider a stellar cluster that, either by its

appearance or by the common motion through space of its member stars, is known to be a self-gravitating group of stars formed at about the same time. In addition, the distance to every cluster member is about the same, so that a plot of apparent magnitude versus color is an H–R diagram. Figure 13–9 shows such a color-magnitude diagram for the Pleiades, a young **open (or galactic) cluster** in the constellation Taurus. Notice (a) the well-defined main sequence, (b) the absence of giants (luminosity classes II to III), (c) the curving up of the early end of the main sequence, and (d) the few subdwarfs (erroneously included because of the problems inherent in determining which stars *actually belong* to the cluster). The stellar spectra exhibit high metal abundance ($Z \approx 0.01$)—they are Population I.

Our Galaxy contains many **globular clusters,** which are extremely compact and spherically symmetric balls of old (12 to 15 billion years) stars (up to 500,000 stars in some). These clusters are found at great distances from the central plane of our Gal-

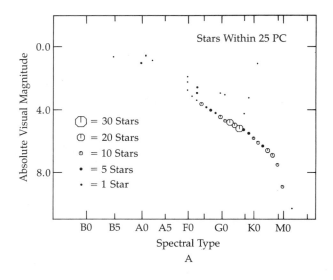

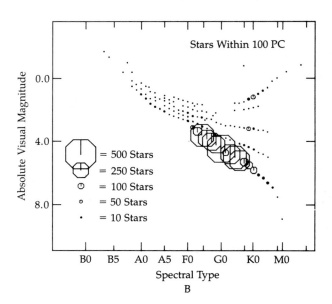

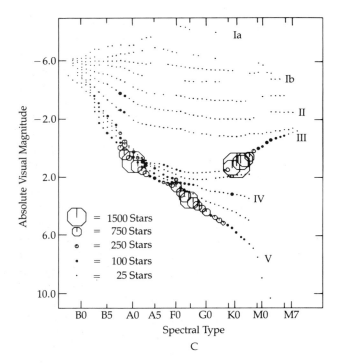

FIGURE 13–8 Hertzsprung-Russell diagrams from the Michigan Spectral Catalog. The size of the symbol represents the number of stars of that spectral type. (A) Stars within 25 pc. (B) Stars within 100 pc. (C) A sample of 36,000 stars. Here we begin to see the true relative proportions in the Milky Way. *(N. Houk)*

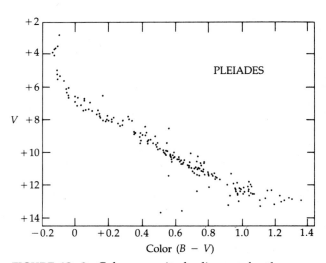

FIGURE 13–9 Color–magnitude diagram for the Pleiades. *(H. L. Johnson and R. I. Mitchell)*

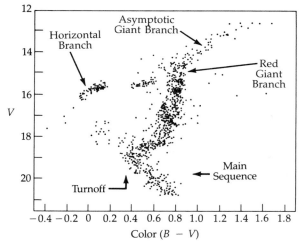

FIGURE 13–10 Color–magnitude diagram for a typical globular cluster. Note the many red giants. *(A. R. Sandage and H. L. Johnson)*

axy. Figure 13–10 is the color–magnitude diagram of the globular cluster M 3 (the third object listed in Charles Messier's catalog of the year 1781). The stellar spectra reveal a very low metal abundance ($Z \leq 0.001$), and so we assign the stars to Population II. Notice the main sequence running from $B - V = 0.8$ up to the turnoff at $B - V \approx 0.4$, the heavily populated giant branch, and the high-luminosity branch running toward the left.

So we have two well-defined stellar populations exemplified by these clusters. Population I stars (open clusters) have a much higher metal abundance (factor ≈ 100) than do Population II stars (globular clusters). As discussed in Chapter 16, the color-magnitude diagram can be understood in terms of stellar evolution in the solar neighborhood: the main sequence represents young (Population I) stars; the giants are older, more-evolved stars (Populations I and II); and the white dwarfs are stars at the end point of stellar evolution. Note that Population II is associated with old stars.

(D) LUMINOSITY CLASSIFICATIONS

So far, we have been discussing *one-dimensional* temperature sequences. As shown in Section 13–1, this description can represent stars only if they all have the same mass, radius, and composition. Chapter 11, however, shows that stars at a given temperature clearly differ in their luminosities, and so a *two-dimensional* (*L, T*) representation is absolutely necessary (Equation 13–16). Over 90% of the stars in the solar neighborhood do define a single band—the **main sequence**—on such an (*L, T*) plot, but many stars do not reside on the main sequence. What other physical parameter determines a star's location on the H–R diagram?

As early as 1897, Antonia Maury at Harvard recognized distinctly different spectra for stars of a given color temperature; certain spectral features were successively sharper (that is, narrower) than for main-sequence stars. In 1905–1907, Hertzsprung confirmed that the narrow-line stars are much more luminous than the corresponding main-sequence stars. Between 1914 and 1935, the Mount Wilson luminosity classification appeared. This scheme, originated by W. S. Adams and A. Kohlschutter, orders stellar spectra (of the same ionization or excitation temperature!) according to the strengths or weaknesses of certain spectral features. After 1937, W. W. Morgan and P. C. Keenan at Yerkes Observatory introduced the currently used **M–K luminosity classification** scheme, which defines six stellar luminosity classes and their subclasses directly in terms of observational criteria. Let's describe the M–K scheme (sometimes called the Yerkes system) in detail.

Before we do so, a word about the philosophy behind the M–K system and its limitations. Morgan and Keenan tried to devise an empirical system based only on the observable features in stellar spectra. They applied their classification to a homogeneous group of stars and did so through a group of standards to which other stellar spectra could be carefully compared. These standards seem to fall into natural groupings. In general, the M–K scheme works best for Population I stars in the solar neighborhood. For other kinds of stars (and for wavelengths outside the visible region of the spectrum), new classification systems will have to be developed. That can be done by using the same process underlying the M–K system but applying them to different stars as standards, different wavelength regimes, and different spectral resolutions.

Six M–K **luminosity classes** are differentiated, with finer subdivisions sometimes indicated, as il-

TABLE 13–2 *The Morgan-Keenan Luminosity Classes*

Class	Subclasses*	Name
I	Ia, Iab, Ib	Supergiant
II	IIa, IIab, IIb	Bright giant
III	IIIa, IIIab, IIIb	Giant
IV	IVa, IVab, IVb	Subgiant
V	Va, Vab, Vb	Dwarf†
VI	VI	Subdwarf

*In a given spectral type, luminosity *decreases* along the sequence a, ab, b.

†These are the **main-sequence** stars.

TABLE 13–3 *The Morgan-Keenan Luminosity Criteria*

Spectral Type	Spectral Criteria*
O	No criteria earlier than O9; near O9, ratio of equivalent widths of lines of He I, He II, C III, O III, and Si IV
B	Ratios of lines of He I, N II, and Si III—especially near B2; after B3, absorption line strengths of hydrogen Balmer lines, especially Hδ and Hλ
A	Until A3, same Balmer strengths; after A2, ratios of lines of Fe II, Mg II, and Ti II. For late-type A, three O I lines in the infrared ($\lambda = 777.1$ to 777.5 nm)
F	Balmer hydrogen strengths ineffective after F5; ratios of lines of hydrogen Balmer, Fe I, and Ca I around F5; in general, line-strength ratios of hydrogen Balmer to Sr II lines
G and K	Strength of molecular G band of CH; enhancement of Hδ and Hλ; relative line intensities of Sr II and Fe I lines; strong blue molecular band of CN and its other absorption bands; line ratios of Fe I, Fe II, Mn I, Ca I
M	Line ratios of Fe I, Cr I, Hδ, Sr II, and Y II; also Fe II, Ni I, Ti I, K I, and Ca I; infrared CN bands

*The luminosity classes are discerned by studying these spectral features, which depend upon temperature and hence upon spectral type.

lustrated in Tables 13–2 and 13–3; see also Figure 13–11A. If we make a two-dimensional plot of absolute visual magnitude M_v versus spectral type, these classes appear as line segments (Figure 13–11B). In this scheme, our Sun is a G2 V star (that is, yellowish main-sequence), and its radius

is much smaller than those of the giants (II to IV) of spectral type G2. Typical designations are B1 III (Beta Centauri), A3 V (Fomalhaut), F0 Ib (Canopus), K1 IVa (37 Librae), and M5 V (Barnard's Star). Note that the subdivision notations are rarely used.

Figure 13–11B is an absolute magnitude-spectral type Hertzsprung-Russell diagram. Let's concentrate on the physical interpretation of the different luminosity classes. For a given spectral type, the equivalent terminology **luminosity effect, surface gravity effect** or **pressure effect** distinguishes the luminosity classes.

Because spectral type corresponds to temperature, Equation 13–13 tells us that classes I to IV represent stars with radii much larger than those of main-sequence stars. For example, a G2 supergiant is about 12.5 magnitudes brighter than our Sun; this then implies a luminosity ratio of 10^5, or a supergiant radius of about $300R_\odot$! Because stellar masses don't exceed $100M_\odot$, this supergiant is about 10^6 times less dense than our Sun, on the average. From Equation 13–3, we see that the **surface gravity** of the supergiant is about 10^{-4} $g_\odot$ so that (by Equation 13–5) the photospheric gas pressure and electron number densities are also about 10,000 times lower than in the Sun's photosphere. Hence, the spectral features of the supergiant are different from those of the Sun—in accordance with the Saha equation—even though both stars are essentially at the *same* temperature. The pressure effect is somewhat less important than the temperature effect for it appears in the equation only linearly, whereas the temperature enters exponentially. A giant will exhibit almost the same spectrum as does a main-sequence star of the same spectral type, as long as the giant's surface temperature is lowered slightly to compensate for its lower electron density (by the Saha equation, the ratio N_{i+1}/N_i will remain the same if both N_e and T decrease appropriately). Even in this case, however, the spectral lines of the giant will be sharper than those of the main-sequence star since the giant's features suffer much less **pressure broadening** (Section 8–5).

To give you some idea of the observed characteristics of stars of different luminosity classes, we present the following data in Table A4–3: absolute visual magnitudes, color indices, effective surface temperatures, bolometric corrections, stellar radii,

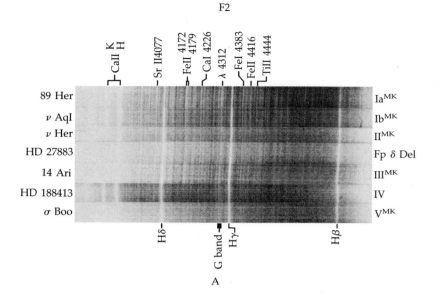

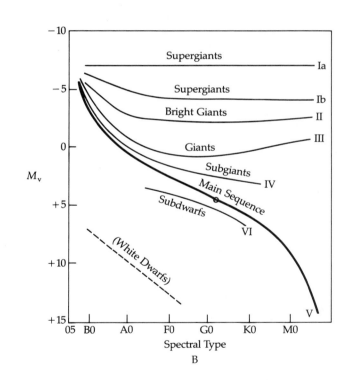

FIGURE 13–11 Stellar luminosity classes. (A) Luminosity classes as indicated by differences in spectral line intensities for spectral type F2. Again, these are negative prints; wavelengths are in Ångstroms. A superscript "MK" at right indicates a Morgan-Keenan standard. Dispersion is 108 Å/mm. *(From An Atlas of Objective Prism Spectra by N. Houk, N.J. Irvine, and D. Rosenbush)* (B) The location of the Morgan–Keenan luminosity classes on an H–R diagram.

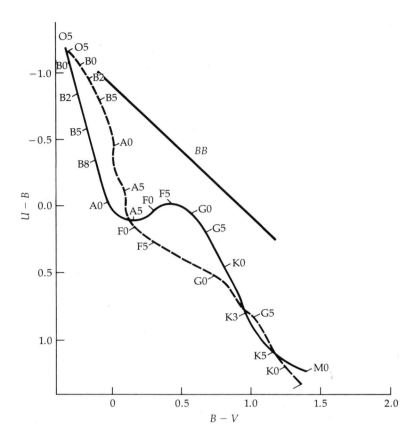

FIGURE 13–12   A color–color diagram of $U - B$ against $B - V$ for main-sequence stars (solid curve) and supergiants (dashed curve); spectral classes are indicated along each curve. The line labeled "*BB*" indicates the colors for pure blackbody radiators.

and stellar masses. Study the values and trends in this table in conjunction with Figures 13–7 and 13–9. Note that the most common stars are G and K giants and A and F main-sequence stars.

(E) COLOR–COLOR DIAGRAMS

Astronomers use a variation of the color-magnitude diagram in which one color index (such as $U - B$) is plotted against another (such as $B - V$). Such plots are called **color–color diagrams** (Figure 13–12), in which the two parameters of magnitude and luminosity are displayed in a different way. (Remember that an astronomical color is actually the ratio of fluxes at two wavelengths.)

If stars radiated exactly as blackbodies, then they would fall along a well-defined straight line in a color–color diagram (such as the one labeled

"*BB*" in Figure 13–12). But stellar atmospheres do *not* radiate as blackbodies; this fact shows up when the actual stellar colors are plotted in a color–color diagram (see the solid line in Figure 13–12 for luminosity class V and the dashed line for luminosity class I). Note the hook in the color–color main-sequence curve between spectral types A5 to G0; it is caused by the nonlinear variation of the Balmer discontinuity with temperature. The Balmer discontinuity falls right into the bandpass of the U filter, and so greatly affects the flux measured through it (Figure 13–13).

(F) ELEMENTAL ABUNDANCE EFFECTS

You may have noticed that we have not yet considered the effects of stellar chemical composition upon spectral classification, nor have we discussed

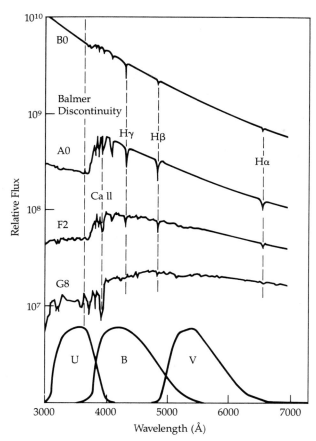

FIGURE 13–13 Profiles of the spectra of various classes of main-sequence stars compared to the *UBV* bandpasses. Note that the Balmer discontinuity falls in the *U* but not the other filters. This nonlinear effect causes the sharp bend in color–color plots.

the subdwarfs. The reason for this is simple. The vast majority of the stars in the solar neighborhood have the same composition: $X \approx 0.70$, $Y \approx 0.28$, $Z \approx 0.02$ (high metal abundance by mass). They belong to the so-called Population I [Section 13–3(c)]. The subdwarfs (Population II, low metal abundance, $Z \approx 0.001$) reside below the main sequence because of their metal deficiency. Fewer heavy elements implies less line blanketing, so that these stars appear to be hotter and bluer (A to F) than they actually are (probably G to K). Let us now consider those rare cases of Population I where elemental abundance effects do appear in the spectra.

Wolf-Rayet Stars In 1867, C. Wolf and G. Rayet discovered three O stars with anomalously strong and wide *emission* lines. Today, only about 200 of these hot (up to 10^6 K), luminous (absolute magnitudes from -4.5 to -6.5) stars are known—the so-called **Wolf-Rayet stars.** As you will see in Chapter 17, the exceedingly wide (a few nanometers) emission features of ionized He, C, N, and O seen in the spectra of these stars arise in an atmospheric envelope expanding from the star at about 2000 km/s. Two abundance branches are distinguished: (a) the **WC stars,** with an apparent overabundance of carbon (spectral features of carbon and oxygen, up to C IV and O VI, are prominent), and (b) the **WN stars,** with an apparent excess of nitrogen (N III to N V lines dominate).

Hot Emission-Line Stars In spectral types O, B, and A, we find the **Of, Be,** and **Ae stars,** with bright emission lines of hydrogen. Similar to the Wolf-Rayet stars, these stars are thought to be slowly losing mass in the form of expanding atmospheric envelopes (where the emission lines arise).

Peculiar A Stars In the spectra of **peculiar A,** or **Ap, stars,** the lines of ionized Si, Cr, Sr, and Eu are selectively enhanced. In many cases, this enhancement is time-varying (the so-called **spectrum variables**), and it appears to be associated with strong magnetic fields (≈ 1 T) at the stellar surface.

Carbon Stars Mixed in with ordinary G, K, and M stars (a temperature range from 4600 to 3100 K), we find the rare **carbon,** or **C, stars,** giants that appear to be overabundant in carbon relative to oxygen. In the early Harvard classification, these were divided into (a) the hotter R stars, distinguished by the bands of C_2 and the bands of cyanogen (CN) and (b) the cooler N stars, exhibiting C_2, CN, and CH bands with little TiO evident.

Heavy-Metal-Oxide Stars Finally, among the M stars we find a significant number of **S stars,** also known to be giants. These stars are distinguished spectroscopically by their enhanced CN absorption bands but more importantly by the presence of the molecular bands of the heavy-metal oxides ZrO, LaO, and YO instead of TiO.

(G) DISTANCE DETERMINATIONS

We end by discussing two methods by which reasonably accurate stellar distances may be determined using H–R diagrams. Both methods depend upon an accurate calibration of the **absolute-magnitude** H–R diagram. [The **moving-cluster method** (Section 15–3) gives the best calibration.]

For individual stars (or clusters in which only one star is observed thoroughly), we employ the method of **spectroscopic parallaxes** (Figure 13–14A). From a star's spectrum, we determine both its spectral type and its luminosity class. These data fix a position in the H–R diagram, from which we can read off the star's absolute magnitude. From the observed apparent magnitude, we compute the distance modulus (Equation 11–6) and then the stellar distance. Observational errors and the scatter of the H–R diagram imply a magnitude uncertainty of about ±1.0, so that the distance is known to within 50%. Spectroscopic parallaxes permit us to probe the vast distances of our Galaxy as well as nearby galaxies.

As an example, consider an MO III star with a measured apparent visual magnitude of +10. What is its distance? From Figure 13–9, note that the absolute visual magnitude of such a star is roughly 0. Then

$$m - M = 5 \log d - 5$$
$$10 - 0 = 5 \log d - 5$$
$$\log d = 15/5 = 3$$
$$d = 10^3 \text{ pc}$$

Greater accuracy in distance determination is possible by using the **main-sequence fitting technique** for an entire *cluster* of stars. Here we plot the color–apparent magnitude diagram of the test cluster and shift this plot up and down (in magnitude) on a calibrated H–R diagram until the two main sequences overlap at the same spectral types (Figure 13–14B). Note that the spectral types (or colors) of both plots *must* line up. The difference between the test-cluster apparent magnitudes and the calibrated absolute magnitudes ($m - M$) is the same for every star in the cluster; this number is the distance modulus of the test cluster. In Figure 13–14B, $m - M$ is 5.5, and so

$$m - M = 5 \log d - 5$$
$$5.5 = 5 \log d - 5$$
$$2.1 = \log d$$
$$d = 126 \text{ pc} = 410 \text{ ly}$$

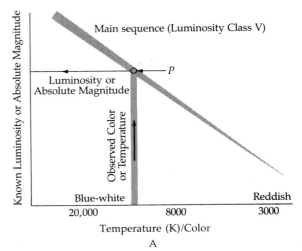

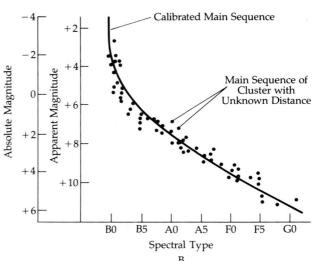

FIGURE 13–14 Distances from spectral types. (A) Estimating luminosity from a calibrated H–R diagram. (B) Matching the main sequence for a cluster whose distance is not known to that of a calibrated H–R diagram.

By using many stars, we can cancel out random errors and achieve good accuracy (±0.2) in determining the distance modulus.

(H) X-RAY EMISSION

In Chapter 10, you saw that X-ray emission from the Sun comes from hot plasma trapped in coronal loops of magnetic flux. The solar dynamo that generates the magnetic field is thought to arise from a complex combination of convection and differential rotation. The interaction of surface magnetic fields and the local plasma governs solar coronal activity. We expect the same for other stars; X-ray observations confounded this expectation. Almost all stars emit X-rays, and most far more energetically than the Sun (Figure 13–15).

Specifically, the X-ray data imply that all main-sequence stars of spectral types F through M emit X-rays with powers ranging from 10^{20} to 10^{25} W. All stars of spectral type earlier than B5 also emit X-rays, with energy outputs that range from 10^{23} to 10^{28} W. In fact, for main-sequence stars, only the narrow range from B8 to A5 shows no evidence for X-ray emission. Finally, most giant and supergiant stars emit X-rays. Those giants that do not range from spectral type A to G; the supergiants that do not are cooler than type G.

These results suggest that almost all stars have hot coronae controlled by large and numerous magnetic flux tubes. For cool stars that means convective zones and differential rotation. For hot stars, the physics is not yet clear; it may relate to vigorous stellar winds.

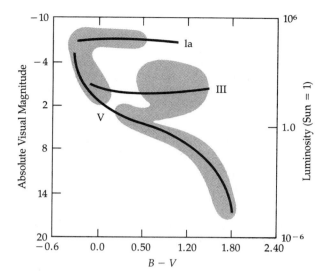

FIGURE 13–15 Regions (shaded areas) on the H–R diagram of stars that are X-ray emitters. (*R. Rosner, L. Golub, and G. S. Viana*)

PROBLEMS ◑

1. The absorption spectra of four stars exhibit the following characteristics. What are the appropriate spectral types?
   (a) The strongest features are titanium oxide bands.
   (b) The strongest lines are those of ionized helium.
   (c) The hydrogen Balmer lines are very strong, and some lines of ionized metals are present.
   (d) There are moderately strong hydrogen lines, and lines of neutral and ionized metals are seen, but the Ca II H and K lines are the strongest in the spectrum.

2. To which spectral types may the following stars be *approximately* assigned if their continuous spectra are of maximum intensity at
   (a) 50 nm   (d) 900 nm
   (b) 300 nm   (e) 1.2 μm
   (c) 600 nm   (f) 1.5 μm
   (*Hint*: Plot λ_{max} from Wien's law as a function of spectral class for the main sequence.)

3. If the parallax of a main-sequence star is in error by 25%, how far and in what direction will this star be displaced from the main sequence in an H–R diagram?

4. Which parameter in the Saha ionization-equilibrium equation is *most* important in explaining the spectral differences between
   (a) giants and dwarfs of spectral type G
   (b) B and A dwarfs

5. (a) What is the ratio of the surface gravities of a K0 V star and a K0 I star?
   (b) If both stars had the same atmospheric temperatures and opacities, what would be the approximate ratio of strengths of the Ca II K lines in their spectra? (The ionization potential of calcium is 6.1 eV.)
   (c) What assumptions did you make to answer part (b)?
   (d) What is the ratio of the mean densities of these two stars?
   (e) If the atmospheric electron densities of these stars were directly proportional to their mean densities, would your answer to (b) remain unchanged? Why?

6. (a) Describe two characteristics that distinguish subdwarfs from ordinary Population I main-sequence stars.
   (b) Why is the H–R diagram of stars in the solar neighborhood (within 500 pc) *not* an unambiguous two-dimensional plot?

7. Why is an absolute magnitude–spectral **type** H–R diagram quite different (in principle) from an apparent magnitude–$(B - V)$ H–R diagram?

8. In determining distances via the main-sequence fitting technique, why must we refrain from comparing the observed H–R diagram of a galactic cluster with the "calibrated" H–R diagram of M 3?

9. **(a)** It is sometimes said that the spectral type of a star depends only upon luminosity and surface temperature; under what conditions is this statement approximately true?
   (b) Give three generic examples that violate the statement that stellar masses are uniquely determined by their colors.

10. In Section 13–1(a), we wrote the mean molecular weight μ for a fully ionized gas as $1/\mu = 2X + (3/4)Y + (1/2)Z$, where, for example,

    $X \equiv$ hydrogen mass fraction
    = mass density of hydrogen/mass density
    of *all* constituents

    Derive this relationship for μ, indicating the assumptions and approximations used at each step.

11. The number 5040 appears in Equations 13–10 and 13–11.
    (a) Show where this number comes from (how it is derived).
    (b) What are the units of this number in these equations?

12. Use hydrostatic equilibrium to compare the central pressure of the Sun and
    (a) a B0 V star,
    (b) a G2 III star,
    (c) a G2 I star.

13. Using Figures 13–7 and 13–9, estimate the distance to an M Ib star of apparent magnitude +1.0.

14. Estimate the distances to the following clusters from their color magnitude figures (use Table A4–3 to convert $B - V$ to spectral type):
    (a) the Pleiades (Figure 13–10)
    (b) M 3 (Figure 13–12)

15. Estimate the radii of both a main sequence M star (M V) and a red supergiant (M I) using information from the H–R diagrams in the text.

16. Using the H–R diagram in Figure 13–7 and the relationship between temperature and spectral type in Figure 13–6, estimate how many times larger is Betelgeuse than
    (a) Antares
    (b) β Crucis
    (c) α Centauri.

17. **(a)** Using the excitation and ionization potentials for calcium, magnesium, helium, and hydrogen (Table 8–3), explain the relative absorption line strengths of neutral and singly ionized atoms of these four elements for stars of different spectral types as shown in Figure 13–6.
    (b) Why do both hot O and cool M stars have weak hydrogen absorption lines in their spectra?

PART THREE
The Milky Way Galaxy

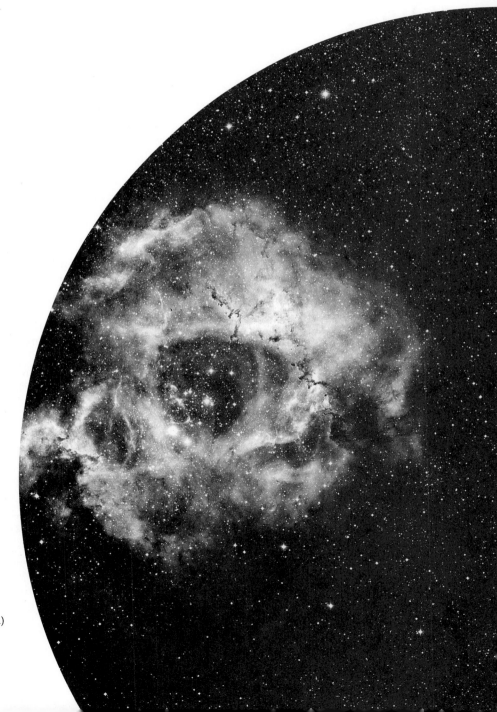

Rosette Nebula in Monoceros. *(NASA)*

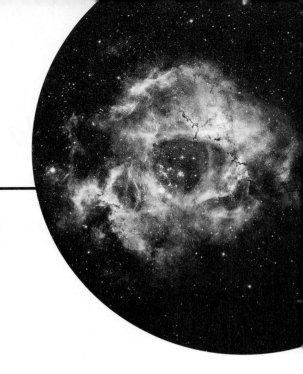

Chapter 14

Our Galaxy: A Preview

We have touched upon some of the basic characteristics of the stars. Now we expand our horizons one step farther (not the last step, by any means; see Part 4) and consider that magnificent whirl of stars—our **Galaxy.** In this chapter, we introduce you to the Galaxy, presenting a broad picture so that Chapters 15 to 18 may be considered in the proper context. In Chapter 20, this context will be reiterated and the material summarized in the framework of the Galaxy's evolution.

14–1 ◐
THE SHAPE OF THE GALAXY

(A) OBSERVATIONAL EVIDENCE

When you get away from brightly lighted urban areas, you can rediscover an irregular band of diffuse light about 10° broad that encircles the celestial sphere approximately along a great circle. We call this band the **Milky Way** (Figure 14–1). The brightest part of the Milky Way lies in the constellation Sagittarius (visible to northern hemisphere observers near the southern horizon during the

summer), and dark lanes of obscuration are evident along the midline of the Milky Way. Directly opposite on the sky, in the constellations Auriga, Perseus, and Orion, the Milky Way appears unspectacular—in fact, the anticenter lies in Auriga.

How can we determine the structure of the Milky Way? Early in this century, the luminous clouds of the Milky Way were called **star clouds** (Figure 14–2); that is, these vague blobs of light were found to be collections of millions of stars. (Galileo discovered that the Milky Way contained individual stars with his telescope in the seventeenth century.) Section 14–2 will show how the stars of our Galaxy (from the Greek **galaktikos,** meaning "milky") are distributed in space.

In 1781, the French astronomer Charles Messier published a catalog (Appendix 1) listing about a hundred fuzzy (nonstellar-appearing) objects he observed while searching for comets; some of these were later found to be glowing clouds of gas in our Galaxy [Section 19–2(b)], open clusters, and distant globular clusters, but some others—the so-called **nebulae** (from the Latin singular **nebula,** meaning "cloud")—remained an enigma until

270

A

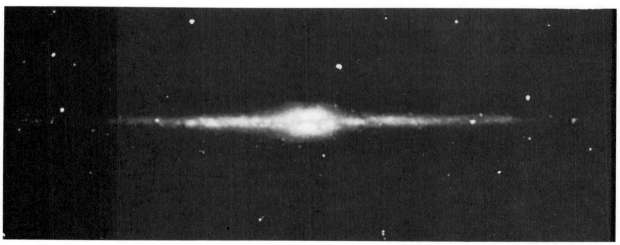

B

FIGURE 14–1 The Milky Way (A) Wide-angle view of the Milky Way viewed from the Earth. The center of the Galaxy lies in the direction of the bright bulge in the upper center. Note the dark lanes through the plane. The three skinny, dark triangles are parts of the camera support. *(A. D. Code and T. E. Houck)* (B) A wide-angle (96°) view of the Milky Way from space. This image from COBE combines views at 1.2 μm, 2.2 μm, and 3.4 μm and reduces the dust obscuration that blocks the optical view in (A). Note the well-defined central bulge. *(NASA)*

FIGURE 14–2 Star clouds in Sagittarius. Note again the dark lanes from dust. (*Hale Observatory, California Institute of Technology*)

well into the twentieth century. Today, we understand that some nebulae are *galaxies*, separate extragalactic entities similar to our own Galaxy (Chapter 21). One of the nearest, the Andromeda galaxy (Messier 31), turns out to be similar in shape and contents to our Galaxy (Figure 14–3).

Our Galaxy is apparently a highly flattened (like a pancake) stellar system that is being viewed (from the Sun) edge-on from a point far from its center. Note that the galactic center lies in the direction of the central **nuclear bulge**, that luminous swelling evident in Figures 14–1. Within this nuclear bulge lies the mysterious **nucleus** of the Galaxy.

(B) THE GALACTIC COORDINATE SYSTEM

To map the Galaxy, we disregard the geocentric coordinates based on the Earth's celestial equator and celestial poles (Appendix 10), as well as the heliocentric ecliptic system, and define a **galactic coordinate system** (Figure 14–4). Here the center line of the Milky Way (more accurately, the mass centroid of the galactic plane) defines a great circle on the sky called the **galactic equator,** along which **galactic longitude,** l, is measured in degrees (0 to 360) eastward from the direction to the center of the Galaxy in Sagittarius. **Galactic latitude,** b, is the angular distance on the celestial sphere (in degrees from 0 to ± 90) either north or south from the galactic equator. Hence, the galactic anticenter is at $l = 180°$, $b = 0°$, the north galactic pole (NGP) at $b = +90°$, and the south galactic pole (SGP) at $b = -90°$.

In terms of right ascension, α, and declination, δ (in the geocentric celestial equatorial system), the (epoch 1950) coordinates of the NGP are roughly $12^h 49^m$, $+27.4°$; hence, the galactic equator is inclined about 63° to the celestial equator and is almost perpendicular to the ecliptic. Figure 14–4

FIGURE 14–3 Messier 31 (M31), the Andromeda galaxy. This is a spiral galaxy similar to our own. Note the central nuclear bulge. *(Lick Observatory)*

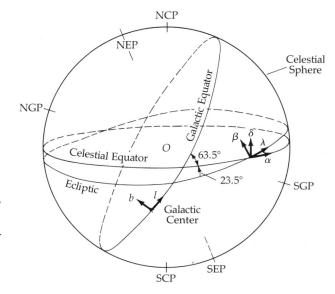

FIGURE 14–4 Coordinate systems. Three coordinate systems are centered on the observer *(O)*: (1) celestial equatorial with the north celestial pole (NCP), south celestial pole (SCP), and equator; (2) ecliptic, inclined 23.5° to the celestial equator, with north ecliptic pole (NEP) and south ecliptic pole (SEP); (3) galactic, with the galactic equator inclined 63.5° to the celestial equator, and a north galactic pole (NGP) and a south galactic pole (SGP).

shows how these three important angular-coordinate systems are related: (a) the celestial equatorial (α, δ) system, (b) the ecliptic (λ, β) system, and (c) the galactic (l, b) system. Note in particular the location of the galactic center $(l = 0°, b = 0°)$ at α (1950) = 17^h 42^m 29^s, δ (1950) = $-28°59'18''$. Observations show the position of the center of the Galaxy, and the galactic coordinate system is defined with this presumed position of the center as the zero point.

14–2 ◗
THE DISTRIBUTION OF STARS

The most direct way to determine the size and shape of our Galaxy is to investigate the spatial distribution of stars. Here you will learn how to "count" the stars, see how misconceptions can arise and have arisen in interpreting the data, and uncover the stellar indications of the Galaxy's size and its spiral-arm features.

(A) STAR-COUNTING

How can we find out the extent and layout of the Galaxy? One technique involves counting stars in different directions in the sky. Assuming a uniform distribution in space, the directions in which more stars are seen are the directions in which the Galaxy extends greater distances. Consider the solid-angular area ω (in units of steradians) on the celestial sphere. At a distance r from the observer, the solid angle subtends an area A,

$$A = \omega r^2$$

The volume contained between r and a distance dr farther out is

$$dV = \omega r^2 \, dr \qquad (14\text{--}1)$$

If $D(r)$ is the number density (number of stars per unit volume) at the distance r, then the number of stars in this volume is

$$n(r) = D(r) \, dV = \omega D(r) r^2 \, dr \qquad (14\text{--}2)$$

$D(r)$ may be treated as constant if one uses the number density in the vicinity of the Sun. Integrating over all volume elements from the observer up to the distance r, we get the total number of stars contained in the solid angle ω within that distance:

$$N(r) = \int_0^r n(r) \, dr = \int_0^r \omega D(r) r^2 \, dr$$
$$N(r) = (1/3)D(r)r^3\omega \qquad (14\text{--}3)$$

To study a large number of stars, however, it is much easier to consider their *apparent magnitudes* rather than their distances. Assume that we consider only stars of the absolute magnitude M (selected by spectral type, for example); then apparent magnitude and distance are related by (Equation 11–6)

$$\log r = (m - M + 5)/5 = 0.2m + \text{constant}$$

or

$$r = 10^{0.2m + \text{constant}} \qquad (14\text{--}4)$$

We now use this in Equation 14–3 to derive an expression for the number of stars of a given absolute magnitude within a certain area of the sky brighter than apparent magnitude m:

$$\log N(m) = 0.6m + C \qquad (14\text{--}5)$$

where the constant C incorporates the dependence on M, ω, and D. This equation tells us that, assuming a uniform density, there should be $10^{0.6} = 3.98$ times as many stars of a given absolute magnitude at apparent magnitude $m + 1$ as at m. A spread in absolute magnitude can be allowed for by an appropriate adjustment in C.

The assumption of uniform star density is the most questionable and has been found to be invalid by comparisons of observed and predicted star counts. Two factors cause this discrepancy: the nonuniform distribution of stars and interstellar absorption.

Interstellar Absorption

We now know that we are *not* located at the center of the Galaxy, but early star-count data indicated that our Sun was at the center. The reason that people believed the Sun was in the center of the Galaxy (wrongly!) was found in 1930 by the work of Robert J. Trumpler, who, while studying star clusters, noted that many appeared strangely faint for their observed angular sizes. By Equation 11–6, these dim clusters would be very distant, but then their linear sizes (in parsecs) would be very large. These problems disappear when we assume, as Trumpler did, that interstellar space is filled with obscuring material (dust) that is concentrated toward the galactic plane. Note the dark lanes of obscuration defining the galactic plane in Figure 14–1.

Starlight is scattered and absorbed by this dust, and so we refer to the phenomenon as **interstellar absorption** (Chapter 19). Because the stars observed at high galactic latitudes (perpendicular to the plane of the Galaxy, at the Sun's position) farther than 500 pc from the plane do not strongly exhibit the characteristic effects of interstellar absorption, we can believe that the disk of the Galaxy is thin, about 1 kpc at the Sun. At low galactic latitudes (near the galactic plane), however, the obscuration becomes severe. We characterize this absorption by a medium of uniform density and opacity (Section 10–2); then Equation 10–2 implies an exponential diminution of starlight with distance traveled through the medium. Because stellar magnitude is proportional to the logarithm of the observed flux (Equation 11–5), the increase in apparent magnitude from interstellar absorption is proportional to the distance to the star. Near the galactic equator, the visual absorption is about 1.0 mag/kpc; in very dense interstellar clouds, it is much higher. So a star 1 kpc distant in the galactic plane has an apparent distance modulus of 11.0 mag even though its actual distance modulus is 10.0 mag; this star appears to be about 1.6 kpc distant. A star at 5 kpc suffers 5.0 mag of absorption, so that not only is it dim, but its distance is overestimated by a factor of 10 if we neglect the effect of interstellar absorption. (Note that 5 mag is a factor of 100 brightness, and $100^{1/2} = 10$, by the inverse-square law for light.)

You can now understand why the Galaxy appeared small and heliocentric to astronomers early in this century; the interstellar obscuration in the galactic plane was the culprit. By including absorption in the analysis of the star-count data, we find that we are roughly midway to the edge of a very large (≈ 50 kpc diameter) stellar system, and the number density of stars increases rapidly toward the galactic center.

Luminosity Function

In addition to star counts, we can characterize the stars in our region of the Galaxy by their **luminosity function:** the number of stars per unit spatial volume with a given absolute magnitude M (or luminosity L). We can infer a star's absolute magnitude by studying its spectrum and placing it ap-

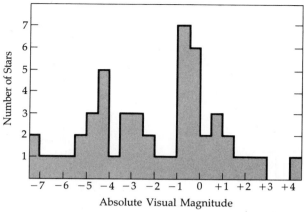

FIGURE 14–5 Luminosity function for the 50 brightest stars, binned in ranges of absolute magnitude.

propriately on a calibrated H–R diagram; that is, we estimate its luminosity from its spectral type.

As a crude, and incorrect, first attempt, we can consider the 50 brightest visible stars, plotting a histogram of the numbers occurring in the absolute magnitude intervals M to $M + 1/2$ (Figure 14–5). We expect that those stars that appear the brightest are also intrinsically bright (note the preponderance at negative M), but Figure 14–5 is biased since very luminous stars can be seen to great distances. A truer picture of the actual luminosity function is given by the histogram of the 50 nearest stars (Figure 14–6); note the overwhelming predominance of intrinsically faint stars ($10.0 \leq M \leq$

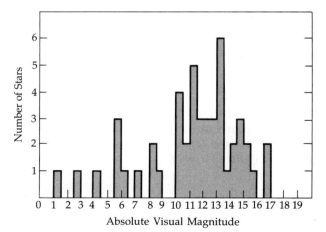

FIGURE 14–6 Luminosity function for the 50 nearest stars, binned in ranges of absolute magnitude. Note that most of the stars have low luminosity.

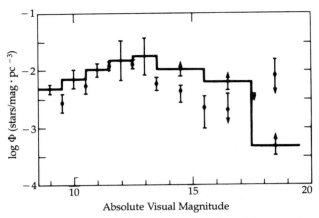

FIGURE 14–7 Local luminosity function. The sample here is complete to an absolute magnitude of 19 (solid circles). The solid line is a step function that fits the data fairly well. *(G. Gilmore, Royal Observatory, Edinburgh)*

15.0) and the dearth of very luminous stars. A stellar sample complete to a limiting magnitude of $M \approx 19.0$ at V band (Figure 14–7) shows that most stars are very dim ($M > 10.0$), with few being brighter than our Sun, which is by no means a very luminous star ($M = 5$). Faint stars are in the overwhelming majority (the distribution has a broad peak near $M \approx 13$), with only a sparse sprinkling of luminous stars. These results apply to the solar neighborhood and to similar regions in the Galaxy. The luminosity function differs in locations, for instance, far above and below the midplane and in the central regions of the nucleus.

(B) LUMINOUS STARS AND STELLAR CLUSTERS

Luminous stars and stellar clusters are discernible to very great distances. For example, if we survey to a limiting magnitude of $m = 15.0$, B stars of absolute magnitude $M = -5.0$ can be seen to a distance of about 100 kpc; interstellar absorption in the galactic plane will, however, limit us to an actual distance of 5 kpc. Stellar type O and B stars are very young, and they tend to occur in small clusters called **associations;** moreover, these hot, luminous stars can ionize the hydrogen gas in which they are imbedded, producing a bright region of ionized hydrogen (Chapter 19). Such asso-

ciations and ionized hydrogen regions give an indication of the spiral-arm structure in the galactic disk. In Chapter 19, you will see that interstellar molecular hydrogen gas best delineates these spiral features.

Recall that a **globular cluster** is a dense spherical cluster of about 10^5 to 10^6 stars; some 200 such globular clusters are known to be associated with our Galaxy, and they are very luminous ($M \approx -4$ to -10). Because the nearest globular cluster is 3 kpc distant, we have a good probe of galactic distances. The distance to a given globular cluster may be found (1) by the main-sequence fitting technique, (2) by the apparent magnitudes of certain well-known types of stars in the cluster, and (3) by the apparent angular diameter of the cluster. When the distance and direction of globular clusters are plotted, we find today that they define a spherical system of radius ≈ 100 kpc, with its center ~ 8 kpc away from the Solar System in Sagittarius (Figure 14–8). Harlow Shapley discovered this effect in 1917, and he correctly concluded that the Sun is far from the galactic center (the center of the globular cluster system). This interpretation receives strong support from the observation of a globular cluster system, of similar size and shape, centered on the middle of the Andromeda galaxy (Figure 14–3). The current conventional standard puts the Sun at about 8.5 kpc (± 1 kpc) from the center, which gives the galactic disk a diameter of at least 50 kpc.

14–3 ◗
STELLAR POPULATIONS

Chapter 13 stated that stars could be assigned to different **stellar populations,** characterized by their observed metal abundances, which are the abundances of all elements (Z values) *except* hydrogen (X) and helium (Y). In our Galaxy, we find that the stars in some globular clusters are extremely metal-deficient ($Z \leq 0.001$); we call these stars **extreme Population II** stars; most of the individual stars seen far from the galactic plane (in what is called the **galactic halo**) are of this population. Within about 500 pc of the galactic plane, the spatial density of stars has increased so markedly that we speak of the pancake-shaped **galactic disk** with its **central bulge** at the galactic center. In the disk, the metal abundance is greater than in Population II stars, and it generally *increases* as we ap-

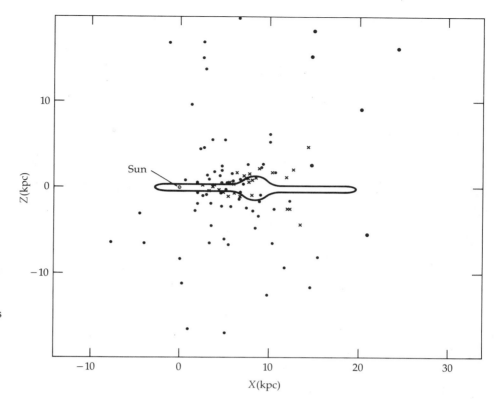

FIGURE 14–8 Space distribution of globular clusters is the *XZ* plane; the Sun's location is indicated. Distance scales are in kpc. The disk and nuclear bulge are outlined. (*W. E. Harris*)

proach the galactic plane. We call these the **disk Population** stars. Both Population II and disk Population stars are very old (billions of years), and their brightest representatives are *red giants*. However, in the galactic plane (especially in the spiral arms) we find the young, luminous blue stars of **Population I** (greatest metal abundance; $Z \geq 0.01$). Population I stars are seen in **open (galactic) clusters,** in O and B associations, and near concentrations of interstellar dust and gas; these turn out to be associated with regions of star formation.

The abundance of heavy elements increases from the halo to the spiral arms buried within the disk. In Chapter 16, you will see that hydrogen and helium are processed into heavier elements in the centers of stars and returned to the interstellar medium upon the demise of these stars. So our Galaxy likely began as a spherical cloud of essentially H and He, from which the metal-poor Population II stars formed, the cloud was enriched with metals as it collapsed toward the galactic plane, and only in the denser regions of the spiral arms is star formation now producing the metal-rich stars of Population I.

In summary, we can roughly divide the stars in the Galaxy into two general groups on the basis of their metal abundances. The point of this scheme is to try to group together stars of similar age, composition, and (as you will see in detail later) kinematics. The key to this choice of collective characteristics is to study the history of the Galaxy as recorded in its stars. Metal abundance and orbital characteristics are fossil properties; they are unchanged by galactic evolution. In essence, we are focusing on stellar *ages*, because they are directly correlated to metal abundances.

14–4 ◗ GALACTIC DYNAMICS: SPIRAL FEATURES

We have described the stellar content of our Galaxy, but what about the dynamics of the system? (Stellar motions are considered in detail in Chapter 15.) We assume that our Galaxy is held together by the mutual gravitational attraction of its stars. The spherical part of the system (the halo with its host of globular clusters) is then analogous to a spherical stellar cluster. The flattened disk, however, clearly implies a rapidly rotating entity that formed *after* the halo of globular clusters.

Every star associated with our Galaxy moves within the gravitational attraction produced by stars and other matter (some of which is invisible). Assume circular stellar orbits about the galactic center in the galactic plane. In particular, our Sun is in such an orbit, with a radius of 8.5 kpc. In Chapters 15 and 20, we show that, near the galactic center, the orbital angular speed ω (radians per second) is approximately constant (**rigid-body rotation**), whereas near the Sun, the circumgalactic speed slowly rises with distance from the center of the Galaxy and the Sun's speed is about 220 km/s toward $l = 90°$. At the Sun's position in the Galaxy, we may approximate the solar motion as a circular Keplerian orbit about a massive central body of mass M_G. Since the centripetal acceleration maintaining this circular orbit is produced by the gravitational attraction between the core (M_G) and the Sun ($M_\odot$), we have

$$v_\odot^2/R_\odot = GM_G/R_\odot^2 \qquad \textbf{(14–6)}$$

where $v_\odot$ is the Sun's circular speed (220 km/s), $R_\odot$ is the distance to the galactic center (8.5 kpc), and G is the constant of universal gravitation. Equation 14–6 may be evaluated to give the mass of our Galaxy within the solar orbit:

$$
\begin{aligned}
M_G &= v_\odot^2 R_\odot/G \\
&= (2.20 \times 10^5 \text{ m/s})^2 (2.6 \times 10^{20} \text{ m})/ \\
&\qquad (6.7 \times 10^{-11} \text{ m}^3/\text{kg} \cdot \text{s}^2) \\
M_G &= 1.9 \times 10^{41} \text{ kg} \approx 10^{11} M_\odot \qquad \textbf{(14–7)}
\end{aligned}
$$

Hence, if an average star has about solar mass, our Galaxy consists of approximately 100 billion stars; the most massive spiral galaxies known have masses about 10 times the Galaxy's mass.

Today we believe that the spiral arms of our Galaxy are also produced by dynamical effects. If this were not the case, these features would wind up and disappear (Figure 14–9) in only a few revolutions of the Galaxy. The Galaxy is about 15×10^9 years old, and at the Sun's location, it rotates once every 240 million years ($\approx 2\pi R_\odot/v_\odot$); it is unlikely that we could observe a spiral structure that was not a steady dynamical feature of the galactic disk. A spiral arm is thought to be a manifestation of a rotating density wave in the galactic disk (Chapter 20). The disk is initially unstable to density perturbations (essentially sound waves), which can grow and gravitationally attract material along spiral

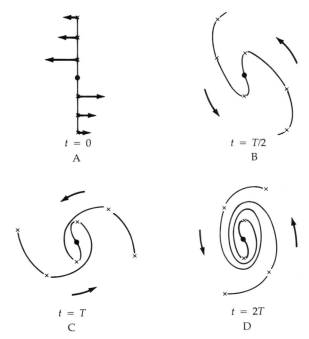

$t = 0$
A

$t = T/2$
B

$t = T$
C

$t = 2T$
D

FIGURE 14–9 The wind-up problem. Here are two hypothetical "spiral arms" made of six stars following Keplerian orbits. After two cycles, the arm configuration has disappeared into a tightly wound pattern.

paths, but these waves rotate only half as fast as the disk, so that material passes through the density pattern in the direction of galactic rotation. Star formation starts as the dust and gas permeating the disk are compressed at the spiral-arm feature; that is why hot, young Population I stars outline spiral structure.

14–5 ◗
A MODEL OF THE GALAXY

We close this chapter by describing a schematic model of our Galaxy that ties together the observations (Table 14–1) and considerations presented above (Figure 14–10). The entire Galaxy is contained within its spherical **halo,** with a diameter of roughly 100 kpc. Here reside the old Population II halo stars and the globular clusters (which contain Population II stars). These objects define the spherical envelope while they move along highly eccentric and inclined orbits about the galactic center with orbital periods near 10^8 years—a close analogy to the spherical comet cloud centered upon our Sun (Section 7–3). The globular clusters

TABLE 14–1   *General Properties of the Galaxy*

Property	Approximate Value
Disk diameter	50 kpc
Halo diameter	100 kpc
Sun's distance from center	8.5 ± 1.0 kpc
Height of Sun above disk	8 pc
Total mass	7.2×10^{11} solar masses
Mass of gas	8×10^9 solar masses
Mass of central bulge	4×10^{10} solar masses
Optical luminosity	3×10^{36} W
Density of stars in solar neighborhood	$0.05\ M_\odot/\text{pc}^3$

Milky Way (Figure 14–11). The halo also includes gas in the shape of an expanded disk with a thickness of no more than 10 kpc. This gas is clumpy, contains both neutral and ionized areas, and corotates with the rest of the Galaxy. Because we can easily see out of the Galaxy, the amount of dust in the halo must be small. The halo also contains dark, as yet unobserved, matter.

Bisecting the spherical halo is the circular **galactic plane.** Inward to the galactic plane, the spatial density of stars increases and their metal abundance rises. We have now reached the thick **galactic disk,** with a perpendicular scale height of about 1 kpc and the large central bulge at its center. About half the mass of the Galaxy ($\approx 10^{11} M_\odot$) interior to the Sun's orbit resides here, in disk Population stars that move in low-eccentricity orbits about the galactic center. The ages of these stars are mostly old.

The central bulge ($R \le 1$ kpc) is very difficult to observe. The metal abundances of the old stars located within the bulge range from lower to greater than that of the Sun. Though the bulge is dominated by old stars, it contains some young stars where it intersects the plane.

At the very center of the Galaxy is the small (< 1 pc diameter) massive (about $10^6 M_\odot$) nucleus. In other spiral galaxies, the nucleus has a star-like appearance, but in our Galaxy it is hidden from our view by the interstellar obscuration in the galactic plane. It makes its presence known by emit-

fall into two subgroups: one with random motions in a spherical distribution with no rotation about the galactic center; the other with a flatter distribution and a mean rotation speed about half the circular speed for objects at this distance.

The outer halo of the Galaxy has no definite boundary. Observations of globular clusters (Figure 14–11) indicate that the halo extends to at least 100 kpc and then includes the Magellanic Clouds and dwarf galaxies that are companions to the

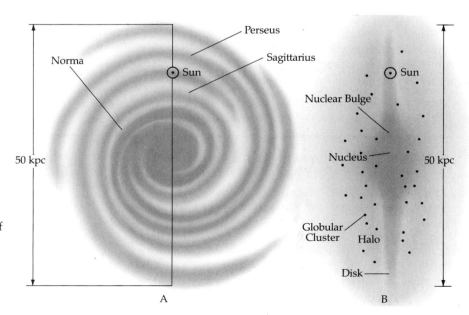

FIGURE 14–10 Overall model of the galaxy. (A) Top view, showing the spiral arm structure and the names of a few nearby arms. (B) Side view, showing the nuclear bulge and halo.

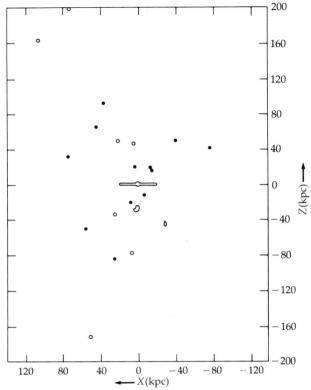

FIGURE 14–11 The local region of space near the Milky Way, showing a cross section (in the *XZ* plane) of the distribution of material in the outer halo of the Galaxy: globular clusters (dots), dwarf galaxies (circles), and the Magellanic Clouds (kidney-shaped). The scale is in kiloparsecs with distances relative to the Galaxy's center. *(Adapted from a diagram by B. Carney)*

ting thermal and nonthermal radio continuum emission, infrared radiation, X-rays, and even gamma rays. In the infrared at wavelengths near 2 μm, the galactic nucleus is 30″ in diameter (1.5 pc in linear dimensions!) with several nearby sources in a sphere of radius 10 pc. The galactic nucleus may be an extremely compact stellar cluster wherein events of great violence are and have been taking place, or it may be a single very massive object, possibly a black hole.

At the midline of the disk is the **galactic plane,** with its veneer of **spiral arms** in a region only about 500 pc thick (consider how thin this really is; the ratio of diameter to thickness is about 100:1). Newborn stars of Population I, as a result of their high luminosities and short lifetimes, signal the presence of the spiral density waves propagating around the galactic plane. Myriad clouds of hydrogen gas and dust occupy the galactic plane, with some tendency to collect near the spiral features. The *interarm* regions contain some disk Population and old Population I stars, with stellar number densities about 10% that in an arm.

Our Sun, a Population I star about 5 billion years old, is located 8.5 kpc from the galactic center, on the inner edge of one spiral arm. Although it is not exactly in the galactic plane, it is only a small distance above it. Carrying the Solar System with it, the Sun orbits the galactic center once every 240 million years in a path of very low eccentricity. As we gaze out from the Earth upon the celestial flywheel that is our Galaxy, we see the neighboring stars participating with us in the galactic rotation toward $l = 90°$ and the breathtaking band of the Milky Way encircling the sky from its center in Sagittarius. The Sun resides in the **thin disk,** which has a vertical scale height of only a few hundred parsecs and contains most of the stars in the Sun's vicinity. These stars have almost circular orbits, metal abundances similar to the Sun, and ages that range from old to newborn. The thin disk also contains the Galaxy's gas and dust, with a scale height smaller than that for the stars.

PROBLEMS ◗

1. Using a simple diagram, explain why our Galaxy appears as the Milky Way in the night sky.

2. Using Figure 14–4, describe the various aspects of the Milky Way on evenings of different seasons for an observer in New Mexico. Use the altitude–azimuth horizon system of coordinates, and mention the tilt of the Milky Way to the horizon and the observability of Sagittarius.

3. The integral count ratio $N(m + 1)/N(m) = 3.98$ with the assumption that stars are uniformly distributed in space. Make a schematic diagram plotting log $N(m)$ versus m and show what effect each of the following would produce:
   (a) a uniformly distributed interstellar obscuration of 1 mag/kpc
   (b) a strongly absorbing interstellar dust cloud localized at the apparent magnitude m_{cloud}

4. What stellar population would you expect to find (and why)
   (a) in the nucleus of our Galaxy

(b) in the spiral arms of the Andromeda galaxy

(c) in the Pleiades star cluster

(d) in intergalactic space (beyond the halo)

(e) in the galactic bulge

5. (a) A globular cluster is in elliptical orbit ($e = 0.9$) about the center of our Galaxy, reaching *apogalacticon* (farthest distance from the center) at the distance 40 kpc. What is the *perigalacticon* (nearest) distance, and how long will this cluster require to complete one orbit?

(b) What is the approximate speed of escape from the Galaxy in the solar neighborhood if the Sun's circular orbital speed about the galactic center is 220 km/s?

6. Consider the center of our Galaxy to be a spherical star cluster of radius R and uniform mass density ρ.

(a) What is the total mass M of this cluster?

(b) What is the mass contained within the sphere of radius $r < R$?

(c) In terms of M, what is the angular speed ω of a star in circular orbit at a distance r ($< R$) from the cluster's center? Note that $\omega = v/r$, where v is the circular orbital speed of the star in kilometers per second.

(d) Using an analogy, explain why we refer to the result of part (c) as *rigid-body rotation*.

7. Assume that the galactic disk may be approximated by a plane-parallel slab 500 pc thick and having the galactic plane at its midline. Take the Sun to be located in the galactic plane.

(a) How many magnitudes of absorption are there at $b = 90°$?

(b) How many magnitudes of absorption are there at the general galactic latitude b?

(c) Explain why the region $b \leq 10°$ is called the "zone of avoidance" (essentially total obscuration in terms of apparent magnitudes).

8. In the direction perpendicular to the galactic plane, approximately how thick (in kiloparsecs) is the galactic bulge?

9. This chapter suggests that stars are formed when disk materials (gas and dust) catch up with a density wave. Assume that the newborn stars continue about the galactic center with a circular speed appropriate to their distance.

(a) How far will an O star move from the spiral arm of its birth in 1 million years?

(b) Are you surprised to find our Sun in a spiral arm (or near one, at least)?

10. If interstellar absorption averages 1 mag/kpc, calculate the optical depth for this absorption in the galactic plane.

11. For each of the bins in Figure 14–6, convert absolute magnitude to luminosity (in solar units). Multiply the luminosity by the number of stars for each bin.

Make a table of the product number of stars × luminosity versus absolute magnitude. If these stars are representative of the Galaxy, at what absolute magnitude (counting upward) is 95% of the Galaxy's luminosity accounted for? Roughly what range of spectral types contributes the bulk of the Galaxy's luminosity?

12. Repeat the previous exercise except convert luminosity to mass by the rough formula $M \approx L^{1/3}$, where M and L are in solar units. Is most of the mass contained in high-, low-, or intermediate-mass stars? Roughly what range of spectral types contributes most of the mass of the Galaxy?

13. The star Deneb, in the constellation Cygnus, is one of the most luminous stars visible with the unaided eye. Stars like Deneb are therefore the most distant stars we can see in our Galaxy.

(a) Given its apparent visual magnitude of 1.3 and distance of 430 pc, calculate the absolute visual magnitude of Deneb (ignore interstellar absorption).

(b) Ignoring interstellar absorption, how distant could a star like Deneb be and still be visible to the unaided eye ($m = 6$)?

(c) Deneb lies close to the galactic plane. Assuming a constant interstellar absorption of 1.0 magnitude per kiloparsec, estimate the absorption towards Deneb in magnitudes.

(d) Assuming interstellar absorption, how distant could a star like Deneb be and still be visible with the unaided eye?

14. Assume that the interstellar absorption in the galactic plane has the roughly constant value of 1 magnitude per kiloparsec.

(a) What is the absolute magnitude of the faintest star that could be detected at the galactic center by a telescope with a limiting magnitude of 23.5?

(b) Compare this result to the value if there were no interstellar extinction.

(c) In each case, what stars on the H–R diagram have the required absolute magnitude?

(d) These stars at the galactic center cannot actually be observed at visual wavelengths. Assuming that the stars do indeed exist, why do you think they are not visible?

15. (a) Assuming a constant interstellar extinction of 1 magnitude per kiloparsec, what is the maximum distance to which we could see a bright globular cluster in our Galaxy using a telescope with a limiting visual magnitude of 23.5?

(b) Globular clusters are observed around the Andromeda Galaxy (M31), a nearby spiral galaxy 960 kpc from our Galaxy. Reconcile this observation with your answer in (a).

Chapter 15

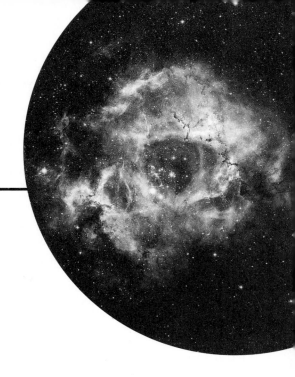

Galactic Rotation: Stellar Motions

O ur Solar System resides in a rotating spiral galaxy known as the Milky Way Galaxy, the Sun one of its many stars. This chapter explains the observable motions of the stars in our Galaxy (especially those in the solar neighborhood) to verify galactic rotation and evaluate its characteristics quantitatively. These results imply a total mass and mass distribution in the Galaxy. We begin with the observational properties of stellar motions. Then we investigate the characteristic rotational velocity of our Sun about the center of the Galaxy and the differential galactic rotation.

15–1 ◐
COMPONENTS OF STELLAR MOTIONS

Before we plunge in, we need to clarify coordinates and reference systems (Appendix 10). The most familiar are **geocentric** (Earth-centered) coordinate systems, such as the system of right ascension and declination, and the **galactic** coordinate system. However, dynamical investigations are best referred to a **heliocentric** (Sun-centered) coordinate system. In practice, astronomers distinguish between apparent celestial positions (the geocentric positions of objects determined from Earth-bound observations) and true celestial positions (the heliocentric positions relative to our Sun). True stellar positions (such as heliocentric right ascension and declination) differ from apparent positions because the latter exhibit periodic displacements, such as geocentric parallax orbits, as a result of the Earth's orbital motion around the Sun. We can correct terrestrial observations to a heliocentric basis because the Earth's orbit is well known. So we discuss only heliocentric celestial positions. This chapter introduces yet another dynamical reference system, called the **local standard of rest,** which is based upon the rotational dynamics of our Galaxy.

Stellar motions arise from the velocities of stars through space. These velocities are vectors, which may be resolved into two perpendicular components: the radial velocity along the line of sight and the tangential velocity in the plane of the sky.

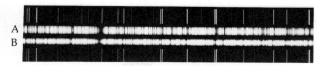

FIGURE 15–1 Doppler shifts of Arcturus. The two spectra, A and B, were taken six months apart. The upper spectrum gives a speed of +18 km/s; the lower one, −32 km/s. The difference in the shifts arises from the Earth's orbital motion. The bright lines above and below are the rest-frame reference spectra. *(Palomar Observatory, California Institute of Technology)*

(A) RADIAL SPEED

The **radial speed**, v_r, of a star is its speed of approach or recession; it is easily obtained from the Doppler shift of stellar spectral lines. To determine this Doppler shift, a laboratory comparison spectrum is photographed adjacent to the stellar spectrum and the relative positions of lines in the two spectra are measured (Figure 15–1). The measured Doppler shift, $\Delta\lambda = \lambda - \lambda_0$, permits us to deduce the radial speed of the star from the Doppler formula (Equation 8–13):

$$v_r = (\Delta\lambda/\lambda_0)c \qquad (15\text{--}1)$$

where c is the speed of light, λ the measured wavelength, and λ_0 the rest wavelength. When $\lambda > \lambda_0$, the spectrum is *redshifted* and v_r is the radial speed of *recession*; when $\lambda < \lambda_0$, the spectrum is *blueshifted* and the star is *approaching* us. To refer these motions to the Sun, we must correct for the component of the Earth's orbital velocity (speed $\approx$ 30 km/s) along the line of sight to the star.

The radial speed may be determined for any star for which a spectrum may be obtained. The distance to the star is irrelevant, for it is only the flux of the star that determines whether it is bright enough to produce a spectrum when observed through a telescope–spectrograph combination. We can now measure stellar radial speeds to a precision of about 0.1 km/s.

(B) PROPER MOTION

The motion of a star in the plane of the celestial sphere is called the **proper motion,** μ, and it is usually expressed in seconds of arc per year. For a given speed perpendicular to the line of sight, the proper motion will be greater the closer the star is to us, an obvious analogy with trigonometric parallax [Section 11–1(a)]. Very distant stars exhibit no measurable proper motion, and they may therefore be used as *reference*, or *background*, stars. Most proper motions are very small (Table 15–1); the largest known is that of Barnard's Star; it is an exceptional 10″/year (Figure 15–2).

Compared with the cyclic nature of parallax orbits, proper motions have the distinct advantage of being cumulative; measurements may be made many years apart, so that small annual angular displacements can accumulate to an easily measured amount. An accuracy of ±0.003″/year is attainable when observations are spaced over decades, but we must exercise great care in the selection of standards of reference. As a result of galactic rotation (Section 15–4), the ideal reference system is based upon distant quasars (Chapter 24). Because observations of proper motions are made many years apart, precautions similar to those taken for paral-

TABLE 15–1 *Selected Stars With Large Proper Motions*

Name	Constellation	V Magnitude	Annual Proper Motion(″)
Groombridge 1830	Ursa Major	6.45	7.05
Lacaille 9352	Piscis Australis	7.34	6.90
61 Cygni	Cygnus	4.84	5.22
Lalande 21185	Ursa Major	7.49	4.77
ϵ Indi	Indus	4.69	4.70
α^1 Cetauri	Centaurus	0.00	3.68
Arcturus	Bootes	−0.04	2.28

FIGURE 15–2 Proper motion of Barnard's Star. Indicated by an arrow, Barnard's Star shows considerable proper motion in the time interval from August 24, 1894 (top) to May 30, 1916 (bottom). *(Yerkes Observatory)*

lax observations are necessary. In particular, corrections must be made for stellar parallax and for the aberration of starlight. Yet proper-motion measurements are well worth the trouble, for they are a key to our knowledge of the structure of the Galaxy.

(C) TANGENTIAL SPEED

The proper motion of a star comes from its **tangential speed** v_t, which is its linear speed in the direction perpendicular to the line of sight. To convert the angular measure of proper motion to a linear

speed (in kilometers per second) transverse to the line of sight, we must know the distance d to the star, for

$$v_t = d \sin \mu \approx \mu d \qquad \textbf{(15–2)}$$

The last part of Equation 15–2 follows because μ is very small (less than 5×10^{-5} rad/year). Be careful to remember the units, for d is normally given in parsecs and μ in seconds per year. Equation 15–2 gives v_t in parsecs per year when μ is in radians per year, but v_t is found in astronomical units per year when μ is in arcseconds per year; we use d in parsecs in either case. By applying the appropriate conversion factors (astronomical units to kilometers and years to seconds), we find that

$$v_t = 4.74\mu''d = 4.74(\mu''/\pi'') \text{ km/s} \qquad \textbf{(15–3)}$$

where the distance d is given in parsecs, the parallax π'' in seconds of arc, and the proper motion μ'' in arcseconds per year. For example, a star whose distance is 100 pc and whose proper motion is 0.1"/year has a tangential speed of 47.4 km/s.

(D) SPACE MOTION

The **space velocity V** of a star with respect to the Sun has been decomposed into two perpendicular components (Figure 15–3): (1) the radial velocity, whose magnitude is v_r (the radial speed), and (2) the tangential velocity, with magnitude v_t (tangential speed). From the Pythagorean theorem (and the law of vector addition), we find

$$V^2 = v_r{}^2 + v_t{}^2 \qquad \textbf{(15–4)}$$

Once again we caution you that all speeds in Equa-

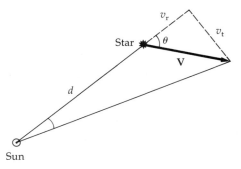

FIGURE 15–3 Components of space velocity. As viewed from the Earth, a star's space velocity, **V**, can be broken down into a radial (v_r) and transverse (v_t) component.

tion 15–4 must be in the same units; this need implies (from Equation 15–3) that we must know the distance to the star. The angle that the space velocity makes with the line of sight is θ, and it is found from

$$\tan \theta = v_t/v_r$$

Hence, these two components can be added (Equation 15–4) to find the total speed and direction of a star in space (relative to the Sun). For stars in the Sun's neighborhood, the magnitude of their space velocities averages 25 km/s.

15–2 ◗
THE LOCAL STANDARD OF REST

Our Galaxy has two physically preferred frames of reference. The first is the **galactocentric** system: centered at the nucleus of the Galaxy, its reference plane is the galactic plane and its reference axis is the galactic rotation axis. We are primarily interested in the second system here—the **local standard of rest (LSR)**.

One way to define the LSR is by the **dynamical LSR:** the reference frame, instantaneously centered upon the Sun, which moves in a circular orbit about the galactic center at the circular speed appropriate to its position in the Galaxy. So all stars in the solar neighborhood that are in circular galactic orbits are essentially at rest in the dynamical LSR. Any deviations from circular motion in the solar neighborhood will appear as stellar peculiar motions with respect to the dynamical LSR.

The Sun's galactic orbit is not perfectly circular. So, with respect to the LSR, a solar motion of 19.5 km/s occurs toward the constellation Hercules ($l = 56°$, $b = 23°$). On the celestial sphere, the Sun is moving toward the **solar apex** and away from the **solar antapex** (Figure 15–4A). The nature and extent of the solar motion were first demonstrated by William Herschel in 1783 using statistical methods; let's paraphrase his analysis in modern terms.

The stars in the solar neighborhood exhibit **peculiar motions** with respect to the LSR; that is, they swarm about like sluggish bees. If the Sun were at rest in the LSR, the average of these peculiar velocities would also be zero with respect to the Sun. If the Sun moves with respect to the LSR, however, each star will (in addition to its peculiar motion) reflect this solar motion to an extent dependent upon its position on the celestial sphere (Figure 15–4B). Stars on the great circle 90° from

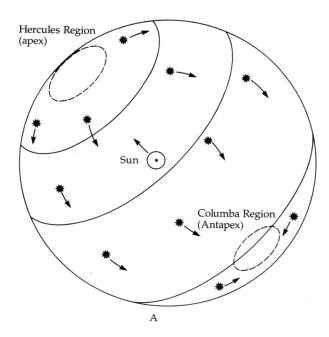

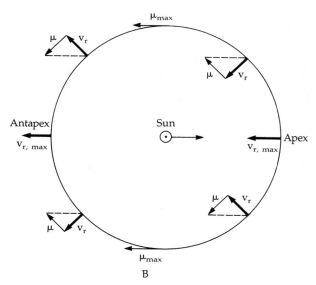

FIGURE 15–4 Solar motions. (A) The Sun's motion relative to local stars in the direction of Hercules, the apex of the motion, makes them appear to stream toward the antapex in Columba. (B) Reflected solar motion. The solar motion with respect to the LSR affects stars around the celestial sphere. Hence, the stars reflect the Sun's motion toward the apex and away from the antapex.

both the apex and the antapex will, on the average, exhibit the largest proper motions toward the antapex; hence, proper-motion averages (out to about 50 pc) reveal the locations of the apex and antapex. We determine the speed of the solar motion by averaging the radial speeds of stars near the apex and stars near the antapex; the average radial speed of approach is greatest at the apex (another way to locate the apex) and that of recession is greatest at the antapex (another way to locate the antapex).

Finally, note that the average proper motion vanishes at both the apex and the antapex. Because the stars sampled for radial speed can be much more distant than those sampled for proper motion, a different apex location will result from each sample; in practice, the two locations are almost identical.

15–3 ◗
MOVING CLUSTERS

We briefly mentioned **moving stellar clusters** in Section 11–1(a). Recall that such a cluster is a gravitationally bound group of stars traveling through the Galaxy. Hence, all the cluster members exhibit the same peculiar motion relative to the LSR because the stellar motions are completely correlated, not random. We can identify the members of a given cluster by this property. Random observational errors permit a small number of stars that are not part of the cluster to slip into our chosen sample and limit us, by Equation 15–3, to those clusters nearer than about 500 pc.

Consider a small, or distant, cluster which subtends a small solid angle on the sky. Its radial speed v_r and angular proper motion μ'', with respect to the LSR, may be obtained by studying a single star (and correcting for the reflected solar motion). To avoid errors from the possible inclusion of stars not part of the cluster, we usually take averages over several stars. The direction of the cluster's space velocity is unknown, however (it is too distant for parallax measurements), so that we cannot find the cluster distance from Equation 15–3. Some other distance criterion must be used in this case (see the following).

On the other hand, the situation improves when we have a nearby cluster that subtends a large solid angle on the sky. Then (Figures 15–5

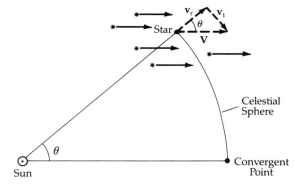

FIGURE 15–5   Moving-cluster geometry. A cluster of stars moves with space velocity **V**, and the proper motions appear directed to the convergent point on the celestial sphere.

and 15–6), because all cluster members are moving in the same spatial direction, their proper motions appear to converge toward (or diverge from) a single point on the celestial sphere. We call this the **convergent point.** This phenomenon occurs from perspective effects that cause parallel motions of stars to appear to radiate from a single point on the sky. Again, we know the radial speed of the moving cluster once we have determined it for a single member star, but now we also know the direction of the cluster's space motion. When we extend the proper motions of the member stars on the sky, they intersect at, for example, the convergent point of the cluster (Figure 15–6). The angular distance θ from the convergent point to a member star is the same as the angle between the line of sight to that star and the star's space velocity vector **V** (Figure 15–5):

$$V = v_r/\cos \theta \qquad (15–5)$$

but we know that $v_t = V \sin \theta$, and so using Equations 15–3 and 15–5 gives us the cluster distance from its derived parallax:

$$\pi'' = 4.74\mu''/v_r \tan \theta \qquad (15–6)$$

These **moving-cluster parallaxes** can apply to individual stars; that is the extraordinary power of this technique.

Radial velocities for about 40 stars in the Hyades have been observed to obtain a cluster radial velocity of 39.1 km/s and a convergent-point position of

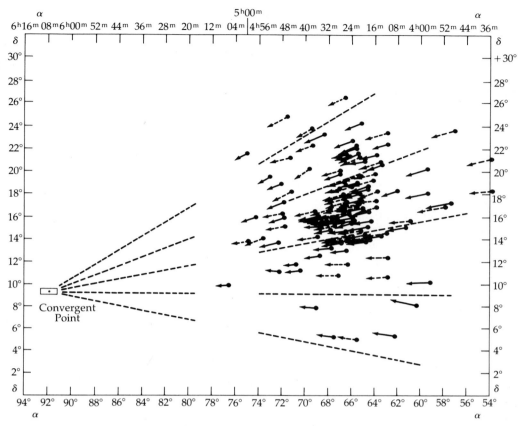

FIGURE 15–6 Hyades motions. This nearby cluster has a well-defined group proper motion that meets at a convergent point (rectangle with dot). Data from 1908 to 1954. *(From O. Struve, B. Lynds, and H. Pillans, Elementary Astronomy, Oxford University Press, 1959)*

RA = 95.3° and DEC = 7.2°. The distance modulus then comes out to $m - M = 3.23$ for a distance of 44.3 pc. This result coincides almost exactly with the distance found from trigonometric parallaxes: $m - M = 3.25$. Hence, we know the distance to the Hyades with an error of a little less than 1 pc, or an accuracy of about 2%, which is amazing for a distance measurement outside of the solar system.

Once we have obtained an accurate cluster distance, we can make an absolute calibration of the H–R diagram of that cluster [Section 13–3(g)]; the Hyades distance of 44.3 pc (obtained by the moving-cluster method) now provides the fundamental distance scale upon which all distances greater than about 100 pc (even to the limits of our Universe!) are based. Distances less than 100 pc are based on trigonometric parallaxes. By using the

Hyades H–R diagram and the main-sequence fitting technique, astronomers have extended the scale to the Praesepe (Beehive) star cluster in Cancer at 159 pc and to the Double Cluster (*h* and *χ* Persei) in Perseus at 2330 pc—with an accuracy of about 10% in the distances. Chapter 22 shows how this distance scale may be further extended to extragalactic objects and the Universe.

15–4 ◖
GALACTIC ROTATION

(A) DIFFERENTIAL GALACTIC ROTATION

Stars in the solar neighborhood orbiting the galactic center in perfectly circular orbits would be at rest in the LSR. This statement implies **rigid-body rotation** (ω = *constant* angular speed about the galactic center) of our region of the Galaxy, however. The particles a rigid body is made up of remain at fixed distances from one another, and every particle moves around the center of the body in the same period. Now $\omega = v/r$, where v is the circular orbital speed and r is the orbital radius; hence, for a rigid body, $v \propto r$ if the stars orbit in such a way that Equation 14–6 holds. Now, the Sun orbits in a roughly Keplerian fashion, but then $v \propto r^{-1/2}$ and $\omega \propto r^{-3/2} \neq$ constant. We say that there is **differential galactic rotation** when the orbital angular speed is a function of distance from the galactic center, $\omega = \omega(r)$.

Let's explicitly show the difference between solid-body and Keplerian motion. Consider most of the Galaxy's mass concentrated at its center. Then Kepler's third law,

$$P^2 = (4\pi^2/GM)R^3$$

where R is the distance to the center at which a mass orbits with circular velocity V and period P such that

$$P = 2\pi R/V$$

gives us

$$4\pi^2 R^2/V^2 = (4\pi^2/GM)R^3$$

and

$$V = (GM/R)^{1/2} \propto R^{-1/2}$$

for Keplerian motion.

In contrast, picture a spherical mass with a uniform density ρ throughout. Then the centripetal acceleration a at a distance R from the center is

$$a = V^2/R = GM(R)/R^2$$

where $M(R)$ is the mass within R, or

$$M(R) = 4\pi R^3\rho/3$$

so that

$$V^2/R = G(4\pi R^3\rho/3)/R^2$$

and

$$V = (4\pi G\rho/3)^{1/2}R \propto R$$

for solid-body rotation (ω = constant).

Let's now follow Jan Oort's brilliant work of 1927 and derive the effects of such differential ga-

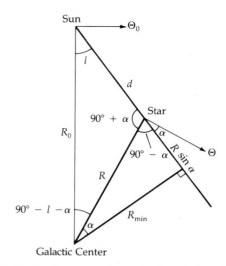

FIGURE 15–7 Geometry for galactic rotation. The Sun, galactic center, and a star define the galactic plane; all motions are assumed to be circular.

lactic rotation (particularly in the solar neighborhood). At the end, you will see that these effects are indeed observable.

First, assume for simplicity *circular* galactic orbits in the galactic plane (Figure 15–7). Here we have defined R as, for example, the star's distance from the galactic center, R_0 as the Sun's distance from the center, d as the Sun–star distance, Θ as the star's circular orbital speed, Θ_0 as the orbital speed of the LSR, l as the galactic longitude of the star, α as the angle between the line of sight to the star and its orbital velocity, ω as the star's galactic *angular* speed, and ω_0 as that of the LSR.

We begin with v_r, the star's radial speed with respect to the LSR. From Figure 15–7, we have

$$v_r = \Theta \cos \alpha - \Theta_0 \sin l \qquad (15\text{–}7)$$

and the law of sines (Appendix 9, "Mathematical Operations") gives

$$\sin (l)/R$$
$$= \sin (90° + \alpha)/R_0 \equiv \cos (\alpha)/R_0 \quad (15\text{–}8)$$

but $\omega = \Theta/R$ (and $\omega_0 = \Theta_0/R_0$), so that using Equations 15–7 and 15–8,

$$v_r = R_0(\omega - \omega_0) \sin l \qquad (15\text{–}9)$$

For a rigid rotation, $\omega \equiv \omega_0$, so that $v_r \equiv 0$; differential galactic rotation implies a finite radial speed for the star.

How about v_t, the star's tangential speed with respect to the LSR? Again, from Figure 15–7:

$$v_t = \Theta \sin \alpha - \Theta_0 \cos l \qquad (15\text{–}10)$$

and now the law of sines gives

$$\sin (l)/R$$
$$= \sin (90° - l - \alpha)/d \equiv [\cos (l + \alpha)]/d$$
$$\equiv (\cos \alpha \cos l - \sin \alpha \sin l)/d \qquad (15\text{–}11)$$

where the last equality follows from the identity $\cos (x + y) = \cos x \cos y - \sin x \sin y$. We solve Equation 15–11 for $\sin \alpha$ and use Equation 15–8 for $\cos \alpha$ to put Equation 15–10 in the form

$$v_t = R_0(\omega - \omega_0) \cos l - d\omega \qquad (15\text{–}12)$$

The **Oort formulas,** Equations 15–9 and 15–12, are for concentric, circular, coplanar orbits; you will see them extensively in Section 15–4(c) and Chapter 20. Here, however, we specialize to the solar neighborhood, where $d \ll R_0$. To do so makes use of an approximation valid for regular functions.

If $f(x)$ is a smoothly varying curve, then we can find the value of this curve near some point $x = x_0$ by using the Taylor expansion:

$$f(x) = f(x_0) + (df/dx)_{x_0}(x - x_0)$$
$$+ (1/2)(d^2f/dx^2)_{x_0}(x - x_0)^2 + \cdots$$

Now $(df/dx)_{x_0}$ is the curve's slope at x_0, and its curvature is $(d^2f/dx^2)_{x_0}$, or the rate of change of the slope with x; recall that $x - x_0$ is small. The first two terms in this expansion, with $(df/dx)_{x_0} = $ constant, approximate $f(x)$ when the curvature is negligible. Near the Sun, ω is approximately equal to ω_0, and so, to a first approximation,

$$\omega - \omega_0 \approx (d\omega/dR)_{R_0}(R - R_0) \qquad (15\text{–}13)$$

where $(d\omega/dR)_{R_0}$ is the rate of change of the orbital angular speed with respect to distance evaluated at the distance $R = R_0$. Historically, we use the **Oort constant** A, defined as

$$A \equiv -(R_0/2)(d\omega/dR)_{R_0} \qquad (15\text{–}14)$$

to write the radial speed (Equation 15–9) as

$$v_r = -2A(R - R_0) \sin l \qquad (15\text{–}15)$$

but you can see that, for small d, Figure 15–7 implies that $R_0 - R \approx d \cos l$, so that Equation 15–15

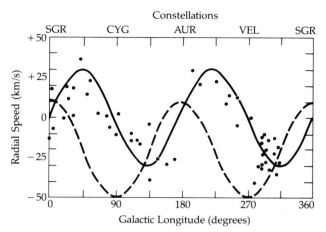

FIGURE 15–8 Observed galactic rotation. The radial velocities of nearby Cepheids (dots) are plotted as a function of galactic longitude. These are motions with respect to the LSR. The dashed curve is the expected motions in the Oort model; the solid curve is a fit to the data.

takes the final form

$$v_r = Ad \sin 2l \qquad (15\text{–}16)$$

where the identity $\sin 2l = 2 \sin l \cos l$ has been used. The tangential speed equation (15–12), if the same approximation (Equation 15–13) and the relation $R_0 - R \approx d \cos l$ are assumed and terms like d^2 or smaller are neglected, becomes

$$v_t = d(A \cos 2l + B) \qquad (15\text{–}17)$$

where the **Oort constant** B is given by

$$B \equiv A - \omega_0 \qquad (15\text{–}18)$$

and where the identity $\cos^2 l = (1/2)(1 + \cos 2l)$ has been used.

If the radial and tangential speeds, given by Equations 15–16 and 15–17, are plotted versus galactic longitude, the resulting curves are termed double-sinusoids (sinusoidal with a period of 180°, not 360°). Figure 15–8 shows observed stellar motions plotted in this way—it is clear that our Galaxy is rotating differentially. Figure 15–9 illustrates the physical cause for the results shown in Figure 15–8. Because A is positive, the angular speed $\omega(R)$ decreases with increasing R in the solar neighborhood. It decreases rapidly enough that the following effects are observable: (a) stars with $R < R_0$

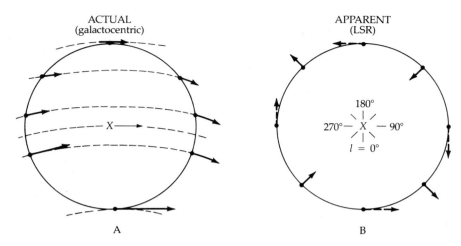

ACTUAL
(galactocentric)

APPARENT
(LSR)

180°

270° — X — 90°

$l = 0°$

A

B

FIGURE 15–9 Differential galactic rotation. (A) Stars and the LSR (X) orbit the galactic center with smaller orbital speeds at greater distances. (B) With respect to the LSR, we expect a double-sinusoidal pattern in stellar radial velocities.

are moving about the galactic center more rapidly than the LSR and (b) stars with $R > R_0$ are orbiting at lower angular speeds than the LSR. So both sets of stars appear to be moving toward larger values of l; those closer in are passing us by, and those farther out are being left behind by the LSR. The resulting apparent stellar motions with respect to the LSR are shown in Figure 15–9, decomposed into radial and tangential parts; compare this with Figure 15–8.

(B) CHARACTERIZING THE LSR

The LSR is specified by its distance R_0 from the center of the Galaxy and its circular orbital speed Θ_0 around the center of the Galaxy. How can we determine these parameters? The stellar effects of differential galactic rotation (Figure 15–8), for example, may be evaluated using Equations 15–16 and 15–17 to find Oort's constants A and B. Since radial velocity data permit us to reach great distances, the value of $A = 15$ km/s · kpc has been found from B stars and Cepheid variables (Section 18–2). Proper-motion data give us the more uncertain value of $B = -10$ km/s · kpc. Using Equation 15–18, we have $\omega_0 = \Theta_0/R_0 = A - B = 25$ km/s · kpc. Clearly, an independent determination of either R_0 or Θ_0 is needed before the other unknown can be found. Recent observations suggest that $A = 14$ and $B = -12$, so that $A - B = 26$ km/s · kpc. A value of $A = -B = 13$ is well within the errors of the observations.

In the next section, you will see that a good value for the combination AR_0 is obtainable from

radio astronomical observations of the neutral atomic hydrogen (H I) and carbon monoxide (CO) in our Galaxy. Using the value for A, we find $R_0 \approx 8.5$ kpc and $\Theta_0 \approx 220$ km/s. If we could somehow determine $\omega(R)$ for our Galaxy and thence $(d\omega/dR)_{R_0}$, Equation 15–14 would give us R_0; unfortunately, we must already know R_0 if we are to obtain $\omega(R)$. The values we use were those recommended to the International Astronomical Union in 1985 to be adopted for R_0 and Θ_0. (The old values were $R_0 = 10$ kpc and $\Theta_0 = 250$ km/s.) Recent observations of motions in the galactic center give a value of 7.1 ± 1.2 kpc; we'll stick with 8.5 kpc.

To better understand differential galactic rotation, let's derive the Oort constants for the LSR in Keplerian orbit about the mass $M_G = 1.5 \times 10^{11} M_\odot$. Then

$$\omega^2 R = GM_G/R^2$$

or

$$\omega(R) = (GM_G/R^3)^{1/2} \qquad \textbf{(15–19)}$$

Differentiating Equation 15–19 with respect to R gives

$$d\omega/dR = -(3/2)(GM_G)^{1/2}R^{-5/2} = -3\omega/2R$$

so that Oort's first constant is (by Equation 15–14)

$$A = (3/4)\omega_0 = 19 \text{ km/s · kpc} \qquad \textbf{(15–20a)}$$

Finally, from Equation (15–18),

$$B = A - \omega_0 = -(1/4)\omega_0 \qquad \textbf{(15–20b)}$$
$$= -6.5 \text{ km/s · kpc}$$

Note that $R_0 = 8.5$ kpc was used in Equation 15–19

to obtain Equations 15–20. The calculated values of A and B assuming Keplerian orbits do *not* agree with the observed values. Why not? Because the Galaxy is not a point mass!

(C) THE ROTATION CURVE OF OUR GALAXY

As a result of differential galactic rotation, $\omega = \omega(R)$, so that $\Theta = \Theta(R)$—the latter relationship is called the **rotation curve** of our Galaxy. The rotation curves for other (nearby) galaxies are found by measuring the radial velocities of H I features in these galaxies, but for our own Galaxy the problem is more intricate.

Return to Figure 15–7 and notice that the maximum radial speed $v_{r,max}$ observed at a given galactic longitude occurs when the line of sight passes closest (R_{min}) to the galactic center. Here the line of sight is tangent to the orbit, and from Figure 15–7 we have

$$R_{min} = R_0 \sin l \qquad \textbf{(15–21)}$$

Now Equation 15–9 is inverted to the general forms

$$\Theta(R_{min}) = v_{r,max} + \Theta_0 \sin l \qquad \textbf{(15–22a)}$$

and

$$\omega(R_{min}) = \omega_0 + (v_{r,max}/R_0 \sin l) \quad \textbf{(15–22b)}$$

where $v_{r,max}$ is observed and R_{min} may be found from Equation 15–21. Note that in Equations 15–22 at least two of the following must be known before we can determine the rotation curve: R_0, Θ_0, and ω_0. If we confine our attention to the radial velocities of nearby objects ($d \ll R_0$), then Equations 15–15 and 15–21 lead to

$$v_{r,max} = 2AR_0 (\sin l)(1 - \sin l) \qquad \textbf{(15–23)}$$

and we obtain the combination AR_0 by observing at galactic longitudes near (but smaller than) 90° and near (but larger than) 270°, because the line of sight can be tangent only to orbits interior to R_0.

Equation 15–23 is useful for distances of a few kiloparsecs, but the interstellar obscuration (Section 19–1) severely impedes stellar studies at greater distances. Fortunately, H I and CO can be seen by radio techniques to vastly greater distances, and then we must use Equations 15–22. Because H I and CO clouds tend to delineate spiral features, only a few places occur where the line of sight is tangent to a spiral arm, and we may evaluate $v_{r,max}$ from these observations, as discussed in Chapter 20. Also, emission from the CO molecule can be used together with data from the associated bright stars to determine the rotation curve beyond the solar circle ($R > R_0$). By combining optical and radio data, we can deduce a rotation curve for our Galaxy (Figure 15–10). From these data, $A = 17.7$ and $B = -8.1$.

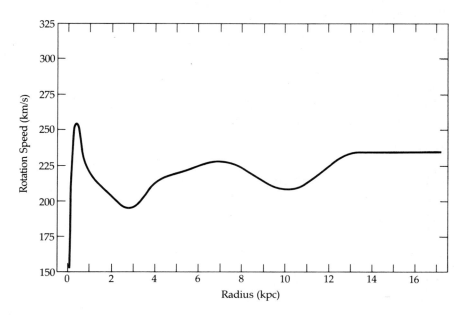

FIGURE 15–10 A rotation curve of the Galaxy. Based primarily on CO observations, this scale is based on a distance of 8.5 kpc from the galactic center. Note how the curve flattens out beyond 14 kpc. *[D. P. Clemens, Astrophysical Journal 295:422 (1985)]*

Note that the actual galactic rotation curve does *not* follow that expected from simple Keplerian motions. From close to the galactic center out to 300 pc, the curve rises steeply, then drops, bottoming out at about 3 kpc. It then rises slowly out to near the position of the Sun. Carbon monoxide observations show the rotation curve in the outer parts of the Galaxy. The curve rises more beyond the Sun, reaching almost 300 km/s at 18 kpc. What does this curve tell us? Because even the outer parts of the Galaxy do not revolve in a Keplerian fashion, much of the Galaxy's material must lie out beyond the Sun's orbit. From the rotation curve out to 18 kpc, the Galaxy's mass is 3.4×10^{11} solar masses (Equation 14–6). So at least as much mass lies exterior to the solar circle as interior to it. Much of this matter is invisible; the Galaxy has a massive halo of nonluminous matter (and so do other galaxies of the same type as ours).

Note that, roughly, the rotation curve is flat outside the central region of the Galaxy. A flat rotation curve requires that $A = -B$, and a value of 13 for each falls within the present error range. However, it is not yet clear that a flat rotation curve applies to the local area in which the rotation constants are derived.

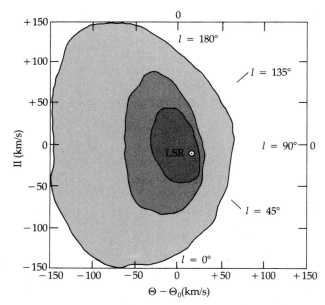

FIGURE 15–11 Stellar motions in the galactic plane. The observed velocities with respect to the LSR are plotted for stars in the solar neighborhood. The darkest shading are A stars; the intermediate shading, older K giants; the lightly shaded area, very old disk and Population II stars.

(D) HIGH-VELOCITY STARS

Now to return to the solar neighborhood to complete our study of stellar motions. Nearby stars are observed to be moving in every possible direction with respect to the LSR (consider the solar motion, discussed in the preceding sections); these stars are in eccentric orbits about the galactic center, and so they cannot be at rest in the LSR. Decompose a star's peculiar motion with respect to the galactic center into three mutually perpendicular components: (1) Π, the speed radially outward (toward $l = 180°$), (2) Θ, the speed toward $l = 90°$ as before, and (3) Z, the speed perpendicular to the galactic plane (positive toward $b = 90°$). Now the stellar-motion components with respect to the LSR are $(\Pi, \Theta - \Theta_0, Z)$; for example, in this notation the solar motion is $(-10.4, 14.8, 7.3)$ km/s. So the Sun is moving inward toward the galactic center and forward toward Cygnus and is rising out of the galactic plane relative to the LSR.

In Figure 15–11, we schematically show the velocity distributions of several types of stars in the galactic plane $(\Pi, \Theta - \Theta_0)$. Here we can clearly see that young stars are almost at rest in the LSR, that older stars are moving faster with respect to the LSR, and that very old Population II stars are in rapid motion relative to the LSR. The Z components of the stellar velocities behave similarly, so that the older a star is, the more rapidly and the farther it moves out of the galactic plane. This phenomenon is associated with the origins of the stars: young stars are born in spiral arms and move in nearly circular orbits in the galactic plane, whereas the older disk and Population II stars were born far from the galactic plane (and even in the galactic halo).

We can compute velocity curves for stellar orbits of different eccentricities and semimajor axes (Figure 15–11). This procedure shows that young stars and our Sun are in nearly circular galactic orbits at $R \approx R_0$ and that very old stars (including subdwarfs) and the globular clusters are in highly eccentric orbits with $R_0/2 \le R \le R_0$ (here R represents the semimajor axis).

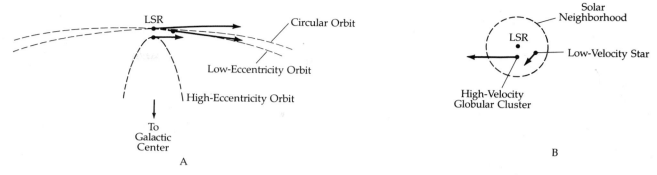

FIGURE 15–12 High-velocity objects. (A) Three galactic orbits are shown: the circular LSR, a low-eccentricity orbit of a nearby A star, and a high-eccentricity globular cluster near apogalacticon. (B) Relative to the LSR, these same objects in the solar neighborhood appear as a low-velocity A star and a high-velocity globular cluster.

Note that the stellar velocities seem to avoid the direction of galactic rotation ($l = 90°$), and few stars move faster than about 65 km/s in this direction. This effect is most noticeable for the so-called **high-velocity stars,** those with in-plane speeds greater than 65 km/s relative to the LSR. (Stars with larger velocities move on very elliptical orbits.) Two separate phenomena are involved here. A local star whose total speed is greater than 450 km/s (the local escape speed) relative to the galactic center will escape from the Galaxy; hence, the deficiency of stars with $\Theta - \Theta_0 \geq 65$ km/s. The high-velocity stars arise not because they are moving extremely rapidly through the Galaxy but because they are moving much more slowly than the LSR. A star moving slowly near **apogalacticon** (farthest distance from the Galaxy's center; at **perigalacticon,** the star is closest to the center) in a highly eccentric orbit (Figure 15–12) appears to be moving rapidly with respect to the LSR. For example, the LSR (moving at $\Theta_0 = 220$ km/s) seems to be receding at 100 km/s from a star at apogalacticon (in the solar neighborhood) with $\Theta = 120$ km/s. Such high-velocity objects also include the RR Lyrae stars and the high-latitude (large b), high-velocity H I clouds.

PROBLEMS

1. The Fe I emission lines (at 441.5 and 444.2 nm) in a comparison spectrum are located 15.00 and 15.43 mm, respectively, from an arbitrary reference point. If a stellar Ca I line (of rest wavelength 442.5 nm) is measured to be at 15.27 mm,
   (a) what is the observed wavelength of the Ca I line?
   (b) what is the radial velocity of this star?

2. By making the appropriate conversions of units, show that Equation 15–3 follows from Equation 15–2.

3. A star located 90° from the solar antapex on the celestial sphere is at rest in the LSR 10 pc from the Sun. As seen from the Sun,
   (a) by what angle (in arcseconds) will this star appear to move on the celestial sphere in ten years?

(b) in what direction will the star appear to move?

4. The star Delta Tauri is a member of the Taurus moving group. It is observed to have a proper motion of 0.115″/year and a radial velocity of 38.6 km/s and to lie 29.1° from the convergent point of the group.
   (a) What is this star's parallax?
   (b) What is its distance in parsecs?
   (c) Another star belonging to the same group lies only 20° from the convergent point. What are *its* proper motion and radial velocity?

5. Refer to the data given in Problem 15–4. Assume that a probable error of ±0.005″/year is associated with the proper motion of Delta Tauri. If we independently measure the trigonometric parallax of this

star (with a probable error of ±0.005″), what are the uncertainties (in parsecs) in the distances determined from these two separate parallaxes?

6. Assume that the mass of our Galaxy is 1.5×10^{11} solar masses and that it is *all* concentrated in a point at the galactic center.
   (a) Plot a rotation curve (Θ versus R), with appropriate units and exemplary values along each axis, for this Keplerian case.
   (b) Indicate the rotation period at $R = 5$, 8.5, and 20 kpc.
   (c) What is the speed of escape from $R = 8.5$ kpc?

7. A star in a galactic orbit of eccentricity 0.8 and semimajor axis 7 kpc moves through the solar neighborhood on its outward journey in the galactic plane. What is the velocity of this star with respect to the LSR? Assume the Galaxy is a point mass (Keplerian motion).

8. Use Figure 15–10 to calculate the Galaxy's interior mass to the distance out to which the curve extends. (*Hint:* The Galaxy is *not* a point mass; the motion is *not* Keplerian!)

9. The star BS 1828 has a proper motion of 0.24″/year along position angle 48° (east of north) and a parallax of 0.012″. The H_β line ($\lambda_0 = 486.1$ nm) appears at $\lambda = 485.9$ nm. What is the magnitude of the star's space velocity and what angle does the velocity make to the line of sight (which points away from the Sun)?

10. Determine the proper motion (relative to the LSR) in arcseconds per year of a star in circular motion about the galactic center 4 kpc from the Sun and at a galactic longitude of 60°. Use the rotation curve given in Figure 15–10. (Expect your answers to be small!)

11. The distance from the convergent point to the center of the Hyades cluster is 29.9°, and the cluster's radial velocity is 39.1 km/s. Use this information and that in the chapter to calculate a moving-cluster distance to the Hyades. If the error in the radial velocity determination is ±0.2 km/s, what is the error in the distance?

12. Derive Equation 15–6 for moving cluster parallaxes from Equations 15–3 and 15–5.

13. Sections 14–5 and 15–4 state that the halo of our galaxy contains nonluminous matter. If we can't see it, how do we know it's there?

14. Barnard's star has a radial speed of −108 km/s, proper motion 10.34″/year, and parallax 0.546″.
    (a) What is the distance to Barnard's star in pc? In km?
    (b) What is the tangential speed of Barnard's star?
    (c) What is the space velocity of Barnard's star and the angle that the space velocity makes with the line of sight?
    (d) In how many years will Barnard's star be its closest to the Sun?
    (e) At its closest approach, how distant (in pc and light years) will Barnard's star be? Compare this to the current distance of α Centauri.
    (f) Barnard's star currently has an apparent visual magnitude of 9.54. What will be its magnitude at closest approach?

Chapter 16

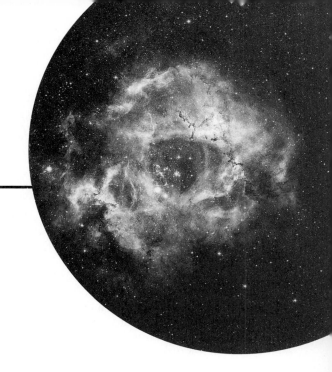

The Evolution of Stars

S tars play the central role in cosmic evolution. The observational characteristics of stars were our primary concern in Chapters 10 through 13; here we discuss the physical laws that govern the structure and evolution of stars. By suitably combining these laws into theoretical stellar models, we can understand the equilibrium configurations of stars, their evolution in time, and the astrophysical basis for cluster H–R diagrams.

We will concentrate upon normal stars, which are in the vast majority, and return later (Chapter 18) to rare peculiar stars with their fascinating variability. These are stars at distinct episodes in their evolution, for a normal star generally exhibits several brief phases of peculiarity during its lifetime. If we lived a billion years, we would see this, but at this instant of cosmic time we happen to catch most stars in their sedate states—only a few are in active or violent phases now.

By its very nature, a normal star is so hot that it must be entirely gaseous. Let's use the physical precepts applicable to stellar-sized masses to see how this comes about.

16–1 ◐
THE PHYSICAL LAWS OF STELLAR STRUCTURE

(A) HYDROSTATIC EQUILIBRIUM

A star is held together by its self-gravitation and supported against collapse by its internal pressures. The simplest stellar model is a static, spherically symmetric ball of matter; any parameter of physical interest depends not upon time or angle but only upon the radial distance r from the star's center. **Hydrostatic equilibrium** ensues when the inward gravitational attraction exactly balances the outward pressure forces at every point (r) within the star (Figure 16–1). As we approach the center of the star, the pressure steadily increases to counterbalance the increasing weight of the material that lies above. The star is in *hydrostatic equilibrium*:

$$dP/dr = -GM(r)\rho(r)/r^2 \qquad \text{(16–1)}$$

Equation 16–1 appears to have three independent variables: $P(r)$, $\rho(r)$, and $M(r)$, but $M(r)$ increases

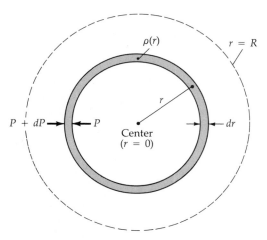

FIGURE 16–1 Hydrostatic equilibrium in a star. A spherical star is divided into a series of shells of thickness dr, each of which is in equilibrium between the outward pressure and weight.

by the amount (Figure 16–1) $dM(r) = 4\pi r^2 \rho(r)\, dr$ when we add the indicated spherical shell (from r to $r + dr$). So $M(r)$ is determined from $\rho(r)$ by

$$dM/dr = 4\pi r^2 \rho(r) \qquad \text{(16–2)}$$

our second fundamental equation, one of *mass continuity*. If the star's outer radius is R, then its total mass is

$$M = \int_{r=0}^{r=R} \rho(r) r^2\, dr \qquad \text{(16–3)}$$

We have added together the masses of all the onion-like shells in the model star.

Hence we need know only $\rho(r)$ to determine first $M(r)$ by Equation 16–2 and then $P(r)$, the pressure profile within the star, by Equation 16–1. Given the star's radius, we also find its total mass using Equation 16–3.

The Sun's Central Pressure
To exercise our physical intuition, let's roughly calculate the pressure at the center of our Sun, using the equation of hydrostatic equilibrium. We know that $G = 6.67 \times 10^{-11}$ N · m²/kg², $M_\odot = 1.989 \times 10^{30}$ kg, and $R_\odot = 6.96 \times 10^8$ m; therefore the mean solar density is $\langle \rho_\odot \rangle = 3M_\odot/4\pi R_\odot^3 = 1410$ kg/m³. Taking the surface pressure as zero, letting $r = dr = R_\odot$, and using $M(r) = M_\odot$ in Equation 16–1 yields

$$P_c \approx GM_\odot \langle \rho_\odot \rangle / R_\odot = 2.7 \times 10^{14} \text{ N/m}^2$$

Since 1 atm = 1.01×10^5 N/m², then the gas pressure $P_c \approx 2.7 \times 10^9$ atm; it supports the weight of the Sun's mass.

Our computed mean solar density is slightly greater than that of water (1000 kg/m³), but since the Sun is centrally condensed, the actual central density is more like 1.6×10^5 kg/m³, and so the actual pressure greater. The tremendous pressures supporting stars imply very high interior temperatures—high enough to maintain the gaseous state.

(B) EQUATIONS OF STATE

To solve our stellar model requires the density profile $\rho(r)$, so the composition and local state of the stellar materials must be examined in detail. In Chapter 17, you will see that quantum liquids and solid lattices of atomic nuclei become important when we consider stellar corpses (such as white dwarfs and neutron stars); in the vast majority of cases, however, stars are gaseous throughout.

For normal stars, we assume that the material is a **perfect gas,** which obeys the *perfect-gas law*:

$$P(r) = n(r)kT(r) \qquad \text{(16–4)}$$

Here the pressure $P(r)$ is related directly to the gas particle number density at r in $n(r)$ (particles/m³), Boltzmann's constant $k = 1.381 \times 10^{-23}$ J/K, and the gas temperature $T(r)$. Now $n(r)$ can be expressed in terms of $\rho(r)$ and the gas chemical composition $\mu(r)$:

$$n(r) = \rho(r)/\mu(r)m_H \qquad \text{(16–5)}$$

where $m_H = 1.67 \times 10^{-27}$ kg is the mass of a hydrogen atom. Recall that, in terms of the *mass fractions* of hydrogen (X), helium (Y), and all heavier elements—"metals"—(Z), the composition expressed as the **mean molecular weight,** μ, is

$$\mu = [2X + (3/4)Y + (1/2)Z]^{-1} \approx 1/2 \quad \text{(16–6)}$$

So the perfect-gas equation of state becomes

$$P(r) = \rho(r)kT(r)/\mu(r)m_H \qquad \text{(16–7)}$$

As you will see, $\mu(r)$ is usually specified for a stellar model, so that only $T(r)$ remains to be determined. In massive stars, the gas pressure is significantly augmented by **radiation pressure:** $P_{rad}(r) =$

$(a/3)T^4(r)$, where $a = 7.564 \times 10^{-16}$ J/m^3 · K^4 is the radiation constant.

The Sun's Central Temperature

By inverting the perfect-gas law of Equation 16–7, using the P_c and $\langle \rho_\odot \rangle$ values found in Section 16–1(a), and making the same crude but simplifying approximations, we estimate the required central temperature T_c of our Sun:

$$T_c \approx P_c \mu m_H / \langle \rho_\odot \rangle k \qquad \textbf{(16–8)}$$

Taking μ as 1/2, we have

$$T_c \approx 12 \times 10^6 \text{ K} \qquad \textbf{(16–9)}$$

which is close to the computer-determined value of about 14.7 million! At these temperatures, the gas is dissociated into ions and electrons—a neutral mixture termed a *plasma*. Our earlier statement that stars are gaseous spheres is consistent and true. The gaseous state persists throughout the star since $\rho(r)$ decreases approximately as rapidly as $T(r)$ outward through the star.

(C) MODES OF ENERGY TRANSPORT

To determine $T(r)$, consider how energy is transported from the stellar interior to the surface, where it radiates away into space. The star's self-gravitation forces the stellar center to be at a much higher temperature (roughly 15 million K for our Sun) than the stellar surface (about 6000 K for the Sun); the heat energy must flow from the higher-temperature regions to the lower-temperature regions (the second law of thermodynamics).

Three processes transport energy: conduction, convection, and radiation. **Conduction** occurs when energetic atoms communicate their agitation to nearby cooler atoms by collisions; this mode works extremely well in solids (especially metals) but poorly in gases because of the low thermal conductivities. (The atoms are too far apart.) **Convection** transports heat energy by means of mass motions in fluids. When $T(r)$ varies rapidly enough with distance (that is, there is a steep temperature gradient dT/dr), the fluid becomes unstable and boils. This process takes place in limited regions of most stars, with hot fluid masses rising, releasing their heat energy, and sinking again to pick up more energy. The top of such a **convective zone** is

observed at the base of the Sun's photosphere (Section 10–2). Unfortunately, no completely adequate mathematical theory of convective transport has yet been devised. However, a useful and reasonable formulation of convection has been applied to stars.

The third mode, **radiative transport,** is an important means of energy flow in sections of most stars. Here high-energy photons flow outward from the star's interior, losing energy by scattering and absorption in the hot plasma of the **radiative zone.** At the extreme temperatures of stellar interiors, the most important sources of such *opacity* (Chapters 10 and 13) are (1) **electron scattering**—the scattering of radiation (photons) by free electrons and (2) **photoionization**—the use of radiant energy to detach electrons from ions (Section 8–3).

Let's now derive the **equation of radiative transport.** At the base of a thin shell, the spherical surface is essentially a blackbody emitter at temperature $T(r)$ so that by Equation 8–40, we have the outward radiative flux (J/m^2 · s), $F(r) = \sigma T^4(r)$, where $\sigma = 5.67 \times 10^{-8}$ J/m^2 · s · K^4 is the Stefan-Boltzmann constant. At $r + dr$, the temperature is $T + dT$, however, and the outward flux is $F + dF = \sigma(T + dT)^4 \approx \sigma(T^4 + 4T^3\, dT)$. Now dT is negative since the exterior of the shell must be cooler than its interior, so that the flux absorbed within the shell is

$$dF = 4\sigma T^3(r)\, dT \qquad \textbf{(16–10a)}$$

This absorption is from the opacity $\kappa(r)$ of the shell's material and from Equation 10–1:

$$dF = -\kappa(r)\rho(r)F(r)\, dr \qquad \textbf{(16–10b)}$$

Combining Equations 16–10 and defining the **luminosity** (J/s) by $L(r) = 4\pi r^2 F(r)$, we find that the total energy flow per second through a thin spherical shell is

$$L(r) = [-16\pi\sigma r^2 T^3(r)/\kappa(r)\rho(r)](dT/dr) \qquad \textbf{(16–11)}$$

The complete machinery of the theory of radiative transfer introduces an additional factor of 4/3 into Equation 16–11, so that the rigorous equation of radiative transport is

$$L(r) = [-64\pi\sigma r^2 T^3(r)]/[3\kappa(r)\rho(r)](dT/dr) \qquad \textbf{(16–12a)}$$

When the opacity is high enough, convection rather than radiation transports the energy through most of the star. The mode that operates depends on which is more efficient. If we let $\gamma =$ the ratio of the specific heats at constant pressure and volume ($=c_p/c_v = 5/3$ for a fully ionized, ideal gas), then for convective energy transport we have

$$dT/dr = (1 - 1/\gamma) \, [T(r)/P(r)] \, dP/dr \quad \textbf{(16–12b)}$$

Either equation is used depending on the physical conditions in the star.

Solar Luminosity from Radiative Transfer

For our Sun, most of the interior conveys energy by radiation. A rough estimate of dT/dr is $-T_c/R_\odot$, which equals a gradient, -2×10^{-2} K/m. Use Equation 16–12 to estimate the solar luminosity. Setting $r = R_\odot$, $T(r) = T_c$, and $\rho(r) = \rho_\odot$, we have

$$L_\odot \approx (-64\pi\sigma R_\odot^2 T_c^3/3\kappa\rho_\odot)(-T_c/R_\odot) \quad \textbf{(16–13)}$$
$$= (9.5 \times 10^{29}/\kappa) \text{ J/s}$$

where we have yet to determine a reasonable opacity (m^2/kg). From its dimensions, κ is the interaction area per gas particle multiplied by the number of particles per kilogram of the stellar material; a mass of 1 kg of completely ionized hydrogen contains 6×10^{26} protons and the same number of electrons. For electron scattering, the interaction area per electron is approximately 10^{-30} m^2; for hydrogen photoionization, this area per atom is near 10^{-20} m^2. In the solar interior, the latter opacity source dominates, so that (very approximately) $10^{-3} \ll \kappa \le 10^7$. Accurate opacities are extremely difficult to determine. Our prediction for the solar luminosity falls in the broad range $10^{22} \le L_\odot \ll 10^{32}$ J/s; a mean value of 10^{27} J/s is very close to the measured value of 3.90×10^{26} J/s, implying an opacity value of about 2.4×10^3.

(D) ENERGY SOURCES

Because stellar luminosity represents energy loss, no star is perfectly static; a stellar model is, however, an excellent approximation for times that are short relative to stellar evolution time. In fact, stars *must* evolve because they lose energy to space. How long will the star remain in essentially steady state, and what energy source maintains this stability? Geological and paleontological evidence indicates that our Sun has been radiating energy at a fairly steady rate for a few billion years; such energy generation takes place in stellar interiors.

The rate of energy production per unit mass of stellar material (J/s · kg) is denoted by $\epsilon(r)$. (In fact, the energy production rate also depends on temperature and density; we write $\epsilon(r)$ as a shorthand for the temperature and density at r.) Now, $\epsilon = 0$ except in stellar cores and in certain localized spherical shells. For our Sun, we estimate the average value of ϵ needed to maintain the solar luminosity to be

$$\epsilon_\odot \approx L_\odot/M_\odot = 2.0 \times 10^{-4} \text{ J/s} \cdot \text{kg}$$

We can find out how such energy generation within our thin spherical shell augments the stellar luminosity (Figure 16–1). The luminosity $L(r)$ enters the bottom of the shell while the greater luminosity $L + dL$ leaves the top—from the energy produced in the shell's mass $4\pi r^2\rho(r)\,dr$. The additional luminosity is

$$dL = 4\pi r^2 \rho(r)\epsilon(r) \, dr \quad \textbf{(16–14)}$$

Equation 16–14 expresses the balance between net energy lost from the shell dL and net energy generated within the shell, an energy, or **thermal equilibrium.**

In a quasistatic gaseous star, energy may be generated only by gravitational contraction and/or thermonuclear fusion reactions. Each process is important at some stage in a star's evolution. Let's consider these energy sources in detail.

Gravitational Contraction

Gravitational potential energy can be transformed to kinetic energy of motion (as when a rock is dropped near the Earth's surface); the bulk form of kinetic energy is *heat*. Consider a very slowly contracting star. The heat energy of its interior provides the pressure—from the random motions of the gas particles—that supports the star against its self-gravitation. When the star contracts to a smaller radius, the self-gravitation increases so that the internal pressures (and hence temperatures and heat energy) must also increase to maintain approximate hydrostatic equilibrium. The gravitational potential energy decreases about twice as fast as the heat energy increases, however, and so to conserve the total energy of the system, approximately half of the potential energy change must be radiated into space—the star's luminosity.

This energy-conversion process may be illustrated by a simple analogy. A small satellite of mass m moves in a circular orbit of radius r at speed v about a greater mass M. From Equation 1–32, the satellite's kinetic energy is $mv^2/2$, and its gravitational potential energy is $-GMm/r$. Since the centripetal acceleration v^2/r maintaining the orbit is provided by the mutual gravitational attractive acceleration $GM/r^2 = v^2/r$, the kinetic energy $mv^2/2 = GMm/2r$, or *half* the magnitude of the potential energy. If we now move the satellite to a smaller (stable) orbit at $r - dr$, the increase in kinetic energy is certainly only half the decrease in potential energy (which becomes more negative). To conserve total energy (potential plus kinetic), the other half of the potential energy change must be transmitted to the agent that alters the satellite's orbit—in the case of a star, this energy is radiated away. This result applies generally and is called the **virial theorem.** It implies that the gravitational contraction of a mass results in the conversion of half of the gravitational potential energy thermal energy and half to radiative energy, that is, $U = KE + PE = -2E_{th}$, where E_{th} is the total thermal energy.

What is the gravitational potential energy for a sphere of mass? Consider a star dispersed to infinity. Bring in one shell of material, mass $dM(r)$, at a time, to add to the mass $M(r)$ at distance r'. Each shell adds to U by the amount

$$dU = \frac{-GM(r)\,dM(r)}{r}$$

so that if we integrate over all shells until the mass, M, is reached:

$$U = -\int_0^M G\,\frac{M(r)\,dM(r)}{r}$$
$$U = -q(GM^2/R)$$

where the value of q depends on the mass distribution in the sphere. For one of uniform density, $q = 3/5$; for most main-sequence stars, $q \approx 1.5$. Let's apply these concepts to our Sun. For each kilogram of solar material, the average gravitational potential energy available for radiation is roughly

$$GM_\odot/2R_\odot = 9.54 \times 10^{10} \text{ J/kg} \quad \textbf{(16–15)}$$

Comparing this with $\epsilon_\odot$, we see that **gravitational contraction** can sustain the Sun at its present luminosity for only 15 million years; some other energy source must be sought if we are to account for bil-

lions of years of sunshine. In Section 16–3, you will see when gravitational contraction is important in stellar evolution.

Thermonuclear Reactions

Only after about 1938 did astronomers understand the **thermonuclear fusion reactions** that provide the long-term energy source for stars. In fusion, light atomic nuclei collide with such violence and frequency in the high-temperature, high-density stellar interior that they fuse into heavier nuclei and release tremendous quantities of energy (such as in a hydrogen bomb). We say that the lighter elements "burn" to form heavier elements in this process of **nucleosynthesis.**

In atomic nuclei (Section 8–2), the strong nuclear force overcomes the electrostatic repulsion of the positively charged protons and binds from one to 260 nucleons (protons and neutrons) in a region about 10^{-15} m in diameter. Two nuclei will fuse to form one larger nucleus if they approach within 10^{-15} m of one another, but their mutual electrostatic repulsion—all nuclei have a positive charge—amounts to a 1-MeV potential barrier. In contrast, at 10^7 K the average thermal energy of a proton is only 1 KeV. Classically, protons cannot fuse because of the strong coulombic barrier. Fusion *does* happen, however, because quantum physics allows the protons to tunnel through the barrier rather than go over it. The easiest fusion reaction involves two protons (hydrogen nuclei); such reactions become significant at temperatures around 10 million K.

The great abundance of hydrogen makes it the key constituent in stellar nuclear reactions. The next stable nucleus is helium, ^{4}He, with atomic weight 4. Since the hydrogen nucleus (one proton) only has atomic weight 1, four protons are required to make one helium nucleus. The atomic weights do not exactly match because the more exact atomic weight of a proton is 1.0078, and four of them add to 4.0312, while the weight of ^{4}He is 4.0026, leaving a **mass defect** of 0.0286. This mass is converted to an amount of energy given by Einstein's equation for the equivalence of mass and energy,

$$E = mc^2 \quad \textbf{(16–16)}$$

where c is the speed of light. Because a unit atomic weight is 1.66×10^{-27} kg, the energy released by the conversion of four ^{1}H nuclei to one ^{4}He nucleus is

$$E = 0.0286(1.66 \times 10^{-27})(9 \times 10^{16}) = 4.3 \times 10^{-12} \text{ J}$$

We can also use Equation 16–16 to determine the total energy store of the Sun if we assume that it originally consisted of pure hydrogen, all of which will eventually be converted to helium. The mass liberated in the form of energy in this thermonuclear conversion is the fraction 0.0286/4.0312 = 0.0071 of the available mass of original hydrogen. Because only in the core are the temperature and pressure high enough to permit nuclear reactions, about 10% of the mass of the Sun is available for energy conversion. So the total thermonuclear energy available in the Sun is

$$\begin{aligned}E_{\text{total}} &= m(4\ ^1\text{H} - ^4\text{He})(c^2)(0.1M_\odot)/m(4\ ^1\text{H}) \\ &= 0.0071(9 \times 10^{16})(0.2 \times 10^{30}) \\ &= 1.28 \times 10^{44} \text{ J}\end{aligned}$$

which, at the present solar luminosity of 3.90×10^{26} J/s, would last about 10 billion years. The best estimates of the age of the Solar System yield figures around 5 billion years, so this reaction will sustain our Sun for another 5 billion years.

Two different fusion processes lead to the conversion of hydrogen to helium: the proton–proton (PP) chain and the carbon (CNO) cycle. The PP chain dominates at temperatures lower than 2×10^7 K, and the CNO cycle is prominent at higher temperatures (Figure 16–2). In the Sun, for instance, both processes take place but the PP chain is the more important. The CNO cycle plays a small role in stars lower on the main sequence than the Sun but predominates in stars hotter than F stars.

The main **proton–proton chain** (called **PP I**) consists of the following reactions (the energy that is released in each step is given in parentheses):

$$\begin{aligned}^1\text{H} + ^1\text{H} &\rightarrow ^2\text{H} + e^+ + \nu \quad &(1.44 \text{ MeV}) \\ ^2\text{H} + ^1\text{H} &\rightarrow ^3\text{He} + \gamma \quad &(5.49 \text{ MeV}) \\ ^3\text{He} + ^3\text{He} &\rightarrow ^4\text{He} + ^1\text{H} + ^1\text{H} \quad &(12.9 \text{ MeV})\end{aligned}$$

where ^{2}H is heavy hydrogen (deuterium), whose nucleus contains one proton and one neutron; e^+ is a positron, ν a neutrino, and γ a photon. A **positron** has the same mass as an electron, but its charge is positive. **Neutrinos** have no (small?) mass or charge, only energy and spin, and are therefore difficult to detect. Conservation of charge is maintained in the first reaction by the emission of the positron. Note that the first two steps must occur twice before the last can take place and that a total of six protons are involved even though two are again released in the final step. Other reactions may occur instead of the last step of this chain—for example,

$$^3\text{He} + ^4\text{He} \rightarrow ^7\text{Be} + \gamma$$

Then there are two possible branches from ^{7}Be, both resulting in ^{4}He. All three chains operate simultaneously in a star, but the PP I is the most important; it occurs 91% of the time in the Sun, according to theoretical models. On the average, the neutrinos carry off 0.26 MeV from each reaction.

The other PP reactions occur less frequently and so contribute a minor amount to the Sun's luminosity. The **PP II chain** is

$$\begin{aligned}^1\text{H} + ^1\text{H} &\rightarrow ^2\text{H} + e^+ + \nu \quad &(1.44 \text{ MeV}) \\ ^1\text{H} + ^2\text{H} &\rightarrow ^3\text{He} + \gamma \quad &(5.49 \text{ MeV}) \\ ^3\text{He} + ^4\text{He} &\rightarrow ^7\text{Be} + \gamma \quad &(1.59 \text{ MeV}) \\ ^7\text{Be} + e^- &\rightarrow ^7\text{Li} + \nu \quad &(0.861 \text{ MeV}) \\ ^7\text{Li} + ^1\text{H} &\rightarrow 2\ ^4\text{He} \quad &(17.3 \text{ MeV})\end{aligned}$$

with the neutrinos taking away as much as 0.86 MeV.

The **PP III chain** involves the same first three steps as PP II and then goes on as

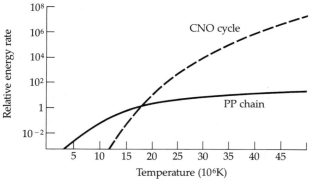

FIGURE 16–2 Energy-generation rates. The relative rates for the PP chain and CNO cycle are compared as a function of temperature for Population I stars. Note the crossover at about 18 million K.

$$^1\text{H} + {}^7\text{Be} \rightarrow {}^8\text{B} + \gamma \quad (0.14 \text{ MeV})$$
$$^8\text{B} \rightarrow {}^8\text{Be} + e^+ + \nu$$
$$^8\text{Be} \rightarrow 2 \ {}^4\text{He} \quad (18.1 \text{ MeV})$$

with the neutrinos escaping with 7.2 MeV.

The neutrinos produced by the PP chain should escape the Sun without interacting with any solar material. Since 1965, Raymond Davis has attempted to detect high-energy solar neutrinos (from PP III) by means of an elaborately shielded tank (located far underground in the Homestake mine in South Dakota) filled with 100,000 gallons of tetrachloroethylene (C_2Cl_4). Although neutrinos react only very weakly with matter, they should transmute some of the ^{37}Cl in the tank to ^{37}A. These radioactive argon atoms can then be collected and counted. The measured neutrino flux, mainly from the boron-8 decays, is only one-third to one-fourth that predicted by current theoretical solar models. This discrepancy is called the *solar neutrino problem*. It has been confirmed by a thousand days of observations by the Kamiokande II detector in Japan, which resulted in 0.46 times the standard model flux.

Two general solutions are: (1) a revision of the standard solar model, or (2) new aspects about neutrinos. The first seems the least likely, for the helioseismology of pressure modes so far gives better support for standard solar models. Hence, the finger appears to point to neutrino physics rather than solar physics. The working concept so far is that neutrinos change their character in the solar interior and transform to a kind of neutrino that current experiments cannot detect.

The **CNO cycle** also converts hydrogen to helium, but it requires a carbon nucleus as a catalyst:

$$^{12}\text{C} + {}^1\text{H} \rightarrow {}^{13}\text{N} + \gamma$$
$$^{13}\text{N} \rightarrow {}^{13}\text{C} + e^+ + \nu$$
$$^{13}\text{C} + {}^1\text{H} \rightarrow {}^{14}\text{N} + \gamma$$
$$^{14}\text{N} + {}^1\text{H} \rightarrow {}^{15}\text{O} + \gamma$$
$$^{15}\text{O} \rightarrow {}^{15}\text{N} + e^+ + \nu$$
$$^{15}\text{N} + {}^1\text{H} \rightarrow {}^{12}\text{C} + {}^4\text{He} \quad \text{(16–17)}$$

Each step in the chain need occur only once to convert four protons to an **alpha particle** (helium nucleus). The second and fifth steps take place because ^{13}N and ^{15}O are unstable isotopes, with half-lives of only a few minutes. (Recall that *half-life* is the time in which half of the original quantity of an isotope has disintegrated into its more stable nuclear form.) The cycle starts with the reaction

between carbon and hydrogen but ends with the release of an identical carbon nucleus; the ^{12}C acts as a catalyst. Although the temperature may be sufficiently high, the CNO cycle cannot operate in a star unless carbon is available.

Higher temperatures are required for the carbon cycle because the coulombic barriers of the carbon and nitrogen nuclei are higher than those of protons and helium nuclei. As a result, temperature dependence goes roughly as T^4 for the PP reaction and as T^{20} for the carbon cycle (Figure 16–2).

Starting at very high temperatures, around 10^8 K, other reactions begin transmuting helium to heavier elements. Three alpha particles (^4He has historically been called an alpha particle) will form carbon:

$$^4\text{He} + {}^4\text{He} \Leftrightarrow {}^8\text{Be} + \gamma \quad \text{(16–18)}$$
$$^8\text{Be} + {}^4\text{He} \rightarrow {}^{12}\text{C} + \gamma$$

This reaction is known as the **triple-alpha process,** the first stage of helium burning. The intermediate beryllium-8 is not very stable, and the back reaction occurs readily. Nevertheless, an equilibrium is established where some ^8Be takes part in the second step. Light elements other than hydrogen, helium, and carbon are rare deep inside stars because such elements (deuterium, lithium, beryllium, and boron) quickly combine with protons at temperatures of only a few million degrees to form one or two helium nuclei—for example,

$$^7\text{Li} + {}^1\text{H} \rightarrow 2 \ {}^4\text{He}$$

The triple-alpha and other helium-burning reactions play a major role in the evolution of stars. The advanced nuclear burning stages, which operate at higher temperatures and densities, involve complex reaction networks. The general sequence is: carbon, neon, oxygen, and magnesium burning. These reactions make it possible to start with stars of pure hydrogen and eventually produce most heavy elements up to iron, which has the highest binding energy per nucleon. Further fusion thus requires the input of energy.

16–2 ◉
THEORETICAL STELLAR MODELS

(A) RECAPITULATING THE PHYSICS

The physical principles basic to stellar structure are hydrostatic equilibrium, the perfect-gas equation of state, the various modes of energy transport, and the gravitational and thermonuclear sources of stellar energy. These are the tools used by astrophysicists in computing **stellar models**— theoretical stars in which physical parameters and their rates of change throughout the star are described. These mutually dependent parameters include temperature $T(r)$, mass $M(r)$, density $\rho(r)$, pressure $P(r)$, luminosity $L(r)$, rate of energy production $\epsilon(r)$, and chemical composition in terms of the mean molecular mass $\mu(r)$. Their interdependence is given by the basic equations of stellar structure:

Hydrostatic equilibrium

$$dP/dr = -GM(r)\rho(r)/r^2 \qquad \textbf{(16–1)}$$

Mass continuity

$$dM/dr = 4\pi r^2\rho(r) \qquad \textbf{(16–2)}$$

Energy transport (radiative and convective)

$$dT/dr = [-3\kappa(r)\rho(r)/64\pi\sigma r^2 T^3(r)]L(r) \quad \textbf{(16–12a)}$$
$$dT/dr = (1 - 1/\gamma)\,[T(r)/P(r)]\,dP/dr \qquad \textbf{(16–12b)}$$

Energy generation (thermal equilibrium)

$$dL/dr = 4\pi r^2\rho(r)\epsilon(r) \qquad \textbf{(16–14)}$$

Equation of state

$$P(r) = k\rho(r)T(r)/\mu(r)m_H \qquad \textbf{(16–7)}$$

These equations describe how the parameters vary through the star *only* if we know their values at some particular points (or shells) in the star, such as at the center and the surface. These values constitute the boundary conditions. At the center, for example, where $r = 0$, the boundary conditions for the mass and luminosity must be $M(r) = 0$ and $L(r) = 0$. So that the theoretical models bear some relationship to real stars, we use observed stellar characteristics for the boundary conditions at the surface. So at $r = R$, the radius of the star, $M(R) = M$, $L(R) = L$, $T(R) = T_{eff}$, the effective surface temperature (or photospheric temperature), and both $\rho(r)$ and $P(r)$ approach zero.

In addition to these equations, we require energy generation relations, $\epsilon(\rho, T)$, which will alternate between gravitational and nuclear; opacity relations, $\kappa(\rho, T)$, which will depend on the chemical composition (and changes with time as the star progresses in its fusion reactions), and the chemi-

cal composition, X (hydrogen), Y (helium), and Z (metals), or expressed as μ. Chemical composition plays a central role in stellar structure. The equation of state shows that $P(r)$ depends on μ; since the hydrostatic equation indicates that $\rho(r)$ is strongly dependent on $P(r)$, it follows that $\rho(r)$ depends on μ. Note that $\rho(r)$ appears in all the other equations. The difficulty lies in knowing how the composition varies through the star.

We must relate $\kappa(\rho, T)$ to the temperature, density, and composition. This relation will be different for different processes that provide the opacity. In general, the calculation of the exact opacity is complex and best done with large computers to handle all the necessary relationships. But one approximate formula, applicable to the range of temperatures and densities found in many main-sequence stars, is known as *Kramers' law* and is given by

$$\kappa = \text{constant} \times Z(1 + X)\,\frac{\rho}{T^{3.5}}$$

where X is the fractional abundance of hydrogen and Z is the fractional abundance of heavy elements.

The quantity ϵ is different for different nuclear reactions, but in general it will depend on temperature, density, and chemical composition. For example, at temperatures around 14 million K, $\epsilon(\rho, T)$ for the PP I chain is given by the expression

$$\epsilon = \text{constant} \times X^2\rho T^4$$

where X is the fractional abundance by mass of hydrogen.

Many theoretical models have been computed for our Sun, for it is the best-observed star and serves as a prototype for others. These models differ in the relative abundances of hydrogen, helium, and the heavy elements assumed for the newly formed Sun and in the degree of mixing of the elements and their participation in thermonuclear reactions in the solar interior. Compositional mixing becomes more thorough as the extent of the convection zone increases. Current solar models that best fit various observations indicate that a large fraction of the hydrogen at the center of the Sun has already been converted to helium (the composition is about 40% H, 60% He), so that there is a central helium-enriched core. Its energy generation, hydrogen burning by the PP chain, still occurs primarily in the core. Energy transport is radiative for most of the interior, but beyond about 0.7–$0.8R_\odot$, the temperature gradient becomes sufficiently steep to maintain convection. The solar granulation is direct evidence for such a convection zone.

(B) THE PHYSICAL BASIS OF THE *M–L* RELATIONSHIP

The empirical mass–luminosity relationship has a physical basis in the equations of stellar structure if they are correct. Start with hydrostatic equilibrium (Equation 16–1) and let $dP \to \Delta P$ and $dr \to \Delta r$; then

$$\Delta P = P_s - P_c = 0 - P_c$$

where P_s = pressure at the surface, P_c = pressure at the center, and $\Delta r = R$, so that

$$P_c \propto M\rho/R$$

For a perfect gas,

$$P \propto \rho T$$

and so

$$\rho T_c \propto M\rho/R$$

and $T_c \propto M/R$. We follow the same approximation with the radiation transport equation (16–12) so that

$$L \propto R^2(T_c^{3}/\kappa\rho)(T_c/R) \propto RT_c^{4}/\kappa\rho$$

Now $\rho \propto M/R^3$, and so

$$L \propto R^4 T_c^{4}/\kappa M$$

into which we substitute T_c from hydrostatic equilibrium:

$$L \propto R^4(M/R)^4/\kappa M \propto M^3/\kappa$$

which is close to the observed relation, $L \propto M^{3.3}$. The difference lies in the dependence of the opacity on temperature and density.

The value for the exponent depends on the mass range covered and the type of stars. If we write the *M–L* relation as $L \propto M^n$, then a general value is 3.3; for stars with mass less than $0.4M_\odot$, $n = 2.3$; for those with greater mass, $n = 4$. These values apply to main-sequence stars *only*.

16–3 ◐ STELLAR EVOLUTION

The study of the physical changes that take place in stars as they alter their composition because of thermonuclear reactions is the heart of **stellar evolution.** Stars follow the same general sequence in their evolution: *protostar, pre-main sequence, main sequence,* and *post-main sequence.* Basically, a star's evolution is determined primarily by its mass. Chemical composition plays a secondary role, so that Population I and Population II stars of the same mass follow somewhat different histories. Stellar evolution aims to understand how the luminosity and surface temperature (two observables) change with time. A plot of the points representing different evolutionary stages with time on an H–R diagram is called a star's **evolutionary track.** This section examines theoretical evolutionary tracks calculated from the basic equations of stellar structure.

(A) THE BIRTH OF STARS: PROTOSTARS AND PMS STARS

Stars are born from the gravitational contraction of interstellar clouds of gas and dust. Chapter 19 provides observational evidence for this process; here we concentrate on the basic physics. Be warned that we do not yet understand the complete process of star formation, but the basic theme is clear: as an interstellar cloud contracts, gravitational potential energy is converted in part (50%) to thermal energy and in part (50%) to radiative energy. Eventually, the core heats up to the ignition temperature of fusion reactions, and a star is truly born. Prior to that event, the star goes through protostar and pre-main-sequence stages. The contracting cloud is a **protostar** before it establishes hydrostatic equilibrium. Between that stage and the ignition of fusion reactions, it is called a **pre-main-sequence (PMS) star.** The track traced on the H–R diagram before the star hits the main sequence is called its **PMS evolutionary track;** before that, its protostar track.

The evolutionary tracks for protostars of different masses differ. Despite detailed differences, the theoretical calculations have the following common features: (1) the collapse starts out in free fall; that is, it is controlled only by gravity (with negligible pressure; "free fall" means that the particles in the cloud do *not* collide as the cloud collapses, so that the internal pressure is zero); (2) it proceeds very unevenly, for the central regions collapse more rapidly than the outer parts and a small condensation in hydrostatic equilibrium forms at the center; (3) once the core forms, it accretes material from the infalling envelope; (4) the star becomes visible to us either by accreting all the surrounding material onto itself or by somehow dissipating it.

Solar-Mass Protostellar Collapse

Now let's take a look at one model for the formation of a sun-like star. Imagine a huge interstellar cloud of dust and gas, mostly in the form of molecular hydrogen (H_2), with sufficient mass to contract gravitationally. Observational data indicate that interstellar dust grains such as those that are part of this cloud are composed of graphite, silicates, and ices (Chapter 19). During the collapse (assumed pressureless, so particles don't collide), material at the cloud's center increases in density faster than at the edge. Because of the density increase, the collapse time at the center is decreased; it collapses faster, grows denser, and so collapses still faster. The rest of the cloud's mass is left behind in a more slowly contracting envelope. This part of the collapse takes place in free fall.

Let's work out free-fall collapse explicitly. Consider a test particle of mass m at the edge of a cloud of mass M, radius R, and initial density ρ_0. Imagine that the particle falls straight into the center, so that it follows an elliptical orbit of semimajor axis $a = (1/2)r$ and $e = 1$. Then

$$M = (4/3)\pi R^3 \rho_0 = (4/3)\pi (2a)^3 \rho_0 = (32/3)\pi a^3 \rho_0$$

so that Kepler's third law,

$$P^2/a^3 = 4\pi^2/GM$$

becomes

$$P = (3\pi/8G\rho_0)^{1/2}$$

The free-fall time t_{ff} equals half this time, P, so that

$$t_{ff} = (3\pi/32G\rho_0)^{1/2}$$

or, if we evaluate the constants

$$t_{ff} = (6.64 \times 10^4)/\rho_0^{1/2} \text{ s} \qquad \textbf{(16–19)}$$

for ρ_0 in kilograms per cubic meter. For the cloud

considered here, $\rho_0 = (10^{10})(3.3 \times 10^{-27} \text{ kg}) = 3.3 \times 10^{-17} \text{ kg/m}^3$, so that

$$t_{ff} = (6.64 \times 10^4)/(3.3 \times 10^{-17})^{1/2} \approx 10^5 \text{ years}$$

Heated by collisions with molecules, the dust grains radiate at infrared wavelengths. As long as this heat radiation can escape into space, the kinetic energy is dissipated, the cloud stays cool, the pressure stays low, and the collapse continues in free fall. At some time, however, the density of the core reaches a critical value, at which point the cloud becomes opaque (optical depth ≥ 1) and traps infrared radiation. Then the core's collapse slows down dramatically as hydrostatic equilibrium is established. The star slowly contracts as a PMS star. The total evolutionary time from the start of collapse to this stage is on the order of 1 million years.

We can trace the evolutionary track of a solar-mass protostar on an H–R diagram (point *A* in Figure 16–3). The pre-main-sequence star has a lower surface temperature than it will on the main sequence, but the radius is much larger, giving a large surface area, so that the luminosity is also higher than it will be when the star reaches the main sequence (recall that $L = 4\pi R^2 \sigma T^4$). The star's temperature is so low that its opacity is relatively high (even though its density is low). Convection rather than radiation transports the energy outward; the star is fully convective, so that it is well mixed. A newly formed pre-main-sequence star is completely convective from center to surface, and the effective transport of energy by convection makes the star very luminous (point *B* in Figure 16–3). The star achieves a high luminosity (about 30 times the sun) for about 10^5 years as matter accretes onto the core.

A PMS star shines by slowly shrinking and accreting, the central temperature rising as the star evolves. As the PMS star shrinks in size, the surface temperature at first does not change very much, and the luminosity decreases. Its point on the H–R diagram moves downward (point *C* in Figure 16–3). Meanwhile the core continues to heat up. As it does, its opacity decreases. Eventually, the opacity drops enough for radiation rather than convection to transport energy most efficiently. The zone of radiative transport starts at the core and slowly creeps outward as the inner layers

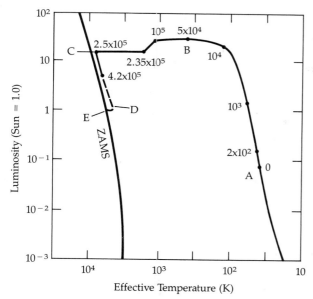

FIGURE 16–3 Pre-main-sequence evolutionary track of a solar-mass star. The times along the track are years since the start at (A), which indicates the formation of the stellar core from an interstellar cloud. The zero-age main sequence (ZAMS) is indicated at the left. *(Based on theoretical calculations by K.-H. Winkler and M. Norman)*

heat up. When a substantial fraction of the star's interior carries energy outward by radiation, its path on the H–R diagram kinks sharply to the left (point *D* in Figure 16–3).

Eventually the core heats up to a few million kelvins, high enough to start thermonuclear reactions. When the PMS star gets most of its energy from thermonuclear reactions (PP reactions in the case of the Sun) rather than gravitational contraction, it achieves full-fledged stardom. It no longer contracts to provide energy; the heat from fusion reactions keeps it in hydrostatic equilibrium. The star is now called a **zero-age main-sequence (ZAMS)** star (point *E* in Figure 16–3). It settles down to the longest stage in its life, calmly converting hydrogen to helium in its core. Most of the interior transports energy by radiation; only the outer region of the envelope is convective. The total time elapsed from initial collapse to arrival as a star on the main sequence is only 20 million years (from point *A* to point *E* in Figure 16–3). When it

hits the ZAMS, the star emits with one solar luminosity and has a radius of about twice that of the present sun.

(B) EVOLUTION ON AND OFF THE MAIN SEQUENCE

A star like the Sun spends about 80% of its total lifetime slowly transforming its hydrogen core to helium, via the PP chain. As the hydrogen abundance decreases, the temperature and density must rise to maintain at least the same rate of nuclear fusion. During this time the temperature in the core increases gradually and the star expands slightly. This results in a greater flow of energy to the surface, and the star's luminosity increases. *Note:* we call the *entire* phase of core hydrogen burning the "main-sequence phase." The zero-age main sequence (ZAMS) is the phase at which a star *first* gets all of its energy from hydrogen fusion, before it has converted any substantial amount of its hydrogen to helium. As hydrogen is converted to helium, the composition and therefore the mean molecular weight change, altering the structure of the star. These changes are gradual at first, and then more rapid. Let's look at a star's lifetime quantitatively. From the overall mass–luminosity relationship for main-sequence stars (Section 12–2b)

$$L*/L_\odot = (M*/M_\odot)^{3.3}$$

A star's lifetime t depends on its store of energy (mass) and on the rate at which it spends that energy (luminosity). So relative to the Sun,

$$t*/t_\odot = (M*/M_\odot)/(L*/L_\odot)$$
$$= (M*/M_\odot)/(M*/M_\odot)^{3.3} = (M*/M_\odot)^{-2.3}$$

So the more massive the star, the shorter its lifetime. The exact power depends on the mass range, but the general statement holds true for all stellar masses.

A Population I Star of 5 Solar Masses

Let's follow the evolution of a Population I star with a mass five times that of the Sun. We have chosen a $5M_\odot$ star as an example because of its more rapid evolution compared to a $1M_\odot$ star; on the main sequence, it will appear as a B-type star.

TABLE 16-1   *Stages of Stellar Evolution (Figure 16-4)*

H–R Position	Stage	Physical Processes
O	ZAMS	Hydrogen burning commences
A, B	Initial evolution on main sequence	Hydrogen consumed in core; some contraction occurs
C	Evolution off main sequence	Hydrogen depleted in core, isothermal helium core and hydrogen-burning shell established
D, E	Evolution to the right in H–R diagram	Core rapidly contracts, envelope expands, hydrogen-burning shell narrows
F, G, H	Red giant	Energy output increases, convective envelope forms, helium burning begins
I, J	Cepheid	Convective shell contracts, core helium burning becomes major energy source
K	Supergiant	Helium-burning shell forms

Table 16–1 summarizes the important characteristics and phases of stellar evolution.

The start of interval *A* at *O* in Figure 16–4 represents the main-sequence position of the star in the H–R diagram at age zero, when hydrogen burning started at the center of the star. Initially, only temperatures at the very center are sufficiently high for the CNO cycle to operate, and fresh hydrogen is supplied by convective mixing throughout a core. Conditions in the core change in a rather complex fashion. When only a small fraction of all the material in the core is hydrogen, first the core and then the whole star contracts (stage *B*).

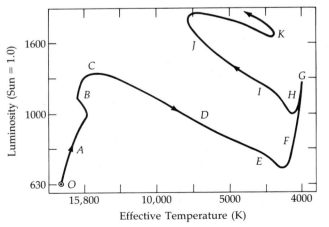

FIGURE 16–4 Evolutionary track for a Population I star of mass $5M_\odot$. Note this is a log–log scale. (*Based on theoretical calculations by I. Iben, Jr.*)

Eventually the core hydrogen is exhausted. As a result of the gravitational contraction of the core, material just beyond the core is pulled into higher-temperature regions and hydrogen burning starts in a shell surrounding the (primarily) helium core. This shell is fairly thick at first (stage *C*), but as a larger fraction of the total mass of the star is concentrated into the core, it narrows (stage *D*). Through stage *C*, the core is so dense that, to a large extent, it supports the weight of material above it. When this equilibrium ends, the core contracts much more rapidly and heats up. When this happens, energy generation in the shell is accelerated and the outer envelope expands. Such an expansion is accompanied by lowering the surface temperature, and the position of the star in the H–R diagram moves to the right (stages *C* to *E*).

During the last part of the expansion phase, convection develops in the envelope and changes the direction of the evolutionary track from decreasing to increasing luminosity (region *E*) by carrying a greater part of the energy outward to the star's surface. This begins the **red-giant** phase of the star's life (stages *F* to *H*). Temperatures in the interior continue to rise owing to core contraction; finally a point is reached (about 10^8 K) where the triple-alpha process can begin at the star's center. Helium burning at this stage is short-lived; the star's surface cools, and it drops from position *G* to position *H*.

Once again gravitational contraction takes over until temperatures are sufficiently high to reignite helium. The balance between helium burning in the core and hydrogen burning in the shell gradu-

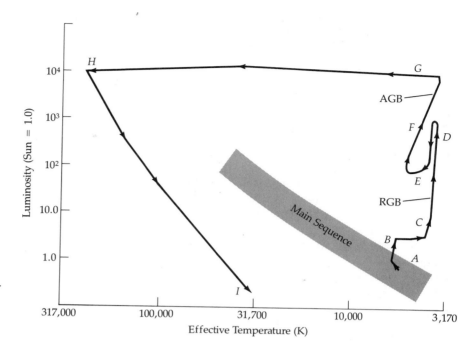

FIGURE 16–5 Evolutionary track for a star of mass $1M_{\odot}$. The main sequence band is indicated; also the locations of the red giant branch (RGB) and the asymptotic giant branch (AGB).

ally shifts in favor of helium burning, and the surface of the star becomes both hotter and brighter (*I, J*). When the helium is exhausted and the core consists of carbon, contraction again occurs and a helium-burning shell is formed; by now the star may be a supergiant (*K*).

A Population I Star of 1 Solar Mass

Now we turn to a star having the same mass as our Sun to see what post-main-sequence evolutionary track the Sun might follow. The star reaches the ZAMS when the PP chain dominates energy production. After about 10 billion years, the main-sequence phase ends when almost all of the hydrogen in the core has been converted to helium. During this time, the temperature in the core increases gradually and the star expands slightly. This results in a greater production of energy, and the star's luminosity increases (*A* to *B* in Figure 16–5; see also Table 16–2).

TABLE 16–2   *Evolution of a Solar Mass Star (Figure 16–5)*

H–R Position	Stage	Physical Processes
A	ZAMS	Core hydrogen burning begins
B	Evolution on main sequence	Core hydrogen burning ceases; shell hydrogen burning begins
C	Evolution off main sequence	Shell hydrogen burning continues; convection dominates energy transport
D	Red giant	Helium flash occurs; core helium burning begins
E	Subgiant	Core helium burning continues along with shell hydrogen burning
F	Red giant again	Thermonuclear reactions in core end; shell helium and hydrogen burning continues
G	Variable star	Expansion and contraction throw off outer layers
H	Planetary nebula	Star enters planetary nebula stage
I	White dwarf	All thermonuclear reactions stop; slow cooling

When the hydrogen in the core is used up, the thermonuclear reactions cease there. However, they keep going in a shell around the core, where fresh hydrogen still exists. With the end of fusion reactions in the core, the core contracts. This heats up the layer of burning hydrogen, and the reactions produce more energy, but the layer of burning hydrogen heats up the surrounding envelope and causes it to expand. So the radius of the star increases, and its surface temperature decreases. The temperature drop increases the opacity, and convection carries the energy outward in the star's envelope. Note that convective energy transport becomes effective at high opacities and that the convective outflow causes the star's structure to change. The radius of the star increases and its surface temperature decreases (points *B–C* in Figure 16–5). The temperature fall increases the opacity, so that at some point convection carries most of the energy outward in the star's envelope. The luminosity then increases greatly (points *C–D* in Figure 16–5); it moves up the **red giant branch** (RGB) on the H–R diagram.

A red giant has a strikingly different structure than a main-sequence star. Most of its mass is concentrated in a dense core only a few Earth radii in size, at temperatures of some 50 million K. The red-giant core is so dense that the electrons in the core become a degenerate gas. In this state, they produce a **degenerate gas pressure,** which depends only on density, not temperature, and this enables the core to attain a pressure sufficient to balance its gravitational force even though no fusion reactions are going on in it.

As the bloated star attains its red-giant status (point *D*), the core temperature—which has been steadily increasing as the core contracts—hits the minimum necessary ($T \approx 10^8$ K) to start helium burning by the triple-alpha process. This helium core is degenerate. Once part of it ignites in the triple-alpha reaction, the heat generated by the fusion spreads rapidly throughout the core by conduction. The rest of the core quickly ignites. If the core were an ordinary gas, this explosive ignition would expand it as a result of the rapid increase in temperature and pressure. The core is degenerate, however, and increased temperature does not increase the pressure in a degenerate gas. So the core does not expand. Instead, the increased temperature increases the rate of the triple-alpha process,

generating more energy, further increasing the temperature, and so on. This out-of-control process in the core is called the **helium flash.** When the core temperature finally reaches about 350 million K, the electrons become nondegenerate. Then the core expands and cools.

Why a helium flash rather than a slow ignition? One property of degenerate matter is that it has a very high thermal conductivity, so that heat flows through it very quickly. Hence, when the helium ignition temperature is reached in one part of the core, the turn-on spreads throughout the core in a flash (by astrophysical standards)—perhaps in just a few minutes!

After the helium flash, the star's radius and luminosity decrease a little and its point on the H–R diagram moves slightly downward and to the left. The star quietly burns helium in the core and hydrogen in a layer around the core (point *E*). This phase is the core-helium-burning analog of the star's main-sequence phase (core hydrogen burning).

Eventually the triple-alpha process converts the core to carbon. The reaction stops in the core but continues in a layer around it. This stage—the core is shut down but thermonuclear reactions are going on in a layer—resembles that when the star first evolved off the main sequence. The physical processes force the same evolution; the burning layer makes the star expand. The star again becomes a red giant (point *F*). The electrons in the core—this time carbon-rich—become degenerate again. The star moves up the **asymptotic giant branch** (AGB) on the H–R diagram.

Because the rate of the triple-alpha reaction is very sensitive to changes in temperature, the helium-burning shell causes the star to become unstable. Here's how: Suppose the star contracts a little. The temperature and energy production in the layer increase; the pressure also increases. However, the increase in pressure more than compensates for gravity, and so the outer parts of the star expand. The expansion leads to decreases in temperature, pressure, and—most dramatically—energy generation rate. The star contracts, the energy generation increases, the star expands, and the cycle repeats. These bursts of triple-alpha energy production are like small thermonuclear explosions in the shell; they have the prosaic name of **thermal pulses** (point *G* in Figure 16–5). The explosions occur about every few thousand years and cause the luminosity of the star to rise and dip rapidly by 20 to 50% in a few years or tens of years! The explosions cause the star to pulsate as well as vary in luminosity. Each blast generates a rush of energy; to move it out efficiently, the region becomes convective and the bubbling gases carry outward elements fused in each explosion.

Meanwhile, the star has developed a very strong outflow of mass from its surface, sometimes called a **superwind** to distinguish it from the normal stellar wind of a red giant. The superwind is triggered by the pulsations of the star and blows in gusts that quickly (in about 1000 years) rip off the envelope of the star. A hot core is left behind (point *H* in Figure 16–5). The expelled material forms an expanding shell of gas heated by the hot core. Astronomers call this a **planetary nebula,** for historical reasons. (It looks like a Jovian planet viewed with a small telescope.) The hot core appears as the central star of the nebula. The nebula keeps expanding until it dissipates in the interstellar medium.

For a star of roughly 1 solar mass or less, the core never reaches the ignition temperature of carbon burning because it has become degenerate and cannot contract and heat up to ignite carbon burning. In about 75,000 years, such a star becomes a white dwarf, composed mostly of carbon (*H* to *I*).

Without energy sources, the white dwarf cools to a black dwarf in a few billion years.

What will happen to the Earth when the Sun becomes a red giant? Recent calculations for a He-core-depleted model give a size just in excess of 1 AU, so that the Earth lies within the red giant envelope. The friction and vaporization will result in an orbital decay timescale of 200 years or less, and the Earth will plunge into the core of the red giant Sun.

Extremely Massive Stars

Let's look at theoretical work on the evolution of really massive stars—50 to 100 solar masses. Mass loss dramatically changes the evolution of certain stars. The Sun loses mass at a rate of about 10^{-14} solar mass per year, by the solar wind. Other stars (Figure 16–6) are known to lose mass at much greater rates in outflows that are called **stellar winds.** Red giants and supergiants blow off their envelopes at rates of 10^{-7} to 10^{-6} solar mass per year. Massive O stars also have stellar winds at about 10^{-7} to 10^{-6} solar mass per year for the strongest winds. Note that an O star will lose a few solar masses of material during its main-sequence lifetime of a few million years.

The mass loss changes the evolutionary tracks of stars with masses of $50M_\odot$ to $100M_\odot$. Such stars lose 50 to 60% of their initial mass by the end of

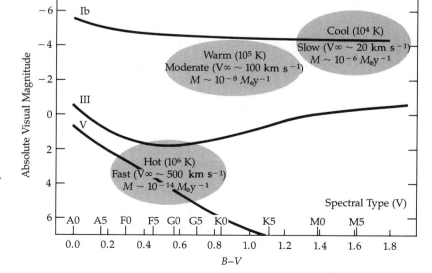

FIGURE 16–6 Locations on the H-R diagram of cool stars with stellar winds. Note that they fall into three major groups: cool, warm, and hot winds. Typical wind speeds and mass-loss rates are given for each. The spectral types along the bottom are only for main-sequence stars. *(A. Dupree)*

their main-sequence lifetime. The outer layers of the stars are stripped off—so much, in fact, that the core is revealed and the products of the CNO cycle (such as nitrogen) lie exposed at the surface. Such stripped cores may never become red giants because the layers above the core, where shell burning would take place, have been removed. (The peculiar objects known as **Wolf–Rayet stars,** which are hot stars with strong emission lines in their spectra, are examples of these stars. They have abnormally large abundances of nitrogen and carbon.)

The evolutionary tracks that result from these calculations (Figure 16–7) start with the stars on the ZAMS, losing mass at observed rates. In general, such stars make large loops in temperature at more or less constant luminosities. For the most

massive stars ($M > 60M_\odot$), the winds strip the outer layers while on the main sequence, leaving a bare He core. Such stars never evolve to red supergiants; they are O stars and then perhaps Wolf–Rayet stars. For mid-range stars ($60M_\odot$ to $25M_\odot$), they rapidly evolve to red supergiants where they undergo core helium burning. These start out as O stars and become blue supergiants and red supergiants; what exactly happens depends on the behavior of the mass loss. The lowest range of masses follow a blue–red–blue supergiant scheme. (Such stars do not have helium flashes because helium ignition takes place before a degenerate core develops.) Finally, these stars blow up in a supernova.

(C) LOW-MASS STARS

The evolution of stars with a mass much lower than the Sun's also has significant differences. First, stars of low enough mass may not get hot enough to burn much helium to carbon before they throw off their envelope. Such stars will end up as *white dwarfs* composed largely of helium.

Second, if the mass of a star is less than about 0.08 solar masses, it will not even reach the main sequence. Gravitational contraction does not heat it very effectively. Before it gets hot enough to start nuclear reactions, the density has risen so high that the matter becomes degenerate. Then the pressure of the degenerate electrons supports the star and keeps it from contracting any further. If gravitational contraction is prevented from heating the star, the nuclear fires can never be lit, and the star simply cools off to become a *brown dwarf.* (More on white and brown dwarfs in Section 17–1.)

(D) CHEMICAL COMPOSITION AND EVOLUTION

Population II stars contain only 0.01% heavy elements; the range is about 0.2 to 0.002%. Do their evolutionary tracks differ much from those of Population I stars with the same mass? The overall sequence is the same, but these stars exhibit a significant difference in position on the H–R diagram during core helium burning. What happens to a star with a heavy-element abundance of 0.01% that

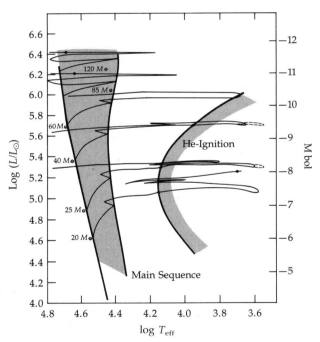

FIGURE 16–7 Evolutionary tracks for very massive stars. The initial masses (20 to $120M_\odot$) are indicated along the main sequence. The decrease in mass from stellar winds is indicated along the tracks. The band of the main sequence is on the left; the region on the right labeled "He-ignition" marks the onset of helium burning. The first dot along the track indicates the start of helium burning for that specific model. *(Adapted from a diagram by C. Chiosi and A. Maeder)*

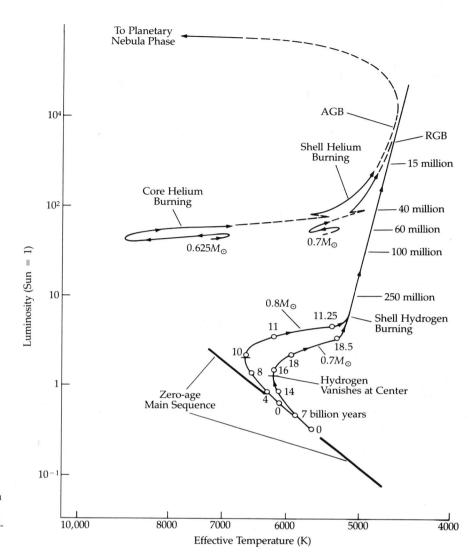

FIGURE 16–8 Evolutionary tracks for stars with low metal abundances. The two tracks are for two stars, one of mass $0.8M_\odot$ and one of mass $0.7M_\odot$, leaving the main sequence. Both stars lose mass through the red-giant phase. *(Based on theoretical calculations by I. Iben Jr.)*

leaves the main sequence with mass of $0.7M_\odot$ (Figure 16–8)? Almost 16 billion years after the start of core hydrogen burning, PP reactions have consumed the core's hydrogen fuel. Shell burning takes over the energy production, and the star rises to the red-giant region, at first slowly and then swiftly.

While a red giant, the star blows off some mass by a strong stellar wind. After the helium flash, the star settles down to core helium burning; it then has a mass of about $0.625M_\odot$ because of the mass lost by a stellar wind. The energy production goes on in a hydrogen-burning shell and a helium-

burning core, where the density is roughly 10^7 kg/m^3 and the temperature about 100 million K. As this star evolves, its luminosity stays roughly constant and its surface temperature changes, first to higher temperatures and then to cooler ones before the core burning stops. The star's evolutionary track makes a zigzag on the H–R diagram.

Stars with fewer heavy elements (only 0.001%) and a range of masses also form a core-helium-burning horizontal branch, but, relative to stars of the same mass yet higher heavy-element composition (0.01% as above), these stars are shifted to the left (higher temperatures) on the H–R diagram. In

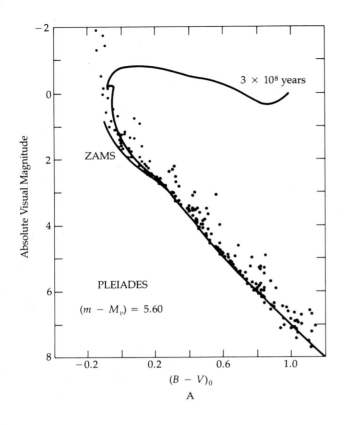

A

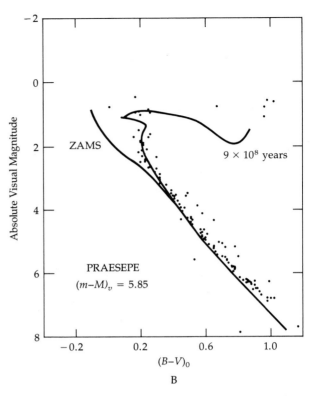

B

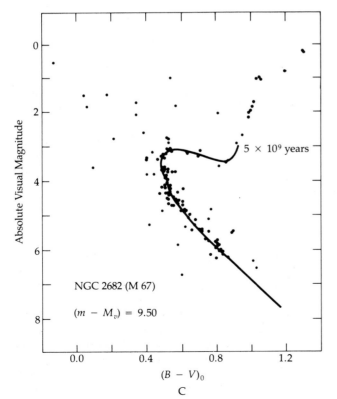

C

FIGURE 16–9 H–R diagrams for clusters. The observed turnoff points give the approximate ages when compared to isochrones from theoretical models (solid lines). (A) H–R diagram for the Pleiades. The solid lines are the theoretical calculations for the ZAMS and for an evolved cluster at age 3×10^8 years. The Pleiades are younger than this age, since the upper end of the main sequence still has massive stars. (B) H–R diagram for the Praesepe cluster with a best-fit theoretical line for an age of 9×10^8 years. (C) H–R diagram for M67. Same comparison as for Praesepe. The apparent distance modulus are given for each cluster. *(Adapted from diagrams by D. A. VandenBerg, Astrophysical Journal Supplement Series 58:711–769 (1985)]*

clusters, the reason for the range of masses along the horizontal branch is the different amounts lost while the stars are red giants. All the horizontal-branch stars in such a group start out with about the same mass and end up with helium cores of the same size, but some lose more mass than others as red giants and end up with very thin hydrogen envelopes around their helium cores. They are very much like main-sequence stars composed of pure helium and lie far to the left on the H–R diagram. Those that lose very little mass as red giants end up with thicker hydrogen envelopes. They are like ordinary red giants, ending up farther to the right on the H–R diagram.

16–4 ◑
INTERPRETING THE H–R DIAGRAMS OF CLUSTERS

If a cluster of stars is formed such that all members contract out of the gas cloud more or less simultaneously, the locations of the stars in the H–R diagram will depend upon the *time* elapsed since the initial formation. In 10^8 years, a cluster has stars on the main sequence up to a luminosity corresponding to $3M_\odot$ or slightly more, say $L/L_\odot \approx 100$ or absolute magnitude ≈ 0. Somewhat more massive stars lie a little to the right of the main sequence, and some stars have reached the giant branch but have not yet gone beyond it.

The **turnoff** from the main sequence reflects the time elapsed since the stars first arrived on the ZAMS. Ages of clusters may be determined by comparing the turnoff points in the theoretical H–R diagram (Figure 16–9), with the scale on the right of the diagram. The observed H–R diagrams of various clusters represent the loci of the ends of the evolutionary tracks for member stars to that particular time since formation. These loci are called **isochrones,** constant-time lines. The fact that computed isochrones closely resemble observed H–R diagrams is one of the major triumphs of modern astrophysics—it demonstrates that the basic physics of stellar models is correct.

The time taken up by the protostar contraction is so short that it can usually be neglected, particularly for well-developed clusters. In many clusters, however, the contraction time for massive stars is so much faster than that for low-mass stars that the massive stars will have already started to evolve off the main sequence by the time the low-mass stars reach it. Such a young cluster is the Pleiades,

which still has massive luminous stars on the main sequence. The turnoff at the very upper end of the H–R diagram of this cluster (Figure 16–9A) and the appearance of the supergiant branch have usually been attributed to the rapid evolution of these massive stars. The Pleiades have an age of 10^7 years. In an intermediate case, the Praesepe (Beehive) cluster has an age of about 9×10^8 years (Figure 16–9B). At the other extreme, we have an old open cluster (Population I) whose stars are highly evolved, M67. It has stars of about $1.25M_\odot$ evolved onto the red-giant branch and stars of $1M_\odot$ just about to leave the main sequence. One estimate gives the age of M67 as 5×10^9 years (Figure 16–9C).

The H–R diagrams of Population II globular clusters (Figure 16–10A) differ from the H–R diagrams typical of both intermediate Population I clusters (like M11) and old Population I clusters (like M67). The globular clusters are older than M67, but because they started with almost pure hydrogen and helium and virtually no heavy elements, their evolutionary tracks took different forms than Population I clusters did. The present H–R diagrams for globular clusters represent the end points of the evolutionary tracks (Figure 16–10B). Differences from one cluster to another are attributed to differences in the initial chemical composition. Globular clusters clearly show the horizontal branch (HB in Figure 16–10B) that results from different mass losses and chemical compositions. Note that a globular cluster's H–R diagram shows stars at all evolutionary phases: main sequence (MS in Figure 16–10B), lasting 10^{10} years with core hydrogen burning; red giants (RGB) for 10^8 years with shell hydrogen burning; horizontal branch (HB) for 10^8 years with core helium burning and shell hydrogen burning; and the asymptotic giant branch (AGB) for 10^7 years with shell double burning.

16–5 ◑
THE SYNTHESIS OF ELEMENTS IN STARS

In order to survive, a star must fuse lighter elements into heavier ones and in this way generate energy. Gravitational contraction provides the initial heat to get fusion reactions going. The more mass a star has, the greater the central temperature produced by gravitational contraction before degeneracy sets in and the heavier the elements it

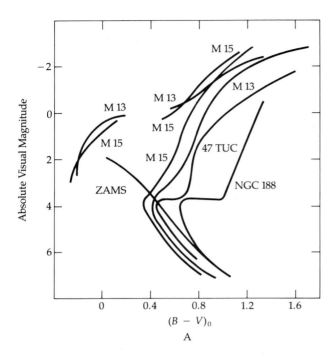

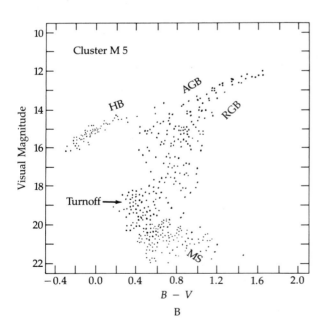

FIGURE 16–10 H–R diagrams for globular clusters. (A) Schematic diagram for a number of clusters. The old galactic cluster NGC 188 has been included for comparison. *(Adapted from a diagram by A. Sandage)* (B) Schematic H–R diagram for one globular cluster with the evolutionary major phases indicated. *(Adapted from a diagram by V. Castellani)*

can fuse. From the ignition temperatures needed for fusion reactions, we can set limits on the heaviest elements that a star of a certain mass can fuse (Table 16–3). For example, our sun can burn helium to carbon but will never get hot enough to fuse carbon. Table 16–3 summarizes the principal stages of nuclear energy generation and nucleosynthesis in stars. Note that the products (or ashes) of one set of reactions usually become the fuel for the next set of reactions, until iron is reached.

Also note in that table that only very massive stars (those with masses greater than about 5 solar masses) can produce elements heavier than oxygen, neon, and sodium. Few stars have this much mass, and so many stars come to the end of their nuclear evolution without having manufactured some important elements. This fact emphasizes the importance of massive stars in the scheme of cosmic evolution—they fuse heavy elements *and*, through the supernova process, throw some back into the interstellar medium.

Red giant stars play a major role in cosmic nucleosynthesis. The thermal pulses in a helium-burning shell as the site for producing certain isotopes, especially those that are rich in neutrons. (A neutron-rich isotope is one that has more neutrons than protons in the nucleus.) This process can occur in two stages for low- and middle-mass stars. Let's see how this happens in a comparison of the post-main-sequence evolutionary tracks for low- and intermediate-mass stars (Figure 16–11).

One stage takes place when a star first becomes a red giant. The convective zone that develops as a result of evolution to a red giant reaches down to the star's core and pulls up elements that have been made with hydrogen burning. At the base of the convection zone, carbon can be converted to nitrogen. The convection brings this processed material up to the surface, so the carbon abundance there goes down while the nitrogen abundance goes up. This whole process, which occurs in every star as it becomes a red giant for the first time, is called the **first dredge-up** (see points marked along the evolutionary tracks in Figure 16–11).

For medium-mass stars, such as one of 5 solar masses, a second phase of nucleosynthesis takes place after a star has burned the helium in its core. Thermal pulses then can convert helium to carbon, carbon to oxygen, nitrogen to magnesium, and iron to certain neutron-rich isotopes of heavier elements. The convective zone brings these to the

TABLE 16–3 *Stages of Thermonuclear Energy Generation in Stars*

Process	Fuel	Major Products	Approximate Temperature (K)	Approximate Minimum Mass (solar masses)
Hydrogen burning	Hydrogen	Helium	$1–3 \times 10^7$	0.1
Helium burning	Helium	Carbon, Oxygen	2×10^8	1
Carbon burning	Carbon	Oxygen, Neon, Sodium, Magnesium	8×10^8	1.4
Neon burning	Neon	Oxygen, Magnesium	1.5×10^9	5
Oxygen burning	Oxygen	Magnesium to sulfur	2×10^9	10
Silicon burning	Magnesium to sulfur	Elements near iron	3×10^9	20

FIGURE 16–11 Theoretical tracks for 1, 5, and 25$M_\odot$ stars. Major phases of fusion burning are indicated (thicker parts of the tracks), as well as important evolutionary stages. Dashed lines shows uncertain phases. *(Adapted from a diagram by I. Iben, Jr.)*

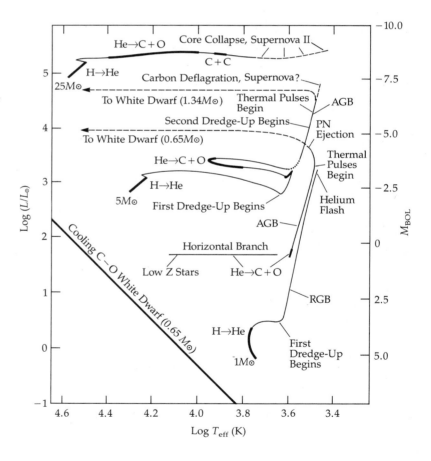

surface, a process called the **second dredge-up** (Figure 16–11).

All these processes would not affect on the rest of the cosmos except for one crucial fact: red giant stars have strong stellar winds. These blow off material from the surfaces of the stars so that the processed material from the first and second dredge-up is sent out into the interstellar medium.

PROBLEMS

1. Verify that about 6×10^{11} kg of hydrogen is converted to helium in our Sun every second.

2. (a) If a star is characterized by $M = 2 \times 10^{32}$ kg and $L = 4 \times 10^{32}$ W, how long can it shine at that luminosity if it is 100% hydrogen and converts all of the H to He?
   (b) Do a similar calculation for a star of mass 10^{30} kg and luminosity 4×10^{25} W.

3. Briefly describe the evolution of the following stars from a cloud of gas and dust to their demise:
   (a) $M = 10M_\odot$
   (b) $M = 0.1M_\odot$
   Clearly indicate which stages of the evolution are highly uncertain.

4. Using Figures 16–7 and 16–9 and the data in Table 16–2, sketch the H–R diagrams for star clusters of ages 10^7, 10^8, and 10^9 years (these are *constant-time* lines!) Clearly label the axes, and comment upon the significance of your results (turnoff points).

5. Although detailed models of stellar structure require the use of complex computer codes, simple scalings can be obtained by making rough approximations. For a variable x, we can substitute $\Delta x/\Delta r$ for dx/dr in order to obtain a crude result. (This method is a rough version of a numerical technique called "finite differences.")
   (a) Use the equation of hydrostatic equilibrium (16–1) to show that the central pressure scales as $P_c \propto M^2/R^4$. Substitute $\Delta P/\Delta r$ for dP/dr and take this difference between $r = 0$ and $r = R/2$, that is, $\Delta P/\Delta r \approx [P(r = R/2) - P_c]/(R/2 - 0)$. You may assume that $P(r = R/2)$ is negligible compared with P_c. Also, substitute the mean density of the star $\langle \rho \rangle$ for $\rho(r)$.
   (b) Now use the radiative transport equation and the same method as in part (a) to approximate dT/dr and obtain the theoretical mass–luminosity relationship, $L \propto M^3$. Assume that $\kappa(r)$ is constant and that $T(r) \propto T_c$.
   (c) Use the same method to obtain a rough relationship between the temperature and mass of a main-sequence star. (No other variables should appear in the proportionality.)
   (d) Combine your answers to parts (b) and (c) to obtain a relationship between T and L. Use the H–R diagrams (Figures 13–7 and 13–9) and Table A4–3 to compare the temperature of a star of $L = 10L_\odot$ with that of the Sun. How does this compare with your theoretical T–L relationship (be quantitative)?

6. Assume that the amount of hydrogen mass available for nuclear reactions in the core of a star is $M_c \approx 0.5M$. Further assume for simplicity that the only energy-generating nuclear reaction is $4\,^1\text{H} \rightarrow\,^4\text{He} +$ energy (ignore the fact that some of the energy is in the form of positrons and neutrinos). Obtain an expression for the hydrogen-burning lifetime of a star in years as a function of mass in solar units. (Assume $L \approx$ constant during the hydrogen-burning phase and use the mass–luminosity relationship $L \approx M^3$ in solar units.)

7. (a) Estimate the central pressure for 0.5, 10, and 50 solar mass stars. Compare these pressures to the central pressure of the Sun.
   (b) Estimate the central temperature for 0.5, 10, and 50 solar mass stars. Compare these temperatures to the central temperature of the Sun.

8. When a 1 solar mass star runs out of hydrogen fuel its outer atmosphere expands into a red giant. At the same time the core collapses. Give a brief physical explanation as to why these two events occur.

9. Assuming that a star radiates as a blackbody during all phases of its evolution, use the Stefan-Boltzmann Law to determine the radius (in units of $R_\odot$) of a $1M_\odot$ star at all nine stages (A–I) labeled in Figure 16–5. [*Hint:* Both the temperature and luminosity axes are logarithmic.]

10. Estimate the energy available and the lifetime for the helium-burning phase in a $1M_\odot$ star:
    (a) Calculate the energy released per net reaction $3\,^4\text{He} \rightarrow\,^{12}\text{C}$ in the triple-alpha process. [*Note:* The weight of ^4He is 4.0026 and the weight of ^{12}C is 12.0000.]
    (b) What fraction of the available mass of 3 helium nuclei is liberated in the form of energy in the triple-alpha reaction? Compare this to the fraction of available mass liberated in the proton–proton reaction.
    (c) Assume that approximately 10% of the original mass of the star is in the form of ^4He in the stellar core during the helium-burning phase. Estimate the total energy available from the triple-alpha process.
    (d) During the helium-core-burning phase, some hydrogen burning is also occurring in a shell. Thus the star's luminosity is not due only to helium burning. Keeping this in mind, assume that the typical luminosity from helium burning is $10^2\,L_\odot$ (see Figure 16–5, stage E). Estimate the lifetime of the helium core burning phase.

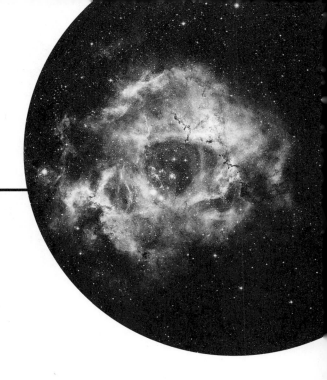

Chapter 17

Star Deaths

We have brought you to the brink of the deaths of stars. We have alluded to the processes of post-main-sequence evolution that result in stellar demises and final states—stellar corpses. This chapter examines the violent deaths of stars in detail and the three types of bizarre corpses that can be left behind: white dwarfs, neutron stars, and black holes. These final states are forever; very little can happen to change them.

The kind of corpse is determined by the *mass at the time of death*, which is *less* than the star's main-sequence mass because of the mass lost as a red giant, and/or in a supernova, or in the formation of a planetary nebula. The three kinds of corpses are white dwarf (mass $< 1.4M_\odot$), neutron star (mass between $1.4M_\odot$ and $3M_\odot$), and black hole (mass $> 3M_\odot$). Their masses while on the main sequence will be higher.

17–1 ◗
WHITE DWARFS AND BROWN DWARFS

White dwarfs evolve from red giant stars, exactly how depending on the mass of the star. For a star

like our Sun [Section 16–3(b)], contraction of the carbon core does not produce temperatures sufficiently high to bring about burning of carbon. The core will contract to a highly compressed state, and the increasing temperatures will accelerate the shell helium-burning rates. The star pulsates until it ejects its outer layers. The envelope becomes separated from the core as a thin shell and expands and cools; this process creates a **planetary nebula** (Figure 17–1 and Chapter 18). The core, having lost its envelope, now stands revealed as a hot, very dense star—a **white dwarf.**

In less massive stars, helium burning may not become appreciable because of low core temperatures. Such stars might continue to contract gravitationally to become white dwarfs even without ejecting an envelope. More massive stars end as white dwarfs by some kind of mass loss by a strong stellar wind, or perhaps by exchange of mass between members of a binary system [Section 12–4(c)]. Many white dwarfs are in fact faint companions to larger, more massive stars; Sirius B and 40 Eridani B are two examples.

Infrared and radio telescopes are revealing new instances of intermediate-mass stars losing mass

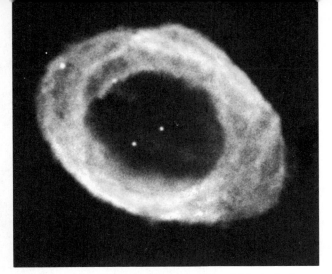

FIGURE 17–1 Messier M57, the Ring Nebula in Lyra, was formed by the expulsion of the atmosphere of a red giant. The star visible in the center will cool to a white dwarf. *(Lick Observatory)*

by strong stellar winds as they evolve from a red giant to a white dwarf. Cool winds from cool stars are dusty, and so would hide the star from optical view. Observations (Figure 17–2) show that such outflows, at least in some cases, tend to be *bipolar*, with material flowing out along an axis in two directions. The star itself is hidden by a disk of dust, which helps to direct the outflows, which move at speeds of about 200 km/s—fast enough to form shock waves in the surrounding medium. This very brief phase then results in the formation of a planetary nebula.

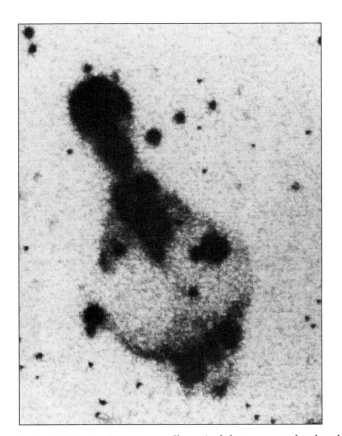

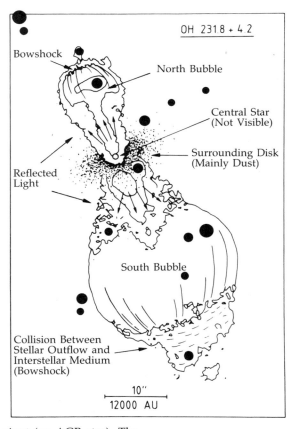

FIGURE 17–2 A strong stellar wind from an evolved red giant (an AGB star). The negative image at left is taken with an Hα filter to show gas that has been excited by the outflow. The diagram on the right is a schematic picture of the interaction of the outflow with the surrounding material. Note the scale at bottom. *(B. Reipurth, ESO)*

Observational evidence indicates that main-sequence stars with $M \leq 7M_\odot$ finish their lives as white dwarfs. Basically, low and intermediate mass stars prefabricate a white dwarf by the growth of a degenerate core while they are on the AGB. Heavy mass loss must then take place to reveal the compact core after the phase of helium thermal flashes.

The kind of white dwarf leftover depends on the mass of the progenitor star. Stars of $0.5M_\odot$ or less will not ignite helium and so form helium white dwarfs; those in the range $0.5–5.0M_\odot$ will not ignite carbon and so leave carbon-oxygen ones; those in the range $5–7M_\odot$ will burn carbon and make oxygen–neon–magnesium white dwarfs.

(A) PHYSICAL PROPERTIES

Because white dwarfs are very dense, the stellar material no longer behaves like an ordinary gas. It becomes so tightly packed that the electrons cannot move completely at random. Hence, their motions are subject to limitations imposed by the proximity of other electrons. Some electrons may still move at very high velocities, but they cannot change their velocities by collisions as in an ordinary gas; the electrons may change velocity only by exchanging orbits with other electrons. The laws of quantum mechanics hold, and the **Pauli exclusion principle** applies: only two electrons with opposite spins can have a given energy in a given volume at one time. Because of the close packing, less space is available and the number of possible velocities or energies permissible for an electron becomes smaller. Such a material is called a **degenerate electron gas.** In such a degenerate gas, the electrons are distributed more or less uniformly throughout the medium, surrounding the nuclei. The nuclei themselves are regularly spaced and become more tightly constrained as the pressures increase, until they are so fixed with respect to each other that they resemble a crystalline lattice. Under such conditions, the material more closely resembles a solid than a gas.

The cause of these high densities lies in the fact that all available nuclear energy has been expended and the star contracts gravitationally until stopped by the pressure of the degenerate electron gas. Only stars of mass smaller than about $1.4M_\odot$ (called the **Chandrasekhar limit**) can be stable white dwarfs because of limitations imposed by

the stellar structure, which in turn depends upon both hydrostatic equilibrium and the nature of the degenerate electron gas. All the peculiar properties of white dwarfs can be traced to the fact that they are made of degenerate material. A main point is the mass–radius relation: *the more massive a white dwarf, the smaller its size.* In contrast, for main-sequence stars, the more massive ones are larger. (Can you reason why?) Let's examine the mass–radius relationship for white dwarfs.

The exact relationship between pressure and density in completely degenerate, **nonrelativistic** matter is

$$P = K\rho^{5/3} \qquad \text{(17–1)}$$

where K is a constant. (For a **relativistic** gas, $P \propto \rho^{4/3}$.) This is the equation of state of such material. Contrast this to the equation of state of an ideal gas,

$$P = nkT \qquad \text{or} \qquad P \propto \rho T$$

Now from hydrostatic equilibrium,

$$P \propto M^2/R^4$$

This result applies to any star. Then, using the above equation of state in the density equation,

$$\rho \propto M/R^3$$

we get

$$P \propto \rho^{5/3} \propto M^{5/3}/R^5$$

Now use

$$P \propto M^2/R^4$$

so that

$$M^2/R^4 \propto M^{5/3}/R^5$$

so

$$R \propto 1/M^{1/3}$$

If we use the equation of state for a degenerate, nonrelativistic gas, we get

$$R = \frac{4\pi K}{G(4/3\pi)^{5/3}\,M^{1/3}} \qquad \text{(17–2)}$$

The mass–radius relation shows that as M gets larger, R gets smaller. This result hints at the idea that white dwarfs may have a maximum mass limit. (They do—the Chandrasekhar limit.)

Generally, most white dwarfs have their masses determined by indirect photometric or spectro-

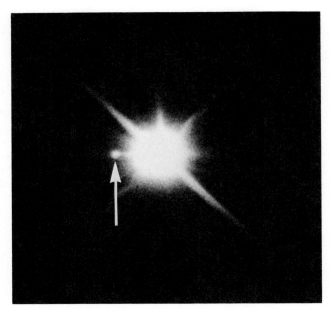

FIGURE 17–3   A white dwarf. Sirius B (arrow), the binary companion to Sirius A. *(Lick Observatory)*

We can make a rough estimate of the cooling time as follows. The total thermal energy is $N(3/2kT)$, where N is the total number of particles in the star. Assume a $0.8M_\odot$ star made of carbon with an average internal temperature of 10^7 K. Then the total thermal energy is about 10^{40} J, and at a constant luminosity of $10^{-3}L_\odot$, the star will have a thermal lifetime of

$$t_{cool} = E_{thermal}/L$$
$$\approx 10^{40} \text{ J}/(10^{-3})(3.8 \times 10^{26} \text{ J/s}) \approx 10^9 \text{ y}$$

So it takes several billion years for a white dwarf to cool to a black dwarf.

(B) OBSERVATIONS

In 1862, the American optician Alvan Clark observed Sirius B (Figure 17–3), the faint companion to Sirius A. Later, this star was found to be a white dwarf. Because Sirius B is part of a binary system,

scopic methods. For stars whose distances are known (say by parallax), we can find the mass from a measurement of the effective temperature, from which we get the radius and then infer the mass from an assumed mass–radius relation (such as Equation 17–2). When distances are not known, we can still use spectrophotometric observations plus model atmospheres to infer the surface gravity, g, from which we arrive at a mass by again applying the mass–radius relation. Overall, the average measured mass of white dwarfs is $\approx 0.6M_\odot$.

Note that white dwarfs shine by the radiation of their thermal energy. The degenerate, isothermal core cools down as its residual heat escapes through the thin, nondegenerate envelope (the star's atmosphere). Any gravitational energy released by compression does *not* contribute to the luminosity; rather, it simply forces degenerate electrons into higher energy levels. Overall, the relation between a white dwarf's age, as defined by its cooling time, and its luminosity is

$$t_{cool} \propto L^{-5/7}$$

So as a white dwarf ages, it cools off at a slower rate.

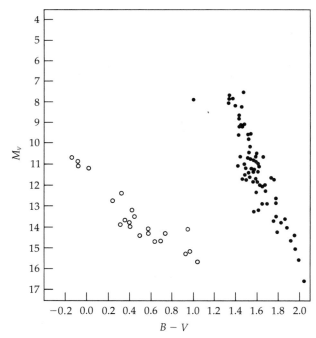

FIGURE 17–4   Absolute visual magnitude and $B - V$ color diagram for stars within 100 pc of the Sun from a survey by the U.S. Naval Observatory. The red dwarf stars fall in the band on the right; degenerate dwarfs are in the band at the left.

its mass can be calculated using Kepler's third law (Section 12–2). This calculation gives a value of about $1.05 \pm 0.03 M_\odot$ for the mass of Sirius B. This star has a low luminosity, about $3 \times 10^{-3} L_\odot$, and a high surface temperature, about 29,500 K and so, from $L = 4\pi R^2 \sigma T^4$, it has a radius of around $7 \times 10^{-3} R_\odot$. Sirius B has an average density of about 3×10^9 kg/m^3. By coincidence, the brightest star in Canis Minor (the Little Dog, near Canis Major), Procyon, also has a white-dwarf companion. The existence of this companion was predicted in 1862 from the motion of Procyon and was observed in 1882. Called Procyon B, it has a mass of about $0.68 M_\odot$. The star 40 Eri B, also in a binary, has a mass of $0.43 \pm 0.02 M_\odot$. It appears that the masses of white dwarfs in binaries are basically the same as those of single stars.

In fact, it turns out, if one examines closely faint stars near the Sun, that a large fraction are degenerate dwarfs. Because these stars are nearby, they also have large proper motions, and show up in surveys. Because white dwarfs are so hot, they are much bluer in color than the common red dwarf stars, and so can be distinguished readily from them (Figure 17–4).

White dwarfs tend to fall into two general categories: those with spectra showing strong hydrogen lines and those with spectra showing strong helium lines. White dwarfs with the strongest hydrogen lines are put in class DA—D for dwarf and A to indicate that the spectra resemble those of A stars (strongest hydrogen Balmer lines). Those showing helium lines resemble stars of class B, and so are assigned spectral type DB. Cooler stars resemble other spectral classes, and some white dwarfs show no lines in their spectra at all; they are called class DC, the C standing for continuous. The spectral lines tell only about the atmosphere of the star. Those stars with strong hydrogen lines may have a thin hydrogen atmosphere, but the interior is still an evolved core of carbon, helium, or other products of nuclear processing.

An H–R diagram of DA white dwarfs for which reliable observational data are available (Figure 17–5) shows that (1) the DA white dwarfs actually fall along a temperature sequence from 6000 to 31,000 K and (2) they lie parallel to lines of constant radius drawn on the H–R diagram. (Recall that $L = 4\pi R^2 \sigma T^4$. If R is kept constant, L will be high where T is high and low where T is low.) The average radius of these stars is $0.013 R_\odot$. Typical values for the physical properties of white dwarfs are mass $0.7 M_\odot$, radius $0.01 R_\odot$ (7×10^6 m), and density 10^9 kg/m^3.

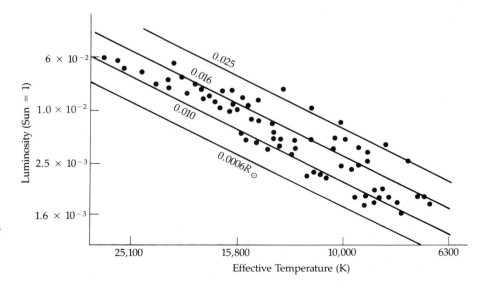

FIGURE 17–5 H–R diagram for DA white dwarfs with well-known temperatures and luminosities. The lines indicate stars of constant radius.
(Adapted from a diagram by V. Weidmann)

(C) WHITE DWARFS AND RELATIVITY

The extremely high densities of white dwarfs provide a test for the general theory of relativity, for their surface gravities are high enough to produce a detectable gravitational redshift in their spectra. A **gravitational redshift** occurs whenever light moves from a stronger to a weaker gravitational field. As it does so, it must do work since a photon has an equivalent mass ($E = mc^2$) and a gravitational field can affect it. In such a situation, an ordinary particle loses kinetic energy (as it gains gravitational potential energy) and slows down. Photons cannot slow down, however; they travel only at the speed of light. Instead of slowing down, a photon's loss of energy shows up as a decrease in its frequency (an increase in its wavelength), that is, as a redshift, because $E = h\nu$.

The gravitational redshift produced by a star depends on its mass-to-radius ratio. The larger this ratio, the larger the gravitational redshift. To calculate the shift, imagine a photon leaving a mass and traveling to infinity. Its total energy is

$$TE = PE + KE = \text{constant}$$

but $PE < 0$ initially and $PE = 0$ at infinity, so that

$$KE_f = KE_i + PE_i < KE_i$$

If we use Newtonian gravitation, then we can imagine that the photon has "lost" kinetic energy and so its frequency changes (since its speed cannot):

$$\Delta KE = \Delta(h\nu) = -GmM/R$$

For a photon,

$$m = E_i/c^2 = h\nu_i/c^2$$

so that

$$h\,\Delta\nu = -G(h\nu_i/c^2)M/R$$

and

$$\Delta\nu/\nu_i = -GM/c^2R \qquad \textbf{(17–3a)}$$

or

$$\Delta\lambda/\lambda_i = GM/c^2R \qquad \textbf{(17–3b)}$$

where $\Delta\lambda = \lambda_f - \lambda_i$. These relations work fine for white dwarfs, because their gravitational fields are relatively weak, $GM/c^2 R \ll 1$. For stronger gravitational fields, we must use general relativity rather than Newton's theory. This application results in

$$\lambda_f/\lambda_i = [1 - 2GM/Rc^2]^{-1/2} \qquad \textbf{(17–4)}$$

where G is Newton's gravitational constant, M is the object's mass, R its radius, and c the speed of light.

The observation is complicated by the motion of the star relative to the Earth because any radial velocity produces a Doppler shift (to the red if the star is receding). Therefore we see both shifts (gravitational and Doppler) together. The two can be separated only if the star's velocity through space can be determined, which is possible for binary systems because their velocities in space can be determined from the spectrum of the primary star. With the velocity known, the Doppler redshift is subtracted from the total redshift to leave any gravitational redshift. For a white dwarf with $M = 0.6M_\odot$ and $R = 0.01R_\odot$, the gravitational redshift amounts to $\approx 10^{-4}$. The measured redshift for Sirius B is 3.0×10^{-4}. The predicted theoretical value is 2.8×10^{-4}. Within experimental error, the redshift observation confirms general relativity.

Note that we can turn this procedure around and infer a white dwarf's mass from its gravitational redshift. Measured values for redshifts range from 20 to 90 km/s; modern CCD spectroscopy can achieve an accuracy of ± 3km/s, which corresponds to an uncertainty in mass of $\pm 0.03M_\odot$. Recent results for a sample of 14 DA stars gives an mean mass of $0.66 \pm 0.05M_\odot$.

(D) MAGNETIC WHITE DWARFS

A few white dwarfs have intense magnetic fields—10^2 to 10^4 T at their surfaces. (Recall that the global magnetic field of the Sun is about 10^{-4} T.) These strong fields are probably relics from the time before the star became a white dwarf. The basic physical concept that supports this idea is called **conservation of magnetic flux**. Consider a star with a magnetic field. The **magnetic flux** is essentially the number of field lines (field strength) times the surface area through which they thread. Imagine compressing the star to a smaller size. The number of field lines remains the same but the surface area decreases, and so the field lines draw closer together. The magnetic field strength increases since the separation of the field lines indicates the intensity of the field.

Because a star's surface area depends on the square of its radius, its magnetic field strength must depend (if flux is conserved) on the inverse square of its radius. An example: Start with a star like the Sun, magnetic field about 10^{-4} T, radius = 7×10^5 km. Imagine collapsing it to the size of a white dwarf, with a radius of 7×10^3 km. What does the conservation of magnetic flux predict for the white dwarf's field strength? We have

$$B_{wd}/B_\odot = (R_\odot/R_{wd})^2$$

where B_{wd} is the white dwarf's field strength, $B_\odot$ the Sun's field strength, R_{wd} the white dwarf's radius, and $R_\odot$ the Sun's radius. Then

$$
\begin{aligned}
B_{wd} &= B_\odot (R_\odot/R_{wd})^2 \\
&= (10^{-4})[(7 \times 10^5)/(7 \times 10^3)]^2 = 1 \text{ T}
\end{aligned}
$$

Although lower than the strongest observed fields, this result shows that the simple collapse idea is plausible.

Polarization observations of white dwarf magnetic fields, which were pioneered by James Kemp, give fields as strong as 10^4 T! Such stars are rare: only a few percent of single stars and about 10% of those in binary systems.

(E) BROWN DWARFS

There is another category of nonnuclear-energy-generating "stars" that must be distinguished from black dwarfs, white dwarfs, and even red dwarfs; these are the **brown dwarfs.** Rather than representing a true star death, these objects more closely resemble gigantic planets than stars. They result from the gravitational collapse and contraction of protostellar nebulae but have insufficient mass to trigger nuclear reactions in their cores. Somewhat arbitrarily, an object is considered a planet if its mass is less than $0.002M_\odot$ and a brown dwarf if its mass lies in the range $0.002M_\odot$ to $0.08M_\odot$. Objects more massive than $0.08M_\odot$ can develop sufficiently high central temperatures to sustain nuclear fusion. The only energy source for a brown dwarf, therefore, is gravitational contraction. Brown dwarfs are cool and have very low luminosities and so are difficult to observe.

Theoretical models show that the boundary between stars and brown dwarfs lies at about 0.07 to $0.08M_\odot$ for a Population I chemical composition. Both kinds of pre-main-sequence objects go through a brief phase of deuterium burning, lasting 10^5 to 10^6 years, when the energy output mostly comes from fusion rather than gravitational contraction. After some 10^{10} years of evolution, a $0.08M_\odot$ star reaches stable hydrogen burning in its core with a power of a mere 4×10^{-5} solar luminosity. Stars with lower mass never reach stable main-sequence hydrogen burning. Note that, because of their slow evolution, brown dwarfs must be very old. Because of their small mass, they are also likely to be fairly common: perhaps some 500 stars in a volume with a 10 pc radius in the Galaxy's disk.

17–2 ◗
NEUTRON STARS

For contracting stellar corpses with masses greater than $1.4M_\odot$, the degenerate electron gas pressure cannot hold off gravity. Matter is crushed to such high densities that inverse-beta decay occurs:

$$p^+ + e^- \rightarrow n + \nu$$

Literally, protons and electrons are squeezed into neutrons; a neutron gas forms. At about 10^{17} kg/m^3, the neutrons become subject to quantum laws and become a degenerate gas. In analogy to the behavior of a degenerate electron gas, a degenerate neutron gas provides internal pressure to form a stable entity: a **neutron star.** Because the equation of state for a degenerate neutron gas is almost the same as for an electron one, neutron stars of greater mass will have smaller radii—and an upper mass limit (thought to be about $3M_\odot$).

(A) PHYSICAL PROPERTIES

Neutron stars have diameters of a few tens of kilometers, depending on their masses. Most of the interior consists of a neutron gas at such high densities that it is a fluid. The outer few kilometers is a mixture of a neutron superfluid and neutron-rich nuclei arranged in a solid lattice. The structure is that of a crystalline solid, similar to the interior structure of a white dwarf. In the outer few meters, where the density falls off quickly, the neutron star has an atmosphere of atoms, electrons, and protons. The atoms here are mostly iron.

Because a neutron star is so dense, it has an enormous surface gravity. For example, a solar-mass neutron star with a radius of 12 km has a surface gravity 10^{11} times greater than that at the Earth's surface. This intense gravitational field results in a huge escape velocity, as much as $0.8c$. Also, objects falling onto a neutron star from a great distance have at least the escape velocity when they hit. That means that even a small mass carries a huge amount of kinetic energy.

The gravitational redshift from a neutron star is substantial. For a solar-mass neutron star about 7 km in radius,

$$\Delta\lambda/\lambda_i \approx GM/c^2R$$
$$= (6.67 \times 10^{-11})(2 \times 10^{30})/(9 \times 10^{16})(7 \times 10^3)$$
$$\approx 0.2$$

This result means that light emitted at 600 nm would be shifted to 720 nm by the time it reached an outside observer.

(B) PULSARS—ROTATING NEUTRON STARS

In 1967, a large radio telescope was developed by Anthony Hewish in Cambridge, England, to study the scintillations of radio sources. Scintillation is the rapid twinkling of a radio source from density fluctuations in the interplanetary plasma (the solar wind) and in the interstellar medium; it is analogous to the twinkling of visible stars (from density fluctuations in the Earth's atmosphere). Almost immediately, weak, precisely periodic radio signals were detected.

Jocelyn Bell Burnell, then a graduate student in charge of preliminary data analysis, noticed a strange signal that suddenly disappeared only to reappear three months later. The Hewish group concentrated on this unusual signal and found radio pulses occurring at a regular rate, once every 1.33730113 s. Excited, they searched the sky for any similar signals and discovered three more objects emitting radio pulses at different rates. They concluded that the objects must be natural phenomena and named them **pulsars.** To date, some 150 pulsars have been studied in detail. The total

TABLE 17–1   *Properties of Selected Pulsars*

Name (PSR)	Period (s)	dP/dt (10^{-9} s/d)	DM(pc/cm^3)
1937 + 214	0.001557	1.07×10^{-5}	71.2
1855 + 09	0.005362	1.8×10^{-6}	13.3
0531 + 21 (Crab)	0.33200	36.5	56.8
0833 − 45 (Vela)	0.089234	10.8	69.1

detected is roughly 500. Table 17–1 gives data on a few selected pulsars.

For a given pulsar, the period between pulses repeats with very high accuracy, better than one part in 10^8. The amount of energy in a pulse, however, varies considerably; sometimes complete pulses are missing from the sequence. Although the intensity and shape vary from pulse to pulse, the average of many pulses from the same pulsar defines a unique shape (Figure 17–6). The average pulse typically lasts for a few tens of milliseconds, with no detectable radio emission between pulses. Individual pulses may be resolved into 20 to 30 subpulses of submillisecond duration, so that the primary pulses are actually the envelopes of these secondary pulses. Pulsars are most readily observed at low frequencies; for instance, the first discoveries were made at 81.5 MHz. The intensity of the pulses diminishes rapidly at higher frequencies, and the pulses become broader and more regular in shape.

For the well-studied pulsars, periods range from 1.6×10^{-3} to 4.0 s, with an average value of 0.65 s. (A few pulsars are known so far to have millisecond pulses: the fastest one with a 1.6-ms interval; these are called **millisecond pulsars.**) In the cases where accurate radio observations have been made, periods have been noted to increase in regular fashion. The rates of change have typical values of about 10^{-8} s/year. Such small increases can be measured only with atomic clocks, whose stability is better than 10^{-10} s/year. Note that, very roughly, the pulse period P divided by its rate of change with time dP/dt gives an estimate of a pulsar's age:

$$t \approx P/(dP/dt)$$

What we have done here is to estimate the time for the pulse rate to decay from its present value to a very large interval. An example: the Crab pulsar

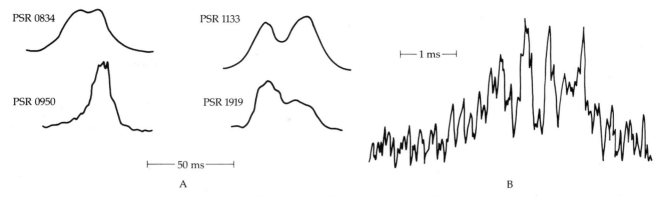

FIGURE 17–6 Pulsars at radio wavelengths. (A) A sample of the variety of pulse shapes. (B) Detailed structure within a pulse from PSR 1133. *(Adapted from a diagram by A. T. Moffet and R. D. Ekers)*

(see below) has $P = 0.03$ s and $dP/dt = 1.2 \times 10^{-13}$ s/s, so that

$$t \approx 0.03/(1.2 \times 10^{-13})$$
$$\approx 10^{11} \ s \approx 10^4 \ \text{years}$$

Approximate distances to pulsars and some properties of the interstellar medium may be deduced directly from pulsar observations. A given pulse arrives at the Earth later as we look at lower frequencies. This phenomenon is called **dispersion,** and it is due to a slowing down of the photon velocity by electrons in the line of sight to the pulsar (analogous to the index of refraction discussed in Section 8–1 and the lower propagation velocity of light in a material medium). Longer wavelengths are slowed down more, and from the observations, we may deduce the mean electron density in the line of sight. Conversely, if we know (or can estimate) the mean electron density, the distance to the pulsar follows immediately. If pulses of two different frequencies f_1 and f_2 are emitted at time t_0, the times at which they arrive at the earth, t_1 and t_2, are different. The times are given by the expressions $t_1 - t_0 = d/v_1$ and $t_2 - t_0 = d/v_2$. We don't know t_0, but we can measure $t_2 - t_1$, which is equal to $(1/v_2 - 1/v_1)d$. The velocities depend on the electron density, so if we know that, we can determine the distance d.

The problem is that the interstellar medium does not have a constant density; it varies along the line of sight to the pulsar. Astronomers define the **dispersion measure** (*DM*) as the integrated electron density, n_e, to a pulsar at distance d as

$$DM = \int_0^d n_e \, dl$$

Then the difference between the pulse arrival times is related to the *DM* by

$$t_2 - t_1 = \frac{e^2}{2\pi m_e c}\left(\frac{1}{f_2^2} - \frac{1}{f_1^2}\right)DM$$

Combined with the observation that most pulsars lie at low galactic latitudes, the distance data imply that pulsars we see are quite local (within a few kiloparsecs) and lie in the galactic disk.

Also, we know that the plane of polarization of linearly polarized radiation (Section 8–1) is rotated when the radiation propagates through a magnetized plasma. This effect, known as **Faraday rotation,** depends upon (1) the mean electron density, (2) the mean magnetic field strength, (3) the square of the wavelength of the radiation, and (4) the distance traveled through the medium. Since pulsar bursts are strongly linearly polarized, we infer that the mean magnetic field in the galactic disk has a strength of about 10^{-10} T. In other words, for a given source, we can measure the angle through which the plane of polarization rotates as a function of wavelength. This gives a value for the product of the electron density and magnetic field strength integrated along the line of sight. Then if we can determine the electron density—and we can, from a measurement of the velocity dispersion—we can infer the mean component of the magnetic field strength along the line of sight.

What mechanism keeps the precise clock of a pulsar? The accepted model is that of a rotating, magnetic neutron star known as the **lighthouse model.** The model has two key components: (1) the neutron star, whose great density and fast rotation insure a large amount of rotational energy and (2) a dipolar magnetic field that transforms the rotational energy to electromagnetic energy.

That neutron stars might possess extremely intense magnetic fields follows from the same conservation-of-flux argument applied earlier to white dwarfs. (Recall that observational evidence supports this argument; some white dwarfs have surface magnetic fields of roughly 10^2 T.) Imagine our Sun collapsed to the size of a neutron star 7 km in radius. Calculating the field strength from conservation of magnetic flux, we have

$$B_{ns} = B_\odot (R_\odot/R_{ns})^2 \approx 10^6 \text{ T}$$

Actual observations indicate that the fields more typically have a strength of 10^8 T. The region close to the neutron star where the magnetic field directly and strongly affects the motions of charged particles is called the pulsar's **magnetosphere.** Here all the energy conversion takes place. The magnetic axis is tilted with respect to the rotational axis.

As the pulsar spins, its 10^8-T magnetic field induces an enormous electric field at its surface. This electric field pulls charged particles (mostly electrons) off the solid crust of iron nuclei and electrons. The electrons flow into the magnetosphere, where they are accelerated by the rotating magnetic field lines. The accelerated electrons emit synchrotron radiation in a tight beam more or less along the field lines.

You can now see how a pulsar emits regular pulses without actually pulsating. If the magnetic axis can fall within our line of sight, each time a pole swings around to view (like the spinning light of a lighthouse), we see a burst of synchrotron emission (Figure 17–7). The time between pulses is the rotation period. The duration of the pulses depends on the size of the radiating region. As the pulsar generates electromagnetic radiation, the torque from accelerating particles in its magnetic field slows down its rotation. This slowdown is observed.

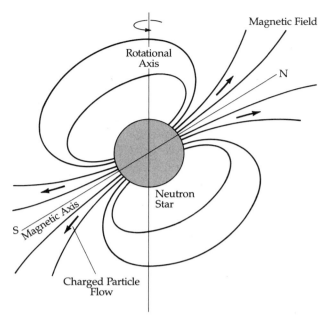

FIGURE 17–7 Generic model for a pulsar. A rapidly rotating, highly magnetic neutron star can emit synchrotron radiation along its dipolar axis. Note that the magnetic axis is inclined with respect to the rotation axis.

Here's a simple argument that shows that fast pulsars must have neutron-star densities. Assume that the clock mechanism is rotation. A sphere can rotate only at a speed such that the centripetal acceleration V^2/R at the equator is equal to or less than the gravitational acceleration GM/R^2:

$$V^2/R = GM/R^2$$

and

$$V = (GM/R)^{1/2}$$

where V is the equatorial velocity of the sphere, R its radius, and M its mass. The period of a rotating sphere is

$$P = 2\pi R/V$$

so that

$$P = 2\pi R/(GM/R)^{1/2} = 2\pi R^{3/2}/(GM)^{1/2}$$

but

$$M = (4/3)\pi R^3 \rho$$

so

$$P = 2\pi R^{3/2}/[G(4/3)\pi\rho R^3]^{1/2}$$
$$= (3.8 \times 10^5)/\rho^{1/2} \text{ s} \qquad \textbf{(17–5)}$$

with the density in kilograms per cubic meter. For a period of, say, 2 ms,

$$2 \times 10^{-3} \text{ s} = 3.8 \times 10^5/\rho^{1/2}$$
$$\rho^{1/2} = 1.9 \times 10^8$$
$$\rho \approx 4 \times 10^{16} \text{ kg/m}^3$$

just the bulk density of a neutron star.

(C) MILLISECOND PULSARS

Because of instrumental limitations, astronomers prior to 1982 had no luck in finding pulsars with periods much shorter than that of the Crab pulsar. Then, while investigating a peculiar radio source in the constellation of Vulpecula, radio astronomers homed in an extremely fast pulsar—a period of 1.558 *milliseconds* (Figure 17–8). Applying the lighthouse model to this pulsar, called PSR 1937+214, requires that it spin 642 times per second (20 times faster than the Crab pulsar) so that its surface rotates at roughly one-tenth the speed of light. That also means that the neutron star lies very close to its breakup speed. Several other millisecond pulsars (those with pulse periods of less than 10 ms) have been discovered in recent years.

One of the curious features of fast pulsars is that their rotation rates are very stable. PSR 1937+214, for instance, loses only 3.2×10^{-12} s/y. This pulsar provides the best time standard available today, even beating out atomic clocks, which are accurate to a few microseconds in a year. This rate contrasts to the very fast spindown rates of ordinary pulsars. One explanation is that the millisecond pulsars have very weak magnetic fields, perhaps one thousand times weaker than typical. Then how did it become an observable pulsar?

The scenario proposed is that of pulsar resurrected in a binary system. Millisecond pulsars may have once been formed in a supernova and aged gracefully. Then, billions of years later, their low-mass companions evolved finally to red giants and matter flowed from them into a disk around the dead pulsar. This material makes a rapidly spinning accretion disk around the neutron star. The magnetic field of the pulsar is entwined with the disk; this linkage spins up the pulsar so that it lives again.

One millisecond pulsar has gained notoriety as the so-called "black widow" pulsar. Its official astronomical name is PSR 1957+20; its special signature is a glowing nebula streaming away from it. This pulsar generates a hot, high-speed wind, with particles traveling close to the speed of light. This wind does damage not only to the surrounding medium but also to the pulsar's companion star. When it went through its red giant phase, this star resurrected the old pulsar by mass transfer and spinup. The pulsar's wind now rams into the companion star, creating a shock wave, heating one side of the star to 5000 K, and stripping off the star's surface material. So in a symbolic sense, the pulsar is gobbling up the star that gave it birth—and so may all millisecond pulsars.

One of those millisecond pulsars, called PSR 1855+09, is a member of a binary system with an

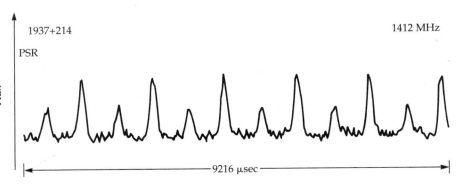

FIGURE 17–8 Early observations at 1412 MHz of PSR 1937 + 214, the first millisecond pulsar to be discovered. Note that the time span of these series of pulses is only about 10 ms. *(D. C. Backer, S. R. Kulkarni, C. Heiles, M. M. Davis, and W. M. Goss)*

PSR 1937+214   1412 MHz

Flux

|← 9216 μsec →|

Time

orbital period of 12.3 days, as indicated by the Doppler-shift variation in its pulse period. The orbit is nearly circular, and from it, we can infer that the companion to the pulsar has a mass in the range of 0.2 to 0.4$M_\odot$. One evolutionary scenario suggests a binary system with a low-mass main-sequence star and a primary of about 3$M_\odot$ that eventually became the pulsar we see today.

(D) BINARY PULSARS

Most stars in the Galaxy are members of binary or multiple-star systems. Even after one member of a binary becomes a supernova, the system usually remains intact. So a radio pulsar could exist in a binary system. The first one observed is called PSR 1913+16 and was discovered by Russell Hulse and Joseph Taylor in July 1974 during a search for new pulsars. This pulsar first attracted attention because its pulse period was only 0.059 s. When Hulse and Taylor reobserved PSR 1913+16 in September 1974, they found that its period went through a large cyclical change in only 7.75 h. They recognized that such regular changes would naturally come about in a binary system consisting of the pulsar and a companion with an orbital period of 7.75 h. What was seen was a Doppler shift in the signal produced by the orbital motion of the system (Figure 17–9). When the pulsar is moving away from us, its pulses are spread out and come at longer intervals. When it is moving toward us, the pulses are pushed together and come at shorter intervals.

PSR 1913+16 lies about 5 kpc from us. Visual and X-ray observations have so far failed to detect either the pulsar or its companion. Radio observations alone indicate an orbital semimajor axis of only 7×10^5 km—a solar radius! The combined mass is 2.83 solar masses, and so if the pulsar has a mass of about 2 solar masses (a typical neutron star), its companion has a mass of about 0.8 solar mass. The companion might be a white dwarf.

Other binary pulsars have been discovered. One called PSR+065564 has a period of 24 h 41 min; its orbit is almost perfectly circular with a radius of only 750,000 km, just a bit larger than the Sun's radius. The companion to the neutron star must be small, perhaps also a white dwarf. In contrast to this system with a circular orbit, the binary

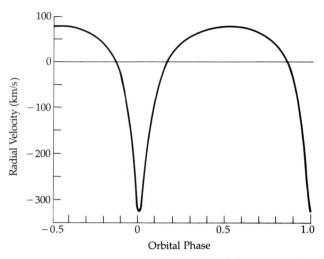

FIGURE 17–9   Doppler shift in the radial velocity of binary pulsar PSR 1913 + 16 from its motion about the center of mass. *[Adapted from a figure by R. Hulse and J. Taylor, Astrophysical Journal (Letters) 191:L59 (1975)]*

pulsar PSR 2303+46 has a highly eccentric orbit: $e = 0.658$. The orbital period is 12.34 days. This system most likely contains two neutron stars of mass ratio around unity, and each star with a mass somewhat larger than 1$M_\odot$.

(E) THE SUPERNOVA CONNECTION

Supernovae are the cataclysmic explosions of stars at the end of their lives (see Chapter 18 for details). These explosions involve the collapse of the core that results in the the detonation of massive stars such that most of the mass is blown away. The core collapse reaches the extreme densities needed to create a neutron star. If this model is correct, we would expect to find neutron stars as pulsars associated with the remnants of supernovae. Two cases are the most famous: the Crab Nebula [Section 18–5(c)], and the Gum Nebula. Several other good candidates have also been discovered.

David H. Staelin and Edward C. Reifenstein discovered the Crab Nebula pulsar (Figure 17–10), called PSR 0531+21. (PSR stands for pulsar, and the numbers 0531+21 refer to its celestial coordinates in the sky—Appendix 10.) PSR 0531+21 has

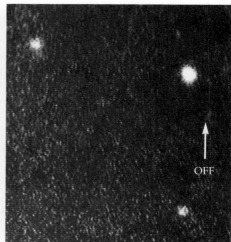

FIGURE 17–10 Optical pulses from the Crab pulsar. At left, we see the peak of an optical pulse; at right, offpeak. *(Lick Observatory)*

a pulse period of 0.033 s, or 30 pulses/s. The Crab pulsar was the first discovered to emit optical pulses as well as radio ones (Figure 17–11); the optical and radio pulses were found to have the same period. Observations of these visible pulses showed a smaller pulse between the main peaks (Figure 17–11B), called an **interpulse.** Remarkably, the star emitting these pulses was picked out by

Walter Baade and R. Minkowski in 1942 as a possible candidate for the stellar remnant of the supernova. Although this star is now known to be the pulsar, astronomers had observed it for years without noticing the optical blinking; a flicker of 30 times/s would be masked in ordinary photographs. Special stroboscopic techniques were used to determine the period of the optical pulses.

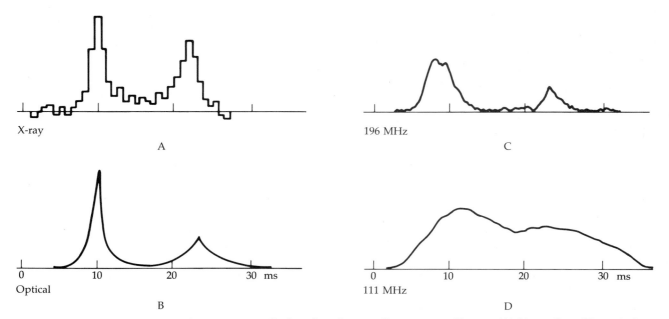

FIGURE 17–11 Crab pulses from radio waves to X-rays. (A) X-ray flux; (B) optical; (C) 196 MHz; and (D) 111 MHz. Note how the relative height and sharpness of the primary and secondary pulse changes with wavelength.

The Crab pulsar is the only one so far observed to pulse in the infrared, radio, optical, X-ray, and gamma-ray regions of the spectrum (Figure 17–8). The total energy emitted in the pulses is about 10^{28} W. The Crab pulsar was one of the first measured to exhibit a slowdown in pulse period, at a rate of about 4×10^{-13} s/s or 10^{-5} s/year.

The discovery of the Crab pulsar solves the energy source of the Crab Nebula. Summing over all wavelengths, the Crab Nebula emits about 10^{31} W. If the pulsar is a rotating neutron star, its slowdown in period gives a change in rotational energy of about 5×10^{31} W. That's enough to power the nebula—if the rotational kinetic energy of the neutron star can somehow be converted to kinetic and radiative energy of the nebula. That's exactly what happens in the highly magnetic, rapidly rotating neutron-star model for pulsars. Let's look at this point in detail.

The rotational kinetic energy of the mass is

$$E_{\text{rot}} = (1/2)I\omega^2$$

where I is the moment of inertia and ω the rotational angular velocity, with

$$\omega = 2\pi/P$$

For a sphere of uniform density,

$$I = (2/5)MR^2$$

where R is the radius. Now suppose some process converts *all* of the rotational energy to radiative energy; then conservation of energy requires

$$dE_{\text{rad}}/dt + dE_{\text{rot}}/dt = 0$$

but

$$\begin{aligned} dE_{\text{rot}}/dt &= (1/2)d/dt(I\omega^2) \\ &= (1/2)(d/dt)[(2/5)MR^2(2\pi/P)^2] \\ &= (4/5)\pi^2 MR^2(d/dt)(1/P^2) \\ &= -(8/5)\pi^2 MR^2 P^{-3}(dP/dt) \end{aligned}$$

Note, however, that

$$L = dE_{\text{rad}}/dt = -dE_{\text{rot}}/dt$$

so that

$$L = (8/5)\pi^2 MR^2 P^{-3}(dP/dt)$$

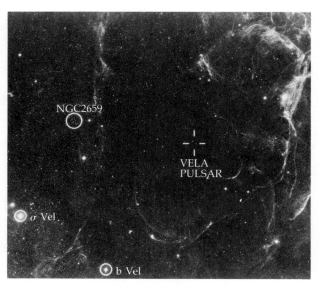

FIGURE 17–12 Gum Nebula, a supernova remnant. The Vela pulsar lies near its center. *(Steward Observatory)*

and

$$dP/dt = (5/8\pi^2)(LP^3/MR^2) \qquad \textbf{(17–6)}$$

which implies that the pulse rate should slow down as the pulsar loses energy. Use 1 solar mass for M, 10 km for R, 10^{31} W for L, and 1 s for P to get, as a rough estimate,

$$\begin{aligned} dP/dt &= (5/8\pi^2)(10^{31})(1^3)/(2 \times 10^{30})(10^4)^2 \\ &= (5/16\pi^2)(10^{31}/10^{38}) \approx 10^{-8} \text{ s/s} \end{aligned}$$

For the Crab pulsar, $P \approx 0.03$ s, and so

$$dP/dt \approx 10^{-13} \text{ s/s}$$

which is the observed rate of slowdown.

If the Crab pulsar were the only one associated with a known supernova remnant, it might be a coincidence, but we know of another one: the pulsar in the constellation Vela, near the center of the Gum Nebula (Figure 17–12), called PSR 0833-45. Its optical pulses come every 80 ms and have two peaks separated by about 22 ms. Gamma-ray telescopes have also detected pulses for Vela (Figure 17–13). The period of the Vela pulsar also slows down but at the slightly different rate of about

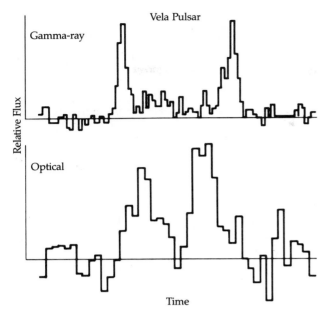

FIGURE 17–13   Optical and gamma-ray pulses from the Vela pulsar.

1.3×10^{-13} s/s (4×10^{-6} s/year). So it, too, provides support for the lighthouse model of pulsars and the supernova–neutron-star connection. So far, searches for the pulsar from SN 1987A have been unsuccessful, though two separate groups at one time thought they had detections.

17–3 ◗
BLACK HOLES

A black hole is a region of spacetime in which gravity is so strong that nothing, not even light, can escape it. Contemporary physics indicates that once a certain minimum mass gets together in a small enough volume, it must eventually become a black hole—collapsing by its own gravity after all its nuclear fuel is exhausted. No known physical force can stop this self-swallowing of mass that makes a black hole. And that minimum mass is not large—about three times the Sun's mass. No material can withstand this final crushing point of matter. The volume will continue to decrease until it reaches zero; the density will increase until it be-

comes infinite. Neither of these events can be exactly true for a real object in this Universe. This theoretical collapse of a nonrotating mass to a singular point of zero volume and infinite density, called a **singularity,** marks a breakdown of the laws of physics as we know them.

(A) BASIC PHYSICS OF BLACK HOLES

Here's a simple way to picture a black hole. Consider an escape velocity such that, when an object leaves with just the escape velocity, it will have zero velocity out at "infinity." There, its total energy ($KE + PE$) is

$$TE = (1/2)mv^2 - GmM/R = 0$$

Since total energy must be conserved, at the moment of launch we require

$$TE = 0 = (1/2)mV^2_{esc} - GmM/R$$
$$(1/2)V^2_{esc} = GM/R$$
$$V_{esc} = (2GM/R)^{1/2}$$

Now no object can travel faster than the speed of light, and so the maximum escape velocity is c. Then the equation for the black-hole radius is

$$R = 2GM/c^2 \qquad \textbf{(17–7a)}$$

In units of solar masses for M,

$$R = 3M \text{ km} \qquad \textbf{(17–7b)}$$

This critical radius is called the **Schwarzschild radius,** after the German astrophysicist Karl Schwarzschild, who worked out this solution shortly after Einstein published his general theory of relativity. For the Sun, the Schwarzschild radius of 3 km would require compression to a density of roughly 10^{19} kg/m^3—about the same density as the nucleus of an atom.

To examine the strange structure of spacetime around a black hole, let us take a theoretical journey into one. Start out from a spaceship orbiting a black hole of mass $10M_{\odot}$ at a distance of 1 AU. The ship orbits the black hole in accordance with Kepler's laws, as it would any ordinary mass. In fact, Kepler's third law and the spaceship's orbit permit you to measure the hole's mass. You will hop in with a laser light and digital watch to send signals back to the spaceship.

For a long time as you fall toward the black hole, nothing strange happens, but as you get closer, stronger and stronger tidal gravitational forces stretch you out from head to toe and squeeze you together at the shoulders. Near a black hole, tidal forces grow enormously because of their inverse-cube dependence on distance. Ordinary human beings would be ripped apart at about 3000 km from a black hole of mass $10M_\odot$. You cross the Schwarzschild radius! But nothing new happens, no signs mark the edge of the black hole. The trip now swiftly ends. About 10^{-5} s after you cross the Schwarzschild radius, you crash into a singularity. Crushed to zero volume, you are destroyed.

What of the view from back in the spaceship? As you drop closer to the black hole, the light from your laser is redshifted, as given by Equation 17–4, a gravitational redshift. The time between laser flashes increases because of the time-dilation effect predicted by general relativity. As you come closer to the Schwarzschild radius, the watches get more and more out of synchronization. In fact, a laser burst sent out just as you cross the Schwarzschild radius would take an *infinite* time to reach the spaceship—even though the light is moving at *c*! It also would suffer an *infinite* redshift (Equation 17–4). The fall seems to grow slower and slower to an outside observer as you get closer to the black hole. Time slows down so much that it seems to be frozen to a distant observer. The light gets more and more redshifted until it can no longer be detected. A black hole practices cosmic censorship; it prevents any outside observer from seeing you fall into it.

(B) THE STRUCTURE OF SPACETIME AROUND A BLACK HOLE

Let's look at the geometry of spacetime outside a black hole. To do so, we will have to examine a spacetime diagram for the geometry there. We get the appropriate spacetime map by solving Einstein's equations of general relativity to find the geometry of spacetime for an empty region of space surrounding a nonrotating, spherical mass (the simplest case). The main point is this: *Spacetime is not static, but dynamic.* You will also see that

spacetime does even stranger things than have been described so far.

The spacetime diagram (Figure 17–14) has coordinates that are not space and time as you experience them. The horizontal axis has properties that are space-like and the vertical one has properties that are time-like, but they are not exactly the same as measured space and measured time. Past is at the bottom of the diagram, future at the top. Light follows a special path in this spacetime diagram; it moves at 45° with respect to the axes. Any object moving slower than light has a path between the time-like axis and the light path; a path between the light line and the space-like axis represents something moving faster than light, which is not normally possible.

The diagram has been divided into four regions separated by **event horizons.** (An event horizon is another term to describe the Schwarzschild radius; it emphasizes the fact that any events taking place inside the Schwarzschild radius are cut off from outside view; they are beyond our visible horizon.) Note that a singularity exists both at the top (future) and bottom (past) of the diagram. Also drawn in the diagram is the path of a spaceship orbiting the black hole (line *O*) and that of a person jumping into the black hole from the spaceship (path *A* to *F*).

With these preliminaries in mind, look at the journey into a black hole, described above, in the spacetime diagram. As the spaceship orbits the black hole, it moves along line *O* from past to future (to the upper right). Your friend jumps out of the spaceship at *A*. Her laser signals are indicated by wavy lines; note that these lie at 45° with respect to the axes. The point at which the wavy line crosses line *O* is the point where and when you see it. The pulse emitted at *B* crosses line *O*, but the pulse emitted at *C*, that is, as your friend crosses the event horizon (Schwarzschild radius), does not intercept *O* until an infinite time has passed. All photons emitted after she crosses the event horizon (at *D* and *E*, for example) eventually get gobbled up in the singularity. At point *F*, your friend plunges into the singularity. You cannot see any events at *C* or beyond.

This example shows that region I is our region of spacetime, that is, our Universe outside the

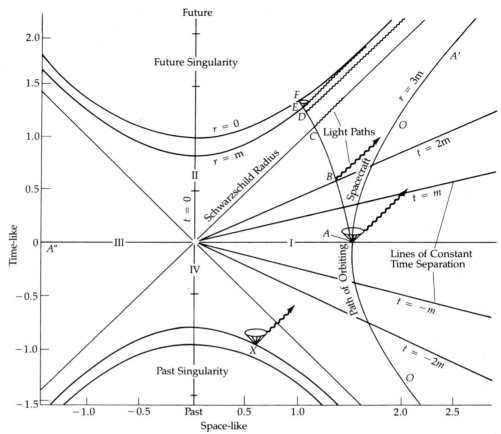

FIGURE 17–14   Spacetime diagram around a black hole. (*Adapted from a diagram by R. Ruffini and J. Wheeler*)

black hole. Region II is that part of spacetime within the Schwarzschild radius and containing the (predicted) singularity. What about region III? It's the mirror image of region I: another realm of spacetime outside of event horizons and singularities. At the bottom of the diagram, region IV, lies a singularity that is the mirror image of the one at the top, that is, a singularity in the past, or a time-reversed black hole. Notice that a permitted photon path (from *X*) can cross the event horizon into region I, our Universe. This we would see as light erupting from an event horizon, a phenomenon sometimes called a **white hole.**

Region III is inaccessible, and so we cannot demonstrate its existence. We are in region I. Suppose we tried to travel to region III along path *AA"*. To go that way means we must *travel faster than light* because such a path makes less than a 45° angle with the space-like axis. In fact, if we examine the figure, no path from region I to III requires less than light speed. Any slower-than-light path crashes into the top singularity. So we have no access to region III. Also, we cannot reach region IV because it lies in the past and we cannot travel back in time. We can get into region II, but we can't get out.

One special solution to Einstein's equations of general relativity indicates one way to avoid the singularity—by making a black hole of *rotating* matter. The angular momentum gives spacetime around the black hole a different character that avoids the singularity problem.

(C) OBSERVING BLACK HOLES

You cannot observe an isolated black hole; you can only detect it by its interactions with other material. Any matter falling toward a black hole gains kinetic energy and heats up, becomes ionized, and emits electromagnetic radiation. If its temperature reaches a few million kelvins or so, the material gives off X-rays. A black hole passing through an interstellar cloud or close to a star can sweep material into it and radiate. If the accreted material has some initial angular momentum, it will form a disk around the black hole—an *accretion disk* that will be the source of X-rays (Figure 17–15). Thus, X-ray sources are good candidates for black holes; you'll find specific ones discussed in detail in Section 18–6.

The general argument goes like this: The luminosities of galactic X-ray sources range from 10^{26} to 10^{31} J/s. To emit strongly at, say 0.3 nm (3×10^{-10} m), the temperature is about 10^7 K (Wien's law; Section 8–6). To produce a luminosity of 10^{30} J/s at this temperature, an object radiating like a blackbody would need a radius of roughly

$$R = (L/4\pi\sigma T^4)^{1/2}$$
$$\approx 10 \text{ km}$$

That's about the radius of a neutron star, or the size of the accretion disk around a black hole.

How rapidly must mass fall onto such an object to produce the X-ray luminosity? Suppose an accretion rate dm/dt falls onto the surface of an object of radius R and mass M each second. The gravitational energy produced is

$$dE_{grav}/dt = (GM/R) \, dm/dt$$

and if all of this energy is converted into radiation (efficiency = 100%), this is the luminosity, L, of the source. Therefore, for $L = 10^{30}$ J/s, $M = 1M_\odot$, and $R = 10$ km

$$L = (GM/R) \, dm/dt$$

or

$$\begin{aligned} dm/dt &= RL/GM \\ &= (10^4)(10^{30})/(6.7 \times 10^{-11})(2 \times 10^{30}) \\ &= 7.5 \times 10^{13} \text{ kg/s} \approx 10^{-9} \, M_\odot/y \end{aligned}$$

a rate of accretion easily obtainable in a close binary system. Of course, 100% conversion will be hard to achieve, but even 50% requires an accretion rate only double the above. The point here is that binary X-ray sources are the best suspects for containing black holes.

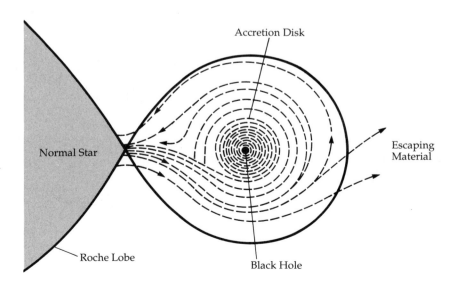

FIGURE 17–15 Schematic view of an accretion disk around a black hole. The disk forms from material from a binary companion star that, in this case, has filled its Roche lobe.

PROBLEMS ◗

1. **(a)** A white dwarf has an apparent magnitude $m_v = 8.5$ and parallax $\pi = 0.2''$. Its bolometric correction is -2.1 mag, and $T_{eff} = 28,000$ K. Calculate the radius of the star. Compare your value with the radius of the Earth.
   (b) A neutron star has $T_{eff} = 5 \times 10^5$ K and a radius of 10 km. What is its luminosity?
   (c) A protostellar cloud starts out with $T_{eff} = 15$ K and $R = 4 \times 10^4 R_\odot$. Determine $L/L_\odot$ and the wavelength of the peak of the Planck curve.

2. Calculate the kinetic energy ($mv^2/2$) for each of the following:
   (a) a nova outburst that accelerates a mass of $10^{-5} M_\odot$ to a velocity of 10^3 km/s
   (b) the formation of a planetary nebula in which a mass of $0.1 M_\odot$ is accelerated to a velocity of 20 km/s
   (c) a supernova outburst that accelerates a mass of $1 M_\odot$ to a velocity of 4×10^3 km/s
   How many years would it take the Sun to radiate away these energies?

3. How hot would a cloud of material become falling into a black hole? If it emitted as a blackbody, at what wavelength would it peak?

4. Calculate the Schwarzschild radius for
   (a) the Earth   **(c)** a globular cluster
   (b) the Sun   **(d)** the Galaxy
   What trend do you notice?

5. Consider stars of mass $1 M_\odot$. Compute the mean mass density for the following:
   (a) our Sun ($R_\odot = 7 \times 10^5$ km)
   (b) a white dwarf ($R = 10^4$ km)
   (c) a neutron star ($R = 10$ km)
   Now consider a ^{12}C nucleus of radius $r = 3 \times 10^{-15}$ m and compute its mean density. Discuss the significance of all these results!

6. The oldest white dwarfs were formed about 10^{10} years ago with initial temperatures of about 10^9 K. Determine the current temperature of an old white dwarf of maximum mass $1.4 M_\odot$, radius 7×10^6 m, and age 1×10^{10} years. Assume for simplicity that the density is constant throughout the star. What is the current wavelength of maximum intensity of this "white" dwarf? (*Hint:* Since the star cools by radiating as a blackbody, set $L = -$volume $\times$ number density of particles $\times k(dT/dt)$, where k is Boltzmann's constant. To solve this equation, separate the variables by putting all terms involving T on the left-hand side; then integrate.)

7. The speed of sound, given by $c_s = (5P/3\rho)^{1/2}$ for a nonrelativistic gas, where P is the pressure and ρ the density, is the speed at which a star will pulsate once oscillations are generated. Determine the speed of sound and the period of pulsations $\approx R/c_s$ as a function of mass (in solar units) for a nonrelativistic white dwarf (equation of state $P \approx 3.2 \times 10^6 \rho^{5/3}$ in SI units). Assume constant density and pressure. How does this time scale compare with the periods of the fastest pulsars? Is it possible for pulsars to be pulsating white dwarfs?

8. The Crab Nebula pulsar radiates at a luminosity of about 1×10^{31} W and has a period of 0.033 s. If $M = 1.4 M_\odot$ and $R = 1.1 \times 10^4$ m, determine the rate at which its period is increasing (dP/dt). How many years will it take for the period to double its present value? (*Hint:* You must integrate after isolating all the terms involving P on the left-hand side for the latter calculation.)

9. Equation 17–4 describes the redshift of electromagnetic waves emitted near a massive, compact object. Since time is in many respects the inverse of frequency, we can express the time-dilation effect by the formula $\Delta t'/\Delta t_{obs} = \nu_f/\nu_i$, where $\Delta t'$ is a time interval between two events (for example, consecutive ticks of a clock) in the source's frame and Δt_{obs} is the time interval between the same two events as measured by the observer. Notice that a clock placed in a strong gravitational field ticks more slowly than normal as observed by a distant observer, while a distant clock ticks more rapidly as observed by an observer in the gravitational field.
   (a) How would a distant observer describe how the timing of events (time between events, how long events last, and so on) changes for an object falling into a black hole? Does the observer ever see the object cross the Schwarzschild radius? Comment.
   (b) Now describe the timing of distant events as observed by an observer falling into a black hole. What does the observer see as she or he crosses the Schwarzschild radius? Is there a paradox here? If so, can you resolve it?

10. A neutron decays into a proton, electron, and antineutrino via the weak nuclear interaction after about 15 min when it is outside the nucleus of an atom. Now imagine that a neutron is freed from its nucleus 3.00 km from the center of a black hole of mass $1 M_\odot$. How long will it take the neutron to decay as measured by a distant observer? (Use the expression for time dilation given in the previous problem.)

11. A star identical to the Sun is in a binary system with a black hole of mass M_H. Assume for simplicity that the density of the star is uniform and that the orbit is circular.
    (a) Use Equation 3–9 to obtain the minimum separation from the black hole the star must have in order *not* to be torn apart by tidal forces.
    (b) At what black-hole mass is this minimum separation less than the Schwarzschild radius? Black holes *more* massive than this can swallow a star whole!

12. Assume a brown dwarf's luminosity derives from gravitational contraction. Its mass is $0.05M_\odot$, and its luminosity is $3 \times 10^{-5}L_\odot$. If we assume that its luminosity has been constant (even when the star had a much larger radius), how long can a star of this type radiate before the contraction is halted by electron degeneracy pressure (when $R \approx 9 \times 10^6 M^{-1/3}$ m, where M is in solar units)?

13. A typical young neutron star has a radius of about 10^4 m and a temperature of roughly 10^6 K.
    (a) What is the blackbody luminosity of such a neutron star?
    (b) What is the maximum distance a neutron star with these properties could have and still be detected by its optical blackbody radiation? Assume a limiting magnitude of 25 for detection and a bolometric correction of zero. Comment.

14. A cloud of hot gas is in a Keplerian circular orbit of radius 4.0×10^4 m about a black hole of mass $10M_\odot$. Plot the ratio of observed to rest frequency or wavelength versus time over one orbital period of an emission line radiated by the cloud, as seen by an observer who views the orbital plane edge-on. Include both the gravitational and normal Doppler effects and use the relativistic formulas.

15. A magnetic dipole rotating with a period P radiates with a luminosity that is proportional to P^{-4}.
    (a) Show that, long after its formation, a pulsar slows down according to the law $P \propto t^{1/2}$ (approximately).
    (b) Using the result from part (a), show that the luminosity of the pulsar decreases approximately as $L \propto t^{-2}$.

16. For the Crab pulsar, calculate the difference between pulse arrival times at 430 and 196 MHz. Then compare this difference to that for the other pulsars listed in Table 17–1.

17. Use the information in Section 17–2(D) and Kepler's Third Law to determine the sum of the masses of the components in the PSR + 065564 binary system.

18. What would be the rotation period of the Sun if it collapsed to a radius of 10 km without losing angular momentum? Compare your result to the rotation periods of known pulsars.

19. Compute the differential tidal force (see Chapter 3) on a person 3000 km from a $10M_\odot$ black hole. (Assume the person's mass is 90 kg and height 2 m.)

Chapter 18

Variable and Violent Stars

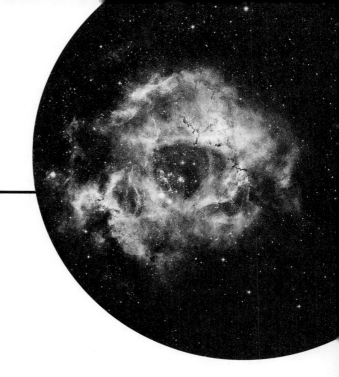

Chapter 16 discussed the structure and evolution of the majority of the observed stars, slowly changing normal stars. We alluded to the existence of rapidly evolving and spectacularly changing stars and mentioned that most normal stars pass through several such phases during their lifetimes. This chapter presents the observational and physical properties of such **variable stars,** which pass through intriguingly peculiar phases of stellar evolution. It illustrates the diverse phenomena of stellar astrophysics.

18–1 ◑
NAMING VARIABLE STARS

The variable stars of primary interest are the **intrinsic variables,** a term used to differentiate these stars from the geometric variables—extrinsic variables (such as eclipsing binary stars). Intrinsic variables may be roughly divided into two categories: (1) **pulsating stars,** whose atmospheres undergo periodic expansion and contraction, and (2) **eruptive,** or **explosive, variables,** which exhibit sudden and dramatic changes. The pulsating stars include Cepheids, RR Lyrae stars, the irregular RV Tauri stars, and the long-period Mira stars. Novae, dwarf novae, and supernovae are some eruptive variables. Variables that do not fit neatly into either category include flare stars, T Tauri stars, spectrum variables, and magnetic stars. Tables 18–1 to 18–3 list the characteristics of the most important types of variable stars; Figure 18–1 indicates their generic positions in the H–R diagram.

A single star may pass through several stages of variability during its evolution. A star of one solar mass may initially be a T Tauri star before settling onto the main sequence; then, much later, it may pulsate as an RR Lyrae star after passing through the red giant stage. In later stages of evolution, massive stars are observed as Cepheid variables, following the giant or supergiant stage. To become a white dwarf, a massive star must lose much of its mass, and it may do this as a red giant, a planetary nebula, or a supergiant.

To designate variable stars (both intrinsic and geometric), astronomers have devised a special system of nomenclature. Each variable star is given a set of capital letters followed by the genitive

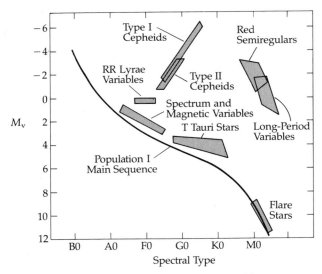

FIGURE 18–1 Locations of selected variable stars on the H–R diagram. *(Adapted from a diagram by J. P. Cox)*

nation have not been renamed. Therefore, we have Beta Lyrae, Delta Cephei, and Algol, all of which are well-known variables and prototypes of their classes.

18–2 ◐
PULSATING STARS

(A) OBSERVATIONS

The most important of the pulsating stars are the **Cepheid variables.** In distinct contrast to the light curves of most eclipsing stars, the light curves of pulsating stars exhibit continual changes in brightness. The spectra of pulsating stars also vary periodically, corresponding to changes in stellar surface temperature that may range over an entire spectral class. The spectral lines show variable Doppler shifts, from which **radial velocity curves** (for the stellar atmosphere) are deduced. From the periodic radial velocities, we find that variable stars alternately expand and contract; we can measure the changes in radii (Figure 18–2).

name of the constellation (Appendix 2) in which the variable occurs. The capital letters follow an alphabetic sequence based on the order of discovery: the first variable discovered in a given constellation is designated R, with the rest of the alphabet through Z used for successively discovered variables. After Z, double letters are used in the order RR, RS to RZ, SS, ST to SZ, . . . , ZZ. Then we return to the beginning of the alphabet (the letter J is omitted to avoid confusion with I): AA, AB to AZ, BB, BC to BZ, . . . , QZ. If there are more variables within a constellation, we resort to numbers starting with V335 (V for variable), since the single and double letters have already accounted for 334 variables. Some examples of variable-star designations are (in sequence!) R Monocerotis, T Tauri, RR Lyrae, UV Ceti, AG Pegasi, BF Cygni, V378 Orionis, and V999 Sagitarii.

In the past, novae were not included in this system; each was simply designated by the constellation name and year of occurrence (such as Nova Aquilae 1918). Since 1925, novae have been given variable-star designations, such as RR Pictoris and DQ Herculis; today, even the earlier novae have been assigned such designations, so that Nova Aquilae 1918 is also known as V603 Aquilae.

Variable stars that are bright enough to have been given a proper name or a Greek-letter desig-

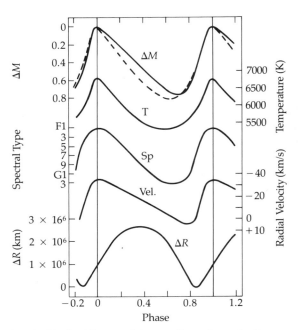

FIGURE 18–2 Observed properties of a pulsating star. Variations with phase of radial velocity curve for changes in magnitude (dashed curve for a constant radius), temperature, spectral type, radial velocity, and radius. *(Adapted from a diagram by W. Beckers)*

TABLE 18–1 *Some Classes of Pulsating Stars*

Type	Prototype	$\overline{M}_v$	Spectral Class	Pulsation Period Range	Characteristic Period
Classical Cepheids	δ Cephei	−0.5 to −6	F6 to K2	1^d to 50^d	5^d to 10^d
Population II Cepheids (W Virgins)	W Virginis	0 to −3	F2 to G6	2^d to 45^d	12^d to 28^d
RR Lyrae stars	RR Lyrae	0.5 to 1	A2 to F6	1.5^h to 24^h	0.5^d
Long-period variables (Mira)	o Ceti	1 to −2	M1 to M6	130^d to 500^d	270^d
RV Tauri stars	RV Tauri	−3	G, K	20^d to 150^d	75^d
Beta Canis Majoris stars	β Canis Majoris	−3	B1, B2	4^h to 6^h	5^h
Semiregular red variables	α Herculis	−1 to −3	K, M, R, N, S	100^d to 200^d	100^d
Dwarf Cepheids	δ Scuti	4 to 2	A to F	1^h to 3^h	2^h

In general, the characteristics described in this section pertain to all pulsating stars. Table 18–1 summarizes the distinguishing features and population characteristics of the several main types of pulsating stars.

The Cepheids and RR Lyrae stars lie in a region of the H–R diagram called the **instability strip.** As low-mass Population II stars ($0.5M_\odot$ to $0.7M_\odot$) traverse this strip during their core helium-burning phase, they become unstable and pulsate; these are the RR Lyrae stars. Population I stars of 3 to $18M_\odot$ also cross the upper region of this strip during their phase of core helium burning; they also pulsate, becoming the Cepheids.

(B) A PULSATION MECHANISM

A star pulsates because it is not in hydrostatic equilibrium; the force of gravity acting on the outer mass of the star is not quite balanced by the interior pressure (Section 16–1). If a star expands as a result of increased gas pressure, the material density (and pressure) decreases until the point of hydrostatic equilibrium is reached and *overshot* (owing to the momentum of the expansion). Then gravity dominates, and the star starts to contract. The momentum of the infalling material carries the contraction beyond the equilibrium point. The pressure is again too high, and the cycle starts anew. Energy is dissipated during such pulsation (analogous to frictional losses), and eventually this loss of energy should result in a *damping* of the

pulsations. The prevalence and regularity of pulsating stars imply that the dissipated energy is replenished in some way.

The rate at which energy is transported outward from the stellar interior can be altered by a damming process. The interior's opacity directly affects the amount of radiative energy absorbed; therefore, a changing opacity will act as a valve. When the stellar atmosphere is transparent, radiation flows freely and the star is bright. When the opacity is greatest and radiation is prevented from escaping, the star is faint. If the star is compressed at the time of greatest opacity, the excess radiation is dammed up and exerts pressure on the outer layers of the star; this process provides the energy necessary to continue the pulsations. The atmospheres of pulsating stars have a zone in which the opacity increases because singly ionized helium absorbs ultraviolet radiation to become doubly ionized. The He^+ ionization region (sometimes called the *He partial ionization zone*) is cooler than the surrounding regions because energy normally used to heat the gas is used to ionize it. The helium ionization zone contributes to the instability of the stellar atmosphere and so perpetuates the pulsations. (Ionization zones of other elements, such as H and C, can function in similar ways.)

Pulsating stars occupy well-defined areas of the H–R diagram (Figure 18–1); this observation can be explained in terms of the depth of the He^+ ionization zone. This depth depends upon the structure of the star, which in turn is a function of the star's stage of development. When the zone lies too deep, the valve action is insufficient to overcome damping. When the zone is shallow, the damming action is inefficient and pulsations are not given the necessary impetus. The period–luminosity law (next section) is explicable in terms of the position in the H–R diagram at which the star becomes unstable to this damming valve mechanism.

Regardless of the pulsation mechanism, we can generally relate a star's period of pulsation to its average density. After maximum expansion, the star's layers free-fall inward. We consider this infall as a special case of orbital motion—along a straight line! So the gas obeys Kepler's law:

$$P^2/R^3 = 4\pi^2/GM$$

where P is the pulsation period, R the star's radius, and M its mass; hence

$$P^2 \propto R^3/M$$

but

$$M \propto \langle\rho\rangle R^3$$

where $\langle\rho\rangle$ is the *average* density. Then

$$P^2 \propto R^3/\langle\rho\rangle^3$$
$$\propto 1/\langle\rho\rangle$$

so that

$$P\langle\rho\rangle^{1/2} = \text{constant}$$

Then the ratio of the pulsation periods for two different Cepheids is inversely proportional to the ratio of the square roots of their average densities:

$$P_A/P_B = (\langle\rho_B\rangle/\langle\rho_A\rangle)^{1/2}$$

which is observed to be approximately the case.

(C) THE PERIOD–LUMINOSITY RELATIONSHIP

Cepheids show an important connection between period and luminosity: the pulsation period of a Cepheid variable is directly related to its median luminosity. This relationship was first discovered from a study of the variables in the Magellanic Clouds, two small nearby companion galaxies to our Galaxy that are visible in the night sky of the southern hemisphere. To a good approximation,

you can consider all stars in each Magellanic Cloud to be at the same distance. Henrietta Leavitt, working at Harvard in 1912, found that the brighter the median apparent magnitude (and so luminosity, since the stars are at the same distance), the longer the period of the Cepheid variable. Harlow Shapley recognized the importance of this **period–luminosity (P–L) relationship** and attempted to find the zero point, for then a knowledge of the period of a Cepheid would immediately indicate its luminosity (absolute magnitude). This calibration was difficult to perform because of the relative scarcity of Cepheids and their large distances. None are sufficiently near to allow a trigonometric parallax to be determined, so Shapley had to depend upon the relatively inaccurate method of statistical parallaxes. His zero point was then used to find the distances to many other galaxies. These distances are revised as new and accurate data become available. Right now, some 20 stars whose distances are known reasonably well (because they are in open clusters) serve as the calibrators for the P–L relationship.

Further work showed that there are *two* types of Cepheids, each with its own separate, almost parallel P–L relationship (Figure 18–3). The classical Cepheids are the more luminous, of Population I, and found in spiral arms. Population II Cepheids, also known as W Virginis stars after their prototype, are found in globular clusters and other Population II systems. Classical Cepheids have periods ranging from one to 50 days (typically five to ten days) and range from F6 to K2 in spectral class. Population II Cepheids vary in period from two to 45 days (typically 12 to 20 days) and range from F2 to G6 in spectral class. Population I and II Cepheids are both **regular,** or **periodic, variables;** their change in luminosity with time follows a regular cycle.

RR Lyrae stars are also periodic variables; they are sometimes called *cluster variables* because of their abundance in globular clusters These stars (named after their prototype, RR Lyrae, whose period is 13.6 h) vary in luminosity with periods of 1.5 to 24 h (typically 12 h). They are Population II (though some have a fairly high metallicity), range in spectral class from A2 to F6, and have about 100 times the Sun's luminosity. About 5000 RR Lyrae stars are known. Note that RR Lyrae stars have a P–L relationship of sorts: they all have essentially

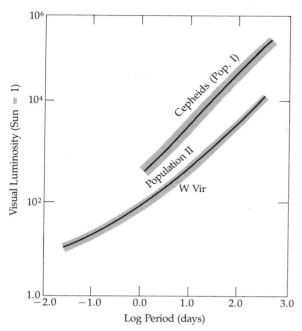

FIGURE 18–3 Period–luminosity relationship for Cepheids. Note that two different relationships exist, one for Population I stars and one for Population II stars.

the same luminosity (absolute visual magnitude of about 0.5) regardless of period.

(D) LONG-PERIOD RED VARIABLES

Long-period variables exhibit a very large change in visible light because they are cool (about 2000 K) and so most of their radiation lies in the infrared. Molecules and dust can exist at these low temperatures, forming a veil over the star's surface; as the temperature increases, these molecules are dissociated and more radiation can penetrate the veil. The **red variables** have irregular cycles of light variation of up to a few magnitudes that last from 100 to 700 days. They contain both Population I and II stars of spectral classes K and M with luminosities roughly 100 times that of the Sun; these are red giants and supergiants. The long-period red variables fall in a region of the H–R diagram in which they undergo shell burning of helium. The cause of their irregular pulsations is yet unknown.

18–3 ◑
NONPULSATING VARIABLES

Some curious variables are not pulsating stars; we briefly mention them here. They include **T Tauri stars,** which are pre-main-sequence, solar-mass stars; **flare stars,** which have arrived on the main sequence but exhibit flaring activity similar to that observed on the Sun; **magnetic variables,** which are in a later phase of evolution, and **RS Canum Venaticorum stars,** which are Sun-like stars evolving off the main sequence.

(A) T TAURI STARS

T Tauri stars are pre-main sequence stars. At this stage of stellar evolution, these low-mass ($0.2 M_\odot$ to $2 M_\odot$) stars have an extensive convection zone and surface magnetic activity (probably driven by convection and rotation) is rampant. (A few T Tauri stars are known to have large, dark active regions in their photospheres from the rotational modulation of their light.) Spectral activity includes emission in the hydrogen Balmer lines as well as emission from ionized calcium and other metals; these probably come from active chromospheres. Some T Tauri spectra also show the forbidden lines typical of gaseous nebulae, indicating the importance of the surrounding nebula. The underlying spectra of these stars are usually of spectral types F to M; a few have spectra resembling those of A0 or B8 stars. The continuum also varies, so that a total light variation occurs together with the sporadic appearance of emission lines. Some T Tauri stars are veiled by circumstellar material; others appear to be naked, so we can see their photospheres directly.

Large fluctuations occur in the ultraviolet; and observations with X-ray telescopes show that some T Tauri stars fluctuate rapidly and violently in X-rays, on the order of a factor of 10 in one day. These outbursts are most likely flares associated with the photospheric active regions.

In many instances, T Tauri stars are found close to and within dusty interstellar clouds, which astronomers call *dark clouds*. Optically visible ones appear at the edges of the clouds, where the extinction is least. Infrared observations can reveal the rest within the cloud. For example, in the ρ Ophiuchi dark cloud, infrared surveys reveal

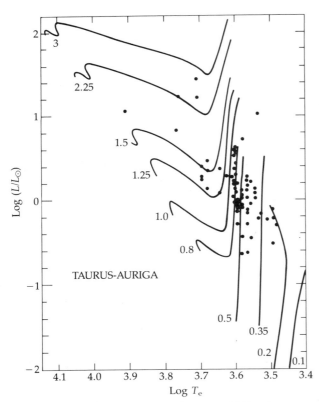

FIGURE 18–4 H–R diagram for some T Tauri stars. Stars observed in the Taurus–Auriga region plotted with theoretical evolutionary tracks labeled by mass. *[M. Cohen and L. V. Kuhi, Astrophysical Journal Supplement Series 41:743 (1979)]*

may well be the sites of planetary formation and formed by an accretion process.

Infrared and radio observations indicate mass outflows (stellar winds) from T Tauri stars of roughly 10^{-7} to 10^{-8} solar mass per year. In fact, mass outflows appear associated with almost all protostellar and pre-main-sequence objects. In many instances, the flows are bipolar along a rotation axis. Such pre-main-sequence stellar objects go under the generic name of *young stellar objects* (YSOs). The winds may well crash into the circumstellar disks, creating a turbulent zone of interaction.

(B) FLARE STARS

Solar flares are the most energetic and spectacular aspect of solar activity. There must be flares on other G stars, but the amount of energy radiated by even the largest solar flares amounts to little compared with the total stellar radiation. However, on a dwarf M star, which radiates very much less than the Sun, a flare with the energy of a large solar flare would lead to a twofold increase in brightness! In fact, several cool main-sequence stars have flared at irregular intervals by brightening several magnitudes in a matter of seconds. The light curves are similar to those of solar flares, for the decline is far slower than the ascent (Figure 18–5). Joint observations by radio and optical ob-

more than 70 stellar objects, many with luminosities in the range from 0.1 to $25L_\odot$.

The position of T Tauri stars on the H–R diagram is just above and to the right of the main sequence, just where we expect to find pre-main-sequence stars (Figure 18–4). T Tauri and related objects radiate very strongly in the infrared region, an excess attributed to a surrounding dust cloud or disk that absorbs much of the star's short-wavelength radiation and then re-emits it at infrared wavelengths. This dust may be either the remainder of the material from which the protostar formed or matter ejected from the star as it collapsed. In many instances, multiwavelength observations show a disk of material, a few hundred AUs in radius, surrounding the T Tauri star. These

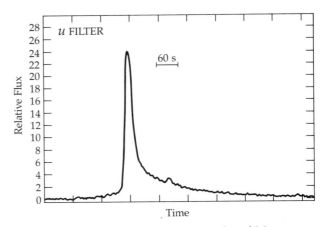

FIGURE 18–5 A stellar flare from a dwarf M star. *U*-band observations of a flare from the star YZ CMi. Note the steep rise and more gradual decline. *[T. J. Moffett, Astrophysical Journal Supplement Series 29:1 (1974)]*

servers have verified that some of these flare stars emit radio bursts simultaneously with the flares.

The total energies of flares from dwarf M stars range from 10^{21} to 10^{27} J for stars whose *B*-band luminosities range from 10^{21} to 10^{26} W. The *B*-band luminosity of the Sun is about 2×10^{26} W. Although stellar flares are more energetic and have a faster rise time than solar flares, the underlying process is probably very similar: a release of magnetically trapped particles into the corona. As with the Sun, most of the energy comes out in X-rays but much more energetically. For example, flares from dwarf M stars hit a maximum luminosity in X-rays of about 10^{24} W (compared with 10^{20} W for the Sun). It may be the case that all stars undergo a flaring phase during pre-main-sequence evolution.

(C) MAGNETIC VARIABLES

Stars that have the basic characteristics of A stars may show spectral peculiarities attributable to abundance abnormalities (Chapter 13). Some of these stars exhibit variable spectra such that the intensity of certain lines varies approximately periodically. Many of the peculiar and metal-lined A stars have strong integrated magnetic fields; these fields range up to a few teslas, but most are from 0.01 to 0.1 T (recall that the strongest magnetic fields on the Sun are about 0.4 T, and these fields are restricted to very small areas in sunspots). Most observed stellar magnetic fields are variable, with some undergoing polarity reversals; sometimes the magnetic variability is coupled with spectrum variability. The light variations of magnetic stars and spectrum variables are very small, amounting to about ±10%.

One possible explanation for magnetic variability is that the magnetic axes are tilted with respect to the rotational axes (like pulsars). This model is referred to as the **oblique rotator.** Perhaps more simply, both the spectrum and magnetic variations arise from large polar spots on the stellar surface that are periodically brought into view by rotation.

(D) RS CANUM VENATICORUM STARS

In this tour of variable stars, we have encountered indications of stellar activity similar to solar magnetic activity but more energetic. For Sun-like stars, hyperactive magnetic activity comes from the **RS Canum Venaticorum** stars, or *RS CVn stars* for short. These are a subclass of chromospherically active stars. The RS CVn stars are in binary systems. A typical orbital period is about seven days; however, periods range from 0.5 day to a few months. In most RS CVn systems, the stars are synchronously locked by tidal forces, so that the rotational period of each is equal to the orbital period. One star is hotter than the other; the hotter star is often main-sequence (luminosity class V), the cooler one a subgiant (luminosity class IV). These sound like ordinary stars, but they are not.

Observations with the VLA have shown that the average radio emission from RS CVn systems is about 10^{20} W. In contrast, the Sun's radio luminosity averages 10^{13} W. Radio flares from RS CVn stars peak at 10^{20} to 10^{21} W, which is 10^{5} to 10^{6} times stronger than the *strongest* solar radio flares. A few superflares have energies 10^{7} times greater than any solar radio flares. Radio observations over a range of wavelengths show that the flares' emissions are highly polarized with nonthermal spectra. These clues point to synchrotron radiation as the source of the flaring radio emissions.

Optically, a light curve of an RS CVn system exhibits waves called *distortion waves* (Figure 18–6). They amount to 1% to 30% of the total light from

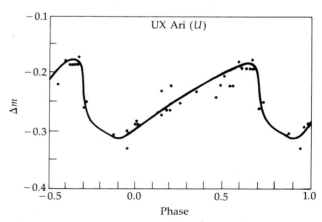

FIGURE 18–6 Light curve for an RS CVn binary, showing the variation of light at *U* band as a function of orbital phase for the system UX Ari. Note the large magnitude changes in this noneclipsing binary. *(Capilla Peak Observatory, University of New Mexico)*

the system. Their amplitude varies from as fast as a few months to over many years. These distortion waves arise from the rotational modulation of large (tens of percent of the total surface area) starspot regions concentrated in longitude and latitude. Such enormous active regions (at sunspot-cycle maximum, the spots cover at most 0.1% of the Sun's surface) indicate that these stars have stronger concentrations of magnetic fields than does the Sun. Associated with these active regions are flares at H_α, optical, ultraviolet, and X-rays. One well-observed optical flare from the system XY UMa in January 1982 lasted for about 30 min (Figure 18–7). Its total energy output amounted to 10^{27} J—10^3 times greater than a solar "white-light" flare.

What drives the hyperactivity of RS CVn stars? From X-ray observations of the Sun, we understand that the solar corona arises from **coronal loops** of magnetic flux tubes. RS CVn stars turn out to be much stronger X-ray emitters than the Sun, so they have more extensive coronae. A loop model for the coronae for RS CVn stars results in 10^3 more loops than for the Sun—essentially a star

completely covered with active regions. The temperature of the gas contained in these loops is a few *tens* of millions K, about ten times hotter than the Sun's coronal gas. Such very hot loops may reach lengths of *tens* of solar radii—larger than the RS CVn stars.

Although we do not yet know the details of how solar flares occur, we do realize that they erupt in the corona, probably initiated by magnetic reconnection, and flow down flux tubes to their bases in the photosphere in a violent rush of high-speed, charged particles. Coronal loops, active regions, flares, and strong X-ray emission are all associated phenomena on the Sun; the same view is supported for RS CVn systems, and similar processes may drive hyperactivity on young stars, such as the T-Tauri class, and the flares from dwarf M stars. What probably drives the dynamo in these systems is their rapid rotation—on the order of a few days.

18–4 ◑
EXTENDED STELLAR ATMOSPHERES: MASS LOSS

(A) AN ATMOSPHERIC MODEL

An extended gaseous envelope around a star (Table 18–2) is deduced from the line profiles of certain spectral lines (Figure 18–8). The portion of the **shell,** or **extended atmosphere,** seen projected against the star's photosphere produces a narrow absorption line, and that part not projected against the disk (the annular region) is seen as an emission line. Normally, the emission is superimposed on the stellar photospheric absorption; the extent to which the underlying absorption profile is distorted by this emission line depends on the strength of the emission, which in turn is a function of the density of the stellar atmosphere. The atmosphere cannot be too dense, however, for then it would affect the continuum radiation as well as the spectral lines. The widths of the several components of the profile depend upon the motions of the contributing regions. For instance, if the atmosphere is turbulent, both the emission and the absorption features will be broad.

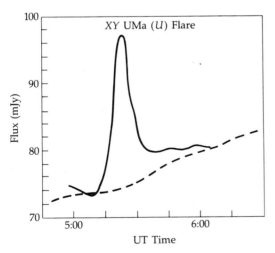

FIGURE 18–7 Flare from an RS CVn system. *U*-band observations of a flare from XY UMa in January 1982. The dashed line shows the underlying light level from the binary system. *(Capilla Peak Observatory, University of New Mexico)*

TABLE 18—2 *Extended-Atmosphere Stars*

Type	Prototype or Example	$\overline{M}_v$	Spectral Class	Expansion Velocity (km/s)	Rate of Mass Loss ($M_\odot$/year)
Be	48 Per	−4	B	—	$\leq 10^{-6}$
Shell star	γ Cas, Pleione	−4	B B	50	$\leq 10^{-7}$
Wolf–Rayet	HD 66811, HD 68273	−4 to −6.8	WN, WC	120 to 2500	10^{-6}
P Cygni	P Cyg	−6?	B	130	?
O and B supergiants	ζ Pup, δ Ori	−7	O and B	1000 to 1800	10^{-6}
M supergiants	α Her, α Ori	−2 to −8	M Ia to II	up to 26	5×10^{-6} to 5×10^{-9}
Planetary nebulae	Ring Nebula in Lyra	0 to +8	W stars? O, B	10 to 30	10^{-4} to 10^{-5}
Beta Lyrae	β Lyr	−6	B	—	10^{-4}

(B) BE AND SHELL STARS

If a star rotates more rapidly than its atmosphere, the underlying stellar absorption profile is widened more than the atmospheric emission line. This results from the fact that the Doppler shifts of the approaching and receding limbs of the photosphere are greater than those of the extended atmosphere (Figure 18–9A). The absorption due to the projected gas remains narrow because the motion is across the line of sight. The **B-emission stars (Be)** and **shell stars** both fit this model, but they differ in that the latter contain more material in the envelope. Sometimes net outward velocities are observed, indicative of expansion and possibly loss of material. The mass lost in this manner is small relative to the mass of the star and to the mass the star must lose if it is to become a white dwarf. Observationally, massive main-sequence stars, giants, and especially supergiants are ejecting material at a substantial rate. The formation of a shell around the Be stars may result from the rapid rotation; the rotational velocities may be so high that material is spun off.

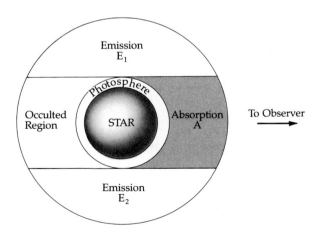

FIGURE 18—8 An extended-atmosphere model. Light from the star's photosphere is absorbed (region A) on its way to the observer and stimulates emission (regions E_1 and E_2) in the atmosphere seen side-on.

(C) MASS LOSS FROM GIANTS AND SUPERGIANTS

In a sense, we should include giants and supergiants in the category of stars with extended atmospheres. For instance, *mass loss* from M-type giants and supergiants indicates that the stellar atmospheres are expanding. The spectra of many M giants and supergiants have narrow absorption lines

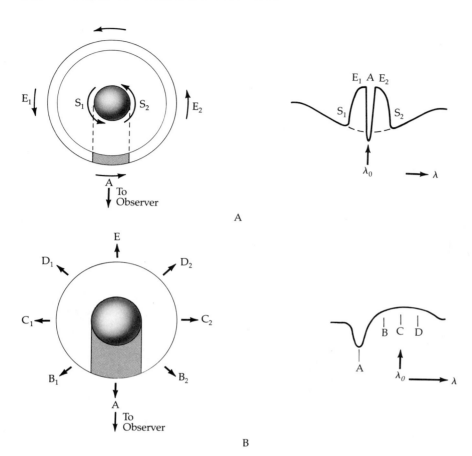

FIGURE 18–9 Expansion of an extended atmosphere. (A) From a rotating star, the line profile shows the absorption, A, of the undisplaced line center and the Doppler-shifted peak (E_1 and E_2). S_1 and S_2 are the absorption lines of the star Doppler-shifted to the blue and red by rotation.
(B) From an expanding atmosphere, the line profile has absorption feature, A, displaced to the blue by the expansion toward the observer. The contribution to the emission, C, is at the undisplaced line center; B and D are Doppler-shifted from the expansion.

superimposed on and shifted with respect to the broad Ca II emission lines of the star itself (Figure 18–9B). These features are interpreted as being due to circumstellar material similar to the shells of the early-type stars discussed earlier. The Doppler velocities of the shifts are of the order of tens of kilometers per second. To convert these velocities to rates of mass loss, we must know the density of the material ejected. Estimates of the density lead to figures for the mass loss ranging from 10^{-6} to 10^{-8} solar mass per year. The more luminous the star and the later the spectral type, the higher the rate of mass loss.

At the hot end of the spectral sequence, O and B supergiants eject mass at high velocities (1000 to 2000 km/s). Spectra in the visible region do not show this phenomenon, for there are no suitable spectral lines. The material being ejected from the star has a very low density; hence, collisional excitation is negligible and most atoms and ions are in

their lowest energy level (ground state). Owing to the high temperature, however, most of the gas is ionized; therefore, it emits at radio wavelengths and can be detected. The ejection velocities are far higher than for the cool supergiants, but the mass loss is only slightly greater because of differences in the gas densities.

(D) WOLF–RAYET STARS

Wolf–Rayet stars are very hot stars ($T_{\text{eff}} \approx$ 30,000 K) whose spectra show strong and wide emission lines of He I, He II, C III, C IV, N III, and N V. That they are young stars of Population I is deduced from their association with OB stars in open clusters, in H II regions, and as binary companions to O or B stars. (Almost all WR stars are in binary systems.) The broad emission lines often have narrower absorption lines superimposed but displaced to the blue. (Such combined blue absorptions superimposed on wide emission lines are often referred to as **P Cygni line profiles** after the prototype star for which such profiles were first observed.) These observations fit an interpretation (Figure 18–9B) of a stellar atmosphere expanding at velocities of 1000 km/s or higher. Wolf–Rayet stars are helium-rich and hydrogen-deficient; carbon dominates the spectrum of some, the WC stars, and nitrogen that of others, the WN stars. Although the true nature of these stars is still not entirely clear, it seems that they represent a particular stage in a star's evolution and that the two types result from real differences in abundances, not from differences in excitation.

As you saw in Section 16–3(b), Wolf–Rayet stars are initially massive stars, with strong stellar winds, that have lost their outer layers, revealing interiors with compositions greatly modified by nuclear reactions. A tantalizing hypothesis to explain the two sequences suggests that, in the WN stars, we see the products of CNO-cycle hydrogen burning, whereas WC stars have already undergone helium burning. We may be seeing either the core of the star or, more likely, intermediate convective layers enriched with core products.

For WR stars in binary systems (about half of the total), we have determined their masses to be roughly 10 to 40 solar masses. In a binary, the mass loss, rather than from a stellar wind, occurs by mass transfer in the red-giant phase of the initially more massive star, which must evolve more rapidly than its companion because of its greater mass. In such a case, the WR star is the remnant, consisting largely of helium and heavier elements. The mass loss must be substantial—tens of solar masses in a few million years—and dramatically affects the evolution of WR stars [Section 16–3(b)]. Theoretical models, starting with a mass of $60 M_\odot$, suggest that WR stars are in a core carbon-burning phase after having been a red supergiant and just before becoming a supernova.

(E) PLANETARY NEBULAE

Planetary nebulae, so named because some resolve into disks reminiscent of planets when seen with a telescope, also fit an expanding-atmosphere model (Figure 18–10). In fact, the atmosphere is

FIGURE 18–10 A planetary nebula. A shell of gas is expanding into space, forming the Helix Nebula. It is illuminated by the central star. *(National Optical Astronomy Observatories)*

really a large shell, large enough and of sufficiently low density that most of the receding portion (E in Figure 18–9B) is visible and the spectral lines are doubled. The velocities of expansion, however, are only some tens of kilometers per second, far lower than those for WR stars.

A substantial fraction of all stars probably go through the planetary-nebula stage after their AGB phase and before they become white dwarfs. The small number of observable planetaries results from the short duration of this stage, which lasts only some 50,000 years. The star develops into a planetary nebula when a nebula of material is ejected from the central star during the contraction that terminates the red-giant stage. The extended envelope includes so much of the former star that what is seen as the central star was formerly the core.

The central star is very hot and so radiates strongly in the ultraviolet; hence, the atoms and ions in the envelope *fluoresce* (absorb ultraviolet radiation to become ionized and emit at longer wavelengths upon recombination and cascade). The central stars generally have spectra that can be classified as type O or WR, but these stars are not identical to normal O and WR stars in either luminosity or mass. Both luminosity and mass are very difficult to establish because planetaries are so far away that we cannot measure their trigonometric parallaxes. We can still make reasonable estimates on the basis of ionization processes operating between star and nebula, however. Most estimated stellar masses range from $0.5M_\odot$ to $0.7M_\odot$, and estimates of the mass contained within the envelopes range from $0.1M_\odot$ (or less) to $0.5M_\odot$.

To ask what place the planetary nebulae occupy in the evolutionary scheme raises the problem of their population characteristics. The planetaries in our Galaxy are primarily concentrated in the disk and center, as are the long-period variables and the RR Lyrae stars. Their galactic orbits are elongated rather than circular, in contrast to the orbits of Population I objects. At least one is known to be a member of a globular cluster—that is, an extreme Population II object. It is probably safe to say that most planetary nebulae belong to the intermediate disk population, but isolated cases are found at both extremes, Populations I and II. Red variable stars, such as Miras, may well be their predecessors.

18–5 ◗
CATACLYSMIC AND ERUPTIVE VARIABLES

This category includes stars that eject matter suddenly and violently, in contrast to the slow expansion and small mass loss of planetary nebulae. These eruptions are accompanied by enormous changes in luminosity, from a few magnitudes for the dwarf novae to over 20 magnitudes in the case of supernovae (Table 18–3).

It is probable that all these variables (except the supernovae) suffer several episodes in their lifetimes. Repeated outbursts of dwarf novae and recurrent novae have been observed, with the interval between outbursts being a function of the amplitude of the eruption. If such a relationship extends to normal novae, their interburst interval must be of the order of 10,000 years. In general, these stars are called **cataclysmic variables.** The common model is that of a Roche lobe filling sec-

TABLE 18–3 *Cataclysmic and Eruptive Variables*

Type		Example	M_{max}	Δm	Energy per Outburst (J)
Supernovae I		Tycho's	−20		10^{44}
				>20	
Supernovae II			−18		10^{43}
Novae:	Fast	GK Per = Nova Per	−8.5 to −9.2	11 to 13	6×10^{37}
	Slow	DQ Her	−5.5 to −7.4	9 to 11	—
	Recurrent	T Cr B	−7.8	8	10^{37}
Dwarf novae		U Gem SS Cyg	+5.5	4	6×10^{31}

Type	Time	Cycle	Mass Ejected per Cycle $(M_\odot)$	Velocity of Ejection (km/s)	Mass of Star $(M_\odot)$
Supernovae I		—	≤1	10,000	1
Supernovae II		—	?	10,000	≥4
Novae:	Fast	10^6? years	10^{-5} to 10^{-3}	500 to 4000	1 to 5
	Slow	—	—	100 to 1500	0.02 to 0.3
	Recurrent	18 to 80 years	5×10^{-6}	60 to 400	2
Dwarf novae		40 to 100 days	10^{-9}	—	≈0.4

FIGURE 18–11 Nova V1500 Cygni 1975. These photographs were taken before (left) and during (right) the outburst. *(Lick Observatory)*

ondary (on or near the main sequence) losing hydrogen-rich material through the inner Lagrangian point onto an accretion disk that surrounds a white dwarf primary. The white dwarf can have a weak or strong magnetic field.

(A) NOVAE

Nova is the Latin word meaning "new"; **novae** (plural) are stars that suddenly become visible in the sky where no star was seen before (Figure 18–11). Actually, the star is usually too faint to be visible prior to its eruption, for a nova characteristically increases in brightness some ten magni-

tudes from the prenova stage to nova maximum. The rise to maximum brilliance is very rapid (Figure 18–12A). The initial rise brings the star to within two magnitudes of the maximum in only two or three days, and the final increase in luminosity takes a day for fast novae and weeks for very slow novae. The decline from maximum is far more gradual; the time spent at maximum is relatively short, generally only a matter of days. Erratic light fluctuations of large amplitude may occur during the decline; occasionally a drop of many magnitudes is observed followed by a practically complete recovery some weeks later (Figure 18–12B). The length of time from the maximum to

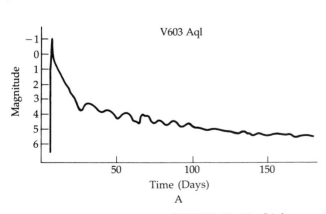

A

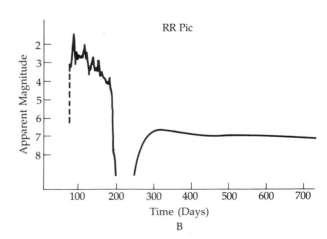

B

FIGURE 18–12 Light curves of novae. Note the rapid rise and more gradual decline, with fluctuations. (A) *V*-band light curve of Nova V1500 Cygni 1975 during its first two months. *(P. Young, H. Corwin, J. Bryan, and G. de Vaucouleurs)* (B) RR Pic, a slow nova. *(Adapted from a diagram by D. B. McLaughlin)*

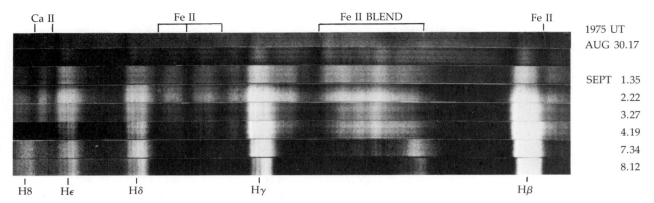

FIGURE 18–13 Spectra of Nova V1500 Cygni 1975. The UT dates are given at the right. These eight spectra show the evolution of the emission lines, especially those of the Balmer series. Note how broad the emission lines appear. *(B. Bohannon)*

the final leveling off of the light curve ranges from months for fast novae to years for slow ones.

Spectra obtained just prior to and at maximum light show that material is ejected from the star at velocities up to 2000 km/s. Complex changes in the spectrum occur during the development of the nova (Figure 18–13); several layers of ejected material are seen as the outer envelope becomes progressively more transparent and exposes the inner layers. The expanding material amounts to some $10^{-5} M_{\odot}$. After maximum, for instance, most nova spectra include the bright forbidden lines characteristic of emission nebulae. The velocities of ejection differ at different nova stages; some velocities are directly related to changes in the light curve, as if there were ejections subsequent to the initial eruption. One nova for which a spectrum was recorded prior to its outburst is Nova Aquilae 1918 (also known as V603 Aquilae). This spectrum resembled that of a hot blue star without any spectral lines. Most novae eventually return to such a state.

Some light fluctuations persist even at minimum. Superimposed on these rapid, erratic variations are others that are best interpreted as stellar eclipses. Such novae clearly must be members of binary systems; in fact, the evidence implies that most novae are in short-period binaries. This conclusion suggests that the companion is a key condition for a star to become a nova.

One binary model consists of a red giant or a star in the process of expanding into the red-giant phase and a white dwarf. As the red star extends, its atmospheric gaseous material crosses the Roche lobe to make a semidetached binary (Section 12–4). Hence, the gas from the red giant's atmosphere escapes and falls onto the white dwarf or forms a gaseous disk around it because of the conservation of angular momentum. The influx of hydrogen-rich gas from this accretion disk onto the degenerate star (Section 17–1) that has used up most of its hydrogen can cause further nuclear reactions at the stellar surface. This new supply of material piles up in a layer until the hydrogen fusion ignition temperature is reached; runaway fusion reactions ensue in the accreted layer. The stellar atmosphere suddenly expands violently—a nova explosion. Repeated explosions may occur if more material flows from the red giant to the white dwarf to build up enough to flash again; that takes about 10^5 years.

The high ejection velocities quoted earlier are based upon measurements of spectral-line profiles—P Cygni-type profiles resembling those of Wolf–Rayet stars. The ejected gas expands as a shell, and sometimes this shell becomes visible as a nebula surrounding the nova. Over the years, the nebula expands perceptibly, and its rate of expansion appears as a proper motion measurable in seconds of arc per year (Section 15–1). Spectra obtained during the same epoch give expansion velocities directly in kilometers per second. If we assume that the velocity of expansion is uniform in all directions, then the observed proper motion corresponds to the same velocity. The geometry of the expansion then allows us to find the distance to the nova. From Equation 15–3, we have

$$d = v_r/4.74 \, \mu'' \qquad \textbf{(18–1)}$$

where v_r is the radial velocity in km/s, 4.74 is a conversion factor that yields d in parsecs, and μ'' is the proper motion in arcsec/year. Nova Persei (GK Persei; Figure 18–14) is a good example of a nova with an expanding nebula. Observations show that the expansion of the shell increases by about 0.5"/year and that the radial velocity of the shell is roughly 1100 km/s. Then the distance is

$$d = 1100/(4.74 \times 0.5) = 460 \text{ pc}$$

The importance of this method of determining the distances of novae is that it is both direct and unambiguous. In this way, we can establish the luminosities of novae at maximum and minimum, and from that information we can find the total amount of energy released in the explosion.

FIGURE 18–14 Expanding shell from Nova Persei. The nova exploded in 1901; this photograph was taken in 1949. *(Palomar Observatory, California Institute of Technology)*

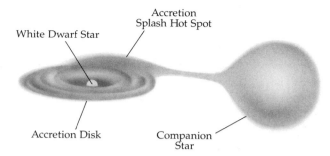

FIGURE 18–15 A schematic model for GK Per, a cataclysmic variable. X-rays come from an accretion disk around the magnetic white dwarf. A hot spot develops there; the infalling matter strikes the accretion disk; it produces stronger X-rays than the disk in general.

A bright nova occurred in August 1975: Nova V1500 Cygni, which peaked at magnitude 1.8 (Figure 18–11). Prenova photographs indicate that the star increased in luminosity at least 16 million times. Nova V1500 Cygni 1975 is a member of a close binary system with an orbital period of about 3 hours. The white dwarf appears to have a very strong magnetic field. The binary system GK Her (Figure 18–15) may be similar. Its overall X-ray emission fluctuates every 351 s; it also pulses. The pulsations may come from a strong magnetic field of the white dwarf, and longer variations from a hot spot on the accretion disk where infalling matter from the companion splashes onto it.

(B) SUPERNOVAE

Supernovae attain absolute magnitudes in the range from −16 to −20. Although supernovae are relatively rare in our Galaxy, some have been noted in historical records. From such records, especially the Chinese chronicles, we find, for example, that the supernova of A.D. 1054 reached an apparent magnitude of −4, bright enough to be seen during the day. Much of what we know about supernovae has been gleaned from studies of galaxies. In the case of small galaxies, the luminosity of a supernova may rival the total brightness of the galaxy. Supernovae exhibit a very rapid rise to maximum and then drop two or three magnitudes within a month before declining more gradually. The total energy output from any supernova is stupendous: 10^{44} J, or approximately as much energy

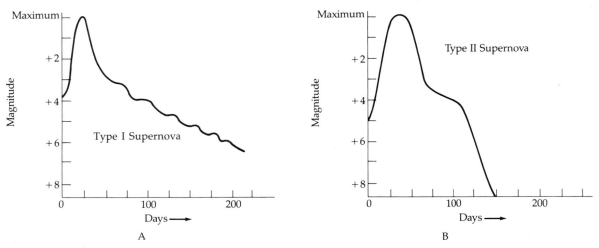

FIGURE 18–16 General supernovae light curves. (A) Type I; note the gradual decline. (B) Type II; note the sharp decline and shoulder.

as the Sun will produce in its entire lifetime of 10 billion years.

Two main types of supernovae occur, differentiated by their spectra and by their light curves (Figure 18–16 and Table 18–3). Type I supernovae appear in both elliptical and spiral galaxies (Chapter 21), and Type II occur only in spirals (especially in the spiral arms). We deduce that Type I supernovae belong to evolved stars of low and intermediate mass, Type II to more massive stars.

At maximum brightness, a Type II supernova shows a nondescript spectrum; the only prominent line is an emission line at 656.3 nm, the H_α line. About a month later, the spectrum has evolved to show more emission lines and a few weak absorption lines. In contrast, Type I supernovae exhibit messy spectra. At maximum light, broad emission lines appear along with some strong, dark lines, the combination having P Cygni-type profiles. Later, four emission lines from Fe II dominate the spectra, along with emission lines of Na I and Ca II. Basically, the difference between Type I and Type II spectra is that Type II show strong hydrogen lines and Type I do not—an indication that Type I involve highly evolved, hydrogen-deficient stars.

Type II supernovae arise from evolved stars much more massive than the Sun (10 to $100M_\odot$)—stars that live their normal lives as O and B stars. Theoretical computer models for Type II supernovae suggest that the explosion occurs in the core of

a red supergiant. The atmosphere of a red supergiant has nearly constant density, so that a shock wave traveling through it moves at almost constant velocity and transmits energy efficiently to the star's surface. The models predict that, at peak brightness, the supernova should have a photospheric temperature of roughly 10,000 K, a surface speed of 5000 km/s, and a maximum radius of some 10^{10} km.

The basic interior model for a Type II supernova involves the collapse of a stellar core to form a neutron star. A violent rebound from this sudden collapse produces the explosion that ejects the outer layers. The stars in which this process occurs fall in the mass range of $10M_\odot$ to $100M_\odot$. Such stars develop carbon–oxygen cores during their normal lives; the carbon fuses to produce neon, magnesium, and finally iron. The core's density eventually reaches a level high enough so that it becomes degenerate, and degeneracy pressure supports the weight of the star on the core. The layer of silicon above the iron core continues to burn, adding to the core until its mass exceeds the Chandrasekhar limit. It collapses; a neutron star forms. The infalling material rebounds off the neutron core to drive a shock wave outward; this triggers violent explosions in the unburned material in the outer layers.

The collapse of the core of a star to nuclear densities easily provides enough energy to power a supernova. The gravitational potential energy of a neutron star of solar mass and radius $R = 15$ km is

$$E_{\text{grav}} \approx \frac{GM^2}{R}$$

$$\approx \frac{(7 \times 10^{-11})(2 \times 10^{30})^2}{1.5 \times 10^4}$$

$$\approx 2 \times 10^{46} \text{ J}$$

much larger than the energy of 10^{44} joules observed in a supernova explosion. But remember, most of the supernova's energy comes out as neutrinos—perhaps 100 times that emitted as electromagnetic radiation.

Because Type I supernovae are associated with stars of roughly the mass of the Sun, they are really a puzzle, for it is hard to see how a solar-mass star can detonate as violently as a supernova. One idea resembles that for binary novae. Imagine a binary system containing a white dwarf and a normal star in which the white dwarf has a mass

very close to the Chandrasekhar limit ($1.4M_\odot$). If enough mass flows onto the white dwarf to push it over the limit, it will collapse violently to a neutron star. This collapse may release sufficient energy to make a supernova. Another model invokes accretion onto a carbon-rich white dwarf. If the accretion reaches high enough temperatures and densities, carbon burning can ignite in a slow sequence called deflagration. This carbon deflagration is still destructive, blowing the star to bits so that no neutron core is left behind. In this model, we have no burst of neutrinos or formation of a neutron star as we have for Type II explosions.

(C) THE CRAB NEBULA—A SPECIAL SUPERNOVA REMNANT

The Crab Nebula (Figure 18–17) is the most intriguing and investigated **supernova remnant.** It

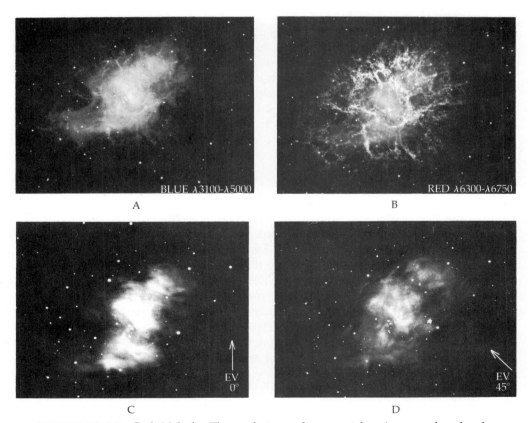

FIGURE 18–17 Crab Nebula. These photographs were taken in normal and polarized light. (A) Blue light, emphasizing the continuum emission. (B) Red light, showing the filamentary structure. (C) Polarized light, electric vector at 0° (arrow). (D) Polarized light, electric vector at 45° (arrow). Note the difference compared to (B). *(Palomar Observatory, California Institute of Technology)*

was the first identified as such and the first connected to a pulsar with the central star, in 1968. Other supernova remnants share a few of the characteristics portrayed by the Crab Nebula, and so it serves as an important prototype.

The distance to the Crab Nebula has been established by the technique outlined for novae (expansion of the gaseous shell); it is about 2000 pc. We can also use the present rate of expansion of the nebula to extrapolate back to the time of outburst; the result corroborates identification with the A.D. 1054 supernova (if we account for some acceleration since the initial expansion). The expansion is not uniform in all directions, and so the central star was not clearly identified until the pulsar was observed optically.

In the visible region of the spectrum, the nebula presents varied aspects, depending upon whether it is photographed in the radiation from one of the emission lines (such as H_α), in the continuum radiation, or through a polarizing filter. The line radiation emanates from clearly defined filamentary features (Figure 18–17B), where the gas density is enhanced over that of the rest of the nebula. Underlying the filaments and more concentrated at the central part of the nebula is the region emitting in the continuum. This region also possesses considerable structure, which can be described as vague wisps or fibers. The appearance and position of these wisps change with time because of motions of the gas or because of compression waves moving through the gas.

An important clue to the nature of the Crab Nebula is the fact that the continuum radiation is strongly linearly polarized (Figure 18–17C and D). Another is that the nebula is a strong radio emitter, for it is identified with the radio source Taurus A (the first radio source to be discovered in the constellation Taurus). The wavelength dependence of the radiation in both the visible and the radio region differs greatly from a Planck blackbody curve: it is *nonthermal radiation*. These characteristics led I. S. Shklovsky to propose (in 1953) that *synchrotron radiation* is the source of both the optical and the radio continuum.

We have already discussed synchrotron radiation in Sections 4–6(c) and 6–1(c), but let us remind you of the basic points. When energetic electrons are accelerated by a magnetic field, they spiral around the magnetic field lines (Figure

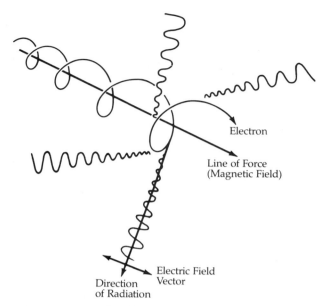

FIGURE 18–18 Synchrotron emission. A relativistic electron spirals along a magnetic line of force. The circular acceleration causes the electron to emit plane-polarized radiation.

18–18). This motion causes them to emit strongly polarized continuous radiation, whose intensity at a given frequency depends upon both the magnetic field strength and the energy of the electrons. The higher the mean electron energy, the higher the frequency at which the intensity is a maximum. Many radio sources have a synchrotron radio spectrum, but few simultaneously emit synchrotron radiation at the higher frequencies corresponding to visible radiation. The Crab Nebula is one of these few. In fact, synchrotron radiation also accounts for the X-ray emission that is observed from the Crab. The nebula as a whole, and not just the pulsar, generates X-rays.

The magnetic field strength in the nebula is estimated at 5×10^{-8} T. This field strength produces copious quantities of synchrotron radiation when the electrons are traveling at relativistic speeds, but where do these relativistic electrons come from? We know the following observational facts about the Crab Nebula: (1) we continue to see it over 900 years after the outburst, (2) the nebular gases exhibit accelerated motions, and (3) phenomenal amounts of synchrotron radiation continue to be emitted. Taken together, these facts demand that a strong energy source lies within the nebula. This energy source is none other than the Crab pulsar, which generates relativistic electrons and synchrotron radiation [Section 17–2(c)]. Recent radio observations show that the overall flux from the nebula has decreased at a rate of 0.17% per year (from 1968 to 1984), a value consistent with pulsar spindown time scales. About half of the energy radiated away is being replenished by the pulsar now.

(D) NUCLEOSYNTHESIS IN SUPERNOVAE

As Chapter 16 pointed out, the most massive stars can fuse elements up to iron; heavier elements require reactions that absorb rather than produce energy. Elements heavier than iron are probably made in the supernova explosions of massive stars (Type II). Here is one scenario of that process.

A star with a mass greater than $10M_\odot$ to $20M_\odot$ will have a layered look at the end of its life: carbon, helium, and hydrogen shells at greater and greater distances from the iron core. This layering results from lack of convection and the outward drop in temperature. The iron core cannot support itself; it contracts and its temperature rises. At about 6×10^9 K, photodisintegration of iron gives $^{56}Fe + \gamma \rightarrow 13\ ^4He + 4n$—an **endothermic** reaction needing about 100 MeV. This robs the core of energy, and so it contracts more rapidly. Once the iron disintegrates,

$$^4He \rightarrow 2\ p + 2\ n$$
$$p + e^- \rightarrow n + \bar{\nu}$$

converts the core to a degenerate neutron gas. Meanwhile, the layers above the core fall rapidly inward and also heat up. They still have fuel for nucleosynthesis. This goes off explosively, blowing off the outer layers. This fusion results in a flood of energetic neutrons, which can be absorbed by heavy nuclei. Now the rapid and slow processes come into play, "rapid" and "slow" referring to how fast the process goes relative to beta decay:

$$n \rightarrow p + e^- + \bar{\nu}$$

which takes about 15 min. In the rapid process, nuclei capture neutrons *faster* than beta decay; this

builds up neutron-rich material. In the slow process, neutrons are captured more *slowly* than beta decay; proton-rich material results. The rapid process is usually abbreviated **r process** and the slow process, **s process.**

Here is a specific example of this type of nucleosynthesis. Starting with ^{56}Fe; then by the r process

$$^{56}Fe + n \rightarrow {}^{57}Fe$$
$$^{57}Fe + n \rightarrow {}^{58}Fe$$
$$^{58}Fe + n \rightarrow {}^{59}Fe$$
$$^{59}Fe + n \rightarrow {}^{60}Fe$$
$$^{60}Fe + n \rightarrow {}^{61}Fe$$

Now, ^{61}Fe is stable for only about 6 *min*, and so if no neutrons are captured in this time, then by the s process,

$$^{61}Fe \rightarrow {}^{61}Co + e^- + \nu$$

In Type II supernovae, time scales are short so that only the r process is effective in nucleosynthesis. The most likely site is in the helium-burning shell as the shock wave plows through it.

We have emphasized nucleosynthesis in supernovae, but red giants also manufacture some heavy elements during their AGB phase by the s process. The s process cannot synthesize the very heavy radioactive elements, however, for the neutron addition is so slow that such nuclei decay by fission before more neutrons can be added. In general, elements made by the s process complement those made in the r process to fill up the periodic table. Nucleosynthesis in red giants makes many of the elements heavier than iron and lighter than lead, but elements heavier than lead, such as uranium and thorium, are produced in Type II supernovae.

(E) SUPERNOVA 1987A

On the night of February 24, 1987, Ian Shelton of the University of Toronto was photographing the Large Magellanic Cloud (abbreviated LMC), a companion Galaxy to our own, from Las Campanas Observatory in Chile. On the photo was a new, bright star in the LMC. By luck, Shelton had taken a photo of the same region 25 hours earlier. A comparison of the photographs dramatically showed the star to be the brightest supernova since Kepler's in A.D. 1604. The supernova was

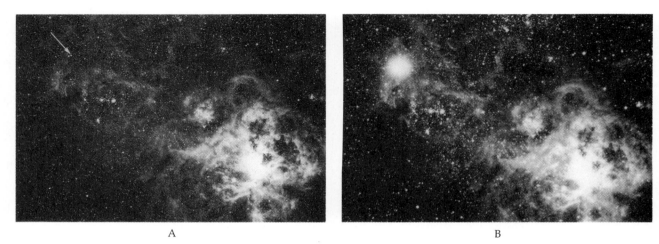

FIGURE 18–19 SN 1987A in the Large Magellanic Cloud. The photo at left shows the progenitor star before its explosion; the one on the right shows the supernova on Feb. 26, 1987, when it had reached an apparent magnitude of 4.4. *(European Southern Observatory)*

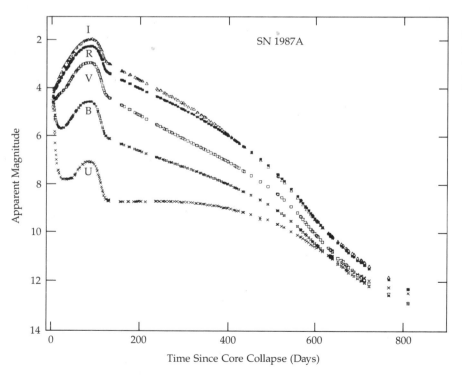

FIGURE 18–20 Light curves for SN 1987A. Data are given in magnitudes at the *UBVRI* bands over the time in days since the core collapse of the progenitor star. The last points are from May 16, 1989. *(CTIO/NOAO)*

given the name SN 1987A ("A" for the first super-nova discovered in 1987). The LMC is 52 kpc away; so the supernova actually exploded some 170,000 years ago. By astronomical standards, the LMC is close by, so we now have the *first* opportunity to study a supernova in detail with modern astronomical equipment (Figure 18–19).

Foremost has been the detection of neutrinos from the explosion. In Kamioka, Japan, a joint experiment of the United States and Japan, called the Kamiokande II, detected a burst of neutrinos at a time about one day before the supernova burst into visibility. Neutrino events were also observed from detectors in a salt mine in Mentor, Ohio. The detection of neutrinos (the first for a source outside of the solar system) strongly implies a Type II supernova—the death of a massive star. But only a few neutrinos could be detected of the estimated 10^{15} per square meter that arrived at the earth from SN 1987A. Kamiokande II detected only 12, six within the first second of the core collapse!

Other observations backed up the idea that this was a Type II explosion. Initial spectra showed strong, broad hydrogen lines, as expected from the hydrogen in the envelope around the exploding core. Doppler shift measurements indicated an expansion rate of about 17,000 km/s, again consistent with a Type II event. But all was not smooth with this Type II scenario. For one thing, the supernova did not get as bright as expected; it was about 100 times fainter than a Type II should reach at maximum. In fact, it stayed pretty much constant for 40 days (Figure 18–20), which is *not* typical behavior for a Type II. For another, observations from the International Ultraviolet Explorer satellite showed a spectrum more like that expected from Type I rather than Type II. IUE observations and those at the European Southern Observatory both indicate that the supernova's progenitor (a star named Sanduleak −69 202) was a *blue* supergiant star, not a red supergiant.

Why a blue star, not a red one, had exploded was initially a puzzle. The answer is that the main-sequence mass of the star that exploded was about 20 solar masses, and it was probably about 10 million years old at the time of its explosion. Because of mass loss, it had swung back to the blue side of the H–R diagram (Figure 18–21).

The visible supernova hit a peak brightness late in May, 1987. By mid-July, light echoes of the ex-

FIGURE 18–21 Evolutionary track for an $18M_\odot$ star evolved off the main sequence (MS) through hydrogen, helium, and carbon burning. The line shows a star that ends up with a final mass of about $16M_\odot$. The "star" shows the position of Sk −69°202, the progenitor star for SN 1987A.

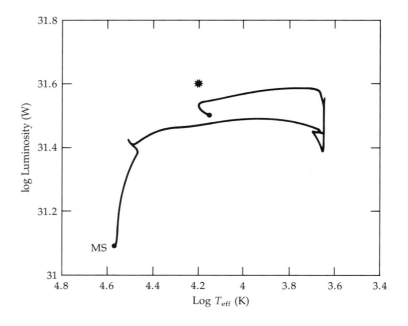

FIGURE 18–22 Light echoes from SN 1987A, visible in Feb. 1989. This specially processed image used a subtraction method to enhance the rings. *(Anglo-Australian Observatory)*

plosion appeared (Figure 18–22); these arose from thin sheets of material 140 and 400 pc in front of the supernova. These reflected the light, which took a longer time to get to the Earth and so arrived after the explosion. From July to November it declined exponentially, with a half-life of 78.7 days (mean lifetime 113.4 days), exactly that of radioactive ^{56}Co, one of the major elements expected to be produced in a supernova explosion. At this stage

the energy of the supernova came from heating of the envelope by the γ-rays produced by the decaying cobalt. By 1988, the opacity of the envelope had decreased to the point where the γ-rays could escape directly to space, so the light started fading more rapidly. The supernova has now faded to below naked-eye visibility, and will continue its descent into obscurity for many years to come.

SN 1987A passed all the major observational tests needed to confirm the main features of supernova models. For the first time neutrinos were observed, and they indicated a total energy of some 10^{46} J generated in the collapse of the star's core—just as predicted. And also for the first time, astronomers have finally observed the actual formation of a neutron star in a supernova event!

18–6 ◗
X-RAY SOURCES: BINARY AND VARIABLE

X-ray telescopes have opened the high-energy Universe (photons with energies ranging from 1 to 100 keV) to our view. The Uhuru and Einstein X-ray instruments in Earth orbit (and now ROSAT) have provided a wealth of data on galactic X-ray sources, many of which are variable and some of which are in binary systems. A few of these sources may contain black holes. Variability seems to be a hallmark of these sources, in part because they are driven by accretion processes.

A number of binary X-ray sources have been seen (Table 18–4). They fall into two general categories: high-mass X-ray binaries (sometime abbreviated *HMXRB*) and low-mass X-ray binaries (*LMXRB*). The "high" or "low" mass refers to the

TABLE 18–4 *Some Binary X-Ray Sources*

Name	Distance (kpc)	Binary Period (days)	X-Ray Luminosity (W) (2–11 keV)	Spectral Type of Visible Star
Cygnus X-1	2.5	5.6	2×10^{30}	O9.7 Iab (HDE226868)
Centaurus X-3	8	2.087	4×10^{30}	O6.5 II–III
Small Magellanic Cloud X-1	65	3.89	6×10^{31}	B0 I
Vela X-1	1.4	8.97	1.4×10^{29}	B0.5 Ib
Hercules X-1	5	1.70	1.0×10^{30}	A9–F0 (HZ Her)
Scorpio X-1	0.7	0.787	2.0×10^{31}	? (V818 Sco)

companion of the X-ray source, not the source itself. The low-mass systems contain a low-mass, late-type star whose optical luminosity is much lower than that of the X-rays; orbital periods range from a few hours to several days. High-mass systems contain early-type giant or supergiant stars and tend to be more concentrated in the galactic plane than the low-mass systems. Their orbital periods range from a few days to hundreds of days.

The sources listed in Table 18–4 have X-ray luminosities in the range from 10^{29} to 10^{31} W. Three of the sources (Hercules X-1, Centaurus X-3, and Small Magellanic Cloud X-1) have short-period X-ray pulses; they are *X-ray pulsars*. These binaries have a magnetized neutron star as the accreting object and X-ray source. The fields channel the accreting material (probably from the stellar wind of the companion) onto a small part of the neutron star's surface; hence, the emission is beamed and modulated by the spin period of the neutron star.

Some systems (Centaurus X-3, Small Magellanic Cloud X-1, Vela X-1, Circinus X-1, and Hercules X-1) exhibit X-ray eclipses; the X-ray source passes behind the normal star as we view the system. Using spectroscopic analysis of the light from the visible star (not the X-ray source), we can observe the changes in Doppler shift and so find the orbital periods, which are typically just a few days. These short periods indicate that the orbits are only a few times larger than those of the primary stars. Then, if we can determine the separation of the two objects, we can ascertain, from Kepler's third law, the sum of the masses (normal star plus X-ray source). With an idea of the mass of the normal star from its luminosity (using the mass–luminosity law), we can also determine the mass of the X-ray source. And if that mass turns out to be large enough (greater than 3 solar masses, the upper limit for a neutron star), the X-ray source is likely a black hole!

For HMXRBs, the neutron stars have masses near $1.4M_\odot$, the Chandrasekhar limit. So these stars may have formed from the collapse of the degenerate core of a highly evolved star or by accretion onto a degenerate dwarf. Most LMXRBs have orbital periods of a few hours, and so probably contain a late-type dwarf companion and an X-ray source of about a solar mass each. The accretor picks up mass from the Roche lobe overflow of the companion. The X-ray source in some cases has a very thick accretion disk, whose geometry

and precession can modulate the X-ray emission in complicated ways.

Let's look in detail at a few of these exotic X-ray sources.

(A) CYGNUS X-1

One likely candidate for a black hole is Cygnus X-1, a strong X-ray source in the constellation Cygnus. Cygnus X-1 emits about 2×10^{30} W in X-rays. Observations have shown that Cygnus X-1 flickers rapidly, in less than 0.001 s. In 1971, radio astronomers discovered radio bursts from Cygnus X-1 and were able to pin down the location better than the X-ray astronomers could. In the most likely place for Cygnus X-1 lies an O supergiant (Figure 18–23), called HDE (Henry Draper Extension) 226868 and is of spectral type O9.7 I, a supergiant with a surface temperature of roughly 31,000 K.

Optical observations show that the dark lines in the spectrum of the blue supergiant go through periodic Doppler shifts in 5.6 days; the star orbits with the X-ray source (a massive but optically invisible companion) about a common center of mass every 5.6 days. The mass of Cygnus X-1 is hard to

FIGURE 18–23   The blue supergiant HDE 226868, which is the visible star about which Cygnus X-1 orbits. *(J. Kristian)*

determine, however, because we do not have enough information to solve for the individual masses in the mass function [Section 12–3(b)]. We can observe the Doppler shift in the spectrum of the visible companion, but we cannot obtain the velocity of the X-ray source. And because Cygnus X-1 has not been found to eclipse, we do not know its orbital inclination. Hence, we cannot determine either individual mass.

We can see this problem as follows. For an X-ray source of mass M_x and a companion of mass M_c, the optical mass function is

$$f(M_x, M_c) = \frac{(M_x \sin i)^3}{(M_x + M_c)^2} = \frac{P^2(V_c \sin i)^3}{2\pi G}$$

where i is the orbital inclination, P is the orbital period, and $V_c \sin i$ the projected velocity of the optical companion. Note that the value of the mass function is the *least* possible value that M_x can have; it corresponds to a system with $M_c = 0$ and $i = 90°$. For Cyg X-1, $f(M_x, M_c) = 0.25 \pm 0.01 M_\odot$, with $V_c \sin i = 76 \pm 1$ km/s. Then if the supergiant has a mass of $33 M_\odot$, Cygnus X-1 has a mass possibly as great as 16 solar masses (Figure 18–24). If so, Cygnus X-1 must be a black hole, if the limit for a neutron star is $3 M_\odot$.

Note that the X-rays come *not* from the black hole itself but from an accretion disk of material around it. Mass from the blue supergiant, blown by a stellar wind, falls toward the black hole. Its angular momentum channels the infall into a disk, which is heated by tidal forces and the conversion of gravitational potential energy into thermal energy. Parts of this disk have temperatures of a few million kelvins; these generate the X-rays.

(B) CENTAURUS X-3

X-ray emission can arise not only from accretion disks around black holes but also from accretion onto neutron stars. An example is Centaurus X-3 (abbreviated Cen X-3). The Uhuru satellite showed that this X-ray source pulses every 4.84 s. Also, long-term observations have revealed that X-ray eclipses take place every 2.087 days and last about 0.5 d. So we know that the orbit of Cen X-3 is tilted so that its plane lies in our line of sight. A faint star at the X-ray source position varies in light with the same period as Cen X-3. The star turns out to be a blue giant about 8 kpc away.

This information all falls into place with a simple model for the Cen X-3 binary system (Figure 18–25). Cen X-3 itself moves in an almost circular orbit around the blue giant at 415 km/s. Its orbit has a radius of about 11×10^6 km. At this close distance, mass flowing from the giant is picked up by the X-ray source. About every two days, the X-ray source orbits behind the giant as seen from the Earth and an X-ray eclipse occurs. These eclipses allow us to estimate the mass of Cen X-3: $1.5 M_\odot$—a low-mass neutron star probably made in a supernova explosion.

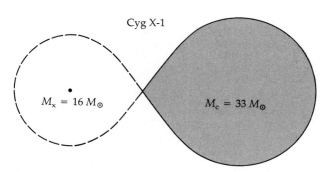

FIGURE 18–24 Schematic drawing of the Cyg X-1 binary system. Its optical companion is shown filling its Roche lobe. The accretion disk is not visible on this scale. *(Adapted from a diagram by J. E. McClintock)*

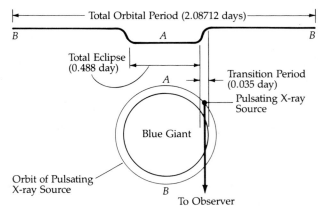

FIGURE 18–25 Model of emission from Centaurus X-3 binary system. *(Adapted from a diagram by H. Gurskey)*

The fact that Cen X-3 is an X-ray pulsar also supports a neutron-star model, in analogy with the model of radio pulsars as magnetic neutron stars. The X-ray pulses might arise from accreting matter channeled into the magnetic polar regions of the neutron star by the intense magnetic field.

(C) SS 433

A binary system with a neutron star can explain the bizarre radio, X-ray, and optical object called SS 433. That name comes from the fact that SS 433 is listed as entry 433 in a catalog of Milky Way objects showing strong emission lines compiled in 1977 by Bruce Stephenson and Nicholas Sanduleak. SS 433 at first appeared to be a nondescript fourteenth-magnitude star in the constellation Aquila. Radio observations showed that it lies near the center of a much larger radio source, called W 50, which may be a supernova remnant. The radio output of SS 433 varies by a factor of 2 to 3 over times of a few hours to days. X-ray observations revealed that the X-ray emission from SS 433 also varies.

These data prompted optical astronomers to take spectra of SS 433. A very strong H_α line appeared, surrounded by less intense but still rather strong lines. They were too strong to be lines emitted by oxygen or any of the heavier elements, and yet they were not at the correct wavelengths to be lines from hydrogen or helium. They turned out to be very highly redshifted and blueshifted lines of hydrogen and helium. Their displacements indicated radial velocities of more than 40,000 km/s, and the Doppler-shifted lines moved about 1% in a few days.

Long-term observations showed that the Doppler-shifted lines move periodically, with the red- and blueshifted lines swinging back and forth in a 164-day cycle (Figure 18–26). The maximum redshift value corresponds to a radial velocity of a little over 50,000 km/s. Very careful spectroscopic observations of the strongest emission lines, which at first seemed stationary, revealed a small Doppler shift (only 70 km/s, about 0.1% of the main moving lines) with a period of 13 days.

A model developed by Bruce Margon and his co-workers solves most of the puzzles presented by SS 433 (Figure 18–27): It is a binary system containing an ordinary (unseen) star and a neutron

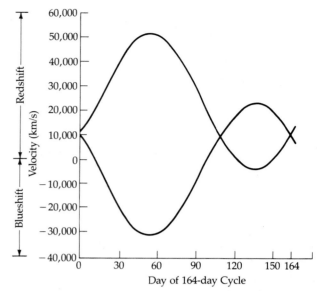

FIGURE 18–26   Cycle of Doppler shifts for SS 433. *(Based on observations by B. Margon)*

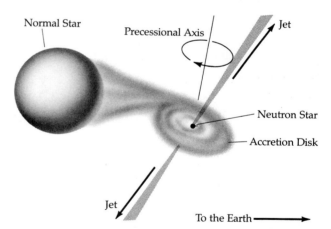

FIGURE 18–27   Model for SS 433. In this binary system, matter from a normal star accretes around a neutron star. Jets pour out along the axis of the accretion disk.

star orbiting every 13 days. The proximity of the two stars channels gas from the ordinary star onto an accretion disk around the neutron star. This gas heats up as it falls in and produces the strongest emission lines. Meanwhile, the accretion disk has two jets of material squirting out of it in opposite directions. The jet blasting away from us generates the highly redshifted lines; the one pointing toward us generates the blueshifted lines. The line of the jets is tilted about 20° to the rotational axis of the neutron star; that, in turn, inclines about 80° with respect to the line of sight from the Earth. Like the Earth's axis, the neutron star's axis precesses in space, with a period of 164 days. As the spin axis slowly whirls in space, it carries the jets around with it. This variation causes the Doppler-shifted lines to fluctuate between maximum and minimum displacements as the inclination of the jets changes along the line of sight.

Radio observations with the VLA (Figure 18–28) imply that the jets consist of small blobs of ionized gas continuously blowing out of the source, like water from a hose. This material shoots out at 78,000 km/s, about 26% of *c*! The jets radiate by synchrotron emission—an idea backed up by the fact that they are highly polarized (15 to 20%). If so, they contain magnetic fields with a strength of about 10^{-6} to 10^{-7} T. X-ray observations of SS 433 confirm this model of relativistic jets. To date, SS 433 is the only *star* system observed to generate relativistic jets.

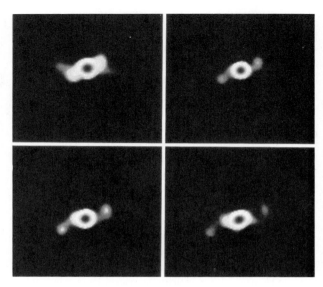

FIGURE 18–28 Radio observations of SS 433. The VLA made these observations over a four-month period. The material in the jets moved outward in this time. *(R. M. Hjellming and K. J. Johnston, National Radio Astronomy Observatory)*

(D) X-RAY AND GAMMA-RAY BURSTERS

X-ray bursters are set apart from other X-ray sources by their emission of brief but powerful bursts of X-rays (Figure 18–29). The bursts can occur in regular intervals of a few hours or a few days. Others fire off in a rapid sequence as a machine gun does, shooting off several thousand bursts in a day. A 10-s burst carries as much X-ray

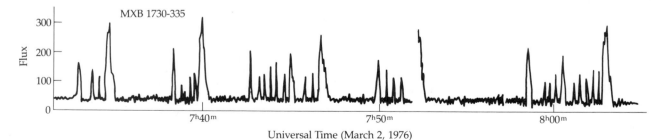

FIGURE 18–29 An X-ray burster. The rapid burster MXB 1730-335 repeats a sequence about every 10 min. [*W. Lewin et al., Astrophysical Journal (Letters) 207:L95 (1976)*]

energy as the Sun gives off in a week at all wave-lengths. Generally, X-ray bursters are LMXRBs, and they tend to cluster toward the galactic center (Figure 18–30A). A few have been observed to have optical bursts that correspond to the X-ray ones.

One X-ray burster is called XB1820-30 ("XB" stands for "X-ray burster"; the other numbers refer to the object's coordinates on the sky). The source lies near the center of the globular cluster NGC 6624. In 1975, Jonathan Grindley and Herbert Gursky discovered a brief burst of X-rays from NGC 6624. Other X-ray observations limited its position as being very close to the core of NGC 6624. X-ray astronomers at MIT then examined some new and old data to find bursts that lasted about 10 s. Remarkably, the bursts seemed to recur every 4.4 h. About 10 bursters are found in globular clusters, about 30% of the total.

To date, it appears likely that the neutron-star model best explains the properties of X-ray bursters: rise to peak in less than 1 s, duration of about 10 s, interval between bursts of about 2 h; blackbody effective temperatures of 3×10^7 K. The bursters are concentrated near the galactic center; if they are about 10 kpc from us, one burst has a luminosity at maximum of a few times 10^{31} W. An effective temperature of 3×10^7 K requires a blackbody of roughly 9-km radius to produce 2 to 3 × 10^{31} W. This hints that bursters are in fact neutron stars in binary systems with a low-mass companion filling its Roche lobe.

Paul Joss has developed a model of thermonuclear helium-burning flashes on an accreting neutron star that accounts for a good number of burster properties. It starts out with a neutron star accreting at a rate of roughly 10^{14} kg/s to build an envelope on it. As matter accretes, the temperature rises to 2×10^9 K at the base of the helium layer. The helium ignites in a flash, producing 2.5×10^{31} W just 0.2 s after the fusion reactions begin. With continuous accretion, the flashes can recur every 15 h. These calculated characteristics are much like those observed. The model does not rely on any special source for the accreted material; it probably comes from a binary companion. This model is similar to that for novae, the differences being that here the material falls onto a neutron star, not a white dwarf, and the temperatures are higher so that the energy comes out as X-rays rather than as visible light. In other words, bursters arise from thermonuclear flashes on weakly magnetized ($\approx 10^4$ T), rapidly accreting ($\approx 10^{-10}$ $M_\odot$/y) neutron stars in binary systems.

Gamma-ray bursters were discovered accidentally by satellites designed to monitor thermonuclear explosions on the Earth. They are transient events, of duration from 0.1–10 s (Figure 18–31)

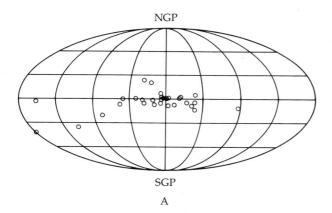

NGP

SGP

A

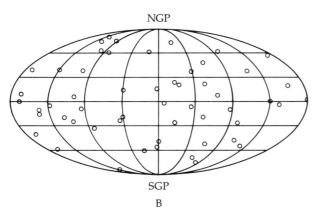

NGP

SGP

B

FIGURE 18–30 Distributions of bursters on the sky in galactic coordinates. (A) X-ray bursters with well-known positions. (B) Gamma-ray bursters. *(Adapted from diagrams by D. Hartmann and S. E. Woosley)*

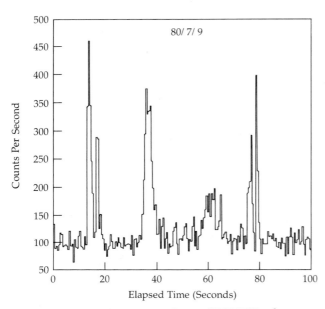

FIGURE 18–31 Gamma-ray burst GB800709, observed by the Pioneer Venus mission. Note the sequence of well-defined, individual spikes over a duration of about a minute.

and occur randomly around the sky (Figure 18–30B); rarely has a source been observed to burst more than once. One such source is associated with a supernova remnant in the Large Magellanic Cloud. In general, the positions of gamma-ray bursters are not well known (positional error boxes contain some 10 arcmin^2), so optical counterparts are hard to pin down.

With so little observational data, models for gamma-ray bursters have tended to derive from those for X-ray bursters. The picture here is for a highly magnetic ($\approx 10^8$ T), slowly accreting ($\approx 10^{-13}$ $M_\odot$/y) neutron star, either isolated or in a binary system. A single neutron star could accrete material from the interstellar medium at a rate large enough to generate a burst every few years or so.

PROBLEMS

1. In a given constellation, the following designations have been assigned to variable stars: V502, SU, II, V956, XY, and AK. List these stars in the order of their discovery.

2. In your own words, describe and correlate the fluctuation profiles of the typical pulsating star illustrated in Figure 18–2.

3. A Cepheid variable in a hypothetical galaxy is observed to pulsate with a period of ten days, and its mean apparent visual magnitude is 18. It is not known whether this is a Population I or Population II Cepheid.
   (a) What are the two possible distances to the galaxy (neglect interstellar absorption)?
   (b) What is the ratio of these distances?
   (c) Would this ratio change if we considered other galaxies? Explain.
   (d) Will this ratio differ for Cepheids of different periods?

4. If our telescope has a limiting magnitude of 22, what is the *maximum distance* to which we can see

   (a) RR Lyrae stars
   (b) classical Cepheids
   (c) W Virginis stars
   (d) ordinary novae
   (e) dwarf novae
   (f) supernovae
   How do these distances compare with the diameter of our Galaxy? (*Hint*: Consult Tables 18–1 to 18–3.)

5. The outburst of Nova Aquilae (V603 Aql) occurred in June 1918, at which time it attained a brightness of −1.1 mag. Spectra showed Doppler-shifted absorption lines corresponding to a velocity of 1700 km/s. By 1926, the star was surrounded by a faint shell 16″ of arc in diameter. Find the distance to Nova Aquilae in parsecs and the absolute magnitude at maximum.

6. If a star becomes a supernova, by what amount does its luminosity change if it originally had an absolute magnitude of 5.0? of 2.0? (The absolute visual magnitude of a supernova at maximum is about −18.0.)

7. Consult Chapter 10 to find the energy output in watts of the most energetic solar flares. Referring to

Chapter 13, compare this with the typical energy outputs of stars of the following spectral types:
(a) F
(b) G
(c) K
(d) M
By what factor is the luminosity of each star increased during such a flare event? Which stars would you consider to be *observable* flare stars?

8. Consider a rotating star with an extended atmosphere. If the stellar atmosphere is both rotating and expanding, draw the observed profile of a given spectral line (Figures 18–9 and 18–10) when the atmosphere rotates more slowly than the star.

9. A white dwarf in a binary system accretes enough material to increase its mass beyond the Chandrasekhar limit ($1.4M_\odot$) and collapse to the radius of a neutron star ($\approx 10^4$ m). Calculate the kinetic energy generated in such a collapse. Compare this with the value of approximately 10^{44} J relevant to a supernova explosion. Comment on the required efficiency of energy conversion.

10. A certain contact-binary system contains a red giant and a neutron star. The neutron star has a mass of $1M_\odot$ and a radius of 10^4 m. The system radiates 10^{31} W in X-rays. Determine the rate of mass flow (in solar masses per year) from the red giant to the neutron star required to produce this luminosity. Assume that half of the change in gravitational potential energy of an accreted gas particle is converted to X-rays and that the separation of the two stars is much greater than the radius of the neutron star.

11. Since the jets of SS 433 are oppositely directed, the point in Figure 18–21 at which the radial velocity curves cross is also the point at which the jets are perpendicular to the line of sight. The velocity at this point is the relativistic transverse Doppler shift, with the equivalent radial velocity given by $1 + v/c = 1/[1 - (v_{\text{jet}}^2/c^2)]^{1/2}$.
(a) Determine the value of v_{jet}.
(b) What angle do the jets make to the line of sight at the point of maximum redshift and blueshift? (*Hint*: Use the relativistic Doppler formula and your answer to part **a**.)

12. Another suggested mechanism for generating a type II supernova is explosive nuclear "burning" of the heavier elements, especially silicon. One model for a massive red supergiant predicts the presence of a shell of mass $\approx 2M_\odot$ and containing mostly ^{28}Si deep within the interior. Once this shell reaches ignition temperature, calculations show that the entire shell undergoes fusion within a fraction of a second. For simplicity, assume that the shell has a mass of $2M_\odot$, with ^{4}He making up half the nuclei and ^{28}Si the other half. The reaction is $^{28}\text{Si} + {}^4\text{He} \rightarrow {}^{32}\text{S} + \gamma$. Calculate the total energy released when the shell ignites, and compare this with the value of about 10^{44} J required for a supernova explosion. (The masses of the nuclei are 27.9769 amu for ^{28}Si, 31.9721 amu for ^{32}S, and 4.0026 amu for ^{4}He.)

13. Sketch the analogs to Figures 18–9 and 18–10 for the case of a cloud of hot gas falling in a spherical shell onto an even hotter star.

14. An O supergiant's mass can range from 15 to $40M_\odot$. Solve the mass function for Cyg X-1 for the mass of the X-ray source for this range of companion masses.

15. (a) Using the information provided in Section 18-5(e) and assuming a constant expansion rate, calculate the proper motion of the expanding nebula from Supernova 1987A.
(b) How long must astronomers wait before the nebula has a diameter of 1″? Approximately what date will this be?

16. (a) Ignoring interstellar absorption, how bright (apparent magnitude) would a Type II supernova be if it exploded 1000 pc from the Earth? Compare this with the apparent magnitude of Venus ($m = -3$).
(b) Ignoring interstellar absorption, how distant could a Type II supernova be and still be visible with the unaided eye ($m = 6$)? Compare this with the diameter of the Galaxy.
(c) Astronomers estimate that a supernova should occur on the average every 25 to 50 years in our Galaxy. Given that the last supernova in our Galaxy visible with the naked eye occurred nearly 400 years ago, comment on your result in (**b**).
(d) Ignoring interstellar absorption, how distant could a Type II supernova be and still be observed visually with a 16-inch telescope ($m = 14$)? Compare this with the distance to the Andromeda Galaxy (0.7 Mpc) and the nearest large cluster of galaxies, the Virgo cluster 15.7 Mpc).

17. Using Appendix 2 for the genitive name of the constellation, what name would astronomers give to:
(a) the third variable star discovered in the constellation Ursa Major?
(b) the thirteenth variable star discovered in the constellation Lupus?
(c) the 290th variable star discovered in the constellation Sextans?

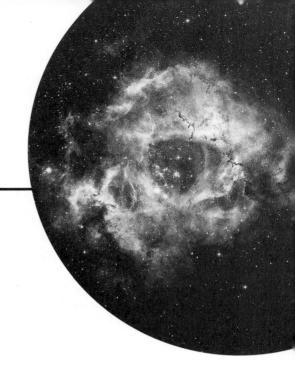

Chapter 19

The Interstellar Medium and Star Birth

We have examined the general form of our Galaxy and the varieties of stars that populate it. Now we will consider the content of the vast regions of space between the stars—the **interstellar medium** (ISM). Just as our own Solar System is pervaded by gas and plasma (the solar wind), magnetic fields, particles, and rocks, so also is interstellar space filled with gas, dust, magnetic fields, and particles. This chapter concentrates on the dust and gas in the galactic disk. Gas and dust make up the bulk of the interstellar medium, and here, from dense molecular clouds, new stars (and planets) are born.

19–1 ◑
INTERSTELLAR DUST

Interstellar dust cohabits the interstellar medium with the gas. On the average, one dust particle exists in every 10^6 m^3, but the dust amounts to about 1% of the total mass of interstellar matter, and it can cut out light from distant objects or from those enshrouded in dense clouds. Piercing the dust veil

has been an important goal of radio and infrared astronomers in revealing the process of star birth.

(A) DARK NEBULAE AND THE GENERAL OBSCURATION

Numerous dark patches stud the Milky Way; some appear only on photographs, but others, such as the Great Rift in Cygnus and the Coal Sack near the Southern Cross (Figure 19–1), are visible to the naked eye. These **dark nebulae** are opaque clouds obscuring the light of the stars behind them. In many cases, dark nebulae lie adjacent to or superimposed upon bright nebulae; an example is the famous Horsehead Nebula in Orion (Figure 19–2). Sometimes very small dark regions called **globules** overlie bright nebulae (Figure 19–3).

Far more difficult to detect is the **general obscuration** caused by dust distributed more uniformly and thinly than in the dark clouds. We have already mentioned this obscuration in our discussion of star counts (Section 14–2). The general absorption from dust requires that the equation for

FIGURE 19–1 A view of the southern Milky Way. The Coal Sack is the dark region near the center. (*Harvard College Observatory*)

FIGURE 19–2 The Horsehead Nebula, a dark cloud, is in Orion. (*Palomar Observatory, California Institute of Technology*)

FIGURE 19–3 Dark globules. These small, dusty regions lie in the Rosette Nebula, an H II region. (*Palomar Observatory, California Institute of Technology*)

the distance modulus should be rewritten from

$$m - M = 5 \log d - 5 \qquad \textbf{(19–1)}$$

to

$$m - M = 5 \log d - 5 + A \qquad \textbf{(19–2)}$$

where A represents the total amount of absorption (in magnitudes at the observed wavelength) for the distance d from our Sun; m is the apparent magnitude, M the absolute. Note that stars appear to be fainter (larger m), and therefore farther away, whenever interstellar obscuration intervenes.

If this general interstellar absorption were truly uniform throughout the Galaxy, A could be expressed as a simple function of distance, $A = kd$. Observations clearly show, however, that the absorption is patchy; the total amount of absorption between us and a star or cluster differs with direction in the sky and with the character of the intervening space.

(B) INTERSTELLAR REDDENING

The absorption A in Equation 19–2 depends upon wavelength. The interstellar dust between us and a star does not dim that star's light identically at all wavelengths; more light is scattered in the blue than in the red. As a result, the light from the star appears redder than in the absence of dust—hence the term **interstellar reddening.** The obscuration is primarily a form of scattering rather than absorption. The reddening arises from **selective scattering,** so that if equal numbers of red and blue photons are incident upon a dust cloud, a greater number of the blue photons are scattered out of the beam. Hence, a proportionately larger number of the red photons penetrate through the cloud and reach an observer (Figure 19–4). In a sense, the light is "deblued" rather than reddened.

Reddening increases the color index observed for a star. We define **color excess** as the difference between the observed and the intrinsic color index [Section 11–4(b)]:

$$CE = CI \text{ (observed)} - CI \text{ (intrinsic)} \qquad \textbf{(19–3)}$$

The intrinsic color index depends upon the spectral type of the star, and it can be established from a spectrum. We also can determine the color excess without taking a spectrogram, by comparing two color indices, such as $(B - V)$ and $(U - B)$.

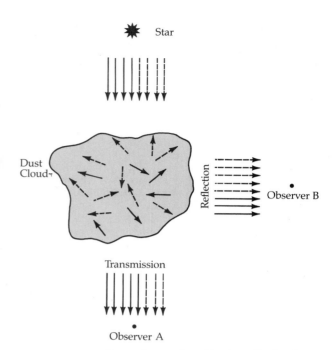

FIGURE 19–4   Scattering and reddening. Grains in a dust cloud scatter blue light more efficiently than red. So observer B sees a bluish reflection nebula, and observer A sees reddened starlight.

The wavelength dependence of the interstellar extinction (absorption) is found by comparing the brightnesses of stars of similar spectral type at a number of wavelengths. By such a comparison of stars reddened by different amounts, astronomers have found that the extinction is proportional to $1/\lambda$ in the visible region. Such data also established that, in most regions of the Galaxy, the absorption in visual magnitudes is approximately three times the color excess (the best current value is 3.2); thus

$$A_v \approx 3(CE) \qquad \textbf{(19–4)}$$

We may use this value of A in Equation 19–2 to find the true distance to the star, providing the star itself does not have a peculiar spectral distribution or affect the nature of dust grains in its immediate vicinity.

Example: Consider a star of spectral type G0 V with $m_v = 13.0$ and $CI = 1.6$. The H–R diagram shows that a star of this spectral type has an intrinsic color index of 0.6, and $M_v = 5$. Therefore, from Equation 19–3, we have $CE = 1.0$ and from Equation 19–4, $A_v \approx 3.0$. Now we substitute in Equation 19–2

$$m_v - M_v = 5 \log d - 5 + A_v$$

to obtain

$$13.0 - 5.0 = 5 \log d - 5 + 3$$
$$\log d = 2$$
$$d = 100 \text{ pc}$$

Although the star appears to be 400 pc distant, its actual distance is only 100 pc. Or, we could say that the true distance modulus is

$$m_v - M_d = m_v - M_v - A_v = 5 \log (d/10)$$
$$= 13 - 5 - 3 = 5$$

Let's quantify the discussion a bit more. Rewrite Equation 19–2 as

$$m_\lambda - M_\lambda = 5 \log d - 5 + A_\lambda$$

where the wavelength (λ) subscript emphasizes that the amount of extinction depends on wavelength. Note that A_λ is the absorption along the line of sight, so that

$$A_\lambda = k_\lambda d$$

where k_λ is the absorption, or extinction, coefficient, which depends on the extinction cross section σ_λ and the number-density distribution of absorbing material n. On the average, k_v equals 1 to 2 mag/kpc.

The amount of extinction can be directly related to the extinction characteristics of the dusty material. Recall (Section 8–7) that, for light of intensity $I(0)$ passing through a uniform slab of thickness L, absorption results in

$$I(L) = I(0) \exp (-\tau_\lambda)$$

where the optical depth

$$\tau_\lambda = \sigma_\lambda \int_0^L n(l) \, dl = \sigma_\lambda n L$$

so that over distance d in the interstellar medium,

$$\tau_\lambda = \sigma_\lambda \int_0^d n(r) \, dr = \sigma_\lambda n d$$

if $n(r)$, the number density of absorbing particles, is uniform along the line of sight. Now, the intensity ratio $I/I(0)$ is related to a difference in magnitudes ΔM by

$$\Delta M_\lambda = -2.5 \log [I/I(0)]$$
$$= -2.5 \log [\exp (-\tau_\lambda)]$$
$$= 2.5 (\log e)\tau_\lambda$$
$$= 2.5(0.434)\tau_\lambda = 1.086\tau_\lambda$$

Since this change in magnitude ΔM_λ is the absorption A_λ,

$$A_\lambda = 1.086\tau_\lambda$$
$$= 1.086\sigma_\lambda n d = k_\lambda d$$

where

$$\sigma_\lambda = \pi a^2 Q_\lambda$$

Here a is the radius of the dust particles and Q_λ is their relative **extinction coefficient,** which can be calculated in the laboratory from the optical properties of the appropriate materials. Note that the absorption in magnitudes approximately equals the optical depth.

(C) INTERSTELLAR POLARIZATION

The light emitted by stars is basically unpolarized. However, observations show starlight to have a polarization of up to 10% at optical wavelengths. (Review Section 8–1 for the nature and properties of polarized light.) The amount of polarization correlates directly with the amount of interstellar reddening: a large polarization is found for stars with large color excesses. Therefore, interstellar dust causes most of the optical polarization of starlight (although some stars do show intrinsic and often variable polarization).

We detect polarization by measuring the intensity of light transmitted through a polarizing filter or other polarization analyzer. Such a filter passes a maximum intensity I in one orientation and a minimum intensity $I_\perp$ when rotated through 90°. We define the **fractional polarization** of light by

$$FP = (I - I_\perp)/(I + I_\perp)$$

For completely unpolarized light, we have $I = I_\perp$ and $FP = 0$, whereas for *linearly* polarized light, $I_\perp = 0$ and $FP = 1$.

Nonspherical particles can polarize light, for light vibrating parallel to the long axis of an elongated particle will be diminished more than that vibrating perpendicular to that axis. Hence, the discovery of interstellar polarization gives clues about the nature of **interstellar dust grains.** Also, even nonspherical particles cannot polarize light if they are oriented at random. On the average, the grains along the line of sight must have a preferential orientation for polarization to occur. Now, under certain conditions, even relatively weak magnetic fields can align particles. One model visualizes elongated particles spinning with their short axes aligned along the magnetic field. So we can use polarization data to map the magnetic field of the Galaxy as seen from the Sun. Polarization will be strong and ordered when the magnetic field is perpendicular to the line of sight and weak and random when we look along the field (down a magnetic flux tube). Observations of interstellar polarization indicate that the galactic magnetic field (on the average) lies along spiral arms. As discussed in Section 20–4(a), radio interstellar polarization confirms the interpretation that the optical polarization is related to the magnetic field.

(D) REFLECTION NEBULAE

When a dust cloud lies to one side of a star rather than between the observer and the star, it scatters light from the star toward the observer. This is the same scattering phenomenon that is responsible for interstellar reddening, but instead of viewing the light that filters through the dust, we see the light that is scattered out of the star-to-cloud direction—a **reflection nebula** (Figure 19–5). Each of the dust particles scatters a bit of the starlight toward us. Since the particles scatter blue light more effectively than red, reflection nebulae appear *bluer* than the incident starlight they scatter. Another type of bright nebula, which appears reddish, is an **emission nebula,** which consists primarily of gas excited by a hot star [Section 19–2(b)]. Although we distinguish between these two types of bright nebulae, they are often found in close proximity to one another. This illustrates that dust and gas are usually closely mingled in the interstellar medium (Section 19–3).

One key observational feature of reflection nebulae is that the light from them is highly polarized—often as much as 20 to 30%. Now, the light emitted by stars is basically unpolarized. When scattered by small particles, light is selectively plane-polarized. The polarization occurs because light is a transverse wave, and so small particles selectively scatter light perpendicular to the direction of travel

FIGURE 19–5 A reflection nebula, caused by a dust cloud around the star Merope in the Pleiades. *(National Optical Astronomy Observatories)*

of the incident light. This polarization differs from that of the general interstellar polarization in that it results from reflection rather than transmission of light. It, too, depends on the nature of the medium but also on the angles of the incident starlight and the line of observation. So again we may deduce some of the properties of the dust grains around stars by observing the polarization of the reflected starlight. Such polarization observations are also possible in the near infrared, so that nonspherical distributions of dust grains (such as disks) can be indirectly detected around stars.

(E) THE NATURE OF THE INTERSTELLAR GRAINS

The observed effects of the interstellar grains—interstellar reddening, extinction, reflection, and polarization—give clues to the nature of the particles involved. Although both theorists and experimentalists have worked diligently to decipher the data, they have not yet found a complete solution. The possibilities that can explain most of the observations include the following:

1. Elongated dirty-ice grains
2. Grains of graphite (carbon)
3. Particles with small cores and large icy mantles
4. Large, complex molecules called polycyclic aromatic hydrocarbons (PAHs)
5. Silicate particles

The strength of the interstellar obscuration and the characteristics of the reddening require solid grains. Interstellar polarization requires *nonspherical* particles that can be aligned by a magnetic field; pure ice particles are excluded because they are not magnetic. Graphite or graphite-core particles could fit, for carbon in the form of graphite readily forms into highly flattened plates or flakes. However, ice and ice-like materials also tend to form flat crystals. There is strong support for silicate particles and for a mixture of silicates and either alternative 2 or 3.

A key clue comes from the average **interstellar extinction curve** (Figure 19–6) in the visible and ultraviolet. Note that the curve rises in the visible, has a bump in the ultraviolet (about 0.2 μm), and then, after a slight dip, rises again into the far ultraviolet (the data are limited at very short wavelengths by our observational techniques). No one type or size of grain can fit the extinction curve; it must be a composite resulting from the interstellar mixture. Calculations show that the bump and the rise in the ultraviolet are caused by very small particles, 0.005 to 0.02 μm in radius. The rise in the visible can be the result of larger grains, 0.05 to 0.2 μm in radius. Note that dust that absorbs ultraviolet will heat up and emit, in equilibrium, in the infrared. This infrared emission has been observed on a large scale from throughout the Milky Way, especially by the IRAS satellite.

The bump at 0.2 μm can be explained by bare, small-radius ($\approx$0.02 μm) graphite (pure carbon)

particles. The bonds between the carbon atoms resonate and absorb at this wavelength. The rise in the ultraviolet must also come from very small particles; silicate particles with a radius of 0.005 to 0.01 μm can play this role. For the visible region of the spectrum, larger particles are needed; their radius is about 0.2 μm. Such particles cannot be composed entirely of silicates, graphite, or pure iron; to account for how much extinction is seen requires some of the abundant hydrogen in the icy materials. To account for the shape and amount of the interstellar extinction curve, astronomers have developed **core–mantle grain models.** The small core, about 0.05 μm in radius, could consist of silicates, iron, or graphite; silicates are plausible. The mantles are made of icy materials, likely some composite of all possible kinds. Note that these grains are much smaller than the dust in your house; in fact, they are smaller than the particles in tobacco smoke.

Infrared observations bolster the idea that silicates and ices (at least water ice) make up part of the interstellar grains. They show absorption bands at 9.7 and 3.07 μm (Figure 19–7). Silicates in terrestrial rocks, meteorites, and lunar rocks have absorption bands at about 10 μm; these involve changes in the energy of vibration in the Si–O bonds. Silicates also have another, but weaker, absorbing band at 18 μm that involves the energy

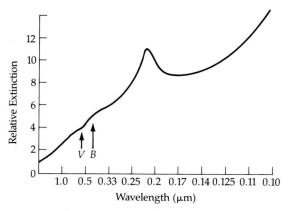

FIGURE 19–6 General interstellar extinction curve. This curve is an average one over many directions in the sky; note the peak near 0.2 μm.

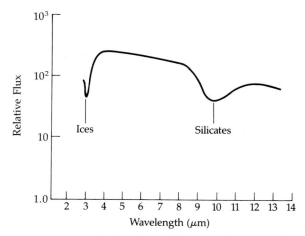

FIGURE 19–7 Infrared spectrum showing absorption features. This spectrum is from the Becklin-Neugebauer source in the Orion Nebula. The absorption bands are probably from ices and silicates. *[F. Gillett and W. Forrest, Astrophysical Journal 179:483 (1973)]*

of bending the O–Si–O bonds; this has been observed, too. The band at 3.07 μm likely occurs from water ice, but the amount of water ice in the grain is not enough to account for all of the extinction. Probably present are other icy substances that have not yet been positively identified because their infrared bands are much weaker than that of water ice.

The PAHs are the newest addition to the grain mixture. They are aromatic hydrocarbons of 20 to 100 carbon atoms in a lattice-like structure, which are much smaller than the conventional grains mentioned so far—with radii of only 10 Å or so. Such structures are extremely stable because of the large binding energies of the carbon atoms. So they can survive high temperatures, greater than 1000 K; some IRAS observations imply emitters at greater than this temperature to explain excess 12-μ emission in infrared cirrus clouds; PAHs may be the source. Also, a number of so-far unidentified IR emission features from 3 to 12 μm may originate from PAHs, especially at 3.3 and 11.3 μm. In addition to the PAHs, the ISM may contain very small grains that bridge the gap in sizes between the PAHs and conventional dust grains. These very small grains (VSGs!) may be made of spherical graphite and have a radius of some 20 to 100 Å.

What is the source of interstellar grains? Basically, they come in two types, ices and refractory materials, and so they must have at least two different sites of origin. Ices solidify at a few hundred kelvins, the refractory materials at a few thousand kelvins. These grains are probably made in the atmospheres of cool supergiants. We know such stars are blowing mass into space at rates of about $10^{-6}M_{\odot}$/y. The surfaces of these M stars have temperatures of only 2500 K or so. As gaseous material streams outward from them, its temperature drops and solids can condense out of the vapor. The spectra of some supergiants show the 9.7-μm silicate feature, indicating that such dust exists around them. In a rarer class of stars, in which carbon is somewhat more abundant than oxygen, graphite-like grains and particles made of silicon carbide can form in the outflowing material. The infrared spectra of these stars indicate a surrounding cloud of carbon particles.

What about the ices that make up grain materials (or perhaps entire grains)? These probably condense on cores in the deep interiors of dense molecular clouds. Here the temperatures are low and

the gas densities high, and so bare grains can grow crusts of ices. A core may have to grow a mantle once every 10^8 years or so, since grains will lose their mantles when in an environment where temperatures range above a few hundred kelvins.

Finally, we note that the Infrared Astronomy Satellite (IRAS) detected emission from interstellar dust all around the sky. This emission is referred to as the **infrared cirrus** because of its resemblance to wispy, high altitude clouds in the earth's atmosphere. This dust emission appears strongest in regions where there is also a concentration of interstellar gas, and the color temperature is higher where the dust is warmed by a nearby star. The cirrus is nearby, within 200 pc, and small in size— 2 to 5 pc. Their masses range from a few to a few hundred solar masses. They may well be the smallest pieces of the ISM that we can observe, perhaps the fluff from larger clouds.

19–2 ◑
INTERSTELLAR GAS

In addition to dust grains, interstellar space contains gas. This section investigates the physical and observational properties of the many species of **interstellar gas,** which makes up most of the interstellar medium. While the dust between the stars makes its presence known by its broad spectral influence upon the continuous light from distant stars, the interstellar gas produces its own characteristic emission and absorption line spectra. The temperature and density of the gas determine these characteristic spectral features. In general, the gas is essentially transparent over a wide spectral range despite the fact that the total mass of the gas in our Galaxy is greater than the total mass of the dust by a factor of about 100.

If there is so much more gas than dust, why is it the dust that is responsible for interstellar obscuration? The number density of dust grains is vastly smaller than the number density of the gas: roughly one part in 10^{12}. To answer the original question, consider the **absorption coefficient** of each material. Dust particles interact very strongly with visible light over a broad range of wavelengths, but the interstellar gas cannot interact with visible light (hydrogen and helium in their ground states cannot absorb visible-light photons). Therefore, at visible wavelengths, the absorption coefficient per particle is much greater for the dust than for the gas.

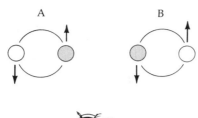

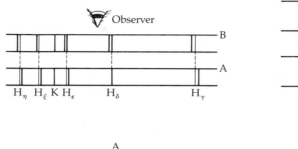

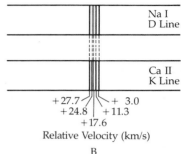

FIGURE 19–8 Optical interstellar absorption lines. (A) Lines from cool clouds appear in absorption toward a binary system, but they do not show any cyclical Doppler shift. (B) The lines can have several Doppler-shifted components, each from a cloud with a different radial velocity.

(A) INTERSTELLAR OPTICAL ABSORPTION LINES

Some stars have in their spectra absorption lines that are quite out of character with the spectral class. For instance, many B stars exhibit sharp, sometimes multiple lines of Ca II. Some spectroscopic binaries show particular spectral lines that remain fixed in wavelength while the rest of the spectral lines shift periodically to the red or blue in response to the binary stellar motions (Figure 19–8A). Clearly, these absorption lines originate in the interstellar medium. Multiple lines arise when there are several absorbing clouds along the line of sight (Figure 19–8B). Optical absorption lines, identified as interstellar in origin, include those from Ca I, Ca II, Ti I, Ti II, Na I, and the molecules CN and CH.

These absorption lines are sharp because thermal Doppler broadening is negligible at the low temperatures that characterize some parts of the interstellar medium. The intensity of a line depends upon the amount of gas lying between the star and the observer; if the gas is distributed uniformly through space, the intensities of interstellar absorption lines depend directly upon the path length traversed by the starlight. Low gas density plays a role in preventing ions from recombining into neutral atoms after photoionization. Sufficiently energetic photons and cosmic rays will occasionally encounter and ionize the widespread gas atoms and molecules. In order to recombine, an ion must capture an electron, but at typical interstellar densities the chance of such a capture is very small.

(B) EMISSION NEBULAE: H II REGIONS

Hydrogen Line Emission

Among the most spectacular objects to be photographed with telescopes are the **emission nebulae**—clouds of gas caused to shine by the intense radiation from a hot star (Figure 19–9). Hot O and B stars emit tremendous amounts of ultraviolet radiation; such energetic photons, with wavelengths less than 91.2 nm, ionize any hydrogen atom they

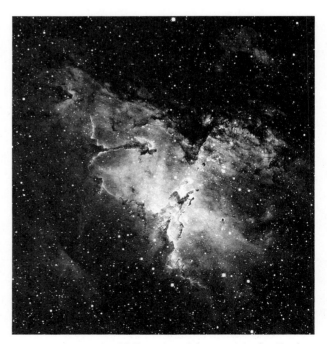

FIGURE 19–9 An H II region, Messier 16, the Eagle nebula. *(KPNO-NOAO)*

encounter. If such a hot star is surrounded by a cloud of gas, the hydrogen atoms close to the star will be ionized and form an **H II region**. Away from the star, the energetic photons have been used up for ionization; eventually none are available to ionize the hydrogen and the H II region sharply terminates (neutral hydrogen H I prevails). Let's now examine the physical processes that determine the structure of an H II region in greater detail.

The hydrogen gas in interstellar space is extremely dilute and cold. Most of the gas is H I (neutral hydrogen) in the ground state because collisional excitation is rare. Therefore, only photons whose wavelengths are less than or equal to 91.2 nm can ionize the gas to H II; note that 91.2 nm corresponds to the Lyman continuum limit (ionization potential) of hydrogen. Imagine a hot star in the midst of this cool H I gas with $T_{eff} \gtrsim$ 20,000 K, which the Planck spectral curves imply will produce ample ultraviolet radiation ($\lambda \gtrsim$ 91.2 nm). If the gas density is reasonably uniform, the ultraviolet radiation from the central star ionizes all of the hydrogen in a roughly spherical volume of space; we term this region the **Strömgren sphere**. (Because the interstellar gas is clumpy, H II regions are rarely spherical.) Equilibrium is established when the rate of recombination (H II + $e^- \rightarrow$ H I) equals the rate of photoionization; the H II region is maintained by the continual reionization of recombined H I atoms due to the flux of ultraviolet photons from the central star.

We can see this process in an idealized case as follows. Consider a single star emitting N_{uv} ionizing photons per second into a uniform medium. Within a volume out to radius R_s, all photons will be absorbed to ionize atomic H. Recombinations will balance ionizations, so that the total number of photons per second will equal the total recombinations per second:

$$N_{uv} = (4\pi/3) \, R_s^3 n_e n_H \alpha(2)$$

where $\alpha(2)$ is the *recombination coefficient* (m^3/s) of H *excluding* the $n = 1$ state. Such captures produce another ionizing photon; captures to $n = 2$ or higher produce photons longward of the Lyman limit. These quickly escape the H II region. So the Strömgren radius is given by

$$R_s = [N_{uv}/(4\pi/3) \, n_e \, n_H \alpha(2)]$$

An O5 star emits about 10^{49} photons/s; at 8000 K (typical of an H II region), $\alpha(2) \approx 10^{-19} \, m^3/s$; $n_e \approx$

$10^9/m^3$; $n_H \approx 10^3/m^3$; so $R_s \approx 100$ pc. For a cluster of stars, we add up the total number of ionizing photons (though one very hot star will dominate).

The outer boundary of an H II region arises from several factors. At greater distances from the star, the inverse-square law diminishes the flux of ultraviolet photons and ionization of the recombined H I atoms is no longer possible. So the ratio of H I to H II rises sharply with increasing distance from the star, and the material quickly becomes opaque to the Lyman continuum, giving rise to the sharp boundary. In addition, most of the H II recombines to an excited state of the neutral H I; the atom then quickly cascades to the ground state, emitting several low-energy ($\lambda > 91.2$ nm) photons in the process. Because the H I atoms spend so little time in the excited states, practically all of these low-energy photons (as well as the star's photons with $\lambda > 91.2$ nm) escape from the H II region. So the H II region **fluoresces** by converting the stellar ultraviolet radiation to lower-energy photons, with the bulk of the radiation escaping as the visible (longer-wavelength) Balmer lines. The H II region is therefore spectacularly visible as a bright reddish *emission nebula*—reddish because most of the Balmer radiation is in H_α. Note that the ionized gas within the H II region is hot (about 10,000 K) because the photons with energies above 13.6 eV expel freed electrons with kinetic energy that they share with the ions by collisions.

Radio line emission at centimeter wavelengths has been observed from very-low-energy electronic transitions between very high excitation levels of H I, such as from level $n = 110$ to $n = 109$ and from $n = 105$ to $n = 104$. The ionized hydrogen has no electrons, and so it cannot radiate spectral lines; nevertheless, radio continuum radiation emanates from the H II region as a result of free–free transitions (next section). Optical fluorescence lines of helium are also strong in the spectra of emission nebulae; together with the radio recombination lines of helium (arising from transitions between high-excitation levels), these lines permit us to (1) study the excitation mechanisms operating in H II regions, (2) investigate the elemental abundances (especially He/H) of the interstellar medium, and (3) probe the spiral structure of our Galaxy.

Continuous Radio Emission

The electrons in an H II region move freely through the gas, sometimes recombining with ions and sometimes, by collisions, exciting atoms or ions (leading to the emission of forbidden lines), but more often interacting with ions in a **free–free transition.** A free electron travels past an ion in a hyperbolic orbit of a given energy. This orbit can be altered by the quantum-mechanical emission of a photon with an energy up to the KE of the electron. When an assembly of electrons and ions (a plasma) is involved, the individual free–free emissions add up to a continuum; because the characteristic kinetic energies are small, this continuum radiation occurs predominantly at infrared and radio wavelengths. In short, an H II region is a source of radio emission characterized by the mean energy of the electrons, by the temperature of the gas. To distinguish this emission from synchrotron radiation, we use the term **thermal radio emission** or **thermal bremsstrahlung.**

Observations at radio wavelengths of thermal bremsstrahlung from H II regions provide a diagnostic of their physical conditions. The emission falls into two regimes: optically thick and optically thin connected by a turnover region. In the optically thin regime (Section 8–7), the flux is a power law:

$$F_\nu \propto \nu^{-\alpha}$$

where $\alpha \approx 0.1$ for a thermal source. On a log–log plot (look back at Figure 8–16), α is just the slope of the spectrum. For the Orion Nebula, for instance, the spectrum turns over at about 1 GHz and α does equal 0.1. In the optically thick regime at low frequencies, the spectrum rises as ν^2, the same as for blackbody radiation at low frequencies.

Following Section 8–7, in the optically thin part of the spectrum,

$$I_\nu = S_\nu \tau_\nu$$

Now, for free–free emission, the observed I_ν is roughly constant and so independent of frequency. If the temperature is roughly constant within the emitting region, then τ is proportional to

$$E_m = \int_0^L n_e^2 \, dl = \langle n_e \rangle^2 L$$

where E_m is called the **emission measure.** Note that L is the distance along the line of sight. The custom is to use n_e in cm^{-3} and L in pc, so E_m has unit pc/cm^6. E_m tells us how many charged particles interact along the line of sight—in this case,

just within the H II region, so L is the diameter of the region. Hence, the measured I_ν of an H II region gives E_m, from which we find $\langle n_e \rangle$. Typically, it ranges from 10^3 to about $10^9/m^3$.

Measurements in the optically thin regime also allow us to compute the size of the H II region, the mass of ionized hydrogen, and the number of ionizing photons per second from the embedded stars and hence to infer the number and spectral type of the stars creating the H II region. These values can be found even if the region is obscured optically by dust surrounding it and along the line of sight.

(C) SUPERNOVA REMNANTS

Material ejected from supernovae certainly becomes part of the interstellar medium. Moreover, the ejected matter sweeps up any surrounding gas and dust as it expands; this produces a shock wave that excites and ionizes the gas, which then becomes visible as an emission nebula. X-rays emitted by supernovae are also instrumental in ionizing nearby gas. Supernova remnants are radio emitters because of their synchrotron radiation (Figure 19–10). The Loop Nebula in Cygnus is such a remnant (Figure 19–11). Note that it looks spherical—a shell produced by the interaction be-

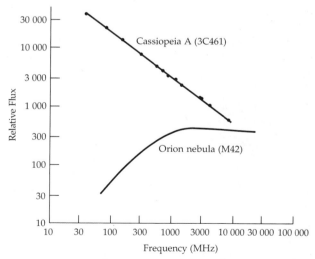

FIGURE 19–10 Comparison of the radio spectra of Cassiopeia A, a supernova remnant with a synchrotron spectrum, and the Orion Nebula, with thermal, free–free emission.

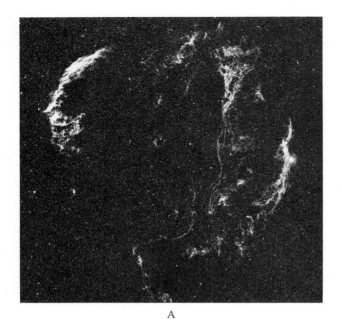

A

B

FIGURE 19–11 Optical images of supernova remnants. (A) Supernova remnant in Cygnus. The Loop Nebula in red light. *(Palomar Observatory, California Institute of Technology)* (B) Vela supernova remnant. A part of the Gum Nebula, these filaments are moving through the interstellar medium at thousands of kilometers per second. *(Royal Observatory, Edinburgh)*

tween the interstellar medium and a supernova shock wave.

A similar nebula in the southern sky—the Gum Nebula (Figure 19–12)—extends over 50° in the sky. The Gum Nebula has a diameter of about 700 pc, its closest edge being only 100 pc from the Sun. This nebula was created by the pulse of ultraviolet radiation and X-rays generated by a supernova 20,000 years ago. An X-ray source named Vela X lies almost in the nebula's center. It is a prime suspect as the supernova site. The discovery of a pulsar at the location of the Vela X source supports its nature as a supernova remnant.

About 150 galactic supernova remnants are cataloged, and more than 30 have been observed with the Einstein X-Ray Observatory. The huge shock waves plow through the interstellar gas and heat it to temperatures of at least a few million kelvins in the zone just behind the wave. This gas emits X-rays by free–free emission because it has such a high temperature. The X-ray pictures of Type I remnants, such as Tycho's SNR (Figure 19–12B) typically show symmetrical shells with variations in brightness around their rims—an indication of the patchy structure of the surrounding interstellar medium. Radio observations (Figure 19–12A) show a similar structure as the expanding material interacts with new material. Note that no source appears at the center—evidence that Type I explosions do *not* produce neutron stars visible as X-ray sources. Tycho's SNR, which is about 400 years old and so one of the youngest supernova remnants, has a diameter of about 13 pc and an L_X (0.2–2 keV) $\approx 4 \times 10^{29}$ W. Only the Crab Nebula has a greater X-ray luminosity.

(D) PLANETARY NEBULAE

Planetary nebulae differ from H II regions in that they are more compact and of higher surface brightness and have a different exciting source. When seen through a telescope, a planetary nebula appears as a round, greenish disk that superficially resembles a planet—hence the name. Closer examination reveals that the nebula is excited by a very hot central star. Gas densities in the nebulae surrounding these stars are higher than in H II regions; hence, collisions between electrons, atoms, and ions occur more frequently. Collisional excitation and de-excitation are therefore significant, so

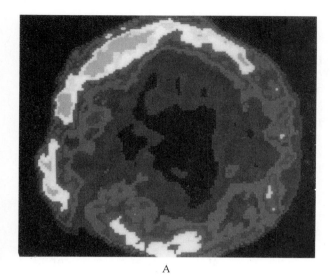

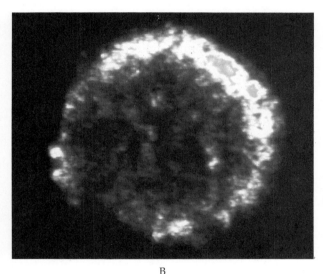

FIGURE 19–12 Tycho's SNR at radio and X-rays. (A) Radio observations at 11 cm. *(D. A. Green and S. F. Gull)* (B) X-ray image of Tycho's supernova remnant by Einstein; energy range is 0.3 to 3.5 keV. Note the spherical shape and overall similarity to the radio image. *(P. Gorenstein and F. Seward, Einstein Data Bank, Center for Astrophysics)*

that spectra of planetaries differ in important ways from those of H II regions.

Although the spectral lines of hydrogen and helium are quite pronounced in the spectra of planetary nebulae, the strongest lines are those of O III, O II, and Ne III. These lines do not correspond to ordinary electronic transitions but instead arise from electronic transitions from an excited metastable state to the ground state, resulting in **forbidden lines.** This situation resembles coronal forbidden lines [Section 10–4(c)] except that in the solar corona the forbidden lines arise from ions that have lost nine or more electrons, whereas in planetary nebulae only one or two electrons have been removed. The physical reason for this difference is the temperatures characterizing the two situations: in the solar corona, temperatures of 2×10^6 K prevail, whereas in planetary nebulae, the temperature is of the order of 20,000 K or less. Forbidden lines occur in both cases because of the low gas density; when an atom is excited to a metastable level, the chance of collisional de-excitation from that level is slight, so that the atom may remain in that level long enough to make the forbidden (low-probability) transition to the ground level. At a given temperature, the forbidden line intensities are a monotonically increasing function of density.

The forbidden nebular lines of [O III] at 500.7 nm and 495.9 nm correspond to the calculated energy differences between the metastable level and two of the three closely spaced ground levels of O III (Figure 19–13A). (The square-bracket notation signifies forbidden transitions.) These lines give planetaries their greenish appearance. A similar pair of forbidden lines arises from [O II], except that in this case the metastable level is double and the ground level is single (Figure 19–13B).

FIGURE 19–13 Forbidden transitions. The metastable state is M, and the ground state is G. (A) For O III, M is a single level and G has three sublevels. Two forbidden lines are seen. (B) For O II, M has two sublevels and G is single. Again, two lines are seen.

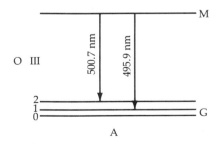

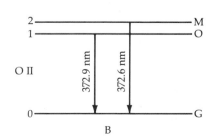

NGC 6720

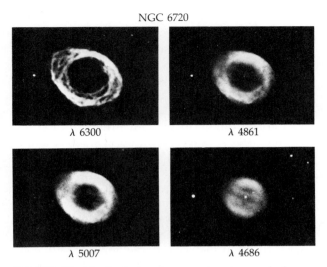

λ 6300 λ 4861

λ 5007 λ 4686

FIGURE 19–14 Emission from a planetary nebula, NGC 6720. Upper left: Hα; upper right: Hβ; lower left: forbidden O III; lower right: He II. *(National Optical Astronomy Observatories)*

Ions such as O II, O III, Ne III, and N II act as cooling agents in gaseous nebulae. Hydrogen atoms require large amounts of energy to become excited (10.15 eV for the first excited state), but most of the free electrons in the nebula do not have this much kinetic energy. The cooling-agent ions, however, all have energy levels near 2 or 3 eV; when an electron collides with one of these ions, it gives up part of its kinetic energy to excite the ion to one of these low (metastable) levels. Within a minute or two (in contrast to the 10^{-8} s for ordinary levels), the ion gives up this energy by emitting a forbidden line that escapes the nebula (Figure 19–14). Collisions must be rare; otherwise, de-excitation would prevent the occurrence of forbidden transitions. On the other hand, collisions must be frequent enough that collisional excitation is fairly common. So these ions extract energy from the electrons, and because the kinetic energy of the electrons is a measure of the temperature of the nebula, the result is a lower temperature for the planetary nebula (or H II region).

(E) INTERSTELLAR RADIO LINES

The Neutral-Hydrogen Line at 21 cm

Where the interstellar gas is cold, hydrogen is neutral and in its ground state. This ground state has two levels separated by a very small energy difference. The reason for this phenomenon lies in the fact that both the proton and the electron have an intrinsic spin. A moving charge produces a magnetic field. Because both the proton and the electron are charged particles, their spin motion generates a dipolar magnetic field (like the field of a tiny bar magnet) that we can characterize by the term **magnetic moment.** The magnetic moment of a spinning particle is represented by a vector, which is proportional to the vector angular momentum of the particle.

Two possible ground-state configurations of the neutral hydrogen atom exist (Figure 19–15). In one configuration, the magnetic-moment vectors of the proton and electron are parallel, or aligned; since vectors add (Appendix 9, "Mathematical Operations"), a high state of magnetic energy is present. Just as two parallel bar magnets will repel one another, so, too, will the proton and electron be less tightly bound to each other in their mutual orbits. If the magnetic-moment vectors are antiparallel, or opposed, we have the second configuration, which is characterized by less magnetic energy and a more tightly bound orbit. So the aligned state lies at a slightly higher energy than the opposed state; we refer to this effect as the **hyperfine splitting** of the ground state of the hydrogen atom. A spontaneous transition from the higher hyperfine state to the lower one can occur, accompanied by a relative spin flip of the electron (from aligned to opposed) and the simultaneous emission of a very-low-energy photon. This emission produces the 21-cm radio spectral line of neutral hydrogen at a frequency of 1.420406 GHz.

When hydrogen atoms collide in the interstellar medium, they generally exchange their electrons; this collisional transfer is the chief mode of changing the hyperfine states of these atoms. If the spin

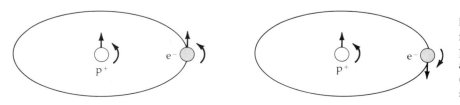

FIGURE 19–15 Spin alignment for hydrogen. The spins of the proton and electron can be either aligned (left) or antiparallel (right), which is the lower energy state.

of the newly acquired electron has the same orientation as the old, no change in energy level occurs; otherwise, there is a change in level (either up or down). In other words, collisions may result in no change, in excitation, or in de-excitation. A change in either direction takes place about once every 400 years for a given interstellar hydrogen atom. On the other hand, an atom in the excited hyperfine state makes a spontaneous downward transition followed by the emission of a 21-cm line quantum only once every few million years (on the average) because this transition is strongly forbidden. Numerous collisional excitations and de-excitations (of the hyperfine levels) occur in the interim. Eventually, an equilibrium is established with an aligned-to-opposed ratio of 3:1, but over distances on the order of a kiloparsec, an enormous number

of hydrogen atoms lie along a line of sight in spite of the exceedingly low gas densities in our Galaxy. Among these atoms, enough downward radiative transitions occur to produce a detectable 21-cm spectral line. The profile of the 21-cm line often has Doppler-shifted several peaks; this indicates that the gas is concentrated in discrete regions, such as spiral arms, rather than distributed smoothly throughout the Galaxy.

Molecular Lines

Interstellar molecules range from simple molecules like CO, CN, and OH to such complex organic molecules as formaldehyde (H_2CO) and methanol (CH_3OH), all found by searching for spectral lines at radio wavelengths. These molecules (Table 19–1) let us probe dense clouds of gas

TABLE 19–1 *Some Interstellar Molecules Observed to Date*

Complexity	Inorganic		Organic	
Diatomic	H_2	hydrogen	CH	methylidyne radical
	HD	deuterized hydrogen	CH^+	methylidyne ion
	OH	hydroxyl radical	CN	cyanogen radical
	SiO	silicon monoxide	CO	carbon monoxide
	SiS	silicon monosulfide	CO^+	carbon monoxide ion
	NS	nitrogen monosulfide	CS	carbon monosulfide
	SO	sulfur monoxide	C_2	carbon
	NO	nitric oxide		
Triatomic	H_2O	water	CCH	ethynyl radical
	HDO	heavy water	HCN	hydrogen cyanide
	N_2H^+	imidyl ion	HNC	hydrogen isocyanide
	H_2S	hydrogen sulfide	DCN	deuterium cyanide
	SO_2	sulfur dioxide	DNC	deuterium isocyanide
	HNO	nitroxyl	HCO	formyl radical
4-atomic	NH_3	ammonia	H_2CO	formaldehyde
			HNCO	hydrocyanic acid
			H_2CS	thioformaldehyde
			HC_2H	acetylene
5-atomic			CH_4	methane
			H_2NCN	cyanamide
			HCOOH	formic acid
			HC_3N	cyanoacetylene
			H_2C_2O	ketene
6-atomic			CH_3OH	methyl alcohol
			CH_3CN	methyl cyanide
			$HCONH_2$	formamide
7-atomic			CH_3NH_2	methylamine
			HC_5N	cyanodiacetylene
8-atomic			$HCOOCH_3$	methyl formate
			CH_3C_3N	methyl cyanoacetylene
9-atomic			CH_3CH_2OH	ethyl alcohol
			HC_7N	cyanotriacetylene
11-atomic			HC_9N	cyanotetraacetylene
13-atomic			$HC_{11}N$	cyanopentaacetylene

and dust, some of which contain protostars. The study of these molecules will eventually lead to a better understanding of the chemistry of the interstellar medium. For many people, however, the most exciting aspect of the molecules concerns their implications with respect to life outside the Solar System. Molecules of H_2O, NH_3, HCN, H_2CO, HC_3N, and $HNCO$ are used in laboratory experiments to synthesize amino acids and nucleotides, the building blocks of life. The fact that these molecules exist in interstellar space indicates that their formation does not require biological conditions.

Although grains make up a very small fraction of the total interstellar medium, they influence the form of the gas. Grains are probably the sites of molecule formation for some of the simpler molecules—at least H_2. Their surfaces act as catalysts by allowing atoms (or simple molecules) to stick to them so that there is time for a second atom to land, interact, and form a molecule that then evaporates back into the gas. Dust grains also shield molecules from dissociating ultraviolet radiation, thus letting the molecule population build up within a cloud.

The first molecule to be detected by radio was the hydroxyl radical, OH, in 1963, after the characteristic spectral frequencies had been firmly established in the laboratory. Four transitions near a wavelength of 18 cm (frequencies of 1612, 1665, 1667, and 1721 MHz) occur because of the splitting of the ground level of the OH molecule. Molecules appear to be connected with dust because OH, H_2CO, and CO lines are fairly widespread and found in large dust clouds. Many of these dense clouds lie in the direction of and are connected to H II regions; the Orion Nebula is a prime example. The number densities in such clouds are estimated as 10^9 to 10^{12} H_2 molecules/m^3; other molecules are, of course, far less abundant though more readily observed. The cloud temperatures are low, usually 10 to 30 K and sometimes as high as 100 K. (Note that we cannot observe H_2 directly by radio because it emits no lines in that wavelength range. Instead, we observe CO and assume that it acts as a good tracer of H_2.)

Although some interstellar molecules, such as carbon monoxide, pop up almost everywhere in the Galaxy, most are concentrated in dark, dense, cold conglomerates called **molecular clouds.** The higher densities here result in more frequent atomic collisions that make molecules, and the dust in these clouds shields out the ultraviolet light that destroys molecules. The net result is a high concentration of many kinds of molecules in dense clouds. Although the chemical reaction networks that create the wide variety of molecules is extremely complex, simple molecules (with 4 atoms or less) form in cool clouds by a sequence of two-body, ion-molecule processes in the gas phase. The initial ionization is provided by interstellar UV for diffuse clouds or by cosmic rays for dense clouds. H II regions often lie near or within these clouds. The Orion Nebula sits at the front of one of the nearest molecular clouds, as we view it from earth. This cloud consists of two parts: a large, low-density cloud (inferred from carbon monoxide emission) surrounding a dense, small core (inferred from formaldehyde, H_2CO, emission). The low-density cloud has an enormous extent: it is at least 10 pc across, has a peak density of 10^9 hydrogen molecules/m^3, and contains at least 10^4 solar masses of material. Its core is only 0.15 pc in size, has a peak density of 10^{11} hydrogen molecules/m^3, and a mass of only 5 $M_\odot$.

Giant Molecular Clouds

The Orion region presents an excellent example of a **giant molecular cloud.** Observations so far indicate that the bulk of the material of the interstellar medium is bound up in complexes of giant molecular clouds. Typical properties are:

1. They consist mostly of molecular hydrogen; many other molecules are present, but make up only a small fraction of the mass.
2. The cloud complexes have average densities of a few hundred million molecules per cubic meter; the individual clouds are slightly denser, with a few billion molecules every cubic meter.
3. They have sizes of a few tens of parsecs.
4. The total masses of the complexes range from 10^4 to 10^7 solar masses; 10^5 solar masses is typical. Masses of individual clouds are about 1000 solar masses.

The cores of these clouds are unusual places compared with the average interstellar medium. Here the temperatures are a frigid 10 K and the densities get as high as 10^{12} molecules per cubic meter. Giant H II regions, which surround young, massive stars, are always found near molecular cloud complexes. This proximity suggests that giant molecular clouds play the essential role in the process of star formation.

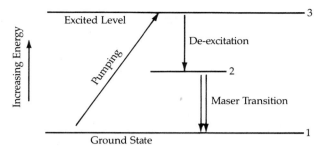

FIGURE 19–16 Energy-level diagram for a hypothetical three-level maser.

Molecular Masers

Most of the radio molecular lines are emission lines from rotational transitions [Section 8–3(c)]. Emission requires that the molecules be excited above the ground state by some mechanism. For example, OH lines appear in emission from parts of bright H II regions. Some, but by no means all, OH sources show strong H_2O emission as well. The H_2O emission is variable, with intensity changes occurring in periods of months or days. Although also variable, the OH radiation changes far less erratically. Superimposed on some H II regions are several groups of OH emission regions separated by distances of only a few astronomical units.

The emissions from these small OH and H_2O regions far exceed that expected from thermal excitation by collisions, which would require temperatures as high as 10^{13} K. The energy levels of the molecules are apparently subject to population inversion, by which we mean that more molecules are in the upper levels than in lower levels; hence, the Boltzmann equation [Section 8–4(a)] is violated and thermal equilibrium does not exist. A **maser action** (molecular laser) is responsible for these inversions. Some mechanism (several have been proposed but none agreed upon) amplifies the energy so as to pump the molecules into the appropriate excited state. Atoms or molecules in a gas are excited to some particular energy state and then stimulated to fall to a lower energy state at a more rapid rate than normal.

We illustrate the maser process with a hypothetical three-level molecular laser (Figure 19–16), with levels numbered 1, 2, and 3 in increasing energy. A molecule in the ground state, 1, is excited to the highest level, 3, either by colliding with another particle or by absorbing radiation. This process is called **pumping** a maser or laser. Assume that from level 3 the most probable transition when the molecule gives off a photon is to level 2, which is relatively stable; the probability of dropping to level 1 is relatively small. The pumping puts many molecules into level 2. Imagine that a photon with energy equal to the difference between levels 1 and 2 comes close to such an excited molecule. It triggers the molecule's drop from level 2 to 1, sending off another photon, which has the same energy as the original photon and is moving in the same direction. The electromagnetic field of the photon promotes emission of a photon equal in energy to the incoming photon. This process is called **stimulated emission.** The two photons now can stimulate two more molecules to radiate. The resulting four photons can trigger four more molecules to radiate, making a total of eight photons, and so on. The chain reaction amplifies the original photon millions of times as the photons travel through the gas. As a result, a maser's light is intense, narrowly directed, and at a single frequency.

Consider this maser process in OH (Figure 19–17). This radical is pumped, probably by infrared photons, to level 5. The possibilities for transitions to lower levels are such that the natural decay in energy leaves most molecules in level 3. The molecules can then be stimulated to drop to level 1 and emit a 1665-MHz photon; another possible, but less likely, drop is to level 2, with the emission of a 1612-MHz photon. This, in fact, is what is seen in sources such as the Orion Nebula. Other cosmic

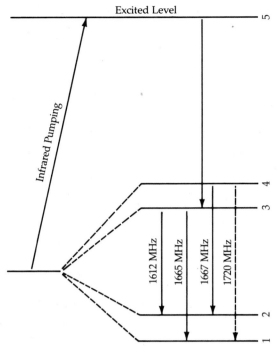

FIGURE 19–17   Energy-level diagram for maser transitions of OH.

masers are water at 22,235 MHz (1.35 cm) and silicon monoxide at 43,122 MHz (6.95 mm) and 86,243 MHz (3.47 mm).

Many hydroxyl and water interstellar masers tend to be found near giant molecular clouds. The regions of maser emission are extremely compact—only a few tens of astronomical units across—and very dense. They turn out to be signposts of incipient star birth.

(F) INTERCLOUD GAS

Optical observations of interstellar absorption lines and 21-cm data indicate that a large fraction of the interstellar gas consists of cool clouds along with denser molecular clouds. The H I clouds have diameters up to a few tens of parsecs, temperatures around 100 K, and densities of 10^6 atoms/m^3. Filling the space between these clouds is an ionized gas, some of which is very hot. Radio observations indicate one partially ionized component at 10,000 K with a mean density of 10^4 ions/m^3, roughly. Ultraviolet and X-ray observations reveal a hotter component at 10^6 K—called the **coronal interstellar gas.** It appears to occupy the largest volume of local interstellar space.

(G) THE EVOLUTION OF THE INTERSTELLAR GAS

Driven by star birth and death, the interstellar gas evolves in various forms. Supernovae play an important role in interstellar gas dynamics. Recall that a supernova blasts a tremendous amount of energy (about 10^{44} J) and material (1 to 50 solar masses) into space. The material blown off by a supernova expands as a shell into the interstellar medium. The expanding shell compresses and heats up the interstellar gas; behind the shell the gas is left hot and rarefied. It is so hot, in fact, that not only is all the hydrogen ionized, but also very highly ionized species such as O VI are formed. The supernova shells are large structures, a few of them hundreds of parsecs (up to 3 kpc) in diameter (Figure 19–18). As these sweep through the interstellar gas, they heat it to about 50,000 K and thin it out to a low density. In the process, they distort and destroy existing cool interstellar clouds. These large, expanding shells may power the evolution of a large fraction of the interstellar gas.

Consider the interstellar medium as containing several major components, which may evolve from one to another as the imbedded stars go through their various life stages:

1. *H II regions.* Zones of glowing, ionized hydrogen surrounding young, hot stars (spectral types O and B); contain a minor amount of the interstellar gas, perhaps 10 million solar masses total in the Galaxy; temperature about 10^4 K; density about 5×10^3 ions/m^3.
2. *H I regions/diffuse neutral clouds.* Clouds of cool, neutral hydrogen roughly 5 pc in diameter and each containing about 50 solar masses of material; total mass in Galaxy may be 3 billion solar masses; temperature about 100 K; density about 5×10^7 atoms/m^3.
3. *Molecular clouds.* Small to huge, containing mostly molecular hydrogen (H_2); total mass of a

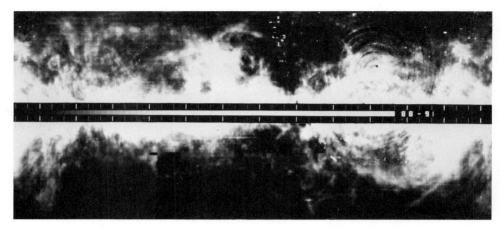

FIGURE 19–18   Filaments and shells of H I in the interstellar medium. This 21-cm map shows the large-scale structure of the interstellar gas, probably from supernova explosions. The bar through the center marks the galactic plane; the view extends 60° above and below the plane. (*C. Heiles*)

few billion solar masses; temperature as low as 10 K; density about 10^9 molecules/m^3 or greater. Although they occupy less than 1% of the interstellar space, they contain a substantial portion of the matter that constitutes the interstellar medium. Stars form out of the dense molecular clouds, some fragments of which develop into H II regions.

4. *Intercloud medium.*   A relatively hot gas composed largely of neutral hydrogen (and therefore observable at 21 cm) plus about 20% ionized gas, including electrons (observable in the radio continuum). This gas surrounds the cooler interstellar clouds and fills about 20% of the volume; temperature from 5000 to 10,000 K; density about 3×10^5 atoms/m^3 and 5×10^4 electrons/m^3.

5. *Coronal gas.*   A very hot (10^6 K), low-density ($<10^4$ particles/m^3), ionized gas that permeates the rest of interstellar space and occupies well over half of it, perhaps as much as 70%.

19–3 ◗
STAR FORMATION

We now turn to the interstellar medium as a star-forming factory to tie together some pieces of the observational picture. We divide the topic into massive (ten or more solar masses) and solar-mass star births. Massive protostars have greater luminosities than solar-mass ones, and once they reach the main sequence, massive stars ionize the gas around them. The ionized gas is detectable by radio telescopes. Since this action takes place cloaked by dust, only infrared and radio observations allow inspection of stellar wombs. Recall, that we call the general class of protostellar and pre-main-sequence stars, visible directly or not, *young stellar objects* (YSOs)

(A) BASIC PHYSICS

In Section 16–3(a), we showed the basics of gravitational collapse and the application to protostars—that is, the time scale for collapse. Here we will review the basic physics of star formation before we address the central issue: *angular momentum* and how a YSO sheds it.

Size Scale for Collapse
Let's apply the virial theorem, $2E_{\text{thermal}} = -U$, where

$$E_{\text{thermal}} = NkT$$

if N is the total number of particles in the cloud. We approximate the gravitational potential energy by

$$U \approx -GMm/R \approx -GM^2/L$$

where the mass attracts itself and L is the size scale

of the collapsing region. If the initial cloud is all molecular hydrogen, then $N = M/2m_H$, and the virial theorem gives

$$2(M/2m_H)kT \approx GM^2/L$$
$$kT/m_H \approx GM/L$$

but for a uniform spherical cloud, $M = (4\pi/3)\rho L^3$, so that

$$kT/m_H \approx G\rho L^2$$

or

$$L \approx (kT/mHG\rho)^{1/2} \propto (T/\rho)^{1/2}$$

If we evaluate the constants

$$L \approx 10^7 \, (T/\rho)^{1/2}$$

where the units are meters. For the coldest regions of a giant molecular cloud, $T \approx 10$ K and $\rho \approx 10^{-15}$ kg/m3, so $L \approx 10^{15}$ m or 0.1 pc. The mass within this volume is roughly $M \approx L^3\rho \approx 10^{30}$ kg $\approx$ solar mass. Hence the size and mass scales are of the right order of magnitude to form a star like the Sun.

Angular Momentum: Collapse with Rotation

Interstellar clouds rotate at least a bit. An isolated rotating cloud must conserve angular momentum. As the cloud collapses, each particle moves closer to the axis of rotation and the cloud must rotate faster. At some point, the rotational velocity may become so high that the centripetal acceleration v^2/r balances the gravitational force per unit mass, and the collapse halts. Note that it is the distance from the axis of rotation, not from the center of the cloud, that determines the angular momentum. A particle that starts out near the rotation axis can fall a large distance toward the center without changing its distance from the axis much at all, and it will be able to move much closer to the center before its rotational velocity is high enough to stop it. Those parts of the cloud originally near the rotation axis will collapse more than those near the equator, and the cloud will flatten into a disk.

Let's apply the conservation of angular momentum to the simple case of a cloud of radius 0.1 pc and mass of one solar mass (as above), with an initial equatorial speed of 1 km/s. We assume no mass loss, so that the initial and final masses are the same. Then

$$m_i v_i r_i = m_f v_f r_f$$

and

$$v_f = v_i(r_i/r_f)$$
$$= 1 \text{ km/s } (3 \times 10^{15} \text{ m/7} \times 10^8 \text{ m})$$
$$\approx 5 \times 10^6 \text{ km/s}$$

which is greater than the speed of light! Therefore, a protostar must shed angular momentum before it becomes a main-sequence star.

The addition of spin to theoretical models of protostar collapse makes the calculations much harder and the results less conclusive. One key point emerges: in some cases, a disk-like core of rings of material results. These rings turn out to be unstable in some instances, and they break up into several blobs (Figure 19–19). Sometimes these

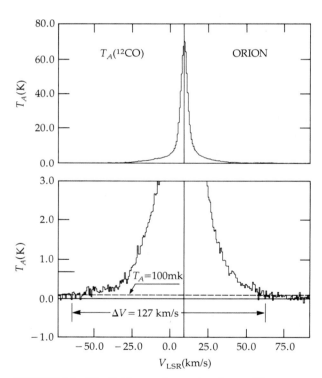

FIGURE 19–19 Evidence for molecular outflows. Observations of the ^{12}CO line from Orion; the vertical scale is the flux, given in units of antenna temperature (T_A). The horizontal scale is the velocity relative to the local standard of rest (V_{LSR}). The top part of the figure gives the complete line profile. The bottom part shows the base of the profile on an expanded scale. At a level of $T_A = 100$ mK, the full width of the line (ΔV) is 127 km/s. *(J. Bally and C. J. Lada)*

blobs coalesce into fewer ones; usually two or three remain. If each blob eventually becomes a star, we then have a natural explanation for the occurrence of multiple star systems.

Magnetic Fields

Now let's apply the conservation of magnetic flux to the collapsing star. We have the magnetic flux, Φ (SI unit: weber)

$$\Phi = \pi r^2 B$$

is conserved, so that the magnitudes are related by

$$B_f/B_i = (r_i/r_f)^2$$

Using the same parameters as before, we include that the average interstellar magnetic field has $B \approx 10^{-10}$ T, so

$$B_f/B_i = (3 \times 10^{15}/7 \times 10^8)^2$$
$$B_f = 10^{-10} \text{ T } (2 \times 10^{15}) \approx 10^5 \text{ T}$$

So we will have a highly magnetic disk at the end of the collapse. Hence, we could solve two problems if we can diffuse the magnetic field and reduce angular momentum together. The bottom line of the basic physics is that starbirth involves a magnetic disk around a YSO that could be dispersed by strong stellar winds collimated by the B-field.

Let's now examine the observational evidence for this concept—evidence that comes primarily from radio and infrared astronomy.

(B) MOLECULAR OUTFLOWS AND STAR BIRTH

Observations of molecules around YSOs have revealed high-speed (30 to 100 km/s) flows of gas. Doppler-shift measurements show that these flows tend to be bipolar: two streams moving in opposite directions—they are bipolar outflows. The bipolar flows carry considerable mass and can span a few parsecs; so enormous amounts of energy push them along. The exact source of that energy and the origin of the flows are a puzzle at the moment. Such outflows do appear to be associated with the birth of stars.

The flows show up as the expanded widths, from Doppler shifts, of molecular lines such as CO (Figure 19–19). The redshifted side is called the red wing; the blueshifted side, the blue wing. For the Orion Nebula, for instance, the total width of the CO line is 127 km/s (Figure 19–19). The total kinetic energy in such an outflow of molecular gas is about 10^{40} J; and the power is some $2600L_\odot$. Hence, these flows are very energetic and pump significant amounts of energy into the surrounding molecular cloud.

One of the cleanest examples of this collimated bipolar flow is associated with a dark cloud called L 1551, which is embedded in a large molecular cloud (Figure 19–20A). An infrared source named IRS 5 lies at the center of a bipolar CO outflow (Figure 19–20B); it has a luminosity a few tens that of the Sun. VLA observations reveal a radio source coincident with IRS 5 (Figure 19–20C). High-resolution radio maps show two jets (Figure 19–20D) aligned with the axis of the molecular bipolar flow. These observations, combined with CCD images, strongly suggest that a disk or torus surrounds the protostar to collimate the outflow (Figure 19–20E), which may arise from a very strong stellar wind from the YSO.

One model to explain the outflows envisions a YSO still accreting material around a central core. Surrounding the star is a dense disk or torus of gas and dust, which is rotating and highly magnetized (Figure 19–21). The gas is ionized out to large distances by UV radiation produced on the core. A hydrodynamic wind results, with the disk naturally channeling the flow of the stellar wind so that it streams out along the rotation axis of the disk, making two streams containing an ionized and a neutral flow. When these two streams push enough material outward, two opposing lobes of gas form as the flows strike the surrounding medium. The outflows can reach 10^{-4} to 10^{-6} $M_\odot/y$ and easily carry away angular momentum to brake the rotation in 10^5 y. The disks around massive YSOs might have densities on the order of $10^{14}/m^3$, rotational speeds of a few kilometers per second, and sizes of about a few parsecs. The disks act as flywheels that convert accretion energy into that of outflows, and so shed the original angular momentum of the system.

The discovery of these two-sided flows strongly hints that disks of material typically form around stars during their formation. It is from such disks that planetary systems might form. So we have a

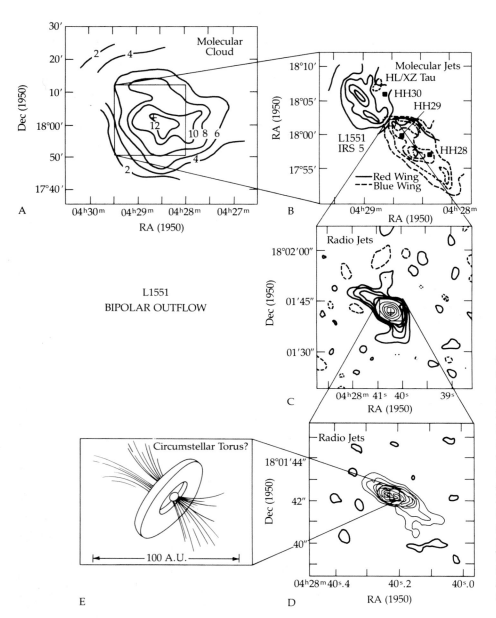

FIGURE 19–20 Bipolar outflow from L 1551, observed at a variety of wavelengths and size scales. (A) Overall emission from the surrounding molecular cloud. (B) The bipolar molecular jets, showing the blue and red wing of the outflows surround IRS 5. Also indicated are the Herbig–Haro objects HH 28, 29, and 30. (C) and (D) Radio continuum emission at better resolutions. (E) A model of the outflow based on a circumstellar torus. *(Adapted from a diagram by R. L. Snell, J. Bally, S. E. Strom, and K. M. Strom)*

clue that the nebular model for planetary formation might actually operate elsewhere in the Galaxy.

(C) THE BIRTH OF MASSIVE STARS

The birth of a massive star has specific hallmarks, in the radio and infrared. First, stars condense from molecular clouds (visible by their emission at millimeter wavelengths). Second, the free-fall col-

lapse, at its early phases, heats the dust to low temperatures, roughly 30 to 50 K. This dust emits infrared radiation that peaks at roughly 10 μm. Third, as the protostar forms, the temperature of the interior dust reaches about 1000 K and therefore it emits with a peak at 3 μm. The exterior dust is cooler, still about 100 K. So the spectrum shows the combination of two blackbody peaks, one near 3 μm and the other near 30 μm. Fourth, as the protostar reaches the main sequence, it ionizes the

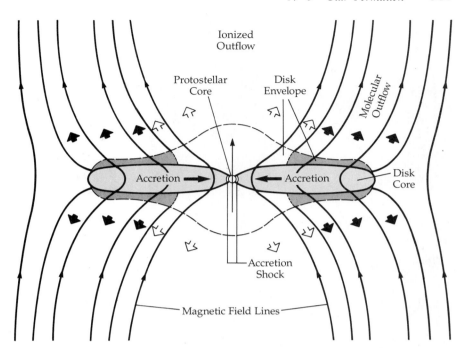

FIGURE 19–21 A model for bipolar outflows from a rotating, magnetized disk around a protostar. This view shows the central region of the disk surrounding the protostar core, which is still accreting material. The ionized part of the disk and surrounding material has magnetic field lines threaded through it. *(R. E. Pudritz)*

hydrogen gas and a compact H II region develops, most easily observable at microwave wavelengths. Fifth, the hot, ionized gas expands. The dust, pushed outward, is cooled; it emits with a peak in the far infrared, but less intensely. When the dust has been pretty much dissipated, the H II region emits a weak continuous spectrum of radio waves. Finally, the H II region expands enough to blow off its dusty cloak, and the star appears to optical view.

With this scenario in mind, consider star formation in the Orion region, a part of the Galaxy especially rich in GMCs. The one we will focus on is the denser part of the GMC behind the Orion Nebula; it is called OMC1. The whole cloud is very clumpy, each about a parsec in size and containing a few tens to hundreds solar masses of material. IRAS observations revealed the overall scene of star formation here (Figure 19–22).

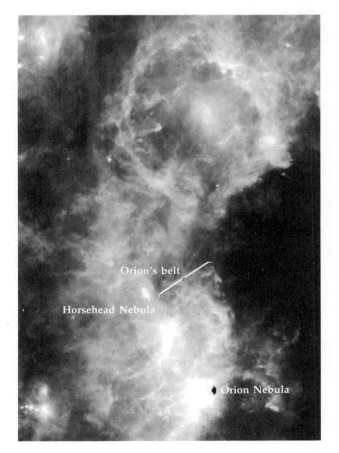

FIGURE 19–22 IRAS composite image of the star-forming regions in the neighborhood of Orion. The strongest regions of infrared emission coincide well with the densest regions of the molecular clouds. Note the strong emission from the Orion Nebula. Infrared cirrus is visible at the top. *(NASA)*

A

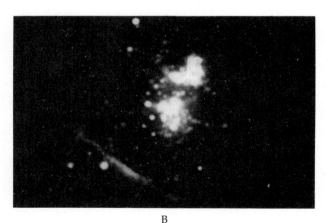

B

FIGURE 19–23 The Trapezium region of the Orion Nebula. (A) Short-exposure optical image, showing the core of the H II region in which the Trapezium cluster is embedded (arrow). *(Lick Observatory)* (B) A mosaic of 2.2-μm images centered on the optical Trapezium stars. Note the clustering of stars in the region; these make up the low-mass component of the Trapezium cluster. *(NASA)*

Let's focus in on the Orion Nebula (Figure 19–23A). The H II region around the Trapezium marks the oldest (most evolved) part of the region; it lies like a hot blister in front of OMC1, and gradually eats away at the molecular cloud. The Trapezium cluster consists of a few hundred stars, optically visible, within 0.3 pc of each other; it is the O and B stars of this group that ionize the gas. These massive stars are no more than 1 million years old; they are the youngest subgroup of the I Orion OB association. The low-mass part of the Trapezium Cluster has been revealed in near-infrared observations (Figure 19–23B). Here we find about 500 stars with masses from 0.5 to $2M_\odot$ and no more than 10^7 years old. In an evolutionary sense, the molecular cloud core (size ≈ 0.05 pc, mass $\approx$ several $M_\odot$) that lies behind the Orion Nebula is the youngest part of the region.

Infrared observations show a strong region emission that has two general parts (Figure 19–24A): the infrared cluster associated with the Becklin–Neugebauer (BN) object and the Kleinmann–Low (KL) nebula. High-resolution infrared observations indicate that at least ten sources lie in the cluster, with separations of a few thousand AUs. In particular, the KL nebula contains IRc2–5 (infrared compact sources 2–5) and surrounding infrared emission. Many maser sources also mark the region, which is the core of OMC1. The infrared emission is highly polarized (up to 50% at 3.8 μm), an indication of an infrared reflection nebula centered on IRc2. Molecular observations show both high and low-velocity outflows here, from 150 to 10 km/s. The flows shock H_2 in the surrounding gas, which then emits at 2 μm; this emission has been observed at the outskirts of the region. IRc2 appears to be the main source of the outflow.

Infrared astronomers have observed the BN object at an infrared line at 4.05 μm, the **Brackett alpha line,** which arises from a transition from $n = 5$ to 4 in hydrogen. This line can arise from recombination within a very small, newly formed H II region around a massive star approaching the main sequence. So the Becklin–Neugebauer object, embedded in the molecular cloud, is, in this interpretation, a B0 star ($\approx 10M_\odot$) just reaching the main sequence, with a circumstellar H II region. It is probably less than 10^6 years old. The hot gas

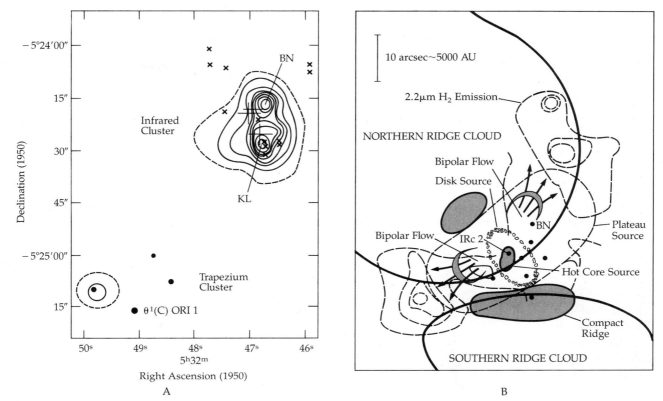

FIGURE 19–24   The Orion Nebula and star formation. (A) Overall view of the infrared emission from the Orion Nebula region. The infrared cluster has two strong
sources: the Kleinmann–Low (KL) source and the Becklin–Neugebauer (BN) object;
they lie in the core of OMC1. Crosses and ''x''-s indicate the positions of maser
sources. The Trapezium Cluster marks the visible part of the H II region. *(Adapted from
a diagram by E. Becklin, G. Neugebauer, and C. Wynn-Williams)* (B) A model of the star formation interactions in the core of OMC1. Powerful outflows have created a cavity in the
molecular cloud around IRc2 and BN. *(W. M. Irvine, P. F. Goldsmith, and A. Hjalmarsson)*

flows out at about 100 km/s at a rate of some
10^{-7} $M_\odot$/y.

Putting these observations into a coherent picture invites some speculation about the intricate
structure of OMC1; here's a working model (Figure 19–24B). An extended ridge cloud lies north
and south of the infrared sources; it has a temperature from 20–70 K and a mass as great as $1000 M_\odot$;
within these are several dense clumps—perhaps
incipient protostars. The energetic core region, the
plateau source, contains several infrared sources,
of which IRc2 appears to drive the powerful outflows. Along with the outflows from BN, the H_2

gas at the edges of the central cavity is excited by
shock fronts. A disk of material surrounds IRc2.
The hot core source may be a remnant from the
collapse that made IRc2 and BN. We see this complexity because OMC1 is so close to us that we can
resolve the different regions.

The situation in Orion and other regions where
H II regions abut giant molecular clouds supports
a scenario of sequential star formation within
them. Massive star formation begins at one end of
the giant molecular cloud (Figure 19–25). (Such
clouds tend to be elongated and cigar-shaped.) A
small group of about ten OB stars forms. They

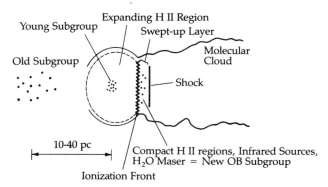

FIGURE 19–25 Sequential model of massive star formation from a giant molecular cloud. *(Adapted from a diagram by B. Elmegreen and C. Lada)*

evolve to the main sequence. Their ultraviolet radiation then dissociates hydrogen molecules around them and ionizes the gas. The H II region, since it is hot, expands, pushing a shock wave into the molecular cloud. The gas behind the shock wave is compressed to densities sufficient to start gravitational collapse. A new group of OB stars is born about 1 million years after the previous one. The process repeats. Small groups of massive stars are born in a sequence of bursts across the molecular cloud.

This model predicts that the fossil remnants of a molecular cloud will be a string of small groups of OB stars, 10 to 30 pc apart and more or less in the same space as the parent molecular cloud. Such loose groupings of OB stars, called **OB associations,** are very young. Since these stars do not live very long (a few tens of millions of years), the OB associations themselves cannot be more than 10^7 years old. Many OB associations, which are 30 to 200 pc across, consist of small clusters of stars called **OB subgroups.** These subgroups contain four to 20 stars (average about ten). Within an association, OB subgroups are lined up in an evolutionary sequence. The most spread out, oldest subgroup lies at one end and the most compact, youngest subgroup at the other. For example, Orion contains a large association that has four OB subgroups; the smallest and youngest subgroup is the Trapezium cluster. Recall that the Trapezium adjoins the southern Orion molecular cloud. Here we see the signposts of massive star formation.

Once star formation starts at one end of a molecular cloud, it propagates through the cloud in a chain reaction. But what starts the first burst of star formation? The answer to that question is not yet known. Perhaps it starts from the collision of molecular clouds or, more likely, from the blast wave of a supernova remnant crushing into an end of a molecular cloud. This idea is aesthetically appealing: a supernova signals the death of a massive star; then the mark of its death ignites the birth of other massive stars.

Note also in this model that giant molecular clouds do not last long once star birth begins—only a few tens of millions of years. Since we see many molecular clouds now, they must form rapidly in order to balance their rapid destruction. Where and how these clouds form most likely has to do with the spiral structure of our Galaxy (Chapter 20).

(D) THE BIRTH OF SOLAR-MASS STARS

The observational picture for the formation of stars like the Sun is skimpy. It does seem clear that, like massive stars, solar-mass stars are also born from molecular clouds. The questions are, in which clouds and how? To date, however, we have no well-confirmed observation of a solar-mass protostar—and not for lack of trying! And solar-mass stars may well form along with more massive ones, as we have seen in the Trapezium cluster. In this picture, most star birth takes place in dark, massive clouds, out of which OB associations form. Thus the Sun may have been born in an OB association, perhaps blasted clean by a supernova.

Solar-mass stars may form within **dark clouds,** which are interstellar clouds containing enough dust to blot out the light of stars within and behind them. These low-density molecular clouds typically have temperatures of 10 K, densities about 10^9 atoms/m^3, and masses from a few tens to a few hundreds of solar masses. Infrared observations reveal a few tens to about 100 candidates for solar-mass YSOs within well-examined dark clouds. The IRAS survey found many warm (70 to 200 K) sources in molecular clouds that may be solar-mass protostars.

The best candidates for solar-mass PMS stars are the T-Tauri stars [Section 18–3(a)]. These lie above the main sequence on an H–R diagram and have ages of 10^5 to 10^7 years; most range from 10^5 to 3×10^6 y. Recently, X-ray observations have identified a new class of T-Tauri stars called naked T-Tauri stars (NTTS), which have large and variable X-ray emission. These stars show a normal photospheric spectrum and active chromospheres. They may well represent the next stage in the evolution of a T-Tauri stars, as the disk around them clumps into smaller agglomerations; that is, the circumstellar environment changes.

From observations so far, we are beginning to suspect that, in general, stars are born from molecular clouds; massive stars from massive clouds and less massive stars (the majority of those in the Galaxy) from less massive clouds. In other words,

high- and low-mass star formation is bimodal in terms of the environment if not the mechanisms. In particular, "cold" (cores T < 10 K) clouds probably are the sites for low-mass stars, and "warm" (cores T > 20 K) clouds develop massive stars.

In summary, we can identify four main stages of star formation from molecular clouds:

1. Formation of a slowly rotating cloud core.
2. Collapse of the core into a protostar and disk, deep within an infalling envelope of material.
3. A stellar wind with collimated jets and a bipolar outflow.
4. Sweeping away of the circumstellar material to reveal the YSO; consolidation of the disk into planets or companion stars.

PROBLEMS

1. An A0 V star has an apparent visual magnitude of 12.5 and an apparent blue magnitude of 13.3.
   (a) What is the color excess for this star?
   (b) What is the visual absorption in front of this star?
   (c) Calculate the distance of the star (in parsecs).
   (d) What error would have been introduced if you had neglected interstellar absorption?

2. (a) How does interstellar reddening alter the Planck spectral energy curve of a star? Sketch approximate curves for an A star (10,000 K) in the visible part of the spectrum, both with and without reddening.
   (b) What effect would interstellar reddening have on the color–magnitude (H–R) diagram of a star cluster? Draw a diagram to support your answer.

3. What are the observational clues to the nature of the interstellar dust grains, and how have these clues been interpreted in terms of models?

4. Several galactic nebulae have been photographed in color. Two different types occur, reddish and bluish.
   (a) Explain the physics of these two types of nebulae.
   (b) Briefly describe the spectra you would expect to observe for these two types of nebulae.

5. Why are none of the hydrogen Balmer lines seen as interstellar absorption lines, even though hydrogen is the most abundant element in the Universe? (*Hint*: Recall the energy level diagram of hydrogen.)

6. (a) Name the two factors that primarily determine the size of an H II region.
   (b) Explain the physical basis for your answer to part (a) in terms of the Strömgren sphere.
   (c) The brightness of an H II region depends only on the gas density of the nebula. Explain this phenomenon in terms of what you know about the hydrogen atom. (*Hint*: Remember the Saha equation.)

7. Forbidden lines are observed both in the solar corona and in gaseous nebulae, but they are not the same forbidden lines.
   (a) How do the lines differ in the two cases?
   (b) Why are nebular forbidden lines not emitted by the solar corona? What about coronal forbidden lines for nebulae?

8. Somewhere in our Galaxy resides a cloud of neutral hydrogen gas with a radius of 10 pc. The gas density is 10^7 atoms/m^3.
   (a) How many 21-cm photons does the cloud emit every second?
   (b) If the cloud is 100 pc from the Sun, what is the energy flux of this radiation (in W/m^2) at the Sun?

9. What are the energies (in electron volts) of the photons that characterize
   (a) the Lyman continuum limit (91.2 nm)
   (b) the nebular line of [O III] at 500.7 nm
   (c) the neutral hydrogen line (21 cm)
   (d) the ammonia (NH_3) emission (1 cm)
   (e) the hydrogen Balmer line, H_α (656.3 nm)

10. Using Bohr's formula for the wavelengths of hydrogen lines, calculate the wavelength and frequency of the radio recombination line with upper level 93 and lower level 92.

11. A pure hydrogen gas cloud of number density $n = 10^7$ atoms/m^3 surrounds an O star that generates 10^{49} photons/s at wavelengths shorter than 91.2 nm. The rate at which such recombinations occur is $\alpha = (2 \times 10^{-19})n^2/m^3 \cdot s$.
    (a) Balance the number of ionizations with the number of recombinations to determine the Strömgren radius of the resultant H II region.
    (b) The Sun produces about 5×10^{23} photons per second with $\lambda < 91.2$ nm. Calculate its Strömgren radius in astronomical units for an interplanetary medium density of 10^9 atoms/m^3.

12. Approximate the Galaxy as a uniform disk of constant thickness. Show that the optical depth of extinction by interstellar dust should approximately obey the law $\tau_\lambda \propto \csc b$ (except for small b), where b is galactic latitude. At what galactic latitudes would an astronomer want to observe other galaxies without having to worry much about extinction?

13. An open cluster of stars is found to contain main-sequence O5 stars with observed color indices $(B - V)$ of 0.4. These O stars are observed to have apparent magnitudes of 10.0.
    (a) Use Table A4–3 to calculate the distance to the cluster. Include the effects of extinction by dust.
    (b) Determine the apparent magnitude of a G0 main-sequence star in the cluster.

14. Consider an extended region filled with neutral hydrogen gas at a density of nm_H and temperature T, where m_H is the mass of the hydrogen atom. Now consider a small spherical volume of radius R inside this region where the density is slightly higher than in the surrounding region.
    (a) Show that the radius of this volume must satisfy the condition $R \geq (3kT/2\pi Gnm_H^2)^{1/2}$ in order for gravitational collapse of the denser region to occur. (*Hint*: Obtain an expression for gravitational pressure by integrating Equation 16–1 over r.)
    (b) Obtain a lower limit for the mass of the collapsing volume.
    (c) For the general interstellar medium, we can take the approximate values $n = 10^6/m^3$ and $T = 100$ K. Obtain a numerical value for the mass limit in solar masses.
    (d) Is a collapsing region of the general interstellar medium likely to produce a single star like the Sun? If not, how would a region which *could* form a single, solar-type star differ from the general interstellar medium?

15. A star count is made on a photograph that contains a dark nebula. Ten times fewer stars are counted in front of the cloud than are found on a region away from the cloud with the same solid angle. For ease of calculation, assume that the dark cloud is completely opaque, that the stars are uniformly distributed in space, that all stars have absolute magnitudes of 5.0, and that the limiting magnitude of the photograph is 15.0. Calculate an approximate distance to the cloud.

16. There is indirect evidence in L 1551–IRS 5 for a roughly solar system size disk surrounding the protostar (Section 19–3, Figure 19–20). Consider the feasibility of directly observing a solar system size disk ($\approx$50 AU) at a distance of 150 pc by estimating the angular size of such a disk. Comment on your result.

17. Using Figure 19–20B and an assumed average outflow velocity of 50 km/s, estimate the age of the molecular outflow associated with L 1551 (L 1551 is 150 pc away).

18. Estimate the initial size of the cloud required to form a $50M_\odot$, $10M_\odot$, and $0.5M_\odot$ star. Compare your result for that of a $1M_\odot$ star and the size of the solar system.

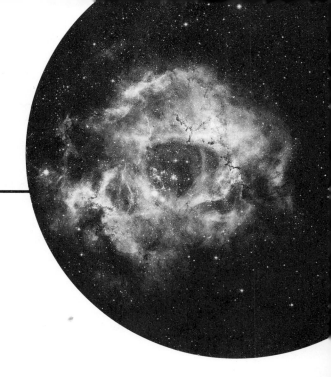

Chapter 20

The Evolution of Our Galaxy

We have so far described the physical characteristics of two main parts of the matter of the Galaxy: the stars and the interstellar medium. This chapter ties them together in the grand design of the structure of the Milky Way Galaxy. You will see that our Galaxy has a spiral layout, with much irregularity imposed on the overall pattern. This spiral imprint must evolve, and its evolution connects to the evolution of the stars, gas, and dust contained within.

20–1 ◖
THE STRUCTURE OF OUR GALAXY FROM RADIO STUDIES

To map out the Galaxy's structure requires a technique that distinguishes the paths of spiral arms from the bulk of the interstellar medium. The traditional mapping technique uses the 21-cm line of H I. As you will see, this technique has serious limitations. Recently, the mapping of molecular clouds in CO has complemented 21-cm observations. In fact, observational evidence so far indi-

cates that giant molecular clouds outline spiral arms more tightly than do H I clouds.

(A) 21-cm DATA AND THE SPIRAL STRUCTURE

The hyperfine transition from neutral hydrogen (H I) at the radio wavelength of 21 cm [Section 19–5(e)], combined with radial velocity variations from differential galactic rotation, allows us to deduce the **spiral-arm structure** in the galactic plane. If the galactic rotation curve is known (and that's a crucial *if!*), the distances to concentrations of neutral hydrogen may be found from the observed Doppler-shifted 21-cm line profiles. Implicit in these distance determinations are two assumptions: (1) differential galactic rotation and (2) circular galactic orbits for the gas near the galactic plane.

Because interstellar absorption is insignificant at the 21-cm wavelength, the line emission is observable throughout our Galaxy. Hence, we can probe galactic regions far beyond the solar neighbor-

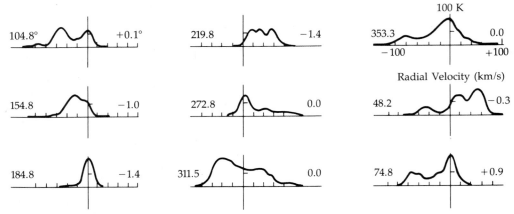

FIGURE 20–1 Line profiles at 21 cm in the galactic plane. These emissions come from regions spanning the galactic equator. Galactic longitude is indicated on the left of each profile; galactic latitude on the right. The flux is calibrated in units of the antenna temperature. *(F. J. Kerr and G. Westerhout)*

hood. The 21-cm line profile for a given line of sight exhibits several Doppler-shifted peaks that are fairly narrow and well defined. The concentration of hydrogen into spiral arms produces the observed forms of the line profiles (Figure 20–1). Line profiles seen near the galactic longitude l have Doppler peaks that shift in a concerted fashion as l varies. Each peak characterizes a spiral arm intersected by the line of sight. If we interpret the Doppler shift in terms of the radial velocity of that section of the arm and apply the rotation formulas of Chapter 15, we find the distance to the arm; since we have assumed circular orbits, the distance is uncertain to the extent that asymmetric motions occur. More realistic results are obtained by including modifications to circular motions.

Let's illustrate this procedure by interpreting one 21-cm profile (Figure 20–2). This profile corresponds to the line of sight at $l = 48°$, and it consists of three Doppler peaks at radial velocities of 55, 15, and −50 km/s. If we denote the Sun's distance from the galactic center by R_0 and note that positive radial velocities correspond to recession, then two of the hydrogen clouds are receding and one is approaching. From the rotation formulas (Section 15–4), we find that recession corresponds to $R < R_0$ and that an approaching cloud must lie at $R > R_0$. Remembering the approximate nature of Equation 15–16, we compute the maximum radial velocity as 57 km/s; therefore cloud A, with a radial velocity of 55 km/s, lies very close to the tangent point. Cloud D, which is then approaching us

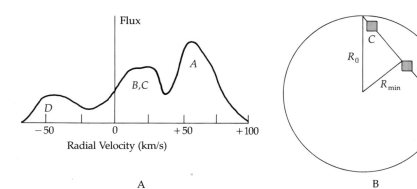

FIGURE 20–2 Line profiles and Doppler shifts. (A) Line profiles from a number of H I clouds at longitude 48°. (B) The line-of-sight geometry for the profiles in A.

(−50 km/s), lies at $R > R_0$, as indicated in Figure 20–2. There is a hint of double structure (clouds B and C) in the 15-km/s peak; latitude scans imply that cloud B lies beyond the tangent point (near $R = R_0$) and cloud C is very close to the Sun.

By combining 21-cm data from both the northern and southern hemispheres, we can construct a schematic picture of the neutral hydrogen distribution in the spiral arms of our Galaxy (Figure 20–3). The spiral structure is poorly determined near $l = 0°$ and $l = 180°$—that is, toward the galactic center and in the diametrically opposed (**anticenter**) direction. Circularly orbiting hydrogen clouds in those directions should exhibit no radial velocity; hence we cannot determine the distances to such clouds, for the line profile is a single peak at 21 cm. A distance ambiguity exists for hydrogen clouds that are closer to the galactic center than is the Sun. Because the maximum radial velocity of recession occurs when the line of sight passes closest to the galactic center (tangent point), a cloud closer to the

Sun than the tangent point may have the same (lower) recession speed as a cloud beyond the tangent point. The ambiguity is difficult to resolve with certainty.

(B) THE GALACTIC DISTRIBUTION OF GAS

Neutral hydrogen is concentrated in the galactic plane. If we define the thickness of the gas layer as the distance from the galactic plane to the half-density point (where the number density falls to half the value found at the galactic plane, or galactic latitude $b = 0°$), then the thickness of the hydrogen layer is observed to range from 80 to 250 pc. The smaller value refers to the region between the Sun and the galactic center. The thickness increases to 250 pc at the spiral arms near the Sun ($R = R_0$), flares out to several hundred parsecs for $R > R_0$, and reaches almost 2 kpc at $R = 30$ kpc. The galactic latitude distribution of neutral hydro-

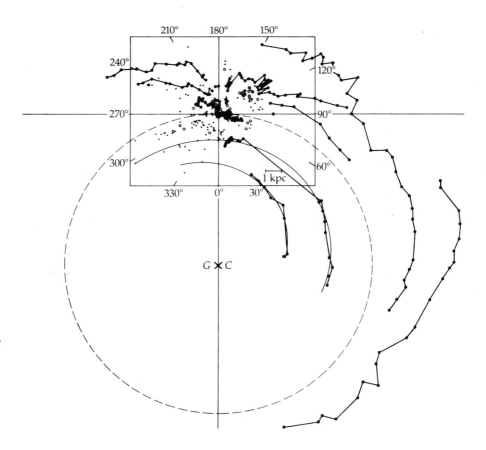

FIGURE 20–3 Galactic structure from 21-cm observations. The lines connecting the dots are from radio data; these trace out a spiral-arm structure. The unconnected dots within the box are optical spiral-arm tracers, such as H II regions. *(Adapted from a diagram by H. Weaver)*

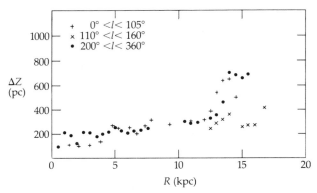

FIGURE 20–4 Warping of gas in the Galaxy's disk. From 21-cm observations, we see that the gas spreads out near the edge of the disk to a height of about 600 pc. *(Adapted from a diagram by H. Van Worden)*

gen (Figure 20–4) has the gas layer very flat (near $b = 0°$) in the region $R < R_0$, while beyond $R = R_0$, the layer is bent in opposite directions (relative to $b = 0°$) at $l = 180°$. This is the **galactic warp,** with a peak-to-peak displacement of some 3 kpc.

Another way to view the H I distribution is to examine its volume density. Outside the solar cir-

cle ($R = R_0$), the density peaks between 12 and 14 kpc and then drops rapidly beyond $R = 20$ kpc (Figure 20–5A). Within the solar circle, the distribution is constant from $R = 4$ kpc. Most of the H I gas lies outside the solar circle—at least 80% of the total gas mass. The CO distribution does not in general follow that for H I. Observations indicate that the CO is densest 6 kpc from the Galaxy's center (Figure 20–5B). Outward from 8 kpc, the density of CO drops but the H I density stays roughly the same. Inside 4 kpc, the CO density also decreases but not as rapidly. The CO layer has a thickness of 125 pc. Remember, the CO indicates the presence of H_2, molecular hydrogen. Within the Sun's distance from the Galaxy's center, about 93% of the hydrogen exists as H_2. In contrast, outside of the solar circle, the hydrogen takes the form of H I.

(C) THE GALACTIC CENTER REGION

If the gas motions were perfectly circular, there would be no observed radial velocities (Doppler-shifted line profiles) near $l = 0°$. The distances to such gas clouds would be indeterminate because the expected 21-cm line profile is a single intense peak at zero radial velocity. Nevertheless, the gas

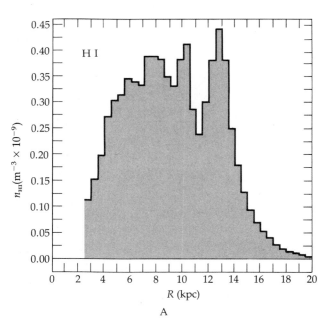

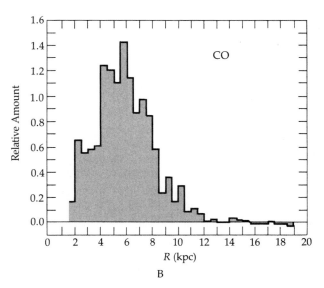

FIGURE 20–5 Gas distribution in the Galaxy's disk. (A) Volume density of H I as a function of distance. (B) Relative density distribution of CO. *(From a diagram by M. Gordon and B. Burton)*

structure in the region may be approximated by extrapolating the features seen at longitudes flanking $l = 0°$.

The 21-cm line profiles observed within a few degrees of the galactic center are extremely complex, largely because of peculiar geometric and velocity perturbations. The major feature seen is a sharp peak at -50 km/s, which reveals material moving outward from the galactic center; this makes up the **expanding,** or **3-kpc, arm.** Within 2000 pc of the center lies a thin gaseous disk tilted 40° with respect to the galactic plane. The disk has a thickness of 100 pc and a maximum rotational velocity of 360 km/s. It contains 10^7 solar masses of H I and has a central bar. In one respect, the molecular distribution here follows that of H I; it shows the same tilted disk structure in the innermost part of the Galaxy. The disk contains 10^9 to 10^{10} $M_\odot$ of H_2. The total mass, determined from the rotation of matter outside the disk, is only a few times 10^{10} $M_\odot$. So this inner part of the Galaxy may be exceptionally rich in gas, rather than mostly stars.

(D) HIGH-VELOCITY HYDROGEN CLOUDS

Chapter 15 showed that stars that move in higher eccentric orbits about the galactic center appear to us as high-velocity stars. The 21-cm profiles at high galactic latitudes ($|b| \geq 10°$) reveal hydrogen structures far from the galactic plane. This gas is distinctly distributed in the form of discrete clouds that fall into three velocity categories: (a) the **high-velocity clouds** with velocities (relative to the Sun) of 70 to 120 km/s, (b) the **intermediate-velocity clouds** with velocities in the range 30 to 70 km/s; and (c) the **very-high-velocity clouds** with velocities greater than 120 km/s in the negative galactic latitudes between $l = 0°$ and $l = 180°$. Nearly all of these clouds exhibit radial velocities of approach (negative), and they appear to be concentrated near $l = 120°$ and $b = 40°$ (they occur in the regions $l = 60$ to 200° and $b = 10$ to 80°).

Interpretations of this fast-moving gas have been the subjects of intense debate, which has yet to be resolved. The most prominent feature of the high-velocity H I is called the **Magellanic Stream.** It envelops the Magellanic Clouds (two companion galaxies to the Milky Way; Chapter 23) and runs as a long filament close to the southern galactic pole

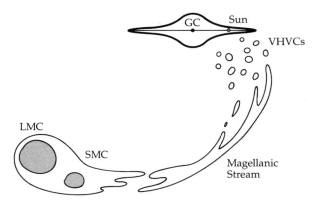

FIGURE 20–6 Schematic drawing of very high velocity clouds (VHVCs). The Magellanic Stream forms a bridge between the LMC and the SMC; it breaks up towards its tip. This region, close to the Galaxy, forms the VHVCs. (*Adapted from a diagram by R. J. Cohen*)

and then into the galactic plane near $l = 90°$. About 30° from the pole, the streams break into cloud fragments—the tip of which may appear as the high-velocity clouds mentioned earlier (Figure 20–6). Most models view the stream as a thread of gas pulled from the Magellanic Clouds by a tidal interaction with our Galaxy.

20–2 ◗
THE DISTRIBUTION OF STARS AND GAS IN OUR GALAXY

We can distinguish populations in the Galaxy in terms of ages, metal abundances, and kinematics. We will stick with the basic division between Population I and II, but keep in mind that this approach is too simple. Unfortunately, the traditional subdivisions are inconsistent for historic reasons. You can find them in Table 20–1 for reference. The astronomical classification of stellar populations is currently undergoing a revolution. We guess that most astronomers use the traditional scheme as given in Table 20–1. We try in this text to move away from it in a gradual manner. The key point is that *metallicity relates to age*—younger objects have a higher metal abundance. The confusion arises because objects inhabiting *similar* parts of the Galaxy are now known to have *different* metallicities.

The metal abundance of stars is an indicator of the degree of heavy-element enrichment of the gas out of which they formed. This metal abundance is

TABLE 20–1 *Characteristics of Traditional Stellar Populations*

	Population Group				
	Extreme Population I	Older Population I	Disk Population II	Intermediate Population II	Halo Population II
Typical objects	Interstellar dust and gas O and B stars Supergiants T Tauri stars Young open clusters Classical Cepheids O associations H II regions	Sun Strong-line stars A stars Me dwarfs Giants Older open clusters	Weak-line stars Planetary nebulae Galactic bulge Novae RR Lyrae stars ($P < 0.4$ day)	High-velocity stars ($Z > 30$ km/s) Long-period variables ($P < 250$ days)	Globular clusters Extremely metal-poor stars (subdwarfs) RR Lyrae stars ($P > 0.4$ day) Population II Cepheids
Characteristics					
$\langle |z| \rangle$, pc	120	160	400	700	2000
$\langle |Z| \rangle$, km/s	8	10	17	25	75
Distribution	Extremely patchy in spiral arms	Patchy	Smooth	Smooth	Smooth
Age (10^9 years)	<0.1	0.1 to 10	3 to 10	≈10	≥10
Brightest stars (M_{vis})	−8	−5	−3	−3	−3
Concentration to galactic center	None	Little	Considerable	Strong	Strong
Galactic orbits	Circular	Almost circular	Slightly eccentric	Eccentric	Highly eccentric

not a smooth function of population, although this was thought to be the case earlier. The stars called Extreme Halo Population II (Table 20–1) do have a very low metal abundance, as do some intermediate Population II stars, but stars in the nucleus and some disk stars are metal-rich, perhaps because of an accelerated birth rate of massive stars early in the life of the Galaxy. ''Metal-rich'' is a relative term. The metal abundance of ordinary Population II stars is about 1% of the Sun's, and for metal-rich Population II stars, this value may go up to 10% or higher.

Recall that we determine ages from isochrones calculated by stellar interior models, and metal abundances from high-dispersion spectroscopy and stellar atmosphere models. Most modern measurements of metallicity are given in terms of the ratio of iron relative to hydrogen compared to Fe/H for the Sun:

$$[\text{Fe/H}] = \log (N_{Fe}/N_H) - \log (N_{Fe}/N_H)_{solar}$$

(A) SPIRAL ARMS: SPIRAL TRACERS

Neutral hydrogen and carbon monoxide emissions enable astronomers to trace out roughly the spiral-arm structure over a large portion of our Galaxy, but spiral arms also include all other objects of a young, metal-rich **Population I.** Because gas and dust are found together, both occur in the spiral arms of our Galaxy. The young T-Tauri stars are usually surrounded by the gas and dust from which they formed. These stars are not very luminous, and even when they occur in groups, they cannot be seen to very great distances. We know that they are good examples of Population I, but they are not good spiral-arm tracers because they have low luminosities. Because early-type stars are Population I, the young **open clusters** that contain these stars are also possible spiral tracers. In these clusters, few stars have evolved away from the upper main sequence.

Massive stars evolve very rapidly and may be seen as main-sequence OB stars; they, too, may still be enveloped by the gas and dust from which they formed. The ultraviolet radiation from such stars ionizes the surrounding gas, rendering it visible as an H II region [Section 19–2(b)]. *H II regions* are usually luminous, for they characteristically include several bright, hot stars whose ultraviolet radiation is converted to visible light. H II regions and the OB associations exciting them are Population I objects and are tracers of spiral structure (Figure 20–7A). The radio recombination lines [Section 19–2(b)] from H II regions allow us to determine their radial velocities and also their distances. Such velocities can then be used in the same manner as the 21-cm line velocities to delineate the spiral structure, making possible a more complete comparison between the distributions of H I and H II regions. The spiral-arm patterns delineated by H I and H II are roughly similar, but a marked difference appears in the large-scale distribution, for H II attains its greatest concentration closer to the galactic center than does the neutral hydrogen.

The best current tracers of spiral arms are the GMCs, from which massive stars are born. Their millimeter-line emission is Doppler-shifted by radial motions, and so their locations can be determined in a way similar to that of H I. Surveys have not yet completed scanning the whole Galaxy, but the results to date strongly delineate some of the Galaxy's major spiral arms (Figure 20–7B).

(B) STELLAR POPULATIONS: GALACTIC DISK AND HALO

Our Sun belongs to an old, metal-rich Population I. Although they are not strictly confined to the spiral arms, old Population I objects still lie fairly close to the galactic plane and have an inhomogeneous distribution throughout the Galaxy; they include clusters whose upper main-sequence stars have evolved to the giant and variable evolutionary stage. The older open clusters are prototypes of this population. These objects (along with gas) make up the Galaxy's *thin disk*.

Metal-rich ($Z \approx 25\%$ solar) Population II stars constitute most of the total mass of our Galaxy, and they form the *thick disk*. Representative disk stars may lie quite far from the galactic plane. The quantity $\langle |z| \rangle$ is the mean stellar distance from the

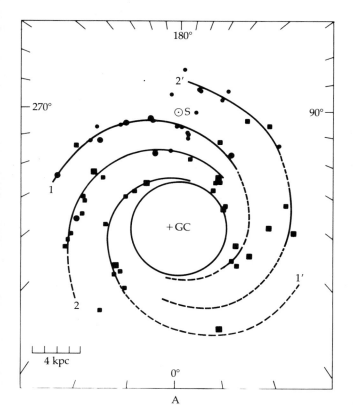

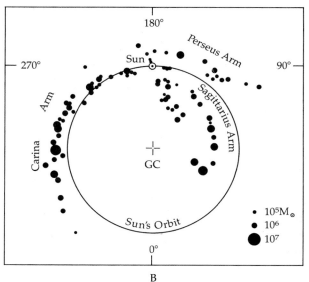

FIGURE 20–7 Spiral arms in the Galaxy. (A) Spiral structure based on Population I spiral tracers. Indicated are the positions of H II regions from optical (circles) and radio (squares) observations. (*Adapted from a diagram by Y. M. Georgelin, Y. P. Georgelin, and J.-P. Sivan*) (B) Spiral arms based on CO observations of GMCs. The mass of each cloud is indicated by the size of the symbol. Note three well-defined arms: Carina, Perseus, and Sagittarius. (*Based on observations by R. S. Cohen, D. A. Grabelsky, J. May, L. Bronfman, H. Alvarez, and P. Thaddeus*)

plane, and the related parameter $\langle|Z|\rangle$ is the mean of the stellar velocity components perpendicular to the galactic plane. The greater the perpendicular velocity component of a star, the greater the likelihood of that star's being far from the galactic plane. For thick disk stars, $\langle|z|\rangle$ ranges from 400 to 800 pc; $\langle|Z|\rangle$ from 20 to 30 km/s. Stars with [Fe/H] = -0.6 can reach as high as 2 kpc about the plane. These objects are distributed fairly smoothly and show no spiral structure.

Enveloping the disk and spiral arms is the **galactic halo,** which extends far above the plane but is still concentrated at the galactic center. Objects that lie in this domain are old, metal-poor and metal-rich **Population II;** these include the globular clusters, field stars, and RR Lyrae stars. The orbits of these objects are highly eccentric, with large velocity components perpendicular to the galactic plane. For instance, the many metal-poor ([Fe/H] < -1) globular clusters form a sphere around the Galaxy's center. Their elliptical orbits bring them out to extreme distances of 10 to 12 kpc from the Galaxy's nucleus. The clusters orbit at speeds of about 100 to 150 km/s, diving into and shooting out of the disk. These passages have helped to wipe globular clusters clear of any gas and dust they once had. In contrast, the metal-rich ([Fe/H] > -1) globulars make up a disk-like distribution with a scale height of ≈ 1 kpc. Overall, the metal-poor stars in the halo have a distribution that is a flattened oblate spheroid with a ratio of major to minor axis of 0.6.

The halo also contains gas, but much less than the disk. The neutral hydrogen shows up as the high-velocity clouds. Otherwise, the halo gas is ionized and extends to greater distances above and below the galactic plane than does the H I. We have observed the ionized gas in two forms: hot (80,000 K) and cool (10,000 to 20,000 K). The hot component has its greatest density at z from 1 to 3 kpc and can be seen as far out as 10 kpc. In contrast, the cool component has been seen only up to 2 kpc. The cool component contains about ten times more mass than the hot, but together they amount to only 1 or 2% of the gaseous mass in the disk.

The halo must also contain other dark or faint objects, currently unobservable, making up the **dark halo.** This invisible component to the halo is needed to give the Galaxy its flattened rotation

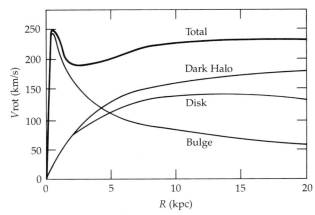

FIGURE 20–8 A model of the mass distribution of the Galaxy, as represented by rotation curves for each component. The Sun's distance is assumed to be 8.5 kpc. The line marked "Total" should match the observed rotation curve. *(Adapted from a diagram by P. C. van der Kruit)*

curve (Figure 20–8) at large distances from the Sun. The total mass in the dark halo may be $2 \times 10^{11}\ M_\odot$. Low-mass main-sequence stars would be very hard to detect. Smaller objects similar to planets and asteroids may also exist. Some astronomers have even suggested numerous low-mass black holes. Recent observations of the rotation curves of other spiral galaxies imply that they have extensive, massive halos of dark matter.

(C) THE CENTRAL BULGE AND GALACTIC NUCLEUS

Wide-angle pictures of our Galaxy (look back at Figure 14–1) show that it has a **central bulge** similar to that of other spiral galaxies. This bulge, which is about 2 kpc in radius, contains a mixture of heavy-element-enriched stars, especially late-type M giants, ordinary Population I K giants, and a few metal-rich RR Lyrae stars. IRAS observations at 12 μm indicate that strong sources here are AGB stars. The bulge has a total mass of some $10^{10}\ M_\odot$ (Figure 20–8).

At the very center of the bulge is the **galactic nucleus,** which is analogous to the stellar-like nucleus in M 31, the Andromeda galaxy. Spectra of the Andromeda galaxy's nucleus show that it, too,

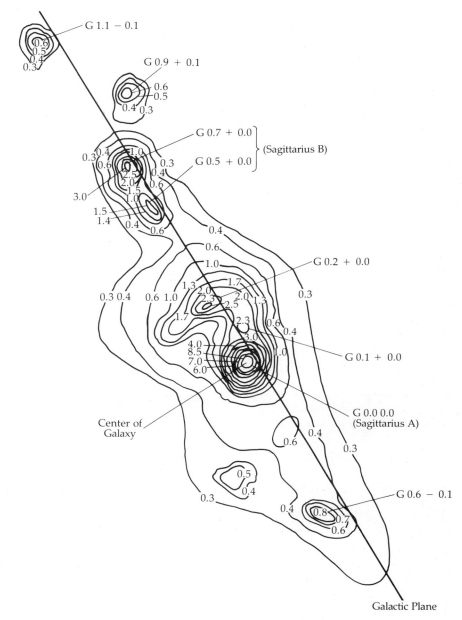

FIGURE 20–9 Radio map of the overall galactic center region; the two main regions are Sgr A and Sgr B. The observations were made at a wavelength of 3.75 cm. Most of this emission is thermal. *(Adapted from a diagram by D. Downes, A. Maxwell, and M. L. Meeks)*

consists of metal-rich giants as well as a large number of low-mass dwarfs. Infrared observations confirm that such stars reside in the Galaxy's nucleus; these and radio data also suggest the presence of very young M and O supergiants. Gas exists as molecular clouds and H II regions. Motions of gas here suggest a high concentration of mass at the center, perhaps a black hole. In the very center of the Galaxy lies a compact radio source less the 140 AU in diameter.

The continuous radio emission from the nucleus shows that an intense radio source lies right in the direction to the center, called **Sagittarius A (Sgr A)**. Clustered around Sgr A and lying more or less along the galactic equator is a string of radio sources (Figure 20–9). When investigated at different radio wavelengths, these sources appear to have characteristics of H II regions. The total extent of this region is about 90 by 260 pc; and the ultraviolet energy output from the OB stars needed to keep the region ionized is at least 2×10^{33} W.

Sgr A consists of a complex of sources: East, West, and A*. **Sgr A East** is a source of nonthermal radiation, usually interpreted as a supernova remnant. **Sgr A West** has a radio spectrum like an H II region. Within Sgr A West lies a nonthermal point-like radio source less than 0.1 arcsec in diameter that may well mark the core of the Galaxy; this is **Sgr A***. The ionized gas here, which amounts to a few million solar masses of material, seems to be rotating at about a few hundred kilometers per second. The thermal emission of Sgr A West, from the inner 3 pc of the Galaxy, shows that the bulk of the radio emission lies along a ridge-like source at the galactic center (Figure 20–10) with a peculiar spiral-like structure. Radio hydrogen recombination lines from the center are very broad. One physical process that commonly broadens spectral lines is the Doppler shift resulting from the rotation of a mass of gas. If Doppler-broadened, the line widths imply rotational speeds of 150 km/s at

a distance of 2 pc from the Galaxy's center. CO observations indicate that this molecular cloud may contain as much as a million solar masses of material.

The galactic center region emits strongly at 2.2 μm from the combined 2.2-μm emissions from all the old Population I stars (probably mostly from K giants) that inhabit the galactic nucleus. The region just around Sgr A is packed with 2.2-μm sources; this infrared cluster coincides with the ridge of radio continuum emission. One near-infrared source, called IRS 16, may be the actual center, rather than Sgr A*. Observations of the same region around 10 μm (Figure 20–11) show the infrared emission from dust that is heated by the radiation from old Population I stars and from high-luminosity O stars; the condensations in the 10-μm map are probably the locations of newly formed O stars. These regions have diameters of less than a few parsecs, the same size as compact H II regions. The combined luminosity from them in the range from 2 to 20 μm is roughly 10^6 $L_\odot$. Far-infrared observations have found that Sgr A emits more intensely at 40 to 300 μm than at 10 μm, about 10^8 $L_\odot$. This emission, too, probably comes from dust heated by OB stars.

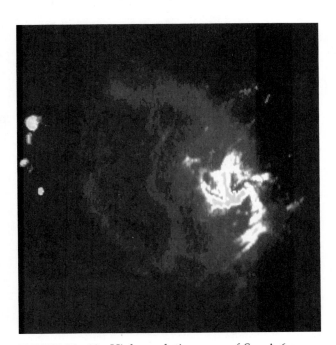

FIGURE 20–10 High-resolution map of Sgr A 6 cm and 20 cm by the VLA. The combination of the two wavelengths brings out details (resolution 1.3 by 2.5 arcsec) and also diffuse emission. Note the spiral shape of Sgr A West, the thermal source, and the shell-like structure of Sgr A East, the nonthermal source. *(W. M. Goss, R. D. Ekers, J. H. van Gorkom, and U. J. Schwarz; NRAO)*

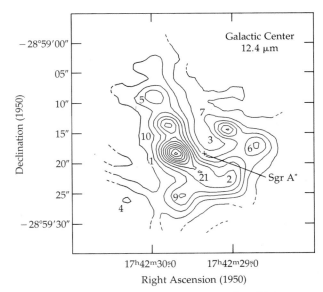

FIGURE 20–11 Galactic center mapped at 12.4 μm. Each individual source is numbered; note that none coincides with the position of Sgr A*. Resolution is 1.3 by 2.3 arcsec. *(D. Y. Gezari, R. Tresch-Fienberg, G. G. Fazio, W. F. Hoffmann, I. Gatley, G. Lamb, P. Shu, and C. McCreight; NASA)*

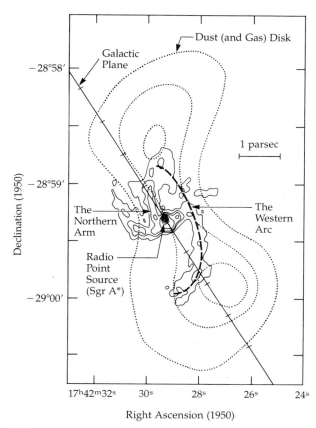

FIGURE 20–12 Schematic diagram of the gas and dust in the central 10 pc of the galactic center. The solid contours show the 2-cm thermal continuum. The dotted contours are those of far-infrared emission from warm dust. The galactic plane has tick marks every parsec. The position of Sgr A* is indicated within Sgr A West. (*Adapted from a diagram by M. K. Crawford, R. Genzel, A. I. Harris, D. T. Jaffe, J. H. Lacy, J. B. Lugten, E. Serabyn, and C. H. Townes*)

The infrared emission line at 12.8 μm produced by [Ne II] comes from ionized gas in the nucleus, and as for radio recombination lines, information about the core can be inferred from the Doppler shift of the line. These observations show that the ionized gas concentrates in a region about 1.5 pc in radius. The Doppler shifts indicate rotational motions of around 200 km/s. These ionized zones are small, less than 0.5 pc in diameter, and contain a few solar masses of ionized material. They appear to orbit around the galactic center on an axis tilted 45° to the main rotational axis of the Galaxy. So the galactic core contains, within the Sgr A molecular cloud complex, a disk of rotating ionized gas (Figure 20–12).

The Einstein X-Ray Observatory has made an X-ray image of the galactic center region. It shows

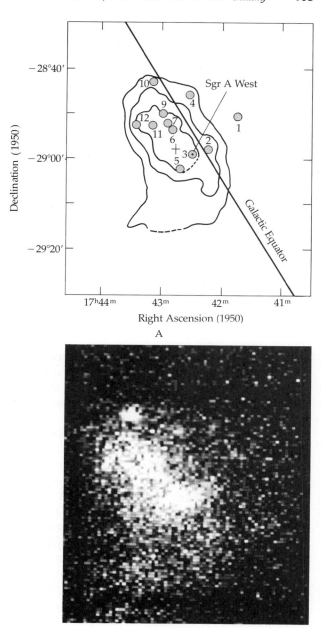

FIGURE 20–13 X-ray emission from the galactic center. (A) This schematic map shows discrete sources (circles) and diffuse emission (contour lines). The cross marks the center of the diffuse emission. (B) A computer-enhanced image of the X-ray emission. [*M. Watson, R. Willingale, J. Grindley, and P. Hertz, Astrophysical Journal 250:142 (1981)*]

modest X-ray emission at wavelengths below 0.6 nm within 100 pc of the galactic center. This emission consists of a complex of weak sources (10^{27} to 10^{28} W) embedded in a weaker halo of diffuse X-ray emission (Figure 20–13). The discrete sources lie along a ridge just south of the galactic equator, in the same location as the cluster of infrared sources. The diffuse emission comes from hot, coronal gas.

Finally, gamma rays from the nucleus have been detected. In particular, a line has been observed with an energy of 511 keV and a luminosity of 10^{31} W, which is attributed to electron–positron annihilation. The source changes luminosity over time leads. The emission appears to come from a source that has a diameter of less than 0.3 pc and is located close to but not at the galactic center.

(D) THE DISTRIBUTION OF MASS IN THE GALAXY

Keep in mind that the neutral and ionized hydrogen gas represents only a small fraction of the total mass of the Galaxy—only 5 to 10%. On the average, approximately 1% of all the hydrogen is ionized, but the percentage varies with distance from the galactic center. For instance, there is a ring of ionized gas beyond the galactic bulge at about 4 kpc, and in this region the concentration of ionized hydrogen is about 10 percent.

The total mass of the Galaxy can be determined, to a first approximation, on the assumption that the Sun moves in a circular Keplerian orbit about a point mass; such calculations lead to a value of $1.5 \times 10^{11}\ M_\odot$. However, recent rotation curves show a rise in the rotational velocity to 300 km/s at 20 kpc. This rotation curve implies a mass of $3.4 \times 10^{11}\ M_\odot$ interior to 20 kpc. So at least as much mass lies exterior to the solar circle as interior to it. The Galaxy has a massive halo of nonluminous matter, the form of which is currently unknown.

The nucleus may hold the densest concentration of mass in the Galaxy (Figure 20–14). Recall that radio and infrared line observations show rapid rotational motions near the Galaxy's core. The rotational velocities increase closer to the core, where they are so high that a huge mass is needed to hold all that rapidly moving gas together. To account for the rapid rotation requires a mass in the core of a few million solar masses—all lumped together in a region only 0.04 pc in diameter.

An approximate calculation shows this point. The infrared line observations imply rotational velocities of about 200 km/s at a radius of 10^{16} m.

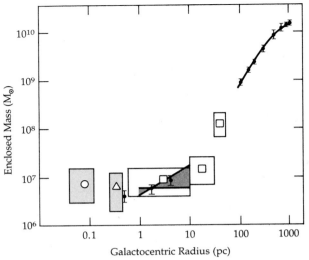

FIGURE 20–14 Mass distribution in the center of the Galaxy, under the assumption of circular rotation. The filled circle at 0.5 pc is for the Northern Arm. The large open circle (within the error box) is an estimate from the width of the He I 2-μm line. Other symbols (outward from center) are: triangle, from [Ne II]; filled circle, highest velocity [Ne II]; filled circles from other line observations, such as [O I], [C II], and CO. Filled circles at distances greater than 100 pc are from 21-cm rotation curve observations. *(Adapted from a diagram by J. B. Lugten, R. Genzel, M. K. Crawford, and C. H. Townes)*

Apply the argument used for pulsars, in which a rotating spherical mass just holds together by it own gravity:

$$V_{eq} = (GM/R)^{1/2}$$

where V_{eq} is the velocity of the object's equator. Solve this equation for the mass:

$$M = RV_{eq}^2/G$$
$$= (10^{16}\ \text{m})(200 \times 10^3\ \text{m/s})^2/$$
$$(7 \times 10^{-11}\ \text{N} \cdot \text{m}^2/\text{kg}^2)$$
$$= 6 \times 10^{36}\ \text{kg} = 3 \times 10^6\ \text{solar masses}$$

What form might this mass have? One possibility is that it is locked up in a central point mass (a supermassive black hole) of 2 to $4 \times 10^6\ M_\odot$. Or, it may be tied up in a very dense star cluster within 2 pc, with about $10^6\ M_\odot$. Indirect support for a black hole comes from observations that a few other galaxies may have a similar mass concentration in their nuclear regions.

20–3 ◗
EVOLUTION OF THE GALAXY'S STRUCTURE

Why do galaxies such as ours exhibit a spiral structure? How stable are the spiral arms? Spiral structure may arise from a perturbation, so that some density irregularity is pulled into a spiral form by the differential galactic rotation. The difficulty with this explanation is that such a feature is expected to persist for only a short time (about 5×10^8 years) before being pulled apart again by the differential rotation. An additional problem is that the initial spiral-like distortion would not extend over the entire galaxy but would occupy only a small part of it. Yet we observe spiral structure throughout the entire plane of a galaxy, and such spiral galaxies are sufficiently common to suggest that they are in fact stable.

(A) DENSITY-WAVE MODEL AND SPIRAL STRUCTURE

One very promising approach to this problem is the **density-wave model** developed by C. C. Lin and Frank Shu. The spiral structure of a galaxy is regarded as a wave pattern resulting from gravitational instabilities. The *density wave* moves through the stellar and interstellar material as a configuration whose shape stays fixed. It rotates at a speed less than that of the material speed of galactic rotation. The presence of a density wave means that the distribution of mass is nonuniform; therefore, the gravitational potential varies over the galactic disk. Stars and gas are concentrated in the regions where the gravitational potential is low, and these mass concentrations in turn influence the orbits of other stars and clouds of gas.

The density wave is a self-sustained phenomenon and is stable. The spiral pattern produced by the density wave is not tied to the matter but instead moves through it. According to this idea, the angular velocity of the pattern may differ appreciably (by a factor of one-half) from that of the material. Hence, stars formed from the gas and dust concentrated at the arms eventually migrate out of the arms. Gas concentrated near the potential minimum largely defines the spiral arms; this is ever-changing gas, for some is consumed in star formation and some is ejected from stars by some mode of mass loss. Stars traveling in orbits that differ greatly from circular orbits come under rapidly varying gravitational attractions. Density waves certainly influence their motions, but not systematically, and therefore no structure will persist for these eccentric-orbit stars.

The density-wave model assumes that a two-armed, spiral density wave sweeps through the galactic plane, but the model does not yet explain the wave's origin or long persistence. The gas in the disk piles up at the back of the wave. The buildup of pressure and density heats up the gas suddenly so that a shock wave forms along the front of the density wave. This shock may initiate the collapse of the clouds to form giant molecular cloud complexes, which in turn form young stars and H II regions. Such squeezing also helps to make dust out of the gas, and a thin dust lane forms along the shock front. The compression of the interstellar medium by the density wave forms the features associated with a spiral arm. During the short lifetimes of the newly formed OB stars, the density wave moves only a short distance. So these stars, while they last, clearly mark the spiral arm. As the density wave moves on, it provokes the formation of more stars. These take the place of the ones that have rapidly faded out. So the spiral arms persist by a continual destruction and creation maintained by the density wave.

How well does the density-wave model describe the observed spiral structure? First, it outlines the grand scheme of the two-armed spiral pattern that we observe in other galaxies and probably in our own. Second, it explains the persistence of the spiral arms in the face of galactic rotation. Third, it predicts the general features of a spiral arm. So the density-wave model succeeds fairly well in explaining the prominent features of spiral structure. However, the model fails on a number of points. It does not explain the origin of the density waves. Nor does it clearly work out what keeps them going. As the density waves ripple through the interstellar medium, they lose energy and should dissipate in about 10^9 years. As evidenced by the abundance of spiral galaxies, however, they must last longer than this. Some mechanism must keep supplying energy to maintain them.

(B) THE GALAXY'S PAST

We will now try to place the Galaxy's present structure in the context of its history. The crucial clues come from the chemical composition of galactic material and its dynamics. The process of galactic evolution links the chemistry with the dynamics.

We stated earlier that the chemical compositions of Population I and Population II stars differ considerably in their abundances of heavy elements. In general, Population II stars contain about 1% of the metal abundance of Population I stars. However, we do not find a simple division of metal abundances into just two groups. Rather, we find a range of abundances, from about 3 to less than 0.1% for the mass ratio of iron to hydrogen. So, though the division into two populations is a useful first approximation, there is really a continuous range of populations. In general, though not for all objects, the lower the metal abundance of an object, the greater its height from the Galaxy's disk.

Interpretation of this observation relies on a basic concept of the recycling of the interstellar medium. First, stars are born from clouds in this medium. Their atmospheric elemental abundances reflect that of the gas from which they formed. Second, stars' orbital motions about the Galaxy are inherited from their parent gas and dust clouds. Third, massive stars evolve quickly and spew back into the interstellar medium material enriched with heavy elements. So as long as new stars, especially massive ones, are born, the abundance of heavy elements in the interstellar medium of the disk gradually increases as the Galaxy ages—basically as a monotonically increasing function of time in the simplest model.

We can estimate the Galaxy's age by finding the oldest stars in the halo. A comparison of theoretical models for metal-poor globular cluster stars with their H–R diagrams indicates an oldest age of 17 billion years. Because globular clusters contain the oldest stars associated with the Galaxy, the halo marks the fossil remains of the Galaxy's birth. Within it, globulars orbit the Galaxy on extremely elongated elliptical paths, moving slowly through the halo at the outer extremes of their orbits and

only briefly whipping in and around the nucleus. These stars exhibit the motions of the cloud from which they were formed. So the Galaxy must have been born from an initially huge gas cloud—at least 100 kpc in radius.

If we apply the virial theorem to the collapse of a cloud to make a galaxy, we have

$$T_{virial} = GMm_p/kR$$

where T_{virial} is the expected virial temperature, M the total mass, R the initial radius, and m_p the mass of a proton (the gas is assumed all hydrogen). In units convenient for galaxies

$$T_{virial} \approx 10^6 \, M_{12}/R_{50}$$

where T_{virial} is in kelvins, R in units of 50 kpc, and M in $10^{12} \, M_\odot$. In these units, $T_{virial} \approx 10^6$ K; in fact, the Galaxy's disk is much colder than this value, so that energy must have been lost in the collapse.

We can estimate the time scale for collapse from the free-fall time (Equation 16–19),

$$t_{ff} = (6.64 \times 10^4)/\rho_0^{1/2} \approx 10^5/\rho_0^{1/2}$$

in seconds for a density in kg/m³. Rapid collapse is that taking place in the free-fall time. To lose the required energy, the cooling time must equal the free-fall time. For $10^{12} \, M_\odot$ with an initial radius of 100 kpc, then

$$t_{ff} \approx 10^5/\rho_0^{1/2} \approx 10^5/(10^{-22} \, \text{kg/m}^3)^{1/2}$$
$$\approx 10^{16} \, \text{s} \approx 10^8 \, \text{y}$$

This proto-Galaxy cloud was probably turbulent, swirling around with random churning currents. Slowly at first, then faster, the cloud's self-gravity pulled it together, with its central regions getting denser faster than its outer parts. Throughout the cloud, turbulent eddies of different sizes formed, broke up, and died away. Shock waves were generated in the collapse, to dissipate energy. Eventually, the eddies became dense enough to contain sufficient mass to hold themselves together. These might be hundreds of parsecs in size—incipient globular clusters. This process happened 17 billion years ago, and it took place quickly—no more than 2 billion years elapsed during the formation of halo stars.

Not all of the gas was consumed in this first burst of globular cluster formation. As the material contracted further, it fell slowly into a disk. Because the original cloud had a little spin, conservation of angular momentum required that it spin faster around its rotational axis as it contracted. As the disk formed, its density increased and more stars formed. Each burst of star birth left behind representative stars at different distances from the present disk. Finally, the remaining gas and dust settled into the narrow layer we see today. Somehow density waves appeared and drove the formation of spiral arms. During this time, massive stars manufactured heavy elements and blew some back into the interstellar medium. So as stars were born in succession, each later type had a greater abundance of heavy elements. That enrichment continues today in the disk of the Galaxy.

20–4 ◗
COSMIC RAYS AND GALACTIC MAGNETIC FIELDS

More than gas and dust fills the interstellar medium. There are also particles, as photons and as matter, traveling at high speeds. The near-relativistic matter makes up the **cosmic rays,** which are intimately bound up in the Galaxy's magnetic field.

(A) OBSERVATIONS OF COSMIC RAYS

Before we can discuss the role of cosmic rays in our Galaxy, we must examine their characteristics. They are not rays at all but rather high-energy charged particles. These particles may be nuclei of atoms whose electrons have been stripped away, or they may be electrons or even positrons. They are called cosmic rays when they travel at essentially the speed of light; in general, they possess large kinetic energies—up to 10^{20} eV. As you saw in Chapter 10, our Sun ejects low-energy (tens to hundreds of MeV) cosmic rays in flares. The Sun

also modulates those cosmic rays coming from outside the Solar System; the interplanetary magnetic field and the solar wind severely distort the orbits of particles with energies of less than 10^9 eV (1 GeV). The amount of modulation varies with solar activity, and the modulation obscures the intrinsic properties of low-energy galactic cosmic rays. We do know, however, that the number of particles increases rapidly with decreasing energy. Most of the observational data on extra–Solar-System cosmic rays pertain to particles having energies in excess of 10^9 eV. Energies as high as 10^{20} eV have been observed for single particles, but these are extremely rare.

The chemical composition of primary cosmic rays (Table 20–2) gives us information on both their source and their journey through space. As in stars, hydrogen nuclei (protons) are by far the most abundant component, constituting about 90% of all cosmic-ray nuclei. Another 9% are helium nuclei, and the remainder is divided among heavier elements. Much interest centers on the so-called **light nuclei**—lithium, beryllium, and boron. These nuclei are virtually absent in stellar atmospheres, where their abundance is 10^{-7} that of helium, while among cosmic rays the light-nuclei-to-helium ratio is about 1:100. The much greater proportion of such nuclei among cosmic rays is attributed to the fact that the original heavy cosmic rays collide with interstellar matter. During such collisions, the heavy nuclei break up into lighter ones; this process is termed **spallation.** Observations of the light, and therefore **secondary,** nuclei enable us to estimate the amount of material traversed by the original cosmic rays. The density of the interstellar medium and the lifetime of cosmic rays are inexorably linked. Current data suggest an average density of interstellar matter roughly 10^5 atoms/m^3 and a mean lifetime of cos-

TABLE 20–2 *Primary Cosmic Rays*

Particle Group	Charge	Average Atomic Weight	Intensity ($\#/m^2$ s sr)
Protons	1	1	1300
Alpha particles (^{3}He, ^{4}He)	2	4	94
Light nuclei (Li, Be, B)	3–5	10	2
Middle nuclei (C, N, O, F)	6–9	14	6.7
Heavy nuclei	≥ 10	31	2
Very heavy nuclei	≥ 20	51	0.5

mic rays from a few million to some tens of millions of years.

From what directions do cosmic rays come? Observationally, they appear to come more or less uniformly from all directions; they are *isotropic*. This phenomenon is not necessarily from a random spatial distribution of the cosmic-ray sources but probably arises because galactic magnetic fields deflect cosmic rays to such an extent that they cannot travel in straight paths. In fact, these high-energy particles travel along the magnetic lines of force in spiral paths, whose size is determined by both the magnetic field strength and the particle energy.

(B) THE SOURCE AND ACCELERATION OF COSMIC RAYS

If cosmic rays arrive at the Solar System isotropically, how can we locate and identify their sources? One clue is the high energy of the particles; we should look for highly energetic phenomena. We have more direct evidence concerning the sources, however, for one component of cosmic rays is electrons, which represent about 1% of all cosmic-ray particles. Cosmic-ray electrons have been observed directly. The fact that positrons (positive electrons) are only one-tenth as numerous indicates that these electrons are primary particles and not secondaries as the light nuclei are. This conclusion is based on the fact that, because of charge conservation, more positrons than electrons are produced when primary cosmic-ray protons collide with interstellar atoms.

We already know that high-energy electrons are observable from a distance, for they emit synchrotron radiation at radio frequencies when they move in a magnetic field. As you saw in Chapter 18, supernova remnants, such as the Crab Nebula, are strong sources of synchrotron radiation from relativistic electrons. Moreover, supernovae generate energy on a scale sufficient to produce cosmic rays. Other sources of cosmic rays may be rotating neutron stars, such as those observed as pulsars.

The lifetime of cosmic rays in the Galaxy is limited; some escape from the Galaxy altogether, and others are consumed by interactions with the interstellar medium. The supply of cosmic rays must be steady and continuous if the energy density and the total energy in them are to remain more or less constant. Supernovae can indeed maintain the supply if supernova explosions occur every 50 years in the Galaxy. Statistics based on the observations of supernovae in other galaxies suggest that this is a reasonable estimate of the frequency of supernova outbursts. A possible additional source of cosmic rays is the galactic nucleus, which appears to be a strong source of synchrotron radiation. The highest-energy cosmic rays (over 10^{18} eV) may well come from outside our Galaxy—that is, from the extragalactic radio galaxies and quasars that are observed to generate such enormous amounts of energy (Chapter 24).

Charged particles must be powerfully accelerated if they are to attain the energies or velocities observed for cosmic rays. Most of this acceleration occurs in the shock wave from a supernova explosion as it drives through the interstellar medium. Regions in interstellar space where magnetic field lines converge are very efficient in accelerating charged particles to produce cosmic rays.

(C) THE GALACTIC MAGNETIC FIELD

The existence of a **galactic magnetic field** is proved by the observation of the Faraday rotation of radiation from radio sources; such radio sources include galactic sources such as pulsars (Chapter 17) and extragalactic sources. **Faraday rotation** is the rotation of the plane of polarization as linearly polarized radiation passes through a magnetized plasma [Section 17–2(b)]. If the electron density and the distance traversed are known, we can solve for the mean magnetic field strength. The best current estimates of the galactic magnetic field strength are about 0.5 nT. Before the observations of Faraday rotation, a galactic magnetic field had already been surmised to exist, both on the basis of interstellar polarization observations (Chapter 19) and from cosmic-ray data.

Another probe of the galactic magnetic field involves the Zeeman effect in 21-cm absorption lines by interstellar clouds along the line of sight to strong radio sources. For hydrogen in the ground state, the magnetic field splits the line into three components, and the frequency difference between two of these (of opposite circular polarization) is directly proportional to the magnetic field strength along the line of sight:

$$\Delta\nu = (2.80 \times 10^2)B \qquad \text{(Hz)}$$

where the field strength is in teslas. Such observations are consistent with those from Faraday rotation and range from about 0.1 to 1.0 nT.

The orientation of the magnetic field with respect to the spiral structure is primarily along the axes of the spiral arms. Superimposed on this mean field is a local field in our vicinity, which dominates most of the observations, making it difficult to ascertain the true nature of the general field. Locally, within 500 pc of the Sun, the field seems to have a turbulent, possibly helical form, with the mean magnetic axis along the spiral arm. Because cosmic rays travel in spiral paths about magnetic field lines, they are closely tied to the galactic magnetic field. Where the field lines converge, the particles are accelerated, as occurs in the Earth's magnetosphere for charged particles from the Sun.

PROBLEMS

1. Shown below is a sketch of the distribution of the neutral hydrogen maxima in the spiral arms of our Galaxy; the position of our Sun is denoted by O. Draw the 21-cm line profiles you would expect to observe in the directions $l = 50°$, $110°$, and $230°$; label the points on these profiles corresponding to the spiral arms. (Do not make any detailed calculations of radial velocity; just make certain that the signs of the velocities and the relative positions of the peaks are correct.)

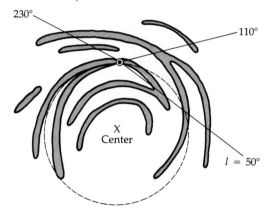

2. Name three important physical characteristics that distinguish Population I stars from Population II stars. Explain these differences in terms of the evolution of our Galaxy.

3. The objects that constitute our Galaxy are usually divided into five major population classes.
   (a) What are these classes?
   (b) Give an example of an object from each class.
   (c) Draw an edge-on view of our Galaxy, indicating the spatial distribution of each of the five classes. Label your diagram carefully.

4. (a) What would be the apparent magnitude of a star like our Sun if it were at the distance of the galactic center (8.5 kpc) from the Sun?
   (b) The Milky Way Galaxy contains 4×10^{11} stars. If we assume that all of these stars are like our Sun ($M_B = 4.7$), what is the absolute magnitude of the whole Galaxy? Compare this result with the apparent magnitude of the Sun.

5. When a particular globular cluster is at its farthest point from the galactic center (apogalacticon), its distance from the center is 10^4 pc. What is its period of galactic revolution? What assumptions must you make to arrive at a unique answer? Can you give any physical justification for your assumptions? (*Hint:* 1 pc = 2×10^5 AU. Assume that the mass of the Galaxy is $10^{12} M_\odot$.)

6. We can deduce the average time between stellar collisions by considering the figure below. If we have identical stars of radius R scattered through space with a mean number density N (stars per unit volume), then a star moving with speed V will sweep out the volume $\pi R^2 V$ per unit time. The average number of stars in this volume is $\pi R^2 VN$, so that in the time $T = 1/\pi R^2 VN$, the star will collide with one other star (on the average)! The average distance between each collision is just $L = VT = 1/\pi R^2 N$; this is called the *mean free path*. In each of the following situations, compute the mean collisional time and the mean free path for
   (a) the solar neighborhood, where $V = 20$ km/s and $N = 0.1/\text{pc}^3$; consider stars of radius $R = R_\odot$
   (b) a galactic nucleus, where $V = 1000$ km/s and where there are 10^9 stars (of radius $R = 10R_\odot$) within a sphere of radius 5 pc. Comment briefly on your results.

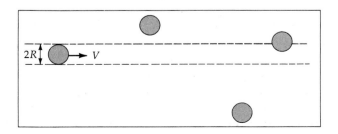

7. The 21-cm line of H I gas has a rest frequency of 1.420406 GHz. A cloud at $b = 0°$, $l = 15°$ has an observed 21-cm emission line frequency of 1.420123 GHz. Use Figure 15–10 to determine, approximately, *two* possible distances to the cloud. Assume only circular motion about the galactic center.

8. The following maximum velocities [relative to the Local Standard of Rest (LSR)] of H I gas are observed using the 21-cm emission along the lines of sight corresponding to the given galactic longitudes: (i) 123 km/s, $l = 15°$; (ii) 95 km/s, $l = 30°$; (iii) 64 km/s, $l = 45°$; (iv) 29 km/s, $l = 60°$; (v) 7.5 km/s, $l = 75°$. Assume circular motions along the galactic center.
    (a) Compute and draw a rough plot of the rotation curve of the inner part of the Galaxy ($R < R_0$) using these data.
    (b) Using the approximation that the mass distribution in the inner part of the Galaxy is spherically symmetric, determine the mass interior to the Sun's orbit, in solar masses.
    (c) Using the same approximation, how does the mass density depend on R between about 6 and 8.5 kpc from the galactic center? (Give a rough proportionality.)

9. A bright radio source has several 21-cm absorption lines caused by neutral hydrogen clouds along the line of sight. Discuss qualitatively how observations of the Doppler shifts of the lines can be used to estimate a lower limit to the distance to the radio source. Under what circumstances would this limit *not* possess the ambiguity caused by two distances along the line of sight having the same radial velocity?

10. Expanding gas rings are thought by some astronomers to be evidence of violent explosions near the center of the Galaxy.
    (a) Calculate the kinetic energy of the 3-kpc arm, which is thought to contain about $10^8 \, M_\odot$ of gas expanding at about 50 km/s.
    (b) Astronomers also believe that there is a ring of molecular gas of mass $10^7 \, M_\odot$ expanding at 150 km/s at a distance of 200 pc from the galactic center. Calculate its kinetic energy.
    (c) Compare the above energies with that of a single supernova explosion. Comment.

11. Confirm the calculation in Section 20–3 for the free-fall time for the proto-Galaxy.

12. Speculate on how one might indirectly observe cosmic-ray electrons in our Galaxy. Do the same for cosmic-ray protons. (*Hint*: Ask how such high-energy particles might produce radiation.) Such measurements indicate that cosmic rays are present throughout the Galaxy and even in the halo!

13. The observed spiral pattern in our Galaxy is *not* produced by differential rotation "winding up" the arms. To demonstrate this, use the rotation curve in Figure 20–8 to estimate how long it would take for differential rotation to smooth out the distinct spiral pattern.
    (a) What is the revolution period for a star at $R = 5$ kpc; for the Sun at $R = 8.5$ kpc; for a star at $R = 20$ kpc?
    (b) How long would it take for the star at $R = 5$ kpc to make one more revolution than the Sun around the center of the galaxy? This section of the spiral arm would thus be "wound up one turn" from its original shape. [*Hint:* see Section 1–1 on the relationship between sidereal and synodic periods. Use a similar approach.]
    (c) How long would it take for the Sun to make one more revolution than the star at $R = 20$ kpc around the center of the galaxy?
    (d) Compare your results in (b) and (c) with the age of the Galaxy. Argue why the spiral pattern must be generated by some mechanism other than differential rotation.

14. Using Figure 20–4, estimate the mass of the Galaxy in the (i) bulge; (ii) disk; (iii) dark halo; and (iv) total, interior to a galactic radius of
    (a) 3 kpc
    (b) 8.5 kpc (the Sun's distance from the galactic center)
    (c) 20 kpc.
    Comment on the mass distribution in the Galaxy, particularly in regard to luminous and dark matter.

15. Section 20–2B describes stars as being metal poor if the metallicity [Fe/H] < -1, and metal rich if [Fe/H] > -1. Calculate the relative abundance N_{Fe}/N_H for the critical value [Fe/H] $= -1$. [*Hint:* see Table 10–1.]

16. (a) The nonthermal radio source Sgr A*, which may mark the Galaxy's core (Section 20–2C), is less than 0.1″ in size. What is its corresponding linear size?
    (b) From the timescale of variation in the gamma-ray luminosity of the galactic nucleus, astronomers estimate the size of the nuclear source to be less than 0.3 pc. What is the angular size of the gamma-ray-emitting source?

PART FOUR
The Universe

The Central region of the Virgo cluster of
galaxies. *(Kitt Peak National Observatory)*

Chapter 21

Galaxies Beyond the Milky Way

Galaxies are the grandest pieces of the Universe. They are the largest objects containing stars, gas, and dust that can still fit in the view of a telescope. Our Milky Way Galaxy (which is a giant) contains 10^{11} stars. A few larger galaxies contain more stars, and dwarf galaxies contain fewer than 1% as many. When we perceive the Universe, we usually picture it as a universe of galaxies. Yet galaxies were recognized only recently (the proof came in 1924) as vast assemblages of stars separate from our Milky Way. This chapter deals with ordinary galaxies in terms of their physical properties, influenced by the distribution of mass within, the relative amounts of gas and dust, and the types of stars they contain. You will often hear that there is no such thing as a "normal" galaxy—perhaps this is strictly true, but one class of galaxies, those with Active Galaxy Nuclei (AGNs), stands out as so different from most others that the term "normal" can be applied to the latter. This chapter concerns the normal galaxies.

Our examination of galaxies will first proceed by presenting the observational data. Most of our in-

formation comes from the optical, radio, infrared, and X-ray windows of the electromagnetic spectrum, and most observations involve either the image or the spectral properties of the galaxy. Once the observations have been presented, we will discuss the basic theoretical interpretations and assess the question of "What is a galaxy?"

21–1 ◗
GALAXIES AS SEEN IN VISIBLE LIGHT

We have certainly learned more about galaxies and the Universe by means of visible light observations than by any other technique. This is partly a result of historical accident; humans have made telescopically aided eye observations of galaxies for well over 200 years, and the first photographs appeared more than 100 years ago. In contrast, the other windows of the electromagnetic spectrum first yielded important results in the 1950s and only reached maturity within the last 10 or 20 years.

The visible light window's importance comes not only from history, however. Starlight, which is the major source of energy emission from most galaxies, lies mostly inside the visible window. In addition, most electronic transitions in atoms have energies on the order of a few eV, which lies within the energy range of visible photons. Therefore, most dilute gases emit line radiation within the visible window. Since stars and gases are the two dominant constituents of galaxies, visible observations form the basis for understanding galaxies.

Having established the priority of visible light, we note that today's astrophysicist must also understand the profound contributions made by different kinds of observations—particularly those in the infrared, radio, and X-ray regions. One example is sufficient to illustrate this point; AGNs are the most powerful phenomena we can observe (short of the origin of the Universe itself). Although some of their manifestations are seen in the visible, their true nature could never be understood without detecting the intricate physical processes seen in such phenomena as their radio jets and extended radio sources or by summing up the phenomenal total energy output from their X-ray and infrared radiation.

(A) VISIBLE LIGHT IMAGING OF GALAXIES

Observations of the images of galaxies provide the most fundamental understanding. Most of our basic terminology (such as elliptical or spiral morphologies) came first from examination of photographic plates. It was also immediately obvious from photographic surveys that galaxies often

come in pairs, triples, etc.—all the way up to very rich clusters.

In recent years, the vast heritage of data from photographs has been augmented by digitized images from CCD cameras. This new technology offers an increase of a factor of approximately 100 in sensitivity, greater dynamic range, and no need for analog-to-digital conversion in order to study the data with computers. Therefore, it is a much simpler process to use a CCD to measure quantities such as colors or magnitudes than it was in the days of photographic emulsions.

Let us now examine the basic data provided by optical imaging. This will include the morphological classification scheme, galaxy photometry, and colors.

The Classification Scheme

Edwin Hubble pioneered the study of galaxies based simply on appearance. Most galaxies may be separated into three major categories: **elliptical, spiral,** and **irregular.** Figure 21–1 shows this classification system in what has come to be termed the "tuning fork" diagram. The splitting of the diagram arises because the barred and ordinary spirals each show a similar progression of structure as one moves from Sa's to Sc's. Hubble thought that the classification sequence probably formed an evolutionary sequence also. For this reason, Sc galaxies have come to be called "late" type spirals and Sa's "early." Although Hubble expected there to be an S0 class, which in his view was the evolutionary bridge from the elliptical stage to the spirals, he did not actually find any. The S0's were

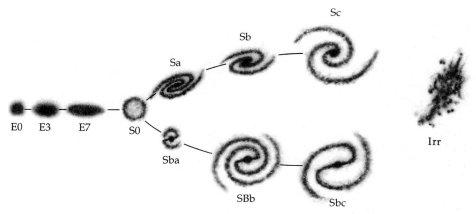

FIGURE 21–1 Hubble's tuning fork diagram. This illustration shows some of Hubble's elliptical classes along with the spirals and irregulars. The spirals separate into parallel "ordinary" and barred sequences based on the tightness of the spiral winding and the bulge/disk ratio. Hubble *incorrectly* thought that the diagram was an evolutionary sequence extending from left to right.

added by Allan Sandage, who also greatly expanded on Hubble's original ideas. The other major additions to the Hubble system came from G. deVaucouleurs who stressed the fact that the system is really a continuum—not just discrete classes—and added spiral classes of type later than Sc—the Sd's and Sm's. (The latter is a ''Magellanic'' type spiral; it often surprises students who have seen photographs of the Large Magellanic Cloud that it has definite, barred spiral rather than irregular structure.) We illustrate several galaxies here; you are urged to browse through *The Hubble Atlas of Galaxies* for many more examples.

Ellipticals   The **elliptical galaxies** (designated by **E**) have the shape of an oblate spheroid (perhaps—there is no consensus as to whether they are oblate, prolate, or triaxial; we'll stick with oblate). They appear in the sky as luminous elliptical disks (Figure 21–2). The distribution of light is smooth, with the surface brightness decreasing outward from the center with an intensity fall-off that varies

FIGURE 21–2   An elliptical galaxy. This example, M32, is type E2 and companion to the Andromeda galaxy. (*National Optical Astronomy Observatories*)

approximately as $\log I(r) \propto r^{-1/4}$. Elliptical galaxies are classified according to the elongation of the apparent projected image; that is, if a and b are the major and minor axes of the apparent ellipse, then $10(a - b)/a$ expresses the observed ellipticity. The true ellipticity cannot be found because the orientation of a particular galaxy cannot be determined. So an E0 galaxy appears circular, while increasingly elliptical ones are given designations from E1 to E7 (the latter is the flattest observed). The E galaxies have *no* axis of rotation; their stars follow orbits with a variety of inclinations. Statistical studies lead to the conclusion that the true ellipticities of these galaxies are represented fairly uniformly from E0 to E7.

cD Galaxies   An additional, somewhat similar class is the **cD** galaxy, which was introduced by W. W. Morgan (*c* historically denotes supergiants in astronomical nomeclature and *D* is for diffuse). These appear superficially to be ellipticals but have greatly extended envelopes and, frequently, multiple nuclei. Their diameters range up to a few megaparsecs (10^6 pc). The most likely origin of cD galaxies is the growth by cannibalism of a supergiant ''normal'' elliptical located at the center of a cluster [see Section 23–1(e)].

Spirals   The **spiral galaxies** are divided into the **ordinary** (designated **S** or **SA**) and the **barred** (**SB**) **spirals.** Both types have spiral-shaped arms, with two arms generally placed symmetrically about the center of the axis of rotation. In the ordinary spirals (Figure 21–3A), the arms emerge directly from the nucleus; in the barred spirals, a bar of material cuts through the center (Figure 21–3B) and the arms originate from the ends of the bar. Both types are classified according to how tightly the arms are wound, how patchy they are, and the relative size of the nucleus. Ordinary spirals of type Sa have smooth, ill-defined arms that are tightly wound about the nucleus; in fact, the arms form almost a circular pattern. The intermediate Sb galaxies have more open arms, which are often partly resolved into patches of H II regions and Population I stellar associations. The nuclei in Sc galaxies are usually quite small, and the spiral arms are extended and well resolved into clumps of stars. Both old and young populations coexist in spiral galaxies, but

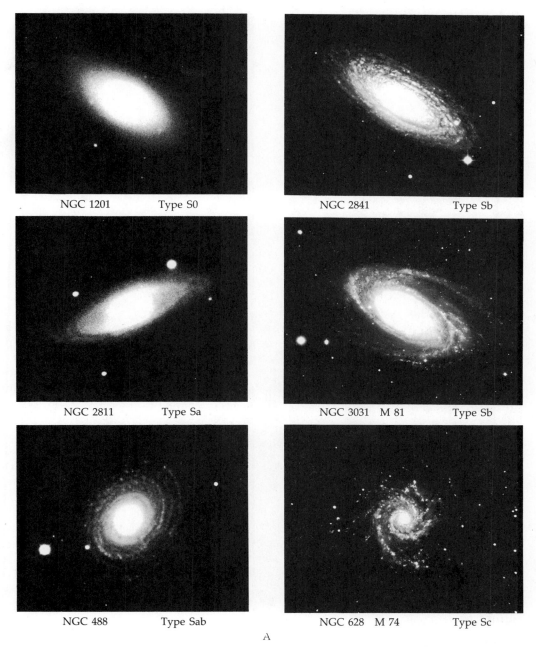

NGC 1201   Type S0

NGC 2841   Type Sb

NGC 2811   Type Sa

NGC 3031   M 81   Type Sb

NGC 488   Type Sab

NGC 628   M 74   Type Sc

A

FIGURE 21–3   Types of spiral galaxies. (A) Normal (*above*). (B) Barred (*see next page*). (*Palomar Observatory, California Institute of Technology*)

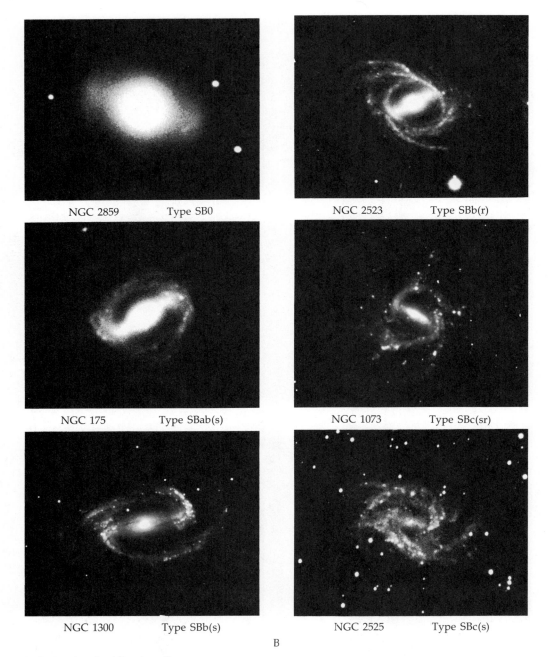

NGC 2859   Type SB0   NGC 2523   Type SBb(r)

NGC 175   Type SBab(s)   NGC 1073   Type SBc(sr)

NGC 1300   Type SBb(s)   NGC 2525   Type SBc(s)

B

FIGURE 21–3 (*Continued*)

the proportion of young Population I objects increases from Sa to Sc. Barred spirals exhibit a parallel sequence of types: SBa, SBb, and SBc.

Why should there be two types of spirals? The most likely explanation came from J. P. Ostriker and P. J. E. Peebles, who studied the dynamics of spiral galaxies with varying halo masses. They found that if a disk galaxy's halo were not sufficiently massive, a bar-like instability would develop. So we see here evidence that some galaxies—the ordinary spirals—possess massive spherical components that make their presence felt by dynamical effects but do not show up directly when the image is viewed in visible light. This is but one piece in a growing puzzle about dark matter—a topic that is becoming increasingly important throughout our discussion of galaxies and the Universe.

The degree to which the spiral arms are developed is related to the luminosity of a galaxy. In analogy to the stellar luminosity classes, those for galaxies are I, II, III, IV, and V, with I the most luminous and V the least. An Sc I galaxy, then, is a very luminous spiral with a small nucleus and extended, well-resolved arms. These galaxian luminosity classes also relate to mass, so that for the same galaxy type, class I galaxies are the most massive and class V the least.

S0 Galaxies The **S0 galaxies** are intermediate between the ellipticals, E7, and the true spirals, Sa. They are flatter than the E7s and also differ from ellipticals in having a thin disk as well as a spheroidal nuclear bulge. Seen edge-on they sometimes have the shape of a convex lens and so are also called **lenticulars**. As opposed to the spheroidal component of galaxies which follow the $r^{-1/4}$ law, the disk components have a slower fall-off of $I(r) = I_0 e^{(-\alpha r)}$. In many respects, S0 galaxies resemble the true spirals but do not contain Population I objects. An S0 galaxy seen *edge-on* is relatively easily distinguished from elliptical galaxies since it is flatter and has a disk, but it is not easily distinguished from an Sa galaxy. On the other hand, an S0 seen *face-on* is relatively easily distinguished from an Sa since the latter has definite spiral structure, but the S0 would be difficult to distinguish from an E0 since neither would have spiral structure. The ability to correctly classify E and S0 galaxies rests on the success of the image at revealing the faint outer disk of the S0. Therefore, detecting S0's is dependent upon the quality of the observational material as well as the distance of the galaxies.

One possible way that S0's could be created is for an early-type spiral to have all of its dust and gas removed by tidal interactions with another galaxy. In even interpenetrating collisions, the stars of the two galaxies would almost never collide, but the gas would tend to be stripped.

Irregulars Other galaxies fall into the **irregular class** since they show no symmetrical or regular structure; nevertheless, even these galaxies may be divided into two distinct groups. Those of type **Irr I** have resolved OB stars and H II regions and clearly a large Population I component. The classification **Irr II** is quite ambiguous and may include galaxies that are simply peculiar. Primarily, however, these galaxies are amorphous and do not re-

FIGURE 21–4 The irregular galaxy M82. *(Lick Observatory)*

solve into stars. Such galaxies do show marked absorption by interstellar dust, and gaseous emission is also observed. The peculiar galaxy M82 is an Irr II (Figure 21–4); it is remarkable in that dusty material extensively blocks out light from its stars in a fashion that makes it appear to be exploding; it is *not*.

Dwarf Galaxies It is important to realize that *the standard classification system does not apply to most galaxies in the Universe*! The galaxies we commonly see illustrated are, naturally, those that are most easily detected. They are the brightest, most outstanding objects, but the number of dwarf galaxies far exceeds that of the familiar giants. The dwarfs have different morphologies. However, the most common type is elliptical in shape and has little gas, so it is called a **dwarf elliptical** or **dE**. The most obvious difference between these and the true ellipticals, aside from size, is the lack of a bright nuclear region in the dE's. This class is numerically the largest in the Universe. The other principal type of dwarf galaxy is the **dIrr** or **dwarf irregular**. Although the dwarfs are numerically superior to the "normal" galaxies, the latter dominate in mass. So it is appropriate that most of our discussions deal with the types of galaxies seen in the *Hubble Atlas of Galaxies*.

Peculiar Galaxies Finally, we note that not all galaxies fall into any of the classes discussed so far, and those that do not are generally called **peculiar**

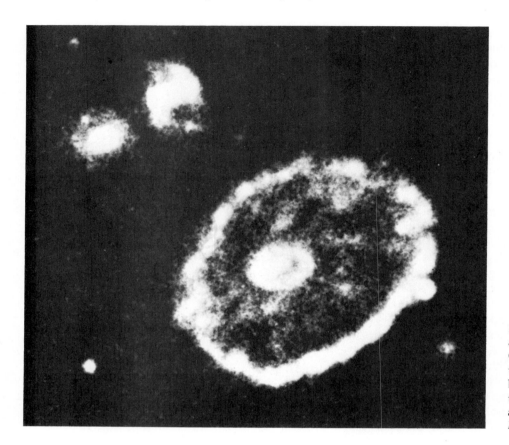

FIGURE 21–5 A ring galaxy. Called the Cartwheel Galaxy, this object may be the result of the collision between a large spiral (now the ring) and two other galaxies. *(Royal Observatory, Edinburgh)*

galaxies. Good examples are the bizarre ring galaxies (Figure 21–5) and other galaxies that may have undergone tidal disruption with another galaxy.

The Morphological Mix   About 77% of *observed* galaxies are spirals, 20% are ellipticals, and 3% are irregulars. This sample is dominated by the luminous spirals, however, which are visible at very great distances. The relative numbers in a given volume of space are quite different. A survey of the region of space out to 9.1 Mpc shows that only 33% of the galaxies in this volume are spirals, 13% are ellipticals, and 54% are irregulars. Many of the irregulars are small galaxies of fairly low luminosity, as are dwarf ellipticals.

(B) PHOTOMETRIC CHARACTERISTICS OF GALAXIES

Integrated Colors

The light from a galaxy arises from all of its stars, with the radiation contribution from the brightest hot stars competing with the light from the fainter (but far more numerous) cool stars. In a gross way, we can use a galaxy's color to infer its stellar component. A direct correlation exists between a galaxy type and its color. Ellipticals tend to be much redder than spirals, and spirals redder than irregulars. Within the spiral group, the galaxies appear redder as their nuclear bulges grow larger and their spiral arms become less extensive. Typical color indices ($B - V$ magnitudes) are listed for different morphological types in Table 21–1, but we note that, especially for spirals, the colors of the outer parts of the galaxies are different from the nuclear bulge regions.

The progression of color from the bluer irregulars to the redder ellipticals reflects a trend in the composition of the galaxies' populations. In general terms, an old Population I predominates in ellipticals and a much younger Population I stands out in irregulars. The mixture in the spirals is determined by the size of the nucleus (old Population I) relative to that of the spiral arms (young Population I). (Population II is probably a minor contributor in all large galaxies, existing mainly in the globular clusters and galactic halo.)

TABLE 21–1 *The Fundamental Characteristics of Galaxies*

	Ellipticals		Spirals			Irregulars I
Mass ($M_\odot$)	10^5 to 10^{13}		10^9 to 4×10^{11}			10^8 to 3×10^{10}
Absolute magnitude	-9 to -23		-15 to -21			-13 to -18
Luminosity ($L_\odot$)	3×10^5 to 10^{11}		10^8 to 2×10^{10}			10^7 to 10^9
M/L ($M_\odot/L_\odot = 1$)	100		2 to 20			1
Diameter (kpc)	1 to 200		5 to 50			1 to 10
Population content	II and old I		I in arms, II and old I overall			I, some II
Presence of dust	Almost none		Yes			Yes
		Sa	Sb	Sc, Sd		
Color index ($B - V$)	$+1.0$	$+0.9$	$+0.4$ to 0.8	$+0.4$ to 0.6		$+0.3$ to $+0.4$
$M_{H\,I}/M_T$(%)	Almost 0	2 ± 2	5 ± 2	10 ± 2		22 ± 4
Spectral type	K	K	F to K	A to F		A to F

Sizes

From simple trigonometry we know that the angular size of an object in the sky coupled with determination of its distance easily yields the linear size of that object. Given the angular diameter, we find the linear diameter from the relationship $a_{\text{rad}} = s/d$, where a_{rad} is the angular diameter in radians, s is the linear diameter, and d is the distance (both in the same units). We will defer discussion of how we find distances until the next chapter. For now we will assume that it is possible to do so.

One hitch in calculating the sizes of galaxies is that the definition of the "edge" of a galaxy is somewhat arbitrary; different definitions result in different diameters. The slow fall-off in intensity in both the disk and spheroidal components means that, for both E and S galaxies, one can "always" look a little fainter and hence find the galaxy a little larger. In general, astronomers judge the edge of a galaxy by using some limiting level of observed brightness. The intensity contour of this designated level—called an **isophotal level**—is drawn around the image of the galaxy. This isophotal level then determines the galaxy's apparent angular size. Such a procedure is relatively easy to carry out today using CCD pictures and computerized image processing.

Dwarf ellipticals and small irregulars tend to be the smallest galaxies, some only 3000 pc in diameter. The typical diameter of galaxies of all types is about 15 kpc. Giant ellipticals can range up to 60 kpc across. The very largest cD galaxies can have diameters up to 2 Mpc—greater than the distance from our Galaxy to the Andromeda galaxy.

Luminosities

If we know distances along with fluxes (measured logarithmically by apparent magnitudes) for galaxies, we can calculate luminosities (absolute magnitudes). However, several considerations must be made in order to ensure that the calculated absolute magnitudes accurately measure the galaxy's power output. One of these is related to the size question raised above. How do we decide where the edge of the galaxy is? We must limit the radius at a given isophote.

There are also three corrections that must be made to a galaxy's absolute magnitude. Here we summarize the corrections as given in the *Second Reference Catalogue of Bright Galaxies*.

The first correction is for *dust obscuration within our own Milky Way*. In the B-magnitude system, the total extinction is $A_B = 0.19$ magnitudes at the North Galactic Pole and $A_B = 0.21$ at the South Pole. The general expressions for the extinction at any given galactic longitude and latitude are the following:

$$A_B = 0.19(1 + S_N \cos b)|C| \quad \text{(for } b > 0)$$

and

$$A_B = 0.21(1 + S_S \cos b)|C| \quad \text{(for } b < 0)$$

Where

$$
\begin{aligned}
S_N(l) = {} & 0.1948 \cos (l) \\
& + 0.0725 \sin (l) + 0.1168 \cos (2l) \\
& - 0.0921 \sin (2l) + 0.1147 \cos (3l) \\
& + 0.0784 \sin (3l) + 0.0479 \cos (4l) \\
& + 0.0847 \sin (4l)
\end{aligned}
$$

and

$$S_S(l) = 0.2090 \cos{(l)}$$
$$- 0.0133 \sin{(l)} + 0.1719 \cos{(2l)}$$
$$- 0.0214 \sin{(2l)} - 0.1071 \cos{(3l)}$$
$$- 0.0014 \sin{(3l)} + 0.0681 \cos{(4l)}$$
$$+ 0.0519 \sin{(4l)}$$

The term C is given for both hemispheres by

$$C = \csc{[b + 0.25° - 1.7° \sin{(l)} - 1.0° \cos{(3l)}]}$$

The second correction is for extinction *internal to the galaxy* in question. Since E galaxies have little dust, this correction is only made for S galaxies. It is a function of the inclination, i, of the image which must be estimated by the observed ellipticity of the image. (We assume that spirals seen pole-on are circular.) The expression is

$$A_B(i) = 0.70 \log{\sec{(i)}}$$

The third is called the *K-correction*; it is needed because the redshift causes the emitted light to be moved out of the rest frame filter band. This correction is morphology dependent. Treating Hubble's classes as units, the standard expression is

$$10^4 \times K_B(cz)$$
$$= 0.15cz$$
$$\text{(for E and S0's)}$$
$$= [0.15 - 0.025 \, (\text{morphology} - \text{S0})]cz$$
$$\text{(for Sa–Sb)}$$
$$= [0.075 - 0.010 \, (\text{morphology} - \text{Sb})]cz$$
$$\text{(for late types)}$$

where cz is the redshift in km/s.

Absolute magnitudes range from -8 (2×10^5 solar luminosities) for dwarf ellipticals to -25 (10^{12} solar luminosities) for supergiant ellipticals. Our Galaxy, viewed from the outside, would have an absolute magnitude of roughly -21 (2.5×10^{10} solar luminosities).

Masses of Galaxies (Round 1)

Let us explore one possible way to estimate the masses of galaxies. Now that we know the energy output, the luminosity, of the galaxy, we can do a simple calculation. If the luminosity is $10^{11}L_\odot$ and if each star, on average, contributes as much light for its mass as the sun does, then the mass of the galaxy ought to be about $10^{11}M_\odot$. A somewhat better estimate would make morphology-dependent corrections for the amount of gas (up to 30%) and dust (up to 5%). The method here is what is important. Note that we are assuming that the contents of galaxies produce about 1 solar unit of luminosity for about 1 solar unit of mass. This method is a luminosity-based calculation, and the results follow the luminosities quite closely. Galaxies, therefore, range from 10^5 to 10^{13} solar units in mass.

Table 21–1 summarizes the fundamental observational data on galaxies. It treats the morphologies as three main types and presents their masses, luminosities, colors, diameters, and population types. Table 21–2 lists the positions of some of the most interesting galaxies.

(C) THE VISIBLE LIGHT SPECTRA OF GALAXIES

We have seen that examination of galaxies at different colors yields important information. Different parts of galaxies emit different amounts of light at various wavelengths. It should not come as a

TABLE 21–2 *Selected Galaxies*

Name	Other ID	$\alpha(2000)$	$\delta(2000)$	Remarks
Andromeda	M31, NGC 224	$0^h 42.7^m$	$+41°16'$	SbI–II
Black Eye	M64	$12^h 56.7^m$	$+21°41'$	SabII
Centaurus A	NGC 5128	$13^h 25.5^m$	$-43°01'$	Radio source
Cygnus A		$19^h 59.4^m$	$+40°43'$	Radio source
Fornax A	NGC 1316	$3^h 22.7^m$	$-37°12'$	Radio source
Hoag's Object		$15^h 17.2^m$	$+21°35'$	Spindle & helix
Maffei I		$2^h 36.3^m$	$+59°39'$	Nearest giant E?
The Mice	NGC 4676a/b	$12^h 46.1^m$	$+30°44'$	Colliding pair
Pinwheel	M33	$1^h 33.9^m$	$+30°39'$	ScII–III
Sombrero	M104	$12^h 39.9^m$	$-11°37'$	Edge-on Sa/b
Stephan's Quintet	NGC 7317-20	$22^h 36.0^m$	$+33°58'$	Odd redshifts?
Whirlpool	M51	$13^h 29.9^m$	$+47°12'$	SbcI–II

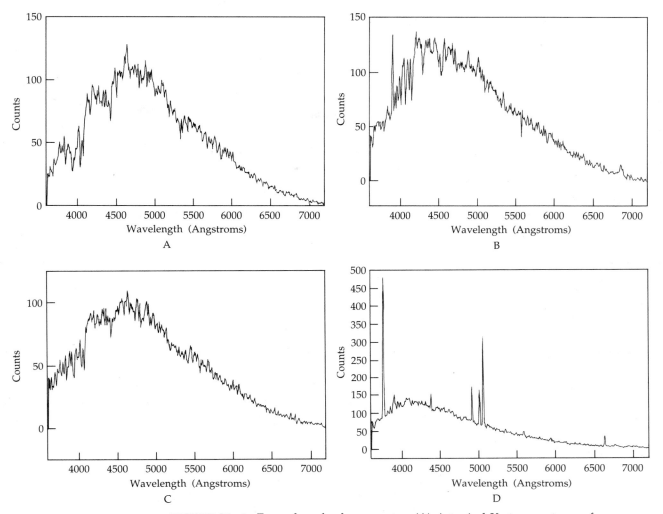

FIGURE 21–6 Examples of galaxy spectra. (A) A typical K star spectrum of an early-type galaxy. Note the Ca II H and K lines, the 4000-Angstrom break, the G band and the Mg I line(s). The dropoff of the spectrum at wavelengths longer than 5000 Angstroms is caused by detector response; the true spectrum is flat in this region. (B) A typical G type spectrum of a later type than seen in (A); note the shallower 4000-Angstrom break. (C) An F-type spectrum of a still later type of galaxy. Note the [O II] 372.7-nm (3727-Angstrom) emission line, which is redshifted to about 3900 Angstroms. (D) A strong emission-line galaxy of late spectral type. The strongest lines are of wavelengths 3727, 4340, 4861, 4959, 5007, and 6563 Angstroms.

surprise then that observations that separate the entire visible window into a spectrum are an essential tool. The history of optical spectroscopy is similar to that of imaging. The first photographs of spectra followed closely the first imaging photographs, and the modern trend is to use CCDs to detect the spectra.

Spectral Types

Most spectroscopic observations treat the galaxy as a single object. These observations measure the nuclear regions only. They can inform us of the gross motions of the galaxy as well as the dynamics and spectral type of the nuclear bulge. A more detailed examination in which spectra of many parts of the galaxy are observed can, in turn, provide us with an understanding of the internal motions within the galaxy and a sampling of quite different populations.

Spectral type is another galaxy property that is dependent upon morphology. W. W. Morgan noted that the characteristic spectral type of a galaxy's nucleus varies with the prominence of the bulge. The spectrum, of course, arises from the integrated light from billions of stars, but from Figure 21–6 one can easily see that some galaxy spectra bear striking resemblance to K-type stellar spectra, while others may resemble G, F, or even A spectral types. This happens when the numbers of luminous giant stars of one particular type numerically dominate in the nuclear regions. Morgan found that there was a smooth progression of K spectral types in early-type morphologies through G and F types in late spirals.

The spectra of galaxies provide much more information than is listed here. In Chapter 22, we will discuss spectroscopically determined distances, and in Chapter 24 we will examine AGN galaxies that exhibit unusual spectral features.

21–2 ◗
GALAXIES AT RADIO WAVELENGTHS

We use radio telescopes in ways similar to those in which we use optical observatories. Broad-band continuum observations are similar to broad-band colors. Aperture synthesis imaging is similar to CCD imaging, and we can scan in radio frequency space to form spectra.

(A) CONTINUUM IMAGING

Continuum observations of galaxies generally measure radiation that was produced by the synchrotron process. From this we immediately know that there must be both a magnetic field and a component of highly energetic particles present in any galaxy that is detected.

Generally, we can separate galaxies into two distinct groups of radio emission. One is the AGN group, which includes Seyfert galaxies, radio galaxies, and quasars, for example. These objects feature powerful, violent phenomena whose origin lies deep within the general nuclear area at the tiny nucleus itself. We will examine these objects in Chapter 24. The current chapter is concerned with more normal galaxies. Although a great many galaxies can be detected with radio telescopes, the term *radio galaxy* is reserved for that special class of AGNs that produce more than 10^{33} J/s of radio power.

In the low-luminosity class of sources, we usually can detect only the later morphological types. This should not be a surprise when one considers the most likely means of producing relativistic

electrons and ordered magnetic fields. The best candidates are all young Population I objects.

(B) LINE RADIATION AND NEUTRAL-HYDROGEN CONTENT

The most widely observed example of radio line radiation is that produced by neutral hydrogen at a wavelength of 21 cm. This provides a great deal of information about the H I content of nearby galaxies, including an estimate of the total amount of hydrogen in the galaxy, the ratio of the mass of the hydrogen gas to the total mass of the galaxy, the distribution of the neutral hydrogen in the system, the rotation curves as a function of distance from the center of the galaxy, and the radial velocity of the galaxy. The most complete data, of course, are for galaxies that are *resolved* by radio telescopes.

As you might expect, the total amount of hydrogen in a galaxy is primarily a function of galaxy size. On the other hand, the ratio of the hydrogen mass to the total galaxy mass ($M_{H\,I}/M_T$) depends on galaxy type (Table 21–1). The percentage of the total mass that is in the form of neutral hydrogen is relevant to our ideas concerning evolution within galaxies. The less hydrogen relative to stars, the more original gas there was that must have already been condensed into stars. Data for a sample of spiral and irregular galaxies indicate that the H I mass makes up a small fraction of the total mass, only 3% for lenticulars and 22% for irregulars. So Irr galaxies have more H I gas, relative to their total mass, than do Sa galaxies.

The present rate of star formation depends on both the amount of hydrogen available and its density. Sc spirals and the irregular galaxies are not necessarily younger than the Sa spirals, but their development has been different. For instance, Population II stars have been observed in irregular galaxies, even though these galaxies have high $M_{H\,I}/M_T$ ratios.

For Sc spirals and irregulars, the extent of the hydrogen in many cases is almost double the optical size of the galaxy. For example, studies of the Magellanic Clouds show that there is hydrogen between these two galaxies, both as a bridge to the Milky Way and as a common envelope surrounding both galaxies.

21–3 ◗
INFRARED OBSERVATIONS
OF GALAXIES

Although infrared observations have been made for over 20 years, this field really came to maturity with the Infrared Astronomy Satellite (IRAS) in the mid 1980s. This satellite's surveys covered over 96% of the sky at 12, 25, 50 and 100 μm wavelengths. Roughly 25,000 galaxies were detected. As is the case with radio continuum and 21-cm observations, IRAS observations of galaxies are also quite sensitive to morphology. Few E and S0 galaxies can be found in the IRAS catalogue.

Although AGNs are often highly luminous in the infrared, the "normal" galaxies are also seen. In the latter the primary emission mechanism is thermal radiation from interstellar dust grains that have been heated by starlight. Therefore, the IRAS observations provide information about the stellar populations as well as the composition and distribution of the dust. A simplified, but useful, model would contain two separate components of the dust. One is the so-called cirrus—wispy, diffuse clouds; the other is a more active component related to H II regions and, therefore, to star formation.

In 1985 G. Helou, B. T. Soifer, and M. Rowan-Robinson found a remarkable correlation between the amount of far-infrared emission and that of the radio synchrotron emission. Figure 21–7 shows that one can predict the radio emission flux density within a factor of 2 from the infrared flux over a range of more than 3 orders of magnitude in luminosity! Since the emission mechanisms are quite different for the two (thermal electrons in one case; nonthermal synchrotron emission from relativistic electrons in the other), this tight correlation is difficult to understand.

21–4 ◗
X-RAY EMISSION FROM
NORMAL GALAXIES

For X-rays, the analogue to the IRAS satellite that advanced infrared studies so far is the HEAO-B satellite—more commonly known as the **Einstein Observatory.** Before this satellite produced X-ray

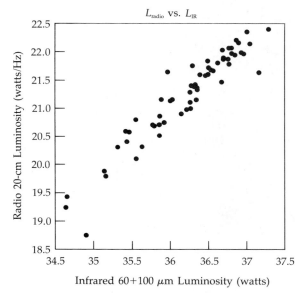

FIGURE 21–7 The radio/infrared luminosity correlation. The far-infrared IRAS luminosity of galaxies is found to be very tightly correlated with the 20-cm radio luminosity. *(N. Duric)*

images in 1978–1980, we generally had to be content with knowing only the total fluxes in objects.

Although the correlation is not so tight as with infrared data, the X-ray and radio flux densities of normal late-type spirals are also found to be related. Again, the emission mechanisms are different. The X-rays are mostly produced by thermal brehmsstrahlung from very hot gas—perhaps accretion disks around compact stars.

Figure 21–8 shows a composite view of the Andromeda galaxy in the visible, radio (21-cm), infrared (60-μm), and X-ray regions of the spectrum. We find that the different classes of photons come from different sources. The blue visible light comes from stars and gas and is quite widely distributed. The X-rays have a very different distribution from all the others; they arise from a relatively small number of discrete sources. The distributions of radio and infrared photons are most similar. Both have a ring of intense, diffuse emission in the disk; but the nucleus also appears prominent in the IRAS data.

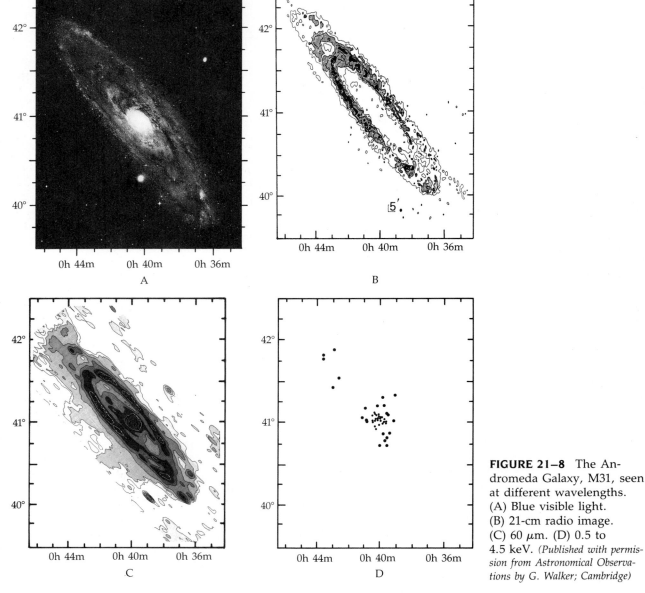

FIGURE 21–8 The Andromeda Galaxy, M31, seen at different wavelengths. (A) Blue visible light. (B) 21-cm radio image. (C) 60 μm. (D) 0.5 to 4.5 keV. *(Published with permission from Astronomical Observations by G. Walker; Cambridge)*

21–5 ◗
SOME BASIC THEORETICAL CONSIDERATIONS

Since galaxies are the most striking elements of the Universe, there must be important physical processes that created them as individuals and have kept them intact during their evolution. The goal of the theorist is to start from a minimal set of basic assumptions and explain the origin and evolution of galaxies in a way that is consistent with both the observations discussed above and the laws of physics.

In this section we will discuss several aspects of the interpretation of observations. First, we will examine the morphological classification scheme from a physical viewpoint and elaborate more on the role of star formation in spiral galaxies. Second, we will introduce the virial theorem and apply it to individual galaxies. (In Chapter 23 it will be applied to clusters.) Finally, we will take another look at galaxy masses. Theories of galaxy formation will be discussed more fully in Chapter 26.

(A) IMPLICATIONS OF THE CLASSIFICATION SCHEME

Is the classification of galaxy types important? In science, morphological distinctions are useful to note only if they represent some underlying physical differences between classes. Let us review three important observations:

1. The color of a galaxy depends strongly on morphological type.
2. The integrated spectral type of the nuclear region depends strongly on morphological type.
3. The spheroidal components of galaxies follow the $r^{-1/4}$ law and the disk components follow an exponential distribution—regardless of the specific morphological type.

The first two points indicate that there is consistency in the evolutionary stage of stellar populations among all galaxies of a given morphological type. Furthermore, there is a regular progression of the population indicators along the morphological sequence. The third point shows that stars in disk regions and those in nuclear bulge regions arrive at consistent dynamical states even though the parent galaxies may follow rather different histories. In other words, the underlying dynamics of all disks are essentially the same, and this also holds for spheroidal systems—whether they be nuclear bulges or elliptical galaxies.

It is quite clear, therefore, that the classification system is enormously useful. The two main classification criteria are *disk/bulge ratio* and the importance and/or *degree of winding of the spiral arms*. We do not understand completely why these two observable quantities so accurately represent both the dynamical state and the dominant evolutionary populations of the component stars within a galaxy. But the fact that there is an important physical basis for separating galaxies by shape is undeniable.

Can we say more about the evolution of galaxies? Why do we not think of the classification scheme as an evolutionary sequence? Although the S galaxies have young stars and the E galaxies have only old stars, the spirals also have old stars. Thus both types are "old." The lack of gas in ellipticals prohibits the formation of new populations.

The form a galaxy takes probably depends mostly upon its **angular momentum** (actually the angular momentum *per unit mass*); the greater the angular momentum, the more flattened the galaxy. This occurs because the gas dissipates its momentum in the direction perpendicular to the plane as gravity collapses gas across the galaxian poles, but the gas in the equatorial plane is rotationally supported. In elliptical galaxies, the condensation of gas into stars was efficient and therefore rapid, which led to a spheroidal distribution of stars and a high concentration of stars at the nucleus. Conversely, in spiral galaxies, star formation occurred more slowly so that the later generations of stars came from an increasingly flattened gas source. A dual type of distribution resulted; the slowly rotating system contains stars distributed spherically, and the rapidly rotating part is a flattened, disk-like system containing stars, dust, and gas. The dust and gas in the plane became subjected to the **density waves** that create spiral arms. The turbulence and magnetic fields may also play important roles in controlling the final shape of a galaxy.

What do the correlated interrelationships among infrared, radio, and X-ray luminosities tell us about normal galaxies? Since only the late-type morphologies are represented, we know that the sources must be part of the young population. How do we relate relativistic electrons, thermal dust sources, and (possibly) accretion disks around collapsed objects?

The canonical answer to these questions shows how important star formation can be to the energetics of disk galaxies. In this scenario, increases in star-formation rates add energy to the interstellar dust, which reradiates the energy from the stellar photons into the infrared—thereby raising the infrared luminosity. This increase in star-formation rates also results in an increase in the number of supernovae, and it is quite likely that the origin of the synchrotron electrons lies in the supernovae. However, the electrons must be accelerated to relativistic velocities, and there are a number of mechanisms in the interstellar medium that can generate shock waves that can effectively accelerate the electrons. Type I supernovae produce collapsed stars, and their associated accretion disks could be the sources of the X-ray luminosity.

Several theorists are currently trying to close the many gaps in this picture, but most agree with its basic plausibility. Somehow star-formation rates seem to regulate the total energy emitted by a galaxy in the infrared, radio, and X-ray windows of the spectrum.

(B) THE ENERGETICS OF GALAXIES— THE VIRIAL THEOREM

The **virial theorem** is used in a very wide range of problems in astrophysics. We have already used it relative to star formation (Section 19–3) and the formation of the Galaxy (Section 20–3). It relates the kinetic energy (KE) of a system to its gravitational potential energy (PE). Its fundamental assumption is that the system in question (star, star cluster, galaxy, galaxy cluster, etc.) is stable—neither collapsing nor flying apart. If the components had too much kinetic energy, then the system would expand; if the gravitational potential energy dominated, then the system would collapse. We write this equation as:

$$2\langle KE \rangle = -\langle PE \rangle \qquad \textbf{(21–1)}$$

For galaxies this reduces to

$$\langle v^2 \rangle = 0.4 GM/r_{\text{h}} \qquad \textbf{(21–2)}$$

where r_{h} is the radius enclosing half the mass; $\langle v^2 \rangle$ is the mean value of the square of the peculiar velocities of the component stars (one form of the velocity dispersion, to be discussed further in Chapter 23), and M is the total galaxy mass.

Since the total energy of a galaxy is $TE = KE + PE$, we find that

$$TE = -KE = 1/2 PE \qquad \textbf{(21–3)}$$

Consider the formation of a galaxy as a widely dispersed cloud of gas initially at rest. In this state, $TE = KE = PE = 0$. As the cloud contracts under gravity, the KE increases and the PE decreases (becomes negative). Equation 21–3 shows that, in its final equilibrium state, the total energy is negative (so the system is bound), and half of the PE released in the collapse has been converted to KE while the other half has been radiated away.

We learn several important things about galaxies from the virial theorem. One is that we can make judgments about which objects belong to a galaxy and which ones are not a permanent feature. For example, if we find a star with unusually high speeds relative to the centroid of a galaxy, we could consider kinetic energy to decide whether the star was a permanent part of the galaxy or a foreground object. Typically, the dispersion of speeds within a galaxy has a width of about 300 km/s. Therefore, if kinetic energy is added to a given component—a star, star cluster, gas cloud, etc.—a likely result is that that component could "evaporate" from the galaxy.

Also note that Equation 21–2 allows estimation of the galaxy mass given the velocity dispersion. This leads us to the next section.

(C) MASSES OF GALAXIES (ROUND 2)

We saw in Section 21–1 a luminosity-based method for estimating galaxy masses. We now introduce some dynamical methods. One is given above in Equation 21–2. A second involves finding the velocity or rotation curve as a function of distance from the center of the galaxy; hence, the method applies only to galaxies that are near enough and bright enough to allow us to obtain spectra at several points. At the telescope the spectrograph slit is placed across the galaxy's diameter, and Doppler shifts are measured at points along it. Knowing the tilt of the galaxy to our line of sight, we can translate the radial velocities to rotational ones.

A simplified view of this method invokes the familiar functional form of Kepler's third law. Although originally defined for the case of a massive sun surrounded by relatively light planets, it can also be applied to any system in which there is a strong central concentration of mass. This certainly seems to apply to the outer reaches of galaxies. Photographs strongly imply that the light is centrally concentrated, so surely(?) the mass is also. We can make predictions about the form of the velocity/radial distance relation, $V(r)$, from Kepler's $P^2 = a^3$. The quantity a is the same as r, and the period P of an orbit is just

$$P = 2\pi r/V$$

Putting V and r into Kepler's equation and rearranging quantities shows that

$$V(r) \propto r^{-1/2} \qquad \textbf{(21–4)}$$

a function that decreases with increasing distance from the nucleus.

Observations show a remarkable feature of the rotation curves for spiral galaxies (Figure 21–9). As one looks outward from the nucleus, the curves initially rise steeply. (This is expected from Keplerian orbits since the bulge is still within the main mass concentration.) The surprising feature is that the curves then flatten out at large distances from

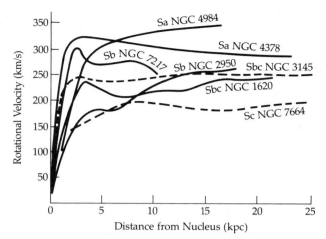

FIGURE 21–9 Rotation curves for spiral galaxies. Note how all flatten out at large distances from their centers. [*V. Rubin, W. K. Ford Jr., and N. Thonnard, Astrophysical Journal (Letters) 225:L107 (1978)*]

velocities. It gave an average mass of 10^{12} solar masses for these spirals (for $H_0 = 75$ km/s · Mpc, see Section 22–2).

We have now estimated galaxy masses by two very different techniques. The first was luminosity based and assumed a mass-to-light ratio of approximately 1 in solar units. This means that we assume that each accumulation of approximately $1M_\odot$ of material in a galaxy contributes approximately $1L_\odot$ of luminosity. Clearly this cannot be true in detail; dark objects such as dust, planets, comets, asteroids, etc. all contribute mass *without* adding to the luminosity. But if our solar system is typical, then the mass contributions of these objects is exceedingly small compared to that of the stars.

The dynamical techniques using velocity dispersions, rotation curves and velocity differences in binaries raise our mass estimates considerably without raising the luminosity estimates. Therefore, the *M/L* values based on dynamical methods rise to values of between 5 and 30 times the solar value. Note that this is consistent with the hypothesis that the difference between ordinary and barred spirals is caused by the presence or lack of a massive but non-luminous halo.

These new estimates of the *M/L* ratio pose serious problems in understanding galaxies. What kind of population of objects can be so numerous as to account for 5 to 30 times the mass of all the stars in the galaxy without producing any light? Could it be dust (no), asteroids, comets, brown dwarfs, black holes? All are unlikely given our understanding of how stars form and evolve.

After all this analysis, can we answer the question "What are galaxies"? They are certainly self-gravitating systems in which we can observe some very familiar phenomena; we see stars move in understandable orbits, many forms of electromagnetic radiation coming from (mostly) understandable sources, and evolutionary patterns of birth and death cycles. For the most part, those aspects of galaxies that are not clearly understood at the moment at least appear to be amenable to comprehension with a little more work from observers and theorists. The most puzzling is understanding the dark matter. In later chapters we will see that the problem only gets worse, and the solution only gets more exotic. Dark matter is truly the central problem in comprehending the Universe.

the nucleus, but they do *not* decrease. This fact implies that a large fraction of the mass lies not in the interior regions but in the halo. Instead of the function $M(r)$ approaching a constant, it seems to have the form $M(r) \propto r$. If the rotation curve of our Galaxy resembles that for other spirals and so continues flat out to 60 kpc, then its total mass is 7×10^{11} solar masses!

Another dynamical method can be used with double galaxies. Just as with visual binary stars, we can apply Newton's form of Kepler's third law [Section 1–4(b)] to determine the mass if we know the distance, the angular size of the orbit, the period, and the position of the center of mass. However, galaxies revolve too slowly for us to see their orbits, periods, and relative centers of mass. So we cannot find the individual masses of binary galaxies. All we can measure are the radial velocities and the separations; we do not know what part of the orbit the galaxies are on or what the inclination is, and so we also do not know what the true orbital velocities are. If we examine a large sample of galaxies, however, and assume that their orbits are nearly circular and randomly oriented to our line of sight, we can estimate from these data the *average* masses of the galaxies sampled. An investigation of 279 binary systems, mostly spirals, used the Doppler shift of the 21-cm line to determine radial

PROBLEMS ◑

1. Our Galaxy and the Andromeda galaxy (M31) are by far the most massive members of the Local Group. If these two giant galaxies form a binary system and move about one another in circular orbits, then calculate
   (a) the distance to the center of mass of the system from our Galaxy
   (b) the orbital period
   Perform the same computations for the orbit of the pair M32 and M31.

2. Use the rotation curves in Figure 21–9 to calculate the masses of NGC 7664 and NGC 4378.

3. The Sun orbits around the Galaxy at a radius of about 8.5 kpc. In a mere 110 million years from now, the Sun will be on the other side of the Galaxy. The nearby galaxies will appear to shift in the sky relative to more distant background galaxies—a galactic parallax. What then would be the number of parsecs in a "galsec," the distance of a galaxy whose galactic parallax equals 1.0"?

4. Based on the photographs and text in this chapter, plot $I(r)$ versus r for a hypothetical (a) E galaxy and (b) S0 galaxy. What kinds of assumptions must you make? How might an Sa galaxy differ from your second plot? Should you use logarithms for either or both axes?

5. Plot the angular diameter vs. distance (in Mpc) for a galaxy with linear diameter of 30,000 pc. At what distance does the galaxy subtend and angle of $d = 4$ arcseconds?

6. Using the data from problem 5, discuss our ability to classify accurately the morphologies of galaxies at distances of 10 Mpc, 100 Mpc, and 1000 Mpc. Assume 1 arcsecond resolution data. Which types are easy to classify? Which are hard?

7. In light of a collapsing gas cloud model of galaxy formation, discuss the observations that halo population stars (Population II) have low metal content and that disk population (Population I) objects with higher metal abundance define thinner disks.

8. Qualitatively discuss why we know that large amounts of dust do not account for the high M/L ratios of galaxies found by the dynamical methods of Section 21–5(C).

9. How many brown dwarfs with $M \approx 10^{-3} M_\odot$ would be needed to give our galaxy an M/L ratio of 10?

10. Assume that the Hubble telescope has recently observed a galaxy that has been given the name Westphal 1. If this galaxy is in the direction $l = 20°$, $b = 30°$, has an observed diameter of $6'' \times 4''$, is of type Sc, has a redshift of $cz = 12,500$ km/s, and has an observed magnitude of $B = 16.5$, what is its absolute magnitude M_B?

11. The Pinwheel Galaxy (M33, also called the Triangulum Galaxy) is a nearby ($D = 690$ kpc) nearly face-on Sc-type spiral galaxy.
    (a) How many parsecs in M33 correspond to an angular size of 1" as seen from the Earth?
    (b) If the rotation curve is like the Sc galaxy NGC 7664 (Figure 21–9), what is the revolution period for a star 10 kpc from the galactic center of M33?
    (c) What is the velocity of the star in (b) in parsecs/year?
    (d) How far in arcseconds would the star appear to move in 100 years?
    (e) If astronomers can measure positions with an accuracy of 0.01", comment on how long it would take before the proper motion of the star in M33 could be readily detected.

12. You observe a star in the field of an elliptical galaxy and want to determine whether the star is a permanent member of the galaxy, an object escaping from the galaxy, or a foreground star in our own Galaxy. You observe the spectrum of the star and find that the Hα line ($\lambda_0 = 656.3$ nm) has an observed wavelength of 656.9 nm. The galaxy has a mass of $10^{12} M_\odot$ and a radius of 100 kpc. What can you conclude?

13. Several possibilities (dust, comets or asteroids, brown dwarfs, black holes) are suggested in the chapter as being unlikely to be the source of the dark matter in galaxies. Explain why each of these classes of objects is unlikely to be the dark matter.

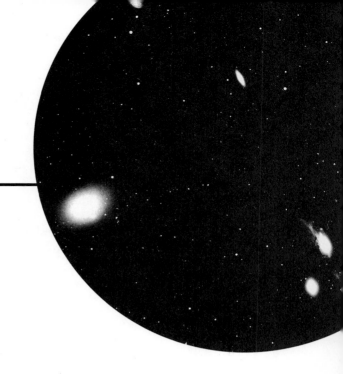

Chapter 22

Hubble's Law and the Distance Scale

We live in a three-dimensional Universe, but astronomers can only measure two dimensions easily. Obviously, these are the east–west and north–south coordinates of an object in the sky. It would not be an exaggeration to state that, historically, the most important work in astrophysics has been the great leaps in understanding brought about when a new method has been found to allow us dramatically to extend measurements of the third dimension, *distance*.

Prior to the twentieth century, the most important distance work came from (1) Copernicus and Kepler who showed us the scale and dynamics of the solar system and (2) Bessel who first measured the trigonometric parallax of stars, which led to the understanding that the Universe was much larger than the volume occupied by our planetary system. We will discuss in the present chapter two more profoundly important methods: (3) the use of the period–luminosity law for Cepheid variable stars, and (4) Hubble's law. These laws have allowed us to appreciate fully the immense size and dynamics of the Universe.

22–1 ◗
THE PERIOD–LUMINOSITY RELATIONSHIP FOR CEPHEIDS

We introduced Cepheids in Chapter 18. Here we will discuss their importance for understanding galaxies and the Universe.

Henrietta Leavitt discovered the **period–luminosity (P–L) relationship** in 1912 for Cepheid variables in the Small Magellanic Cloud. We now recognize that the SMC is a small, nearby galaxy that is separate from the Milky Way and in orbit around it, but, of course, she did not know this at the time. However, she did assume correctly that the stars in the SMC were all at roughly the same distance from the Earth, and this is critical to the discovery. For the period–luminosity relationship is also a period–apparent magnitude relationship when all the variable stars are at roughly the same distance. Harlow Shapley made the P–L relation useful by calibrating it and then applied it in the analysis of the size and shape of the Milky Way.

Using the Mt. Wilson 100-inch telescope, Edwin Hubble searched for Cepheid variables in the great nebula in Andromeda. He announced the results in 1924. This was the first of Hubble's great discoveries, for he found that the Cepheids in Andromeda were so much farther away than those in any portion of the Milky Way or in the LMC that Andromeda must be a separate galaxy. The long debate about the nature of the spiral nebulae was over—they were not components of the Milky Way as Shapley had previously argued. Immediately, the vast scale of the Universe was clear. Spirals similar to Andromeda could be seen in photographs all the way down to the faintest detectable images. Our perception of the Universe changed from one that was at most a million light years across to one that was at least billions of light years in scale!

22–2 ◗
HUBBLE'S LAW

In the last chapter, we examined some important results of optical spectroscopy, but we saved by far

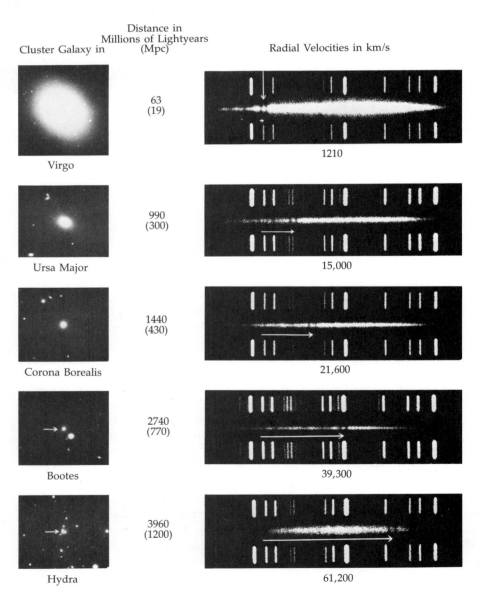

FIGURE 22–1 Redshifts and distances for galaxies. White arrows in the spectra on the right represent the redshift for calcium H and K lines. Distances are based on a Hubble constant of 50 km/s · Mpc. *(Palomar Observatory, California Institute of Technology)*

the most important result, Hubble's law, for this chapter, since its implications are quite profound. Like most discoveries, Hubble's was based on the work of earlier astronomers; in this case V. M. Slipher and K. Lundmark. By 1912 Slipher had found that the Doppler shifts seen in the spectra of spiral nebulae were often larger than the 300 km/s maximum found for individual stars in the Milky Way. Prior to Hubble's demonstration that the spiral nebulae were galaxies, the large size of the radial velocities was one of the major points used in debates about the nature of the spirals.

The 1929 paper in which Hubble announced the relationship that we now call **Hubble's law** followed a long tradition advocated by Lundmark in which the spiral galaxies were used to try to find the sun's motion with respect to an assumed rest frame. The solar motion solutions included not only terms dependent on direction but also terms dealing with distance. Hubble's discovery amounted to finding that the coefficient of the distance term was decidedly nonzero.

Consider a given galaxy at a distance d. If this galaxy emits a spectral line of wavelength λ_0 and we detect this line at a (greater) wavelength λ, then the redshift is defined as

$$z = (\lambda - \lambda_0)/\lambda_0 = \Delta\lambda/\lambda_0 \qquad \textbf{(22–1)}$$

(Note that $\Delta\lambda$ would be negative if the object were approaching; this would be a blueshift.) If the change in wavelength is interpreted as a velocity Doppler shift, then the speed of recession of the observed galaxy is

$$v = c(\Delta\lambda/\lambda_0) = cz \qquad \textbf{(22–2)}$$

where c is the speed of light. By using the apparent magnitude of the galaxy to measure its relative distance, Hubble discovered the correlation

$$cz = Hd \qquad \textbf{(22–3a)}$$

where H is the **Hubble constant;** comparing Equation 22–3a with Equation 22–2, we find an alternative form of Hubble's law:

$$v = Hd \qquad \textbf{(22–3b)}$$

Hubble's law is commonly illustrated in a variety of ways. Figure 22–1 shows photographs of galaxy images along with their spectra for objects spanning a wide range of redshifts. Figure 22–2 directly

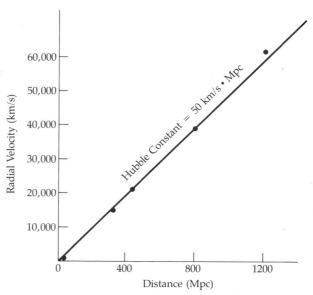

FIGURE 22–2 Hubble's law. Plotted here are the galaxies shown in Figure 22–1.

plots v versus d for the same galaxies. Notice that the slope of the line is Hubble's constant, H.

(A) REDSHIFT, DISTANCE, AND THE AGE OF THE UNIVERSE

Why is this simple mathematical statement (Equation 22–3b) so important? There are many answers. First, it is a new distance-determining method. Hubble used distances found by other means to discover the law and to calibrate it, but once the calibration is known it can be inverted to give a distance estimate based on a galaxy's redshift. Let us rearrange Equation 22–3a to obtain

$$d = cz/H \qquad \textbf{(22–3c)}$$

Then, for example, if $z = 0.1$,

$$d = (0.1)(3 \times 10^5 \text{ km/s})/(50 \text{ km/s} \cdot \text{Mpc})$$
$$= 600 \text{ Mpc}$$

or 300 Mpc if $H = 100$ km/s · Mpc. The second important aspect of Hubble's law is that almost all galaxies have redshifted spectra. (The only blueshifts arise in a few nearby galaxies from either small random motions or because the Earth's rest frame is itself in orbital motion about the center of the Milky Way.) Therefore, the galaxies are flying away from each other—*the Universe is expanding.* The third aspect is related; galaxies at greater distances are moving away faster than those that are near us. This effect is not acceleration but rather *uniform expansion.*

Prior to Hubble's discovery, essentially all philosophical thought about the dynamical state of the Universe held that it was static—neither expanding nor contracting. Indeed, Einstein's 1916 theory of General Relativity was "doctored" with the inclusion of the famous *cosmological constant* [see Section 25–2(b)] in order to "save" the theory. Einstein realized that his equations best described a dynamically active Universe but could not at that time believe that it was possible.

We emphasize that the simple relationship of Equation 22–3c holds only for nearby galaxies with small z. For values of z greater than about 0.8, cosmological effects become important for converting redshifts to distances. For a flat Universe, the proper relationship is

$$d = cz(1 + z/2)/H(1 + z)^2 \qquad \textbf{(22–3d)}$$

Now we will compare Hubble's law, Equation 22–3b, with the distance–time relationship for motion at a constant speed. This relationship is

$$d = Vt$$

so that

$$t = d/V$$

Compare this with Equation 22–3b

$$1/H = d/V$$

which implies that

$$t = 1/H \qquad \textbf{(22–4)}$$

This t is the time since the expansion started, an "age" of the Universe—it is commonly called the **Hubble time.** Using a value of 50 km/s · Mpc for H and conversion factors to get everything in the same units, we obtain the value of $t = 1/H$ in years:

$$\begin{aligned}
t = 1/H &= 1/(50 \text{ km/s} \cdot \text{Mpc}) \\
&= [1/(50 \text{ km/s} \cdot \text{Mpc})] \times (10^6 \text{ pc/Mpc}) \\
&\qquad \times (3 \times 10^{13} \text{ km/pc}) \\
&= 6 \times 10^{17} \text{ s} \\
&= (6 \times 10^{17} \text{ s})/(3 \times 10^7 \text{ s/year}) \\
&= 2 \times 10^{10} \text{ years}
\end{aligned}$$

Our very simple method of estimating the age of the Universe is not quite correct. It does not account for deceleration. One can obtain more accurate estimates for given models (see Chapter 25). A typical calculation would give an age estimate that is about 60% to 70% of the value in Equation 22–4.

(B) PARAMETRIZING EQUATIONS WITH H

Currently, there is considerable uncertainty among various evaluations of Hubble's constant, so many of the equations that are frequently used have a corresponding uncertainty. Therefore, many astronomers write their equations in parametrized form using the definition

$$H = 100 \text{ h km/s} \cdot \text{Mpc}$$

The h is now a parameter having a value in the range $0.5 < h < 1.0$.

Let us look at a couple of examples. Equation 22–3c becomes

$$d = cz/H = 0.01vh \text{ Mpc}, \qquad \textbf{(22–5)}$$

where $v = cz$ is the redshift in symbolic velocity units of 10^3 km/s.

Similarly, Equation 22–4 becomes

$$t = 1/H = 1 \ h^{-1} \times 10^{10} \text{ years} \qquad \textbf{(22–6)}$$

These new forms of the equations allow simple evaluations. Equation 22–6 is particularly revealing. Note that for $h = 1$ ($H = 100$ km/s · Mpc), t is only 1×10^{10} years, less than the inferred age of globular clusters.

(C) THE PHYSICAL MEANING OF THE COSMIC EXPANSION

The Hubble time just calculated is the time in the past when all galaxies, if no acceleration had occurred in their motions, were jammed together at the beginning of the expansion. We call this event the **Big Bang.** Note that the Big Bang took place at a certain time, not at a certain place, because all parts of the Universe were close together at the Big Bang. Since then, the Universe has been expanding—but not into "empty" space: space itself expands as time passes. Galaxies simply serve as luminous markers of this expansion. Hence, the redshifts of galaxies, although commonly called

TABLE 22–1   *Distance Indicators*

Object	M_v	Population	Method	Basis of Calibration
Nearby stars	$>0^m$	Disk	Trigonometric parallax	Radar determination of AU
Galactic clusters	—	I	Main-sequence fitting	Trigonometric parallaxes
				Moving clusters
				Main sequence of nearby stars
Lower main-sequence stars (A–M)	$>0^m$	Disk	Spectroscopic parallax	Trigonometric parallaxes
O and B stars	0^m to -6^m	I	Spectroscopic parallax	Galactic-cluster C–M diagrams
				Galactic-cluster C–M diagrams
				Statistical parallaxes
Supergiants	-6^m, -7^m	I	Spectroscopic parallax	Galactic-cluster C–M diagrams
RR Lyrae stars	$+0.^m5$	II	Period from light curve	Statistical parallaxes
				Globular clusters
Classical Cepheids	$-0.^m5$ to -6^m	I	P–L law	Statistical parallaxes
				Galactic-cluster C–M diagrams
W Virginis stars (Population II Cepheids)	0^m to -3^m	II	P–L law	Statistical parallaxes
				Globular-cluster C–M diagrams
Globular clusters	-5^m to -9^m	II	Integrated magnitude	RR Lyrae stars
				C–M diagrams
Novae	-8^m	Disk	Maximum light	Expansion rate of shell
H II regions	-9^m	I	Angular size	Nearby galaxies
Supernovae	-16^m to -20^m	I and II	Maximum light	Nearby galaxies
Brightest galaxies in clusters of galaxies	-21^m	I and II	Integrated magnitude (a) Brightest galaxy (b) Fifth brightest (c) Mean of ten brightest	Nearby galaxies
Galaxies	—	I and II	Hubble constant of redshift (expansion of Universe)	Doppler shifts of nearest clusters of galaxies

C–M = color–magnitude (= H–R); P–L = period–luminosity.
Statistical parallaxes are averaged over a group of stars, using one component of their proper motions, chosen so that they are *not* affected by solar motion and their peculiar velocities are random.

Doppler shifts, are not really so. They arise from the different distances of our cosmic markers, galaxies, at different times in the history of the Universe. Neither does the observed expansion imply that we lie at the center of the Universe. If the expansion is uniform, then an observer on another galaxy would observe the same Hubble's law.

Finally, recognize that H is not really a constant; it must change with time because of the gravitational effects the galaxies have on each other. In general, H decreases as the Universe ages. The values of H mentioned in this chapter are for *now*, our epoch; they are usually written as H_0.

(D) EVALUATING HUBBLE'S CONSTANT

The problem in finding H is to find the distance d accurately because redshifts z can be measured well. (Typical uncertainties in the determination of a galaxy's redshift range from ≈100 km/s for a low surface brightness object observed at optical wavelengths, down to ≈1 km/s for late-type objects observed in the 21-cm line by a radio telescope.) So what is the value of H? Basically, we don't know yet to within a factor of 2! Estimates in the last ten years range from 50 to 100 km/s · Mpc. In the next section we will examine two techniques that give results at either end of this range. Compare them in light of Table 22–1.

22–3 ◗
DISTANCES TO GALAXIES— THE DISTANCE SCALE

(A) BUILDING UP THE SCALE

Now that we have seen the importance of the distance scale, let's examine in more detail how it is established and what implications its uncertainties have upon our knowledge of the Universe. The distance scale is bootstrapped from direct measurements of planets and nearby stars through a variety of techniques all the way out to some quite tenuous methods that are used on the most distant galaxies. For galaxies, we must generally use indirect methods related to the brightness of objects and the inverse-square law of radiative intensity. The goal here is to calibrate the most luminous objects possible, those that can be seen at the greatest distances.

The calibration process starts locally in the Solar System and progresses by steps to span our Galaxy (Table 22–1 and Figure 22–3), the galaxies nearest to ours [called the **Local Group;** Section 23–1(b)], and more distant galaxies. Objects that are brighter than absolute magnitude zero can be used as distance indicators for galaxies within the Local Group. Galactic-cluster, main-sequence fitting is

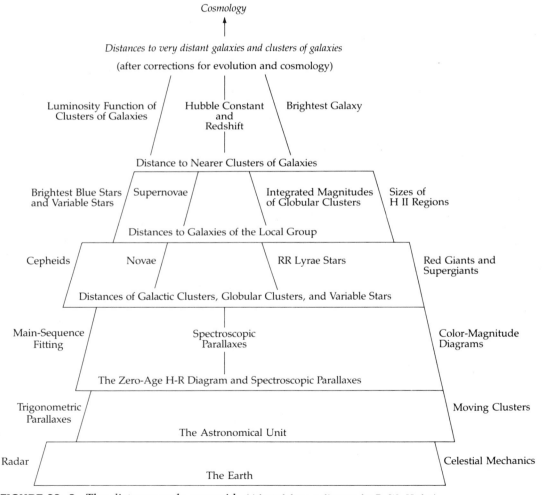

FIGURE 22–3 The distance-scale pyramid. *(Adapted from a diagram by P. W. Hodge)*

not useful even for nearby galaxies, but this method allows calibration of bright stars, such as OB stars and classical Cepheids. Although the methods listed in the second part of Table 22–1 permit one to find distances to galaxies with large distance moduli, these distance indicators are substantially less reliable than those in the first part of the table. The power of the methods listed in the second part of the table lies in the very great brightness of the objects and in the subsequently large limiting distance moduli. The bright objects can give distances to galaxies well beyond the Local Group to other large clusters of galaxies, such as the Virgo cluster. Statistics from these clusters tell us about the luminosities (absolute magnitudes) of the brightest galaxies in a cluster and about the luminosity function within a cluster. Although there is some variation from cluster to cluster, the statistics are good enough to give a first approximation of distance. Distinct groupings appear at certain luminosities; for instance, the brightest galaxy in a large cluster is nearly always a giant elliptical.

What kinds of galaxies serve as useful standards? The candidates are often supergiant spiral galaxies with small nuclei and spread-out spiral arms, that is, Sc I galaxies. (M101 is such a galaxy.) Several teams have calibrated the absolute magnitude of such galaxies as about −21.2, or 25×10^9 times the Sun's luminosity. Seen in distant clusters, these galaxies are relatively easy to identify because of their high luminosity and distinctive shape. With contemporary telescopes, they serve as standard candles to distances of roughly 400 Mpc. Giant elliptical galaxies are also used. Again, their luminosities must be calibrated by reference to some nearby galaxy or cluster whose distance is known by other techniques. Note that the lower levels of the distance pyramid (Figure 22–3) affect the outcome most strongly.

(B) THE SANDAGE–TAMMANN PROCESS TO OBTAIN H

Here are the steps in a contemporary procedure followed by Allan Sandage and Gustav Tammann that yields a value of 50 km/s · Mpc (±10 percent) for *H*:

1. Measure the astronomical unit using radar reflection from Venus and then measure the distances to nearby stars using trigonometric parallax with the astronomical unit as the baseline.
2. Determine the distance to the Hyades, a galactic cluster, using the moving-cluster method. Check for consistency by comparing brightnesses of Hyades stars with those of nearby stars with distances known by trigonometric parallax.
3. Find the distances to Cepheids in our Galaxy by searching out galactic clusters that contain Cepheids, comparing the brightness of the stars in these clusters with those of stars in the Hyades cluster, and so finding the distances to the clusters with Cepheids. This calibrates the Cepheid P–L relationship.
4. Use Cepheids to determine the distances to nearby galaxies by the P–L relationship.
5. Measure the angular sizes of H II regions in these nearby galaxies. Find that the sizes of H II regions in spiral and irregular galaxies depend on the luminosity of the galaxy. Use nearby galaxies to calibrate the relationship between size of H II region and luminosity.
6. Extend this calibration to Sc I galaxies, for which the nearest, M101, has its distance determined by different methods as a check. We now have the sizes of H II regions in Sc I galaxies.
7. Use the size–luminosity relationship for H II regions to find the distances of galaxies (about 60) to the limit of this method. Now we know the absolute magnitudes of these galaxies and their relationship to luminosity classes.
8. Look at Sc I galaxies, the objects we can see distinctly at the greatest distances. Use the absolute-magnitude calibration (Step 7) to find their distances. Measure their redshifts to get their radial velocities; divide their radial velocities by their distances to get the Hubble constant.

Note that this procedure relies strongly on Cepheids as distance indicators. If the Hubble Space Telescope can eventually be made to work as designed, astronomers should be able to resolve Cepheids in galaxies about ten times farther away than the present ground-based limit. This telescope would then increase direct measurements of distances to galaxies by a factor of 10—and the number of galaxies by a factor of 1000.

(C) THE AARONSON–HUCHRA–MOULD WAY TO H

Marc Aaronson, John Huchra, and Jeremy Mould have developed a new technique—independent of the Sandage–Tammann steps—that gives a value of $H = 90$ km/s · Mpc (with a claimed error of 5%). Let's briefly look at their procedure, which relies on radio and infrared observations to infer the luminosities of galaxies.

The technique rests on a relationship, derived by R. Brent Tully and J. Richard Fisher, between the absolute magnitudes (in blue light) of spiral galaxies and the spreads in frequency of their 21-cm H I emissions. The wider the 21-cm line, the greater a galaxy's luminosity. The 21-cm emission comes from the neutral gas in a spiral galaxy's disk, which is broadened by rotational motion. Most galaxies have maximum rotational velocities of 100 to 300 km/s. So the Tully–Fisher relationship is really one between rotational velocity and luminosity of spiral galaxies. Note that the rotational velocity at a given distance from the center relates to the mass of a galaxy, and the total luminosity of the stars in a galaxy is also related to the total number of stars and hence the total galaxy mass. The relationship shows relatively little variation from one galaxy to the next, at least among galaxies of the same type.

If properly calibrated, the Tully–Fisher relationship provides a galaxy's luminosity from a measurement of the width of its 21-cm line. The main problem with the method, as originally developed, lay in its use of blue magnitudes. Dust scatters blue light well, so that blue-light observations of another galaxy suffer from an unknown amount of extinction produced by dust in our Galaxy and in the disk of the other galaxy. Aaronson, Huchra, and Mould noted that infrared light suffers virtually no extinction from galactic or extragalactic dust. So they made infrared observations of nearby galaxies (such as M31 and M33) whose distances are known from standard candles, such as the Cepheids. This gives the absolute infrared magnitudes (essentially the infrared luminosities at a wavelength of 1.6 μm). These infrared absolute magnitudes connect well with the widths of the 21-cm lines of local galaxies. Hence, we can infer the infrared luminosities of galaxies from the widths of their 21-cm lines (Figure 22–4). Compare

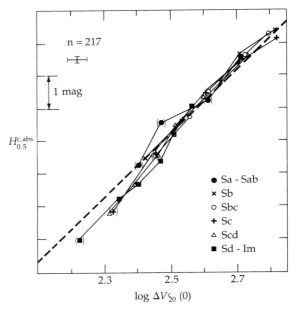

FIGURE 22–4   The infrared Tully–Fisher relationship. Plotted here are infrared absolute magnitude versus rotational velocity for more than 200 spiral galaxies. The dashed line is the average of all values. (*Adapted from a diagram by M. Aaronson*)

these with their infrared fluxes, and we can infer the distances and so a value for H. This technique yields a best result of 90 km/s · Mpc for observations of 11 clusters of galaxies (Figure 22–5).

Who's right? Both the Sandage–Tammann method and the Aaronson–Huchra–Mould method probably contain unknown systematic errors, but we can say confidently that H lies in the range 50 to 100 km/s · Mpc. We note from Equation 22–6 that there are problems with the higher values of H in regard to the age of the Universe being in conflict with the ages of globular clusters, but this does not necessarily mean that the larger values are excluded—there might be other solutions to this problem. We also do not mean to indicate that we have discussed the only methods available for determining H. Several other investigations also give values in the same range. A very recent one using supernovae as standard candles concluded that H is 68 to 76 km/s · Mpc (Figure 22–6).

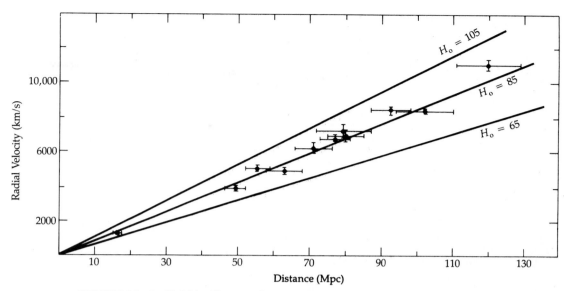

FIGURE 22–5 Hubble diagram for 11 galaxy clusters. These results are based on the infrared Tully–Fisher relationship. Shown is a range of values for the Hubble constant. *(Adapted from a diagram by M. Aaronson)*

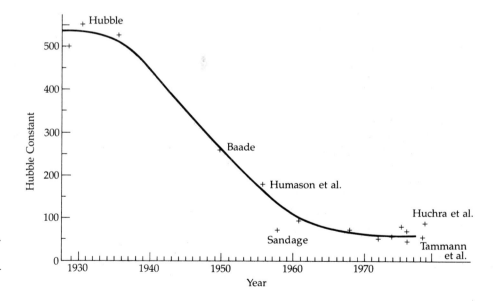

FIGURE 22–6 Hubble constant versus year of measurement. Astronomers are beginning to come close to the actual value. *(Adapted from a diagram by D. N. Schramm)*

PROBLEMS ◗

1. An approximate distance for the separation between our Galaxy and the Magellanic Clouds is 50 kpc. What will be the observed apparent magnitude for the following stars in the Magellanic Clouds:
   (a) an RR Lyrae variable with a period of 0.5 day
   (b) a classical Cepheid with a pulsation period of 100 days
   (c) a Population II Cepheid with a ten-day period

2. What methods of distance determination are most useful for finding the distance to
   (a) the Pleiades
   (b) a globular cluster in our Galaxy
   (c) the Large Magellanic Cloud
   (d) M31, the Andromeda galaxy
   (e) the Virgo cluster of galaxies
   (f) the Hercules cluster of galaxies
   (g) our Sun
   (h) the nucleus of our Galaxy

3. The Ca II K stellar absorption line has a rest wavelength of 393.3 nm. In a particular galaxy, the Ca II K line is observed at a wavelength of 410.0 nm. What is the distance to the galaxy, assuming
   (a) $H = 50$ km/s · Mpc
   (b) $H = 100$ km/s · Mpc

4. How would the derived age of the Universe (Equation 22–4) change if
   (a) the expansion is decelerating (H decreasing with time)
   (b) the expansion is accelerating (H increasing with time)
   (c) What observations could be made to determine if either of these possibilities is correct?

5. Most of the nearest galaxies do not obey Hubble's law. Explain why.

6. Devise a lecture demonstration of Hubble's law using some easily attainable elastic materials.

7. Plot Equation 22–3d for redshifts up to $z = 5$.

8. We often use the equation $m - M = 5 \log (d) - 5$ to obtain absolute magnitudes for distant galaxies. The distance d is usually estimated by Hubble's law. Parametrize this equation by using h from Section 22–2(b).

9. Table 22–1 lists a number of types of objects used as distance indicators. What is the maximum distance of detectability for each of these indicators for a telescope with a limiting magnitude of +25? (The Hubble Space Telescope will be able to detect objects of this brightness.)

10. Show that for small values of z, Equation 22–3d reduces to Equation 22–3c.

11. For relatively nearby galaxies, the hydrogen Balmer-alpha line can be redshifted out of the visible part of the spectrum.
    (a) For what redshifts would the Balmer-alpha line of hydrogen be shifted out of the visible into the infrared portion ($\lambda > 720$ nm) of the electromagnetic spectrum?
    (b) To what distances does this correspond? Express your answer in terms of the Hubble parameter h.
    (c) To what distances do these correspond for a Hubble constant $H = 100$ km/s · Mpc; for $H = 50$ km/s · Mpc?

12. For relatively distant galaxies (and quasars—see Chapter 24), the hydrogen Lyman-alpha line can be redshifted into the visible part of the spectrum.
    (a) For what redshifts would the Lyman-alpha line of hydrogen be shifted out of the ultraviolet into of the visible portion ($\lambda \approx 390$ nm to 720 nm) of the electromagnetic spectrum?
    (b) For a flat Universe, to what range in distance do these redshifts correspond? Express your answer in terms of the Hubble parameter h.
    (c) To what distances do these correspond for a Hubble constant $H = 100$ km/s · Mpc; for $H = 50$ km/s · Mpc? Compare these distances to the age of the Universe.

Chapter 23

Large-Scale Structure in the Universe

So far we have only hinted at the large-scale layout of matter in the Universe. Recent observations have revolutionized the picture, changing it from one in which galaxies congregated in spherical clusters imbedded in a uniform background of galaxies to one in which chain-like superclusters snake among vast voids of space. This chapter deals with that structure, which may be the undistorted fossil of an earlier stage in the Universe's evolution.

23–1 ◐
CLUSTERS OF GALAXIES

Most galaxies—maybe all—are members of some type of cluster. The Milky Way Galaxy is one of the dominant members of the Local Group of galaxies, which in turn is part of a local supercluster that also includes the rich Virgo cluster.

(A) TYPES OF CLUSTERS

There are different ways of defining categories of galaxy clusters, but the simplest is that of George Abell, who separates rich clusters into regulars and irregulars. **Regular clusters** tend to be giant systems with spherical symmetry and a high degree of central condensation; they frequently contain many thousands of member galaxies, of which perhaps 1000 are brighter than absolute magnitude -15. Almost all members of regular clusters are either elliptical or S0 galaxies, whereas **irregular clusters** contain a mixture of all types of galaxies. Among the irregular clusters of galaxies are included (1) small groups, such as our own Local Group, (2) loose aggregates of subgroups with several centers of condensation, and (3) fairly large but diffuse clusters. The **luminosity function for galaxies** in clusters (the number of galaxies per integrated magnitude interval) indicates that there is a great preponderance of faint galaxies.

(B) THE LOCAL GROUP OF GALAXIES

Our Milky Way Galaxy and the Andromeda galaxy (M31) dominate the small group of galaxies referred to as the Local Group, which contains at

TABLE 23–1   *Selected Members of the Local Group*

Name	Type	M_v	Distance (kpc)	$m - M$	Mass $(M_\odot)$	Radial Velocity (km/s)
M31 = NGC224	Sb	−21.1	690	24.6	3×10^{11}	−267
Galaxy	Sb or Sc	−21	8.5	14.7	4×10^{11}	—
M33 = NGC598	Sc	−18.9	690	24.6	4×10^{10}	−190
LMC*	Irr I	−18.5	50	18.6	6×10^{9}	+275
SMC*	Irr I	−16.8	60	19.1	1.5×10^{9}	+163
NGC205	E6p	−16.4	690	24.6	—	−239
M32 = NGC221	E2	−16.4	690	24.6	2×10^{9}	−220
NGC6822	Irr I	−15.7	460	24.2	1.4×10^{9}	−34
NGC185	dE0	−15.1	690	24.5	—	−270
NGC147	dE4	−14.8	690	24.5	—	—
IC1613	Irr I	−14.8	740	24.5	4×10^{8}	−235
Fornax	dE3	−13.0	188	21.4	2×10^{7}	−73
Sculptor	dE3	−11.7	84	19.7	3×10^{6}	—
Leo I	dE3	−11.0	220	21.8	3×10^{6}	—
Leo II	dE0	−9.4	220	21.8	10^{6}	—
Ursa Minor	dE6	−8.8	67	19.5	10^{5}	—
Draco	dE3	−8.6	67	19.6	10^{5}	—

*LMC = Large Magellanic Cloud; SMC = Small Magellanic Cloud.

least 20 members (Table 23–1). The other members are fainter and less massive and to some extent seem to be concentrated around one or the other of the two largest spiral galaxies. A wide range of galaxy types is included in the group, from the three major spirals (our Galaxy, M31, and M33) to the dwarf ellipticals and irregulars. Most of the galaxies are dwarfs with low mass and luminosity. The Local Group spans about 1 Mpc along its largest dimension (Figure 23–1).

The importance of the Local Group lies not only in the fact that it is the closest cluster but also that the study of its individual galaxies enables us to learn a great deal about the characteristics of galaxies; we may then use this knowledge to extrapolate to the more distant galaxies. Local Group members also serve as critical calibrators for the distance scale. Finally, we point out that, in comparison with other clusters, the Local Group does not contain a large number of galaxies, nor are many of these very massive.

(C) OTHER CLUSTERS OF GALAXIES

Other clusters range from compact groups to rather loose collections. Several are listed in Table 23–2. The Fornax cluster, one relatively close to us,

contains many types of galaxies even though the total number is only 16. The huge Coma cluster spreads over at least 7 Mpc of space and contains thousands of galaxies. From these observations, we know that a typical cluster contains about 100 galaxies brighter than $M = -16$ and is separated by tens of millions of lightyears from its neighboring clusters. Under the Abell classification, the Coma cluster (Figure 23–2) is a regular cluster. Two large, bright elliptical galaxies lie near the center, about which the others seem to concentrate; other rich, regular clusters, such as A2199, are dominated by cD galaxies, which are supergiant, elliptically shaped galaxies with extensive halos. Examples of irregular clusters are the Local Group, the Hercules cluster (Figure 23–3), and the Virgo cluster. Of the 205 brightest galaxies in the Virgo cluster, the four brightest are giant ellipticals, but ellipticals make up only 19% of the total, whereas 68% are spirals. The Virgo cluster covers about 7° in the sky, which with its distance of 15.7 Mpc, implies that its diameter is some 3 Mpc.

The most widely used catalog of clusters is the one compiled by Abell. From his examination of the Palomar Sky Survey, he found 2712 clusters. His richness classification is based on the number of galaxies within 2 magnitudes of the third bright-

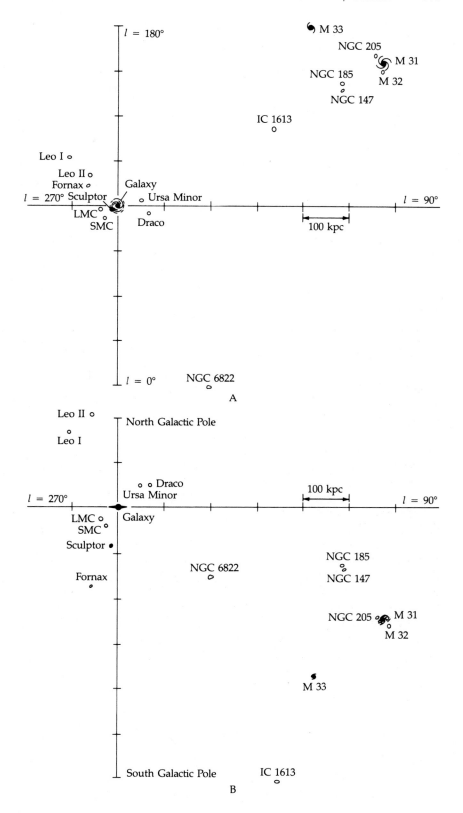

FIGURE 23–1 The Local Group. (A) A top view projected on the plane of our Galaxy. (B) A side view. Note the concentrations around our Galaxy and M31.

TABLE 23–2   Seclected Clusters of Galaxies

Name	Abell #	$\alpha(2000)$	$\delta(2000)$	Redshift (cz, km/s)	Remarks (SC = Supercluster)
Perseus	426	3^h 18.6^m	$+41°$ $32'$	5460	Perseus SC
Hydra	1060	10^h 36.9^m	$-27°$ $32'$	3000	Hydra-Centaurus SC
Virgo		12^h 30^m	$+12°$ $23'$	1200	Local SC
Centaurus		12^h 50^m	$-41°$ $18'$	3200	Hydra-Centaurus SC
Coma	1656	12^h 59.8^m	$+27°$ $59'$	6647	Coma/ A1367 SC
Corona Borealis	2065	15^h 22.7^m	$+27°$ $43'$	21,600	
Hercules	2151 + 2152	16^h 05.2^m	$+17°$ $43'$	11,200	Hercules SC
	2199	16^h 28.6^m	$+39°$ $31'$	9200	A 2197/ 2199 SC

FIGURE 23–2 Central region of the Coma cluster. *(National Optical Astronomy Observatories)*

FIGURE 23–3 Central region of the Hercules cluster. *(Palomar Observatory, California Institute of Technology)*

est cluster member. Richness class 1 clusters have 50–79 galaxies satisfying the criterion, while richness class 5 clusters (only one is known) have at least 300. All counts were made within an angular distance that was adjusted to the cluster's estimated distance (judged from the tenth brightest cluster member), so the same physical volume should have been studied for all clusters. Although many clusters known prior to this work have proper names—such as Coma, Hercules, Perseus, and Corona Borealis—references in the literature are often made to the Abell number. For example, Coma is A1656. We note in passing that the Virgo cluster does not appear in Abell's catalogue. This is partly because it is too close; Abell excluded very near clusters that had large angular extent, but Virgo also may not be rich enough!

(D) CLUSTERS AND THE GALAXIAN LUMINOSITY FUNCTION

The local stars have a luminosity function; that is, a certain number of stars can be found in a given range of luminosity or absolute magnitude. The basic trend is that there are many fewer extremely luminous (OB) stars than low-luminosity (M V) stars [Section 14–2(a)]. A similar hierarchy exists for galaxies in clusters. In the Local Group, only three galaxies are highly luminous (the Milky Way, M31, and M33); most galaxies in the Local Group are dwarf, low-luminosity galaxies. So galaxies have a luminosity function somewhat like that of stars.

Clusters provide a direct way of determining the luminosity function of galaxies, for you can see a wide range all at once. You can count the number of galaxies in specific apparent-magnitude ranges, from the brightest to the faintest. Because every galaxy in a cluster lies at about the same distance, a plot of number versus *apparent* magnitude translates into a plot of number versus *absolute* magnitude once you know the distance to the cluster. (Abell thought that this process might be reversable; perhaps fitting luminosity functions could lead to a new method of distance determination. It does not work well, however.) The main drawback here is that dwarf galaxies are undercounted because they are so faint at great distances. Abell determined the integrated luminosity functions of a

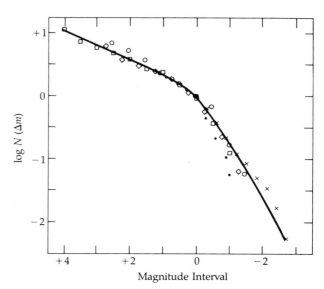

FIGURE 23–4 Luminosity function of galaxies. Number of galaxies in a magnitude interval versus magnitude interval, with an arbitrary zero. (*Adapted from a diagram by G. Abell*)

few clusters and found that the number of bright galaxies falls off very rapidly as the luminosity increases (Figure 23–4). This behavior tells us that there are many more low-luminosity galaxies in a cluster than high-luminosity ones.

Let's examine the concept of luminosity function (l.f.) in a bit more detail. We do so because if we know the l.f. for clusters, then we can determine (1) the (unobserved) dwarf galaxy populations in clusters, (2) the mass distributions in clusters, and (3) perhaps the distances to clusters. A basic luminosity function gives the number of galaxies per luminosity (or magnitude) interval per unit volume of space. The exact expression of the function can take many forms. Paul Schechter has proposed an analytical form for the differential l.f. that seems useful and appropriate to a wide variety of samples of galaxies:

$$\Phi(L)dL = \Phi^*(L/L^*)^\alpha \exp\,(-L/L^*)\,d(L/L^*) \quad \textbf{(23–1)}$$

where $\Phi(L)$ is the density of galaxies in the range from L to $L + dL$, Φ^* is the normalization parameter, L^* is a characteristic luminosity at which the slope of the function changes rapidly (at $M_B = -19.4$), and α is the slope of a plot of log Φ ver-

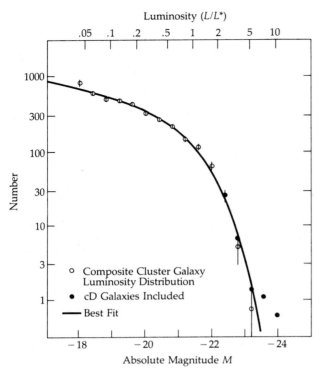

FIGURE 23–5 Schechter luminosity function.
[P. Schechter, Astrophysical Journal 203:297 (1976)]

sus log L (for $L < L^*$). Typically, for rich clusters, $\alpha = -5/4$ and

$$\Phi^* = 0.005(H_0/50)^3/\text{Mpc}^3$$

See Figure 23–5 for a graph of such a function.

It's difficult to estimate the masses of these clusters. Not all the material in them can be seen, and so adding up all the observable galaxies gives a lower limit to a cluster's mass. On the other hand, if the cluster is assumed to be gravitationally bound, the motions of its members establish an upper limit on its mass. Masses range from 10^9 to 10^{15} solar masses. However, it is not possible to tell if all clusters are bound and stable or if they are unstable and expanding. If they are *not* stable, the mass estimates fall into the lower end of the range.

(E) GALAXIAN CANNIBALISM

A remarkable fact about clusters is that—relative to the sizes of the galaxies in them—the galaxies are spaced close together. Compare galaxian spacings with those of planets and stars. In the Solar System, planets are spaced out about 10^5 times their diameters. In the Galaxy, stars are spaced about 10^6 times their diameters. In a cluster, however, the spacing is only 100 times a typical galaxy's diameter. Now, if in relative terms galaxies are so close, consider also that the most massive galaxies (supergiant ellipticals) are at least 10^7 times more massive than the least massive ones (irregulars and dwarf ellipticals). Tidal forces could cause the largest galaxies to disrupt the smaller ones strongly enough to destroy their structure and then pull the pieces in. This devouring of a smaller galaxy by a larger one has been called *galaxian cannibalism*. (Current usage of *galactic* refers to the Milky Way; *galaxian* is the adjective used for galaxies in general.)

What observations support this idea? Some point to supergiant cD galaxies as a special class of galaxies. Their peculiar properties include (1) extensive halos, up to 1 Mpc in diameter, (2) multiple nuclei (sometimes), and (3) location at the center of clusters. These observed properties plus theoretical calculations of the motions of these galaxies in clusters suggest that the cD galaxies result from cannibalism, that is, from close encounters at the centers of clusters or the infall of material tidally stripped from other cluster members. Dynamical friction may also play a role. These processes all assume that the growing galaxies lie in the center of the gravitational potential well of a cluster, so that material freed from other galaxies congregates there.

Some observations support this notion: (1) cD galaxies *do* lie at the centers of clusters—within 200 kpc, which is about the diameter of a cD galaxy, and (2) photometry of the inner parts of cD galaxies indicates that they have the same properties as E galaxies. This observation backs up the scenario that cD galaxies were once E galaxies that added material to create their extensive halos.

At least 50% of cD galaxies have more than one nucleus. Might these be the leftovers from cannibalism? Recent observations suggest so. Doppler-shift data indicate that the nuclei within a cD move at relative speeds of about 1000 km/s. In contrast, the stars within a cD galaxy move in orbits at about 300 km/s. Hence, the nuclei do not share the stellar dynamics in a cD galaxy. They are moving much faster, and their velocities carry them far away from the center of the galaxy. So the nuclei may be normal elliptical galaxies that have come close to and passed through the cD galaxy. Dy-

FIGURE 23—6   Tidally interacting galaxies. NGC4038 and 4039 are interacting by tidal forces, which result in the streaming tails. *(Palomar Observatory, California Institute of Technology)*

FIGURE 23—7   The galaxies M82 (top) and M81 (bottom). The activity of M82 may be the result of a tidal interaction. *(National Optical Astronomy Observatories)*

namical friction has decreased their orbital energies so that they are now bound to the cD galaxy.

Although galaxies may not actually merge, they certainly undergo close encounters and interact by tidal forces. Such interactions would have some general effects. First, as illustrated by the Earth's tidal bulges, matter would be pulled out in bulges on both sides of each galaxy. Second, because galaxies rotate, their material would conserve angular momentum after a tidal encounter and move off in arc-shaped streams. So we expect that tidal bridges may join two tidally interacting galaxies and tails may flow away from each in opposite directions. Have such interacting galaxies been seen? Many galaxies with peculiar shapes—those that do not fall into the standard Hubble form—show some of the characteristics of tidal interactions. An excellent example is the pair NGC4038 and 4039 (Figure 23–6). Here is visible a bridge of material between the galaxies and tails heading off in opposite directions. Computer simulations of such an encounter show that similar structures should result from the gravitational interactions. The peculiar galaxy M82 likely had its star formation triggered by a tidal interaction with M81 (Figure 23–7).

(F) CLUSTER REDSHIFTS AND VELOCITY DISPERSIONS

Redshifts are the key to understanding the structure and dynamics of the Universe. We use these observations in galaxy and cluster work to locate the object in three dimensions, gauge its kinetic energy by means of the velocity dispersion, and isolate its members by using large redshift differences to reject foreground or background objects. At first sight, the mean redshift and the standard deviation about that mean might be expected to provide the desired statistical parameters. It's not that simple!

Velocity dispersions are particularly troublesome, and this comes partly from the fact that extragalactic astronomers use them in two rather different arenas. For galaxies, we cannot measure velocities for the individual components (except for very nearby objects), so the velocity dispersion must be measured by the broadening of the galaxy's spectral features. The position of the line center gives us the galaxy redshift, while the line width gives us the velocity dispersion, σ_s, of the stellar populations inside the galaxy.

One sophisticated technique for quantifying σ_s comes from cross correlating a galaxy's spectrum with that of a standard star of similar overall spectral type. The width of the lines in the star's spectrum are the minimum values consistent with the resolution of the spectrograph. The cross correlation technique employs the Fourier transforms of the two spectra to find the best match. Think of it as translating (the redshift) and broadening (σ_s) the star's spectrum to match the galaxy's spectrum.

In contrast, when we measure the velocity dispersion, σ_c, of a cluster, we can observe the individual components. Since the cosmological implications of σ_c are profound, let's examine one method of calculating it that is unbiased and adds redshifts correctly; it also accounts for observational uncertainties. We follow the formulae of L. Danese, G. De Zotti, and G. di Tullio, and their method is applicable for clusters with redshifts $z < 1.0$.

First, let us review the usual definition of the standard deviation.

$$\sigma = \sqrt{\frac{\Sigma(x - \langle x \rangle)^2}{n}} \qquad \text{(23–2)}$$

This expression is inconvenient for computing and is biased if only a subset of the cluster is sampled. A better expression is:

$$\sigma = \sqrt{\frac{n}{n-1}} \sqrt{\frac{\Sigma x^2}{n} - \left(\frac{\Sigma x}{n}\right)^2} \qquad \text{(23–3)}$$

Next, we turn to the mean redshift of the cluster. Redshifts are complicated by the fact that we do not directly observe the cosmological component, z_r. Instead, we see the sum of the line-of-sight components of z_r plus the peculiar motions of the individual galaxies with respect to the cluster center and our own peculiar velocity. The latter is surprisingly complex. The easiest parts of our peculiar motion are earth's rotational and orbital motions. Somewhat less certain is the sun's motion relative to the center of the Milky Way; this has traditionally been corrected for by

$$v_{corr} = 300 \sin (l) \cos (b)$$

where l and b are the galactic longitude and latitude, respectively, for the galaxy in question. In Section 23–2(b) we will find that the Milky Way itself has important peculiar motions, but we will ignore these for the moment. We can intuitively write

$$v_{obs} = v_{cluster} + v_{us} + v_{galaxy} \qquad \text{(23–4)}$$

Here $v_{cluster}$ is the mean redshift of the cosmological component of the cluster, i.e., $v_{cluster} = cz_r$. In addition, v_{obs} is the observed redshift; v_{us} is the peculiar motion of the Earth, and v_{galaxy} is the individual motion of the galaxy relative to its cluster center. Although Equation 23–4 is usually used in practice, the relativistically correct expression is

$$(1 + z_{obs}) = (1 + z_r)(1 + z_{us})(1 + z_{galaxy}) \qquad \text{(23–5)}$$

If we now take averages (quantities enclosed in angle brackets) and assume $\langle z_{galaxy} \rangle = 0$, then

$$\langle z_r \rangle = \frac{(\langle z_{obs} \rangle - \langle z_{us} \rangle)}{(1 + \langle z_{us} \rangle)} \qquad \text{(23–6)}$$

The denominator in Equation 23–6 is not intuitively obvious and is often overlooked. For statistical purposes $\langle z_r \rangle$ is not sufficient, we also need its uncertainty, $\sigma \langle z_r \rangle$. If we have a homogeneous data set with uniform uncertainty in an individual redshift of ζ (typically $c\zeta = 30$ to 100 km/s in the optical), then

$$\sigma \langle z_r \rangle = \sqrt{\frac{\sigma^2 + \zeta^2}{n}} \qquad \text{(23–7)}$$

where σ is obtained from Equation 23–3.

Now we turn to the velocity dispersion. In the spirit of Equation 23–6, we write

$$v_p = \left(\frac{V_p - \langle V_p \rangle}{1 + \langle V_p \rangle / c}\right) \qquad \text{(23–8)}$$

Here v_p is that component parallel to the line of sight of the velocity of a galaxy with respect to the cluster center. V_p is the observed redshift of the galaxy *uncorrected* for the peculiar motion of the observer, and $\langle V_p \rangle$ is the mean value of the V_p's. We now can determine the line-of-sight velocity dispersion, σ_p.

$$\sigma_p = \sqrt{\frac{\Sigma v_p{}^2}{(n-1)} - \frac{(c\zeta)^2}{(1 - v_p/c)^2}} \qquad \text{(23–9)}$$

In some calculations, we want the *physical* velocity dispersion, which can only be estimated from σ_p using a model. Under the assumptions of large n and randomly oriented orbits, we can write

$$\sigma = \sqrt{3} \, \sigma_p$$

Typical velocity dispersions for rich Abell clusters fall in the range $400 < \sigma_p < 1500$ km/s.

23–2 ◗
SUPERCLUSTERS

Does the Universe have a higher level of organization than clusters of galaxies? Are there clusters of clusters—**superclusters**? For some years, astronomers were extremely skeptical about the reality of superclusters. Before observations confirmed that reality, the standard picture was that of mostly spherical clusters imbedded in a very uniform distribution of isolated, noncluster galaxies. Recently that view has rapidly changed to one in which superclusters have a strung-out, filamentary structure hundreds of megaparsecs long. Between them lie vast voids, empty of luminous matter such as galaxies. The superclusters may be interconnected—the fundamental network of the Universe.

(A) DISCOVERY

Once Hubble had shown in 1924 that the Universe is filled with galaxies, clusters were also known to exist—they are that *obvious*! Some are so rich that many hundreds appeared on a single photographic plate as Hubble was amassing his great collection. Several contemporaries speculated that there might exist larger structures—*clouds* of clusters in an older terminology. (Today, we use *superclusters* to refer to systems containing multiple galaxy clusters.) For example, Harlow Shapley

pointed out that there were more galaxies and clusters seen in the North Galactic Hemisphere than in the South. Also, while searching for Pluto, Clyde Tombaugh noted many possibly associated concentrations of clusters in what is now know as the Perseus supercluster. Still, Hubble thought that his photographic surveys established large-scale homogeneity, and his reputation was such that most astronomers held the same view.

Starting in the 1950s new, wide field studies with telescopes such as the Palomar Schmidt (leading to the **Palomar Observatory Sky Survey**—a crucially important tool for studying clusters) led the way to further two-dimensional analyses (few redshifts being known at the time). George Abell and Fritz Zwicky compiled catalogues of clusters, and Donald Shane and Carl Wirtanen examined the statistics of galaxy counts. They counted galaxies brighter than magnitude 19 in $1/6°$ squares in the northern sky—some million galaxies in a survey that took 12 years. The map (Figure 23–8)

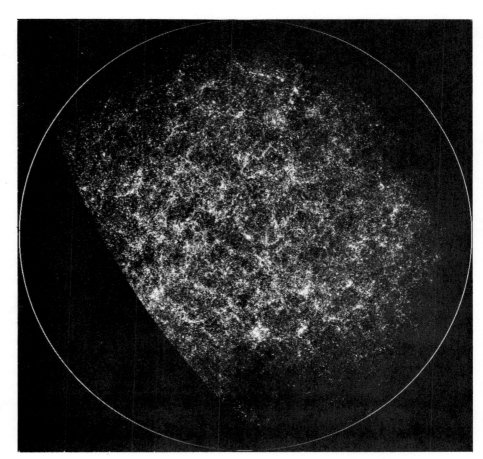

FIGURE 23–8 Galaxy clustering. This plot of more than one million galaxies shows filamentary structures in two dimensions. *(P. J. E. Peebles)*

shows that clusters of galaxies interconnect in a chain-like fashion—the explosive imprint of the Big Bang. (But note that this map is a two-dimensional projection of three-dimensional structure, so that not all of the chains are real.)

All of these analyses except Zwicky's concluded that larger structures existed, but Zwicky's clusters often had multiple concentrations. Many astronomers believed that superclusters existed, and the implicit assumption was that they would have *spherical symmetry*. A *core/halo model* was suggested with a rich cluster surrounded by groups and isolated galaxies.

One reason for this model was the persistent work of Gerard deVaucouleurs who described the **Local Supercluster** in similar terms. The dominating cluster was Virgo, and the Local Group was one of the outlying components. Strict spherical structure was known not to hold, since deVaucouleurs had defined a plane (the supergalactic plane) along which the galaxies concentrate.

The next major step in the search for very large structures came in the 1970s with the widespread introduction of image intensifying tubes. These devices amplified faint light signals by factors of 10^3 to 10^5. Several groups employed these devices in a new technique of making statistically complete redshift surveys in relatively large-area regions of the sky. S. Gregory, L. Thompson, and W. Tifft were the first to actually demonstrate with three-dimensional positions the existence of *external superclusters*. They showed that the Coma and A1367 clusters (separated by about 20° in the sky) were joined together by a bridge of galaxies and small clusters. In their work, they also surveyed the foreground and background galaxies since they had no *a priori* knowledge of membership. This led to the serendipitous discovery that a large region in the foreground was remarkably empty. Quickly, other superclusters were found; G. Chincarini, H. Rood, and M. Tarenghi joined in demonstrating the second—in Hercules. Another was soon found in Perseus and another in the South in the Hydra/Centaurus region. All of these supercluster studies also found the empty regions, and the term *void* was universally adopted.

In the following decade, many more examples of voids and superclusters were found. R. Kirshner, A. Oemler, P. Schechter and S. Shectman

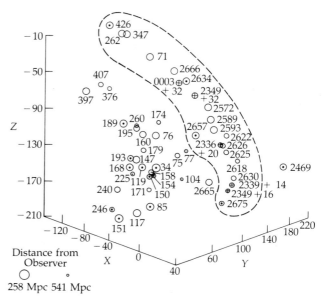

FIGURE 23–9 Perseus-Pegasus supercluster filament. Circles show the location of Abell clusters projected against the sky. The dashed contour outlines the filament. *(J. Burns and D. Batuski)*

found a much larger void (10^5 to 10^6 Mpc^3) in the direction of Bootes. D. Batuski and J. Burns, analyzing the Abell clusters, compiled a catalogue of possible superclusters and voids. One of their voids covered much of the North Galactic Hemisphere and included the **Bootes void** as a small corner! They also found the largest scale structure presently known (Figure 23–9). This supercluster has a length of approximately 1 billion light years and includes the Perseus supercluster as one small part.

R. B. Tully and R. Fisher have thoroughly explored the 3 dimensional structure (see Figure 23–10) of the Local Supercluster by obtaining 21-cm redshifts of a large number (more than 2200) of late-type galaxies. They discovered a rich, convoluted structure that breaks into two main clouds with streamers—thin, cigar-shaped clouds—emerging above and below the central plane. Most of the Supercluster is empty space; 98% of the visible galaxies are contained in just 11 clouds that fill a mere 5% of the overall volume. Yet, the clouds do delineate a disk structure, with a width about 10 times its thickness—a cosmic pancake. The Virgo cluster is the densest concentration of galax-

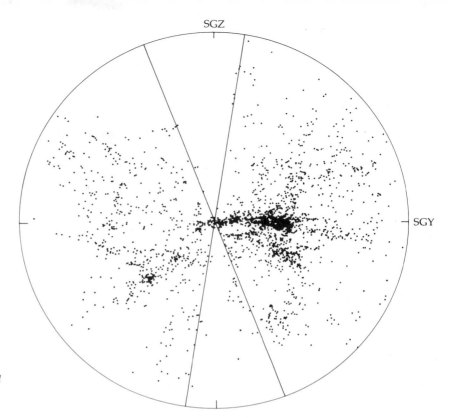

FIGURE 23–10 The Local Super-cluster. This map is a top view that roughly corresponds to the plane of our Galaxy. Each dot represents one galaxy. The wedges indicate regions of the sky obscured by dust in our Galaxy. *[R. B. Tully, Astrophysical Journal 257:389 (1983)]*

ies in Figure 23–10. Note that it is *not* obviously the center of the structure although the figure is misleading since it is centered on our position. Virgo may not dominate the Local Supercluster.

Figures 23–11 and 23–12 illustrate **wedge diagrams** or **cone diagrams** (if you think in 3-D) for galaxies in the Coma and Hercules regions, respectively. This type of diagram uses redshifts as a distance axis, and voids and superclusters appear as empty regions or clumpy regions in redshift space.

One tool for studying the statistics of large-scale structure is by *n*-point correlation functions. In

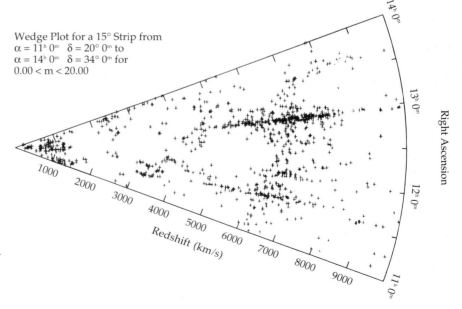

Wedge Plot for a 15° Strip from
$\alpha = 11^h\,0^m$ $\delta = 20°\,0^m$ to
$\alpha = 14^h\,0^m$ $\delta = 34°\,0^m$ for
$0.00 < m < 20.00$

FIGURE 23–11 The Coma super-cluster. This wedge diagram includes parts of the Local Supercluster (clumped objects near the vertex) and the Coma/A1367 supercluster (at a redshift of about 7000 km/s). Note the empty void regions and the elongated appearance of rich clusters caused by high-velocity dispersions within the clusters.

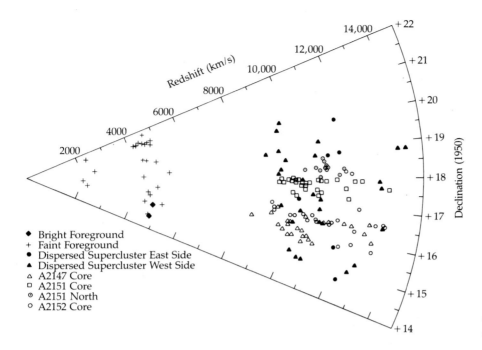

FIGURE 23–12 The Hercules supercluster. This wedge diagram shows the large void between the supercluster and foreground galaxies. [*M. Tarenghi, W. G. Tifft, G. Chincarini, H. J. Rood, and L. A. Thompson, Astrophysical Journal 234:793 (1979)*]

practice, higher-order functions than the 2-point correlation are rarely used. The defining equation is

$$\delta P = n[1 + \zeta(r)] \, \delta V$$

where n is the mean volume number density of galaxies. Then $\zeta(r)$ (the **two-point correlation function**) gives the excess probability (compared to random chance) of finding a galaxy in the volume δV. Generally, $\zeta(r)$ has the form

$$\zeta(r) = 4.7 \, h^{-1} \, r^{-1.8}$$

This is generally valid for r in the range of 2 to 25 Mpc. The smoothness of the functional form of the correlation function shows that there is no preferred length scale, and the large range in distances shows nonrandom distributions at scale lengths much larger than clusters. The correlation function can be found for various cosmological samples such as quasars, radio galaxies, rich clusters, etc. in order to examine their clustering tendencies. We will see in Chapter 26 that $\zeta(r)$ can be used to discriminate among theoretical models of galaxy formation.

Based on a decade of observations let's list some of the features of large-scale structures.

1. Superclusters are *not* spherical. They are dominated by flattened structures—mostly gently curving filaments, but some have a pancake shape.
2. All rich clusters lie in superclusters.
3. At least 95–99% (possibly 100%) of all galaxies lie in superclusters.
4. Voids *are* predominantly spherical.
5. Voids are empty of (at least) bright ($M < M^*$) galaxies.

(B) PECULIAR MOTIONS AND THE GREAT ATTRACTOR

With a growing awareness of the significantly large masses that might be concentrated in large-scale structures, several groups have questioned the possibility that superclusters could cause detectable **peculiar motions**. In order to make observational tests, we must find distance-determining methods that do not employ the redshift. If such methods are found to be reliable at the large distances encountered with superclusters, then any systematic differences between the observed red-shift and that predicted by Hubble's law would be caused by peculiar motions.

One method was introduced by S. Faber and R. Jackson. They found that the absolute magnitude of a galaxy could be predicted from observations of its internal velocity dispersion. This **Faber-Jackson relation** is usually written in luminosity units as $L_B \propto \sigma^4$—implying that the luminosity (hence mass) is proportional to the fourth power of the galaxy's internal velocity dispersion. Proper calibration yields M_B which, of course, leads to the distance.

The group consisting of D. Burstein, R. Davies, A. Dressler, S. Faber, D. Lyndon-Bell, R. Ter-levich, and G. Wegner have found another correlation involving the internal galaxy velocity dispersion. This $\sigma - D_n$ relationship relates σ to the diameter of a circle within which the integrated surface brightness of the galaxy is 20.75 magnitudes per arcsecond in the B band. If D_n refers to the same physical location independently of velocity or distance, then proper calibration will yield confident distances.

These groups, along with M. Aaronson and J. Mould, have found significant systematic motions. It appears that the Local Group is falling into the central regions of the Local Supercluster [toward $l = 284°$, $b = 74°$ (near Virgo)] at about 250 km/s. The whole Local Supercluster is also moving toward $l = 307°$, $b = 9°$, which is in the direction of the Hydra-Centaurus supercluster. The total Local Group motion is about 570 km/s. The most recent results suggest that the cause of all this bulk motion—named the **Great Attractor**—has about $5 \times 10^{16} M_\odot$ but is somewhat more distant than the Hydra-Centaurus supercluster. Unfortunately, at $b = 9°$, there is too much galactic obscuration to have a clear view of this region in the optical window.

This bulk motion result is consistent with one other means of measuring the peculiar motion of our local group with respect to the average of the matter in the Universe. There is a significant dipole character to the cosmic microwave background radiation (c.f. Section 25–3). In the direction toward which we are moving, the background radiation is somewhat brighter and has a higher characteristic temperature than in the opposite direction. Both the speed and direction of our peculiar motion determined by the cosmic microwave background method agree with the bulk motion result.

(C) WHAT IS A VOID?

Certainly, observations of supercluster dynamics and their detailed structures will continue to improve. But several basic questions about voids make them seem exceptionally mysterious. The first obvious question is "Are they really empty?" Such a seemingly simple question is not easy to answer observationally. By "empty" does one mean that the voids just lack galaxies? Do they contain dark matter? In deep surveys of the Bootes void, about a dozen galaxies have been found and all are peculiar to the extent of having emission lines in their spectra. However, similar studies of the Coma void find no galaxies. This difference might be attributable to small number statistics and different volume sizes.

Searches have also been made for a gaseous void component. This possible gas could conceivably be detected either through emission lines (optical or 21-cm) or by absorption lines in the light of distant galaxies or quasars. None of these searches has yet been conclusive.

Another interesting question concerns the nature of void topologies. One possibility is that of sponges. In this case, empty regions are connected to each other and the percolation length (the distance one could traverse without encountering a boundary) is quite long. A second possibility is that of bubbles in which each empty region is self-contained with a continuous, built-up boundary. This point has not yet been settled, but, since different physical processes lead to the two different topologies, this is potentially an important observational test of formation theories.

23–3 ◗ INTERGALACTIC MATTER

Is intergalactic space empty, or is there an intergalactic medium similar to the interstellar medium? If an intergalactic medium is present, it may contain both gas and dust. The gas (probably hydrogen) may be neutral or ionized. We can look for the intergalactic medium in two locations: *between* clusters of galaxies and *within* clusters.

Consider the first possible location. Let's examine the possibility of intergalactic dust. Such dust, if it resembles the interstellar dust in our Galaxy, would extinguish and redden the light from distant galaxies. This extinction and reddening effect has been searched for but not found. It is less than 4×10^{-4} magnitudes per megaparsec. So intergalactic space cannot contain very much dust; the density must be less than 4×10^{-30} kg/m³. In other words, intergalactic dust is optically thin over the distances between clusters of galaxies.

How to detect neutral hydrogen? Hydrogen atoms absorb ultraviolet radiation well, especially at 121.6 nm, the Lyman alpha absorption. Such ultraviolet absorption has been sought in the spectra of distant objects at both small and large redshifts. It has not been detected. This lack of ultraviolet absorption implies that neutral hydrogen cannot have a density greater than about 10^{-9} atom/m³. So if hydrogen is there, it must be ionized because H II is far more transparent than H I. These observations probe the Universe's past, and so they also imply that any intergalactic gas must have remained highly ionized for most of the history of the Universe.

These arguments leave ionized hydrogen (H II) as the most likely candidate for the intergalactic medium. Because intergalactic material would not have a high density, ionized hydrogen would take a very long time to find an electron and recombine. Unfortunately, detecting a low-density ionized gas is difficult. If it is hot (a few tens of millions of kelvins), you can expect X-ray or ultraviolet emission. X-ray observations of local superclusters show 15 sources that are probably clustered in seven superclusters. The sources consist of spots centered in rich clusters; this implies that hot gas in the superclusters is highly clumped.

Recent X-ray observations support this idea (Figure 23–13). To date, at least 40 clusters of galaxies are known to emit X-rays. The X-ray luminosities of clusters range from 10^{36} to 10^{38} W. The sizes of the X-ray-emitting cores range from 50 kpc to 1.5 Mpc. The richer clusters tend to be the more luminous in X-rays—specifically it looks as if $L_x \propto \sigma^4$, where σ is the cluster velocity dispersion. A reasonably confirmed model for this X-ray emission is that it comes from hot, ionized gas. This model requires typical temperatures of 10 to 100 million K and densities of about 1000 ions/m³ to explain the X-ray observations. So we have evidence of intergalactic gas in clusters, about equal

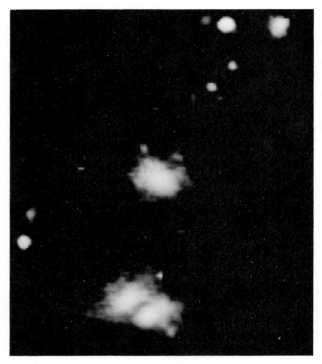

FIGURE 23–13 X-ray emission from superclusters. The cluster at the center is about 6 Mpc from the double cluster below it. *(C. Jones and W. Foreman)*

to the amount of mass in the galaxies themselves. We have no evidence so far for much gas in *between* clusters.

Finally, we note that for plasmas with temperatures between 10^4 and 10^7 K, trace elements dominate the emission and produce emission lines, which can be used to deduce the physical properties of the emitting gas. In particular, inner-shell transitions of neon, silicon, iron, sulfur, and argon can be very strong. X-ray spectroscopy of the emission of the gas from clusters shows highly ionized iron lines in almost all cases (mainly in a line at 7-kev); the iron lines confirm that the emission mechanism is thermal brehmsstrahlung from an extremely hot gas rather than synchrotron radiation. Theoretical models of the emission require iron abundances (relative to hydrogen) about half that of the Sun. Hence, the intergalactic gas must have been processed through stars and then removed from the galaxies (perhaps by supernova explosions).

23–4 ◐
MASSES—ROUND 3: THE MISSING? MASS

In Chapter 21 we introduced the virial theorem as applied to individual galaxies. It has, perhaps, an even more astounding application to clusters. Here the assumptions are the same as in almost all applications of the virial theorem in astrophysics. We must suppose that the cluster is in a steady state, neither expanding nor contracting. When we apply this theorem and use the velocity dispersions discussed in Section 23–1(f), we find that cluster masses are surprisingly high.

As always, the most useful quantity in examining the missing mass is the mass-to-light ratio M/L, where both are in solar units. The M/L ratios for individual galaxies can be as high as $M/L = 50$, and for double galaxies $M/L = 100$. When we apply the virial theorem to clusters, however, we often find that the M/L ratios range from 300 to 500! This provides strong evidence that as R_c (the characteristic scale of the system) increases, the M/L value obtained from dynamical methods correspondingly increases.

23–5 ◐
SUMMARY

What does all this mean? The study of clusters, voids, and superclusters affords us unique opportunities to study the formation and evolution of galaxies.

The high M/L ratios found from virial studies indicate that there is an important constituent to the gravitational properties of these structures that we cannot detect directly. Apparently, the dark matter (dm) can coexist with normal galaxies; it suffuses them, causing the M/L ratio to be somewhat larger than 1. Yet the increase of M/L with R_c indicates that the dm also lies between the galaxies in binary systems, small groups, and rich clusters. So it does not clump together quite the same way as does luminous matter. Does the dark matter collect around superclusters? Does it fill voids? If we could see the dark matter, would it trace out the same large-scale structures that we see in luminous matter, or would it form other structures? Or is it, perhaps, smooth? No matter how the dark matter is distributed, it is clear that there is much more of it than there is of normal matter.

Another problem is the iron lines in the intracluster X-ray gas. Only inside of stars can the Universe manufacture iron. This shows that stars within the cluster galaxies have contributed a large amount of gas—perhaps equal in mass to that of the stars themselves. Somehow this gas has left the individual galaxies and collected near the cluster center with sufficient energy input to raise the kinetic temperature to around 10^7 kelvins. It is difficult to understand these processes.

How is progress made in using the observations as input into improved theories? An important use of the 2-point correlation function is to test models of galaxy formation. For example, some theories can successfully predict the qualitative formation of superclusters and voids but cannot match the observed functional form of $\zeta(r)$. In Chapter 26, we will try to sketch out the most plausible current model. In order to be successful, it must match the current observations of clusters, voids, and superclusters.

Perhaps the most exciting aspect of large-scale structure studies is that, finally, at the scale of voids and superclusters, we see matter distributed as it was in its primordial state. At all smaller scale lengths, mixing has occurred. For example, a galaxy moving at 500 km/s will travel about 5 Mpc in a Hubble time. Therefore, since cluster cores are smaller than this number, the galaxy could have crossed the cluster a few times. Hence we have no direct association of its present position within the cluster to its original position. In contrast, superclusters are larger than the mixing scale length. Therefore, the position of a galaxy within the supercluster is approximately that of its origin. In other words, we are viewing the "fossils" of the primordial matter distribution. If we find galaxy properties to vary across a supercluster, for example, then that variation has its roots in the physical properties of the formation process.

Are all of the structures we see important? You might at first think of voids as a necessary consequence of the existence of superclusters, but this is not the case. The Universe might have constructed itself so that superclusters were separated by a homogeneous sea of galaxies or perhaps a population of small groups and clusters. (A mixture of these ideas was for a long time the standard operational assumption.) The reality is that both voids and superclusters are vital clues in our search for understanding of the origin and evolution of galaxies.

PROBLEMS ◗

1. Calculate the crossing time for the Hercules super-cluster and compare it with the age of the Universe ($\approx 15 \times 10^9$ years).

2. (a) Demonstrate that a low-density hydrogen gas at 10^7 K produces X-rays. What is the peak wavelength of emission?
   (b) Compare the mass of the hot gas in a cluster with the mass contained in galaxies.

3. We state in the chapter that intergalactic space must have a dust density of less than 4×10^{-30} kg/m^3. Assume that ten times this amount really exists. What would the intergalactic extinction be, in magnitudes per megaparsec? How would that affect our ability to see the Virgo cluster?

4. How close does a spiral galaxy like the Milky Way have to get to a cD galaxy to be tidally disrupted?

5. Estimate the free-fall time for a cluster of galaxies and compare your result with the age of the Universe ($\approx 15 \times 10^9$ years). What do you conclude?

6. An approximation to the luminosity function of galaxies in rich clusters is $\Phi(L) = \Phi^*(L/L^*)^{-5/4}$ for $L < L^*$ and $\Phi(L) = 0$ for $L > L^*$, where Φ^* is 0.005/Mpc^3 for $H = 50$ km/s · Mpc, $L^* \approx (10^{10})L_\odot$, and $L_{min} \approx (10^3)L_\odot$. The number of galaxies per cubic megaparsec is

$$N_{gal} = \int_{L_{min}}^{L_{max}} \Phi(L)dL$$

and the total luminosity per cubic megaparsec is

$$l_{gal} = \int_{L_{min}}^{L_{max}} \Phi(L)L \, dL$$

   (a) Show that most galaxies have luminosities near the low end of the range.
   (b) Show that most of the luminosity of a cluster is produced in the higher mass galaxies.

7. Only the radial velocity is observable for galaxies that are members of clusters. By integrating to obtain the mean value, determine how v^2 is related to v_r^2.

8. Discuss how the existence of superclusters affects the Hubble diagram. What are the dangers in deriving the Hubble constant using galaxies within the Local (Virgo) Supercluster?

9. Discuss what happens to the kinetic energy of a galaxy that is cannibalized by a central cD galaxy. Relate this to the extended haloes of cD galaxies.

10. Discuss the evidence that M/L increases as the char-acteristic scale of the system increases. What does this imply about the distribution of dark matter?

11. Below we list the redshifts obtained for 10 galaxies in the A2199 cluster. They are given in km/s, and have already been referred to a galactocentric frame by the addition of 193 km/s. Evaluate Equations 23–2, 23–3, 23–6, 23–7, 23–8, 23–9, and 23–10 for this set of data. Assume that the observational uncertainties are all 50 km/s.

9473	8369
8667	10250
10083	8311
7723	8773
7953	10733

12. From your answers to Problem 11, estimate the total kinetic energy in the cluster. What assumptions must you make? What is your best estimate of the potential energy?

13. Imagine that an astronomer in the Virgo Cluster of galaxies ($D = 15.7$ Mpc) wanted to observe our Local Group cluster of galaxies.
   (a) What angular size would the Local Group subtend in the sky?
   (b) How large would the Milky Way Galaxy appear?
   (c) What would be the apparent visual magnitude of the Milky Way Galaxy? Compare this with our naked-eye limiting magnitude of $m = +6$.
   (d) Assuming that our observer could build a telescope to detect $m = +20$ objects, could she detect all the Local Group galaxies listed in Table 23–1?
   (e) Based upon your answer to (d), comment on how observational selection effects may influence our understanding of distant galaxy clusters.

14. Observations of the maximum light from novae and supernovae in other galaxies are important as distance indicators (see Figure 23–3 and Table 22–1).
   (a) Could a nova or supernova be observed in a galaxy in the Virgo Cluster ($D = 15.7$ Mpc) using a modest size Earth-based telescope with limiting magnitude $m = +23$?
   (b) Could a nova or supernova be observed in a galaxy in the Corona Borealis Cluster (assume a cosmological redshift: $cz = 21,600$ km/s) using this telescope? Assume $H = 50$ km/s · Mpc.

15. Using Figure 23–11, estimate the linear extent of the void region near $cz = 5500$ km/s (assume $H = 50$ km/s · Mpc). Compare this size to the size of the Local Group, to the size of the Virgo Cluster, and to the distance between the Local Group and the Virgo Cluster.

Chapter 24

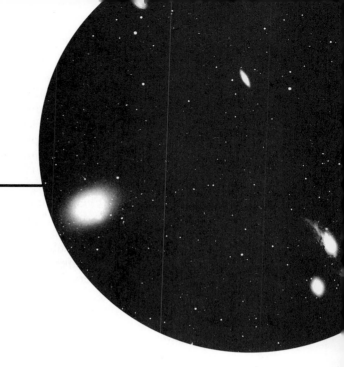

Active Galaxies and Quasars

This chapter examines the power and the glory of active galaxies and quasars in the distant Universe—galaxies that may contain supermassive black holes in their cores to supply their enormous energy outputs. We will see several classes of unusual galaxies with different names, but keep in mind as you examine each one that observations more closely support a small number of basic models with a continuous range of activity.

In certain classes, most of the activity is spread over relatively large parts of the galaxy's disk. In the most powerful classes the activity arises from a tiny region deep within the nucleus. Could it be that the nuclei of most galaxies contain low-power models of radio galaxies and quasars?

24–1 ◗
RADIATION MECHANISMS

The radiation from normal galaxies is dominated by thermal processes. These include starlight, thermal radio emission, and infrared radiation from heated interstellar dust. In contrast, the active galaxies feature either nonthermal processes such as synchrotron radiation or thermal processes with unusually large energies. An important indi-

cation of activity is the presence of a strong emission line component to the galaxy's spectrum.

(A) EMISSION LINES

As discussed in Chapters 8 and 19, emission lines come from bound–bound atomic transitions in which the electron drops from an excited state. Often the emission lines arise from an initially ionized atom in which the electron re-combines and cascades downward through a series of levels. A major goal of interpreting emission line spectra is to understand the mechanism for exciting or ionizing the gas since most astrophysical gases remain in or close to the ground state.

Among the excitation/ionization mechanisms we can identify two classes. One is collisional; here we need to look at cloud–cloud collisions or a variety of ways of producing interstellar shock waves. The other broad category includes several radiative processes. Generally, this demands a source of "hard" photons whose energies meet or exceed the excitation or ionization energy of the transition in question. Synchrotron radiation is a common source of hard photons; but, by Wien's law (Chapter 8), we know that thermal radiation from blackbodies at very high temperature can also produce large numbers of photoionizing UV photons.

Since the elements found in the interstellar medium of galaxies are also abundant on Earth, we have laboratory examples of many astrophysically important emission lines. However, an important class of lines is not observed in the lab. These are the **forbidden lines** that arise from metastable excited states [Section 19–2(d)]. Even quite good laboratory vacuum conditions on Earth have much higher densities than those found in astrophysics, and in the high densities the metastable states are always collisionally de-excited rather than radiating away their energies via observable photons—hence the term *forbidden*; they are, of course, not forbidden in most astrophysical settings.

In Table 24–1, we list some of the most commonly observed emission lines. Those that are enclosed in square brackets are forbidden (from electric quadrupole, magnetic dipole, or magnetic quadrupole transitions), and those with a single square bracket are only partially forbidden (electric dipole). An absence of brackets indicates permitted lines. We include several useful lines in the ultraviolet since they are observable by the IUE satellite, the Hubble Space Telescope, and by ground telescopes in high redshift systems such as quasars. We provide wavelengths in units of nanometers; but, for historical reasons, most astronomers employ Angstrom units in everyday usage. Thus, for example, [O III] 500.7 would be termed 5007.

TABLE 24–1   *Important Emission Lines*

Line	λ_0 (nm)
H I (Ly-α)	121.6
N V	124.0
C IV	154.9
C III]	190.9
Mg II	279.8
[O II]	372.7
[Ne III]	386.8
Hδ	410.2
Hγ	434.1
Hβ	486.1
[O III]	495.9
[O III]	500.7
[N II]	654.8
Hα	656.3
[N II]	658.4
[S II]	671.7
[S II]	673.1

(B) SYNCHROTRON RADIATION

We introduced synchrotron radiation in Chapters 18 and 19. The necessary conditions for this process are a magnetic field and a source of relativistic electrons. The major mystery in understanding active galaxies is to find a convincing mechanism for the acceleration of electrons by collapsed objects in the galaxy nucleus.

In general, the flux, F, of nonthermal emission has the spectral form

$$F(\nu) = F_0 \nu^{-\alpha}$$

so that

$$\log F(\nu) = -\alpha \log \nu + \text{constant}$$

Hence, on a log–log plot of the spectrum, the spectral index α is simply the slope.

Radio galaxies and quasars have similar synchrotron properties. Their spectra come in two types. The extended emission has a spectral index α with values between 0.7 and 1.2. At low frequencies the spectrum often is seen to turn down. This is called *synchrotron self-absorption* and is just a result of the source becoming optically thick at these frequencies. In compact sources, the spectra are much flatter with $\alpha \approx 0.4$. Because of the slower fall-off with frequency, compact sources are more readily found in high frequency surveys. See Section 24–3(c) for a more detailed discussion.

24–2 ●
MODERATELY ACTIVE GALAXIES

(A) PRELUDES TO ACTIVITY

Although absorption lines dominate the spectra of most galaxies surveyed, the Hα and [O II] 372.7 emission lines can often be found in the nuclear spectrum. Also, spectrograph slits often include some light from the disk in spiral galaxies, and therefore H II regions may contribute light to the spectrum—leading to a characteristic emission line component.

It is sometimes difficult to dissociate nuclear and disk emission. One, somewhat more active type of galaxy than the quiescent types, has a very strong H II region–like spectrum that overwhelms the starlight component. Since H II regions are stellar nurseries, this type of galaxy seems to be experiencing an increased star-formation episode. Another type of strong emission spectrum that looks superficially like the H II region class was recognized by T. Heckman who termed the class *LINERs* (Low Ionization Nuclear Emission Region). As the name implies, these emission lines do not arise in the disk, and the ionization energies are relatively small. Hence [O II] lines are relatively stronger than the [O III] lines, for example.

(B) STARBURST GALAXIES

Appearances can be deceiving in astronomy—especially for distant objects of which we get a two-dimensional view projected on the sky. The nearby galaxy M82 (NGC 3034) makes this point most dramatically, for it was once believed to be an exploding galaxy because of its optical appearance (Figure 21–4). An Irregular II galaxy, M82 is a radio source but not a strong one, radiating only 10^{32} W. Photographs of this galaxy in Hα show filaments extending above the plane of the galaxy to a distance of 3 to 4 kpc. Spectra of this galaxy also show material expanding outward from the center at velocity differences of about 100 km/s.

The nuclear region of M82 is much larger than that for our Galaxy and a strong infrared emitter: about 10^{37} W. Star clusters and giant H II regions dominate the nuclear region; within it lies a small, nonthermal source. Clumps and filaments of dust here block out the optical view, a fact that misled people into believing that the galaxy's appearance resulted from explosive, high-velocity outflows.

X-ray observations show that the nucleus has a clumpy structure with a total power of 3×10^{33} W in soft X-rays. The most likely candidates for the emission source are massive Population I objects, such as X-ray binaries (like Cyg X-1), OB stars, and supernova remnants. These and other observations lead to the interpretation that the nuclear region underwent a huge burst of star formation 10^7 to 10^8 years ago and that this burst is continuing today. M82 lies close to the spiral galaxy M81 (Figure 23–7). Tidal interactions between them could trigger the star formation in M82, which is rich in the gas and dust needed to give birth to stars.

24–3 ◑ AGNs

As a group, the most active galaxies show all or most of the following characteristics: (1) high luminosity, greater than 10^{37} W; (2) nonthermal emission, with excessive ultraviolet, infrared, radio, and X-ray flux (compared with normal galaxies); (3) a small region of rapid variability (a few lightmonths across at most); (4) high contrast of brightness between the nucleus and large-scale structures; (5) explosive appearance or jet-like protuberances; and (6) broad emission lines (sometimes). The nucleus of the Milky Way Galaxy has some of these characteristics, but it does not generate as much energy in total (some 10^{35} W) as do the nuclei of active galaxies. As a class those galaxies with active nuclei are called **Active Galaxy Nuclei (AGNs).** Several of the best known AGNs are listed in Table 24–2.

(A) SEYFERT GALAXIES

In 1943, Carl Seyfert identified six spiral galaxies with unusual, broad emission lines in their spectra. When viewed optically, such galaxies have

TABLE 24–2 *Selected* **AGNs**

Name	α(1950)	δ(1950)	z	m_V	Remarks
I Zwicky 1	$00^h\ 51^m\ 00^s$	$+12°25'$	0.061	14.3	Seyfert 1–1.5
NGC 1052	$02^h\ 38^m\ 37^s$	$-08°28'$	0.005	13.5	LINER
NGC 1068	$02^h\ 40^m\ 07^s$	$-00°14'$	0.003	10.5	Seyfert 2
3C 120	$04^h\ 30^m\ 32^s$	$+05°15'$	0.033	14.6	Radio galaxy
Mrk 79	$07^h\ 38^m\ 47^s$	$+49°56'$	0.020	13.4	Seyfert 1–1.5
Ark 160	$08^h\ 17^m\ 52^s$	$+19°31'$	0.019	17.5	LINER
OJ 287	$08^h\ 51^m\ 57^s$	$+20°18'$	—	14.0	BL Lac
NGC 4151	$12^h\ 08^m\ 01^s$	$+39°41'$	0.003	12.0	Seyfert 1–1.5
3C 273	$12^h\ 26^m\ 33^s$	$+02°20'$	0.158	12.8	Quasar
3C 351	$17^h\ 04^m\ 03^s$	$+60°49'$	0.371	15.3	Quasar
BL Lac	$22^h\ 00^m\ 40^s$	$+42°02'$	0.069	14.5	BL Lac
3C 445	$22^h\ 21^m\ 15^s$	$-02°21'$	0.057	15.4	Radio galaxy

FIGURE 24–1   The Seyfert galaxy NGC 1275 in Perseus. *(National Optical Astronomy Observatories)*

unusually bright nuclei (Figure 24–1). They are now called **Seyfert galaxies;** about 90 are known to date. Seyfert galaxies, some of which are close, provide a possible clue to the nature of active galaxies—a clue that comes from their broad emission lines and their partially nonthermal, synchrotron continuum.

Photographs of Seyfert galaxies show that they are almost always spirals. A detailed survey of 80 Seyferts finds that only 5 to 10% might be ellipticals. (The small angular size of some Seyferts makes it hard to classify them.) Compare this with the fact that extended radio galaxies [Section 24–3(c)] are all ellipticals. Overall, about 1% of all spiral galaxies (ordinary and barred) are Seyferts. Perhaps all spiral galaxies go through a Seyfert phase, with only 1% active at any time.

It is in their spectral properties that Seyferts really distinguish themselves from normal galaxies. We now recognize that Carl Seyfert's original group was not homogeneous. Daniel Weedman suggested a useful classification scheme. His Type 1 represents the kind that most people think of as a Seyfert; these objects have permitted lines (such as the Balmer series of hydrogen) with *extreme* widths. If the cause is Doppler broadening it corresponds to 5000–10,000 km/s! The forbidden lines (from O II, O III, N II, S II, etc.) are only moderately broadened—typically 200–400 km/s.

Type 2 Seyferts have only "narrow" lines present. The characterization as narrow is only relative to the broad lines of Type 1's, since the Type 2 profiles (which range from 200–400 and in one case up to 1000 km/s) are broader than in most normal galaxies. Interestingly, the permitted lines in a Seyfert 2 have about the same widths as the forbidden lines.

The standard model for Seyferts is that of a central, tiny source of photoionizing photons that encounter two distinctly different gaseous regions as they flow outward from the active nucleus. The inner one is the **Broad Line Region** (BLR). We can confidently place an upper limit of about 10^{14} m as the scale size of the BLR from variability studies [see Section 24–4(d)]; dramatic changes in the broad lines and continuum have been detected over a period of a few weeks or months. Outside of the BLR is the **Narrow Line Region** (NLR) with a scale size that is perhaps 10^2–10^3 times larger. We have not seen much evidence for variability in the narrow lines.

The BLR must have a density greater than a certain critical value since there are no broad forbidden lines. This implies densities around 10^{13}–10^{15} ions/m³. Estimates of the total gas content in the BLR are in the range of 30–50$M_\odot$, which must be in a very chaotic state given the observed Doppler shifts.

In the NLR there is nothing to prevent the production of permitted line photons, so there is generally a narrow component to them as well as the narrow forbidden lines. A modern addition to the Seyfert classification system is to call those objects with both broad and narrow components to the permitted lines Type 1.5. In this system there are few Type 1.0 Seyferts. Figure 24–2 illustrates the spectrum of a Type 1.5 Seyfert in the Hβ / [O III] region where there are both broad and narrow components. Slight differences in the widths of the forbidden lines show that the density of the NLR decreases with radial distance from the nucleus. The critical densities of forbidden lines from differ-

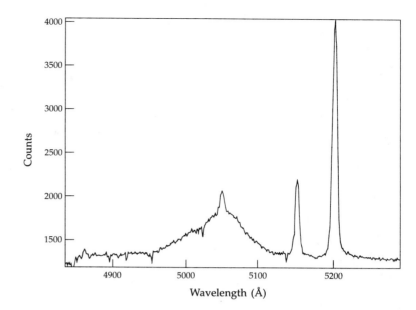

FIGURE 24–2 The spectrum of the Type 1.5 Seyfert galaxy Markarian 79. Note the extreme width of Hβ along with narrow [OIII] lines.

ent atomic species vary, and those lines coming from the least dense regions exhibit somewhat narrower profiles.

Both the BLR and NLR are probably more complex than a simple two-region model implies. As an example of some of this complexity, four AGN galaxies have been seen to change from Type 1.5 to Type 2 in the time scales of from 1 to 4 years. How might this happen? The simplest explanation is that a dust cloud interior to the NLR moved across our line of sight to the BLR—possible, given the small scale size of the BLR. The dust model is strengthened by the fact that in both cases the broad component at Hα did not completely disappear. Most Seyfert spectral classification is made from the appearances of Hβ and the [O III] lines, and dust would extinguish broad Hβ photons more effectively than the redder Hα photons.

In addition, the continuous spectra of Seyferts have a combination of stellar, nonthermal, and infrared (from dust) radiation. The total energy output is 10^{37} to 10^{38} W. Their luminosities vary, sometimes by large amounts, over time spans of a few days to a few months. Seyferts tend *not* to be strong radio sources; sensitive radio surveys have detected only about half of the Seyferts examined.

What is the ultimate cause of all this activity? It must be very small to fit inside the BLR and yet must be very powerful to accelerate the gas in the BLR to very high speeds as well as to produce the total luminosity of the AGN. Further constraints on the size are imposed by recent observations showing light fluctuations that happen on time scales of hours or minutes. We will return to this subject in Section 24–5(a), when we discuss the power sources for quasars.

A recent survey of spiral galaxies reached a curious conclusion with respect to Seyferts: most of them tend to be in close, binary galactic systems. Tidal interactions may then induce the Seyfert phenomena for some short period of time.

In galaxy activity, we have been careful to distinguish between nuclear and disk activity. However, one galaxy, Arp 220, may have both! This object is probably the most luminous galaxy known. It was included in a catalogue of peculiar objects compiled by H. Arp and appears quite bright in the IRAS catalogue for its distance. It is likely that the nucleus is an AGN while at the same time there is a starburst in the disk. Could they be triggered by the same physical mechanism?

(B) BL LACERTAE OBJECTS

One other type of active galaxy is named after its prototype, BL Lacertae, and so these galaxies are

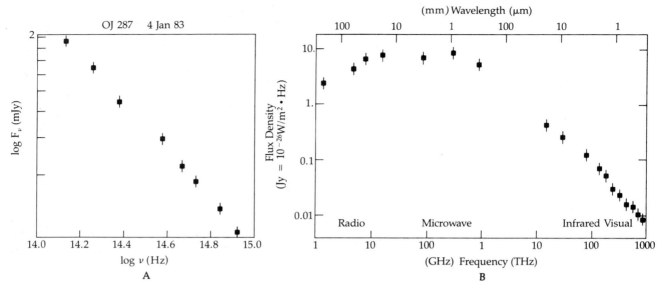

FIGURE 24–3 Spectra of OJ 287. (A) Infrared-optical continuum. *(P. Smith)* (B) Radio-optical spectrum in March 1983. *(T. Balonek)*

called **BL Lac objects.** As a group, the BL Lac objects have most of the following characteristics: (1) rapid variability at radio, infrared, and visual wavelengths, (2) no emission lines, (3) nonthermal continuous radiation (Figure 24–3) with most of the energy emitted in the infrared, and (4) strong and rapidly varying polarization. Also, BL Lac objects generally have a star-like appearance—structure is rarely visible. The radio emission from many BL Lac objects is compact or only slightly extended. The extended radio structure is weak (only a few percent of the total) in contrast to the intense emission from the nucleus.

The greatest difference between BL Lac objects and other active galaxies is that the emissions of the former vary so frequently and erratically (Figure 24–4). For example, BL Lac itself, recorded on a long sequence of Harvard Observatory photographs, fluctuates between visual magnitudes 14 and 16 with occasional bursts to brighter than 13. These fluctuations mean that BL Lac's optical emissions vary by a factor of 20 times or so. Observers have noted night-to-night luminosity variations of 10 to 30 percent. A few BL Lac objects have changed their luminosities by as much as a factor of 100.

A great puzzle about the BL Lac objects is that their energy variations take place in objects that

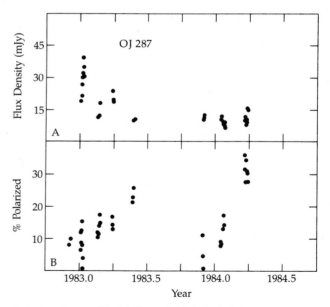

FIGURE 24–4 Variability of OJ 287. (A) Intensity variations; note the large outburst at the beginning of 1983. (B) Polarization variations of the linear polarization. *(P. Smith)*

show almost no emission lines in their spectra. As discussed above, the standard model for active galaxies pictures synchrotron emission output in the ultraviolet (and even the electrons themselves) as ionizing any gas near the nucleus and producing emission lines through recombination. But where are the BL Lac's emission lines if they are powered the same way? For a few BL Lac objects that have weak continuum outputs, the emission lines *are* visible and are probably powered by nonthermal radiation from the nuclear source. The hint here is that the emission lines are usually present but often submerged in the continuum emission.

Some 40 BL Lac objects have been classified to date. They may actually not all be the same kind of beast. A few are possibly the nuclei of galaxies. Some, like BL Lac itself, have a faint surrounding "fuzz" that might be a galaxy. Others look point-like without a hint of enveloping material. A number of BL Lac objects are found in clusters of galaxies—indirect evidence that they are also galaxies.

Finally, we must note that we do not have good distance determinations for very many BL Lac objects. Clearly, redshift-derived distances are hard for a class whose main characteristic is a lack of spectral lines! A redshift of 0.07 appears in the weak absorption features of the nebulosity around BL Lac, which corresponds to a radial velocity of 2.1×10^4 km/s. The nebulosity has a spectrum like that of a luminous elliptical galaxy. If the redshift is cosmological (from the expansion of the universe), this radial velocity corresponds to a distance of 430 Mpc (for $H_0 = 50$ km/s · Mpc).

(C) RADIO GALAXIES

A majority of the spiral galaxies of apparent visual magnitude 11 or brighter radiate in the radio continuum; the typical amount of this radiation appears to be at about 10^{33} W. A few extragalactic radio sources are exceptionally strong radio emitters, with some of them generating energies in excess of 10^{37} W. We use the term **radio galaxy** for those galaxies with a radio luminosity greater than 10^{33} W. This is not to say that the "weak" sources are all normal, but rather that normal galaxies are relatively weak radio sources.

Radio galaxies fall into two types: compact and extended. **Extended radio galaxies** have radio

FIGURE 24–5 The Active elliptical galaxy M87, a nearby radio galaxy. (*National Optical Astronomy Observatories*)

emission that is larger than the optical image of the galaxy; **compact radio galaxies** have radio emission same size or smaller. Compact radio galaxies often display very small (usually nuclear) radio sources, often no more than a few lightyears in diameter. Extended radio sources, in contrast, sometimes show a double structure of two gigantic lobes separated by distances of megaparsecs and symmetrically placed on opposite sides of the nucleus. In such sources, often called *classical doubles*, the two radio components are well separated from the optical galaxy and are much larger. As a rule, these sources lie on a line, with the galaxy at about the center of the radio-emission pattern. Typically, the nucleus is also a radio source.

M87 is a typical radio galaxy (Figure 24–5). A giant elliptical galaxy, it lies near the center of the Virgo cluster of galaxies and so lies about 20 Mpc away. One radio source only 1.5 lightmonths in diameter appears in the core of M87 along with a group of other compact radio sources. Poking out from the core, an optically visible jet (Figure 24–6A) extends over a length of some 6000 lightyears. This jet has a luminosity of roughly 10^{34} W; its emission is polarized. A detailed photograph shows that the jet contains at least six blobs

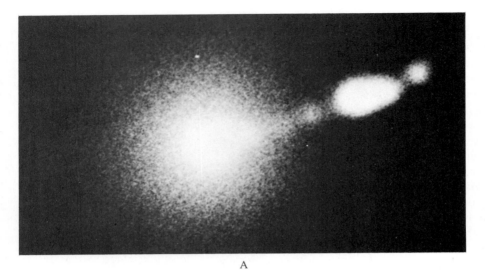

A

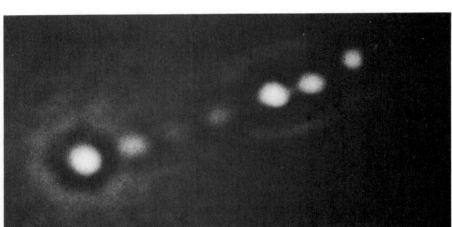

B

FIGURE 24–6 The jet of M87. (A) Photograph of galaxy and jet exposed for maximum detail (negative print). (B) Computer-processed image to bring out the structure in the jet. *(H. Arp and J. Lorre)*

of material, each no more than a few tens of lightyears across (Figure 24–6B). Over 22 years, the blobs have changed significantly in intensity and polarization.

M87 also emits X-rays with about 50 times more energy than its optical emission, about 5×10^{35} W in X-rays from the whole galaxy. The jet itself is also known to emit X-rays. High-resolution observations with the Einstein Observatory show that the jet contains knots that emit X-rays strongly. The VLA has mapped the M87 jet in detail and confirmed that its radio emission coincides with the optical and X-ray emissions and extends into one radio lobe (Figure 24–7). So the jet overall emits over a wide range of frequencies, from radio to X-rays, and each knot of the jet generates this same spectrum of energies. Recent infrared observations have demonstrated a break in the spectrum's slope at about 600 nm. The emission is nonthermal both before and after the break.

For the M87 knots, $\alpha \approx 0.6$ before the break (in the radio) but 1.7 after the break (in the optical). For synchrotron emission from relativistic electrons, such a sharp change in spectral index probably results from a sudden change in the electron energy distribution (of the accelerated electrons). The average magnetic field strength in the knots is about 10^{-7} T.

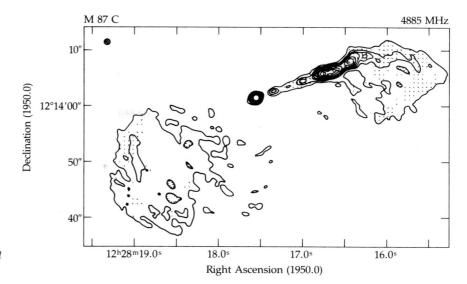

FIGURE 24–7 Radio emission from M87. A VLA map made at 5 GHz; note the jet at upper right. *(F. Owen and P. Hardee, National Radio Astronomy Observatory)*

Other elliptical galaxies also possess nuclear jets. In fact, radio jets are common—almost all radio galaxies in the lowest luminosity range have them.

Extended radio galaxies are commonly double, with the lobes lined up with the galaxy's center. These radio clouds are huge: most are 50 to 1000 kpc in diameter. When classified by structure, extended radio galaxies fall into three main groups: (1) classical doubles (example, Cygnus A), with highest luminosities, lobes aligned through center of galaxy, and bright hot spots at ends;

(2) wide-angle tails, or bent tails (example, Centaurus A), with intermediate luminosities, a bend through the nucleus, and tail-like protrusions; (3) narrow-tail sources (example, NGC1265), with lowest luminosities, U shapes, and rapidly moving galaxies in a cluster.

Cygnus A, one of the strongest radio sources in the sky and one of the first discovered, provides an excellent example of the classical double structure typical of an extended radio galaxy. Its radio output, 10^{38} W, comes from two gigantic lobes set on opposite sides of the optical galaxy (Figure 24–8).

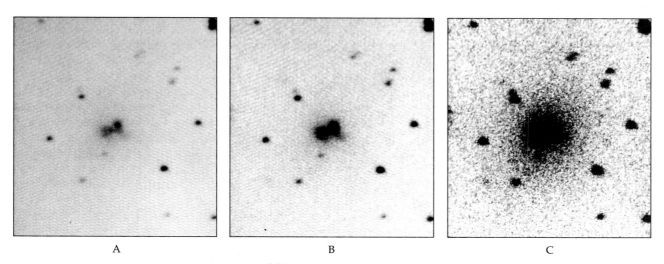

A   B   C

FIGURE 24–8 CCD negative images of Cygnus A. Made under conditions of excellent seeing, these images show (A) brighter features in the core, (B) lower flux levels resembling previous photographs, and (C) lowest flux showing the galaxy's halo. *(L. A. Thompson)*

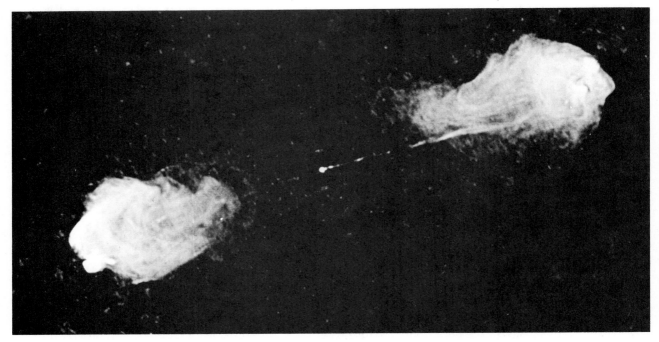

FIGURE 24–9   Radio map of Cygnus A. This extraordinary image shows the filamentary structure in the lobes and the thin jet to the right lob. Map made with the VLA. *(R. A. Perley, J. W. Dreher, and J. J. Cowan, National Radio Astronomy Observatory)*

Each lobe has a diameter of 17 kpc, and they extend roughly 50 kpc away from the central galaxy. Each lobe contains a cloud of energetic electrons and magnetic fields that stores the 10^{53} J needed to account for the radio luminosity lasting 10^7 to 10^9 years. The central galaxy of Cygnus A is an elliptical with a dust lane down its middle. It has an active nuclear region (perhaps with two nuclei), with a spectrum showing emission lines and a synchrotron continuum. Beyond 8 kpc from the center, the spectrum is just that of a mix of stars.

The VLA has produced a remarkable radio image of Cyg A (Figure 24–9) that shows a needle-like jet connecting the elliptical galaxy to one of its lobes. The radio map details a wispy structure in the lobes with small, bright spots of emission. Bright patches appear along the jet, which clearly holds its structure through most of the lobe. The filamentary structure of the lobes implies that the emitting gas occupies only 3 to 30% of their overall volumes. Within the lobes lie "hot spots" of intense radio emission.

Centaurus A (NGC 5128) is another extended radio source, somewhat similar to Cygnus A. It is an E2 galaxy bisected by an irregular dust lane. At a distance of 4 Mpc, Centaurus A is the closest active galaxy; it almost outshines M33 visually. Viewed with a radio telescope, Centaurus A has two huge outer lobes, 200 kpc and 400 kpc in diameter. The lobes form a bend through the nucleus and have tail-like protrusions. Closer in, another pair of radio lobes sits on the edges of the optical galaxy; each lobe is about 10 kpc in diameter. The inner and outer lobes have almost but not exactly the same alignment.

The nucleus of Centaurus A has a direct connection to one of the inner radio lobes. The Einstein X-Ray Observatory detected an X-ray-emitting jet streaming northeast from the nucleus and consisting of at least seven distinct blobs. A VLA map at a wavelength of 20 cm shows radio emission along a jet that extends to one of the nuclear radio lobes (Figure 24–10). The jet has a blob-like structure that coincides with the X-ray emitting blobs. So

FIGURE 24–10 Nuclear region of Centaurus A. A VLA map showing the nuclear emissions. The inner lobes with a jet extending to the left-hand lobe is superimposed on an optical photograph. *(J. Burns, National Radio Astronomy Observatory)*

about 1 nT) and high-speed electrons. Then a typical lobe contains more than 10^{52} J. The jets are the channels through which the nucleus conducts particles and energy to the lobes. The lobes and jets emit nonthermally, which suggests that the synchrotron process is operating. The nucleus provides the high-energy electrons. These are expelled either as a fairly constant beam of particles or as a sequence of ionized blobs thrown out along a magnetic field so that half the particles fly in one direction and the other half fly in the opposite direction. If the nuclear machine is active and stable, extended lobes of ionized material build up at the end of the jets. Repeated bursts from the nucleus can account for variability. The jets imply that high-speed electrons are channeled, perhaps in bursts, from the nucleus into the circumgalactic medium, where they pile up to form a lobe.

Centaurus A and M87 look similar in that both have nuclear jets that emit radio waves and X-rays. New high-resolution (24 pc by 7 pc beam size) VLA observations of Cen A have revealed new structural features in the inner 700 pc of the jet (Figure 24–11). The radio emission shows limb brightening on alternating sides—in contrast to the center brightening of the M87 jet. Thin filaments of radio emission emanate from the knots; they appear to point "downstream" in the direction of presumed matter flow along the jet (from the nucleus out to the lobe).

Many radio galaxies have emission in the form of lobes or streams that extend far beyond the visible galaxy. The lobes may be a few million lightyears apart and thousands of lightyears across. The vexing problem with these extended radio lobes is the vast amount of energy they contain: a typical lobe luminosity is 10^{36} to 10^{37} W, whereas the visible elliptical galaxy with which it is associated may emit only 10^{35} W. If the emission is synchrotron radiation, the lobes must be energy reservoirs of tangled magnetic fields (strengths

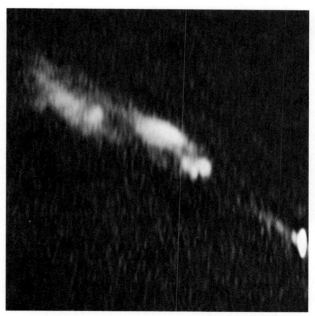

FIGURE 24–11 High-resolution image of the Cen A jet. This map shows the nucleus (right) and all 700 pc of the jet extending to the left. Note the limb brightening on alternating sides. *(J. Burns, D. Clarke, E. D. Feigelson, and E. J. Schreier, National Radio Astronomy Observatory)*

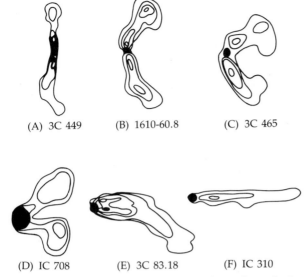

(A) 3C 449 (B) 1610-60.8 (C) 3C 465

(D) IC 708 (E) 3C 83.18 (F) IC 310

FIGURE 24–12 Bending sequence for radio galaxies. The position of the tails is influenced by the speed of the galaxy through the local intergalactic medium and by the density of that medium. *(Adapted from a diagram by G. Miley)*

Extended radio galaxies show a bending sequence (Figure 24–12) from linear classical doubles to nuclear emission bunched up at one end of a tail. This sequence strongly implies that clusters of galaxies contain a hot, ionized gas. Imagine that a galaxy moving rapidly through this medium shoots out material (high-speed electrons, for instance) in a jet. The material flowing out of the galaxy is decelerated by the intracluster medium, and the moving galaxy leaves it behind. As the galaxy travels along, it leaves behind a radio-visible trail—a fossil record of where it's been (Figure 24–13). The bending sequence shows that a dense gas exists in clusters. The discovery of head–tail galaxies prompted the acceptance of intracluster gas before X-ray observations confirmed its existence (at a density of 10^{-24} to 10^{-27} kg/m^3) and showed it to be very hot (about 10^7 K).

To sum up: as more extensive observations are made of radio galaxies, it appears that at least 50% of the classical doubles (which also have relatively high luminosities) show jets, which tend to be one-

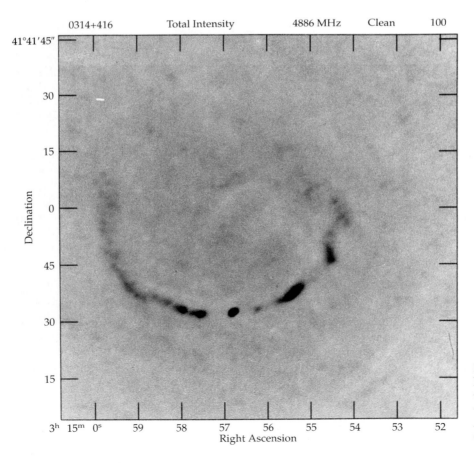

FIGURE 24–13 VLA map of the head-tail galaxy NGC 1265. Note that the darkest areas represent the largest flux. *(F. Owen, J. Burns, L. Rudnick, National Radio Astronomy Observatory)*

sided. About 80 to 90% of lower-luminosity radio galaxies exhibit jets—perhaps all of them do. The jets are typically two-sided, clumpy, and well-aligned for distances up to kiloparsecs. The physics of these cosmic radio jets is a critical, unsolved problem in modern astrophysics.

24–4 ◗
QUASARS: DISCOVERY AND DESCRIPTION

In the late 1950s, radio astronomers compiled catalogs replete with radio sources that were not identified with any familiar visible objects. Hunting for possible associations of radio and optical sources, Thomas Matthews and Allan Sandage in 1960 discovered a faint, 16th-magnitude, star-like object (hence the name **quasi-stellar object,** or **quasar**) at the position of radio object 3C 48. (3C means the Third Cambridge Catalog.) This object had a spectrum of broad emission lines that could not be identified, and it emitted more ultraviolet light than an ordinary main-sequence star.

3C 48 remained a unique object until 1963, when the strong radio source 3C 273 was identified with a 13th-magnitude, star-like object (Figure 24–14). The emission lines of 3C 273 were just as puzzling as the emission lines from 3C 48: they coincided with no known atomic lines. What was happening?

(A) EMISSION-LINE CHARACTERISTICS

The first quasi-stellar object whose spectral redshift was identified was the radio source 3C 273.

FIGURE 24–14   The quasar 3C 273. Note the jet. *(Palomar Observatory, California Institute of Technology)*

The emission lines in the spectrum of 3C 273 show a familiar regularity reminiscent of the hydrogen Balmer lines, but they seem to be greatly displaced to the red of the normal Balmer lines (Figure 24–15). Maarten Schmidt did identify them as the Balmer series, and he calculated a redshift of $z = 0.158$ (where z is $\Delta\lambda/\lambda_0$). With the acceptance of the concept of large values of z, other lines could be identified. Most of the emission features, all of which are *broad* lines, are similar to those found in gaseous nebulae and active galaxies.

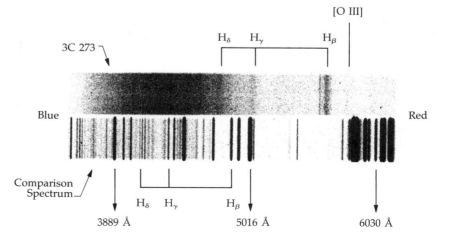

FIGURE 24–15   Spectrum of 3C 273. The comparison spectrum below serves as a wavelength standard. Note this is a negative print, so emission lines appear darker. *(M. Schmidt)*

The spectra of most quasi-stellar objects show very broad emission lines similar to those found in the spectrum of 3C 273. The outstanding feature of these lines is the fact that they are all very greatly shifted to the red, so that the redshift corresponds to values of z from 0.06 to greater than 4.0. For example, in several cases the Lyman line, normally 121.6 nm, is shifted into the visible part of the spectrum.

Note that redshifts greater than unity require the use of the **relativistic Doppler-shift** formula rather than the classical one,

$$z = \Delta\lambda/\lambda_0 = v/c$$

which for $z > 1$ would imply that the source is moving faster than the speed of light! One form of the relativistic Doppler shift is

$$z = \Delta\lambda/\lambda_0 = [(1 + v/c)/(1 - v/c)]^{1/2} - 1 \quad \textbf{(24–1)}$$

where $\Delta\lambda = \lambda - \lambda_0$, λ_0 is the original (laboratory) wavelength, v the radial velocity, and c the speed of light. Note that for $v/c \ll 1$, this equation reduces to the classical result.

Now we will use the relativistic Doppler shift to find v for quasars. Say $z = 2$; then

$$z = 2 = \Delta\lambda/\lambda_0 = [(1 + v/c)/(1 - v/c)]^{1/2} - 1$$

so that

$$(1 + v/c)/(1 - v/c) = 3^2 = 9$$
$$v/c = 8/10 = 0.8$$

Once identified, the emission lines from quasars can be analyzed in the same way as those from H II regions. This process indicates that a strong flux of ultraviolet and X-ray photons ionizes a low-density, transparent gas. Within that gas are clouds or filaments moving at high speeds—usually greater than 1000 km/s—to explain the width of the emission lines.

(B) THE ABSORPTION-LINE SPECTRA

Although the emission lines of quasars first gained them notice and remain perhaps their most distinguishing characteristic, their absorption line spectra also receive a great deal of observational attention. Most, perhaps all, quasars with emission-line redshifts greater than 2.2 also have strong absorption lines in their spectra; those with lesser redshifts typically do not have absorption lines. The absorption-line redshifts are almost always less than or equal to the emission-line redshift.

In order to detect absorption lines, a very specific set of conditions must be met. First, there must be a source of continuum emission. Second, located between the continuum source and the observer there must be a medium capable of absorbing some of the continuum photons. The absorption lines provide much more information about the absorbing medium than they do about the source. Since quasars are probably located at great distances, the spatial position of the absorbing medium could be anywhere from quite near the quasar to deep in intergalactic space.

In astrophysics, we frequently encounter velocity differences between the source and the absorbing medium. The best example of this occurs when the medium is moving outward with respect to the source—an expanding shell model. This situation causes spectral lines to have the well-known P-Cygni profile in which an absorption feature is found on the blue side of an emission feature (see Figure 18–10). Many quasar absorption lines have a P-Cygni character.

R. J. Weymann has proposed a classification system for absorption line systems in quasars. Figure 24–16 illustrates three of the 4 classes in the spectrum of the quasar 0135.0–4001. The classes are described below:

Type A: Broad Absorption Line (BAL) quasars. In these spectra we see very wide troughs of absorption. Inferred ejection velocities range up to about $0.1c$.

Type B: Low Velocity Sharp Line Systems. In these spectra we see velocity differences between the absorbing and emitting regions of up to 3000 km/s. The C IV lines are most commonly seen.

Type C: Sharp Metallic Lines. Type C systems have a velocity difference up to 30,000 km/s.

Type D: The Lyman-alpha Forest. These systems show sharp Ly-α lines with velocity differences again up to 30,000 km/s. Often many systems with different redshifts can be seen.

How are we to interpret the absorption-line classes? Models for the BAL systems seemingly must put the absorbing gas in proximity to the quasars. One model proposes pre-existing interstellar clouds within about 1 kpc of the quasar nucleus, which is imbedded in an edge-on spiral galaxy. A high-velocity, low-density wind accelerates

FIGURE 24–16 C IV absorption lines in the quasar 0135.0–4001. Three of the four kinds of absorption lines are seen here. Type A is the broad trough seen at an observed wavelength of 4230 Angstroms. Type B is represented by the narrow pair at $z \cong 1.86$, and there are four type C systems at $z = 1.62$, 1.76, 1.78, and 1.83. The type D system (the Ly-α forest) would only be seen at wavelengths shorter than the Ly-α line (redshifted here to 3470 Angstroms). *(R. Weymann)*

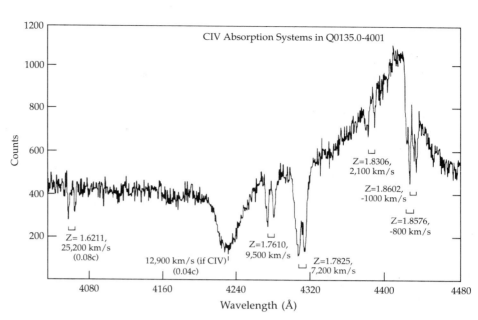

the clouds. Thus there are atoms flowing outward at a continuous range of velocities and capable of absorbing photons at a wide range of wavelengths.

Types B and C both feature narrow lines of the metals. Since elements heavier than hydrogen and helium are only created in stars, these lines must come from material originating in galaxies. The lowest velocity difference systems probably come from clouds within the parent galaxy or from galaxies in the same cluster.

A most interesting controversy has arisen over one aspect of the sharp metallic lines. Some believe that there is evidence for a process termed **line locking**. Imagine that an atom is accelerated outward by succesive absorptions of photons from one of the emission lines. If the velocity becomes too great, the atom stops accelerating because the emission-line photons have been Doppler shifted out of the acceptable range of wavelengths for absorption by that atom. If the atom slows its outward flow, then it would again be susceptible to accelerating absorptions. Hence the outward flow is "locked" in velocity by the line absorption and emission properties of the gas. The controversy comes from the fact that not all specialists believe line locking to be very effective.

The Lyman-alpha forest features lines of very low velocity dispersion—perhaps less than 60 km/s. Thus the clouds responsible for the absorptions must have temperatures less than about 10^5 K. The high-velocity differences indicate that most of these clouds must be completely indepen-

dent of the parent galaxy or even its cluster. For example, the spectrum of the quasar PHL 938 has an emission redshift of 1.955 and absorption redshifts of 1.949, 1.945, and 0.613. It has been calculated that light from a quasar with redshift of $z = 2.0$ might cross 10 or 20 superclusters in traversing the whole path between the quasar and the Earth. The outskirts of the intervening galaxies seem the most likely means of accounting for so many different absorbing systems.

(C) CONTINUOUS EMISSION

Like radio galaxies, when observed over a span of wavelengths from radio to X-rays, the continuous emission from quasars is nonthermal. Similarly, we find two distinct categories of spectral index— flat and steep, with the division at $\alpha \approx 0.5$. The spectral indices range from 0.0 to 1.6. So synchrotron emission is most likely the source of the continuous radiation.

Quasars also split into two groups when their polarization is examined. Most quasars have low polarizations (less than 3%). Only 1% or so of optically bright quasars have polarizations greater than 3% (up to 35%). The highly polarized quasars are compact radio sources, have flat radio spectra and steep optical ones, and exhibit rapid (days to years), large-amplitude variability at optical wavelengths. Hence, high-polarization quasars share many characteristics with BL Lac objects.

(D) OPTICAL APPEARANCE

As suggested by the name, the optical appearance of quasars is stellar; that is, their angular diameters are less than 1 arcsec. Some, however, have faint nebulosities associated with them. The optical variations of several quasars have been well established; in some cases, changes of as much as 1 to 2 magnitudes over a few months have been observed. Some of these variations are long-term, with time scales of years; in other cases, the brightness changes have been very rapid, with small changes (of the order of a few percent) occurring in a matter of 15 min. Other changes occur more slowly, over a matter of several days or a month or two. One phenomenal case, 3C 446, changed its luminosity by a factor of 20 within a year and continued to show sizable variations thereafter. Quasars that vary appreciably at radio frequencies are also optical variables. A strong correlation exists between optical variability and polarization. In fact, the brightness changes are often accompanied by a variation in polarization, indicating that it is the polarized component that is responsible for the light fluctuations. About 20% of known quasars have rapid (days to weeks) variations at optical wavelengths (and also in the radio region).

Optical variations are important in setting limits on the size of quasars. If an object varies with a period t, the radius of the object must be equal to or smaller than t times the speed of light—$R \leq ct$. Outbursts at different places within the source would average out to only slight overall changes if this limit did not hold. A total variation by any appreciable factor requires that the outbursts be synchronized; this means that the signal from one region must travel the distance to all other regions within the period of variation. Moreover, if an object with radius larger than ct varied as a whole, the light travel time would smooth out the time variations. If ct is 1 lightmonth but the object is 1 lightyear in diameter, radiation from the farthest point is delayed by 1 year relative to that from the side nearest the observer, masking the monthly variations.

The fact that the variations in quasars have been observed to take place with time scales of less than 1 month (even days) suggests that the radius of the object must be about 1 lightday, or 10^{13} m. The entire quasar need not participate in the variation, but the fact that variations by a factor of 2 and even more have been observed does mean that a very sizable portion of the total radiation from the object participates in this variation; so the region from which this radiation arises is restricted by this size limitation.

24–5 ●
PROBLEMS WITH QUASARS

The main feature of quasars is their very large redshifts. The most natural explanation of these redshifts is a cosmological one: quasars participate in the Universe's expansion. If so, their enormous redshifts indicate that they are very far from us and must expend vast amounts of energy. For example, the redshift of $0.16c$ for 3C 273, if due to the expansion of the Universe, implies a distance of 770 Mpc for $H_0 = 50$ km/s $\cdot$ Mpc. At this distance, to appear at its observed apparent magnitude of 13, 3C 273 must emit about 10^{40} W, or about 40 times as much as the most luminous galaxies. A typical quasar produces about 1000 times as much power as an ordinary spiral galaxy, most of it emitted in the infrared.

Not only do quasars emit energy at enormous rates, but that energy comes from relatively small regions of space in the quasar centers—from light-hours or lightdays to no more than a few lightyears in diameter. Two pieces of evidence point to small energy-emitting volumes. First, the variation of light output over days to years. The size of the region that emits the energy can be no more than the light travel time across it, and so the regions cannot be larger than a few lightyears. Second, VLBI observations have revealed in some quasars radio structures that are no more than a few tens of lightyears in diameter (for a cosmological redshift).

This then defines the energy problem for quasars: how to generate about 100 times the energy of a galaxy in a region only a few lightyears across!

(A) ENERGY SOURCES

Almost all of a quasar's continuous spectrum comes from synchrotron emission: high-speed electrons gyrating in a magnetic field. As these electrons emit electromagnetic radiation, they lose energy and move more slowly. So they emit radiation with lower and lower energy. This loss of speedy electrons implies that the supply of high-energy ones must be replenished at least about every year or so. The central energy source of a quasar must yearly blast out clouds of high-energy electrons containing a total of at least 10^{43} J. The rest of the quasar acts like a transformation machine, trapping the energy of electrons and converting it to other forms. What energy source lies at the heart of a quasar?

The most developed quasar model to date involves supermassive black holes, objects of mass about 10^7 to 10^9 solar masses. This model originates from that for binary X-ray sources (Chapter 17), in which material from a normal star forms an accretion disk around a black hole before the material falls into it. In the quasar model, a supermassive black hole in a dense galactic nucleus is fueled by the tidal disruption of passing stars. The stellar material forms an accretion disk and radiates as it spirals into the black hole, powering the quasar. Outflows of ionized gas may be generated perpendicular to the spin axis of the disk; these may be visible as jets from the nucleus [Section 24–3(c)].

The model calculations show that luminosities of 10^{12} solar luminosities, about that of bright quasars, are possible with an infall of 1 solar mass or less of material per year. One aspect of this model that works is that the supermassive black hole can easily generate the level of quasar luminosity in a region of space only a few lightyears in diameter (the Schwarzschild radius of a $10^8 M_\odot$ black hole is only 3×10^8 km, or about 2 AU). And it does the energy conversion (from gravitational to radiative) with high efficiency.

You can see this quantitatively by recalling that the potential energy of mass m brought in from infinity to distance R next to mass M is

$$PE = -GMm/R$$

so for a 1-kg mass moved to the Schwarzschild radius of a $10^8 M_\odot$ black hole, $PE \approx 4 \times 10^{16}$ J/kg. Then to provide $10^{12} L_\odot = 4 \times 10^{38}$ J/s requires the infall of

$$L/E = (4 \times 10^{38} \text{ J/s})/(4 \times 10^{16} \text{ J/kg})$$
$$\approx 10^{22} \text{ kg/s} \approx 10^{29} \text{ kg/year}$$

with the assumption that the conversion from potential energy to light will be 100% efficient. (It won't be, so the amount of mass will be greater.)

(B) SUPERLUMINAL MOTIONS

VLBI radio observations have also presented a new puzzle with several quasars. Parts of these objects appear to be moving with speeds greater than c! This apparent faster-than-light motion is called **superluminal motion**.

A good example is 3C 273. Recall that it has an optical jet about 100 pc long. Its radio emission comes mainly from the body of the quasar (called 3C 273B) and from near the end of the jet (called 3C 273A). So 3C 273 has a core with a radio jet protruding from it. A series of radio observations from 1977 to 1980 showed that a knot in the source has steadily moved away from the central peak (Figure 24–17). The total separation has increased in the three years by 2 milliarcseconds. The expansion seems to have happened at a constant rate over this time. Now, if 3C 273 is at a distance given by its redshift, the observed angular separation rate corresponds to a transverse velocity of almost *six times the speed of light*!

Sources that show superluminal motions all have a common feature: a radio structure consisting of a strong central source with a weaker jet out one side. So they resemble nearby radio galaxies with single jets, such as Centaurus A. These jets are thought to be electrons flowing outward at close to the speed of light from a nuclear source; they may be relativistic jets. The same process may occur in the superluminal sources. In 3C 273, the jet is pointing almost directly at us (tilted only about 10° away from a perfect alignment). The moving knot is a blob of material streaming out along the jet. The knot is *not* moving faster than the speed of light, however—it only *appears* to be. The apparent superluminal speed is an optical illusion caused by the almost head-on orientation of the relativistic jet and the finite speed of light.

To understand this effect, consider a jet emitting blobs of material moving at close to c. Suppose the jet opens at some small angle, say 8° with respect to our line of sight (Figure 24–18). Imagine that a blob ejected by the nucleus (point N) gets to point A in 101 years. Suppose the light emitted from the blob when it was at N reaches point B after 100 years. The separation between A and B is 14 lightyears (for an 8° angle), but the light at B is one year ahead of that emitted by the blob when it reaches A. (It has taken 100 years for the light from N to reach B, 101 years for the blob to reach A.) Many years later, the light that was at B reaches us; only one year later, that emitted at A reaches us. The source seems to have moved from B to A—14 lightyears—in only one year. It seems to have a transverse speed of $14c$. Yet no such physical mo-

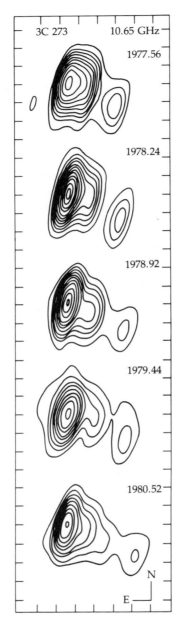

FIGURE 24–17 Motions in 3C 273. High-resolution radio maps at 10.65 GHz show a blob of material moving away from the nucleus. *(T. J. Pearson, S. C. Unwin, M. H. Cohen, R. P. Linfield, A. C. S. Readhead, G. A. Seielstad, R. S. Simon, and R. C. Walker, Owens Valley Radio Observatory and the National Radio Astronomy Observatory)*

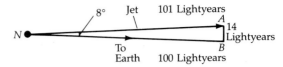

FIGURE 24–18 Geometry for superluminal motions.

tion has occurred. The superluminal velocity is only an apparent motion. And the smaller the angle of the jet to our line of sight, the faster the apparently superluminal motion will appear.

The main point is that bulk relativistic linear motion aligned close to the line of sight provides a simple explanation for rapid flux density outbursts, the superluminal motions, and the strength of the emission in the jets. Physically, this model requires that the speeds in the beam be very close to light ($0.99c$ or greater) and that the Lorentz factor γ, defined by

$$\gamma = (1 - \beta^2)^{-1/2}$$

where $\beta = v/c$, is on the order of 10. For $\beta = 1$, for small angles (θ) to the line of sight, the apparent transverse velocity is

$$v = 2c/\sin \theta$$

so for $\theta = 8°$, as in the example above, the apparent transverse velocity is increased by a factor of 7. Also, relativistic effects concentrate the flux in a narrow forward beam—an effect known as **relativistic beaming**—that boosts the observed flux. For a beam that is head-on with $\gamma > 1$, the flux is increased in proportion to γ^3, so that if $\gamma = 10$, the flux goes up by 10^3 relative to an unbeamed signal.

(C) DOUBLE QUASARS AND GRAVITATIONAL LENSES

Another optical illusion related to quasars derives from the bending of light in strong gravitational fields—an effect predicted by general relativity and confirmed in the Solar System by the bending of starlight as rays pass close to the Sun. Large masses can image light, usually imperfectly. Recently, observations of quasars have confirmed this gravitational-lens effect on a cosmological scale.

Quasars are rarely close together. Two called 0957 + 561A and 0957 + 561B are only 6 arcsec apart. Even more surprising, their spectra are

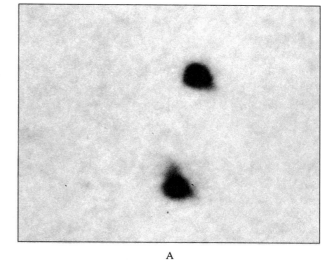

A

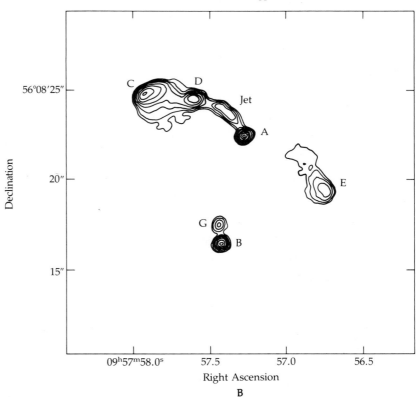

B

FIGURE 24–19 The double quasar. (A) Optical negative image on a night of excellent seeing; note the fuzz sticking up from the lower image. *(A. Stockton)* (B) Radio map at 6 cm done with the VLA. The double quasar (A and B) is seen; note that A has a jet flowing into two lobes (C and D). G has properties similar to those of an active elliptical galaxy. *(D. H. Roberts, P. E. Greenfield, J. N. Hewitt, B. F. Burke, and A. K. Dupree, National Radio Astronomy Observatory)*

nearly identical (in line positions and intensities), and their redshifts are the same: $z = 1.41$. It turns out that these quasars are not twins but rather optical images of the same quasar made by an intervening galaxy. Photographs made on a night of exceptionally good seeing show that quasar B has a little piece of "fuzz" sticking out of it (Figure 24–19A). This fuzz turns out to be the poorly re-

solved image of a faint galaxy—the gravitational lens! Since the galaxy is an extended mass, however, it acts like an imperfect lens and produces a complex pattern of up to three images. By a quirk of placement, we see two of the three images formed by a gravitational lens: an intervening, elliptical (probably cD, with mass $\approx 10^{12} M_\odot$) galaxy between us and the quasar (Figure 24–19B).

This discovery has three important implications: (1) it provides another confirmation of general relativity; (2) it proves in this case that the quasar is more distant than the galaxy, and so the quasar's redshift is cosmological; (3) cool gas around the galaxy creates the quasar's absorption-line spectrum; this situation may well be the case of other quasars, too.

(D) QUASARS COMPARED WITH ACTIVE GALAXIES

Quasars and active galaxies share some observed characteristics. First, consider the radio galaxies. Most have only absorption lines in their spectra except for the common presence of [O II] 372.7. Those that have emission lines come in two types. One type has narrow emission lines. Cygnus A is a good example; it has strong emission in the lines of [O III] at 500.7 nm and [N II] at 658.3 nm and other lines from such ionic states as [S II], [Ne V], and [Fe X]. These lines are all narrow; the Doppler widths are roughly 500 km/s. Other narrow-line radio galaxies have emission-line widths of 400 to 800 km/s. Narrow-line radio galaxies make up about two-thirds of the total. The other one-third have broad lines of hydrogen and helium, some as wide as 10^4 km/s. Other emission lines for these galaxies tend to be narrow. Quasars with low redshifts have optical spectra that resemble those of the broad-line radio galaxies in terms of the emission lines present, their widths, and the shape of the optical continuous spectrum. For instance, the hydrogen and helium lines of such quasars have widths of 3000 to 6000 km/s.

Quasars also resemble Seyferts in their emission-line spectra. The Seyfert 1 galaxies, just like the broad-line radio galaxies, look like low-redshift quasars in terms of their emission spectra. So the physical conditions in the regions producing the spectra may be basically the same in Seyfert 1 galaxies and in low-redshift quasars. Note that if Seyferts were more distant than they are, we would see not the galaxy but only the bright core. A distant Seyfert would look a lot like a quasar.

Comparing quasars and BL Lac objects, we note a crucial difference: BL Lac objects do not have strong emission lines. However, in both BL Lacs and quasars, the nonthermal nature of the contin-uous emission stands out more clearly than in other active galaxies. BL Lac objects and about 15% of radio-bright quasars show wide variations in optical output over periods of days, weeks, and months. These swings in luminosity often occur very abruptly.

One other possible connection relates to radio structure. Quasars have high luminosities and tend to exhibit symmetrical double structures. The higher luminosity active galaxies have similar radio shapes. So the physical processes responsible for both may be the same. The nuclear jets imply that this structure is somehow tied to violent activity in the nuclei of luminous active galaxies and quasars.

The observations suggest that quasars are somewhat physically similar to active galaxies. One popular idea views the two as objects at different stages of evolution, that is, the quasar phenomenon signals very violent activity in the nucleus of a galaxy at a very early stage in its life. So quasars may be hyperactive galaxies. This evolutionary connection is a nice idea and implies that the energy sources in quasars relate directly to those in active galaxies. If the sources are black holes, then all (or most) galaxies need to have supermassive black holes in their nuclei.

If we accept a black-hole model, then the nucleus of a quasar should show an intense, point-like source of light (from a concentration of stars around the black hole), stars orbiting the center should have high velocities, and emission lines with high Doppler shifts might be visible from the infalling matter. Observations of such effects have been reported for a high-luminosity point in the nucleus of M87, including higher velocities for stars in the nucleus than for those outside. A model consistent with these observations is one of a $5 \times 10^9 M_\odot$ black hole hiding in the inner 100 pc of the nucleus. (An alternative model uses a dense concentration of ordinary stars.)

Given that a quasar may simply be the hyperactive nucleus of a very distant galaxy, what about the rest of the galaxy? At cosmological distances, the disk of a quasar's parent galaxy would be too small and faint to see easily. For example, if the Andromeda galaxy were placed at a distance of 900 Mpc—equivalent to a redshift of 0.2—its angular diameter would be a mere 4″. Remember that the Earth's atmosphere limits seeing to 1″ at best.

So a distant galaxy would make an image that is difficult to distinguish from that of a faint star.

Recent CCD observations of the quasar 1059 + 730 ($z = 0.089$) have noted that the bright nucleus is surrounded by an elongated, fuzzy structure about 9″ by 16″ in size that resembles an edge-on disk. In May 1983, a stellar object with an apparent visual magnitude of 19.6 appeared within the fuzz. This outburst was most likely a supernova explosion. If so, then 1059 + 730 lies at a cosmological distance, as inferred from its redshift, and the fuzz surrounding it must contain stars—the stellar disk of the underlying galaxy.

We might also look for a quasar in a cluster of galaxies—a good sign that we're not looking at individual stars. An example is the quasar 3C 206, which is surrounded by some 20 faint galaxies (Figure 24–20). The quasar's redshift is 0.206; that of a nearby pair of galaxies is 0.203 (and the same is presumed for the cluster). This result argues in favor of the quasar's residing in the cluster.

Other observations are beginning to reveal that quasars are surrounded by very faint envelopes that may well be the hard-to-detect disks of galaxies. For example, 3C 273, the closest of the high-luminosity quasars, has a fuzzy appearance in computer-processed photographs. The faint fuzz

has emission lines in its spectrum with the same redshift as the quasar. Another good example is Markarian 1014, a low-redshift (0.163) quasar whose fuzz has been examined spectroscopically. Stellar absorption lines in the spectra show a mixture of early and late-type stars with the same redshift as the quasar. The luminous matter surrounding it appears asymmetric and spiral-like and has a linear diameter of about 100 kpc. This seems to be a galaxy with a "quasar" nucleus.

(E) NONCOSMOLOGICAL REDSHIFTS

Our discussion has presumed that the redshifts of quasars are cosmological—in other words, that they arise from the expansion of the Universe—but a small, persistent band of astronomers have argued that all or part of the redshifts come from noncosmological causes. Such effects require that quasars be closer so that the energy problem is alleviated somewhat.

The proponents of noncosmological redshifts have looked for evidence that some quasars are associated with galaxies—in the sense of proximity on the sky—and then searched for a possible *physical* connection between the galaxy and the quasar. The arguments rely heavily on statistical inference to demonstrate that the chances of proximity are small. (On the average, two quasars appear per square degree of sky down to a limiting magnitude of $V = 19$.)

We give two representative examples, both from the work of Halton Arp. In one, a quasar-like object ($z = 0.044$) lies in *front* of an elliptical galaxy ($z = 0.009$), based on the fact that a dark ring rims the quasar—material perhaps absorbing light from the galaxy. Arp concludes that the object was ejected by the galaxy. He finds another curious case in the cluster of galaxies Abell 1367, where two quasars lie close to a galaxy (within about 1′)—an association that Arp argues has only 7×10^{-6} probability of happening by chance. A physical connection has yet to be established in either case.

Also troubling to the standard view is the work of W. Tifft who has found evidence for the quantization of redshifts of normal galaxies. Whereas Arp's methods tend toward finding peculiar associations of high and low redshift objects, Tifft's method examines the statistical properties of large

FIGURE 24–20 Quasar 3C 206 in a small cluster of some 20 galaxies. (*S. Wyckoff, P. Wehinger, H. Spinrad, and A. Boksenberg*)

numbers of redshifts. One example is Tifft's study of the redshift differences between the two galaxies in a complete sample of binary galaxies. Because of the wide range of orbital eccentricities and viewing angles, one would expect the differences to vary smoothly with a difference of zero being quite likely (because many systems would be seen pole-on). Tifft found that zero was excluded in the observations and that several regularly separated (24 km/s spacing) peaks were seen in the distribution.

Most astronomers believe that the evidence to date more strongly supports a cosmological rather than a noncosmological interpretation. It seems useful for most workers to continue to push the "standard" picture as far as possible, but it is also important for a few brave souls to question the foundations of our science. Perhaps there is another Copernicus waiting in the wings.

24–6 ◑
ASTROPHYSICAL JETS

High-resolution radio observations in recent years have revealed that long, narrow flows of material—loosely called **astrophysical jets**—are a common and colorful phenomenon. Within the Galaxy, we see bipolar flows from protostars (Chapter 19), a possible outflow from the Galaxy's nucleus (Chapter 20), the relativistic flow from SS 433 (Chapter 18), and surprising flows from the Crab Nebula and some planetary nebulae. This chapter describes the nuclear jets from active galaxies and quasars. These jets range in size from a few parsecs to megaparsecs across, with lengths typically 100 times the diameters. In just about all cases (such as Cen A and M87), the jets show internal knots, the emission from which is likely synchrotron radiation from electrons in magnetic fields. In most cases, the knotty beam ends in an amorphous, wispy lobe that surrounds a hot spot. Some jets twist and bend before they reach the limits of their flows.

The observed properties of astrophysical jets appear similar to those of supersonic gaseous jets that have been studied in laboratories and in theoretical computation for well over a century, pioneered by Ernst Mach and others. A key property of a supersonic flow is the **Mach number** M, which is the ratio of the flow speed v to the local sound velocity s:

$$M = v/s$$

where

$$s = (\gamma P/\rho)^{1/2}$$

and γ is the adiabatic index, which equals 5/3 for an ideal monotonic gas. When $M > 1$, the flow is supersonic, that is, greater than the local sound speed.

Supersonic flows generate shocks (such as the sonic boom created by a jet airplane that crosses Mach 1 in our atmosphere). Laboratory experiments show that artificial jets with high Mach numbers produce internal, reflected shocks and, as a result, high-pressure knots. Astrophysical jets exhibit knotty structures of higher pressures, temperatures, and densities that may very well be caused by internal shocks. These shocks do not disrupt the beam's flow, which does end in a strong shock at the head of the beam. The termination is surrounded by a cocoon of low-density material (the radio lobe).

Michael Norman, Larry Smarr, and Karl-Heinz Winkler have used a supercomputer (Cray-1) at Los Alamos National Laboratory to investigate the properties of computational supersonic jets and to compare them with the observed astrophysical ones. We will outline their basic results here for jets whose internal pressures initially just match those of the surrounding media. (Indirect evidence suggests that astrophysical jets are kept collimated by the external gas pressures of the surrounding media.)

As a jet penetrates the ambient medium, it drives a shock before it (Figure 24–21). The discontinuity forms at the boundary of the jet and ambient gas. A shock forms behind the contact boundary; here the gas in the beam stops and is deflected around in a return flow that creates a cocoon around the jet. This cocoon contains vortices within itself and waves along its surface. The extent of the cocoon depends on the Mach number and density of the jet's gas relative to the medium. Greater Mach numbers generate larger cocoons, and lower-density jets also produce relatively large cocoons. Note that the cocoon works to dissipate the flow of the jets. In real astrophysical jets, shocks within the cocoons might generate the syn-

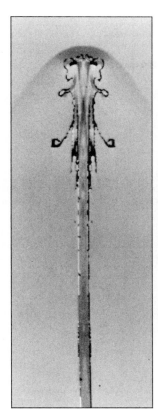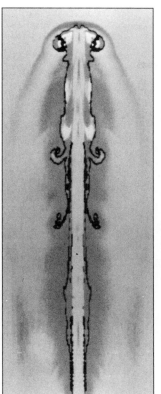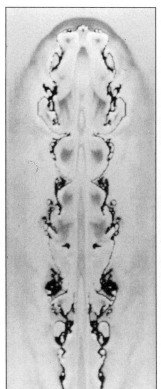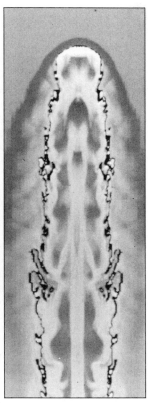

FIGURE 24–21 Shock wave in front of a jet. These theoretical calculations show the envelope of a shock around the front of a supersonic jet. *(M. Norman and K.-H. Winkler)*

chrotron emission that is observed. Eventually, the flow becomes unstable and develops kinks and wiggles far out from the source. Here the jet also hits a high-density piece of the surrounding medium that brings the flow to an end. Overall, the

model of knotty supersonic flows describes some of the key features of astrophysical jets, and more computations are in progress to gain a better match between theoretical models and the observed properties of jets.

PROBLEMS

1. The special theory of relativity says that no material object can move faster than the speed of light. The classical Doppler formula says that when the redshift z is greater than unity, we have $v > c$ (which is impossible).
   (a) Referring to Chapter 8, derive the *exact relativistic relation* between v and z. Express your result in the form $v = f(z)$.
   (b) Make a table with three columns: column one for z, where you should enter five exemplary values

   of z from 0 to 3.0; the second column headed $(v/c)_{cla}$, where you compute the classical *nonrelativistic* result for your five values of z; and the last column headed $(v/c)_{rel}$, where you use your formula from part **(a)**.

2. Assume that the optical galaxy associated with the radio source Centaurus A is the same size as our Galaxy and draw a scale diagram of the radio and optical emission regions of Centaurus A. Clearly

indicate the dimensions and the relative positions of the various components.

3. The radio galaxy Cygnus A has an observed radio *flux density* of 2.18×10^{-23} W/m$^2 \cdot$ Hz at a frequency of 10^3 MHz. (Note that the unit of bandwidth $\Delta\nu$ is 1 Hz.) The observed redshift of the galaxy is $\Delta\lambda/\lambda_0 = z = 0.170$.
   (a) If the radiation is received at 10^3 MHz, at what (rest) frequency was it emitted by Cygnus A?
   (b) What is the distance to Cygnus A? (Use a Hubble constant of $H_0 = 50$ km/s $\cdot$ Mpc.)
   (c) What is the *radio luminosity* (W/Hz) of this radio source at 10^3 MHz?
   (d) To find the total radio luminosity of Cygnus A, we must multiply the result of part (c) by the *bandwidth* $\Delta\nu$ of our detector. Assume $\Delta\nu = 10^4$ Hz and compute the energy radiated per second at radio frequencies.
   (e) What is the minimum mass of hydrogen (in solar masses) that must be converted to helium during each second to provide this luminosity?
   (f) If Cygnus A continues to radiate at this rate for 10^8 years, how many solar masses of hydrogen must be converted to He? Express this result in terms of the mass of our Galaxy ($\approx 10^{12} M_\odot$).

4. 3C 9 is a quasi-stellar object that has a redshift of 2.0 and an apparent visual magnitude of 18.2. Answer the following questions using the cosmological interpretation of the redshift.
   (a) What is the speed of recession?
   (b) What is the distance to 3C 9?
   (c) What is the intrinsic luminosity relative to that of our Galaxy?
   (d) What is the maximum size of the emitting region if 3C 9 exhibits luminosity variations on a time scale of two months?

5. (a) Quasar 3C 273 has a redshift of 0.16. What is its distance?
   (b) The V magnitude of 3C 273 is 12.8. What is its flux density at V band? Its luminosity at V band?
   (c) The fuzz around 3C 273 has a diameter of 15″. What is its linear size?
   (d) The absolute magnitude of the fuzz is -25. What is its luminosity?

6. The quasar PKS 1402 + 044 has a redshift of 3.2. What is its distance? Note that $z > 1$!

7. Observations of the radio jet in Cen A indicate that the spectral index is about 0.5. The 20-cm flux density of the strongest blob of the jet is 2.3 Jy (janskies, 1 Jy $= 10^{-26}$ W/m$^2 \cdot$ Hz). If this emission is synchrotron, what should the flux density be at 2.2 μm?

8. (a) How close must a star pass a black hole of mass $10^6 M_\odot$ to be disrupted tidally?
   (b) What *yearly* rate of matter infall is needed to power a quasar at 10^{39} W if a black hole of mass $10^6 M_\odot$ lies in its core?

9. In Cen A, the radio jet and the innermost lobe are separated by a distance of 4 arcmin.
   (a) What is their physical separation?
   (b) How long would it take relativistic electrons to travel from the jet to the lobe?

10. The quasar 1059 + 730 has a redshift of 0.089.
    (a) What is the distance to the quasar if the redshift is cosmological? What is the range, given the uncertainty in H_0?
    (b) The fuzz around the quasar has an angular size of 9″ × 16″. What is its physical size? How does this compare with the size of a typical spiral galaxy?
    (c) The supernova observed in 1059 + 730 had an apparent V magnitude of 19.6. Use this value to estimate the distance to the quasar. Compare this value with your results from (a).

11. The argument is often made that the dimensions of an object with variable brightness cannot exceed the speed of light times the time scale of the variations. Critically examine this argument by considering
    (a) two nonspherical geometries
    (b) special relativistic effects (note that observed time intervals are inversely proportional to observed frequencies)

12. The quasar NRAO 140 ($z = 1.26$) has a compact radio component that moves at an angular speed of 0.15 milliarcseconds/year. What is that apparent velocity of the component as a percentage of c? (Multiply your answer by $1 + z$ to correct for relativistic time dilation. Use a Hubble constant of 50 km/s $\cdot$ Mpc.)

13. A quasar has an emission line, identified as Ly-α of hydrogen ($\lambda_0 = 121.6$ nm), that is observed at 581.2 nm. Calculate the redshift and distance to the quasar for a Hubble constant of 50 km/s $\cdot$ Mpc. How fast is the quasar moving away from us?

14. At what redshift is the Ly-α line brought into a visible light detector that is sensitive to photons of wavelength greater than 370 nm?

15. Discuss the evidence that Seyfert 1, Seyfert 2, LINERs, radio galaxies, BL Lac objects, and quasars are all related and should be lumped together into one class called AGNs. What kinds of physical phenomena or conditions make us detect these objects as *different* classes when we observe them?

16. How fast would a dust cloud have to be moving to occult a typical Seyfert galaxy BLR in the time scale of one year? Assume that the cloud is in circular orbit around the galaxy nucleus.

17. To appreciate the observational difficulties associated with trying to detect a galaxy's stellar emission around a distant quasar, calculate the apparent magnitude (ignore the k-correction) and angular size that a large, luminous galaxy ($M = -21; R = 50$ kpc) would have at a redshift of: (Use $H = 50$ km/s · Mpc.)
    (a) $z = 0.1$,
    (b) $z = 1.0$.

18. Using Figure 24–15, determine the redshift of the quasar 3C 273. How well does your measurement compare with the accepted value of $z = 0.158$?

19. Which of the emission lines listed in Table 24–1 would be observable in the visible portion of the electromagnetic spectrum (from 390 nm to 720 nm) for a quasar with the following redshifts?
    (a) $z = 0.1$;
    (b) $z = 1.0$;
    (c) $z = 4.0$?

20. (a) Calculate the Schwarzschild radius of a $10^7 \, M_\odot$ black hole.
    (b) What is the average density inside the Schwarzschild radius?

Chapter 25

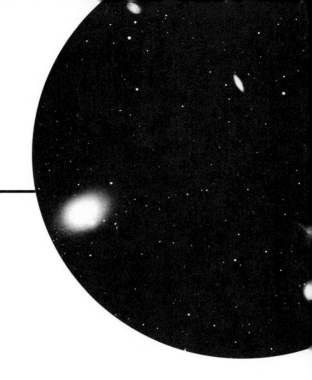

Cosmology: The Big Bang and Beyond

In this chapter we will look at the two most important new concepts about our Universe to come after Hubble's great discoveries of the 1920s. One of these is theoretical—the development of models of the Universe based on general relativity. The second is observational—the discovery of the microwave background radiation that showed that the origin of the Universe was an inferno of unimaginable heat and fury.

25–1 ◗
STEPS TOWARD GENERAL RELATIVITY

Cosmology means the science of the cosmos, or Universe. A cosmologist is not very concerned with details like comets, planets, stars, etc. Even such large objects as galaxies are thought of primarily as markers that indicate where matter lies, allowing us to find their motions, distributions, and numbers. We'll first concentrate on theoretical models of the Universe. Then at the end of the chapter we will discuss how these theories stand up to the observations.

We will focus on **relativistic** cosmology—models that rely on Einstein's theory of relativity.

With the appearance of Einstein's special (1905) and general (1915) theories of **relativity,** theoretical cosmology became a true science. In the special theory, Einstein united space and time in a four-dimensional spacetime continuum and deduced the correct kinematic theory for all objects in uniform motion. The *general theory (GR) is a theory of gravitation* that supplants Newton's laws. In its geometry, spacetime is curved and the trajectories of objects are determined by this curvature; the curvature, in turn, is produced by the material and energy content of the Universe. In contrast to Newton's view, Einstein saw gravity not as a force but as a manifestation of the curvature of spacetime. Although general relativity is certainly an improvement, calculations of particle motions by means of Newtonian mechanics work well in all but the most extreme gravitational fields.

(A) SPECIAL RELATIVITY

One of the great advances of special relativity can best be expressed by the following equation:

$$dl^2 = c\,dt^2 - dx^2 - dy^2 - dz^2 \qquad \textbf{(25–1)}$$

On the right side, we see infinitesimals of the three standard cartesian coordinates (polar, cylindrical, etc. could be substituted), but additionally we see an infinitesimal of the time coordinate playing a role equal to that of the spatial coordinates. Note the similarity of this equation to the Pythagorean theorem. In this light, dl is seen to be the separation of two points in a four-dimensional space. We have given this four-dimensional construct the name **spacetime**. Special relativity assumes that, for any two events, dl^2 is the same for all observers travelling in **inertial frames** (unaccelerated motion). Note that the difference in signs on the right side indicates that the time coordinate is not precisely the same as the spatial coordinates. If we took a square root, the time and space coordinates would differ by a factor of the imaginary number i. Nevertheless, with this theory, time has taken its place as a component of our 4-dimensional Universe that is on the same level of importance as space.

We will not attempt to give a full discussion of special relativity, but its well-known phenomena of *length contraction, time dilation,* and the *addition of velocities* all are used by some areas of astrophysics. For example, muons, which are created as secondary particles by the impact of cosmic rays high in the atmosphere, could not be detected at the Earth's surface without time dilation. Given their short lifetimes, they could not pass through the whole distance of the atmosphere even at $v = c$; but with their internal clock slowed down by time dilation, they survive the journey and are easily detected.

In other experiments as well, special relativity has been successfully verified. However, its usefulness to cosmology is limited because it only applies to unaccelerated objects. Therefore, a force such as gravity, which dominates the dynamics of the Universe, would accelerate particles and be beyond the pale of special relativity. For this reason, we need a relativistic theory that includes gravitation, and this is the realm of general relativity.

(B) NEWTONIAN COSMOLOGY

Before we develop the equations of motion for a relativistic Universe, we will do so for a Newtonian model. These equations have a direct analogy to the relativistic ones, and they are physically a bit easier to intuit. Assume here that the Universe is infinite and homogeneous, that the speed of light is infinite, and that, as a consequence, a universal time (*absolute time*) applies to all observers.

Newton proved that only the matter *interior* to some point affects the motion at that point *if* the distribution of matter is homogeneous (which we assume). Consider the matter in the Universe to be a noninteracting gas (so that pressure = 0); then the equation of motion for a test particle is

$$d^2R/dt^2 = -GM(R)/R^2$$
$$\text{(spherical symmetry!)} \quad \textbf{(25–2)}$$

where $M(R)$ is the mass interior to R:

$$M(R) = 4\pi \int_0^R \rho(r)r^2\, dr$$
$$= (4/3)\pi\rho R^3$$

since $\rho(r) = \rho = $ constant. Multiply Equation 25–2 by dR/dt to get

$$(dR/dt)(d^2R/dt^2) = -[GM(R)/R^2](dR/dt) \quad \textbf{(25–3)}$$

Integrate this equation with respect to t:

$$\int_0^t (dR/dt)(d^2R/dt^2) + \int_0^t [GM(R)/R^2](dR/dt) \equiv 0$$

To do this integration, note that

$$d/dt[(dR/dt)^2/2] = 2[(dR/dt)/2](d^2R/dt^2)$$
$$= (dR/dt)(d^2R/dt^2)$$

so that we get

$$(dR/dt)^2/2 - GM(R)/R = k = \text{constant}$$

but $M(R) = (4/3)\pi\rho R^3$, and so

$$(dR/dt)^2/2 - (4/3)\pi G\rho R^2 = k$$

Dividing this equation by R^2 and multiplying by 2, we have

$$[(dR/dt)/R]^2 - (8\pi/3)G\rho = 2k/R^2$$
$$[(dR/dt)/R]^2 = 8\pi G\rho/3 + 2k/R^2$$
$$(dR/dt)/R = (8\pi G\rho/3 + 2k/R^2)^{1/2} \quad \textbf{(25–4)}$$

What does this equation mean? Note that $R \geq 0$. So if we start (initial condition) with $(dR/dt)/R > 0$ (expansion), then

$k = 0$ means $(dR/dt)/R$ is always greater than 0
$k > 0$ means $(dR/dt)/R$ is always greater than 0
$k < 0$ means $(dR/dt)/R$ eventually equals 0
and expansion "turns around"

Equation 25–4 in essence is the energy form of the escape-velocity equation. Recall that

$$V_{esc} = (2GM/R)^{1/2}$$

so that

$$V_{esc}/R = [(8/3)\pi G\rho]^{1/2}$$

which is Equation 25–4 for the case $k = 0$ (total energy equals zero at infinity). We see here a classically derived description of a *dynamical Universe*. If ρ is small, this model expands forever; but, if ρ is larger than a critical value, expansion speed might be less than V_{esc} and the Universe eventually will collapse.

25–2 ◑
EINSTEIN'S THEORY OF GENERAL RELATIVITY

With his general theory of relativity in 1915, Einstein replaced the force of gravitation by a coupling between this geometric spacetime and the material content of the Universe. He noted that Newton defined mass by two different operations: Newton's second law and the law of gravitation. Suppose you apply a known force to an object and measure its acceleration. When the force and acceleration are known, Newton's second law gives the object's mass—its **inertial mass.** Now take the same object and weigh it. Weight is a force, the amount of gravitational force acting on the object. Mass measured this way is called **gravitational mass.** Newton believed that an object's inertial mass and its gravitational mass were the same. He knew this from Galileo's experiments with falling bodies as well from his own careful experiments. These results demonstrated that, near the Earth, all masses fall with the same acceleration.

Experimentally, this equality of gravitational and inertial mass holds true to a very high degree of accuracy. To the limits of sensitivity of the testing methods, no difference has ever been detected. The best experiment to date, done by V. B. Braginsky and V. I. Panov at Moscow University, found the inertial and gravitational masses of gold and platinum to be the same within one part in 10^{12}. Einstein felt that *the equality of gravitational and inertial mass was no accident.* He saw it as a fundamental fact about the Universe and gave it a special

place in the general theory as the **principle of equivalence.** Here is Einstein's own example of the principle of equivalence. Imagine that you are on the Earth in a spacecraft. You drop objects in the spacecraft, measure their accelerations, and find that they all fall with the same acceleration, 9.8 m/s². Now suppose that, without your knowledge, you and the spacecraft are instantly transported out into space and constantly accelerated at 9.8 m/s². As you continue your experiment, no change occurs in the acceleration of the falling objects. It remains 9.8 m/s². Only by looking out a window could you tell that you had left the Earth. You cannot, by your experiments, distinguish between gravitational mass and inertial mass.

The principle of equivalence provides a way to cancel out gravity locally. Put yourself in an elevator in a tall building. Let the elevator free-fall. You find yourself weightless; gravity has vanished! Transport yourself into space far away from any large masses. Your condition is the same—you are weightless without gravity. You might object that in this example of a free-fall, gravity will reappear quite dramatically when the elevator hits the ground. Imagine, however, a long tunnel drilled through the Earth, so that the elevator never hits the ground. Then the elevator would free-fall back and forth from one side of the Earth to the other, completing a cycle in about 84 min. Throughout this swing, no acceleration would be felt inside the elevator, even though it's passing through the Earth! Note that if we dropped two test particles—one from each pole—through the tunnel, they would execute simple harmonic motion with a period of 84 min. Relative to each other, their paths would be curved (Figure 25–1). So this experiment demonstrates that *spacetime within the Earth is curved.* In Einstein's general theory of relativity, the distribution of mass (and energy) determines the geometry of spacetime. A massive object produces a curvature of nearby spacetime. And that curvature shows itself by accelerated motion, which Newton would say was caused by gravitational forces. In other words, spacetime curvature is proportional to mass–energy density. Therefore, cosmological models may be constructed by making reasonable estimates of the type and amount of material content of the Universe; the equations of general relativity then yield the *evolutionary* behavior of the model.

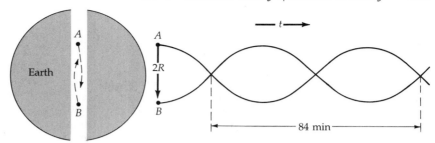

FIGURE 25–1 Motion of two masses falling through the Earth. One oscillation takes about 84 min.

(A) GEOMETRICAL RESULTS

Although the basic theory is attributable to Einstein, he did not follow through in examining the consequences of dynamical models of the Universe. Instead, this work was initiated in relative obscurity by the Russian mathematician A. Friedmann and only brought to the attention of the rest of the world after Hubble's observations of an expanding Universe.

Let's examine general relativity (GR) in more detail. In the spirit of Equation 25–1, we can write a generalization as follows:

$$dl^2 = \Sigma^4_{\mu=1} \Sigma^4_{\nu=1} \, g_{\mu\nu} \, dx^\mu \, dx^\nu \qquad \textbf{(25–5)}$$

This is a *tensor* equation with the matrix of $g_{\mu\nu}$ called the *metric* of spacetime. You can see that Equation 25–1 can be obtained from Equation 25–5 if $g_{\mu\nu} = 0$ for $\mu \neq \nu$ and if $g_{11} = g_{22} = g_{33} = -1$, and $g_{44} = +1$. Deviations from the flat metric of special relativity represent curved spacetime. As an example, near the surface of a neutron star, the components of the metric deviate from the flat values by about 10%.

In order to pursue Friedmann's models, we must assume that the Universe is isotropic and homogeneous; this assumption is so crucial that it is given the name of the **cosmological principle.** Essentially no theoretical progress is possible without starting at this point. We must also assume that the speed of light is finite and constant (no universal time).

We now can write a specific version of the metric of the Universe. This expression is quite general under these assumptions and is called the *Robertson–Walker metric.*

$$dl^2 = c^2 \, dt^2 - R(t)^2 \left[\frac{dr^2}{1 - kr^2} + r^2(d\theta^2 + \sin^2 \theta \, d\phi^2) \right]$$

$$\textbf{(25–6)}$$

In this equation r, θ, and ϕ are *comoving coordinates*. They are fixed on galaxies and are dimensionless. Also k is formally a free parameter and can have the values -1, 0, or $+1$, which describe positive, zero, and negative curvature spaces, respectively. The function $R(t)$ is the scale size, having the dimension of length, and it represents a dynamical Universe if it changes with time.

Distances and Coordinates
The quantity r in Equation 25–6 is not the measurable distance between two points. In order to find distances, we need to integrate the following:

$$l = \int_0^r dl = R(t) \int_0^r \frac{dr}{\sqrt{1 - kr^2}}$$

$$= \begin{cases} R \sin^{-1} r; & k = +1 \\ Rr; & k = 0 \\ R \sinh^{-1} r; & k = -1 \end{cases} \qquad \textbf{(25–7)}$$

Here we have rotated the coordinate system into the $\theta = \phi = 0$ plane but have not lost any generality.

Next, consider the paths of photons. They follow *null geodesics*, which are defined by $dl^2 = 0$. From Equation 25–6 we see that

$$\int_0^r \frac{dr}{\sqrt{1 - kr^2}} = c \int_{t_1}^{t_0} \frac{dt}{R(t)} \qquad \textbf{(25–8)}$$

Using Equation 25–7, we find

$$\begin{Bmatrix} \sin^{-1} r \\ r \\ \sinh^{-1} r \end{Bmatrix} = c \int_{t_1}^{t_0} \frac{dt}{R(t)} \qquad \textbf{(25–9)}$$

where t_0 is the time of photon detection and t_1 is the time of emission.

Next we introduce two important expressions. The first is known as the Lemaitre Equation

$$1 + z = R_0/R_1 \qquad \textbf{(25–10)}$$

where again the subscripts refer to the epoch of emission and reception of the photons. The second expresses the geometrical dynamics of the Universe; it is (with a dot representing the derivative with respect to time; two dots for the second derivative)

$$\left(\frac{\dot{R}}{R}\right)^2 + \frac{2\ddot{R}}{R} = -\frac{kc^2}{R^2} \qquad \textbf{(25–11)}$$

We are now at a position where we could investigate world models in detail. For example, integrating Equation (25–11) gives $R(t)$, which can be related to the observable redshift and the coordinate r by means of Equations 25–10 and 25–9.

Let's look at one specific case—the flat space solution to Equation 25–11. Here $k = 0$, and we find that

$$R(t) = R(t_0)(t_1/t_0)^{2/3} \qquad \textbf{(25–12)}$$

This is a very important result. It shows that the expansion of this model Universe follows a *power law form*. Figure 25–2 illustrates $R(t)$ for the three models, and one can see that power law expansion approximately holds for the other two models during the early phases of the Universe (probably up to the present) as well.

Let's look at one more purely geometric result. Notice that all three possible expressions for the

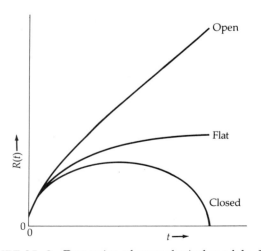

FIGURE 25–2 Dynamics of cosmological models. For isotropic and homogeneous models, distances vary with time depending on the overall geometry (open, flat, or closed).

right hand side of Equation 25–7 approximate Rr for $r \ll 1$. Therefore,

$$l = R(t)r$$

In words, the measurable distance equals the scale factor of the Universe multiplied by the coordinate distance. Now speed is just the time derivative of the distance, so

$$v = \dot{l} = \dot{R}r$$

but

$$\dot{R}r = \frac{\dot{R}}{R}l$$

so

$$v = \frac{\dot{R}}{R}l \qquad \textbf{(25–13)}$$

Examine the quantity $\dot{R}/R$; it has the dimensions of t^{-1}. These are the same dimensions as Hubble's constant H. Indeed,

$$H = \dot{R}/R \qquad \textbf{(25–14)}$$

So Equation 25–13 is a statement of Hubble's law that was derived by assuming only homogeneity and isotropy! Note that it was derived by approximating $r \ll 1$, so at large distances the simple form of Hubble's law might not hold; this is a potentially important way of discriminating among models.

(B) SOLVING EINSTEIN'S FIELD EQUATIONS

The coupling that we have mentioned between spacetime and the matter that it contains comes from *Einstein's field equations*. These equate a tensor that is constructed from the $g_{\mu\nu}$ and their derivatives to another tensor, the *stress-energy tensor*, which represents the matter. In simple terms, this means that the presence of mass curves spacetime, and we interpret the curved motions of particles in this curved space as accelerations caused by a force—in this case gravity.

If the Robertson–Walker metric is used in the field equations, and we use only those equations for which $\mu = \nu$, then the three space-like equations yield

$$\frac{8\pi GP(t)}{c^4}$$

$$= -\frac{k}{R_0{}^2}\left(\frac{R_0}{R}\right) - \frac{2}{c^2}\left(\frac{\ddot{R}}{R}\right) - \frac{1}{c^2}\left(\frac{\dot{R}}{R}\right)^2 + \Lambda \qquad \textbf{(25–15)}$$

and the time-like equation gives

$$\frac{8\pi G U(t)}{c^4} = \frac{3k}{R_0{}^2}\left(\frac{R_0}{R}\right)^2 + \frac{3}{c^2}\left(\frac{\dot{R}}{R}\right)^2 - \Lambda \quad \textbf{(25-16)}$$

Here $P(t)$ and $U(t)$ are the relevant components, the physically measurable pressure and the internal energy of the Universe respectively, of the stress-energy tensor; R_0 is the present value of R, and Λ is the *cosmological constant*. Formally, Λ is a constant of integration resulting from solving the differential field equations. Einstein introduced Λ in order to allow for static solutions since he did not know about Hubble's law at the time of the introduction of GR; it appears that Λ has a value of zero.

Through algebra, Equations 25–15 and 25–16 become

$$\frac{\ddot{R}}{R} = -\frac{4\pi G}{c^2}\left[P + \frac{U}{3}\right] + \Lambda\frac{c^2}{3} \quad \textbf{(25-17)}$$

and

$$\frac{d}{dt}(UR^3) = -P\frac{d(R^3)}{dt} \quad \textbf{(25-18)}$$

Notice that Equation 25–18 is a form of the first law of thermodynamics. If we simplify Equations 25–17 and 25–18 by the assumptions that $\Lambda = 0$, $P = 0$, and $U = \rho c^2$ (all are reasonable), then we find

$$\frac{\ddot{R}}{R} = -\frac{4\pi G\rho}{3c^2} \quad \textbf{(25-19)}$$

and

$$\frac{d}{dt}(\rho R^3) = 0 \Rightarrow \rho_0 R_0{}^3 = \rho R^3 \quad \textbf{(25-20)}$$

Note the similarity of Equation 25–19 to the Newtonian results of Section 25–1(b). In both cases the ballistic expansion is slowed down by ρ and stopped if ρ is large enough. Also note that Equation 25–20 describes the **conservation of mass** for the Universe as a whole.

(C) WORLD MODELS

We have seen that there are a variety of possible theoretical models for how the Universe behaves. Some of the properties of these *world models* come from geometric arguments alone; others arise when we couple the geometry to the matter content.

The biggest difference in geometry is determined by the value of k. This quantity algebraically governs the shape of the Universe. For $k = -1$, the geometry is locally similar to that of a sphere. For example, the sums of the interior angles of triangles would be greater than 180°. Whereas for $k = +1$, the geometry is locally that of a saddle point, with the angles of a triangle summing to less than 180°. If $k = 0$, then the Universe is flat. Figure 25–2 illustrates the function $R(t)$ for the three classes of possible Universes. The open Universe expands forever; the closed will eventually stop expanding and collapse into the **Big Crunch.** The flat Universe is balanced between the other two. Note that, in the early phases of expansion, the three models behave very similarly. All expand in a power-law fashion that would be hard to discriminate. At the present cosmic time, we seem to be in the situation of early expansion; it is difficult to decide among world models.

We suggest that you not try to visualize the global properties of an open Universe that has a saddle shape locally, but the closed Universe is not too hard to imagine. Consider a balloon that is being blown up. If you ignore the opening through which the air comes, and consider the two-dimensional surface of the balloon as it is curved into the third dimension, then this is an analogue to a three-dimensional sphere in a curved 4-D space-time. There would be no definable center to the balloon's surface, and uniform expansion would mimic Hubble's law for imaginary galaxies painted on its surface. A traveler could walk forever on the surface and never leave it, but that wanderer might, from a different direction, chance upon the point from which the journey started. Also note that triangles drawn on a balloon have angles that sum to more than 180°. In addition, consider a longitude–latitude system inscribed on the balloon. If an observer at the North Pole measured the angle subtended by a meter stick quite close to the pole and then moved it farther away, the angle would, of course decrease. But this would only continue until the meter stick reached the equator; then, as the stick was moved closer to the South Pole, the angle would *increase*.

The balloon analogy carries over to the real Universe in many ways. Perhaps the hardest to comprehend is the lack of a center to the Universe. You cannot identify any single point in space from which all else sprang. Because spacetime is curved and the Universe expands spacetime itself, every point within the observable Universe came from the initial singularity. Note that, in principle, we could make purely geometrical measurements that would allow the shape of the Universe to be determined. Doing this in fact has proven to be very hard. Tests have been devised to examine the magnitude of galaxies, diameters, surface brightnesses, etc. as a function of redshift. If galaxies had uniform properties, this method of discriminating among world models might work, but galaxies are wonderfully rich in variety, and we do not know much about how their properties vary with time. So looking back in cosmic time, as we do when we look to high redshift objects, only confuses the picture as far as geometrical measures go.

Finally, let us reiterate that H_0 can be related to geometrical properties by Equation 25–14. We know that H is not a constant, so there must also be a means of describing the deceleration of the Universe. In purely geometrical terms, this deceleration parameter q_0 is defined by the following:

$$q_0 = -\frac{\ddot{R}_0}{R_0 H_0{}^2} \qquad (25\text{–}21)$$

Physical Properties

Equations 25–15 through 25–20 relate the geometrical properties of the Universe to such important physical properties as pressure, density, and energy content. Therefore, again in principle, we could determine valid world models if we could make accurate measurements of, say, the mean density of the Universe.

The way is to write Equations similar to 25–14 and 25–21 but including density. We do not have tight theoretical constraints on how fast the original expansion was; therefore, the present value of Hubble's constant is not tightly constrained either. However, we know that the Universe is not too far from critical density, $\rho_c = 5 \times 10^{-27}$ kg/m³ (which would make the Universe flat), in which case we can write

$$H = [(8\pi G/3)\rho_c]^{1/2} \qquad (25\text{–}22)$$

In contrast, we *can* write a specific expression for the deceleration parameter.

$$q_0 = 4\pi G\rho/3H^2 \qquad (25\text{–}23)$$

Clearly, if we could measure H_0 and q_0 accurately, then we would know the mean density of the Universe. If $\rho_c < 1$, then the Universe would be open, if $\rho_c > 1$ closed, and if $\rho_c = 1$ the Universe would be flat.

What is the difficulty with this procedure? Obviously, the answer is dark matter. The density is often expressed by the quantity $\Omega = 2q_0$. Visible matter accounts for $\Omega = 0.02$ to 0.2, but there are reasons for believing that $\Omega = 1$ (see Chapter 26). An accurate direct assessment of the contribution of dark matter to the mean density is quite difficult.

25–3 ◗
THE PRIMEVAL FIREBALL

In 1964, Arno Penzias and Robert Wilson, scientists with the Bell Telephone Laboratories in New Jersey, began a sensitive study of the radio emission from the Milky Way. They detected an annoying excess radiation in their special low-noise radio antenna. They had tuned their radio receiver to 7.35 cm (4080 MHz), where the radio noise from the Galaxy is very small. Still they picked up the static. They further discovered that the noise did not change in intensity with direction in the sky, time of day, or season. The excess noise had an intensity equivalent to that of a blackbody at 3.5 K. What could it be?

At the same time, a group at Princeton was pondering the consequences of the expansion of the Universe from a *hot* dense state. Photons from such a time would be highly redshifted by now and would permeate the cosmos. Also, if the early Universe were so dense that it was opaque to the photons, then they would have had a blackbody spectral distribution. Subsequent redshifting by expansion would lower the temperature of the blackbody spectrum but not change its shape.

Penzias and Wilson got in touch with the Princeton group and concluded that their excess noise could very well be redshifted radiation from a hot Big Bang. That conclusion was solidified when the Princeton group and others confirmed the existence of the background radiation and found that its spectrum (Figure 25–3) matched that of a blackbody at 2.7 K. We will call it the *3-K back-*

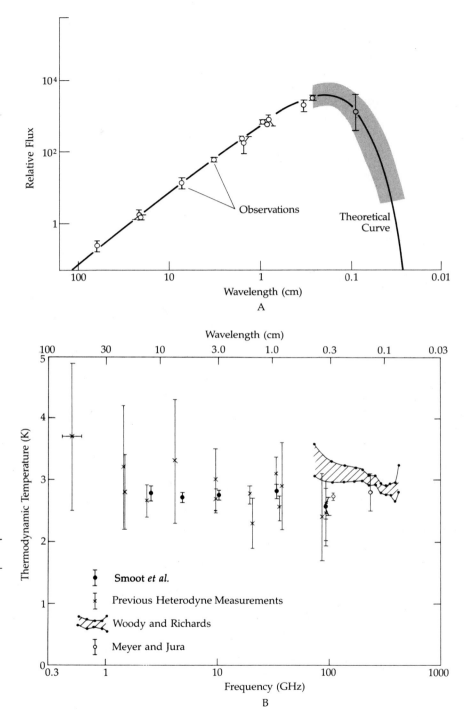

FIGURE 25–3 Spectrum of the cosmic background radiation. (A) Spectral observations; the shaded areas are infrared observations. The solid line is a 3-K blackbody curve. *(Adapted from a diagram by P. J. E. Peebles)* (B) Comparison of measurements of the temperature of the background radiation in the radio spectral region. *[G. F. Smoot, G. De Amici, S. D. Friedman, C. Witebsky, G. Sironi, G. Bonelli, N. Mandolesi, S. Cortiglioni, G. Morigi, R. B. Partridge, L. Danese, and G. DeZotti, Astrophysical Journal (Letters) 291:L23 (1985)]*

ground radiation. Its discovery supports a hot Big Bang model (sometimes called the **primeval fireball**), which is the standard relativistic model (see Section 25–4) accepted by most astronomers today.

Now, the energy density u for blackbody radiation is

$$u = aT^4 \qquad (25\text{-}24)$$

and from $E = mc^2$, we can convert this energy density to an equivalent mass density ρ:

$$m = E/c^2$$
$$\rho_r = aT^4/c^2 \qquad (25\text{-}25)$$

where a, the radiation density constant, is

$$a = 4\sigma/c$$

with σ, the Stefan–Boltzmann constant, equal to 5.6697×10^{-8} W/m$^2 \cdot$ K^4, so that

$$\begin{aligned} a &= (4)(5.6687 \\ &\times 10^{-8} \text{ W/m}^2 \cdot \text{K}^4)/(2.998 \times 10^8 \text{ m/s}) \\ &= 7.564 \times 10^{-16} \text{ W/m}^3 \cdot \text{K}^4 \end{aligned}$$

Then from Equation 25–25, the radiation density is

$$\begin{aligned} \rho_r &= (7.564 \times 10^{-16})(2.7^4)/(2.998 \times 10^8)^2 \\ &= (7.564 \times 10^{-16})(53)/(8.988 \times 10^{16}) \\ &= 4.5 \times 10^{-31} \text{ kg/m}^3 \end{aligned}$$

Note that this density is much less than that of luminous matter, $\rho_m \approx 4 \times 10^{-28}$ kg/m^3. Hence, we say that the Universe is now matter-dominated and in the **matter era.**

This was not always the case, however. Consider shrinking the Universe. Then the matter density, since lengths scale as $R(t)$, increases as

$$\rho_m \propto R^{-3} \qquad (25\text{-}26)$$

In contrast, the radiation density, from Equation 25–25, goes as T^4. Now, the wavelength of a photon is proportional to $R(t)$, so that

$$\lambda \propto R$$

and because a photon's energy is $E = h\nu = hc/\lambda$,

$$E = h\nu \propto R^{-1}$$

and for blackbody radiation,

$$T \propto R^{-1}$$

so that

$$\rho_r \propto R^{-4} \qquad (25\text{-}27)$$

At some time in the past, the **radiation era,** the energy density of radiation exceeded that of matter and the Universe was radiation-dominated.

Another way to compare the cosmic radiation with matter is to compute the ratio of number of photons to number of protons or neutrons. Blackbody radiation peaks at a wavelength

$$\lambda_{\max} = 2.9 \times 10^{-3}/T$$

The energy of a photon of wavelength λ is

$$E_{\text{ph}} = hc/\lambda$$

and so for photons of wavelength $\lambda = \lambda_{\max}$,

$$E_{\text{ph}} = hcT/(2.9 \times 10^{-3})$$

With $h = 6.6 \times 10^{-34}$, $c = 3 \times 10^8$, and $T = 2.7$ K, the energy of the photons at the peak of the blackbody curve is

$$\begin{aligned} E_{\text{ph}} &= (6.6 \times 10^{-34})(3 \times 10^8)(2.7)/(2.9 \times 10^{-3}) \\ &= 2.0 \times 10^{-22} \text{ J/photon} \end{aligned}$$

The photons making up the blackbody radiation have a variety of energies, but if we take this peak energy as typical, the number of photons needed to produce the 4×10^{-14} J of radiation contained in every cubic meter of the cosmic radiation at 2.7 K is

$$\begin{aligned} n_{\text{ph}} &= (4 \times 10^{-14} \text{ J/m}^3)/(2.0 \times 10^{-22} \text{ J/photon}) \\ &\approx 2 \times 10^8 \text{ photons/m}^3 \end{aligned}$$

Now consider the matter. The mass of one nucleon (a proton or neutron) is 1.7×10^{-27} kg. The 4×10^{-28} kg of matter contained in 1 m^3 is equivalent to

$$\begin{aligned} n_{\text{nu}} &= (4 \times 10^{-28} \text{ kg/m}^3)/(1.7 \times 10^{-27} \text{ kg/nucleon}) \\ &\approx 2 \times 10^{-1} \text{ nucleons/m}^3 \end{aligned}$$

The ratio of these two numbers is

$$n_{\text{ph}}/n_{\text{nu}} = 2 \times 10^8/2 \times 10^{-1} = 10^9$$

So there are 1 billion times as many photons in the Universe as nucleons.

One of the characteristics of cosmic blackbody radiation is that the total number of photons remains the same if account is taken of the expanding volume. Similarly for matter: the total number of nucleons remains the same. So the above ratio of photons to nucleons will stay constant as the Universe expands.

Now to recast our equations of motion for the matter-dominated and radiation-dominated cases. To do so, we need to use the conservation of energy for a sample volume V. We do so in the form of the first law of thermodynamics:

$$dE + P\,dV = 0$$

where P is the pressure and E is the matter–energy density in V, so that $E = \rho c^2$. Now $V \propto R^3(t)$, so that

$$dE/dt + P(dV/dt) = 0$$

gives (with dots representing the derivative with respect to time)

$$d/dt\,(\rho c^2 R^3) + P(d/dt)(R^3) = 0$$
$$c^2 R^3 \dot{\rho} + 3\rho c^2 R^2 \dot{R} + 3PR^2\dot{R} = 0$$
$$\dot{\rho} = -3(\rho + P/c^2)(\dot{R}/R)$$
$$\dot{\rho} = -3(\rho + P/c^2)H \qquad \textbf{(25–28)}$$

Let $P = 0$ (noninteracting matter) and assume that matter dominates $(E_{mat} = \rho_{mat}c^2 > E_{rad} = \rho_{rad}c^2)$, so that we have

$$\dot{\rho} = -3\rho H$$

For a flat geometry, however (which is the transitional case),

$$H = [(8\pi G/3)\rho]^{1/2}$$

so that

$$\dot{\rho} = -3\rho[(8\pi G/3)\rho]^{1/2} = 3\rho^{3/2}(8\pi G/3)^{1/2}$$
$$\rho^{-3/2}\dot{\rho} = -(25\pi G)^{1/2}$$

Now integrate this equation with respect to time:

$$\rho^{-3/2}\,d\rho = (25\pi G)^{1/2}\,dt$$
$$2\rho^{-1/2} = (25\pi G)^{1/2}t$$
$$t = (1/6G\pi\rho)^{1/2} \quad \text{(matter-dominated)} \qquad \textbf{(25–29)}$$

This equation gives the relationship of time and density for a matter-dominated, zero-pressure model, that is, it gives $\rho(t)$.

Now we will take Equation 25–28 to see what happens in the radiation-dominated case. Then

$$P = (1/3)(\text{energy density}) = (1/3)u = (1/3)\rho c^2$$

where T, ρ, and P are the temperature, density, and pressure of the radiation. Equation 25–28 becomes

$$\dot{\rho} = -3[\rho + (1/3)\rho]H = -4H\rho$$
$$= -4(8\pi G\rho/3)^{1/2}\rho$$

so that we have

$$\rho^{-3/2}\dot{\rho} = -(128\pi G/3)^{1/2}$$

and, as before, we integrate with respect to t:

$$2\rho^{-1/2} = (128\pi G/3)^{1/2}t$$
$$t = (3/32\pi G\rho)^{1/2} \quad \text{(radiation-dominated)} \qquad \textbf{(25–30)}$$

Now that we have time and density related, we can substitute temperature for radiation density because

$$P = (1/3)\rho c^2 = E/3 = aT^4/3$$

so that

$$\rho = aT^4/c^2$$

and Equation 25–30 becomes

$$T = (3c^2/32\pi Gat^2)^{1/4} \qquad \textbf{(25–31)}$$

or, if we substitute for the constants,

$$T(\text{K}) \approx 1.5 \times 10^{10}t^{-1/2} \qquad \textbf{(25–32)}$$

You should see that this equation relates every time t in a radiation-dominated Universe to a temperature T. So we can trace out the thermal history of the Universe as it evolves.

At the end of 1989, a new satellite was launched for the purpose of improving our knowledge of both the spectrum and uniformity of the background radiation. The first results from the Cosmic Background Explorer (COBE) dramatically confirmed previous earth-based observations. (See Figures 25–4 and 25–5.) The spectrum is indistinguishable from that of a blackbody with a temperature of 2.74 K, and the intensity distribution is remarkably smooth. At both large scales (1 radian) and small scales (10^{-3} radian), the measured value of $\Delta T/T$ is less than 3×10^{-5} after we subtract the dipole effect caused by our bulk motion [Section 23–2(b)].

The implication of COBE's confirmation of the spectrum lies mainly in giving us confidence that we understand the essence of a hot Big Bang. There had been hints of deviations from a purely thermal form, but these have now been set aside. Equally profound is the smoothness of the radiation. We will fully examine this question in the next chapter.

A   3.3 mm

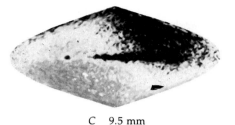

C   9.5 mm

FIGURE 25–4 COBE's view of the microwave background. Note the smooth texture of the photon distribution. Apart from the dipole effect caused by the peculiar motion of the Earth, no anisotropies are found.

B   5.7 mm

25–4 ◐
THE STANDARD BIG BANG MODEL

Can we assemble the concepts introduced in this chapter into a coherent picture of the origin and evolution of the Universe? The answer is generally positive, although you will see in the next chapter that the model is incomplete.

Modeling the origin of the Universe as a Big Bang explosion from a highly condensed initial state seems unavoidable. The existence of the microwave background demands a very hot early Universe, and the observed expansion demands a compact early Universe.

At the very earliest stages, the picture presented in this chapter is not very complete, but we have seen that the expansion probably follows a power law form with $R(t) \propto t^{2/3}$ or, at least, close to this. During the radiation-dominated era, the temperature of the matter content was forced to be the same as that of the radiation. This happened because the electrons and nuclei had not yet com-

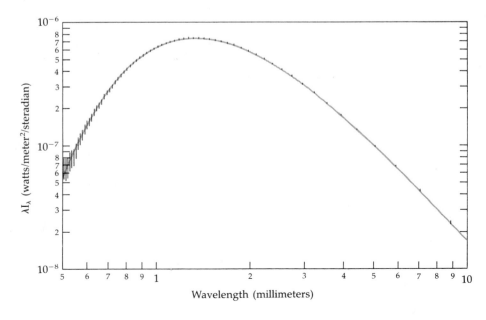

FIGURE 25–5 COBE's measurements of the blackbody spectrum. The blackbody nature of the background is seen at $T = 2.785 \pm 0.005$ K.

bined to form neutral atoms. The free electrons easily scattered the background photons, and this constant interaction kept the two components coupled. However, as T dropped (according to Equation 25–32), it eventually reached a value of about 3000 K. Suddenly (in cosmic time scales), the electrons and nuclei combined to form atoms, and the coupling between radiation and matter ceased since hydrogen atoms interact with photons at a much lower rate than with free electrons.

Our present view of the microwave photons has two aspects. One is the expected continual blackbody energy distribution of the photons—with, however, a characteristic temperature that decreased with time. The second is the spatial distribution of the photons, which represents the location of the matter at the time of the last scattering

off of the electrons. (In the next chapter you will see that the spatial distribution causes problems.)

When did the decoupling occur? We have the ratio between the temperature at the decoupling era and the present temperature:

$$T_d/T_0 = 3000 \text{ K}/3 \text{ K} = 1000$$

A characteristic of the expanding radiation is that

$$T \propto 1/R$$

Therefore,

$$R_0/R_d = 1000$$

So the Universe was about 1/1000th its present size at the time of decoupling. This corresponds to a cosmic time of about 700,000 years.

PROBLEMS ◗

1. What is the approximate volume of our Galaxy (express your answer in cubic kiloparsecs)? By what scale factor must the dimensions of our Universe shrink if there is to be no empty space between the galaxies? Is this stage of cosmic expansion a reasonable time for galaxy formation?

2. Planck's law for the intensity of blackbody radiation (Chapter 8) is

$$I_\lambda = (2hc^2/\lambda^5)(e^{hc/\lambda kT} - 1)^{-1}$$

As the Universe expands with a scale factor (radius) $R(t)$, the intensity varies as $I_\lambda \propto R^{-5}$ while the wavelength goes as $\lambda \propto R$.
   (a) Show that $T \propto R^{-1}$ if the blackbody formula is to remain valid.
   (b) At what wavelength does the blackbody curve reach a maximum for the observed 2.7 K background radiation?

3. If the Hubble constant is observed to be $H_0 = 50 \pm 5$ km/s · Mpc, what are the permissible ranges for the Hubble time ($t_0 \approx H_0^{-1}$), the size of the Universe ($r_{\text{hor}} \approx cH_0^{-1}$), and the critical mass density ($\rho_0 \propto H_0^2$)?

4. The diagram at right illustrates the famous expanding-balloon analogy for our Universe. All of space is represented by the surface of the spherical balloon, and clusters of galaxies are represented by spots painted on this surface. The radius of the balloon corresponds to $R(t)$—the radius of the Universe.
   (a) As the balloon expands, the spots remain at constant angular separations (θ) from one another.

Let the balloon expand at a constant rate and verify that

$$\Delta s/\Delta t = (1/R)(\Delta R/\Delta t)s$$

where s is the separation between any two spots on the surface and $\Delta s/\Delta t$ is the speed of recession of one spot from another. (Note that this is Hubble's law.)
   (b) The photons from distant galaxies may be represented by ants crawling along the balloon's surface at speed 1. Show that, for uniform cosmic expansion ($\Delta R/\Delta t = $ constant), there is a distance s from beyond which these ants cannot ever reach our Galaxy (this distance is called the *horizon*).
   (c) Discuss the effects that take place if the balloon's expansion is decelerated [the increase of $R(t)$ is slowed down].

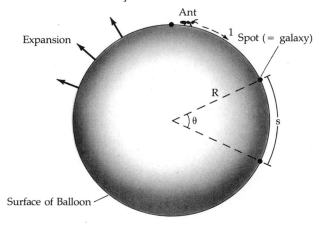

5. Consider an expanding gaseous sphere of uniform mass density ρ, total mass M, and radius $R(t)$. A gas particle at the surface of this sphere will move radially outward in accordance with the vis-viva equation (Chapter 2):

$$v^2/2 = GM/R + \text{constant}$$

where $v = \Delta R/\Delta t$ is the radial speed.

(a) Show that this equation may be written in the form

$$[(1/R)(\Delta R/\Delta t)]^2 = 8\pi G\rho/3 + 2(\text{constant})/R^2$$

Note that this is the equation that governs the expansion of our Universe and leads to the three cosmological models discussed in this chapter.

(b) From your knowledge of the vis-viva equation, show that the constant can be either positive, zero, or negative; illustrate the evolution of $R(t)$ in each case by drawing an approximate graph of R versus t. Comment upon your results.

6. Demonstrate that the Universe is now matter-dominated. Argue that in the past it must at some time have been radiation-dominated. (*Hint:* Background radiation at about 3 K.)

7. For a flat ($k = 0$) Universe, show that

$$t_0 = (2/3)H_0^{-1}$$

where t_0 is the age of the Universe. Evaluate t for the uncertainty in H_0.

8. Follow the algebra in deriving Equation 25–17 from Equation 25–15.

9. Follow the algebra in deriving Equation 25–18 from Equation 25–16.

10. Evaluate Equations 25–22 and 25–23 using the value of the critical density given in the text.

11. A friend of yours plans to give a lesson on cosmology at a local middle school. He wants to demonstrate the Big Bang expansion of the universe by placing a handful of marbles, representing galaxies, in the center of a large bandana, representing the space between galaxies, and then disrupting the marbles. The outward motion of the marbles then illustrates the expansion of the Universe. Comment on this demonstration. Is it a good analogy? Is there a better way? Has your friend lost his marbles?

12. At what times in the history of the universe does the radiation temperature of the Universe correspond to the temperature of the Sun's core? Surface?

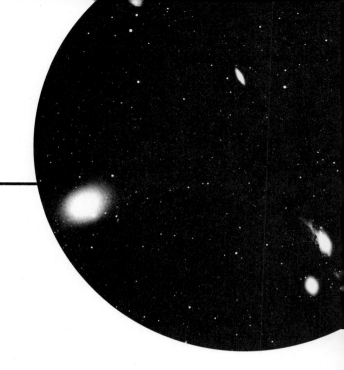

Chapter 26

The New Cosmology

The study of cosmology has radically changed in the last decade or two. A marriage of particle physics and large-scale structure shows that the Universe must have had a period of extraordinarily rapid expansion. This inflationary model explains several problems that are otherwise insoluble.

Astronomy offers to humans perhaps the most profound philosophical questions. These include "How did life arise?" and "How did stars and planets form?" The branch of astronomy called **cosmology** extends these questions to "How did the Universe form?" and "How will the Universe end?" Until recently, the last two questions boiled down to rather simple statements about a hot Big Bang as the origin and an evaluation of H_0 and q_0, the Hubble "constant" and the deceleration parameter, to decide whether the Universe was open or closed.

In this chapter, two themes dominate. One of these is the rich complexity of the physical processes that influenced the early stages. Such exotic concepts as quantum gravity, extra dimensions, phase transitions, free quarks, and symmetry breaking mark the first stages and may have left

observable consequences. We also find that some of these processes have a random character to them; "domains" or "bubbles" probably formed in different locations. *Our observable Universe may not be unique*, and other "Universes" might be quite different in character.

The second theme adds a profoundly important element to philosophy. We find that the Universe is comprehensible only if we appreciate the intimate relationship between the largest structures and the microscopic world of subatomic particles. As an example, the shapes and scale sizes of superclusters and voids are determined in models by the properties of the particles that make up the dark matter—such as whether the particles are "hot" or "cold."

26–1 ◐
PROBLEMS WITH THE EXISTING MODEL

Let's look at five aspects of the Universe that pose problems. In each case, current observations cannot be reconciled with cosmological models that we have so far discussed.

1. *Isotropy and Homogeneity*

The first problem is one of causality. Because of the finite speed of light, two regions separated at a cosmic time t by a distance greater than ct cannot have communicated with each other. Regions within a diameter of ct are said to be within the *horizon*; those farther away than ct are outside of the horizon. If we project the current power law expansion of the Universe back to times prior to the epoch of recombination, then one horizon diameter should correspond to about 1 radian in the microwave background. This means that we should see several "domains" in the microwave radiation. Each of these regions should have a different temperature since they could not have communicated their original internal temperatures with one another prior to the last time the background photons were scattered. Therefore, the microwave sky should be patchy—there should be large-scale anisotropies. Yet it is not; the microwave sky is remarkably free of large-scale anisotropies.

How did the Universe grow from a region small enough to have a uniform temperature?

2. *Galaxy Formation*

The view we obtain by looking at the luminous matter is one of a very clumpy Universe. In contrast, the Universe appears to have been exceedingly smooth at the time of decoupling of the matter and radiation. Indeed, if we project any realistic growth pattern that would result in galaxies, clusters, and superclusters back to the decoupling era, then we should see the "seeds" of galaxy formation. There should be small-scale anisotropies in the microwave background that correspond to the small density perturbations that would grow to become structures, but there are no small-scale anisotropies in the microwave sky.

Why don't we see the seeds of galaxy formation in the microwave background?

3. *Flatness*

The density parameter, Ω, can have an infinite range of values from zero to positive infinity. The only "magic" number in this range is the precise value of unity. A Universe with this value is at **critical density.** It is balanced between open and closed—between eternal expansion and eventual contraction.

What is the current value of the density? Optical observations of the luminous matter show Ω to lie between 0.02 and 0.20. The high M/L ratios of galaxy clusters raise the density to values close to 1 when dark matter is included. All values in this range are remarkably close to unity. This becomes evident when you consider that the expansion of the Universe tends to force Ω away from unity. So if the density is close to 1 now, it used to be much closer. Calculations show that if Ω differed from unity by one part in 10^{55} at a cosmic age of 10^{-35} s, then either the Universe would have already collapsed or the galaxies would have receded to such great distances that the density would be an order of magnitude less than the current observational limits allow.

Therefore, the Universe is known to be so close to critical density that there must be some reason for this to occur.

Why is the Universe at critical density?

4. *Net Baryon Number*

Compared to the energies of particle collisions at the time of the Big Bang, our present Universe involves very low energies. Even our largest particle accelerators do not provide sufficiently powerful collisions to come close to those whose remnants govern our observations. In our low-energy world, there is a high degree of symmetry between particles and antiparticles. *If the energy density is high enough to create particles, then both a particle and antiparticle are created. Similarly, in order to completely annihilate a particle, one must annihilate its antiparticle also.* If this symmetry existed in the early Universe, then it should consist of equal numbers of particles and antiparticles.

Observations clearly show that the background of gamma rays is very low (as contrasted to microwaves, for example). However, if the number of antiparticles equalled the number of particles, then continual annihilations of pairs throughout the Universe would create an easily detectable background of extremely high-energy photons.

A related item that needs explaining is the very small value for the ratio of nucleons to photons (see Section 25–3); the value is about 10^{-9}. Most of the photons in the Universe are part of the background radiation, and, compared to these, nucleons are rare.

How did the Universe become asymmetric between particles and antiparticles and why are nucleons so rare?

5. Population III Stars

In our galaxy, we see two populations of stars. Population I is the younger; these stars have a relatively high percentage of heavy elements. The older Population II stars have a lower metal abundance, but it is not zero. Since cosmic nucleosynthesis (see Section 26–2) does not make appreciable amounts of any elements heavier than helium (these are only produced by stellar nucleosynthesis), the first generation of stars *must* have zero metal abundance.

Why don't we see any Population III stars?

These five problems will be discussed in the remainder of this chapter. We do not yet have the solution to all, but the most recent forms of inflation solve most of them.

26–2 ◗
COSMIC NUCLEOSYNTHESIS

As a preface to this section, let us remark about the extreme presumption of humankind trying to comprehend the early Universe. Does it seem difficult for you to reconcile the fact that, although we cannot distinguish between 10 or 20 billion years as the Hubble age of the Universe, we will try to convince you that we know what the Universe was like at an age of 1 second? In this section we will show you the most successful "postdictions" (It is hard to *pre*dict the formation of the Universe!) we have about primordial times. The close correspondence between theory and observation will lend confidence about extending theories backward toward earlier, more energetic times.

In astrophysics, nuclear reactions are primarily confined to two arenas. One of these is continual; it is the fusion process that powers stars and is discussed in Chapter 16. The second concerns a few brief moments at the beginning of the Uni-verse before stars existed when fusion reactions raged throughout the Universe. Since fusion builds heavier elements out of lighter ones, we term these two processes **stellar nucleosynthesis** and **cosmic nucleosynthesis,** respectively.

Qualitatively, fusion requires both high temperatures and high densities in order for the kinetic energies of the nucleons to overcome the *coulomb barrier*. If the particles come sufficiently close [on the order of 1 fermi (f) which is 10^{-15} meter] then the attractive nature of the strong nuclear force overwhelms the repulsive nature of the electromagnetic force. Certainly, the early Universe satisified both criteria. Interestingly, there were conditions when the Universe was *too* hot and dense for fusion to take place. For example, in the process of fusing single protons (hydrogen nuclei) to make alpha particles (helium nuclei), deuterons must maintain stability. The latter are the nuclei of a heavy isotope (**deuterium**) of hydrogen consisting of a proton and a neutron. In successive stages two more protons could collide with the deuteron forming ^3He and then ^4He. But deuterons are easily broken up by high-energy gamma rays—a process termed **photodissociation.** So even though there were protons available for fusion before an age of about 1 second, when two of them fused to form a deuteron, a positron, and a photon, the deuteron was destroyed before a more stable ^4He nucleus could form. Hence fusion could not progress until the Universe *cooled* to well below 10^{10} K.

By the time the Universe was 10^2 seconds old, the temperature had dropped to about 10^9 K, and there were very few photons energetic enough to photodissociate deuterons. At this stage the cosmic nucleosynthesis began in earnest. All isotopes of hydrogen and helium with an atomic number of 4 or less were produced, but bottlenecks prevented significant buildup of heavier elements. These bottlenecks occurred because there are no stable nuclei with atomic number 5 or 8. Only small traces of ^7Li and ^7Be were produced.

As the Universe continued to cool and expand, the temperatures and densities continued to drop. So even though some nuclei heavier than hydrogen were built up and hence were potential fuel for further nucleosynthesis, the falling densities and temperatures could no longer supply particles with energies higher than the coulomb barrier. By an age of 10^3 seconds, the Universe had ceased

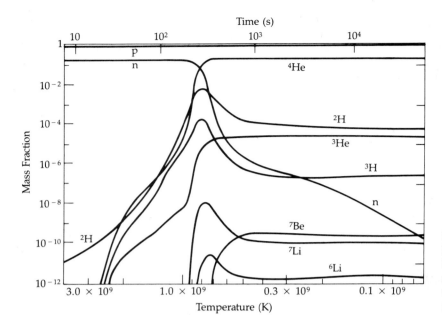

FIGURE 26-1 Nucleosynthesis in the Big Bang. The top axis gives the age of the Universe; the bottom gives temperature; the vertical axis is abundance in terms of the fraction of total mass. *(Adapted from a diagram by R. V. Wagoner)*

producing fusion reactions. Figure 26–1 illustrates the accumulation of deuterium, ^{3}He, ^{4}He, ^{7}Li, and ^{7}Be (all with respect to the total mass density) with time during the cosmic nucleosynthesis phase.

There is very little disagreement among the theoretical calculations. ^{4}He should have a mass fraction of about 25%, deuterium and ^{3}He should be about 10^{-4} as abundant as hydrogen and both lithium and beryllium should be about 10^{-9} to 10^{-10} as abundant as hydrogen.

Observations of the abundances of these nucleons is complicated by the addition of the effects of stellar nucleosynthesis. ^{4}He is the main product of stellar fusion, and both deuterium and ^{3}He are not only produced but also burned in the varying conditions of stellar envelopes and interiors. However, when these effects are sorted out by theoretical modeling, the results are quite close to those predicted by the models of the primordial Universe and are not very sensitive to uncertain parameters.

We see from the close agreement between theory and the observations of cosmic abundances that we probably do know a great deal about the state of the Universe during the era spanning 1 to 10^3 seconds. Thus to refute the seeming contradiction of not knowing *precisely* when the Big Bang happened, we can state that we know *accurately*

much about the processes that occurred. It seems as though we can extend our knowledge of the laws of physics to very early times. Our successes, of course, are achieved because we still have not reached a time early enough to exceed the energies that are accessible in our studies of nuclear energy. As we try to examine progressively earlier and more energetic times throughout the rest of this chapter, we will also move toward less certainty. Again, we see the unity between large and small. If we could build larger, more powerful accelerators on Earth and peer deeper into the inner workings of elementary particles, we could better understand the Universe as a whole. At some stage, we can no longer build bigger machines, and we must, in the end, accept the Universe as the ultimate particle accelerator!

26–3 ◗
PARTICLE PHYSICS

At the conditions of the present day Universe, particles are influenced by four forces of Nature. These are the **strong nuclear, electromagnetic, weak nuclear,** and **gravitational** forces. Their important properties are listed in Table 26–1. The nuclear forces are quite strong but of very limited

TABLE 26–1   *The Four Forces of Nature*

Force	Coupling Strength	Range	Carriers of Force	Particles Acted on
Strong nuclear	1	10^{-15} m	Gluons (total of 8 in 3 colors)	Quarks
Electromagnetic	1/137	Infinite	Photons	Any particle with charge
Weak nuclear	1×10^{-5}	10^{-18} m	W^+, W^-, Z^0	Quarks and leptons
Gravitational	10^{-40}	Infinite	Graviton (not detected yet)	Any particle with mass

range—only about the size of an atomic nucleus. Although gravitation is the weakest of the forces, it dominates the structure and dynamics of the Universe since its range is infinite and it has no negative charges. We now know that the electromagnetic and weak nuclear forces are unified at higher energies. It is only at the relatively cool, low-energy conditions of the present Universe that the two forces act differently. One can in a sense predict some of the future of particle physics by understanding the great efforts being made to unify the electroweak force with the strong nuclear force and, ultimately, with gravity also.

(A) KNOWN PARTICLES AND THE STANDARD MODEL

We know of many particles. **Fundamental particles** are those that are not composed of yet smaller particles. Some of the most important particles, the protons and neutrons in particular, are not fundamental. They are composites of smaller particles.

The known particles are categorized by a variety of means. One of the most important is a classification based on the spin angular momentum of the particle. Those with integral (0, 1, or 2) spin are called **bosons,** and those with half-integral spin (1/2, 3/2, etc.) are called **fermions.** The statistical properties of the two groups are quite different.

Atoms are made up of electrons and nucleons. The two types of particles are very different. Although the particles of both groups have spin 1/2, they obey different forces. We will discuss the two groups separately and mention the other known members of the groups.

Leptons

There are three generations of leptons—the **electron, muon,** and **tau** types. Their characteristics are that they all have spin 1/2 and participate in the weak, electromagnetic, and gravitational interactions but not the strong. Table 26–2 lists the 12 known leptons. The six on the right side are **antiparticles** to those on the left. The best known antiparticle is the **positron,** which only differs from the electron in its charge. For each of the electron, muon, and tau, there is an associated **neutrino.**

The neutrinos are extremely important for astrophysics. For example, supernovae emit most of their energy in the form of neutrinos. Only about 10^{-4} of the energy comes out as photons. (It is quite impressive to realize that, although the visible light output of a star during a supernova explosion can rival the total luminosity of the rest of the galaxy, this increase in luminosity is just the "tip of the iceberg.") Since the neutrinos have no electrical charge, the only force they feel is the weak nu-

TABLE 26–2   *Leptons*

Leptons Symbol/name	Charge	Antileptons Symbol	Charge
e^- (electron)	−1	e^+ (positron)	+1
ν_e (electron neutrino)	0	$\bar{\nu}_e$	0
m^- (muon)	−1	m^+	+1
ν_m (muon neutrino)	0	$\bar{\nu}_m$	0
τ^- (tau minus)	−1	τ^+	+1
ν_τ (tau neutrino)	0	$\bar{\nu}_\tau$	0

clear force, and they therefore do not interact well with matter. For this reason, they are highly penetrating. For example, the average neutrino can pass through about 1 lightyear of lead before being absorbed. In cosmology, neutrinos may be the ultimate answer to many questions. Although neutrinos have very low mass (and may join photons as massless particles), there are so many neutrinos in the Universe that their total mass may dominate—*neutrinos may be the major component of dark matter*!

Particles Composed of Quarks

Although the leptons appear to be truly fundamental, the nucleons are certainly made up of other particles. These component particles are called **quarks** (with spin 1/2), and in the present Universe they cannot exist except in combinations (with a force mediated by spin 1 particles called **gluons**). Quarks are influenced by all four forces. The particles that are made up of combinations of quarks are termed **hadrons**. They also come in two types; when 3 quarks combine, we have half-integral spin particles called **baryons,** combinations of 2 quarks form integral spin particles called **mesons.** Like leptons, there are three known generations of quarks and antiquarks. These are listed in Table 26–3. Note the fractional charges of quarks. As an example of how quarks combine, protons consist of two *u*'s and one *d* and neutrons of one *u* and two *d*'s.

The standard model explains the known particles and describes their interactions through a set of equations. It is a *gauge field theory*, which means that the equations of motion are invariant under a group of transformations that can be performed at any point. In this type of theory, the force is communicated between particles by means of bosons. The standard model explains all the known parti-

cles and their interactions, as observed in our most powerful accelerators, very well. Yet particle physicists are working both in theory and experiment to take the next step beyond the standard model— why? The answer is that the standard model is too complex and has too many input parameters. Physics has so far shown us that the underlying phenomena are simple and generally symmetric, but a typical version of the standard model has about 20 unspecified parameters. This seems too many for a fundamental theory.

(B) GRAND UNIFIED THEORIES AND SUPERSYMMETRY

Grand Unified Theories (GUTs) eliminate several of the arbitrary features of the standard model by unifying the strong nuclear force with the already unified electroweak force. (Remember that these forces would only appear the same at very high energies.) In these models quarks, antiquarks, leptons, and antileptons can be tranformed into each other. Successes from GUTs include the easy explanation of why electrical charges are quantized. (This is one of the free parameters in the standard model.)

Most GUTs make some exceedingly important predictions about cosmology. The first is that protons probably decay with a half-life of about 10^{32} years. Since the Universe is only about 10^{10} years old, this does not have immediate consequence to humankind! Philosophically, however, this implies that the elements that make up matter as we know it are only a passing phase. (One problem with GUTs is that no experiments have yet detected proton decay, and the tests have been sensitive enough to rule out some GUTs on this basis.)

A second prediction, more immediately to the point, is that (again most) GUTs predict that the neutrinos are not massless. Estimates of the neutrino rest masses range downward from 30 eV (meaning the rest mass energy, $mc^2 = 30$ eV). We have reason to believe that there are vast numbers of neutrinos. So if their masses are in the few eV to a few tens of eV range, then they could easily account for the missing mass or dark matter problems.

The final prediction of interest is that GUTs demand the existence of vast numbers of **magnetic**

TABLE 26–3 Quarks

Name	Quarks		Antiquarks	
	Symbol	Charge	Symbol	Charge
up	u	+2/3	$\bar{u}$	−2/3
down	d	−1/3	$\bar{d}$	+1/3
charmed	c	+2/3	$\bar{c}$	−2/3
strange	s	−1/3	$\bar{s}$	+1/3
top	t	+2/3	$\bar{t}$	−2/3
bottom	b	−1/3	$\bar{b}$	+1/3

monopoles. These particles would be quite massive and therefore might also contibute to the dark matter. What is a magnetic monopole? In human experience all magnetic phenomena begin with a dipole—a pairing of north and south poles. In a simple bar magnet with both poles, the action of cutting the bar in two does not isolate the poles but rather makes two separate bars each of which has two poles. Particles with magnetic properties can be thought of as tiny, spinning bar magnets that cannot be cut. So if there are magnetic monopoles, their properties would be quite different from anything else yet known. A problem that could now be added to the five introduced in Section 26–1 is that GUTs predict the existence of so many magnetic monopoles that we should not only have already detected them easily (we haven't yet) but also that we should be overwhelmed by them. *Why have we not detected magnetic monopoles?*

One more particle that should be mentioned is the *Higgs boson*. It has not been detected and its theoretically predicted properties are not universally agreed upon. The Higgs particle is responsible for spontaneous symmetry breaking, which we will see later is responsible for inflation. It also seems that it is the interaction of the Higgs particle and all others that leads to the most fundamental property of all—**mass.**

Another extension of particle theories is termed *supersymmetry* and holds if there is a pairing of bosons and fermions. The names of the supersymmetric particles of bosons are constructed by adding *-ino* to the end of the boson name. For example, the *photino* would be the hypothetical supersymmetric fermion to the photon. Others are called *zinos*, *winos*(!), *gluinos*, and *Higgsinos*. The names of the supersymmetric particles of fermions are made by adding an *-s* to the beginning of the name—e.g., *squarks* and *sleptons*. We have not yet experimentally found any supersymmetric particle.

26–4 ◗
INFLATION THEORY

Fortunately, we now have a remarkable theory that explains most of the problems in cosmology that have been presented in this chapter. It was originally developed by A. Guth who called it the **inflation theory** (for reasons soon to be made

clear, having nothing to do with economics!). Important improvements were later added by A. Albrecht, P. Steinhardt, and A. Linde.

Before explaining inflation, let's stop for a moment and consider symmetries and phase changes. These phenomena are esoteric arguments in modern physics—including inflation. Symmetry is so important that the existence of some particles can be confidently predicted on symmetry arguments alone. An example is the *top quark*, which has not been detected in any way but "must" exist in order to preserve the symmetry of the *bottom quark* and the three generations of both quarks and leptons.

An analogy may make these concepts easier to understand. Our analogy is to imagine we start with a hot universe made up only of hydrogen and oxygen atoms that will eventually become water molecules. At the earliest epochs in our model, the temperature is high enough that the phase is gaseous and any molecules that form are rapidly dissociated again. At this stage the model universe is highly symmetric; there are no preferred directions. As the universe cools, molecules become stable. First the molecules are in the gaseous state, but they undergo a phase change when $T = 373$ K (100°C). Below this temperature, the phase is liquid, and a second phase transition, of course, occurs at another 100°C lower temperature. This last state, ice, is particularly useful as an analogy since ice can have a variety of asymmetries, such as molecular rotational symmetries about 60-degree angles, and different refracting properties in different directions; ice also has a variety of fractures. These can be thought of as topological defects and include bubbles (nearly point-like) and both linear and planar cracks.

Turning from the analogy to the inflationary Universe, the important phase transition happened at about $t = 10^{-35}$ s, when the strong nuclear force separated from the electroweak. This symmetry breaking was generated by the Higgs particle. At the critical temperature, "bubbles" (to become separate "Universes"?) of the new phase started and grew within the old. This growth, however, was exponential rather than power law as during the rest of the expansion.

During the exponential inflation, the Universe doubled in size about every 10^{-35} s and grew from a diameter of about 10^{-23} cm to about 10 cm by the time inflation was over at about 10^{-32} s. Space itself grew much faster than the speed of light during this period (not a violation of causality since information was not passed from point to point in spacetime at a rate faster than c). So the thermal properties of one small "domain" were spread throughout very many regions of size ct—thereby solving the problem of large-scale homogeneity in the microwave background.

Inflation theory solves the magnetic monopole problem with GUTs by applying the fact that the monopoles only form at the walls between domains. So there might be only on the order of one monopole in our whole observable Universe. The monopoles appear as defects in the metric of spacetime (similar to the defects in ice).

The inflationary model also solves the problem of why the Universe is so flat (why the density parameter is so close to unity) in a very natural manner with rapid expansion. Here you need to appreciate the interrelationship between density and the curvature of the Universe. The more the density deviates from the critical value, the greater the curvature of spacetime. Or phrased differently, the flatter the Universe, the closer the density is to critical. Now consider another analogy—that of a balloon being blown up. At first the balloon is quite small, and a given area of the surface shows significant curvature. Then, as the balloon expands, the same surface area becomes less and less curved. In a similar manner, the Universe is forced to become flat by the enormous factor of expansion caused by inflation.

We also need to consider the thermal history of the Universe during the period of inflation. During the brief interval before inflation, the Universe was certainly very hot—around 10^{27} K. As temperatures dropped, the little bubble of the new phase that would become our observable Universe became supercooled. (Here again the water analogy is appropriate since water can sometimes remain liquid at temperature below 0°C before freezing, and when it does freeze it releases latent heat since the moving liquid molecules have more energy than the slowly vibrating ice molecules at the same temperature.) At the end of inflation, the latent heat was then released as a overwhelming flood of hot matter and radiation. Indeed, it is this fireball whose remnant is the microwave background seen today. All of the matter and radiation that is observable originated in this event.

26–5 ◗
GALAXY FORMATION

Let's now examine how structures formed to produce the clumps of matter that we see today as galaxies, clusters, and superclusters. In a hot Big Bang model for the Universe, the only material available to start with was an expanding gas made up of hydrogen and helium. At times prior to the decoupling era, the matter (gas) was intimately tied to the radiation field. This occurred mainly by Thomson scattering of the background photons off free electrons. Once the temperature dropped to around 3000 K, the electrons combined with protons and alpha particles to form neutral hydrogen and helium atoms, which have much lower cross sections for interactions with the background photons. So, from this point, the matter was free of the dominance of the radiation field.

(A) PERTURBATIONS

Our main goal in following galaxy formation is to understand how small clumps originated in the gas and then grew in density. The clumps, called *perturbations*, behaved as waves passing through the medium. A wave contains complementary regions of higher and lower than average density. We normalize the density in a region by using the **density contrast** $\Delta\rho/\rho$. Under certain conditions, the regions of slightly higher than average density will expand more slowly because of gravitational effects, and the density contrast grows with time. For regions of lower than average density, the expansion will be at a higher than average rate, so the density drops faster than the average. Therefore, the effects of gravitation in the expanding Universe are to amplify the primordial density contrasts.

Under what conditions do the perturbations grow? The general problem of collapsing gas clouds was described early in this century by James Jeans. The question is when pressure gradients within a gas cloud are dominated by gravitational forces. A scale length, termed the **Jeans length,** l_J, denotes a wavelength in which the total mass has a critical value. For perturbations with a wavelength $l > l_J$ gravitation dominates the dynamics, while for $l < l_J$ pressure dominates and the perturbation just behaves like a sound wave.

Another important scale length is l_D, the damping scale. For the smallest wavelengths, those with $l < l_D$, the waves lose energy to viscosity and thermal conductivity and are damped out.

We must distinguish between two different kinds of perturbations. In one, the **adiabatic perturbations,** both the matter density and radiation temperature fluctuate. The two would be related by $\Delta\rho/\rho = 3\Delta T/T$. A second type, the **isothermal perturbations,** derives its name from the fact that there are no fluctuations in the radiation (hence T is constant) to complement the matter fluctuations.

If isothermal fluctuations dominated, then the history of galaxy formation was as follows: During the radiation era the perturbations neither dissipated nor amplified. At decoupling, the Jeans length corresponded to a scale that encompassed about $10^{5-6}M_\odot$. Perturbations of this size and larger then grew as $\Delta\rho/\rho \propto t^{2/3}$. This value of l_J corresponds to the masses of globular clusters. The simplest scenario has galaxies forming by gathering globular clusters through gravity and galaxy clusters forming by gathering galaxies. This is an example of a *bottom-up* theory in which the largest structures formed after smaller structures.

For adiabatic perturbations the history is a little more complicated. The longest scale length perturbations always had $l > l_J$ and they grew throughout both the radiation- and matter-dominated eras, but intermediate-scale lengths could have grown at first, then oscillated and then grown again. This happened because l_J grew during the radiation era but then dropped quickly at decoupling. Since the largest adiabatic modes grew throughout the history of the Universe, they represent models in which the largest gas clouds became well separated from each other before smaller-scale gas clouds collapsed to form galaxies. So these are *top-down* models.

(B) THE IMPLICATIONS OF LARGE-SCALE STRUCTURE ON THEORY

Observations of large-scale structures have a certain "gee whiz" aspect to them. Fame and fortune (not very likely!) can come to the astronomers who find the largest superclusters or voids. Yet the scientific payoff of such observations transcends fleeting frivolity because extreme objects in important classes push theories to the limits and can discriminate against incorrect ideas. Nowhere is this seen more readily than in cosmology. For example, a model that can make reasonable galaxy clusters but cannot make superclusters must be discarded.

An important task for observers, then, is to identify not only the typical scale lengths for superclusters and voids but also the largest examples. Similarly, we need to find out whether all galaxies belong to larger systems. Now, it looks that way. If there are any galaxies that are truly isolated, their numbers and luminosities cannot be very large. This means that the physical processes that create large-scale structures must be very efficient.

Let's now look at how the observations constrain current theories. Although parts of the theoretical models are subject to analytical methods, computer modeling based on many particles or zones is essential for so complex a problem. Currently, most codes represent galaxies as collisionless particles; these are *n*-body simulations. Hydrodynamic (fluid) and magneto-hydrodynamic simulations will surely follow.

Since there are important theoretical reasons for believing that $\Omega = 1$, roughly 90% of the Universe must be dark matter. The simulations must make basic assumptions about the dark matter. One that is usually made is that the dm is *nonbaryonic*. Since all familiar matter is made up of nucleons (baryons), this rules out rocks, gas, dust, faint stars, ices, etc. One of the major reasons for believing that the dark matter is nonbaryonic is the small-scale isotropy of the microwave background. The implication of not seeing the seeds of galaxy clusters and superclusters in the relic photons is that the dark matter and photons were smoothly distributed and the density fluctuations only occurred in the minor baryonic component.

Another choice the simulations must make is whether the dark matter is hot (hdm) or cold (cdm). Hot particles would be those that were moving at relativistic speeds when they entered the horizon. The only candidate that we currently have is the neutrino. A variety of supersymmetric particles could form the cdm.

In hdm models the particles can stream to great distances. Therefore, it is relatively easy for them to build up large structures such as filamentary superclusters or to evacuate large voids. These models have more difficulty in making galaxies, and most theorists favor cdm models.

Since the dark matter in cdm models moves sluggishly, it may be gravitationally bound to clusters or individual galaxy halos. One variant of the general cdm class hypothesizes *biased galaxy formation*. Up to now, we have not mentioned the spectrum of the density fluctuation amplitudes. The majority, of course, would lie within 1σ of the mean. Biased formation would occur if only the most extreme perturbations formed the luminous giant galaxies. These 3 or 4σ perturbations are also likely to have been correlated—to be located nearer to each other than the average. A test of this model would be to observe the distribution of a large number of dwarf galaxies. These would have come from the 1σ perturbations; they would have collapsed more slowly, and they would be more widely distributed. Perhaps dwarfs are relatively common in voids.

Another manner in which observations might constrain theories comes from an alternative way of creating voids. Gravity can amplify density contrasts, thereby making voids less dense and larger with time. An alternative possibility is to envision pre-galactic, supermassive Population III objects exploding and evacuating a spherical region that would then grow. This is an explosive amplification of the density contrast. Most versions of the explosive models have trouble making voids whose diameters exceed about 10 Mpc. Here again, finding the sizes of the largest voids will be helpful in discriminating among theories.

We caution the reader that today's problems might be overcome by the theorists of tomorrow. We do not necessarily advocate any of these models. Our intent is to show you something about the current status and direction of models of galaxy formation.

26–6 ◗
HISTORY OF THE UNIVERSE

We are now prepared to sketch out our current best ideas about how the Universe evolved. This history is complex, with chapters in which the strong nuclear and weak nuclear forces command the stage as far as which component particles exist and what interactions they are participating in at the time. The electromagnetic force became important after expansion made interparticle distances much greater than the effective scale lengths of the short-ranged nuclear forces. Of course, in the background it was the weakest of the four forces, gravitation, that commanded the curvature and deceleration of the Universe as a whole. This happened because gravitation alone has an infinite effective range and always causes attraction; there are no negative gravitational charges whose effect would cancel out attraction.

We will follow this history in terms of time units. We could just as easily follow it by using temperature, mass density, energy density, or scale size of the Universe as the independent parameter. Remember that it is more significant to think in terms of exponents than linear units. A time of 10^{-35} s is, in this sense, farther removed from the epoch of cosmic nucleosynthesis (10^{2-3} s) than we are 10^{17-18} s (10–20 billion years) later. You can see this by considering how many times the Universe doubled its size before the nucleosynthesis era compared to how many times it doubled afterward.

10^{-45} s   We have no real understanding of what physics would be like at the energies prevailing before this time. Probably, the gravitational force was quantized, and there may have been extra dimensions beyond the four that we experience as spacetime. It may be that the perturbations that would eventually become galaxies had their origins in quantum fluctuations right at the beginning when gravity was quantized.

10^{-35} s   This marked the end of Grand Unification. The strong nuclear force separated from the electroweak force, thereby initiating inflation. Before this epoch the numbers of quarks (and antiquarks) and photons were equal, but a variety of processes that violated charge-parity conservation resulted in the annihilation of most of the quarks. Since the current ratio of the number of photons to the number of baryons (made out of quarks) is 10^{9-10}, only about 1 in one billion quarks survived this period of annihilation. No antiquarks survived. This is the origin of the baryon/antibaryon asymmetry and is also the reason why baryons are so rare compared to photons.

10^{-32} s   Inflation ended. The "bubble" of phase that is our observable Universe had grown from about 10^{-23} cm to perhaps 10 cm—which would subsequently expand to much larger than we can presently see (about 10^{28} cm). The principal components of the Universe were photons, quarks, antiquarks, and colored gluons. Notice that protons were not stable, so at this stage there were no elements—not even hydrogen.

10^{-12} s The weak nuclear and electromagnetic forces separated. Notice that the Universe at this point was 10^{20} times older than when the last major event happened. Very little activity occurred in this period, and it is often referred to as a "desert."

10^2–10^3 s This was the period of cosmic nucleosynthesis. Nucleons were fused, leaving about 25% of the baryonic matter in the Universe in the form of helium. Essentially all the rest of the baryons were hydrogen nuclei (protons).

10^{11} s This marked the era of decoupling between photons and baryons. Before this point, the energy density of radiation was much higher than that of the matter. After this, the matter dominated. Since the decoupling accompanied the combination of free electrons with nuclei, this marked the beginning of atoms—the form of matter that we are most familiar with.

10^{16} s Galaxies, stars, and planets began to form.

10^{18} s The present. As time passes from this stage, the galaxies continue to recede from each other, but the Hubble "constant" decreases. The temperature of the Universe will continue to drop, asymptotically approaching absolute zero. Since $\Omega = 1$, the expansion will continue for an indefinite period of time.

10^{40} s Protons decay (perhaps). Atomic matter ceases to exist.

Figure 26–2 illustrates several aspects of this history up to the present. Each of the horizontal lines should be thought of as coincident, but we could not annotate one line and retain clarity; the history of the Universe is too complex for one simple illustration.

26–7 ◑ SUMMARY

The recent progress that has been made in both astronomical observations of large-scale structures and the understanding of particle physics has shown that several problems in understanding the

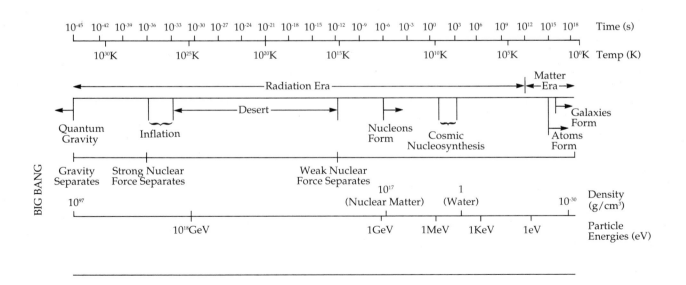

FIGURE 26–2 History of the Universe. All the horizontal axes are actually coincident. The history proceeds from the Big Bang on the left to today on the right.

Universe can be solved. The theory of inflation has been wonderfully successful at explaining several of the problems. It shows us why the whole Universe has a common background temperature, why magnetic monopoles are not common, and why the density is astoundingly close to critical. We also now understand that the present-day symmetry between matter and antimatter was violated in the early Universe in a manner that annihilated most of the matter and all of the antimatter. Rather than solving the problem of the small-scale isotropy of the microwave background, we use this observations to constrain our theories and believe that it tells us much about the dark matter.

We do not know if these solutions are unique, and we have not solved all of the problems that have been raised. In particular, we cannot claim to understand cosmology without significantly improved knowledge of the dark matter. Actually detecting these particles could be exceedingly difficult. This is particularly true if the dm is dominated by supersymmetric particles whose interactions with baryonic matter are unknown. How would we trap an ''ino'' in the laboratory? Similarly, we cannot claim to know the history very well until Population III stars have been identified. Cosmic nucleosynthesis and stellar nucleosynthesis cannot be reconciled without Population III. Finally, we have made progress but do not yet know with any definitiveness how large-scale structures were formed. What exactly were the processes that made superclusters and voids?

PROBLEMS

1. Assume that dust grains, whose characteristic size is 1 μm, are uniformly distributed throughout intergalactic space at a mean mass density of 10^{-27} kg/m^3 (the critical mass density).
   (a) What is the number density (number per cubic meter) of this dust, and what is the average separation between grains?
   (b) Compare your answers to (a) with the number density and separation of the interstellar dust grains in our Galaxy.
   (c) Show that this hypothetical intergalactic dust will drastically *redden* the stars observed in the Andromeda galaxy (such reddening is *not* observed in practice).

2. If intergalactic space is filled with H II at a temperature of 10^6 K (a plasma),
   (a) What are the mean speed and mean kinetic energy (per particle) of these protons?
   (b) To what wavelength of electromagnetic radiation does this individual kinetic energy correspond?

   Could such radiation be detected from the surface of the Earth?

3. Electrons are the lightest stable particles made in the Big Bang. What is the *latest* time they could have been formed?

4. Use Equation 25–21 and Figure 25–2 to verify our statement in this chapter that Ω tends to evolve away from unity.

5. If we use 10^{10} K as the critical temperature for photodissociation of deuterium, what is the binding energy of the deuteron?

6. Section 26–6 outlines various stages in the history of the universe in time units. Estimate the corresponding density and radiation temperature for each of these times. Comment on the accuracy of your results.

7. The GUTS and inflation models help explain many of the problems of the standard Big Bang model. What problems still exist?

APPENDIX 1
THE MESSIER CATALOG

TABLE A1–1

Messier No.	NGC	RA (2000)	DEC (2000)	Const.	Size (arcmin)	V (mag)	Type	Name
1	1952	5ʰ 34ᵐ5	+22°01′	Tau	6 × 4	8.4:	Di	Crab Nebula
2	7089	21 33.5	−0 49	Aqr	13	6.5	Gb	
3	5272	13 42.2	+28 23	CVn	16	6.4	Gb	
4	6121	16 23.6	−26 32	Sco	26	5.9	Gb	
5	5904	15 18.6	+2 05	Ser	17	5.8	Gb	
6	6405	17 40.1	−32 13	Sco	15	4.2	OC	
7	6475	17 53.9	−34 49	Sco	80	3.3	OC	
8	6523	18 03.8	−24 23	Sgr	90 × 40	5.8:	Di	Lagoon Nebula
9	6333	17 19.2	−18 31	Oph	9	7.9:	Gb	
10	6254	16 57.1	−4 06	Oph	15	6.6	Gb	
11	6705	18 51.1	−6 16	Sct	14	5.8	OC	
12	6218	16 47.2	−1 57	Oph	14	6.6	Gb	
13	6205	16 41.7	+36 28	Her	17	5.9	Gb	Hercules Cluster
14	6402	17 37.6	−3 15	Oph	12	7.6	Gb	
15	7078	21 30.0	+12 10	Peg	12	6.4	Gb	
16	6611	18 18.8	−13 47	Ser	7	6.0	OC	
17	6618	18 20.8	−16 11	Sgr	46 × 37	7:	Di	Omega Nebula
18	6613	18 19.9	−17 08	Sgr	9	6.9	OC	
19	6273	17 02.6	−26 16	Oph	14	7.2	Gb	
20	6514	18 02.6	−23 02	Sgr	29 × 27	8.5:	Di	Trifid Nebula
21	6531	18 04.6	−22 30	Sgr	13	5.9	OC	
22	6656	18 36.4	−23 54	Sgr	24	5.1	Gb	
23	6494	17 56.8	−19 01	Sgr	27	5.5	OC	
24		18 16.9	−18 29	Sgr	90	4.5:		
25	IC 4725	18 31.6	−19 15	Sgr	32	4.6	OC	
26	6694	18 45.2	−9 24	Sct	15	8.0	OC	
27	6853	19 59.6	+22 43	Vul	8 × 4	8.1:	Pl	Dumbbell Nebula
28	6626	18 24.5	−24 52	Sgr	11	6.9:	Gb	
29	6913	20 23.9	+38 32	Cyg	7	6.6	OC	
30	7099	21 40.4	−23 11	Cap	11	7.5	Gb	
31	224	0 42.7	+41 16	And	178 × 63	3.4	S	Andromeda Galaxy
32	221	0 42.7	+40 52	And	8 × 6	8.2	E	
33	598	1 33.9	+30 39	Tri	62 × 39	5.7	S	
34	1039	2 42.0	+42 47	Per	35	5.2	OC	
35	2168	6 08.9	+24 20	Gem	28	5.1	OC	
36	1960	5 36.1	+34 08	Aur	12	6.0	OC	

TABLE A1–1   (*continued*)

Messier No.	NGC	RA (2000)	DEC (2000)	Const.	Size (arcmin)	V (mag)	Type	Name
37	2099	5^h 52^{m}4	+32°33′	Aur	24	5.6	OC	
38	1912	5 28.7	+35 50	Aur	21	6.4	OC	
39	7092	21 32.2	+48 26	Cyg	32	4.6	OC	
40		12 22.4	+58 05	UMa		8:		
41	2287	6 47.0	−20 44	CMa	38	4.5	OC	
42	1976	5 35.4	−5 27	Ori	66 × 60	4:	Di	Orion Nebula
43	1982	5 35.6	−5 16	Ori	20 × 15	9:	Di	
44	2632	8 40.1	+19 59	Cnc	95	3.1	OC	Praesepe
45		3 47.0	+24 07	Tau	110	1.2	OC	Pleiades
46	2437	7 41.8	−14 49	Pup	27	6.1	OC	
47	2422	7 36.6	−14 30	Pup	30	4.4	OC	
48	2548	8 13.8	−5 48	Hya	54	5.8	OC	
49	4472	12 29.8	+8 00	Vir	9 × 7	8.4	E	
50	2323	7 03.2	−8 20	Mon	16	5.9	OC	
51	5194 − 5	13 29.9	+47 12	CVn	11 × 8	8.1	S	Whirlpool Galaxy
52	7654	23 24.2	+61 35	Cas	13	6.9	OC	
53	5024	13 12.9	+18 10	Com	13	7.7	Gb	
54	6715	18 55.1	−30 29	Sgr	9	7.7	Gb	
55	6809	19 40.0	−30 58	Sgr	19	7.0	Gb	
56	6779	19 16.6	+30 11	Lyr	7	8.2	Gb	
57	6720	18 53.6	+33 02	Lyr	1	9.0:	Pl	Ring Nebula
58	4579	12 37.7	+11 49	Vir	5 × 4	9.8	S	
59	4621	12 42.0	+11 39	Vir	5 × 3	9.8	E	
60	4649	12 43.7	+11 33	Vir	7 × 6	8.8	E	
61	4303	12 21.9	+4 28	Vir	6 × 5	9.7	S	
62	6266	17 01.2	−30 07	Oph	14	6.6	Gb	
63	5055	13 15.8	+42 02	CVn	12 × 8	8.6	S	
64	4826	12 56.7	+21 41	Com	9 × 5	8.5	S	
65	3623	11 18.9	+13 05	Leo	10 × 3	9.3	S	
66	3627	11 20.2	+12 59	Leo	9 × 4	9.0	S	
67	2682	8 50.4	+11 49	Cnc	30	6.9	OC	
68	4590	12 39.5	−26 45	Hya	12	8.2	Gb	
69	6637	18 31.4	−32 21	Sgr	7	7.7	Gb	
70	6681	18 43.2	−32 18	Sgr	8	8.1	Gb	
71	6838	19 53.8	+18 47	Sge	7	8.3	Gb	
72	6981	20 53.5	−12 32	Aqr	6	9.4	Gb	
73	6994	20 58.9	−12 38	Aqr				
74	628	1 36.7	+15 47	Psc	10 × 9	9.2	S	
75	6864	20 06.1	−21 55	Sgr	6	8.6	Gb	
76	650 − 1	1 42.4	+51 34	Per	2 × 1	11.5:	Pl	
77	1068	2 42.7	−0 01	Cet	7 × 6	8.8	S	
78	2068	5 46.7	+0 03	Ori	8 × 6	8:	Di	
79	1904	5 24.5	−24 33	Lep	9	8.0	Gb	
80	6093	16 17.0	−22 59	Sco	9	7.2	Gb	
81	3031	9 55.6	+69 04	UMa	26 × 14	6.8	S	
82	3034	9 55.8	+69 41	UMa	11 × 5	8.4	Ir	
83	5236	13 37.0	−29 52	Hya	11 × 10	7.6:	S	
84	4374	12 25.1	+12 53	Vir	5 × 4	9.3	E	
85	4382	12 25.4	+18 11	Com	7 × 5	9.2	E	
86	4406	12 26.2	+12 57	Vir	7 × 6	9.2	E	
87	4486	12 30.8	+12 24	Vir	7	8.6	E	Virgo A
88	4501	12 32.0	+14 25	Com	7 × 4	9.5	S	
89	4552	12 35.7	+12 33	Vir	4	9.8	E	
90	4569	12 36.8	+13 10	Vir	10 × 5	9.5	S	
91	4548	12 35.4	+14 30	Com	5 × 4	10.2	S	

TABLE A1–1   (*continued*)

Messier No.	NGC	RA (2000)	DEC (2000)	Const.	Size (arcmin)	V (mag)	Type	Name
92	6341	17^h 17^{m}1	+43°08′	Her	11	6.5	Gb	
93	2447	7 44.6	−23 52	Pup	22	6.2:	OC	
94	4736	12 50.9	+41 07	CVn	11 × 9	8.1	S	
95	3351	10 44.0	+11 42	Leo	7 × 5	9.7	S	
96	3368	10 46.8	+11 49	Leo	7 × 5	9.2	S	
97	3587	11 14.8	+55 01	UMa	3	11.2:	Pl	Owl Nebula
98	4192	12 13.8	+14 54	Com	10 × 3	10.1	S	
99	4254	12 18.8	+14 25	Com	5	9.8	S	
100	4321	12 22.9	+15 49	Com	7 × 6	9.4	S	
101	5457	14 03.2	+54 21	UMa	27 × 26	7.7	S	
102								M101 reobser-vation
103	581	1 33.2	+60 42	Cas	6	7.4:	OC	
104	4594	12 40.0	−11 37	Vir	9 × 4	8.3	S	Sombrero Galaxy
105	3379	10 47.8	+12 35	Leo	4 × 4	9.3	E	
106	4258	12 19.0	+47 18	CVn	18 × 8	8.3	S	
107	6171	16 32.5	−13 03	Oph	10	8.1	Gb	
108	3556	11 11.5	+55 40	UMa	8 × 2	10.0	S	
109	3992	11 57.6	+53 23	UMa	8 × 5	9.8	S	
110	205	0 40.4	+41 41	And	17 × 10	8.0	E?	

: denotes approximate value.
Types: diffuse nebula (Di), globular cluster (Gb), open cluster (OC), planetary nebula (Pl), or galaxy (E for ellipitcal, Ir for irregular, S for spiral).
(Adapted from *Sky Catalogue 2000.0*, vol. 2, Sky Publishing Corp., 1985.)

APPENDIX 2
CONSTELLATIONS

TABLE A2–1 *The Constellations*

Name	Genitive Form of Name	Abbreviation	Equatorial Coordinates α (h)	Equatorial Coordinates δ (°)	Galactic Coordinates l (°)	Galactic Coordinates b (°)
Andromeda	Andromedae	And	1	+40	135	−25
Antlia	Antilae	Ant	10	−35	270	+15
Apus	Apodis	Aps	16	−75	315	−15
Aquarius	Aquarii	Aqr	23	−15	50	−60
Aquila	Aquilae	Aql	20	+5	45	−15
Ara	Arae	Ara	17	−55	335	−10
Aries	Arietis	Ari	3	+20	160	−35
Auriga	Aurigae	Aur	6	+40	175	+10
Boötes	Boötis	Boo	15	+30	45	+65
Caelum	Caeli	Cae	5	−40	245	−35
Camelopardalis	Camelopardalis	Cam	6	+70	145	+20
Cancer	Cancri	Cnc	9	+20	210	+35
Canes Venatici	Canum Venaticorum	CVn	13	+40	110	+80
Canis Major	Canis Majoris	CMa	7	−20	230	−10
Canis Minor	Canis Minoris	CMi	8	+5	215	+20
Capricornus	Capricorni	Cap	21	−20	30	−40
Carina	Carinae	Car	9	−60	270	−10
Cassiopeia	Cassiopeiae	Cas	1	+60	125	−5
Centaurus	Centauri	Cen	13	−50	305	+10
Cepheus	Cephei	Cep	22	+70	110	+10
Cetus	Ceti	Cet	2	−10	170	−65
Chamaeleon	Chamaelontis	Cha	11	−80	300	−20
Circinus	Circini	Cir	15	−60	320	0
Columba	Columbae	Col	6	−35	240	−25
Coma Berenices	Comae Berenices	Com	13	+20	320	+85
Corona Australis	Coronae Australis	CrA	19	−40	355	−20
Corona Borealis	Coronae Borealis	CrB	16	+30	50	+50
Corvus	Corvi	Crv	12	−20	290	+40
Crater	Crateris	Crt	11	−15	270	+40

TABLE A2–1 *(continued)*

Name	Genitive Form of Name	Abbreviation	Approximate Position			
			Equatorial Coordinates		Galactic Coordinates	
			α	δ	l	b
			h	°	°	°
Crux	Crucis	Cru	12	−60	295	0
Cygnus	Cygni	Cyg	21	+40	85	−5
Delphinus	Delphini	Del	21	+10	60	−25
Dorado	Doradus	Dor	5	−65	275	−35
Draco	Draconis	Dra	17	+65	95	+35
Equuleus	Equulei	Equ	21	+5	55	−30
Eridanus	Eridani	Eri	3	−20	205	−60
Fornax	Fornacis	For	3	−30	225	−60
Gemini	Geminorum	Gem	7	+20	195	+10
Grus	Gruis	Gru	22	−45	355	−55
Hercules	Herculis	Her	17	+30	50	+35
Horologium	Horologii	Hor	3	−60	380	−50
Hydra	Hydrae	Hya	10	−20	260	+25
Hydrus	Hydri	Hyi	2	−75	300	−40
Indus	Indi	Ind	21	−55	340	−40
Lacerta	Lacertae	Lac	22	+45	100	0
Leo	Leonis	Leo	11	+15	230	+60
Leo Minor	Leonis Minoris	LMi	10	+35	190	+55
Lepus	Leporis	Lep	6	−20	225	−20
Libra	Librae	Lib	15	−15	345	+35
Lupus	Lupi	Lup	15	−45	325	+10
Lynx	Lyncis	Lyn	8	+45	175	+30
Lyra	Lyrae	Lyr	19	+40	70	+15
Mensa	Mensae	Men	5	−80	290	−30
Microscopium	Microscopii	Mic	21	−35	10	−40
Monoceros	Monocerotis	Mon	7	−5	210	0
Musca	Muscae	Mus	12	−70	300	−10
Norma	Normae	Nor	16	−50	330	0
Octans	Octantis	Oct	22	−85	305	−30
Ophiuchus	Ophiuchi	Oph	17	0	30	+15
Orion	Orionis	Ori	5	+5	195	−15
Pavo	Pavonis	Pav	20	−65	330	−30
Pegasus	Pegasi	Peg	22	+20	80	−25
Perseus	Persei	Per	3	+45	145	−10
Phoenix	Phoenicis	Phe	1	−50	300	−70
Pictor	Pictoris	Pic	6	−55	260	−30
Pisces	Piscium	Psc	1	+15	125	−45
Piscis Austrinus	Piscis Austrini	PsA	22	−30	20	−50
Puppis	Puppis	Pup	8	−40	255	−5
Pyxis	Pyxidis	Pyx	9	−30	255	+10
Reticulum	Reticuli	Ret	4	−60	270	−45
Sagitta	Sagittae	Sge	20	+10	50	0
Sagittarius	Sagittarii	Sgr	19	−25	10	−15
Scorpius	Scorpii	Sco	17	−40	345	0
Sculptor	Sculptoris	Scl	0	−30	10	−80
Scutum	Scuti	Sct	19	−10	25	−5
Serpens	Serpentis	Ser	17	0	20	+5
Sextans	Sextantis	Sex	10	0	240	+40
Taurus	Tauri	Tau	4	+15	180	−30

TABLE A2—1   (*continued*)

Name	Genitive Form of Name	Abbreviation	Equatorial Coordinates		Galactic Coordinates	
			α	δ	l	b
			h	°	°	°
Telescopium	Telescopii	Tel	19	−50	350	−20
Triangulum	Trianguli	Tri	2	+30	140	−30
Triangulum Australe	Trianguli Australis	TrA	16	−65	320	−10
Tucana	Tucanae	Tuc	0	−65	310	−50
Ursa Major	Ursae Majoris	UMa	11	+50	160	+60
Ursa Minor	Ursae Minoris	UMi	15	+70	110	+45
Vela	Velorum	Vel	9	−50	260	0
Virgo	Virginis	Vir	13	0	310	+65
Volans	Volantis	Vol	8	−70	280	−20
Vulpecula	Vulpeculae	Vul	20	+25	65	−5

APPENDIX 3
SOLAR SYSTEM DATA

TABLE A3–1 Planetary Orbits

Planet	Symbol	Synodic Period (Days)	Sidereal Period Tropical Years	Sidereal Period Days	Semimajor Axis Au	Semimajor Axis 10^6 km	Eccentricity	Inclination to Ecliptic
Mercury	☿	115.9	0.241	87.96	0.387	57.9	0.206	7.00°
Venus	♀	583.9	0.615	224.70	0.723	108.2	0.007	3.39
Earth	⊕	—	1.000	365.26	1.000	149.6	0.017	0.00
Mars	♂	779.9	1.881	686.98	1.524	228.0	0.093	1.85
Jupiter	♃	398.9	11.86	4333	5.203	778.3	0.048	1.31
Saturn	♄	378.1	29.46	10,759	9.54	1427	0.056	2.49
Uranus	♅	369.7	84.01	30,685	19.18	2871	0.047	0.77
Neptune	♆	367.5	164.8	60,188	30.06	4497	0.009	1.77
Pluto	♇	366.7	248.6	90,700	39.44	5913	0.249	17.15

TABLE A3–2 Planetary Rotation

Planet	Sidereal Rotation Period	Oblateness	Obliquity*
Mercury	58.65 days	0	0.0°
Venus	243 days	0	177.4
Earth	23^h 56^m 4.1^s	0.0034	23.5
(Moon)	27.3 days	0.0006	6.7
Mars	24^h 37^m 22.6^s	0.0052	25.2
Jupiter	9^h 50.5^m	0.062	3.1
Saturn	10^h 14^m	0.096	26.7
Uranus	17^h 14^m	0.06	98
Neptune	16^h 3^m	0.02	29
Pluto	6.439 days	?	65

*Obliquity is defined as the inclination of the equator to the orbital plane. Obliquities greater than 90° imply retrograde rotation.

TABLE A3–3 *Planetary Physical Data*

Planet	Mass 10^{24} kg	Mass $\oplus = 1$	Equatorial Radius km	Equatorial Radius $\oplus = 1$	Average Density (kg/m^3)	Surface Gravity $(\oplus = 1)$	Albedo	Escape Speed (km/s)	Equilibrium Blackbody	Observed	Subsolar Blackbody
Terrestrial											
Mercury	0.33	0.056	2,439	0.38	5.4×10^3	0.38	0.06	4.2	445	100–700	633
Venus	4.87	0.815	6,052	0.95	5.2	0.91	0.76	10.3	325	700	464
Earth	5.97	1.000	6,378	1.00	5.52	1.00	0.4	11.2	277	250–300	395
Moon	0.07	0.012	1,738	0.27	3.34	0.16	0.07	2.4	277	120–390	395
Mars	0.64	0.107	3,393	0.53	3.9	0.39	0.16	5.1	225	210–300	319
Pluto	0.01	0.0018	1,140	0.18	0.5(?)	0.03	0.5	2.1	44	40	63
Jovian											
Jupiter	1900	318	71,398	11.19	1.40	2.74	0.51	61	122	110–150	173
Saturn	569	95	60,000	9.41	0.69	1.17	0.50	36	90	95	127
Uranus	87	14.5	25,559	4.01	1.19	0.94	0.66	21	63	58	90
Neptune	103	17.2	24,800	3.89	1.66	1.15	0.62	24	50	56	72

TABLE A3–4 *Satellites of Terrestrial Planets*

Planet	Moon	Distance from Planet $(10^3$ km)	Sideral Period (days)	Orbital Eccentricity	Orbital Inclination (Degrees)	Radius (km)	Mass (Planet = 1)	Bulk Density (kg/m^3)
Earth	Moon	384	27.32	0.055	5.1	1738	0.012	3300
Mars	Phobos	9.37	0.32	0.021	1.1	$14 \times 11 \times 9$	1.5×10^{-8}	1900
	Deimos	23.52	1.26	0.003	1.6	$8 \times 6 \times 5$	3.1×10^{-9}	2100

TABLE A3–5 *Satellites of Jupiter*

Name	Number	Distance from Jupiter 10^3 km	Distance from Jupiter Jupiter radii	Orbital Period (days)	Radius (km)	Mass (planet = 1)	Bulk Density (kg/m^3)
Metis	J16	128	1.79	0.29	10	5×10^{-11}	—
Andrastea	J14	129	1.80	0.30	20	1×10^{-11}	—
Almathea	J5	181	2.55	0.50	130×80	2×10^{-9}	3000
Thebe	J15	222	3.11	0.67	45	4×10^{-10}	—
Io	J1	422	5.95	1.77	1820	4.7×10^{-5}	3530
Europa	J2	671	9.47	3.55	1570	2.6×10^{-5}	3030
Ganymede	J3	1070	15.10	7.16	2630	7.8×10^{-5}	1930
Callisto	J4	1883	26.60	16.69	2400	5.7×10^{-5}	1790
Leda	J13	11,094	156	239	≈8	3×10^{-12}	—
Himalia	J6	11,480	161	251	85	5.0×10^{-9}	1000
Lysithea	J10	11,720	164	259	≈20	4×10^{-11}	—
Elara	J7	11,737	165	260	≈40	4×10^{-10}	—
Ananke	J12	21,200	291	631(R)	15	2×10^{-11}	—
Carme	J11	22,600	314	692(R)	20	5×10^{-11}	—
Pasiphae	J8	23,500	327	735(R)	20	1×10^{-10}	—
Sinope	J9	23,700	333	758(R)	20	4×10^{-11}	—

(R) indicates retrograde orbit

TABLE A3–6   *Major Satellites of Saturn*

Name	Distance from Saturn		Orbital Period (days)	Radius (km)	Mass (planet = 1)	Bulk Density (kg/m^3)
	10^3 km	Saturn radii				
Atlas	137.67	2.28	0.602	20×10	—	—
Prometheus	139.35	2.31	0.613	$70 \times 50 \times 40$	—	—
Pandora	141.70	2.35	0.629	$55 \times 45 \times 35$	—	—
Epimetheus	151.42	2.51	0.694	$70 \times 60 \times 50$	—	—
Janus	151.47	2.51	0.695	$110 \times 100 \times 80$	—	—
Mimas	185.54	3.08	0.942	197	8.0×10^{-8}	1200
Enceladus	238.04	3.95	1.370	250	1.3×10^{-7}	1200
Tethys	294.66	4.88	1.888	524	1.3×10^{-6}	1200
Telesto	294.67	4.88	1.888	$17 \times 14 \times 13$	—	—
Calypso	294.67	4.88	1.888	$17 \times 11 \times 11$	—	—
Dione	377.42	6.26	2.737	560	1.85×10^{-6}	1400
1980 S6	377.42	6.26	2.737	$18 \times 16 \times 15$	—	—
Rhea	527.07	8.74	4.518	765	4.4×10^{-6}	1300
Titan	1221.86	20.25	15.945	2575	2.36×10^{-4}	1880
Hyperion	1481.00	24.55	21.277	$205 \times 130 \times 110$	3×10^{-8}	—
Iapetus	3560.80	59.02	79.33	718	3.3×10^{-6}	1200
Phoebe	12,954	214.7	550.45	≈ 110	7×10^{-10}	—

TABLE A3–7   *Major Satellites of Uranus, Neptune, and Pluto*

Planet	Satellite	Distance from Center of Planet (10^3 km)	Orbital Period (days)	Radius (km)	Mass (planet = 1)	Bulk Density (kg/m^3)
Uranus	Ariel	191.8	2.52038	580	1.8×10^{-5}	—
	Umbriel	267.3	4.14418	595	1.2×10^{-5}	—
	Titania	438.7	8.70588	805	6.8×10^{-5}	~1500
	Oberon	586.6	13.46326	775	6.9×10^{-5}	~1500
	Miranda	130.1	1.414	160	2×10^{-6}	—
Neptune	Triton	354.3	5.87683	1430	13×10^{-3}	—
	Nereid	5510	359	470	2.0×10^{-7}	—
Pluto	Charon	19.7	6.39	600	0.1	2100

APPENDIX 4
STELLAR DATA

TABLE A4–1 *The Nearest Stars (within 5 parsecs)*

Star	α (1975)	δ (1975)	Distance (parsecs)	Proper Motion ("/yr)	A m_v	A M_v	A Spectral Type	B m_v	B M_v	B Spectral Type	C m_v	C M_v	C Spectral Type
α Centauri	14^h 38^{m}0	$-60°44'$	1.31	3.68	0.01	4.4	G2 V	1.4	5.8	K5 V	10.7	15	M5e V
Barnard's Star	17 56.6	+4 37	1.83	10.34	9.54	13.2	M5 V	Unseen companion					
Wolf 359	10 55.3	+7 10	2.32	4.71	13.66	16.8	M6e V						
BD + 36° 2147	11 2.0	+36 8	2.49	4.78	7.47	10.5	M2 V	Unseen companion					
Sirius	6 44.0	-16 41	2.65	1.32	-1.47	1.4	A1 V	8.7	11.5	WD			
Luyten 726-8	1 37.7	-18 5	2.74	3.35	12.5	15.4	M5.5e V	12.9	15.8	M6e V			
Ross 154	18 48.3	-23 52	2.90	0.72	10.6	13.3	M4.5e V						
Ross 248	23 40.7	+44 3	3.16	1.60	12.24	14.7	M5.5e V						
ϵ Eridani	3 31.8	-9 33	3.28	0.97	3.73	6.1	K2 V						
Luyten 789-6	22 37.2	-15 27	3.31	3.25	12.58	14.9	M5.5e V						
Ross 128	11 46.4	+0 57	3.32	1.40	11.13	13.5	M5 V						
61 Cygni	21 5.8	+38 37	3.43	5.22	5.19	7.5	K5 V	6.02	8.3	K7 V	Unseen companion		
ϵ Indi	22 1.4	-56 53	3.44	4.69	4.73	7.0	K5 V						
Procyon	7 38.0	+5 17	3.48	1.25	0.34	2.7	F5 IV$-$V	10.7	13.0	WD			
BD + 59° 1915	18 42.5	+59 35	3.52	2.29	8.90	11.1	M4 V	9.69	11.9	M5 V			
BD + 43° 44	0 16.9	+43 53	3.55	2.91	8.07	10.3	M2.5e V	11.04	13.2	M4e V			
CD $-$ 36° 15693	23 4.2	-36 0	3.59	6.90	7.39	9.6	M2 V						
τ Ceti	1 42.9	-16 4	3.67	1.92	3.50	5.7	G8 Vp	Unseen companion					
BD + 5° 1668	7 26.1	+5 18	3.76	3.73	9.82	11.9	M4 V						
CD $-$ 39° 14192	21 15.8	-38 58	3.85	3.46	6.72	8.7	M10 V						
CD $-$ 45° 1841	5 10.6	-45 0	3.91	8.72	8.81	10.8	M0						
Kruger 60	22 27.1	+57 34	3.94	0.87	9.77	11.8	M3 V	11.43	13.4	M4.5e V			
Ross 614	6 28.1	-2 48	4.02	1.00	11.13	13.1	M4.5e V	14.8	16.8	?			
BD $-$ 12° 4523	16 28.9	-12 36	4.02	1.18	10.13	12.0	M4.5e V						
v. Maanen's Star	0 47.9	+5 16	4.28	2.98	12.36	14.3	WD						
Wolf 424	12 32.1	+9 10	4.37	1.78	12.7	14.4	M5.5e V	12.7	14.4	M6e V			
CD $-$ 37° 15492	0 3.9	-37 29	4.45	6.11	8.59	10.3	M3 V						
BD + 50° 1725	10 9.9	+49 35	4.61	1.45	6.59	8.3	M0 V						
CD $-$ 46° 11540	17 26.8	-46 52	4.63	1.06	9.34	11.3	M4 V						
CD $-$ 49° 13515	21 31.9	-49 7	4.67	0.81	9	11	Me V						
CD $-$ 44° 11909	17 35.3	-44 18	4.69	1.14	11.2	12.8	M5 V						
Luyten 1159-16	1 58.7	+12 58	4.72	2.08	12.3	13.9	M7?						

TABLE A4–1   The Nearest Stars (within 5 parsecs)   (continued)

Star	α (1975)	δ (1975)	Distance (parsecs)	Proper Motion ("/yr)	A m_v	A M_v	A Spectral Type	B m_v	B M_v	B Spectral Type	C m_v	C M_v	C Spectral Type
BD + 15° 2620	13ʰ 44ᵐ5	+15° 2'	4.80	2.30	8.6	10.2	M2 V						
BD + 68° 946	17 36.6	+68 22	4.83	1.31	9.1	10.7	M3 V	Unseen companion					
Luyten 145-141	11 44.2	−64 42	4.85	2.69	11	12.5	WD						
Ross 780	22 51.9	−14 24	4.85	1.12	10.2	11.8	M5 V						
o² Eridani	4 14.2	−7 42	4.87	4.08	4.5	6.0	K0 V	9.2	10.7	WD	11.0	12.5	M5e V
BD + 20° 2465	10 18.2	+20 0	4.95	0.49	9.4	10.9	M4.5 V	Unseen companion					

BD refers to Bonner Durchmusterung
CD refers to Cordoba Durchmusterung

TABLE A4–2   The 25 Brightest Stars

Star	α (1975)	δ (1975)	m_v	Distance (parsecs)	Proper Motion ("/yr)	Spectral Type	M_v
Sirius, α CMa	6ʰ 44ᵐ0	−16°41'	−1.5*	2.7	1.32	A1 V	+1.4
Canopus, α Car	6 23.6	−52 41	−0.7	55	0.02	F0 Ib	−3.1
α Centauri	14 38.0	−60 44	−0.3*	1.3	3.68	G2 V	+4.4
Arcturus, α Boo	14 14.5	+19 19	−0.1	11	2.28	K2 III	−0.3
Vega, α Lyr	18 36.0	+38 46	0.0	8.1	0.34	A0 V	+0.5
Capella, α Aur	5 14.8	+45 52	0.0*	14	0.44	G2 III	−0.7
Rigel, β Ori	5 13.3	−8 14	0.1*	250	0.00	B8 Ia	−6.8
Procyon, α CMi	7 38.0	+5 17	0.3*	3.5	1.25	F5 IV–V	+2.7
Achernar, α Eri	1 37.8	−57 22	0.5	20	0.10	B5 V	−1.0
β Centauri	14 02.1	−60 15	0.6*	90	0.04	B1 III	−4.1
Altair, α Aql	19 49.5	+8 48	0.8	5.1	0.66	A7 IV–V	+2.2
Betelgeuse, α Ori	5 53.8	+7 24	0.8†	150	0.03	M2 Iab	−5.5
Aldebaran, α Tau	4 34.0	+16 28	0.9*	16	0.20	K2 III	−0.2
α Crucis	12 25.2	−63 00	0.9*	120	0.04	B1 IV	−4.0
Spica, α Vir	13 23.9	−11 01	1.0†	80	0.05	B1 V	−3.6
Antares, α Sco	16 27.8	−26 22	1.0*†	120	0.03	M1 Ib	−4.5
Pollux, β Gem	7 43.8	+28 05	1.2	12	0.62	K0 III	+0.8
Formalhaut, α PsA	22 56.2	−29 45	1.2	7	0.37	A3 V	+2.0
Deneb, α Cyg	20 40.6	+45 11	1.3	430	0.00	A2 Ia	−6.9
β Crucis	12 46.2	−59 33	1.3	150	0.05	B0.5 IV	−4.6
Regulus, α Leo	10 7.0	+12 5	1.4*	26	0.25	B7 V	−0.6
ϵ Canis Majoris	6 57.7	−28 56	1.5	240	0.00	B2 II	−5.4
Castor, α Gem	7 33.0	+31 56	1.6	14	0.20	A1 V	+0.9
λ Scorpii	17 31.8	−37 5	1.6	96	0.03	B2 IV	−3.3
Bellatrix, γ Ori	5 23.8	+6 20	1.6	210	0.02	B2 III	−3.6

*Multiple-star apparent magnitude is integrated magnitude, other data are brightest component.
†The star is a variable.
Distances for more distant stars are from spectroscopic parallaxes.

TABLE A4–3 Stellar Characteristics by Spectral Type and Luminosity Class

Spectral Type	M_v			$B - V$			T_{eff}(K)			BC	$R/R_\odot$			$M/M_\odot$		
	V	III	Ib*	V	III	I	V	III	I	V	V	III	I	V	III	I
O5	−6.0			−0.32	−0.32	−0.32	50,000			−4.30	18			40		100
B0	−4.1	−5.0	−6.2	−0.30	−0.30	−0.24	27,000			−3.17	7.6	16	20	17		50
B5	−1.1	−2.2	−5.7	−0.16	−0.16	−0.09	16,000			−1.39	4.0	10	32	7		25
A0	+0.6	−0.6	−4.9	0.00	0.00	+0.01	10,400			−0.40	2.6	6.3	40	3.6		16
A5	+2.1	+0.3	−4.5	+0.15	+0.15	+0.07	8200			−0.15	1.8		50	2.2		13
F0	+2.6	+0.6	−4.5	+0.30	+0.30	+0.24	7200			−0.08	1.3		63	1.8		13
F5	+3.4	+0.7	−4.5	+0.45	+0.45	+0.45	6700	6500	6200	−0.04	1.2	4.0	80	1.4		10
G0	+4.4	+0.6	−4.5	+0.60	+0.65	+0.76	6000	5500	5050	−0.06	1.04	6.3	100	1.1	2.5	10
G5	+5.2	+0.3	−4.5	+0.65	+0.86	+1.06	5500	4800	4500	−0.10	0.93	10	126	0.9	3	13
K0	+5.9	+0.2	−4.5	+0.81	+1.01	+1.42	5100	4400	4100	−0.19	0.85	16	200	0.8	4	13
K5	+8.0	−0.3	−4.5	+1.18	+1.52	+1.71	4300	3700	3500	−0.71	0.74	25	400	0.7	5	16
M0	+9.2	−0.4	−4.5	+1.39	+1.65	+1.94	3700	3500	3300	−1.20	0.63		500	0.5	6	16
M5	+12.3	−0.5	−4.5	+1.69	+1.85	+2.15	3000	2700		−2.10	0.32			0.2		

*All class Ia stars have an absolute visual magnitude of −7.0.
BC is bolometric correction.

APPENDIX 5
ATOMIC ELEMENTS

TABLE A5–1 *The Periodic Table*

Element	Symbol	Atomic Number	Atomic Weight*	Element	Symbol	Atomic Number	Atomic Weight*
Hydrogen	H	1	1.008	Krypton	Kr	36	83.8
Helium	He	2	4.003	Rubidium	Rb	37	85.5
Lithium	Li	3	6.9	Strontium	Sr	38	87.6
Beryllium	Be	4	9.0	Yttrium	Y	39	88.9
Boron	B	5	10.8	Zirconium	Zr	40	91.2
Carbon	C	6	12.0	Niobium	Nb	41	92.9
Nitrogen	N	7	14.0	Molybdenum	Mo	42	96.0
Oxygen	O	8	16.0	Technetium	Tc	43	(99)
Fluorine	F	9	19.0	Ruthenium	Ru	44	101.1
Neon	Ne	10	20.2	Rhodium	Rh	45	102.9
Sodium	Na	11	23.0	Palladium	Pd	46	106.4
Magnesium	Mg	12	24.3	Silver	Ag	47	107.9
Aluminum	Al	13	27.0	Cadmium	Cd	48	112.4
Silicon	Si	14	28.1	Indium	In	49	114.8
Phosphorus	P	15	31.0	Tin	Sn	50	118.7
Sulfur	S	16	32.1	Antimony	Sb	51	121.8
Chlorine	Cl	17	35.5	Tellurium	Te	52	127.6
Argon	A	18	39.9	Iodine	I	53	126.9
Potassium	K	19	39.1	Xenon	Xe	54	131.3
Calcium	Ca	20	40.1	Cesium	Cs	55	132.9
Scandium	Sc	21	45.0	Barium	Ba	56	137.4
Titanium	Ti	22	47.9	Lanthanum	La	57	138.9
Vanadium	V	23	51.0	Cerium	Ce	58	140.1
Chromium	Cr	24	52.0	Praseodymium	Pr	59	140.9
Manganese	Mn	25	54.9	Neodymium	Nd	60	144.3
Iron	Fe	26	55.9	Promethium	Pm	61	(147)
Cobalt	Co	27	58.9	Samarium	Sm	62	150.4
Nickel	Ni	28	58.7	Europium	Eu	63	152.0
Copper	Cu	29	63.5	Gadolinium	Gd	64	157.3
Zinc	Zn	30	65.4	Terbium	Tb	65	158.9
Gallium	Ga	31	69.7	Dysprosium	Dy	66	162.5
Germanium	Ge	32	72.6	Holmium	Ho	67	164.9
Arsenic	As	33	74.9	Erbium	Er	68	167.3
Selenium	Se	34	79.0	Thulium	Tm	69	168.9
Bromine	Br	35	79.9	Ytterbium	Yb	70	173.0

TABLE A5–1 (*continued*)

Element	Symbol	Atomic Number	Atomic Weight*	Element	Symbol	Atomic Number	Atomic Weight*
Lutetium	Lu	71	175.0	Actinium	Ac	89	(227)
Hafnium	Hf	72	178.5	Thorium	Th	90	232.1
Tantalum	Ta	73	181.0	Protoactinium	Pa	91	(231)
Tungsten	W	74	183.9	Uranium	U	92	238.1
Rhenium	Re	75	186.2	Neptunium	Np	93	(237)
Osmium	Os	76	190.2	Plutonium	Pu	94	(244)
Iridium	Ir	77	192.2	Americium	Am	95	(243)
Platinum	Pt	78	195.1	Curium	Cm	96	(248)
Gold	Au	79	197.0	Berkelium	Bk	97	(247)
Mercury	Hg	80	200.6	Californium	Cf	98	(251)
Thallium	Tl	81	204.4	Einsteinium	E	99	(254)
Lead	Pb	82	207.2	Fermium	Fm	100	(253)
Bismuth	Bi	83	209.0	Mendeleevium	Md	101	(256)
Polonium	Po	84	(209)	Nobelium	No	102	(253)
Astatine	At	85	(210)	Lawrencium	Lw	103	(256)
Radon	Rn	86	(222)	Rutherfordium	Rf	104	(261)
Francium	Fr	87	(223)	Hahnium	Ha	105	(260)
Radium	Ra	88	226.1				

*Where mean atomic weights have not been well determined, the atomic mass numbers of the most stable isotopes are given in parentheses.

APPENDIX 6
CONVERSION OF UNITS

Astronomers have traditionally used a cgs system of units, while physicists have pretty much adopted SI units. Here we give the basic SI units and some useful conversions to cgs and English units. SI stands for Système International, the International System of units.

SI Basic Units

Length: meter (m)
Time: second (s)
Mass: kilogram (kg)
Current: ampere (A)
Temperature: kelvin (K)
Luminous intensity: candela (cd)

SI Derived Units

Force: newton (N)	$1 \text{ N} = 1 \text{ kg} \cdot \text{m/s}^2$
Work and energy: joule (J)	$1 \text{ J} = 1 \text{ N} \cdot \text{m}$
Power: watt (W)	$1 \text{ W} = 1 \text{ J/s}$
Frequency: hertz (Hz)	$1 \text{ Hz} = \text{s}^{-1}$
Charge: coulomb (C)	$1 \text{ C} = 1 \text{ A} \cdot \text{s}$
Magnetic induction: tesla (T)	$1 \text{ T} = 1 \text{ N/A} \cdot \text{m}$
Pressure: pascal (Pa)	$1 \text{ Pa} = 1 \text{ N/m}^2$

Conversion

Length

1 km = 0.6215 mi
1 mi = 1.609 km
1 m = 1.0936 yd = 3.281 ft = 39.37 in.
1 in. = 2.54 cm
1 ft = 12 in. = 30.48 cm
1 yd = 3 ft = 91.44 cm
1 lightyear = 9.461×10^{15} m
1 Å = 0.1 nm

Area

$1 \text{ m}^2 = 10^4 \text{ cm}^2$
$1 \text{ km}^2 = 0.3861 \text{ mi}^2$
$1 \text{ in}^2 = 6.4516 \text{ cm}^2$
$1 \text{ ft}^2 = 9.29 \times 10^{-2} \text{ m}^2$
$1 \text{ m}^2 = 10.76 \text{ ft}^2$

Volume

$1 \text{ m}^3 = 10^6 \text{ cm}^3$
$1 \text{ L} = 1000 \text{ cm}^2 = 10^{-3} \text{ m}^3$
$1 \text{ gal} = 3.786 \text{ L} = 231 \text{ in}^3$

Time

1 h = 60 min = 3.6 ks
1 day = 24 h = 1440 min = 86.4 ks
1 year = 365.24 days = 31.56 Ms

Speed

1 km/h = 0.2778 m/s = 0.6215 mi/h
1 m/h = 0.4470 m/s = 1.609 km/h

Angle and Angular Speed

π rad = 180°
1 rad = 57.30°
1° = 1.745×10^{-2} rad
1 rev/min = 0.1047 rad/s
1 rad/s = 9.549 rev/min

Mass

1 g = 0.035 oz
1 kg = 1000 g
1 tonne = 1000 kg = 1 Mg

Conversion *(continued)*

Density

1 g/cm^3 = 1000 kg/m^3 = 1 kg/L

Force

1 N = 0.2248 lb = 10^5 dyn
1 lb = 4.4482 N

Pressure

1 Pa = 1 N/m^2
1 atm = 101.325 kPa = 1.01325 bars
1 atm = 14.7 lb/in^2 = 760 mmHg
1 torr = 1 mmHg = 133.32 Pa
1 bar = 100 kPa

Energy

1 kW · h = 3.6 MJ
1 Btu = 778 ft · lb = 252 cal = 1054.35 J
1 eV = 1.602 × 10^{-19} J
1 erg = 10^{-7} J

Power

1 hp = 550 ft · lb/s = 745.7 W
1 Btu/min = 17.58 W
1 W = 1.341 × 10^{-3} hp

Magnetic Induction

1 G = 10^{-4} T
1 T = 10^4 G

TABLE A7–1   Astronomical Constants

Astronomical unit	$AU = 1.496 \times 10^{11}$ m
Parsec	$pc = 206{,}265$ AU
	$= 3.26$ ly
	$= 3.086 \times 10^{16}$ m
Lightyear	$ly = 6.324 \times 10^4$ AU
	$= 0.307$ pc
	$= 9.46 \times 10^{15}$ m
Sidereal year	1 yr $= 365.26$ days
	$= 3.16 \times 10^7$ s
Mass of Earth	$M_\oplus = 5.98 \times 10^{24}$ kg
Radius of Earth at equator	$R_\oplus = 6378$ km
Orbital velocity of Earth	$V_\oplus = 30$ km/s
Mass of Sun	$M_\odot = 1.99 \times 10^{30}$ kg
Radius of Sun	$R_\odot = 6.96 \times 10^5$ km
Luminosity of Sun	$L_\odot = 3.90 \times 10^{26}$ W
Effective temperature of Sun	$T_{\text{eff}} = 5780$ K
Mass of Moon	$M_\mathrm{D} = 7.3 \times 10^2$ kg $= 0.0123 M_\oplus$
Radius of Moon	$R_\mathrm{D} = 1738$ km $= 0.273 R_\oplus$
Radius of Moon's orbit	$d_\mathrm{D} = 3.84 \times 10^5$ km
Sidereal month	$P_\mathrm{D} = 27.3$ days
Synodic month	$= 29.5$ days
Distance of Sun from center of Galaxy	$R_\odot = 8.5$ kpc
Velocity of Sun about galactic center	$V_\odot = 220$ km/s
Diameter of Galaxy	$= 120$ kpc
Mass of Galaxy	$M = 7 \times 10^{11} M_\odot$

TABLE A7–2   Physical and Mathematical Constants

Velocity of light	$c = 3.00 \times 10^8$ m/s
Constant of gravitation	$G = 6.67 \times 10^{-11}$ N $\cdot$ m²/kg²
Planck constant	$h = 6.623 \times 10^{-34}$ joule $\cdot$ s
Boltzmann constant	$k = 1.38 \times 10^{-23}$ joule/K
Rydberg constant	$R = 1.097 \times 10^7$/m
Stefan–Boltzmann constant	$\sigma = 5.67 \times 10^{-8}$ W/m² K^{-4}
Wien's law constant	$\lambda_{\max} T = 2.898 \times 10^7$ Å K
Mass of hydrogen atom	$m_\mathrm{H} = 1.67 \times 10^{-27}$ kg
Mass of electron	$m_\mathrm{e} = 9.11 \times 10^{-31}$ kg
Charge of electron	$e = 1.60 \times 10^{-19}$ C
Electron volt	1 eV $= 1.602 \times 10^{-19}$ J
Wavelength equivalence of eV	1 eV $\rightarrow 1.24 \times 10^4$ Å

$\pi = 3.1416$

$e = 2.7183$; $\log_{10} e = 0.4343$

TABLE A7–3   Units and Conversions

tera (T)	$= 10^{12}$
giga (G)	$= 10^9$
mega (M)	$= 10^6$
kilo (k)	$= 10^3$
hecto (h)	$= 10^2$
deca (da)	$= 10$
deci (d)	$= 10^{-1}$
centi (c)	$= 10^{-2}$
milli (m)	$= 10^{-3}$
micro (μ)	$= 10^{-6}$
nano (n)	$= 10^{-9}$
pico (p)	$= 10^{-12}$

Temperature

$K = °C + 273$

$°F = (5/9)°C + 32$

Angular Measure; Degrees and Time

$360° = 24^\mathrm{h} = 2\pi$ rad; 1 rad $= 57°17'45'' = 206{,}264.8''$

$1° = 60' = 3600''$; $1° = 0.01745$ rad

$\quad = 4^\mathrm{m}$ $1'' = 4.848 \times 10^{-6}$ rad

$15° = 1^\mathrm{h}$ Solid angle: sphere $= 4\pi$ sr

APPENDIX 8
THE GREEK ALPHABET

Alpha	A	α	Nu	N	ν
Beta	B	β	Xi	Ξ	ξ
Gamma	Γ	γ	Omicron	O	o
Delta	Δ	δ	Pi	Π	π, ϖ
Epsilon	E	ϵ	Rho	P	ρ
Zeta	Z	ζ	Sigma	Σ	σ
Eta	H	η	Tau	T	τ
Theta	Θ	θ	Upsilon	Υ	υ
Iota	I	ι	Phi	Φ	ϕ, φ
Kappa	K	$\kappa, \varkappa$	Chi	X	χ
Lambda	Λ	λ	Psi	Ψ	ψ
Mu	M	μ	Omega	Ω	ω

APPENDIX 9
MATHEMATICAL OPERATIONS

In the following seven sections, we briefly review the basic mathematical methods used in astronomy and astrophysics: trigonometry, exponential notation, analytical geometry, vector analysis, series, the calculus, and mensuration formulas. The most useful results are placed in boxes and in tables for handy references.

A9–1 ◑
TRIGONOMETRY

(A) ANGULAR MEASURE

Figure A–1 depicts a circle of unit radius. The angular measure θ may be specified in three ways. The most ancient and familiar procedure is to divide the circle's circumference into 360 equal parts and to term that θ corresponding to one of these parts an *arc-degree* (°). Each arc-degree is further subdivided into 60 *arc-minutes* (') and each arc-minute into 60 *arc-seconds* ("). Hence, there are $360 \times 60 \times 60 = 1,296,000''$ in the full circle.

Astronomically, one rotation of the Earth requires 24 *hours* (^{h}) of time; we are accustomed to dividing the hour into 60 *minutes* (^{m}) and each minute into 60 *seconds* (^{s}). Hence, there are $24 \times 60 \times 60 = 86,400^s$ per rotation. A complete rotation (24^h) corresponds to the full circle (360°), however, so that we may say $1^h = 15°$, $1^m = 15'$, and $1^s = 15''$.

Finally, we may define a *radian* (rad) as that angle θ corresponding to a unit distance along the circumference of our unit circle. Since the entire circumference is 2π units in length ($\pi = 3.141593 . . .$), there are 2π rad in the full 360°. Therefore: $1 \text{ rad} = 360°/2\pi = 57.2958° = 206,264.81.''$ [Radian measure is extended to *angular areas* by noting that the surface area of a sphere of unit radius is 4π square units, that is, 4π *steradians* (sr). Since a steradian is one square radian, there are 41,252.96 square arc-degrees on the sphere.]

(B) THE RIGHT TRIANGLE

The triangle *OHA* in Figure A–2 is a *right triangle* since the angle at vertex *H* is 90°. With respect to the angle θ, the three sides of this triangle are la-

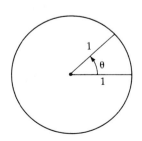

FIGURE A–1

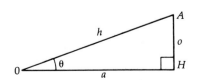

FIGURE A–2

TABLE A9–1

Region	sin	cos	tan
0°–90°	+	+	+
90°–180°	+	−	−
180°–270°	−	−	+
270°–360°	−	+	−

TABLE A9–2

Angle					
Arc-Degrees	Rad	sin	cos	tan	cot
0	0	0	1	0	∞
30	$\pi/6$	1/2	$\sqrt{3}/2$	$\sqrt{3}/3$	$\sqrt{3}$
45	$\pi/4$	$\sqrt{2}/2$	$\sqrt{2}/2$	1	1
60	$\pi/3$	$\sqrt{3}/2$	1/2	$\sqrt{3}$	$\sqrt{3}/3$
90	$\pi/2$	1	0	∞	0

beled a (adjacent), o (opposite), and h (hypotenuse). The fundamental trigonometric functions, *sine* (sin) and *cosine* (cos), are defined as

$$\sin \theta = o/h \qquad \cos \theta = a/h$$

A dependent function, the *tangent* (tan), then follows as

$$\tan \theta = o/a = (o/h)/(a/h) = \sin \theta/\cos \theta$$

The trigonometric functions may be extended to the full circle ($0° \leq \theta \leq 360°$) by using the signs given in Table A9–1, the special values listed in Table A9–2, and the values for every arc-degree from 0° to 90°. The following practical identities are needed in this extension:

$$\sin \theta = +\cos(\theta - 90°) = -\sin(\theta - 180°) = -\cos(\theta - 270°)$$
$$\cos \theta = -\sin(\theta - 90°) = -\cos(\theta - 180°) = +\sin(\theta - 270°)$$
$$\tan \theta = -\cot(\theta - 90°) = +\tan(\theta - 180°) = -\cot(\theta - 270°)$$

Also occasionally encountered are the three reciprocal functions:

$$\text{cosecant} \rightarrow \csc \theta = h/o = 1/\sin \theta$$
$$\text{secant} \rightarrow \sec \theta = h/a = 1/\cos \theta$$
$$\text{cotangent} \rightarrow \cot \theta = a/o = 1/\tan \theta$$

The following trigonometric identities are extremely useful:

Pythagorean	$\sin^2 \theta + \cos^2 \theta = 1 \qquad 1 + \tan^2 \theta = \sec^2 \theta$
Sum-and-difference	$\begin{cases} \sin(\theta \pm \phi) = \sin \theta \cos \phi \pm \cos \theta \sin \phi \\ \cos(\theta \pm \phi) = \cos \theta \cos \phi \mp \sin \theta \sin \phi \end{cases}$
Double-angle	$\begin{cases} \sin 2\theta = 2 \sin \theta \cos \theta \\ \sin^2 \theta = (1/2)(1 - \cos 2\theta) \qquad \cos^2 \theta = (1/2)(1 + \cos 2\theta) \end{cases}$

(C) THE PLANAR TRIANGLE

Figure A–3 illustrates the general *planar triangle* ABC, with vertex angles A, B, and C and corresponding opposite sides a, b, and c. For any such triangle, the following formulas obtain:

Area	$= \sqrt{s(s-a)(s-b)(s-c)}$, where $s = \frac{1}{2}(a + b + c)$
Law of sines	$\begin{cases} \dfrac{a}{\sin A} = \dfrac{b}{\sin B} = \dfrac{c}{\sin C} \end{cases}$
Law of cosines	$\begin{cases} a^2 = b^2 + c^2 - 2bc \cos A \\ b^2 = c^2 + a^2 - 2ca \cos B \\ c^2 = a^2 + b^2 - 2ab \cos C \end{cases}$

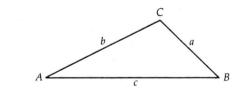

FIGURE A–3

A9–2 ◑
EXPONENTIAL NOTATION

(A) POWERS AND ROOTS

When a positive number a is multiplied against itself an integer m number of times, the result is the *mth power of a:*

$$a \times a \times a \times \cdots (m \text{ times}) \cdots \times a = a^m$$

When several powers of the same number are multiplied together, their *exponents* add: $a^m a^n = a^{m+n}$. If we define $a^0 \equiv 1$, then negative exponents are admitted and are called *reciprocals:* $a^{-m} = 1/a^m \Rightarrow a^m a^{-m} = a^{m-n} = a^0 = 1$.

Similarly, we term $a^{1/m}$ the *mth root of a* since we recover a when its root is raised to the *m*th power: $(a^{1/m})^m = a^{m/m} = a^1 = a$. Note that when a power or root is raised to a power, the two exponents involved multiply. These results are readily generalized to *any real exponent* (not necessarily an integer or a rational fraction) by the following formulas:

$$
\begin{array}{ll}
a^0 = 1 & a^{-m} = 1/a^m \\
\multicolumn{2}{c}{(ab)^m = a^m b^m} \\
a^m a^n = a^{m+n} & (a^m)^n = a^{mn}
\end{array}
$$

We define the *factorial* of an integer n as the product of n with all smaller integers (down to 1): $n! \equiv n(n-1)(n-2) \cdots (3)(2)(1)$. It is conventional to also define $0! \equiv 1$.

The following simple examples illustrate these manipulations:

$$
\begin{aligned}
3^4 &= 3 \times 3 \times 3 \times 3 = 81 \\
2^{-3} &= 1/2^3 = 1/(2 \times 2 \times 2) = 1/8 \\
15^2 &= (3 \times 5)^2 = 3^2 \times 5^2 = 9 \times 25 = 225 \\
6^2 \times 6^3 &= 6^{2+3} = 6^5 \\
(\sqrt{2})^3 &= (2^{1/2})^3 = 2^{3/2} \\
4! &= 4 \times 3 \times 2 \times 1 = 24
\end{aligned}
$$

(B) EXPONENTIALS AND LOGARITHMS

When the *base a* is given, the *exponential* formula

$$y = a^x = \text{the base } a \text{ to the power } x$$

yields a value of y for every value of x (*exponent*) we choose. However, if we know both a and y and desire to learn x, we must invert this relationship to obtain the *logarithmic* formula

$$x = \log_a y = \text{the exponent of } a \text{ that yields } y$$

For example, given $8 = 2^x$, we know that $x = 3$, since $2^3 = 2 \times 2 \times 2 = 8$; hence, $\log_2 8 = 3$.

The general properties of powers and roots lead to the following useful relationships for logarithms:

Product	$\log_a (xy) = \log_a x + \log_a y$
Quotient	$\log_a (x/y) = \log_a x - \log_a y$
Power	$\log_a (y^n) = n \log_a y$
Change of base	$\log_a y = (\log_a b)(\log_b y)$

In this text, we most frequently encounter the *decimal* base, $a = 10$; logarithms with respect to this base are termed *common* logarithms (written "log"). Every common logarithm consists of two parts: an integer (the *characteristic*) and an "endless" decimal (the *mantissa*). For example,

$$\log 33.7 = \log 10^{1.5276} = 1.5276$$

characteristic ——⌐ ⌐—— mantissa

When we use the power-of-ten notation, $33.7 = 3.37 \times 10^1$, the characteristic 1 is immediately evident.

Important in the calculus (Section A9–6), although infrequently encountered in this text, are exponentials to the base $e \equiv 2.71828 \ldots$. The associated *natural*, or *Naperian*, logarithms are denoted "ln." In all practical computations, we will make a change of base to the decimal system (common logarithms) using the relationships:

$$
\begin{aligned}
e^x &= 10^{0.4343x} \\
\ln x &= (2.3026) \log x
\end{aligned}
$$

A9–3 ◐
ANALYTICAL GEOMETRY

(A) COORDINATE SYSTEMS

There are three common coordinate systems used to locate a point in three-dimensional space. The most familiar system is *rectangular Cartesian* coordinates (x, y, z). Beginning at the *origin* O $(x = 0, y = 0, z = 0)$, we move out the x axis x units, across parallel to the y axis y units, and up parallel to the z axis z units (Figure A–4).

In *cylindrical polar* coordinates (ρ, ϕ, z), the point is found by moving out from the origin in the xy plane the distance ρ at the angle ϕ to the x axis, then up z units parallel to the z axis (Figure A–4). These coordinates are clearly related to Cartesian coordinates by

$$x = \rho \cos \phi \qquad y = \rho \sin \phi \qquad z = z$$

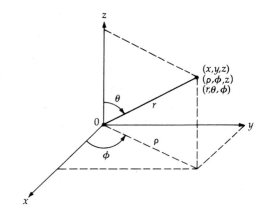

FIGURE A–4

Finally, in *spherical* coordinates (r, θ, ϕ), we move the distance r from the origin at the angle θ to the z axis; the projection of this motion on the xy plane is inclined the angle ϕ to the x axis and has the length $\rho = r \sin \theta$ (Figure A–4). The connection to Cartesian coordinates is therefore given by

$$x = r \sin \theta \cos \phi \qquad y = r \sin \theta \sin \phi \qquad z = r \cos \theta$$

(B) GRAPHS

We define "y as a *function* of x" by the algebraic equation $y = y(x)$. Therefore, for each value of x, the function yields a value of y; we have an (x, y) pair. To better illustrate the properties of the function, let us *graph* every (x, y) pair as a point in a two-dimensional Cartesian coordinate system; the result is a *curve*.

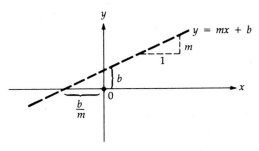

FIGURE A–5

Consider the *linear* equation $y = mx + b$, where m and b are constants. When $x = 0$, $y = b$. When $x = -b/m$, $y = 0$. And for every unit increase of x, y "increases" by m units; we say that the *slope* is m. The graph of this function is the *straight line* shown in Figure A–5.

Now consider the *quadratic* equation $y = ax^2 + bx + c$, where a, b, and c are constants. When $x = 0$, $y = c$; the two zeroes of the equation (where $y = 0$) are given by the *quadratic formula:*

$$x = \frac{-b \pm \sqrt{b^2 - 4ac}}{2a}$$

The graph of this function (Figure A–6) is a *parabolic* curve.

The usefulness of graphs is most evident when we consider more complicated functions. Figure A–7 shows the trigonometric functions $\sin x$,

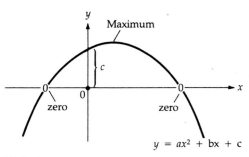

FIGURE A–6

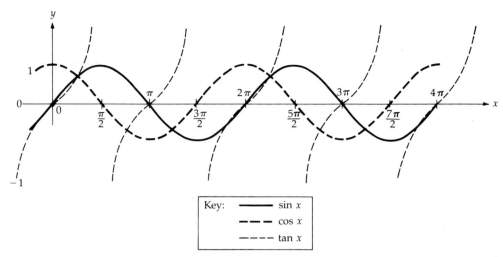

Key: —— sin x
- - - cos x
- - - - tan x

FIGURE A-7

cos x, and tan x. Figure A–8 depicts the exponential function $y = a^x$; the logarithmic function $x = \log_a y$ may be seen by rotating the diagram 90° counterclockwise.

(C) THE CONIC FUNCTIONS

In planar polar coordinates (ρ, ϕ), all gravitational orbits may be described by the single equation $\rho = d(1 + e)/(1 + e \cos \phi)$, where $\rho = d$ is the distance of *closest approach* to the origin (at $\phi = 0°$). The graph of this function yields a variety of curves called the *conic sections* (Figure A–9).

When $e = 0$, we have a *circle* of radius d. When e lies in the range $0 < e < 1$, the curve is an *ellipse*; we usually write $d = a(1 - e)$, so that the *major axis* (longest dimension) of the ellipse is $2a$. When $e = 1$, the curve is a *parabola* that is "open" to the left at $\phi = 180°$. Finally, we speak of a *hyperbola* when $e > 1$; this curve exhibits $\rho \to \infty$ at the two angles where $\cos \phi = -1/e$ (along lines called *asymptotes*).

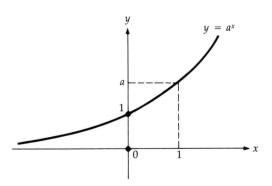

FIGURE A-8

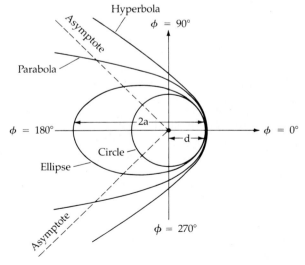

FIGURE A-9

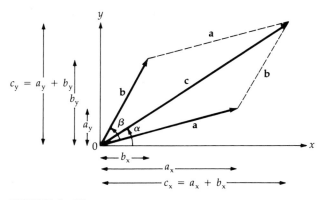

FIGURE A–10

A9–4 ◗
VECTOR ANALYSIS

(A) VECTORS

A *vector* is like an *arrow*, for it has both a *magnitude* (length) and a *direction*. The magnitude is a *scalar*, a simple number without a direction (like temperature or mass). We denote a vector by a letter printed in boldface, **c,** and its magnitude by the same letter printed in italic type, *c*.

Two vectors are added, $\mathbf{c} = \mathbf{a} + \mathbf{b}$, using the *parallelogram rule of vector addition* illustrated in Figure A–10. Conversely, a vector may always be decomposed into two *component* vectors. For convenience, we decompose along the coordinate axes and write the vector as $\mathbf{c} = (c_x, c_y)$. Now the rule of vector addition may be stated in terms of components as

$$c_x = a_x + b_x \qquad c_y = a_y + b_y$$

From the Pythagorean theorem and Figure A–10, it is clear that the magnitude of **c** is just $c = (c_x^2 + c_y^2)^{1/2}$. In terms of the angle α between **c** and the *x* axis, the direction of **c** is given by $\tan \alpha = c_y/c_x$. Finally, as a consequence of vector addition, the magnitude of **c** may be written as

$$
\begin{aligned}
c &= [c_x^2 + c_y^2]^{1/2} \\
&= [(a_x + b_x)^2 + (a_y + b_y)^2]^{1/2} \\
&= [(a_x^2 + a_y^2) + (b_x^2 + b_y^2) + 2(a_x b_x + a_y b_y)]^{1/2} \\
&= [a^2 + b^2 + 2(\mathbf{a} \cdot \mathbf{b})]^{1/2}
\end{aligned}
$$

(see the vector dot product in the following section) or from the law of cosines

$$c^2 = a^2 + b^2 + 2ab \cos \beta$$

where β is the smallest angle between **a** and **b.**

As an example, consider the vectors $\mathbf{a} = (1, 1)$ and $\mathbf{b} = (3, -4)$. Their magnitudes are

$$
\begin{aligned}
a &= (a_x^2 + a_y^2)^{1/2} = (1^2 + 1^2)^{1/2} \\
&= (1 + 1)^{1/2} = (2)^{1/2} = \sqrt{2} \\
b &= (3^2 + 4^2)^{1/2} = (9 + 16)^{1/2} = 5
\end{aligned}
$$

Their vector sum is

$$
\begin{aligned}
\mathbf{c} = \mathbf{a} + \mathbf{b} &= (a_x + b_x, a_y + b_y) \\
&= (1 + 3, 1 - 4) = (4, -3) = (c_x, c_y)
\end{aligned}
$$

with the magnitude

$$c = (4^2 + 3^2)^{1/2} = 5$$

(Construct a diagram like Figure A–10 for these vectors and show that **a, b, c** form an *isoceles* triangle!)

(B) DOT PRODUCT

In three-dimensional Cartesian coordinates, the *vector dot product* of **a** and **b** is defined as the scalar

$$\boxed{\mathbf{a} \cdot \mathbf{b} \equiv a_x b_x + a_y b_y + a_z b_z}$$

If ψ is the smallest angle between **a** and **b**, then we can easily show that

$$\boxed{\mathbf{a} \cdot \mathbf{b} \equiv ab \cos \psi}$$

Therefore, the dot product is a measure of the component of **a** *in the direction of* **b** (or vice versa), and $\mathbf{a} \cdot \mathbf{b} = 0$ when the two vectors are perpendicular ($\psi = 90°$; Figure A–11).

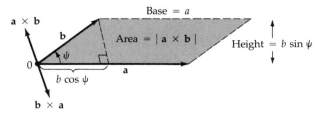

FIGURE A–11

Consider the example from the previous section, $\mathbf{a} = (1, 1)$ and $\mathbf{b} = (3, -4)$. Now, $\mathbf{a} \cdot \mathbf{b} = a_x b_x + a_y b_y = (1)(3) + (1)(-4) = 3 - 4 = -1$. The angle ψ satisfies

$$\cos \psi = \mathbf{a} \cdot \mathbf{b}/ab = (-1)/(\sqrt{2})(5)$$
$$= -\sqrt{2}/10 = -0.1414$$

so that Table A9–1 implies that $\psi \approx 98°$.

(C) CROSS PRODUCT

The *vector cross product* of $\mathbf{a}$ and $\mathbf{b}$, denoted $\mathbf{a} \times \mathbf{b}$, is *another vector* and is perpendicular to both $\mathbf{a}$ and $\mathbf{b}$. The direction of the resultant vector is given by the *right-hand rule:* "Align the fingers of your right hand along $\mathbf{a}$, and then rotate this hand through the smallest angle (ψ) between $\mathbf{a}$ and $\mathbf{b}$ toward $\mathbf{b}$; your thumb will point in the direction of the cross-product vector." In terms of components, the cross product is defined by

$$\mathbf{a} \times \mathbf{b} \equiv (a_y b_z - a_z b_y, \; a_z b_x - a_x b_z, \; a_x b_y - a_y b_x)$$

The cross product is essentially a measure of the component of $\mathbf{a}$ *perpendicular to* $\mathbf{b}$ (or vice versa), and it is therefore also given by

$$|\mathbf{a} \times \mathbf{b}| \equiv ab \sin \psi$$

Note that when $\mathbf{a}$ and $\mathbf{b}$ are parallel (or antiparallel), $\mathbf{a} \times \mathbf{b} = 0$; also, it is true that $\mathbf{b} \times \mathbf{a} = -\mathbf{a} \times \mathbf{b}$ (check this by using the right-hand rule and the component definition of cross product).

Figure A–11 illustrates some of the properties of the dot product and the cross product of two vectors $\mathbf{a}$ and $\mathbf{b}$.

We conclude with the calculation of the cross product of $\mathbf{a} = (1, 1)$ and $\mathbf{b} = (3, -4)$ (see the preceding sections):

$$\mathbf{a} \times \mathbf{b} = (0, 0, -4 - 3)$$
$$= (0, 0, -7) \quad \text{since } a_z = 0 = b_z$$

Therefore, $\mathbf{a} \times \mathbf{b}$ is directed in the negative z direction (perpendicular to both $\mathbf{a}$ and $\mathbf{b}$, which lie in the xy plane) and the area of the parallelogram in Figure A–11 is $|\mathbf{a} \times \mathbf{b}| = 7$. An alternate method for finding $\mathbf{a} \times \mathbf{b}$ is the following. First discover its direction by using the right-hand rule; then find its magnitude via $|\mathbf{a} \times \mathbf{b}| = ab \sin \psi$:

$$|\mathbf{a} \times \mathbf{b}| = (\sqrt{2})(5) \sin 98° = (1.414)(5)(0.99)$$
$$= 6.999 = 7$$

A9–5 ◗
SERIES

In the functional relationship $y = y(x)$, we term x the *argument*. In many practical applications of astronomy and astrophysics (and particularly in the calculus; Section A9–6), we need to know the behavior of certain functions for very small values of the argument ($0 \leq x \ll 1$). Hence, we expand the function in a *series* of powers of x; useful series expansions are listed below (together with the precise range of applicable x values):

Binomial	$(1 \pm x)^n = 1 \pm nx + (1/2)n(n-1)x^2 \pm (1/6)n(n-1)(n-2)x^3 + \cdots$	$(x^2 < 1; \text{ all } n)$
Trigonometric	$\sin x = x - (1/6)x^3 + (1/120)x^5 - \cdots$ $\cos x = 1 - (1/2)x^2 + (1/24)x^4 - \cdots$ $\tan x = x + (1/3)x^3 + (2/5)x^5 + \cdots$	$(x^2 < 1)$ $(x^2 < 1)$ $(x^2 < \pi^2/4)$
Exponential	$e^x = 1 + x + (1/2)x^2 + (1/6)x^3 + \cdots$	$(x^2 < 1)$
Logarithmic	$\ln (1 + x) = x - (1/2)x^2 + (1/3)x^3 - (1/4)x^4 + \cdots$	$(x^2 < 1)$

Three simple examples illustrate the use of these series. First, let us evaluate $\sqrt{e}$. We have (approximately)

$$e^{1/2} = 1 + (1/2) + (1/2)(1/2)^2$$
$$+ (1/6)(1/2)^3 + \cdots$$
$$= 1 + 1/2 + 1/8 + 1/48 + \cdots$$
$$= 79/48 + \cdots \approx 1.65$$

Second, consider the very narrow triangle used in stellar parallax (Chapter 12), with short side = 1 AU, adjacent side = d AU, and included angle at the star = π(rad) $\ll$ 1. Then we compute

$$(1\ \text{AU})/(d\ \text{AU}) = \tan \pi(\text{rad}) \approx \pi(\text{rad}) \Rightarrow$$
$$d(\text{pc}) \approx 1/\pi''$$

since there are 206,265" per radian and 1 pc = 206,265 AU.

Finally, when computing tidal accelerations, we seek the very small difference between two large quantities: $GM/[r \pm (d/2)]^2$. Extract the r^2 in the denominator, and use the binomial series on the remaining denominator (since $d \ll r \Rightarrow x = d/2r \ll 1$):

$$\left(1 \pm \frac{d}{2r}\right)^{-2} = 1 \mp 2\left(\frac{d}{2r}\right) + 3\left(\frac{d}{2r}\right)^2 \cdots$$

$$\approx 1 \mp \frac{d}{r} + \cdots$$

Therefore, we find

$$\frac{GM}{\left(r - \dfrac{d}{2}\right)^2} - \frac{GM}{\left(r + \dfrac{d}{2}\right)^2}$$

$$= \frac{GM}{r^2}\left[\left(1 - \frac{d}{2r}\right)^{-2} - \left(1 + \frac{d}{2r}\right)^{-2}\right]$$

$$= \frac{GM}{r^2}\left[\left(1 + \frac{d}{r} + \cdots\right) - \left(1 - \frac{d}{r} + \cdots\right)\right]$$

$$\approx 2GMd/r^3$$

A9–6 ◗
THE CALCULUS

(A) DERIVATIVES

We seek the *derivative* (or *instantaneous slope)* of the function $y(x)$ at the point x. As illustrated in Figure A–12, we select a nearby point $x + \Delta x$, evaluate

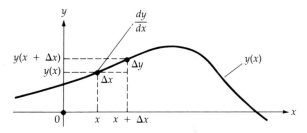

FIGURE A–12

$y(x + \Delta x)$, and in the limit as Δx becomes infinitesimally small $\left(\lim\limits_{\Delta x \to 0}\right)$, we *define* the derivative as

$$\boxed{\frac{dy}{dx} \equiv \lim_{\Delta x \to 0} \frac{y(x + \Delta x) - y(x)}{\Delta x}}$$

Let us use this definition to derive two simple derivatives. Consider first $y(x) = x^2$; then

$$y(x + \Delta x) = (x + \Delta x)^2 = x^2 + 2x(\Delta x) + (\Delta x)^2$$

Therefore

$$\frac{dy}{dx} = \lim_{\Delta x \to 0} \frac{x^2 + 2x(\Delta x) + (\Delta x)^2 - x^2}{\Delta x}$$

$$= \lim_{\Delta x \to 0} 2x + \Delta x = 2x$$

Hence the derivative of x^2 at x is $2x$.

Second, consider $y(x) = \sin x$. Then (using the sum-of-angles identity) $y(x + \Delta x) = \sin(x + \Delta x) = \sin x \cos \Delta x + \cos x \sin \Delta x$. Since Δx becomes very small, however, we may use the series expansions for $\sin \Delta x \approx \Delta x$ and $\cos \Delta x \approx 1$. Therefore,

$$\frac{dy}{dx} = \lim_{\Delta x \to 0} \frac{\sin x + (\Delta x) \cos x - \sin x}{\Delta x} = \cos x$$

the sought-for derivative of $\sin x$.

Proceeding in just this fashion, you may verify the following useful formulas for derivatives [where a and n are constants and $u = u(x)$ and $v = v(x)$]:

Definitions	$da/dx = 0$ $dx/dx = 1$
Linearity	$\begin{cases} d(au)/dx = a(du/dx) \\ d(u + v)/dx = (du/dx) + (dv/dx) \end{cases}$
"Chain rule"	$\begin{aligned} d(uv)/dx &= u(dv/dx) \\ &\quad + v(du/dx) \end{aligned}$
Powers	$d(u^n)/dx = nu^{n-1}(du/dx)$
Trigono-metric	$\begin{cases} d(\sin u)/dx = \cos u(du/dx) \\ d(\cos u)/dx = -\sin u(du/dx) \\ d(\tan u)/dx = \sec^2 u(du/dx) \end{cases}$
Exponential	$\begin{cases} d(a^u)/dx = a^u(\ln a)(du/dx) \\ d(e^u)/dx = e^u(du/dx) \end{cases}$
Logarithmic	$d(\ln u)/dx = (1/u)(du/dx)$

For example, here are the steps in finding $d(x \sin x)^2/dx$:

1. Note that this is in the form of $d(u^n)/dx$:

$$\frac{d}{dx}(x \sin x)^2 = 2(x \sin x)\frac{d}{dx}(x \sin x)$$

2. Apply the chain rule:

$$= 2(x \sin x)\left[x\frac{d(\sin x)}{dx} + \sin x\left(\frac{dx}{dx}\right)\right]$$

3. Note that $\dfrac{dx}{dx} = 1$ and $\dfrac{d(\sin x)}{dx} = \cos x$:

$$= 2(x \sin x)(x \cos x + \sin x)$$

4. Expanding:

$$= 2x^2 \sin x \cos x + 2x \sin^2 x$$

[Note that the derivative of a vector is defined in terms of the derivatives of its components: $d\mathbf{a}/dx = (da_x/dx, da_y/dx, da_z/dx)$.]

(B) INTEGRALS

The *integral* of the function $y(x)$ may be either indefinite or definite. The *indefinite integral*, denoted by $\int y(x)\,dx$, is to be thought of as "that function of x whose derivative is $y(x)$." Hence, it is clear that $\int \cos x\,dx = \sin x$, since $d(\sin x)/dx = \cos x$. Therefore, the indefinite integral is the antiderivative, in the sense that $\int [dy(x)/dx]\,dx = y(x)$.

The *definite integral*, denoted by $\int_a^b y(x)\,dx$, is the net area under the curve $y(x)$ between $x = a$ and

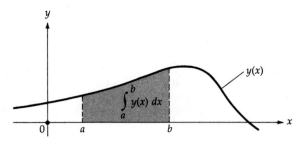

FIGURE A–13

$x = b$ (Figure A-13). If we have $y(x) = df(x)/dx$, then by definition it follows that

$$\int_a^b y(x)\,dx = \int_a^b (df/dx)\,dx = f(x)\Big]_a^b$$

$$= f(b) - f(a)$$

In general, indefinite integrals are found by trial and error, but we can tabulate some useful known results (see the list of derivatives above):

Linearity	$\begin{cases} \displaystyle\int ay(x)\,dx = a\int y(x)\,dx \\[2mm] \displaystyle\int (u + v)\,dx = \int u\,dx + \int v\,dx \end{cases}$
"By parts"	$\displaystyle\int u\,dv = uv - \int v\,du$
Powers	$\displaystyle\int x^n\,dx = x^{n+1}/(n + 1)$ (except for $n = -1$)
Trigono-metric	$\begin{cases} \displaystyle\int \sin x\,dx = -\cos x \\[2mm] \displaystyle\int \cos x\,dx = \sin x \\[2mm] \displaystyle\int \sec^2 x\,dx = \tan x \end{cases}$
Exponential	$\displaystyle\int e^{ax}\,dx = e^{ax}/a$
Logarith-mic	$\begin{cases} \displaystyle\int [(dy/dx)/y(x)]\,dx = \ln y(x) \\[2mm] \displaystyle\int (1/x)\,dx = \ln x \end{cases}$

For a vastly more extensive tabulation, look in any standard *table of integrals*.

We may illustrate the usefulness of this brief table of integrals by considering $\int (\sin^2 x) \cos x \, dx$. Let $u = \sin x$; then $du = \cos x \, dx$ and our integral is $\int u^2 \, du = u^3/3$. Substituting back in for $u = \sin x$, we have the answer: $(1/3) \sin^3 x$. If this had been the definite integral, $\int_0^{\pi/2} (\sin^2 x) \cos x \, dx$, we would find

$$\int_0^{\pi/2} (\sin^2 x) \cos x \, dx = [(1/3) \sin^3 x]_0^{\pi/2}$$

$$= (1/3)[\sin^3 (\pi/2) - \sin^3 (0)]$$
$$= (1/3)(1^3 - 0^3) = 1/3$$

[Note that the integral of a vector is another vector, defined in terms of the components $\int \mathbf{a} \, dx = \int (a_x(x), a_y(x), a_z(x)) \, dx = (\int a_x \, dx, \int a_y \, dx, \int a_z \, dx)$.]

A9–7 ◖

MENSURATION FORMULAS

Such things as lengths, areas, and volumes are given by *mensuration formulas*; a typical example is the area of a circle of radius R: $A = \pi R^2$. In Section A9–7(a), we show how to derive these formulas using the integral calculus; those interested only in the answers should proceed at once to Section A9–7(b).

(A) MULTIPLE INTEGRALS

At a given point in a coordinate system, we make infinitesimal changes in the three coordinates and define (a) infinitesimal lengths, (b) infinitesimal surface areas, and (c) infinitesimal volumes. By appropriately summing (that is, integrating) these, we obtain finite lengths, areas, and volumes. In general, we will be dealing with *multiple integrals*.

In rectangular Cartesian coordinates (x, y, z), the infinitesimal extensions are (dx, dy, dz). The distance along the x axis from $x = 0$ to $x = L$ is then just $\int_0^L dx = x]_0^L = L$. The infinitesimal surface areas are $dx \, dy$ [in the xy plane at (x, y, z)], $dy \, dz$, and $dz \, dx$. Therefore, the area in the xy plane bounded between $0 \le x \le L$ and $0 \le y \le W$ is $\int_0^L dx \int_0^W dy = x]_0^L y]_0^W = LW$. Finally, at (x, y, z), the infinitesimal volume is $dx \, dy \, dz$. The volume of a rectangular parallelopiped of dimensions $L \times W \times H$ is clearly $\int_0^L dx \int_0^W dy \int_0^H dz = LWH$.

In cylindrical polar coordinates, the elementary lengths are $(d\rho, \rho \, d\phi, dz)$, the elementary areas are $(\rho \, d\rho \, d\phi, \rho \, d\phi \, dz,$ and $d\rho \, dz)$, and the elementary volume is $\rho \, d\rho \, d\phi \, dz$. Therefore, the *circumference* of a circle of radius $\rho = R$ is $\int_0^{2\pi} R \, d\phi = R \int_0^{2\pi} d\phi = R\phi]_0^{2\pi} = 2\pi R$; the *area* of this circle is $\int_0^R \rho \, d\rho \int_0^{2\pi} d\phi = (1/2)\rho^2]_0^R \phi]_0^{2\pi} = (R^2/2)(2\pi) = \pi R^2$; and the *volume* of a right cylinder (of radius $\rho = R$ and height $z = H$) is $\int_0^R \rho \, d\rho \int_0^{2\pi} d\phi \int_0^H dz = \pi R^2 \int_0^H dz = \pi R^2 H$.

In spherical coordinates, the basic lengths are $(dr, r \, d\theta, r \sin \theta \, d\phi)$, the basic areas are $(r \, dr \, d\theta, r^2 \sin \theta \, d\theta \, d\phi,$ and $r \, dr \sin \theta \, d\phi)$, and the basic volume element is $r^2 \, dr \sin \theta \, d\theta \, d\phi$. Therefore, the *surface area* of a sphere of radius $r = R$ is $\int_0^\pi R^2 \sin \theta \, d\theta \int_0^{2\pi} d\phi = 2\pi R^2 \int_0^\pi \sin \theta \, d\theta = 2\pi R^2[-\cos \theta]_0^\pi = 4\pi R^2$, and the volume is $\int_0^R r^2 \, dr \int_0^\pi \sin \theta \, d\theta \int_0^{2\pi} d\phi = [(1/3)r^3]_0^R[-\cos \theta]_0^\pi[\phi]_0^{2\pi} = 4\pi R^3/3$.

A final example illustrates how these techniques are extended to more complex cases. Suppose we want to know the surface area of a sphere of radius $r = R$ in the range $0 \le \theta \le \theta_0$. The appropriate multiple integral is $R^2 \int_0^{\theta_0} \sin \theta \, d\theta \int_0^{2\pi} d\phi = 2\pi R^2[-\cos \theta]_0^{\theta_0} = 2\pi R^2(1 - \cos \theta_0)$. Note that the area is $2\pi R^2$ when $\theta_0 = \pi/2$ (half the sphere's surface) and $4\pi R^2$ when $\theta_0 = \pi$ (the entire surface), as should be the case.

(B) USEFUL MENSURATION FOMULAS

Using methods similar to those shown in Section A9–7(a), it is relatively straightforward to verify the following handy formulas:

Planar
  Arbitrary Triangle
    Area = (1/2)(base length) × (vertical height)
        $= \sqrt{s(s-a)(s-b)(s-c)}$ $\begin{cases}\text{where } s = (12)(a+b+c) \text{ and} \\ \text{the sides have lengths, } a, b, c\end{cases}$
  Parallelogram and Rhombus
    Area = (base length) × (vertical height)

  Trapezoid
    Area = (1/2)(a + b) × (vertical height) $\begin{cases}\text{where } a \text{ and } b \text{ are the} \\ \text{lengths of top and bottom}\end{cases}$

  Circle
    Circumference = 2π(radius) = π(diameter)
    Area = π(radius)^2 = (π/4)(diameter)^2
    Area of segment = (1/2)(radius)^2(θ − sin θ) $\begin{cases}\text{where } \theta \text{ is the central} \\ \text{angle in radians}\end{cases}$
        Area of sector = (1/2)(radius)^{2}θ
      Area of thin annulus = 2πR(ΔR) $\begin{cases}\text{where } R \text{ is the radius and} \\ \Delta R \text{ the radial thickness}\end{cases}$

  Ellipse
    Area = πab   (a = semimajor axis, b = semiminor axis)
Solid
  Rectangular Parallelopiped
    Volume = abc   (the sides have lengths a, b, c)
  Pyramid and Cone
    Volume = (1/3)(base area) × (vertical height)
  Right Cylinder
    Volume = πR^{2}H   (R = radius, H = height)
  Sphere
    Surface area = 4π(radius)^2 = π(diameter)^2
    Volume = (4π/3)(radius)^3 = (π/6)(diameter)^3
    Surface area of segment = 2π(radius) × (height of segment)
    Volume of segment = (π/3)(height)^2 × (3 radius − height)

  Ellipsoid
    Volume = (4π/3)abc $\begin{cases}\text{where } a, b, c \text{ are the lengths} \\ \text{of the three semiaxes}\end{cases}$

APPENDIX 10
THE CELESTIAL SPHERE

To map the sky, we assign locations to each of the celestial phenomena we study. We specify the three-dimensional spatial position of an event by its Cartesian (rectangular), polar, and spherical coordinates. Because angular positions are our primary interest in positional astronomy, we discuss almost exclusively *spherical coordinate systems*. On the surface of a sphere, the circumference and radius of a circle are not related by 2π and the sum of the interior angles of a triangle is always greater than 180°. Hence, the familiar planar geometry and trigonometry are not applicable; they must be replaced with spherical geometry and trigonometry, for which the most important formulas are given in Appendix 9.

To begin, consider the surface of a sphere of arbitrary radius. Any plane passing through the center of the sphere intersects the surface in a *great circle*. We select one plane—usually the one perpendicular to an axis of rotation—and designate its great circle the *primary circle*. All great circles intersecting the primary circle perpendicularly are called *secondary circles;* all the secondary circles meet at only two points, the *poles*. We define one intersection point of the primary circle and a given secondary circle (the reference circle) as the *point of origin*. A coordinate system may now be set up on the spherical surface as follows: the position of a point A is specified by (1) the angular distance in a conventional direction along the primary circle from the point of origin to the point of intersection closest to A of the secondary circle passing through A and (2) the shortest angular distance along this secondary circle from the primary circle to point A. Before proceeding to celestial coordinates, let us illustrate these ideas using the surface of the Earth.

A10–1 ◗
LONGITUDE AND LATITUDE ON THE EARTH

Figure A–14 shows the familiar longitude–latitude system of terrestrial coordinates. The *equator* is the primary circle, defined by the central plane perpendicular to the Earth's axis of rotation; the rotational axis intersects the surface of the Earth at the north and south poles. Secondary circles pass through the poles, and each semicircle terminating at both poles is termed a *meridian*. The reference semicircle, the *prime meridian*, passes through Greenwich, England, and meets the equator at the point of origin (0° longitude). *Longitude* is the shortest angular distance along the equator from the prime meridian to a given meridian; it is mea-

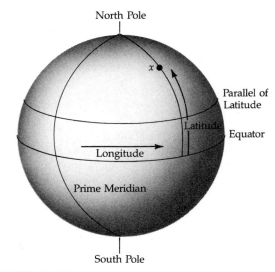

FIGURE A–14

sured eastward or westward from 0° to 180°. The International Date Line is located essentially at longitude 180° E (or W). *Latitude* is the angular distance north or south from the equator, measured along a meridian in degrees from 0° (the equator) to 90° (the poles). Notice that planes parallel to the equator slice the Earth's surface in small circles—the *parallels* of latitude. Some examples of approximate locations specified by this system are New York City (73°58'W, 40°40'N) and Sydney, Australia (151°17'E, 33°55'S).

A10–2 ◐
THE HORIZON SYSTEM

Primary observations are clearly location-dependent, and so let us outline the observer-based *horizon*, or *azimuth–altitude*, system of coordinates. On the celestial sphere, we construct a spherical coordinate system with the observer at its center (Figure A–15). The point vertically overhead is termed the *zenith*, and the opposite point (directly underfoot) is the *nadir*. Together these two points define an axis. The plane passing through the observer perpendicular to this axis meets the sky at the *celestial horizon*, which is everywhere 90° distance from both zenith and nadir. Because of natural and artificial obstructions, the actual horizon is seldom the celestial horizon; the closest approximation occurs for a sea-level observer in the middle of a calm ocean. Planes parallel to the axis and passing through the observer cut the celestial sphere in great circles called *vertical circles*. The reference circle is that vertical circle which includes the observer's zenith and the north and south points on her or his horizon—this circle we designate the observer's *celestial meridian*. The north point on the horizon is the point of origin, and east and west lie on the horizon midway between the north and south points.

The position of a celestial phenomenon is specified in the horizon system by giving its instantaneous azimuth and altitude. *Azimuth* is the angular distance along the horizon measured eastward from the north point to the foot of the vertical circle containing the phenomenon; the foot closest to the phenomenon is the one intended, and the azimuth ranges from 0° to 360°. *Altitude* is the shortest angular distance upward along this vertical circle from the horizon to the phenomenon, and it ranges from 0° (the horizon) to 90° (the zenith). The complement of a body's altitude is its *zenith distance* (90° minus the altitude). Two events on the observer's celestial meridian are of particular interest: A celestial body is said to be in *upper transit* when it crosses the celestial meridian moving westward and in *lower transit* when it crosses moving eastward. In the cases of meteors and artificial satellites, which may rise in the west and set in the east, another criterion for upper transit is necessary: Usually upper transit is that crossing of the celestial meridian visible to the observer.

A10–3 ◐
CELESTIAL EQUATORIAL COORDINATES

We come now to the most important astronomical coordinate system—the *celestial equatorial system*. Recall that the *celestial sphere* is centered upon the Earth's center and that its radius is indefinitely large. This last stipulation follows because we intend to map the entire sky onto the surface of this sphere and because the lines of sight to any star (other than our Sun) are essentially perfectly parallel for any two earthbound observers. There are many ways to project a spherical coordinate system onto the interior surface of the celestial sphere; the Earth's rotation is the basis of the present method. Though much of the terminology is different, the celestial equatorial coordinate system is almost completely analogous to terrestrial longitude and latitude.

To a given observer, the apparent rotation of the celestial sphere causes all stars to circumnavigate the heavens once each day; hence, the azimuth and altitude of each star are constaly changing with time. By transforming to a spherical coordinate system that rotates with the celestial sphere—the celestial equatorial system—we can obtain positions that are *fixed* to an accuracy of one part in 10^4 per year. The chief cause of the remaining variations is the Earth's precession, which results in

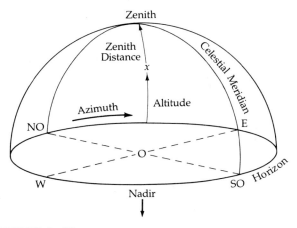

FIGURE A–15

the westward precession of the equinoxes by about 50″ per year.

Visualize a stationary celestial sphere at the center of which the Earth rotates eastward upon its axis once each day. Imagine halting the Earth's rotation and projecting the terrestrial longitude–latitude coordinate mesh onto the surface of the celestial sphere, so that the Earth's equatorial plane cuts the celestial sphere in the great circle of the *celestial equator* and extending the Earth's rotational axis so that it intersects the sphere at the north and south *celestial poles* (Figure A–16). Meridians of longitude are mapped into *hour circles* on the celestial sphere, and parallels of latitude appear as small circles concentric to the poles. If we now permit the Earth to resume its rotation, this celestial equatorial coordinate mesh fixed to the celestial sphere appears, to any observer on the Earth's surface, to rotate westward as an entity once each day.

In the celestial equatorial coordinate system, the celestial equator is the primary circle and the hour circles are the secondary circles. The position of a celestial body is specified by its declination (DEC or δ) and its right ascension (RA or α). *Declination,* the analog of terrestrial latitude, is the smallest angular distance (measured in degrees, minutes, and seconds of arc) from the celestial equator to the body along the hour circle passing through the body. Positions between the celestial equator and the north celestial pole have positive declination by convention, and those between the celestial equator and the south celestial pole have negative declination; hence, declination ranges from 0° (the celestial equator) to 90° (the north celestial pole) or −90° (the south celestial pole). *Right ascension,* the analog of terrestrial longitude, is the angular distance (measured in hours, minutes, and seconds of *time,* or ^{hms}) eastward along the celestial equator from the prime hour circle (see below) to the hour circle containing the body. Right ascension ranges from $0^h0^m0^s$ to $23^h59^m59^s$. To understand how the time units of right ascension come about, consider the prime hour circle to exactly coincide with an observer's local celestial meridian. Since the Earth rotates through 360° in 24 h, the observer's celestial meridian will lie 15° east of the prime hour circle after 1 h. Therefore, we call this right ascension $1^h0^m0^s$; a rotation of 1° corresponds to 4 min of time, of 1 arc-min to 4 s of time, and of 1 arc-sec to 1/15 s of time.

The point of origin of right ascension is the *vernal equinox* (♈), historically known as the first point of Aries. This origin, fixed on the celestial sphere, is defined by the celestial equator and the ecliptic. The *ecliptic,* the apparent annual path of the Sun in the sky, is the great circle where the orbital plane of the Earth intersects the celestial sphere. Figure A–17 is a map of the celestial sphere showing the celestial equator and the sinuous ecliptic. These two great circles are inclined at the *obliquity* angle of 23°26.5′ to one another, so that they intersect at only two points—the *equinoxes.* As the Sun progresses eastward along the ecliptic, it crosses the celestial equator moving northward at the vernal equinox (spring) and again six months later moving southward at the *autumnal equinox* (fall). By definition, the right ascension–declination of the vernal equinox is (0^h, $0°$).

For observational purposes, we sometimes use the celestial equatorial system but take as our reference circle the local celestial meridian. Declination measures north–south angular positions, whereas *hour angle* tells us how far west of the celestial meridian a celestial body is in units of time. The hour angle of an astronomical object depends both on time and on the observer's location, being $0^h0^m0^s$ at upper transit and $2^h30^m0^s$, 2.5 h later.

Table A10–1 gives the transformation equations between the horizon and equatorial systems.

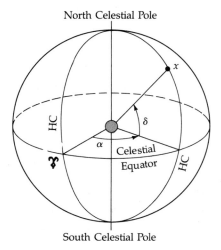

FIGURE A–16

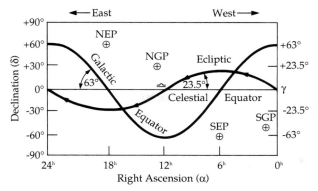

FIGURE A–17

A10–4 ◑
ECLIPTIC AND GALACTIC COORDINATES

For describing the motions of bodies within the Solar System, the *ecliptic coordinate system* is extremely useful. Here the ecliptic is the primary circle and the poles are called the *north ecliptic pole* (that pole closest to the north celestial pole) and the *south ecliptic pole*. *Celestial longitude* (λ) is the angular distance (from 0° to 360°) eastward along the ecliptic from the vernal equinox (the point of origin). *Celestial latitude* (β) is the angle from the ecliptic, measured positively toward the north ecliptic pole and negatively toward the south ecliptic pole; it ranges from 0° to ±90°. Hence, when the Sun's center lies at the autumnal equinox, its coordinates are (λ, β) = (180°, 0°). Table A10–1 gives the relations between the ecliptic and equatorial systems.

When we discuss phenomena associated with our Galaxy, we find it convenient to employ modern *galactic coordinates*. Here the primary circle is defined by the central plane of the Milky Way and is called the *galactic equator*. The center of the Galaxy (in Sagittarius), which lies on the galactic equator, is the point of origin. *Galactic longitude* (l or l^{II}) is measured eastward along the galactic equator from the direction to the galactic center and ranges from 0° to 360°. As viewed from the *north galactic pole*, galactic longitude increases in the counterclockwise direction. *Galactic latitude* (b or b^{II}) is the angle from the galactic equator toward the north or south galactic pole and ranges from 0° to ±90°. To avoid confusion, it should be noted that prior to August 1958, a different system of galactic coordinates (l^{I}, b^{I}) was in use; we shall have no occasion to refer to these obsolete coordinates. Table A10–1 provides the transformations between the galactic and equatorial systems.

TABLE A10–1   *Astronomical Coordinate Transformations*

I. *Horizon–equatorial (celestial) systems*
$$\cos a \sin A = + \cos \delta \sin h,$$
$$\cos a \cos A = -\sin \delta \cos \phi + \cos \delta \cos h \sin \phi,$$
$$\sin a = \sin \delta \sin \phi + \cos \delta \cos h \cos \phi,$$
$$\cos \delta \sin h = \cos a \sin A,$$
$$\cos \delta \cos h = \sin a \cos \phi + \cos a \cos A \sin \phi$$
$$\sin \delta = \sin a \sin \phi - \cos a \cos A \cos \phi,$$
h = local sidereal time − α,
A = azimuth, toward west from south,
a = altitude,
ϕ = observer's latitude,
h = local hour angle,
α = right ascension,
δ = declination.

II. *Ecliptic–equatorial (celestial) systems*
$$\cos \delta \cos \alpha = \cos \beta \cos \lambda,$$
$$\cos \delta \sin \alpha = \cos \beta \sin \lambda \cos \epsilon - \sin \beta \sin \epsilon,$$
$$\sin \delta = \cos \beta \sin \lambda \sin \epsilon + \sin \beta \cos \epsilon,$$
$$\cos \beta \cos \lambda = \cos \delta \cos \alpha,$$
$$\cos \beta \sin \lambda = \cos \delta \sin \alpha \cos \epsilon + \sin \delta \sin \epsilon,$$
$$\sin \beta = \sin \delta \cos \epsilon - \cos \delta \sin \alpha \sin \epsilon,$$
α = right ascension,
δ = declination,
λ = ecliptic longitude,
β = ecliptic latitude,
ϵ = obliquity of the ecliptic
   = 23°27′8″.26 − 46″.845 T
   − 0″.0059 T^2 + 0″.00181 T^3
   where T is the time in centuries from 1900.

III. *Galactic–equatorial (celestial) systems*
$$\cos b^{\mathrm{II}} \cos(l^{\mathrm{II}} - 33°) = \cos \delta \cos(\alpha - 282.25°),$$
$$\cos b^{\mathrm{II}} \sin(l^{\mathrm{II}} - 33°) = \cos \delta \sin(\alpha - 282.25°) \cos 62.6° + \sin \delta \sin 62.6°,$$
$$\sin b^{\mathrm{II}} = \sin \delta \cos 62.6° - \cos \delta \sin(\alpha - 282.25°) \sin 62.6°,$$
$$\cos \delta \sin(\alpha - 282.25°) = \cos b^{\mathrm{II}} \sin(l^{\mathrm{II}} - 33°) \cos 62.6° - \sin b^{\mathrm{II}} \sin 62.6°,$$
$$\sin \delta = \cos b^{\mathrm{II}} \sin(l^{\mathrm{II}} - 33°) \sin 62.6° + \sin b^{\mathrm{II}} \cos 62.6°,$$
l^{II} = new galactic longitude,
b^{II} = new galactic latitude,
α = right descension (1950.0),
δ = declination (1950.0),
For, $l^{\mathrm{II}} = b^{\mathrm{II}} = 0$: $\alpha = 17^{\mathrm{h}}\ 42^{\mathrm{m}}.4$, $\delta = -28°55'$ (1950.0);
$b^{\mathrm{II}} = +90.0$, galactic north pole: $\alpha = 12^{\mathrm{h}}\ 49^{\mathrm{m}}$, $\delta = +27°.4$ (1950.0).

In Figure A–17 we have also indicated the poles and equators of the ecliptic and galactic coordinate systems. There we see that the great circle of the galactic equator is inclined 63° to the celestial equator. Hence, we already know of three different coordinate systems, each of which completely covers the celestial sphere.

Glossary

absolute magnitude A measure of the brightness a star would have if it were to be placed at a standard distance of 10 pc from the Sun.

absorption (dark) lines Colors missing in a continuous spectrum because of the absorption of those colors by atoms.

absorption-line spectrum Dark lines superimposed on a continuous spectrum.

acceleration The rate of change of velocity with time.

accretion The colliding and sticking together of small particles to make larger masses.

accretion disk A disk made by infalling material around a mass; the conservation of angular momentum results in the disk shape.

active galaxies Galaxies characterized by a nonthermal spectrum and a large energy output compared to a normal galaxy.

active galaxy nucleus (AGN) The tiny central engine (possibly a supermassive black hole) that drives the active galaxy phenomenon.

active regions Areas on the photosphere of the Sun (and other stars) where magnetic field lines are concentrated; these generate sunspots and flares.

adiabatic perturbations Waves in the primordial radiation/matter mixture in which both the radiation and matter have small changes in amplitude.

albedo A measure of an object's reflecting power; the ratio of reflected light to incoming light for a solid surface, where complete reflection gives an albedo of 1.

Alpha Centauri The closest star to the Sun, a triple system; component A happens to have almost the same luminosity and surface temperature as the Sun.

alpha particle A helium nucleus emitted in the radioactive decay of heavy elements.

Andromeda galaxy (M31) The closest spiral galaxy to the Milky Way Galaxy at a distance of 680 kpc; it has a diameter of about 50 kpc.

Ångstrom A unit of length, equal to 10^{-10} m, often used to measure the wavelength of visible light.

angular diameter The apparent diameter of an object in angular measure; the angular separation of two points on opposite sides of the object.

angular distance The apparent angular spacing between two objects in the sky.

angular momentum The tendency for bodies, because of their inertia, to keep spinning or orbiting.

angular separation The observed angular distance between two celestial objects, measured in degrees, minutes, and seconds of angular measure.

angular speed The rate of change of angular position of a celestial object viewed in the sky.

anorthosite A basaltic mineral composed of calcium and sodium with aluminum silicate; the predominant mineral of the lunar highlands.

antapex The direction in the sky from which the Sun appears to be moving relative to local stars; located in the constellation Columba.

antiparticles Subatomic particles that annihilate with so-called "normal" particles when they collide; an example is the positron, which is the antiparticle of an electron.

aperture synthesis Interferometric techniques in which an array of radio antennae are used to attain high resolutions characteristic of their widest spacings.

apex The direction in the sky toward which the Sun appears to be moving relative to local stars; located in the constellation Hercules.

aphelion For a body orbiting the Sun, the point on its orbit that is farthest from the Sun.

Aphrodite Terra A large highland region on Venus.

apogee The point in its orbit where an earth satellite is farthest from the earth.

apparent magnitude The brightness of a star (or any other celestial object) on the magnitude scale as seen from the earth; an astronomical measure of the object's flux.

arroyo A channel carved in the ground by sporadic water flows.

asteroid (Minor planet) one of several thousand very small members of the solar system that revolve around the Sun, generally between the orbits of Mars and Jupiter.

asteroid belt The region lying between the orbits of Mars and Jupiter, containing the majority of asteroids.

astronomical unit (AU) The semimajor axis of the earth's orbit and average distance between the earth and the Sun; 149.6 million km or 8.3 light minutes.

astrophysical jets Collimated beams of material (usually ions and electrons) expelled from astrophysical objects, such as the nuclei of active galaxies.

atmosphere A gaseous envelope surrounding a planet, or the visible layers of a star; also a unit of pressure (abbreviated *atm*) equal to the pressure of air at sea level on the earth's surface.

atmospheric escape The process by which particles at the exosphere with greater than escape velocity and unhindered by collisions leave a planet.

atmospheric extinction The decrease in light caused by passage through the atmosphere.

atmospheric reddening Preferential scattering of blue light over red by air particles so that an object appears redder than it actually is.

atom The smallest particle of an element that exhibits the chemical properties of the element.

AU Abbreviation for astronomical unit.

aurora Visible light emission from atmospheric atoms and molecules excited by collisions with energetic, charged particles from the magnetosphere.

axis One of two or more reference lines in a coordinate system; also, the straight line, through the poles, about which a body rotates.

Balmer jump or edge The large change in opacity at 365 nm due to bound-free transitions from the second energy level of hydrogen atoms.

Balmer series The set of transitions of electrons in a hydrogen atom between the second energy level and higher levels; also the set of absorption or emission lines corresponding to these transitions that lies in the visible part of the spectrum, the first of which is the H-alpha line.

bandwidth The range of frequencies which a radio receiver (or another electromagnetic sensor) detects at the same time.

barred spirals A subclass of spiral galaxies that have a bar across the nuclear region.

baryons Half-integral spin particles that are composites of 3 quarks.

basalt An igneous rock, composed of olivine and feldspar, that makes up much of the earth's lower crust.

basins Large, shallow lowland areas in the crusts of terrestrial planets created by asteroidal impact or plate tectonics.

Becklin–Neugebauer (B–N) object An infrared source associated with the Orion Nebula; probably a very young, massive star.

belts Regions of downflow and so low pressure in the atmosphere of a Jovian planet.

beta decay A process of radioactive decay in which a neutron disintegrates into a proton, an electron, and a neutrino.

beta particle An electron or positron emitted by a nucleus in radioactive decay.

Beta Regio A highland region on Venus containing at least two large shield volcanoes.

Betelgeuse A red supergiant star with a visual luminosity roughly 2×10^4 times that of the Sun.

Big Bang model A picture of the evolution of the Universe that postulates its origin, in an event called the Big Bang, from a hot, dense state that rapidly expanded to cooler, less dense states.

binary-accretion model A model for the origin of the moon in which the moon and earth form by accretion of material from the same cloud of gas and dust.

binary galaxies Two galaxies bound by gravity and orbiting a common center of mass.

binary stars Two stars bound together by gravity that revolve around a common center of mass.

binary X-ray sources A binary system containing an X-ray emitter, which is usually a collapsed object surrounded by a hot accretion disk giving off X-rays; fall into two general classes: high mass and low mass, depending on the mass of their companion.

bipolar outflows High-speed outflows of gas in opposite directions from a young stellar object; probably the result of a magnetized accretion disk around the object.

blackbody A (hypothetical) perfect radiator of light that absorbs and re-emits all radiation incident upon it; its flux depends only on its temperature and is described by Planck's law.

blackbody spectrum The continuous spectrum emitted by a blackbody; the flux at each wavelength is given by Planck's law.

black dwarf The cold remains of a white dwarf after all its thermal energy is exhausted.

black hole A mass that has collapsed to such a degree that the escape velocity from its surface is greater than the speed of light, so that light is trapped by the intense gravitational field.

BL Lacertae (BL Lac) objects A type of active galaxy whose nonthermal emission varies rapidly (one day or so) and is highly polarized.

blue compact dwarf galaxy Small galaxies that are probably undergoing star formation; there may be many more dwarfs with primordial gas that has not had an episode of star formation.

blue shift A decrease in the wavelength of the radiation emitted by an approaching celestial body as a consequence of the Doppler effect; a shift toward the short-wavelength (blue) end of the spectrum.

Bohr model of the atom A simple picture of atomic structure in which electrons have well-defined orbits about the nucleus of the atom.

bolometer A detector of infrared radiation, usually a small chip of semiconductor material cooled to a few kelvins; absorption of infrared radiation causes a change in its resistance, which can be measured in an electronic circuit.

bolometric magnitude The magnitude of an object measured over all wavelengths of electromagnetic radiation.

bolometric luminosity The complete energy output per second at all wavelengths of an astronomical object.

Boltzmann's constant The number that relates pressure and temperature, or kinetic energy and temperature in a gas; the gas constant per molecule.

boson A particle with integral spin.

bound-bound transition A transition of an electron between two bound energy states of an atom or ion.

bound-free transition A transition of an electron between a bound and an unbound (free) state.

breccias Rock and mineral fragments cemented together; a common part of the lunar surface.

bright-line spectrum See *emission-line spectrum*.

bright nebula See *diffuse nebula*.

brightness An ambiguous term, usually meaning the energy per unit area received from or emitted by an object—its flux—but sometimes used to refer to an object's luminosity.

broad absorption line (BAL) quasars High redshift quasars in which very broad troughs of absorption are found.

broad line region (BLR) The gas immediately surrounding an AGN that contributes very broad emission line radiation to the galaxy's spectrum.

brown dwarf A very low mass object (roughly 0.01 to 0.08 solar mass) of low temperature and luminosity that never becomes hot enough in its core to ignite thermonuclear reactions.

C-type asteroids Dark asteroids with low albedos (around 0.04) that probably have a large percentage of carbon materials.

canali (Italian, "channels") term used by Giovanni Schiaparelli to describe dark linear features seen on the surface of Mars.

capture model A model of the moon's origin—proposed about 1955—that pictures the moon as captured by the earth's gravity, after which it spiraled in toward the earth, reversed orbital direction, and spiraled outward.

carbonaceous chondrites A class of meteorites that contain chondrules imbedded in a material with a large percentage of carbon (about 4%).

carbon-nitrogen-oxygen (CNO) cycle A series of thermonuclear reactions taking place in a star's core, in which carbon, nitrogen, and oxygen aid the fusion of hydrogen into helium; it is a secondary energy-production process in the Sun, but the major process in high-mass, main-sequence stars.

Cassegrain reflector A design of a reflecting telescope where the secondary mirror directs the beam to a focus through a hole in the center of the primary mirror.

Cassini's division A gap about 2000 km wide in Saturn's rings, discovered in 1675 by Giovanni Cassini; now known to contain many small ringlets.

catastrophic models Models for the origin of the Solar System in which an improbable event involving a large mass (usually collision with another star) led to the collection of gaseous materials that became the planets.

causally connected A description of events in space-time that can be affected by a given event; all those events within the light cone of a given event are causally connected to it.

cD galaxies See *supergiant elliptical galaxies*.

celestial equator An imaginary projection of the earth's equator onto the celestial sphere; declination is zero along the celestial equator.

celestial pole An imaginary projection of the earth's pole onto the celestial sphere; a point about which the apparent daily rotation of the stars takes place.

celestial sphere An imaginary sphere of very large radius centered on the earth on which the celestial bodies appear fastened and against which their motions are charted.

Centaurus X-3 A high-mass binary X-ray source in the constellation Centaurus; contains a low-mass neutron star (the X-ray source) orbiting a blue giant star.

center of mass The balance point of a set of interacting or connected bodies.

central force A force directed along a line connecting the centers of two objects.

centripetal acceleration The acceleration of a body toward the center of a circular path.

centripetal force A force required to divert a body from a straight path into a curved one, directed toward the center of the curve.

cepheid variables (cepheids) Stars that vary in brightness as a result of a regular variation in size and temperature; a class of variable stars for which the star Delta Cephei is the prototype.

Ceres The first observed asteroid, discovered by Father Giuseppe Piazzi in 1801.

Chandrasekhar limit The maximum amount of mass for a white dwarf star, about 1.4 solar masses; this amount leads to the highest density and smallest radius for a star made of a degenerate electron gas; more than this and the star collapses gravitationally.

charge-coupled device (CCD) A small chip of semiconductor material that emits electrons when it absorbs light; the electrons are trapped in small regions called pixels; the pattern of charges is read out in a way to preserve the image striking the chip.

Charon Moon of Pluto, with a diameter of about 1200 km.

chemical condensation sequence The sequence of chemical reactions and condensation of solids that occur in a low-density gas as it cools at specific pressures and temperatures.

chondrite A stony meteorite characterized by the presence of small, round silicate granules (chondrules).

chondrules Round silicate granules lacking volatile elements; found in chondritic meteorites, or chondrites, they are believed to be primitive solar-system materials.

chromosphere The part of the Sun's atmosphere just above the photosphere; hotter and less dense than the photosphere; it creates the flash spectrum seen during eclipses.

circular velocity The speed at which an object must travel to maintain uniform circular motion around a gravitating body.

circumpolar stars For an observer north of the equator, those stars that are continually above the northern horizon and never set; for a southern observer, those stars that never set below the southern horizon.

closed geometry See *spherical geometry*.

CNO cycle See *carbon-nitrogen-oxygen cycle*.

collisional deexcitation Loss of energy by an electron of an atom in a collision so that the electron drops to a lower energy level.

collisional excitation Forcing an electron of an atom to a higher energy level by a collision.

color excess The difference between the actual color (of a star) and its observed color; usually redder because of interstellar dust along the line of sight.

color index The difference in the magnitudes of an object measured at two different wavelengths; a measure of the color and hence the temperature of a star.

color temperature Temperature inferred from color, usually by fitting a Planck function to the continuous spectrum of a star at two wavelengths.

coma The bright, visible head of a comet.

compact radio galaxies Active galaxies that have a small, strong radio source in their nuclei.

comets Bodies of small mass that revolve around the Sun, usually in highly elliptical orbits, and consist, in the dirty-snowball model, of small, solid particles of rocky material imbedded in frozen gases.

compound A substance composed of the atoms of two or more elements bound together by chemical forces.

condensation The growth of small particles by the sticking together of atoms and molecules.

conduction Transfer of thermal energy by particles colliding into one another.

conjunction The time at which two celestial objects appear closest together in the sky.

conservation of angular momentum The principle stating that, with no torques, the total angular momentum of an isolated system is constant.

conservation of energy A fundamental principle in physics that states that the total energy of an isolated system remains constant regardless of whatever internal changes may occur.

conservation of magnetic flux The physical principle stating that, under certain circumstances, the number of magnetic field lines passing through an area remains constant.

conservation of momentum The physical principle stating that, with no outside net forces, the total momentum of an isolated system is constant.

constellation An apparent arrangement of stars on the celestial sphere, usually named after ancient gods, heroes, animals, or mythological beings; now an agreed-upon region of the sky containing a group of stars.

continental drift The model that the present continents were at one time a joined landmass that fragmented and drifted apart.

continuous spectrum A spectrum showing emission at all wavelengths, unbroken by either absorption lines or emission lines.

contour map A diagram showing how the intensity of some kind of radiation varies over a region of the sky; lines on such a map connect points of equal intensity; closely spaced lines mean that the intensity changes rapidly over a small distance, widely spaced lines mean it changes more slowly.

convection The transfer of energy by the moving currents of a fluid.

coplanar Lying in the same plane.

core (of the earth) The central region of the earth; it has a high density, is in part liquid, and is believed to be composed of iron and iron alloys.

core (of the Sun) The inner 25% of the Sun's radius, where the temperature is great enough for thermonuclear reactions to take place.

core (of the Galaxy)   The inner few parsecs of the nucleus, which contains a small, nonthermal radio source and fast-moving clouds of ionized gas.

core-mantle grains   Interstellar dust particles with cores of dense materials (such as silicates) surrounded by a mantle of icy materials (such as water).

corona   The outermost region of the Sun's atmosphere, consisting of thin, ionized gases at a temperature of about 10^6 K.

coronal holes   Regions in the Sun's corona that lack a concentration of high-temperature plasma; here magnetic field lines extend out into interplanetary space and mark the source of the solar wind.

coronal interstellar gas   High-temperature interstellar plasma made visible by its X-ray emission.

Cosmic Background Explorer (COBE)   A satellite launched in 1989 to measure the spectrum and intensity distribution of the microwave background radiation.

cosmic blackbody microwave radiation   Radiation with a blackbody spectrum at a temperature of about 2.7 K permeating the Universe; believed to be the remains of the primeval fireball in which the Universe was created.

cosmic nucleosynthesis   The production of the lightest few nuclei during the first 100 seconds of the Big Bang.

cosmic rays   Charged atomic particles moving in space with very high energies (the particles travel close to the speed of light); most originate beyond the Solar System, but some of low energy are produced in solar flares.

cosmological principle   The statement that the Universe, averaged over a large enough volume, appears the same from any location.

cosmology   The study of the nature and evolution of the physical universe.

cosmos   The Universe considered as an orderly and harmonious system.

Crab nebula   A supernova remnant, located in the constellation Taurus, produced by the supernova explosion visible from earth in 1054 A.D.; a pulsar in the nebula marks the neutron-star corpse of the exploded star.

crater   A circular depression of any size, usually caused by the impact of a solid body or by a surface eruption.

cratered terrain   Landscape with an abundance of craters, which implies that it is old and unevolved.

crescent   The phase of a moon or planet when it has less than half of its visible surface illuminated.

critical density   In cosmology, the density that marks the transition from an open to a closed universe; the density that provides enough gravity to just bring the expansion to a stop after infinite time.

crust   The thin, outermost surface layer of a planet; on the earth, it is composed of basaltic and granitic rocks.

cyclone   Spiral flows in a planet's atmosphere produced by the planet's rotation.

Cygnus arm   A segment of one of the spiral arms in the outer part of our Galaxy, about 14 kpc from the center.

Cygnus X-1   Binary X-ray source in the constellation Cygnus; it contains a probable black hole orbiting a blue supergiant star.

dark cloud   An interstellar cloud of gas and dust that contains enough dust to blot out the light of stars behind it (as seen from the earth).

dark-line spectrum   See *absorption-line spectrum*.

dark matter   The probable dominant form of matter in the Universe; it may be non-baryonic and does not form stars or galaxies—hence is "dark."

decoupling   The time in the Universe's history when the density became low enough so that matter and light stopped interacting.

deferent   An ancient geometric device used to account for the apparent eastward motion of the planets; a large circle, usually centered on the earth, that carries around a planet's epicycle.

degenerate electron gas   An ionized gas in which nuclei and electrons are packed together as much as possible, filling all possible low energy states, so that the perfect gas law relating pressure, temperature, and density no longer applies.

degenerate gas pressure   A force exerted by very dense, compacted matter that depends mostly on how dense the matter is and very little on its temperature.

degenerate neutron gas   Matter made up of neutrons packed together as tightly as possible.

Deimos   The smaller of the two moons of Mars; it has a dark, cratered surface.

density   The amount of mass per volume in an object or region of space.

density-wave model   A model for the generation of spiral structure in galaxies, which pictures density waves (similar to sound waves) plowing through the interstellar matter and sparking star formation.

differential rotation   The tendency of a fluid, spherical body to rotate faster at the equator than at the poles.

dipole field   A magnetic field configuration like that of a bar magnet, with opposed north and south poles.

diffuse (bright) nebula   A cloud of ionized gas, mostly hydrogen, with an emission-line spectrum.

dirty-snowball comet model   A model for comets that pictures the nucleus as a compact solid body of frozen materials, mixed with pieces of rocky matter, that turns into gases as a comet nears the Sun, creating the head and tail.

disk (of a galaxy)   The flattened wheel of stars, gas, and dust outside the nucleus of a galaxy.

dispersion The effect that causes pulses of radiation at different frequencies emitted simultaneously to arrive at different times after traversing the interstellar medium.

D-lines A pair of dark lines in the yellow region of the spectrum, produced by sodium.

double quasar Two images on the sky of a single quasar, produced by a gravitational lens.

Doppler shift A change in the wavelength of waves from a source reaching an observer when the source and the observer are moving with respect to each other along the line of sight; the wavelength increases (red shift) or decreases (blue shift) according to whether the motion is away from or toward the observer.

dust tail The part of a comet's tail containing dust particles, pushed out by radiation pressure from the Sun.

dwarf A star of relatively low light output and relatively small size; a main-sequence star of luminosity class V.

dynamo model A model for the generation of a planet's (or star's) magnetic field by the organized circulation of conducting fluids in its core.

eccentric An ancient geometric device used to account for nonuniform planetary motion; a point offset from the center of circular motion.

eccentricity The ratio of the distance of a focus from the center of an ellipse to its semimajor axis.

eclipse The phenomenon of one body passing in front of another, cutting off its light.

eclipsing binary system Two stars that revolve around a common center of mass, the orbits lying edge-on to the line of sight, so that each star periodically passes in front of the other.

ecliptic From the earth, the apparent yearly path on the celestial sphere of the Sun with respect to the stars; also, the plane of the earth's orbit.

Eddington luminosity For a given mass of an object, the luminosity that would just provide sufficient radiation pressure on surrounding material to prevent gravity from pulling it in.

Eddington mass For a given luminosity of an object, the mass that would produce enough gravity to just overcome radiation pressure on surrounding material and allow it to fall inward.

effective temperature The temperature a body would have if it were a blackbody of the same size radiating the same luminosity.

Einstein Observatory The HEAO-B satellite that was capable of imaging X-rays.

electromagnetic force One of the four forces of nature; particles with electromagnetic charge either attract or repel each other depending upon whether the two charges are opposite or identical.

electromagnetic radiation A self-propagating electric and magnetic wave, such as light, radio, ultraviolet, or infrared radiation; all types travel at the same speed and can be differentiated by wavelength or frequency.

electromagnetic spectrum The range of all wavelengths of electromagnetic radiation.

electron A lightweight, negatively charged subatomic particle.

element A substance that is made of atoms with the same chemical properties and cannot be decomposed chemically into simpler substances.

ellipse A plane curve drawn so that the sum of the distances from a point on the curve to two fixed points is constant.

elliptical galaxy A gravitationally bound system of stars that has rotational symmetry but no spiral structure and that contains mainly old stars and little gas or dust.

elongation The angular separation of an object from the Sun as seen in the sky.

emission (bright) lines Light of specific wavelengths or colors emitted by atoms; sharp energy peaks in a spectrum caused by downward electron transitions from a discrete quantum state to another discrete state.

emission-line spectrum A spectrum containing only emission lines.

emission nebula See *diffuse nebula*; a hot cloud of hydrogen gas whose visible spectrum is dominated by emission lines.

empirical Derived from experiment or observation.

energy The ability to do work.

energy level One of the possible quantum states of an atom, with a specific value of energy.

ephemeris (pl., ephemerides) A table that gives the positions of celestial objects at various times.

epicycle A small circle whose center lies on a larger one (the deferent) used by ancient astronomers, such as Ptolemy, to account for the westward retrograde motion and other irregular motions of the planets.

equant An ancient geometrical device invented by Ptolemy to account for variations in planetary motion; essentially an eccentric in which the center of the circle is not the center of uniform motion.

equation of state A relationship that describes the conditions in a gas, such as an equation relating the pressure, temperature, and density of a gas.

equatorial bulge The excess diameter, about 43 km, of the earth through its equator compared with the diameter through its poles.

equilibrium A state of a physical system in which there is no overall change.

equilibrium temperature The temperature achieved by a body when it emits the same energy per second that it absorbs.

equinox Time of year of equal length of day and night; the two times of the year when the Sun crosses the celestial equator; spring (vernal) equinox occurs about March 21, and fall (autumnal) equinox about September 23.

escape temperature The temperature that particles in an atmosphere need in order to escape into space with the escape velocity.

escape velocity The speed a body must achieve to break away from the gravity of another body and never return to it.

Euclidean (flat) geometry Geometry in which only one parallel line can be drawn through a point near another line; the sum of the angles in a triangle drawn on a flat surface is always 180°.

event A point in four-dimensional spacetime.

evolutionary track On a temperature-luminosity diagram, the path made by the points that describe how the temperature and luminosity of a star changes with time.

excitation The process of raising an atom to a higher energy level.

exosphere The topmost region of a planet's atmosphere, from which particles in the atmosphere can escape into space.

expanding (3 kpc) arm A segment of spiral-arm structure encircling the center of our Galaxy at a distance of about 3 kpc; it appears to be moving toward us and away from the Galaxy's center.

exponential decay A process, such as radioactive decay, for which the rate of change is directly proportional to the quantity present, so that the quantity remaining is given by an exponential function of time.

extended radio galaxies Active galaxies that show extended radio emission, usually in the form of two lobes on either side of the nucleus.

extinction The dimming of light when it passes through some medium, such as the earth's atmosphere or interstellar material.

extragalactic Outside of the Milky Way Galaxy.

eyepiece A magnifying lens used to view the image produced by the main light-gathering lens of a telescope.

Faber-Jackson relation The correlation between a galaxy's luminosity and the fourth power of its internal velocity dispersion.

fermion A particle with half-integral spin.

first dredge-up Convection acting to bring up material processed by hydrogen burning to a star's surface during its first episode as a red giant.

first quarter The phase of the moon one quarter of the way around its orbit from new moon, when it looks half illuminated when viewed from the earth.

fission See *nuclear fission*.

fission model The earliest of the major models for the origin of the moon, suggesting that a young, rapidly spinning, molten earth lost a piece that spiraled out into orbit and cooled down to form the moon.

flash spectrum The spectrum that appears immediately before the totality of a solar eclipse as the normal absorption spectrum is replaced briefly by the chromosphere's own emission spectrum.

flatness problem In the Big Bang model, the fact that the geometry of the Universe is flat (or very close to it) and remains flat as the Universe evolves.

flocculent galaxies Spiral galaxies that show a puffy structure in their disks rather than well-defined spiral arms.

fluorescence The process by which a high-energy photon is absorbed by an atom and re-emitted as two or more photons of lower energy.

flux The amount of energy flowing through a given area in a given time.

focal length The distance from a lens (or mirror) to the point where it brings light to a focus for a distant object.

focus (pl., foci) The point at which light is gathered in a telescope.

forbidden line An emission line from an atom produced by a transition with a low probability of occurrence.

forced motion Any motion under the action of a net force.

frame of reference A set of axes with respect to which the position or motion of something can be described or physical laws can be formulated.

Fraunhofer lines The name given to absorption lines in the spectrum of a star, especially the Sun.

free-bound transition A transition of an electron between a free energy state and one bound to an atom; results in the atom adding an electron with the emission of a photon; the reverse process is a bound-free transition.

free-fall Gravitational collapse under the condition of no resisting internal pressure caused by collisions among the particles.

free-fall collapse time The time for a cloud to collapse gravitationally under free-fall conditions (no internal pressure); a function only of the initial density of the cloud.

free-free transition A transition of an electron between two different free states; if energy is lost by the electron, the process is free-free emission; if gained, it is free-free absorption.

frequency The number of waves that pass a particular point in some time interval (usually a second); usually given in units of hertz, one cycle per second.

full moon The phase of the moon when it is opposite

the Sun in the sky, looking fully illuminated as seen from the earth.

fusion See *nuclear fusion*.

galactic cannibalism A model for galaxy interaction where more massive galaxies strip material, by tidal forces, from less massive galaxies.

galactic center (core) The innermost part of the Galaxy's nuclear bulge.

galactic (open) cluster A small group, about ten to a few hundred, of gravitationally bound stars of Population I, found in or near the plane of the Galaxy.

galactic equator The great circle along the line of the Milky Way, marking the central plane of the Galaxy.

galactic latitude The angular distance north or south of the galactic equator.

galactic longitude The angular distance along the galactic equator from a zero point in the direction of the galactic center.

galactic rotation curve A description of how fast an object some distance from the center of a galaxy revolves around it.

galaxy A huge assembly of stars (between 10^6 and 10^2), plus gas and dust, that is held together by gravity; the Galaxy, our own galaxy, containing the Sun.

Galilean moons The four largest satellites of Jupiter (Io, Europa, Ganymede, Callisto), discovered by Galileo with his telescope.

gamma ray A very high energy photon with a wavelength shorter than that of X-rays.

gas tail The part of a comet's tail that consists of ions and molecules; it is shaped by its interaction with the solar wind.

gauss A physical unit measuring magnetic field strength (not the SI unit, which is the tesla, but commonly used by astronomers).

general theory of relativity The idea developed by Albert Einstein that mass and energy determine the geometry of spacetime and that any curvature of this spacetime shows itself by what we commonly call gravitational forces; Einstein's theory of gravity.

geocentric Centered on the earth.

geomagnetic axis The axis that connects the earth's magnetic poles; it is inclined about 12° from the geographic spin axis and does not pass through the earth's center.

giant-impact model A scenario for the moon's origin in which a Mars-size object strikes the young earth with a glancing blow; material from the colliding object and the earth form a disk around the earth out of which the moon accretes.

giant molecular clouds Large interstellar clouds, with sizes up to tens of parsecs and containing 100,000 solar masses of material; found in the spiral arms of the Galaxy, giant molecular clouds are the sites of massive star formation.

gibbous The phase of the moon occurring between first quarter and full moon, when the moon is more than half illuminated as seen from the earth.

globular cluster A gravitationally bound group of about 10^5 to 10^6 Population II stars (of roughly solar mass), symmetrically shaped, found in the halo of the Galaxy.

gluon The boson that carries the strong nuclear force.

grand design spirals Spiral galaxies that have a well-defined spiral arm structure.

grand unification theories (GUTs) Physical theories that attempt to unite the elementary particles and the four forces in nature as the actions of one particle and one force.

granite The type of rock making up the continental regions of the earth's crust.

granule Brief-lived (3 to 10 min) bright spots that appear as a rough texture on the solar photosphere.

gravitation In Newtonian terms, a force between masses that is characterized by their acceleration toward each other; the size of the force depends directly on the product of the masses and inversely on the square of the distance between them; in Einstein's terms, the curvature of spacetime.

gravitational bending of light The effect of gravity on the usually straight path of a photon.

gravitational collapse The unhindered contraction of any mass from its own gravity.

gravitational field The property of space having the potential for producing gravitational force on objects within it; characterized by the acceleration of free masses.

gravitational focusing The directing of the paths of small masses by a larger one so that their paths cross, which enhances their accretion onto the larger mass.

gravitational force The weakest of the four forces of nature; all particles with nonzero mass attract each other.

gravitational instability The tendency for a disturbed region in a gas to undergo gravitational collapse.

gravitational lens The bending effect of a large mass on light rays so that they form an image of the source of light.

gravitational mass The mass of an object as determined by the gravitational force it exerts on another object.

gravitational potential energy Potential energy related to a body's position in a gravitational field.

gravitational redshift The change to longer wavelengths that marks the loss of energy by a photon that moves from a stronger to a weaker gravitational field.

great circle Shortest distance between two points on a sphere.

Great Red Spot A large, long-lived, high-pressure storm in Jupiter's atmosphere.

greenhouse effect The effect producing an increased

equilibrium temperature at the surface of a planet due to the opacity of its atmosphere in the infrared, trapping outgoing heat radiation.

ground state The lowest energy level of an atom.

GUTs See *grand unification theories*.

H I region A region of neutral hydrogen in interstellar space.

H II region A zone of ionized hydrogen in interstellar space; it usually forms a bright nebula around a hot, young star or cluster of hot stars.

Hadley cells Large regions of convective flow in the earth's atmosphere.

hadron Any particle composed of quarks; hadrons are divided into the baryon and meson groups.

half-life The time required for half of the radioactive atoms in a sample to disintegrate.

Halley's Comet The periodic comet (orbital period about 76 years) whose orbit was first worked out by Edmund Halley from Newton's laws; has a small nucleus (about 10 km diameter), dark, irregular in shape and emitting jets of gas and dust.

halo (of a galaxy) The spherical region around a galaxy, not including the disk or the nucleus, containing globular clusters, some gas, and probably a few stray stars.

H-alpha line The first line of the Balmer series, the set of transitions in a hydrogen atom between the second energy level and levels with higher energy; it lies in the red part of the visible spectrum.

head–tail radio galaxy An active galaxy whose radio lobes have been swept back to form a tail because of interaction with the surrounding medium.

heavy-particle era In the hot Big Bang model, the time up to 0.001 sec, when gamma rays collided to make high-mass particles, such as protons.

heliocentric Centered on the Sun.

heliocentric parallax An apparent shift in the positions of nearby stars (relative to more distant ones) from the changing position of the earth in its orbit around the Sun; the size of the shift can be used to measure the distances to close stars; see *trigonometric parallax*.

helium burning Fusion of helium into carbon by the triple-alpha process.

helium flash The rapid burst of energy generation with which a star initiates helium burning by the triple-alpha process in the degenerate core of a low-mass red giant star.

hertz A physical unit of frequency equal to 1 cycle/s.

Hertzsprung-Russell (H–R) diagram A graphic representation of the classification of stars according to their spectral class (or color or surface temperature) and luminosity (or absolute magnitude); the physical properties of a star are correlated with its position on the diagram, so a star's evolution can be described by

its change of position on the diagram with time (see *evolutionary track*).

Higgs boson A predicted but not yet experimentally verified particle; it may be responsible for giving other particles mass and may have triggered cosmic inflation.

high-velocity clouds Clouds of gas associated with the Galaxy, moving at speeds of hundreds of kilometers per second.

high-velocity stars Stars in the Galaxy with velocities greater than 60 km/s relative to the Sun; they have orbits with high eccentricities, often at large angles with respect to the galactic plane.

homogeneous Having a consistent and even distribution of matter, the same in all parts.

horizon The intersection with the sky of a plane tangent to the earth at the location of the observer.

horizon distance In cosmology, the maximum distance that light can travel in some epoch of the Universe.

horizon problem In the Big Bang model, the problem that arises from the rapid expansion of the early Universe so that different regions cannot communicate with each other.

horizontal branch A portion of the Hertzsprung-Russell diagram reached by Population II stars of low mass after the red giant stage and typically found in a globular cluster; it ranges from yellow to red stars all having the same luminosity (about 100 times that of the Sun).

H–R diagram A Hertzsprung-Russell diagram.

Hubble's law A description of the expansion of the Universe, such that the more distant a galaxy lies from us, the faster it is moving away; the relation, $v = Hd$, between the expansion velocity (v) and distance (d) of a galaxy, where H is the Hubble constant.

Hubble constant The proportionality constant relating velocity and distance in the Hubble law; the value, now around 75 km/s/Mpc, changes with time as the Universe expands.

Hubble time Numerically the inverse of the Hubble constant; it represents, in order of magnitude, the age of the Universe.

hydrogen burning Any fusion reaction that converts hydrogen (protons) to heavier elements.

hydrogen corona A shell of hydrogen gas around the earth, coming from the atmospheric escape of hydrogen produced by the dissociation of water.

hydrostatic equilibrium An equilibrium characterized by the absence of mass motions, when pressure balances gravity.

hyperbola A curve produced by the intersection of a plane with a cone; the shape of the orbit of a body with more than escape velocity.

hyperbolic geometry An alternative to Euclidean geometry, constructed by N. I. Lobachevski on the premise that more than one parallel line can be drawn through a point near a straight line; the sum of the angles of a triangle drawn on a hyperbolic surface is always less than 180°.

igneous rock Rock formed by the cooling of molten lava.

image Light rays gathered at the focus of a lens or mirror in the same relative alignment as the real object.

image processing The computer manipulation of digitized images to enhance specific aspects of them.

impact model The idea that craters are formed by the impact of solid objects onto a surface.

inertia The resistance of an object to a force acting on it because of its mass.

inertial frame Any system of reference in special relativity that is not accelerated.

inertial mass Mass determined by subjecting an object to a known force (not gravity) and measuring the acceleration that results.

inferior conjunction For a planet orbiting interior to another, the alignment with the Sun when the interior planet lies on same side of the Sun as the outer planet.

inflation theory A modification of the Big Bang model in which the Universe undergoes a brief interval of rapid expansion.

Infrared Astronomy Satellite (IRAS) A satellite that surveyed the sky at wavelengths of 12, 25, 60, and 100 μm.

infrared telescope A telescope, optimized for use in the infrared part of the spectrum, fitted with an infrared detector.

infrared cirrus Patches of interstellar dust, emitting infrared radiation, which look like cirrus clouds on the images of the sky produced by the Infrared Astronomy Satellite.

intergalactic medium The gas and dust found between the galaxies.

instability strip The region on the H–R diagram where cepheid variable stars are found; these stars are burning helium in their cores.

interferometer See *radio interferometer*.

interstellar dust Small (micrometers in diameter), solid particles in the interstellar medium.

interstellar extinction curve The amount of extinction from interstellar dust as a function of wavelength.

interstellar gas Atoms, molecules, and ions in the interstellar medium.

interstellar medium All the gas and dust found between stars.

inverse beta decay The process in which electrons and protons are forced together to form neutrons and neutrinos; the reverse process of neutron decay.

inverse-square law for light The decrease of the flux of light with the inverse square of the distance from the source.

ion An atom that has become electrically charged by the gain or loss of one or more electrons.

ionization The process by which an atom loses or gains electrons.

ionized gas A gas that has been ionized so that it contains free electrons and charged ions; a plasma if it is electrically neutral overall.

ionosphere A layer of the earth's atmosphere ranging from about 100 to 700 kilometers above the surface where oxygen and nitrogen are ionized by sunlight, producing free electrons.

iron meteorites One of the three main types of meteorites, typically made of about 90% iron and 9% nickel, with a trace of other elements.

irregular galaxy A galaxy without spiral structure or rotational symmetry, containing mostly Population I stars and abundant gas and dust.

Ishtar Terra A large upland plateau in the northern hemisphere of Venus, bounded on the east by Maxwell Montes, a shield volcano.

isophote A contour of constant intensity in a digital image.

isothermal perturbation A fluctuation of the matter component of the early Universe with no accompanying fluctuation of the radiation field.

isotope Atoms with the same number of protons but different numbers of neutrons.

isotropic Having no preferred direction in space.

Jeans length The minimum size a disturbance in a gas must have to result in gravitational contraction; it depends on the pressure, temperature, and density of the medium.

Jeans mass The mass contained in the Jeans length; the minimum mass a disturbance must have in a gas in order to grow larger.

jet streams Latitudinal, coherent flows at high speed in the upper atmosphere of the earth or another planet.

joule A physical unit of work and energy.

Jovian planets Planets with physical characteristics similar to Jupiter: large mass and radius, low density, mostly liquid interior.

K correction A correction to the apparent magnitude of a galaxy necessitated by the Galaxy redshift.

Kepler's laws Kepler's three laws of planetary motion that describe the properties of elliptical orbits with an inverse-square force law.

Keplerian motion Orbital motion that follows Kepler's laws.

kiloparsec One thousand parsecs.

kinematic distance In our Galaxy, the distance to an object inferred from its Doppler shift, direction, and the Galaxy's rotation curve.

kinetic energy   The ability to do work because of motion.

Kirchhoff's rules   Empirical descriptions of the physical conditions under which the main types of spectra originate.

Kleinmann-Low nebula   A diffuse dust cloud with a temperature of about 70 K near the Orion nebula emitting strongly at far infrared wavelengths.

KREEP   A lunar material composed of potassium (K), rare-earth elements (REE), and phosphorus (P).

Large Magellanic Cloud (LMC)   A small galaxy, irregular in shape, about 50 kpc from the Milky Way.

last quarter   The phase of the moon when it is three-quarters of the way around its orbit from new moon, when it looks half-illuminated when viewed from the earth.

lens   A curved piece of glass designed to bring light rays to a focus.

lenticular galaxies   Galaxies of Hubble type S0, with a disk like a spiral galaxy but with no spiral arms, and no gas or dust.

lepton   An elementary particle that participates in the weak nuclear force (the electron, for example).

light curve   A graph of a star's changing brightness with time.

light gathering power (LGP)   The ability of a telescope to collect light as measured by the area of its objective.

lighthouse model   For a pulsar, a rapidly rotating neutron star with a strong magnetic field; the rotation provides the pulse period and the magnetic field generates the electromagnetic radiation.

light-particle era   In the hot Big Bang model, the interval from 0.0001 to 4 s when gamma rays can collide to make low-mass particles, such as electrons.

light rays   Imaginary lines in the direction of propagation of a light wave.

lightyear   The distance light travels in a year, about 3.09×10^{13} kilometers.

limb darkening   The apparent darkening of the Sun along its edge caused by an optical depth effect.

line profile   The variation of a spectral line's intensity as a function of wavelength.

Local Group   A gravitationally bound group of about 20 galaxies to which our Milky Way Galaxy belongs.

Local Supercluster   The supercluster of galaxies in which the Local Group is located; spread over 10^7 pc, it contains the Virgo and Coma clusters.

longitudinal wave   A sound wave that moves in a push-pull motion through solids, liquids, and gases with a velocity that depends on the density of the medium.

long-period comets   Comets with orbital periods greater than 200 years.

low-ionization nuclear emission region (LINER)   A type of emission line galaxy that may be the lowest level of activity of AGN galaxies.

low-velocity stars   Stars with close to circular orbits in the plane of the Galaxy; they travel at less than 60 km/s with respect to the Sun.

luminosity   The total rate at which radiative energy is given off by a celestial body, over all wavelengths; the Sun's luminosity is about 4×10^{26} watts.

luminosity class   The categorization of stars that have the same surface temperatures but different sizes, resulting in different luminosities; based on the widths of dark lines in a star's spectrum, giant stars having narrower lines than dwarf stars.

luminosity function   (of stars) The number of stars in a region of the Galaxy as a function of their luminosity.

luminosity function   (of galaxies) The number of galaxies per unit volume as a function of their luminosity.

lunar eclipse   The cutoff of sunlight from the moon, when the moon lies on the line between the earth and Sun so that it passes through the earth's shadow; a lunar eclipse can occur only at full moon, only at those times of the year when the full moon lies very close to the ecliptic plane.

lunar occultations   The passage of the moon in front of a star or planet.

lunar soil   The fine particles created by the bombardment of the lunar surface by meteorites that, with larger rock fragments, compose the lunar soil.

Lyman series   All transitions in a hydrogen atom to and from the lowest energy level; they involve large energy changes, corresponding to wavelengths in the ultraviolet part of the spectrum; also, the set of absorption or emission lines corresponding to these transitions.

M-type asteroids   Asteroids with albedos of about 10% and reflective properties that resemble metals.

Mach number   The ratio of an object's speed in a medium to the speed of sound in that medium.

Magellanic Clouds   Two neighboring galaxies, the Large Magellanic Cloud (LMC) and the Small Magellanic Cloud (SMC), visible in the Southern Hemisphere to the unaided eye; companions to our Galaxy.

magnetic field   The property of space having the potential of exerting magnetic forces on bodies within it.

magnetic field lines   A graphic representation of a magnetic field showing its direction and, by the degree of packing of the lines, its intensity.

magnetic flux   The number of magnetic field lines passing through an area.

magnetic monopoles   A particle predicted by GUTs that would have only one type of magnetic pole.

magnetic reconnection   The sudden connection of magnetic field lines of opposite polarity.

magnetometer A device to measure the strength of a magnetic field.

magnetosphere The region around a planet where particles from the solar wind are trapped by the planet's magnetic field.

magnifying power The ability of a telescope to increase the apparent angular size of a celestial object.

magnitude An astronomical measurement of an object's brightness; larger magnitudes represent fainter objects.

main sequence The principal series of stars in the Hertzsprung-Russell diagram; such stars are converting hydrogen to helium in their cores by the proton–proton process or by the carbon-nitrogen-oxygen cycle; this is the longest stage of a star's active life.

major axis The larger of the two axes of an ellipse.

mantle The major portion of the earth's interior below the crust, made of a plastic rock probably composed of olivine.

mare (pl., maria) (Latin for "sea") A lowland area on the moon that appears darker and smoother than the highland regions, probably formed by lava that solidified into basaltic rock about 3 to 3.5×10^9 years ago.

mare basalts Basaltic rocks found on the surface of the lunar maria; tend to be the youngest (most recently formed) rocks on the lunar surface, with ages around 3.2×10^9 years.

mascons Abnormal concentrations of mass beneath the lunar maria; they have been detected by their effect on the orbits of moon-orbiting satellites.

mass A measure of an object's resistance to change in its motion (inertial mass); a measure of the strength of gravitational force an object can produce (gravitational mass).

mass loss The rate at which a star loses mass, usually by a stellar wind, per year.

mass-luminosity ratio For galaxies, the ratio of the total mass to the luminosity; a rough measure of the kind of stars in a galaxy.

mass-luminosity relation An empirical relation, for main-sequence stars, between a star's mass and its luminosity, roughly proportional to the third power of the mass.

matter era In the Big Bang model, the time interval from about one million years after the Big Bang to now, when matter dominates the Universe.

maximum elongation The greatest angular distance of an object from the Sun.

mean lifetime The time it takes for the number of atoms of a radioactive substance to decrease by one factor of e.

mechanics A branch of physics that deals with forces and their effects on bodies.

megaparsec 1 million parsecs.

megaton An explosive force equal to that of 1 million tons of TNT (about 4×10^{15} J).

meridian (celestial) An imaginary line drawn through the north and south points on the horizon and through the zenith.

meson Particles that are composites of quarks having integral spin.

mesosphere Region of the earth's atmosphere between 50 and 100 km where the temperature falls rapidly.

Messier 17 (M17) A nearby H II region that is the site of recent massive star formation at one end of a giant molecular cloud.

metallic hydrogen A state of hydrogen, reached at high pressures, where it is able to conduct electricity.

metal-poor stars Stars with metal abundances much less than the Sun's.

metal-rich stars Stars with metal abundances like that of the Sun (about 1 to 2% of the total mass).

meteor The bright streak of light that occurs when a solid particle (a meteoroid) from space enters the earth's atmosphere and is heated by friction with atmospheric particles; sometimes called a falling star.

meteorite A solid body from space that survives a passage through the earth's atmosphere and falls to the ground.

meteorite fall A meteorite seen in the sky and recovered on the ground.

meteorite find A recovered meteorite that was not seen to fall.

meteoroid A very small solid body moving through space in orbit around the Sun.

meteor shower A rapid influx of meteors that appear to come out of a small region of the sky, called the *radiant*.

meteor stream A uniform distribution of meteoroids along an orbit around the Sun.

meteor swarm Meteoroids grouped in a localized region of an orbit around the Sun; the source of meteor showers.

meteor trail The visible path of a meteor through the atmosphere created by ionization of the air and vaporization of the meteoroid.

micrometeorite Very small meteorites (about one-tenth a micrometer in diameter) that cool off and solidify before they hit the ground.

microwave background radiation A universal bath of low-energy photons having a blackbody spectrum with temperature of about 2.7 K.

midoceanic ridge An almost continuous submarine mountain chain that extends some 64,000 km through the earth's ocean basins.

Milky Way The band of light that encircles the sky, caused by the the blending of light from the many stars lying near the plane of the Galaxy; also sometimes used to refer to the Galaxy to which the Sun belongs.

millisecond pulsar Generic name given to any pulsar with a pulse period of a few milliseconds.

minimum resolvable angle The smallest angle a telescope can clearly show.

minute of arc 1/60 of a degree.

molecular cloud Large, dense, massive clouds in the plane of a spiral galaxy; they contain dust and a large fraction of gas in molecular form.

molecular maser Microwave amplification by stimulated emission of radiation from a molecule.

molecule A combination of two or more atoms bound together electrically; the smallest part of a compound that has the properties of that substance.

momentum The product of an object's mass and velocity.

moving-cluster method A method for finding a distance to a cluster of stars by determining their radial velocities and the convergent point of their proper motions.

muon One of three generations of leptons (along with the electron and tau generations).

nanometer 10^{-9} of a meter; common unit of wavelength measurement for light.

narrow-line region (NLR) The gas surrounding an AGN which contributes relatively narrow forbidden emission lines.

narrow-tailed radio galaxies Radio galaxies that show a U-shaped tail behind the nucleus; they are fast-moving galaxies in a cluster of galaxies.

natural motion Motion without forces.

nebula (Latin for "cloud") A cloud of interstellar gas and dust.

nebular model A model for the origin of the Solar System, in which an interstellar cloud of gas and dust collapsed gravitationally to form a flattened disk out of which the planets formed by accretion.

neutrino An elementary particle (lepton) with no (or very little) mass and no electric charge that travels at the speed of light and carries energy away during certain types of nuclear reactions.

neutron A subatomic particle with about the mass of a proton and no electric charge; one of the main constituents of an atomic nucleus; the union of a proton and an electron.

neutron star A star of extremely high density and small size that is composed mainly of very tightly packed neutrons; cannot have a mass greater than about 3 solar masses.

new moon The phase of the moon when it is in the same direction as the Sun in the sky, appearing almost completely unilluminated as seen from the earth.

newton The SI unit of force.

Newtonian reflector A reflecting telescope designed so that a small mirror at a 45° angle in the center of the tube brings the focus outside the tube.

nonthermal radiation Emitted energy that is not characterized by a blackbody spectrum; usually used to refer to synchrotron radiation.

noon Midday; the time halfway between sunrise and sunset when the Sun reaches its highest point in the sky with respect to the horizon.

Norma arm A segment of a spiral arm of our Galaxy, about 4000 pc from the Sun toward the center of the Galaxy in the direction of the constellation Norma.

north magnetic pole One of the two points on a star or planet from which magnetic lines of force emanate and to which the north pole of a compass points.

nova (Latin, "new") A star that has a sudden outburst of energy, temporarily increasing its brightness by hundreds to thousands of times; now believed to be the outburst of a degenerate star in a binary system; also used in the past to refer to some stellar outbursts that modern astronomers now call supernovas.

nuclear bulge The central region of a spiral galaxy, containing old Population I stars.

nuclear fission A process that releases energy from matter; in it, a heavy nucleus hit by a high-energy particle splits into two or more lighter nuclei whose combined mass is less than the original, the missing mass being converted into energy.

nuclear fusion A process that releases energy from matter by the joining of nuclei of lighter elements to make heavier ones; the combined mass is less than that of the constituents, the difference appearing as energy.

nucleosynthesis The chain of thermonuclear fusion processes by which hydrogen is converted to helium, helium to carbon, and so on through all the elements of the periodic table.

nucleus (of an atom) The massive central part of an atom, containing neutrons and protons, about which the electrons orbit.

nucleus (of a comet) Small, bright, starlike point in the head of a comet; a solid, compact (diameter a few tens of kilometers) mass of frozen gases with some rocky material as dust embedded in it.

nucleus (of a galaxy) The central portion of a galaxy, composed of old Population I stars, some gas and dust, and, for many galaxies, a concentrated source of nonthermal radiation.

number density The number of particles per unit volume.

OB association Loose groupings of O and B stars in small subgroups; they are not bound by gravity and so dissipate in a few times 10^7 years.

OB subgroup A small collection of about ten O and B

stars, a few tens of parsecs across, within an OB association.

objective   The main light-gathering lens or mirror of a telescope.

observable universe   The parts of the Universe that can be detected by the light they emit.

occultation   The eclipse of a star or planet by the moon or another planet.

Olbers' paradox   The statement that if there were an infinite number of stars distributed uniformly in an infinite space, the night sky would be as bright as the surface of a star, in obvious contrast to what is observed.

Olympus Mons   A large shield volcano on the surface of Mars in the Tharsis ridge region.

Oort's cloud   A cloud of comet nuclei in orbit around the Solar System, formed at the time the Solar System formed; the reservoir for new comets.

Oort's constant   In Oort's equations describing differential galactic rotation, the number relating the Sun's distance from the center of the Galaxy and the change in angular speed as a function of radius; current value is 15 km/s/kpc.

opacity   The property of a substance that hinders (by absorption or scattering) light passing through it; opposite of transparency.

open cluster   Same as galactic cluster.

open geometry   See *hyperbolic geometry*.

opposition   The time at which a celestial body lies opposite the Sun in the sky as seen from the earth; the time at which it has an elongation of 180°.

optics   The manipulation of light by reflection or refraction.

orbital angular momentum   The angular momentum of a revolving body; the product of a body's mass, orbital velocity, and the distance from the system's center of mass.

orbital inclination   The angle between the orbital plane of a body and some reference plane; in the case of a planet in the Solar System, the reference plane is that of the earth's orbit; in the case of a satellite, the reference is usually the equatorial plane of the planet; for a double star, it is the plane perpendicular to the line of sight.

Orion arm   A segment of the Galaxy's larger spiral arm structure; the Solar System lies within it.

Orion Nebula   A hot cloud of ionized gas that is a nearby region of recent star formation, located in the sword of the constellation of Orion; also called Messier 42 (M42).

outgassing   Release of gasses from nongaseous materials; extrusion of gasses from the body of a planet after its formation.

ozone layer (ozonosphere)   A layer of the earth's atmosphere about 40 to 60 kilometers above the surface, characterized by a high content of ozone, O_3.

parallax   The change in an object's apparent position when viewed from two different locations; specifically, half the angular shift of a star's apparent position as seen from opposite ends of the earth's orbit.

parabola   A geometric figure that describes the shape of an escape-velocity orbit.

parent meteorite bodies   Small solid bodies, a few hundreds or thousands of kilometers in size, believed to be the source of nickel-iron meteorites; formed early in the history of the Solar System and then broken up through collisions.

parsec (pc)   The distance an object would have to be from the earth so that its heliocentric parallax would be 1 second of arc; equal to 3.26 lightyears; a kiloparsec is 1000 parsecs, and a megaparsec is 10^6 parsecs.

pascal   Unit of pressure in the SI system.

Pauli exclusion principle   The statement from quantum theory that no two electrons can be in the same quantum state at the same time.

P-Cygni profile   A characteristic spectral line profile of an expanding shell; the emission lines have an absorption component on the blue side of the profile.

perfect cosmological principle   The statement that the Universe appears the same to an observer at all locations and at all times.

perigee   The point in its orbit at which an earth satellite is closest to the earth.

perihelion   The point at which a body orbiting the Sun is nearest to it.

period   The time interval for some regular event to take place; for example, the time required for one complete revolution of a body around another.

periodic comets   Comets that have relatively small elliptical orbits around the Sun, with periods of less than 200 years.

periodic (regular) variables   Stars whose light varies with time in a regular fashion from various causes.

period-luminosity relationship   For cepheid variables, a relation between the average luminosity and the time period over which the luminosity varies; the greater the luminosity, the longer the period.

Perseus arm   A segment of a spiral arm that lies about 3 kpc from the Sun in the direction of the constellation Perseus.

phases of the moon   The monthly cycle of the changes in the moon's appearance as seen from the earth; at new, the moon is in line with the Sun and so not visible; at full it is in opposition to the Sun and we see a completely illuminated surface.

Phobos   The larger of the two moons of Mars.

photodissociation   The breakup of a molecule by the absorption of light with enough energy to break the molecular bonds.

photometer A light-sensitive detector placed at the focus of a telescope; it is used to make accurate measurements of small photon fluxes.

photometry Measurement of the intensity of light.

photon A discrete amount of light energy; the energy of a photon is related to the frequency f of the light by the relation $E = hf$, where h is Planck's constant.

photon excitation Raising an electron of an atom to a higher energy level by the absorption of a photon.

photosphere The visible surface of the Sun; the region of the solar atmosphere from which visible light escapes into space.

physical universe The parts of the Universe that can be seen directly plus those that can be inferred from the laws of physics.

pitch angle The angle between a spiral arm's direction and the direction of circular motion about the Galaxy.

pixel The smallest picture element in a two-dimensional detector.

Planck curve The continuous spectrum of a blackbody radiator.

Planck's constant The number that relates the energy and frequency of light; it has a value of 6.63×10^{-34} $J \cdot s$.

planet From the Greek word for "wanderer"; any of the nine (so far known) large bodies that revolve around the Sun; traditionally, any heavenly object that moved with respect to the stars (in this sense, the Sun and the moon were also considered planets).

planetary nebula A thick shell of gas ejected from and moving out from an extremely hot star; thought to be the outer layers of a red giant star thrown out into space, the core of which eventually becomes a white dwarf.

planetesimals Asteroid-sized bodies that, in the formation of the Solar System, combined with each other to form the protoplanets.

plasma A gas consisting of equal numbers of ionized atoms and electrons.

plate tectonics A model for the evolution of the earth's surface that pictures the interaction of crustal plates driven by convection currents in the mantle.

Polaris The present north pole star; the outermost star in the handle of the Little Dipper.

polarization A lining up of the planes of vibration of light waves.

polarized light Light waves whose planes of oscillation are all the same.

polodial magnetic field Magnetic field configuration with two poles and field lines running along meridians.

Population I stars Stars found in the disk of a spiral galaxy, especially in the spiral arms, including the most luminous, hot, and young stars, with a heavy element abundance similar to that of the Sun (about 2% of the total); an old Population I is found in the nucleus of spiral galaxies and in elliptical galaxies.

Population II stars Stars found in globular clusters and the halo of a galaxy; may be older than any Population I stars, and contain a smaller abundance of heavy elements.

positron An antimatter electron; essentially an electron with a positive charge.

potential energy The ability to do work because of position; it is storable and can later be converted into other forms of energy.

PP chain See *proton–proton chain.*

precession of the equinoxes The slow westward motion of the equinox points on the sky relative to the stars of the zodiac because of the wobbling of the earth's spin axis.

precession of Mercury's orbit The turning, with respect to the stars, of the major axis of Mercury's orbit at a rate of 43 arcsec per century.

pre-main-sequence (PMS) star The evolutionary phase of a star just before it reaches the main sequence and starts hydrogen core burning.

pressure Force per unit area.

pressure gradient The rate of change of pressure along a direction.

primary The brighter of the two stars in a binary system.

primeval fireball The hot, dense beginning of the Universe in the Big Bang model, when most of the energy was in the form of high-energy light.

principle of equivalence The fundamental idea in Einstein's general theory of relativity; the statement that one cannot distinguish between gravitational accelerations and other kinds of acceleration, or, equivalently, a statement about the equality of inertial mass and gravitational mass; a consequence is that gravitational forces can be made to vanish in a small region of spacetime by choosing an appropriate accelerated frame of reference.

prominences Cool clouds of hydrogen gas above the Sun's photosphere in the corona; they are shaped by the local magnetic fields of active regions.

proper motion The angular displacement of a star on the sky from its motion through space.

protogalaxies Clouds with enough mass that they are destined to collapse gravitationally into galaxies.

proton A massive, positively charged elementary particle; one of the main constituents of the nucleus of an atom.

proton–proton (PP) chain A series of thermonuclear reactions that occur in the interiors of stars, by which four hydrogen nuclei are fused into helium; this process is believed to be the primary mode of energy production in the Sun.

protoplanet A large mass formed by the accretion of planetesimals; the final stage of formation of the planets from the solar nebula.

protostar A collapsing mass of gas and dust out of which a star will be born (when thermonuclear reactions turn on) whose energy comes from gravitational contraction.

pulsar A radio source that emits signals in very short, regular bursts; thought to be a highly magnetic, rotating neutron star.

quantum (pl., quanta) A discrete packet of energy.

quantum number In quantum theory, one of the four special numbers that determine the energy structure and quantum state of atoms.

quantum state The quantum description of the arrangement of electrons in an atom; allowed quantum states are filled starting with those of lowest energy first.

quark An elementary particle with third-integral charge that makes up others, such as protons and neutrons.

quasar or quasi-stellar object (QSO) An intense, point-like source of light and radio waves that is characterized by large red shifts of the emission lines in its visible spectrum.

radar mapping The surveying of the geographic features of a planet's surface by the reflection of radio waves from the surface.

radial velocity The component of relative velocity that lies along the line of sight.

radian (rad) A unit of angular measurement; 1 rad equals 57.3°; 2π radians equals 360°.

radiant The point in the sky from which a meteor shower appears to come.

radiation belts In a planet's magnetosphere, regions with a high density of trapped solar wind particles.

radiation Usually refers to electromagnetic waves, such as light, radio, infrared, X-rays, ultraviolet; also sometimes used to refer to atomic particles of high energy, such as electrons (beta-radiation), helium nuclei (alpha-radiation), and so on.

radiation era In the Big Bang model, the time in the Universe's history in which the energy in the Universe was dominated by radiation.

radiative energy The capacity to do work that is carried by electromagnetic waves.

radioactive decay The process by which an element fissions into lighter elements.

radio galaxies Galaxies that emit large amounts of radio energy by the synchrotron process, generally characterized by two giant lobes of emission situated on opposite ends of a line drawn through the nucleus; they are divided into two types, compact and extended.

radio interferometer A radio telescope that achieves high angular resolution by combining signals from at least two widely separated antennas.

radiometric dating A process that determines the age of an object by the rate of decay of radioactive elements within the object.

radio recombination line emission Sharp energy peaks at radio wavelengths caused by low-energy transitions in atoms from one very high energy level to another nearby level following recombination of an electron with an ion.

radio telescope A telescope designed to collect and detect radio emissions from celestial objects.

rapid process (r-process) Formation of very heavy elements by the rapid addition of neutrons to a nucleus followed by beta decay.

ray On the moon or other satellite, a bright streak of material ejected from an impact crater.

recombination The joining of an electron to an ion; the reverse of ionization.

recombination line Emission line from an electron following the process of recombination.

reddening The preferential scattering or absorption of blue light by small particles, allowing more red light to pass directly through.

red giant A large, cool star with a high luminosity and a low surface temperature (about 2500 K), which is largely convective and has fusion reactions going on in shells.

red shift An increase in the wavelength of the radiation received from a receding celestial body as a consequence of the Doppler effect; a shift toward the long-wavelength (red) end of the spectrum.

red variables A class of cool stars variable in light output.

reference frame A set of coordinates by which position and motion may be specified.

reflecting telescope A telescope that has a uniformly curved mirror as a primary light gatherer.

reflection The return of a light wave at the interface between two media.

reflection nebula A bright cloud of gas and dust that is visible because of the reflection of starlight by the dust.

refracting telescope A telescope that uses glass lenses to gather light.

refraction Bending of the direction of a light wave at the interface between two media, such as air and glass.

regular cluster (of galaxies) A cluster of galaxies with definite symmetry and a well-defined core.

relativistic Doppler shift Wavelength shift from the radial velocity of a source as calculated in special relativity, so that very large red shifts do not imply that the source moves faster than light.

relativistic jet   A beam of particles moving at speeds close to that of light.

relativity   Two theories proposed by A. Einstein; the special theory describes the motion of nonaccelerated objects, and general relativity is a theory of gravitation.

resolving power   The ability of a telescope to separate close stars or to pick out fine details of celestial objects.

retrograde motion   The apparent anomalous westward motion of a planet with respect to the stars, which occurs near the time of opposition (for an outer planet) or inferior conjunction (for an inner planet).

retrograde rotation   Rotation from east to west.

revolution   The motion of a body in orbit around another body or a common center of mass.

rift valley   A depression in the surface of a planet created by the separation of crustal masses.

Roche lobe   In a binary star system, the region in the space around them where their gravitational fields provide a path from one star to another.

rotation   The turning of a body, such as a planet, on its axis.

rotation curve   The relation between rotational velocity of objects in a galaxy and their distance from its center.

RR Lyrae stars   A class of giant, pulsating variable stars with periods of less than one day; they are Population II objects and commonly found in globular clusters.

runaway accretion   The process by which a planetesimal that starts out with an escape velocity greater than its neighbors grows very rapidly.

Rydberg constant   A number relating to the spacing of the energy levels in a hydrogen atom.

S0 galaxy   A type of galaxy intermediate between ellipticals and spirals; they have a disk but no spiral arms.

S-type asteroids   Asteroids whose albedos indicate a surface made of silicates.

Sagittarius A (Sgr A)   Radio sources at the center of the Galaxy; Sgr A West is a thermal radio source (H II region), Sgr A East a nonthermal source, and Sgr A* a pointlike source that may mark the Galaxy's core.

Sagittarius arm   A portion of spiral-arm structure of the Galaxy that lies about 2000 pc from the center of the Galaxy in the direction of the constellation Sagittarius.

scarp   A long, vertical wall running across a flat plain.

scattering (of light)   The change in the paths of photons without absorption or change in wavelength.

Schwarzschild radius   The critical size that a mass must reach to be dense enough to trap light by its gravity, that is, to become a black hole.

scientific model   A mental image of how the natural world works, based on physical, mathematical, and aesthetic ideas.

secondary   The fainter of the two stars in a binary system.

second dredge-up   The process by which convection brings the products of helium burning to the surface of a massive star during the second time it becomes a red giant.

second of arc   1/3600 of a degree, or 1/60 of a minute of arc.

secular parallax   A method of determining the average distance of a group of stars by examining the components of their proper motions produced by the straight-line motion of the Sun through space.

seeing   The unsteadiness of the earth's atmosphere that blurs telescopic images.

seismic waves   Sound waves traveling through and across the earth that are produced by earthquakes.

seismometer   An instrument used to detect earthquakes and moonquakes.

semimajor axis   Half of the major axis of an ellipse; distance from the center of an ellipse to its farthest point.

Seyfert galaxies   A type of AGN galaxy; the nuclear spectrum shows intense emission lines with either narrow (type 2) or both broad and narrow (type 1) components; the host galaxy is usually a spiral.

shepherd satellites   Small moons that confine a planet's ring in a narrow band; one is located on the inside edge of the ring, one on the outside.

sexagesimal system   A counting system based on the number 60, such as 60 minutes in an hour, or 60 minutes of arc in one degree.

shield volcano   A large volcano with gentle slopes formed by the slow outflow of magma.

shock wave   A discontinuity in a medium created when an object travels through it at a speed greater than the local sound speed.

short period comets   Comets with orbital periods less than 200 years, probably captured from longer period orbits by an encounter with a major planet.

sidereal month   The period of the moon's revolution around the earth with respect to a fixed direction in space or a fixed star; about 27.3 days.

sidereal period   The time interval needed by a celestial body to complete one revolution around another with respect to the background stars.

signs of the zodiac   The twelve equal angular divisions of 30° each into which the ecliptic is divided; each corresponds to a zodiacal constellation.

silicate   A compound of silicon and oxygen with other elements, very common in rocks at the earth's surface.

singularity   A theoretical point of zero volume and infinite density to which any mass that becomes a black hole must collapse, according to the general theory of relativity.

slow process (s-process) Formation of very heavy elements by the slow addition of neutrons to the nucleus followed by beta decay.

Small Magellanic Cloud (SMC) The smaller of the two companion galaxies to the Milky Way; it is an irregular galaxy containing about 2×10^9 solar masses.

solar core Region of the Sun's interior where temperatures and densities are high enough for fusion reactions to take place.

solar cosmic rays Low-energy cosmic rays generated in solar flares.

solar day The interval of time from noon to noon.

solar eclipse An eclipse of the Sun by the moon, caused by the passage of the moon in front of the Sun, so the moon's phase must be new.

solar flare Sudden burst of electromagnetic energy and particles from a magnetic loop in an active region.

solar mass The amount of mass in the Sun, about 2×10^{30} kg.

solar nebula The disk of gas and dust, around the young Sun, out of which the planets formed.

solar wind A stream of charged particles, mostly protons and electrons, that escapes into the Sun's outer atmosphere at high speeds and streams out into the Solar System.

solstice The time at which the day or the night is the longest; in the Northern Hemisphere, the summer solstice (around June 21), the time of the longest day; and the winter solstice (around December 21), the time of the shortest day; the dates are opposite in the Southern Hemisphere.

south magnetic pole A point on a star or planet from which the magnetic lines of force emanate and to which the south pole of a compass points.

space A three-dimensional region in which objects move and events occur and have relative direction and position.

spacetime A 4-dimensional Universe with space and time unified; a continuous system of one time coordinate and three space coordinates by which events can be located and described.

spacetime curvature The bending of a region of spacetime because of the presence of mass and energy.

spacetime diagram A diagram with one axis representing the three dimensions of space and the other of time; it shows the relation of events and worldlines.

space velocity The total velocity of an object through space, combining the components of radial and transverse velocities.

special theory of relativity Einstein's theory describing the relations between measurements of physical phenomena as viewed by observers who are in relative motion at constant velocities.

spectral line A particular wavelength of light corresponding to an energy transition in an atom.

spectral sequence A classification scheme for stars based on the strength of various lines in their spectra; the sequence runs O–B–A–F–G–K–M, from hottest to coolest.

spectral type (or class) The designation of the type of a star based on the relative strengths of various spectral lines.

spectroscope An instrument for examining spectra; also a spectrometer or spectrograph if the spectrum is recorded and measured.

spectroscopic binary Two stars revolving around a common center of mass that can be identified by periodic variations in the Doppler shift of the lines of their spectra.

spectroscopic parallax A technique for measuring distance by comparing the brightnesses of stars with their actual luminosities, as determined by their spectra.

spectroscopy The analysis of light by separating it by wavelengths (colors).

spectrum (pl., spectra) The array of colors or wavelengths obtained when light is dispersed, as by a prism; the amount of energy given off by an object at every different wavelength.

speed The rate of change of position with time.

spherical (closed) geometry An alternative to Euclidean geometry, constructed by G. F. B. Riemann on the premise that no parallel lines can be drawn through a point near a straight line; the sum of the angles of a triangle drawn on a spherical surface is always greater than 180°.

spicules Spears of hot gas that reach up from the Sun's photosphere into the chromosphere.

spin angular momentum The angular momentum of a rotating body; the product of a body's mass distribution, rotational velocity, and radius.

spiral arm A structure, part of a spiral pattern in a galaxy, composed of gas, dust, and young stars, that winds out from near the galaxy's center.

spiral galaxy A galaxy with spiral arms; the presumed shape of our Milky Way Galaxy.

spiral nebula An older term for a spiral galaxy as it appeared visually through a telescope.

spiral tracers Objects that are commonly found in spiral arms and so are used to trace spiral structure; for example, Population I cepheids, H II regions, and OB-stars.

spontaneous emission The emission of a photon by an excited atom in which an electron falls to a lower energy level.

sporadic meteor A meteor that occurs at random and so is not associated with a shower.

stadium (pl., stadia) An ancient Greek unit of length, probably about 0.2 km.

standard candle An astronomical object of known luminosity used to estimate distances to galaxies.

star counting A technique to measure the extent of the Galaxy, first used by William Herschel, in which it is assumed that the directions in space in which more stars are found (in a specific area) mark regions of greater extent of the Galaxy.

statistical parallax A parallax, and so a distance, determined from the average proper motions of selected groups of stars.

steady-state model A theory of the Universe based on the perfect cosmological principle, in which the Universe looks basically the same to all observers at all times.

Stefan-Boltzmann law The relation for a blackbody radiator between temperature and energy emitted per unit area of surface.

stellar interior model A table of values of the physical characteristics (such as temperature, density, and pressure) as a function of position within a star for a specified mass, chemical composition, and age, calculated from theoretical ideas of the basic physics of stars.

stellar nucleosynthesis A process in which nuclear fusion builds up heavier nuclei while supplying the energy by which stars shine.

stimulated emission Radiation produced by the effect of a photon stimulating an atom in an excited state to emit another photon of the same wavelength.

stochastic star formation A model for the generation of spiral arms by the random formation of stars in a molecular cloud, triggered by a supernova explosion, which then are drawn out into a spiral pattern by differential galactic rotation.

stony-iron meteorite A type of meteorite that is a blend of nickel-iron and silicate materials.

stony meteorite A type of meteorite made of light silicate materials.

straight line. The shortest distance between two points in any geometry.

stratosphere A layer in the earth's atmosphere in which temperature changes with altitude are small and clouds are rare.

strong nuclear force One of the four forces of nature; the strong force acts over short distances to keep the nuclei of atoms together.

summer solstice See *solstice*.

sunspot A temporary cool region in the Sun's photosphere, associated with an active region, with a magnetic field intensity of a few 0.1 T.

sunspot cycle The 11-year number cycle and a 22-year magnetic polarity cycle of the formation of sunspots.

superclusters A system containing multiple clusters of galaxies.

supergiant A massive star of large size and high luminosity.

supergiant elliptical (cD) galaxies Largest and most massive elliptical galaxies, sometimes with more than one nucleus; found at the core of a rich cluster of galaxies.

supergravity A model which combines quantum ideas with gravity in an attempt to describe the unification of all forces of nature during the first 10^{-43} second of the Universe's history.

superior conjunction A planetary configuration in which an inner planet lies in the same direction as the Sun but on the opposite side of the sun as viewed from the outer planet.

superluminal motion Motion apparently faster than the speed of light.

supermassive black hole A black hole with a mass of 10^6 solar masses or more; probably powers active galaxies and quasars.

supernova A stupendous explosion of a massive star, which increases its brightness hundreds of millions of times in a few days.

supernova remnant Expanding gas cloud from the outer layers of a star blown off in a supernova explosion; detectable at radio wavelengths; moves through the interstellar medium at high speeds.

superwind A very strong stellar wind.

synchrotron radiation Radiation from an accelerating charged particle (usually an electron) in a magnetic field; the wavelength of the emitted radiation depends on the strength of the magnetic field and the energy of the charged particles.

synodic month The time interval between similar configurations of the moon and the Sun; for example, between full moon and the next full moon; about 29.5 days.

synodic period The interval between successive similar lineups of a celestial body with the Sun, for example, between oppositions.

tau One of three generations of leptons (along with the electron and muon generations).

temperature A measure of the average random speeds of the microscopic particles in a substance.

temperature gradient The change in temperature over a unit change in distance.

terrestrial planets Planets similar in composition and size to the earth; Mercury, Venus, Mars, and the moon.

tesla In the SI system, a unit of measure of magnetic flux.

Tharsis ridge A highland region on Mars containing a cluster of volcanoes, including Olympus Mons.

thermal equilibrium Steady-state situation characterized by no large-scale temperature changes.

thermal pulses Bursts of energy generation from the triple-alpha process in the shell of a red giant star.

thermal radiation Electromagnetic radiation due to the fact that a body is hot; often characterized by a blackbody spectrum.

thermal speed The average speed of the random motion of particles in a gas.

thermosphere A layer in the earth's atmosphere, above the mesosphere, heated by X-rays and ultraviolet radiation from the Sun.

three-degree cosmic blackbody microwave radiation The relic radiation from the Big Bang; has a blackbody spectrum and a current temperature near 3 K.

threshold temperature The temperature at which photons have enough energy to create a given type of particle and antiparticle.

tidal force The difference in gravitational force between two points in a body caused by a second body, which may cause the deformation of the second body.

time A measure of the flow of events.

Titan Saturn's largest satellite, the first satellite detected to have an atmosphere.

Titius–Bode law A nonphysical formula that gives the approximate distances of the planets from the Sun in AUs.

ton (metric) 1000 kilograms.

torodial magnetic field A magnetic field configuration in which the field lines run parallel to the equator.

torque A twisting force.

transition (in an atom) A change in the electron arrangements in an atom, which involves a change in energy.

transition region In the Sun's atmosphere, the region between the chromosphere and corona where the temperature rises rapidly.

transverse fault A crack in a solid surface where the ground has moved sideways.

transverse velocity The component of an object's velocity that is perpendicular to the line of sight.

transverse wave A wave in which the oscillatory motion is perpendicular to the direction of propagation; such waves cannot travel through liquids.

Trapezium cluster A small cluster of young massive stars located in the Orion Nebula.

trigonometric parallax A method of determining distances by measuring the angular position of an object as seen from the ends of a baseline having a known length; see *heliocentric parallax*.

triple-alpha reaction A thermonuclear process in which three helium atoms (alpha particles) are fused into one carbon nucleus.

Triton Largest moon of Neptune; it has a thin atmosphere.

tropopause The boundary in the earth's atmosphere between the troposphere and the stratosphere.

troposphere The lowest level of the earth's atmosphere, reaching 10 kilometers from the surface; the area in which most of the weather takes place.

Tully–Fisher relation The relation between the luminosity of a galaxy and the width of its 21-cm emission line.

turbulence Irregular and sometimes violent convective motion.

turbulent viscosity The property of a gas (or any fluid) by which turbulent flow in one part affects the flow of a nearby part; an important effect in the transfer of angular momentum outward in the solar nebula.

turnoff point The point on the H–R diagram of a cluster at which the main sequence appears to terminate at the high-luminosity end.

T-Tauri stars Newly formed stars of about 1 solar mass; usually associated with dark clouds; some show evidence of flares and starspots.

21-cm line The emission line, at a wavelength of 21.11 cm, from neutral hydrogen gas; it is produced by atoms in which the direction of spin of their proton and electron change from parallel to opposed.

Type I, Type II supernovae Classification of supernovae by their light curves and spectral characteristics; Type I show a sharp maximum and slow decline with no hydrogen lines; Type II have a broader peak and a very sharp decline after 100 days with strong hydrogen lines in the spectrum.

two-sphere universe The basic premise of the celestial coordinate systems that the Universe is composed of two concentric spheres, the earth and the celestial sphere.

uniformity of nature The assumption that astronomical objects of the same type are the same throughout the Universe.

universal law of gravitation Newton's law of gravitation; see *gravitation*.

universality of physical laws The assumption, borne out by some evidence, that the physical laws understood locally apply throughout the Universe and perhaps to the Universe as a whole.

Universe The totality of all space and time; all that is, has been, and will be.

upland plateaus Large highland masses on the surface of Venus.

Valles Marineris Extensive canyonlands region near the equator of Mars.

Van Allen radiation belts Belts of charged particles

(from the Sun) trapped in the earth's lower magnetosphere.

variable star Any star whose luminosity changes over a short period of time.

vector A quantity that expresses magnitude and direction; for example, forces and accelerations are vector quantities.

velocity The rate and direction in which distance is covered over some interval of time.

velocity dispersion The range of velocities around an average velocity for a group of objects, such as a cluster of stars or galaxies.

vernal equinox The spring equinox; see *equinox*.

vertical circle On the celestial sphere, any great circle through the zenith.

Very Large Array (VLA) A radio interferometer located in New Mexico; it consists of 27 antennas spread over a **Y**-shaped pattern.

Very Large Baseline Array (VLBA) A radio interferometer with antennas spread across the United States; the processing and control center is in New Mexico.

Virgo cluster of galaxies The nearest large cluster of galaxies; it appears to lie in the direction of the constellation of Virgo.

virial theorem The statement that the gravitational potential energy is twice the negative of the kinetic energy of a system of particles in equilibrium.

visual binary Two stars that revolve around a common center of mass, both of which can be seen through a telescope so that their orbits can be plotted.

visual flux The flux from a celestial object measured across the visual part of the electromagnetic spectrum.

visual luminosity The luminosity from a celestial object measured across the visual part of the electromagnetic spectrum.

void (cosmic void) A large region of space empty of visible galaxies.

volatiles Materials, such as helium or methane, that vaporize at low temperatures.

volcanic model The formation of craters as cones left over from lava eruptions.

W Virginis stars Pulsating variable stars; Population II cepheids.

watt A unit of power; 1 joule expended per second.

wavelength The distance between two successive peaks or troughs of a wave.

weak nuclear force A short-range force that operates in radioactive decay and governs leptons.

weight The total force on some mass produced by gravity.

weightlessness The condition of apparent zero weight, produced when a body is allowed to fall freely in a gravitational field; in general relativity, weightlessness signifies motion on a straight line in spacetime.

white dwarf A small, dense star that has exhausted its nuclear fuel and shines from residual heat; such stars have an upper mass limit of 1.4 solar masses, and their interior is a degenerate electron gas.

Widmanstätten figures Large crystal patterns that appear on the surfaces of iron meteorites when they are polished and etched.

Wien's law The relation between the wavelength of maximum emission in a blackbody's spectrum and its temperature; the higher the temperature, the shorter the wavelength at which the peak occurs.

winter solstice See *solstice*.

worldline A series of events in spacetime.

X-rays High-energy electromagnetic radiation with a wavelength of about 10^{-10} meter.

X-ray burster An X-ray source that emits brief, powerful bursts of X rays; probably occurs from accretion onto a neutron star in a binary system.

young Population I stars A class of stars found in the disk of a spiral galaxy, especially in the spiral arms; they have a metal abundance similar to the Sun's and are the youngest stars in the Galaxy.

ZAMS Acronym for zero-age main sequence.

zenith The point on the celestial sphere that is located directly above the observer at 90° angular distance from the horizon.

zero-age main sequence (ZAMS) The position on the H–R diagram reached by a protostar once it derives most of its energy from thermonuclear reactions rather than from gravitational contraction.

Zodiac The twelve constellations through which the Sun travels in its yearly motion, as seen from the earth.

zodiacal light Sunlight reflected off of dust in the plane of the ecliptic.

zone A region of high pressure in the atmosphere of a Jovian planet.

zone of avoidance A region near the plane of the Galaxy where very few other galaxies are visible because of obscuration by dust.

Index

Key:

- Page numbers for figures are *italicized*.
  This often results in listing the same page twice, once for text, once for figure.
- Page numbers for table are followed by t
- Occasional references to material in a problem are followed by p.
- There are no references to items in the Glossary.
- References to material in the Appendix are simply given as Appendix, without page number.

Aaronson, Marc, 436, 451
Aaronson-Huchra-Mould method, 436
Abell catalogue of clusters, 447, 448
Abell classification, 440
Abell, George, 439, 447
Aberration of starlight, 42, 284
Absorption, *see* Interstellar absorption
Absolute magnitude, 226, 256
Absorption coefficient, 369, *see also* extinction coefficient
Absorption lines, 160, 161, 163 *see also* Fraunhofer absorption spectrum; Spectral lines
  profile, 164
  quasars, 468, 475
  stellar atmospheres, 251
  temperature dependence, 166
Abundance of elements, *see also* Metal abundances
  asteroids, 133
  galaxy, 407
  globular clusters, 261
  intergalactic matter, 452
  interstellar gas, 374
  magnetic variables, 343
  Population I, 265
  stars, 264
  stellar populations, 259
  sun, 203, 204t
Acceleration, 10–14
  centripetal, *12*

Accretion
  impact cross section, 144, *144*
  neutron stars, 363
  planetary, 143–145, 146, 147
  solar nebula, 141–145
  star formation, 385
  supernova, 353
Accretion disk(s), 334, 351
  Cygnus X-1, 360
  jets, 361
  neutron stars, 360
  novae, 350
  SS 433, 361
  X-ray binaries, 359
Acoustic waves (sun), 199, 203
Active galaxies, 455–467, 474
  jets, 463, 476
Active Galaxy Nuclei (AGN), 412, 413, 422, 423, 457, 457t, 459
Active optics, 179
Active regions, 214–221; *see also* Solar activity
Adams, J. C.
  discovery of Neptune, 113
Adams, W. S., 261
Adaptive optics, 194
Adiabatic index, 476
Adiabatic perturbations, 501
AGB (Asymptotic giant branch) stars, 308, 319, 347, 400
AGN, (Active Galaxy Nuclei), 412,

413, 422, 423, 457, 457t, 459
Air mass, 230
Albedo, 24
  asteroids, 132, 133, 145
  Mercury, 82
  Phobos and Deimos, 120
  planets, Appendix 3
  Venus, 87
Albrecht, A., 499
Aldebaran
  diameter, 246
Alpha Centauri, 43, 226
Alphonsus (crater), 64
Amalthea, 125
Andromeda Galaxy (M31),272, *273*, 400, 423, *424*, 431, 439
Angular momentum, 15
  gravitational collapse, 384
  solar system, 33
  spin angular momentum, 50p
  star formation, 383
Angular resolution, 154
Anomalistic month, 54, 56
Anorthosites (lunar), 66, *66*
Antarctic Circle, 38
Antiparticles, 497
Aperture synthesis, 182, 194
Aphelion, 8, 20
Apogalacticon, 293
Apogee, 20
Apollo Program, 53, 58, 64, 71

Apparent magnitude, 225
    distribution of stars, 274
Apparent solar time, 35, *35*
Appenine Mountains (lunar), 64
Arctic Circle, 38
Ariel, 129
Arp, Halton C., 459, 475
Artificial satellites, *see also* Space
        astronomy, Space missions by
        name
    apogee, 31
    energy equation, 31
    escape speed, 31
    orbits, 31
    perigee, 31
Associations(OB), 276, 277
Asteroid belt, 27
Asteroidal moons, 124–125
Asteroids, 27, 120, 132–134, 145
    albedo, 132, 133, 145
    Apollo, 27
    Ceres, 132, 134
    composition, 133
    origin, 145
    physical characteristics, 132–134
    Trojans, 27
Astrometry, 224
Astronomical seeing, *see* Seeing
Astronomical unit, 6, 223, 435,
        Appendix 7
Astrophysical jets, *see* Jets
Asymptotic giant branch (AGB), 308,
        319, 347, 400
Atmosphere(s),
    Earth, 67–71, *see also* Earth,
        atmosphere
    ionosphere, 69, 180
    planets, 25-27, *see also* individual
        planets
    stars, *see* Stellar atmospheres
    transmission of electromagnetic
        radiation, 153, 180, *180*
    unit, 69, Appendix 6
Atmospheric extinction, 229
Atmospheric refraction, 70, 71, 224
Atmospheric seeing, 178
Atmospheric transmission, 153, 180
Atomic nuclei, 299
Atomic number, 157, Appendix 5
Atomic particles, 157t
Atomic structure, 156–162
Atomic transitions, 159–162
    excitation potentials, 162
    forbidden, 161
    ionization potentials, 162
Aurorae, 74–75, *75*
    Jupiter, 107
Azimuth, Appendix 10

B stars, 155
    evolution, 305, 345, 345t, 346
    Oort constant A, 290
B-emission stars, 345, 345t
Baade, Walter, 329
Background radiation, 451, 486–489,
        487, 490, 494, 500
Background stars, 283
Bailly (lunar crater), 62

Balmer lines, 159, *159*, 160, *160*
    chromosphere, 204, 205
    discontinuity, 264
    excitation/ionization, 165, 166, *166*
    H II regions, 374
    spectral sequence, 255, 262t
    sun, 203–205, 207
    T Tauri stars, 341
    white dwarfs, 321
Bandpass, 227, 227t
Barnard's Star, 262, 283
Barometric equation, 69, 250
Barred spirals (SB), 416; *see also* Spiral
        galaxies
Barringer Meteor Crater, 29
Barycenter, 53, *53*, 235, *253*
Baryons, 498
Batuski, D., 448
Becklin-Neugebauer objects, 388
Bessel, Friedrich Wilhelm, 43, 223
Beta Canis Majoris stars, 339t
Beta Lyrae stars, 243, 345t
Beta Regio (Venus), 88, 89
Betelgeuse, 246–247
Biermann, Ludwig
    solar wind, 135
Big Bang, 432, 448, 486, 490, 494, 496,
        500
Big Crunch, 485
Binary pulsars, 328
Binary system(s), 233–245, 243t
    apparent binary, 233
    astrometric binary, 233
    black holes, 334
    center of mass, 235
    contact binaries, 244
        Roche lobes, 245
    data, 243t
    eclipsing binaries, 234, 240–245, *see
        also* Eclipsing binaries
    inclination of orbit, 235
    light curve, 240
    mass function, 239–240
    novae, 350
    orbits, 234, *234*, 235, *235*, *238*, *239*
    pulsars, 328, 327, 328
    RS CVn stars, 343
    Sirius, 320
    spectroscopic binary, 234, 237–240
    spectrum binary, 234
    SS 433, 361
    stellar data obtained from, 243t
    stellar masses, 233, 235–240, 243t
    stellar radii, 240–243, 243t
    supernovae, 353
    tidal effects, 242
    velocity curves, 237, 238, *238, 239*
    visual, 233, 234–237, *234*
    Wolf-Rayet stars, 346
    X-ray binaries, 358–362, 358t
    X-ray bursters, 328
Bipolar Magnetic Regions (BMR), 216
    flare model, 219
    prominences, 216
    solar, 215
    solar cycle model, 220
Bipolar outflow, 385, *386, 387*
BL Lacerta objects, 459–461, 474
    spectra, *460*
Black dwarf, 309, 320

Black hole(s), 331–334
    accretion disk, 334, *334*
    binary, 334
    Cygnus X-1, 359, 360, *360*
    galactic center, 280, 401
    mass, 331, 332
    observations, 334
    quasars, 471
    Schwarzchild radius, 474
    singularity, 331
    spacetime diagram, *333*
    x-ray sources, 334, 359
Blackbody radiation, 168–171, *169*
    3-K background, 486–489, *487, 490*
    binary systems, 241
    color index, 229
    color-color diagram, 264
    cosmic, 486–489, *487, 490*
    effective temperature, 230
    Jovian planets, 104, 104t
    Planck radiation law, 168, *169*
    planets, 24
    radio corona, 209
    solar photosphere, 201
    spectral distribution, 168, *169*
BMR, *see* Bipolar Magnetic Regions
Bode's law, 19, 27
Bode, Johann, 19
Bohr model, 157–159, *159*
Bohr, Niels, 157
Bolometer, 183, 184
Bolometric correction (BC), 230
Bolometric magnitudes, 230
Boltzmann constant, 164
Boltzmann's equation, 164–167, 171t,
        252
    chromospheric temperature, 204
    molecular clouds, 381
    stellar atmospheres, 249, 255
Boltzmann-Saha equations, 165–167,
        255
Bombardment, *see* Cratering
Bootes void, 448, 451
Bosons, 497
    Higgs bosons, 499
Brackett alpha line, 388
Bradley, James, 42
    nutation, 48
Braginsky, V.B., 482
Brahe, Tycho, 223
Breccia, 66
    meteorites, 140
Bremsstrahlung
    galaxies, 423
    intergalactic gas, 452
    H II regions, 375
Broad line radio galaxies, 474
Broadening (spectral lines), 167
Brown dwarfs, 310, 323
Burnell, Jocelyn Bell, 324
Burns, J., 448
Burstein, D., 451

Ca II lines, 166, 205, 346
Callisto, 123–124, *124*, 457A
Caloris Basin (Mercury), 83, *83*, 85
Cannon, Annie J., 255
Capilla Peak Observatory, 177

Capture model (moon), 146
Carbon burning, 314
  supernovae, 352
Carbon (CNO) cycle, *see* CNO cycle
Carbon dioxide (CO_2)
  comets, 135, 137
  Earth's atmosphere, 67t, 69
  Mars, 93
  Venus, 86
Carbon monoxide (CO)
  density distribution, 396
  galactic center, 402
  galactic rotation curve, 291, *291*
  interstellar, 379, 380
  Oort constant determination, 290
  spiral arm structure, 398
  star formation, 385
Carbonaceous chondrites, 139
Carbon-burning, 315t, *315*, 316
  supernovae, 352, 353
  Wolf-Rayet stars, 347
Cassini division, 127, 128
Cataclysmic variables, 337, 348–358,
  348t
Cavendish experiment, 14, *14*
Cavendish, Henry, 14
CCD (Charge-coupled devices), 188,
  189, 192, 193
cD galaxies, 414, 440, 444
  gravitational lenses, 474
Celestial coordinates, 34, 35,
  Appendix 10
Centaurus A, 464, *465*
  jets, 471, 476
Centaurus X-3, 359, 360, *360*
  eclipses, 360
  orbit, 360
Center of mass, 16
  binary systems, 235, *235*
  Earth-Moon system, 53
Central bulge (Galaxy), 272, 279, 400
Centripetal force, 12–13, *13*, 31, 41
Cepheid variables, 290, 308, 338, 339,
  339t, 340, 429
  classical Cepheids, 340
  distance scale, 435
  instability strip, 339
  period-luminosity relationship, 340,
  341
  W Virginis stars, 340
Ceres, 132, 134
  discovery, 27
Chandrasekhar limit, 319, 352
  supernovae, 352
  white dwarfs, 319, 353
  X-ray binaries, 359
Charge-coupled devices (CCD), 188,
  189, 192, 193
Charon, 115, 116, *see also* Pluto
Chemical composition, 302, *see also*
  Abundance of elements, Mean
  molecular weight
  cosmic rays, 407
  stellar models, 302
Chincarini, G., 448
Chinese chronicles, 351
Chlorofluorocarbons, 70
Chondrites, 139
  origin, 145
  Chondrules, 139, 145

Christy, James
  discovery of Charon, 115
Chromosphere, 198, 204–207
  fine structure, 205
  flares, 218
  HeII lines, 204
  Hydrogen Balmer lines, 204
  Lyman alpha, 207
  network, 205, 206
  plages, 216
  spectrum, 204
  temperature, 201
Circumstellar dust, 370, 388
Circumstellar material, *see* Mass loss
Clark, Alvan, 320
Clusters, 270, 276–277, 313; *see also*
  Globular clusters, Open clusters
  ages of, 313
  Hertzsprung-Russell (H-R)
  diagrams, 313
Clusters of galaxies, 439–450, 442t,
  447, *448*, 500, *see also*
  Superclusters
  Abell classification, 439
  active galaxies, 466
  intergalactic matter, 451, 452
  irregular, 439
  luminosity function, 443
  mass distribution, 443
  mass-luminosity (M/L) ratio, 453
  masses, 453, 444
  quasars, 475
  redshifts, 445
  regular, 439
  tidal effects, 445
  velocity dispersions, 445
CNO cycle, 300, *300*, 301, 306
  massive stars, 310
  WN stars, 347
CO *see* Carbon monoxide
Coal Sack, 390
COBE (Cosmic Background Explorer),
  489
Collisional broadening, 167
Collisional excitation
  neutral-hydrogen 21-cm line, 379
  planetary nebulae, 378
  solar transition region, 207
Color, *see* Effective temperature
Color excess, 229, 368, *see also*
  Interstellar dust, Color index
Color index, 228–230, *228*, 255, 259,
  264
  interstellar reddening, 368
  stellar temperature, 171t, 250, 255
Color-color diagram, 264, *264*
Color-magnitude diagram, 259, *260*,
  261, *see also* Hertzsprung-Russell
  diagram
  dwarf stars, *320*
Coma (comet), 134, 135, 137
Coma cluster, 448, 440
Comet Arend-Roland, *28*
Comet Giacobini-Zinner, 135
Comet Halley, 28, *136*, 136–137, *138*
Comet Kohoutek, *134*
Comets, 28, 134–137, 138
  albedo, 137
  brightness, 134, 135–136, 137
  coma, 134, 135, 137

composition, 135, 135t, 137
dirty iceberg cometary model, 136,
  137
escape from solar system, 50
fluorescence, 135
formation, 145
Giotto spacecraft, 137
halo, 134, 136, 137
ICE (International Cometary
  Explorer), 135
ion tail, 135, *134*, *135*, 137
jets, 137
magnetic field, 137
mass, 137
meteoroid material, 139
nucleus, 134, 135–136, 137, *134*, *138*
Oort cloud, 28, 136, 145
orbits, 28, 136
size, 137
solar wind, 135, 137
spectrum, 135
structure, 134–136, 137, *134*, *135*
tail, 28, 135, 136, *134*, *135*
Comparative planetology, 79
  internal heat, 105t
  magnetic fields, 105t
  Mars, 93
  Mercury, 84–85
  terrestrial planets, 98–99
  Venus, 87, 88, 89, 90–92
Composition, *see* Abundances of
  elements
Conduction, 297
Conic sections, 9, *9*,
Conjunction, 4
Conservation of energy, 16–18, 489
Conservation of mass, 485
Constellations, Appendix 2
Contact binaries, 244
Continental drift, 58
Continuous radio emission, 374–375
Continuum, 164, 168–171, *see also*
  Blackbody radiation,
  Electromagnetic spectrum
  Planck curves, *169*
  radio, 173, 374–375
  stellar atmospheres, 250
  sun, 201–203, 207
    K corona, 208
    radio, 208–209
Contraction, *see* Gravitational
  contraction,
Convection, 200, 297, 303, 306, *see also*
  Sun, photosphere
  Earth's atmosphere, 41
  sun, 200
  stars, 297, 303, 306, 314
  T Tauri stars, 341
  Wolf-Rayet stars, 347
Convergent point, 286
Coordinate systems, Appendix 10
  galactic, 282
  galactocentric system, 285
  geocentric, 282
  heliocentric, 282
  terrestrial, 34–38
Copernicus, Nicolaus, 2
  heliocentric model, 2, 43, 223
Copernicus (lunar crater), *62*, 63, *63*,
  64

Core(s) (stellar), 298, 299, 302–303
  helium burning, 308, 310
  hydrogen burning, 305–307
  novae, 350
  planetary nebula, 347
  supernovae, 355
  Wolf-Rayet stars, 310
Core-mantle grain models, 371
Coriolis effect, 39–40, *39*
Coriolis, Gaspard Gustave de, 39
Corona, 198, 207–210, 266–267
  coronal streamers, 208, 216
  electron scattering, 208
  emission lines, 209
  F corona, 208
  forbidden lines, 209
    Fe X, 209
    Fe XIV, 209
  free-free transitions, 208
  holes, 210
  K corona, 208
  magnetic fields, 209
  mass ejections, 216
  radio emission, 208, 218
  RS CVn stars, 343, 344
  solar wind, 210
  temperature, 208, 210
  transition region, 207
  ultraviolet lines, 209
  X-rays, 218, 266–267
Coronal holes, 210
Coronal interstellar gas, 382, 383
Coronal loops, 210, 344
  RS CVn stars, 344
Coronal streamers, 208, 216
Coronium lines, 209
Correlation function, 449, 450, 453
Cosmic background radiation, 486–489, *487*, *490*
Cosmic nucleosynthesis, *see* Nucleosynthesis
Cosmic rays, 218, 407–408
  galactic, 407, 408
  interstellar gas, 373
  interstellar medium, 407
  molecular clouds, 380
  solar, 216, 218
Cosmic time, 491
Cosmological constant, 432, 485
Cosmological effects, 432
Cosmological principle, 483
Cosmological redshifts
  quasars, 470
Cosmology, 480–491, 493–504, *see also* Universe
  Grand Unified Theories (GUT's), 498–499
  inflation theory, 499–500
  models, *484*
  Newtonian, 481
  relativistic models, 483–486
Coulomb (unit), 73, Appendix 6
Coulomb barrier, 299, 496
Crab Nebula, 328, 324, 353, *353*, 354, 376
  cosmic rays, 408
  jets, 476
  pulsar, 324, 328, *329*, 330, 354
Cratering, *see also* individual planets and moons

ejecta, 64, *63*
Mercury, 85
meteoric impact, 62–64
Moon, 62–64 *99*
terrestrial planets, 99
Venus, 87–89
Critical density, 494
Curvature of space, 482
Curvature of Universe, 484, 500
Cyclone, 40
Cyclotron frequency radiation
  Jupiter, 107
Cygnus A, 463, *464*
Cygnus X-1, 359, 360, *360*; *see also* X-ray binaries

Danese, L., 446
Dark clouds, 341, *367*, 390, *see also* Interstellar dust
Dark halo, 400
Dark matter, 400, 416, 427, 451, 453, 486, 493, 498, 499, 500, 504
Dating by radioactive decay, 60–61
Daughter atoms, 60
Davies, R., 451
Davis, Robert, 301
De Zotti, G., 446
De-excitation, *see* Excitation of atoms
Decay constant (radioactivity), 61
Deceleration parameter q, 486
Declination, Appendix 10
Degenerate gas, 308
  neutron star, 323, 355
  red giants, 308
  supernova, 352, 355
  white dwarf, 310, 319
Deimos, 119–120
Density waves, 278, 405, 407, 425
Descartes, Rene, 11
Detectors, 187–190
Deuterium, 157
  early Universe, 496
  proton-proton chain, 300
deVaucouleurs, Gerard, 414, 448
di Tullio, G., 446
Differential forces, 44–49; *see also* Tidal forces
Differential galactic rotation, 288, 288, 289, 405, *see also* Galactic rotation
Differential rotation
  Jupiter, 102
  Saturn, 108
  sun, 214
  Uranus, 113
Diffraction, 153–154, *154*, 178, 194
Diffraction grating, 190–191, *191*
Dione, 126
Disk, *see* Galactic disk
Disk Population, 399
Dispersion, 153, 190
  pulsars, 325
Distance determinations, 266, 394; *see also* Distance modulus, Distance scale
  galaxies, 431, 434
  geometric methods, 223–225
  luminosity methods, 225

moving clusters, 286
novae, 351
Distance modulus, 226, 266, 275, 368, 369
Distance scale, 287, 433, 433t, 434, *434*, 435, 440
Doppler broadening, 167
  corona, 208
  Seyfert galaxies, 458
Doppler effect, 154, 155, *155*, 283; *see also* Mass loss
  binary pulsars, 328
  binary systems, 237
  Earth's revolution, 44
  expanding atmospheres, 265, 345–346
  galactic center, 402
  galaxies, 426, 431
  Mercury
    radar mapping, 80
    rotation, 80
  moons of Saturn, 125–127
  planetary nebulae, 347
  planetary rotation, 20
  pulsating stars, 338, *338*
  quasars, 470
  radar, 21, 80–81, *80*
  relativistic, 155
  rings of Saturn, 129
  shell stars, 345–346
  solar oscillations, 199, 203
  SS 433, 361
  star formation 385
  thermal broadening, 167
  21-cm line, 379, 393–395, *394*
  Venus, 85
  Wolf-Rayet stars, 347
  X-ray binaries, 359
  Young stellar objects (YSO's), 385
Doppler, C. J., 154
Draconic month, 54
Draper, Henry, 255
Dressler, A., 451
Dust, *see also* Interstellar dust
  galactic center, 402
  intergalactic, 452
  stellar winds, 318
  T Tauri stars, 341
Dwarf galaxies, 417

Early universe, *see* Cosmology and Universe
Earth
  accretion, 146
  age, 60–61
  atmosphere, 67–71, 147
    carbon dioxide, 69
    chlorofluorocarbons, 70
    composition, 67t
    evolution, 76, 147
    global warming, 69
    human activity, 69
    hydrostatic equilibrium, 67
    ionosphere, 25, 69, 180, 216–217
    ozone, 69, *69*, 70
    refraction, *71*
    scale height, 69
    scattering of light, 70–71
    secondary atmosphere, 67, 76

Early universe (Cont.)
    structure, 67–69, *68*
    transparency, 70–71,
atmospheric seeing, *see* Seeing
aurorae, 74–75, *75*
central pressure, 58
composition, 57
core, 57, 58, 71
crust, 59, *60*
density, 56–58
eccentricity of orbit, 35, 37
equatorial bulge, 41
evolution, 76–77, 99
formation, 146–147
global warming, 69
hydrosphere, 59
hydrostatic equilibrium, 58, 67
interior, 56–58, *57*
internal heat, 76
lithosphere, 59
magnetic field, 71, 105t
magnetosphere, 71–72, *72*
mantle, 57, *57*
nutation, 47
oblateness, 41, *41*
orbit, 37,41, 47, 53, 54, 233,
    Appendix 3
physical characteristics, Appendix 3,
    Appendix 7
plate tectonics, 59, *60*, 147
precession, 47, *47*, 54
revolution, 41–44
rotation, 38–41
    tidal evolution, 46
seasons, 37–38, *37*
surface, 59–60, *60*
tidal forces, 44–46, *45*
wind patterns, *40*
Earth-Moon system, 46, *53*,
    angular momentum, 46
    dimensions, 52, 53
    evolution, 75–76
Eccentricity of planetary orbits, 8
Echelle grating, 191
Eclipses, 55–56
Eclipsing binaries, 240–245, 242
    ellipticity effect, 242
    Mass radius correlation, 243
    orbits, 239, 242
    partial, 242
    reflection effect, 242
    spectroscopic, 240, 242
    X-ray binaries, 359
Ecliptic, 4, *3*, 37,
Ecliptic coordinate system, Appendix 10
Eddington, Arthur S., 236
Eddington-Barbier relation, 172
Eddy, John, 212
Edlen, B., 209
Effective temperature, 230, 255, 171t,
    302
    binary systems, 241
    stellar atmospheres, 250
Effective wavelength, 227, 227t
Einstein X-ray Observatory, 184, 358,
    403, *see also* X-ray emission
    Centaurus A, 464
    galaxies, 423
    M87, 462
    supernova remnants, 376

Einstein's field equations, 484
Einstein, Albert, 482
    Doppler effect, 155
    general relativity, 432
    Schwarzschild radius, 331
    theory of relativity, 480
Ejecta blanket, 64
Electric field vector **E**, 151
Electromagnetic force, 496, 497t
Electromagnetic radiation, 150–155
    Doppler effect, 154–155
    flux, 156
    intensity, 155
    Planck blackbody radiation, 168–171
    spectrum, 152, 152t
    ultraviolet, *see* Ultraviolet radiation
        radiation
    wavelengths, 152t
Electromagnetic wave, 150–152, *152*
Electron, 497, Appendix 7
Electron degeneracy, 308, 310, 319
Electron density
    interstellar, 325
Electron pressure, 252
Electron scattering, 297
    corona, 208
    Sun, 298
Electron volt, 159
Elementary particles, 157t
Elements, 156, 157, Appendix 5, *see
    also* Abundance of elements
    synthesis, 313–316, 315t, *see also*
        Nucleosynthesis
Ellipse, 8–9, *8*, Appendix 9
Elliptical galaxies, 414, *414*, 419t; *see
    also* Galaxies
Elongation, 4
Emission coefficient, 172
Emission lines, 158, 161, *163*
    active galaxies, 455, 456t
    Be stars, 265, 345
    H II regions, 374
    novae, 350–351
    P Cygni profiles, *346*, 347; *see also*
        P Cygni profiles
    planetary nebulae, 347, 377–378
    quasars, 467
    radio, 379–380
    radio galaxies, 474
    shell stars, 344
    sun, 203
    supernovae, 352, 357
    temperature dependence, 166
    21-cm line, 378–379; *see also* 21-cm
        line
    Wolf-Rayet stars, 346
Emission measure, 375
Emission nebula(e), 370, 373, 374, 390;
    see also H II regions
Emission-line stars, 265, 345
Energy, 17–18
    radiation, 168
    units, Appendix 6
Energy generation, 298, 302
Energy level(s), 159, *160*
    Hydrogen, 159
    molecules, 163
Energy transport, 297, 298
    pulsating stars, 339
Ephemeris time, 36

Equation of state, 68, 249, 302
    degenerate gas, 319
    Earth's atmosphere
    neutron star, 324
    stellar atmospheres, 252
    stellar structure, 302
    white dwarf, 319
Equation of time, 36
Equatorial coordinates, 35,
Equinoxes, 37,
Equivalent width, 164, *164*, *251*, 251
Eratosthenes, 52
Escape velocity
    atmospheric gas, 26, *27*
    black holes, 331
    neutron star, 324
    spacecraft, 31, *31*
Europa, 122–123, *123*
European Southern Observatory
    (ESO), 180
    New generation telescope, 194
Evening star, 2
Event horizon, 332
Evolution, 277, 303–313; *see also*
    Evolutionary tracks
    carbon burning, 314
    helium burning, 314
Evolutionary track(s), 303, 304, 308,
    310, *315*
    clusters, 313
    low metal abundance stars, *311*
    massive stars, *310*
    Population I, *306*
    pre-main-sequence, *305*
    solar mass star, *307*
Excitation/De-excitation of atoms, 159,
    160, 252
    collisional, 161
    forbidden transitions, 161
    potential(s), 161, 162t
    radiative, 160, 168
Exosphere, 67, 68
Expanding arm, 397
Expanding atmosphere(s), 265
    novae, 350, 351
    planetary nebula, 347
    supernovae, 352
    Wolf-Rayet stars, 347
Expansion of Universe, 432, *see also*
    Big Bang, Cosmology
Explosive variables, 337 *see also*
    Cataclysmic variables
Extended atmosphere stars, 344, 345t,
    345, 346
Extinction,
    Earth's atmosphere, 70, 229
    interstellar, 369
    stellar, 171 *see also* Optical depth
Extragalactic radio galaxies, 408

F corona, 208
F ratio, 176
Faber, S., 451
Faber-Jackson relation, 451
Faculae, 215, *see also* plages
Faraday rotation, 408
    pulsar, 325
Filaments (solar), 205, 216
Fireball, 29

Fisher, J. Richard, 436, 448
Fizeau, Armand
    Doppler effect, 155
Flare stars, 341, 342
Flares, 216–219
    energy, 218
    magnetic fields, 210
    model, 219
    prominences disruption, 216
    radio bursts, 217
    RS CVn stars, 343, 344
    solar activity, 210
    solar cosmic rays, 218
    solar wind, 211, 218
    stellar, 342, *342, 344*
    X-ray bursts, 217
Fluorescence, 347, 374
    comets, 135
Flux, 156, *156,* 201, *see also* Apparent
        magnitude
    apparent magnitude, 225
    bolometric, 230
    monochromatic, 227
    observed, 227, 264
    ratio, 228
    solar, 201
    stellar, 230
    visual, 230
Focal length, 176
Focus, 176
Forbidden lines, 161, 209, 456t
    active galaxies, 456, 456t
    corona, 209
    interstellar, 377, *377*
    novae, 350
    planetary nebulae, 377
    Seyfert galaxies, 458
    T Tauri stars, 341
Force(s), 11–12, *11*
    early universe, 503
    four forces of Nature, 496, 497t, 498
Foucault, Bernard Leon, 40
Foucault's pendulum, 40, *41*
Fourier analysis, 445
    speckle interferometry, 246
Fraunhofer absorption spectrum, 202–
        203
    corona, 208
Fraunhofer, Joseph von, 202
Free-fall collapse, 304, *see also*
        Gravitational collapse
Free-free emission, 208, 375
    spectrum, *375*
Friedmann, A., 483
Fundamental particles, 497
Fusion, 299, 303, 355; *see also*
        Nucleosynthesis

Galactic, *see also* entries under Galaxy
        and galaxies
Galactic bulge, 276
Galactic cannibalism, 444
Galactic center, 274, 288, 395, 396,
        400, 402, 403, *403, see also*
        Galactic nucleus
    black hole, 280
    distance, 276
    mass distribution, *400,* 404
Galactic clusters, *see* Open clusters

Galactic coordinate system, 274,
        Appendix 10
Galactic disk, 276, 279, 399
    formation, 406–407
    gas distribution, 395–396, *396*
Galactic dynamics, 277, 278
Galactic equator, 274
Galactic halo, 276, 399, 400, 404
Galactic magnetic field, 408, 409
Galactic nuclei, *see* Active Galaxy
        Nuclei (AGN)
Galactic nucleus, 400, 408, 476, *see also*
        Galactic center
Galactic plane, 274, 279, 280, 399
Galactic rotation, 278, 287–293, 393,
        404, 405
Galactic structure, 393–404, *395*
Galactic warp, 396, *396*
Galactocentric system, 285
Galaxies, 272, 412–427, 419t; *see also*
        specific classes of galaxies, e.g.,
        Elliptical and Spiral galaxies
    absolute magnitudes, 420
    active, 455–467, 474, *see also* Active
        galaxies
    angular momentum, 425
    cannibalism, 414
    characteristics, 419t
    classification, 413–418, *415, 416,*
        419t, 424, 425
    clusters, *see* Clusters of galaxies
    colors, 418, 425
    dark matter, 427
    density waves, 425
    disk/bulge ratio, 425
    distances, 436
    double, 427
    evolution, 425, 474
    formation, 494, 500
    halos, 416
        clusters of galaxies, 444
    H I content, 422
    interstellar dust, 423
    isophotal level, 419
    luminosities, 419, 420, 425
    luminosity function, 439
    mass distribution, 417
    mass-luminosity (M/L) ratio, 420,
        427
    masses, 420, 426, 427
    morphology, 413, 418, 425
    peculiar motions, 446
    radio observations, 422; *see also*
        Radio galaxies
    redshifts, *430, 431,* 475
    rotation, 400, 426, *427*
        Tully-Fisher relationship, 436
    size, 417, 419
    spectral type, 421, 425, 445
    spiral arms, 414, 425
    star formation, 425
    supernovae, 408, 425
    virial theorem, 425, 426
Galaxy, 270, 277, 278, 285, 368; *see also*
        entries under Galactic
    age, 406
    anticenter, 272
    center, 272, 279
    chemical composition, 406
    coordinate system, 272, *273*

density wave, 278, 405
dimensions, 279t
disk, 274, 278
distribution of hydrogen, 395
distribution of mass, 405
distribution of stars, 274
dynamics, 277, 278, 287–293
evolution, 406
gravitational collapse, 406
halo, 278, 292, 406
heliocentric view, 274, 275
magnetic field, 370, 408
mass, 278, 279t, 288, 292, 373, 404
mass distribution, *400,* 404, *404*
model, 279, *279, 395*
nucleus, 272, 280, 404
poles, 272
rotation, 278, 287–293, 393–395, 405
rotation curve, 291–292, *291*
size, 279t, 406
spiral arms, 278, 280, 390, 393, 399,
        405
star counting, 274
stellar evolution, 277
structure, 278–280, 393–406, *395,*
        399
tidal interaction, 397, *397*
Galaxy clusters, *see* Clusters of
        galaxies
Galilean moons, 121–124, 121t, *121,*
        145, 457A
Galilei, Galileo, 6, 10, 193
    discovery of Neptune, 113
    gravity experiments, 433
    lunar craters, 61
    lunar librations, 54
    moons of Jupiter, 121
    star clouds, 270
Galle, Johann G.
    discovery of Neptune, 113
Gamma rays, 184
    early Universe, 494, 496
    galactic center, 404
    pulsars, 330
Gamma-ray bursters, 363
Ganymede, 123, *124,*
Gas law, 25,
Gaseous nebulae, 173, 373–378, *see*
        also H II regions and Planetary
        nebulae
Gauge field theory, 498
General relativity, 432, *see also*
        Relativity
    black hole, 331–334
    gravitational lenses, 472, 473
General theory of relativity, 322
Giant molecular clouds (GMC), 380,
        405; *see also* Molecular clouds
    density wave, 405
    spiral structure, 399, *399*
Giants, 261–264, 265
Gibbous phase, 6
Giotto (ESA) spacecraft, 137
Global warming, 69
Globular clusters, 259, 261, 276, 278,
        400, 406, 407
    color-magnitude diagram, 261, *261*
    distribution, 277, 400
    metal abundance, 400
    orbits, 279, 406

Globular clusters (Cont.)
  system of, 276
  X-ray bursters, 363
Globules, 366, *367*
GMC, *see* Giant molecular clouds and
    Molecular clouds
Grains, *see* Interstellar grains
Grand Unified Theories (GUTs), 498–
    499, 502
Gravitation, 13–15, 481, 496, 497t, 502
  constant of, 13–14 322
  differential forces, 44–49
  in hydrostatic equilibrium, 67–69,
    250
  relativity, 480
Gravitational collapse, 304
  angular momentum, 384
  black holes, 331, 334
  galaxies, 425
  Galaxy, 406
  massive stars, 390
  supernovae, 352
Gravitational constant, 322
Gravitational contraction, 145, 303, 306
  Jupiter, 105, 145
  pulsating stars, 339
  stellar, 298
  White dwarf, 317
Gravitational impact parameter, 144,
    144
Gravitational lenses, 472, 473
Gravitational mass, 482
Gravitational potential energy, 299
  neutron star, 352–353
Gravitational redshift, 322
  black hole, 332
  neutron star, 324
  white dwarfs, 322
Gravity, *see* gravitation
Gravity effect, 262
Great Attractor, 451
Great Rift in Cygnus, 390
Greenhouse effect, 86
  Earth, 69–70
  Mars, 93
  Venus, 86
Greenwich mean time, 36
Gregorian calendar, 37
Gregory, S., 448
Grindlay, Jonathan, 363
Grotian, W., 209
Gum Nebula, 328, 330, *330*, 376
Gursky, Herbert, 363
Guth, A., 499
GUTs (Grand Unified Theories), 498–
    499, 502

H I (neutral hydrogen gas); *see also*
    21-cm line
  clouds, 382
  density distribution, 396
  galactic rotation curve, 291
  interstellar gas, 374, 382, 399
  21-cm transition, 378–379
H II (ionized hydrogen gas)
  intergalactic, 452
H II regions, 173, 374, 382, 380, 388,
    398, 401, 405; *see also* Planetary
    nebulae

distance scale, 435
galactic center, 402
galaxies, 423, 456
M82, 457
molecular clouds, 381
planetary nebulae, 376
protostar, 388
in spiral galaxies, 414
H alpha line, 159; *see also* Balmer lines
  chromosphere, 205
H$^-$ (negative hydrogen ion), 201–203
H-R diagram, *see* Hertzsprung-Russell
    diagram
Hadley cells, 40
  Venus, 87
Hale telescope, 177
Hale, George E., 214
Half-life, 61, *61*
Hall, Asaph, 119
Halley's comet, 28, *136*, 136–137, *138*,
    188
Halo, 404, 400
Halo population, *see* Population II
Harvard spectral sequence, 255, 256t
Hawaiian volcanoes, 96
He$^+$ ionization, 339
HEAO, *see* Einstein X-ray Observatory
Heavy element composition, *see* Metal
    abundances
Heckman, T., 457
Heliocentric model, 2, 43
Heliocentric parallax, 223–224
  Helioseismology, 199, 301
Helium
  early Universe, 496
  nuclear fusion, 300
  sun, 204
Helium burning, 308, 314, 339
  long period red variables, 341
  supernovae, 355
  WC stars, 347
  X-ray bursters, 363
Helium flash, 308
Helium ionization region, 339
Helou, G., 423
Henderson, T., 43
Henry Draper Catalog, 255
Henry Draper Extension (HDE), 359
Hercules cluster, 440
Herschel, William, 233, 285
  magnitude scale, 225
Hertzsprung, Ejnar, 256
  narrow line stars, 261
Hertzsprung-Russell (H-R) diagram,
    256–267, 304, 308
  asymptotic giant branch, 308, *314*
  clusters, 259–261, *312*, 313, *314*
  distance determination, 266, *266*
  evolutionary tracks, 303–316, *311*
  instability strip, 339
  long period red variables, 341
  Population I, 306
  Population II, 311
  pulsating stars, 339
  red giants, 308
  stellar wind stars, *309*
  T Tauri stars, 342, *342*
  variable stars, *338*, 339
  white dwarfs, 321, *321*
Hewish, Anthony, 324

Hey, John S., 181
Higgs particle, 499
High-velocity clouds, 400
High-velocity stars, 292, 293, 400
Hipparchus
  magnitude scale, 225
Hipparcos mission, 224
Homestake mine, 301
Horizon, 494
Horsehead nebula, 390
Houk, Nancy, 255
Hour angle, 35
Hubble Space Telescope (HST), 184,
    185, *185*, 193, 224, 435, 456
  limiting magnitude, 225
  parallax measurements, 224
  spherical aberration, 185
Hubble tuning fork diagram, 413, *413*
Hubble's constant, 431, 432, 433, 484,
    486, 503
Hubble's law, 430–433, *431*, 451, 484,
    485, 486
  parametrized form, 432
Hubble's time, 432
Hubble, Edwin, 413, 431, 447
Huchra, John, 436
Hulse, Russell, 328, 328
Hyades cluster, 261, 286, 287, 435
  moving cluster method, 225
Hydrogen, *see also* H I and H II
  atom, 157, 158
  Balmer lines, 159, *159*, 160, *160*; *see
    also* Balmer lines
  excitation/ionization, 165, 166, *166*
  Bohr model of atom, 157, 158
  Bracket series, 388
  distribution in galaxy, 395
  energy level diagram, 159, *160*
  excitation potential, 162
  intergalactic, 452
  interstellar, *see* 21-cm line,
    Hydrogen molecule
  ionization potential, 161, 162t
  Jupiter, 103–104
  Lyman lines, 159
  metallic hydrogen, 104
  molecule, 380, 382
  negative ion (H$^-$), 201
  nuclear fusion, 300
  Saturn, 109, 110
  spectral series, 159
  spin, *378*
  21-cm transition, 378
Hydrogen burning, 300, 302, 305
Hydrogen molecule (interstellar), 380,
    382
Hydrostatic equilibrium, 58, 250, 295,
    296, 302
  Earth, 58, 67
  Jupiter, 100
  Mercury, 82
  protostar, 303–305
  pulsating stars, 339
  stellar atmosphere, 250, 252
  stellar structure, 295, 302, 303
  white dwarf, 319
Hydroxyl (OH) molecule
  comets, 134, 135, 135t
  interstellar, 379–381, 379t
Hyperfine splitting, 378
Hyperbola, 9, Appendix 9

Image, 176
Image processing, 188, 186
Impact parameter, 144
Impacts, see cratering
Index of refraction, 190
  Earth's atmosphere, 70
Inertia, 11
Inertial frames, 481
Inertial mass, 482
Inferior conjunction, 4
  moon, 54
  inferior planets, 4
Inflation theory, 499, 502
Infrared astronomical satellite (IRAS),
  see IRAS
Infrared radiation, 152t, 153, 180, 183,
    184
  bolometer, 183
  cirrus, 372
  galactic nucleus, 401
  galaxies, 423, 459
  interstellar dust, 183, 371, 372
  M82, 457
  M87, 462
  M stars, 255
  molecular clouds, 390
  Orion Nebula, 388
  protostars, 304, 386, 390
  T Tauri stars, 341, 342
  Wien's law, 183
  YSO, 388, 390
Insolation, 38
Instability limit, 48, 49
Instability strip, 308, 339
Intensity, 155, 156, 171
  blackbody, 168
  radiation, 154, 155–156, 168
  spectral lines, 163–167
Intercloud medium, 382, 383
Interference, 153, 154
Interferometry, 182, 182
  aperture synthesis, 182
  lunar occultation, 245
  speckle, 246
  stellar, 245, 246, 246t
Intergalactic matter, 451, 452
  clouds, 469
  hot gas, 466
  jets, 476
  quasar absorption lines, 470
Intermediate Population II, see
  Population II
International Astronomical Union
    (IAU), 290
International Cometary Explorer (ICE)
  Comet Giacobini-Zinner, 135
International Ultraviolet Explorer
    (IUE), 184, 456
  Supernova 1987, 357
Interplanetary medium, 29, 141, see
    also Solar wind
Interstellar absorption, 274, 369, see
    also Interstellar dust
  color excess, 229, 368
  galactic disk, 275
Interstellar absorption lines, 373, 373
Interstellar clouds, see Dark clouds,
    Molecular clouds, Nebulae,
    Protostars

Interstellar dust, 274–275, 372, see also
    Color excess
  color excess, 229, 368
  dark clouds, 341, 390
  in galaxies, 423
  grains, 303, 304, 369–372
    absorption bands, 372
    extinction curve, 371, 371
    molecule formation, 380
    polarization, 369
    properties, 371
  obscuration
    latitude effect, 419
  reddening, 368
  reflection nebulae, 370
  spiral arm structure, 398
  T Tauri stars, 341
Interstellar Gas, 372–383, 388, see also
    H I, H II regions, 21-cm line
  absorption lines, 374
  distribution, 374
  galactic center, 401
  galactic structure, 393
  spiral arm structure, 398
Interstellar medium, 366–383, 387–390,
    406, 408, see also other entries
    under Interstellar, Molecular
    clouds
  cosmic rays, 408
  dispersion measure, 325
  electron density, 325
  galactic disk, 407
Interstellar molecules, 379, 379t, 380,
    397, see also Molecular clouds
Interstellar polarization, 369, 408
  dust grains, 371
  radio, 370, 408
  reflection nebulae, 370
Interstellar radio lines, 378–382
Interstellar reddening, 229, 368, 368,
    369
Inverse square law of light, 156
  measure of distances, 225
  stellar distance, 226
Io, 121–122, 122, 123
  atmosphere, 121
  sodium cloud, 121
  volcanoes, 122
Ionization, 161–163
  active galaxies, 455
  equilibrium, 165
  potential(s), 162, 162t, 165
  stage of ionization, 165, 166
Ionosphere, 25, 69, 180
  solar flares, effect of, 216
IRAS (Infrared Astronomical satellite),
    184, 371, 372, 387, 390, 423, 459
Iron lines
  clusters of galaxies, 453
  solar corona, 209
Irregular galaxies, 417, 417, 419t
Isochrones, 313
Isoelectronic series, 162
Isophotes (of galaxies), 419
Isothermal perturbations, 500
Isotopes, 157
  nucleosynthesis, 355
Isotropy, 494
IUE (International Ultraviolet
    Explorer), 184, 357, 456

Jackson, R., 451
Jansky, Karl, 180
Jeans length, 500
Jets, 476, 477
  active galaxies, 466
  Centaurus A, 465, 465
  comet, 136, 137
  Cygnus A, 464
  M87, 461–462, 461, 462
  quasars, 471, 474
  radio galaxies, 462, 465
Johnson, H. L., 227
Jolly, Phillip von, 14
Joss, Paul, 363
Jovian planets, 98–110, Appendix 3;
    see also individual planets
  atmospheric composition, 25
  formation, 145
  internal heat, 105t
  orbits, Appendix 3
  physical data, Appendix 3
Jupiter, 101–108, 101, 102, see also
    Jovian planets
  albedo, 101
  asteroidal moons, 124–125
  atmosphere, 102, 103, 102
  aurorae, 107
  blackbody radiation, 104
  composition, 102, 103
  core, 104, 105
  cyclotron frequency radiation, 107
  decameter radiation, 106
  density, 23, 102, 104
  differential rotation, 102
  excess energy, 104
  formation, 145
  Galilean moons, 121–124
  gravitational contraction, 105
  Great Red Spot, 103, 104, 103
  interior, 104, 105, 104
  internal heat, 104, 105, 105t
  Io influence on magnetic field, 106
  lightning, 106
  magnetic field, 105–108, 105t
  magnetosphere, 106–107, 106
  mass, 102
  metallic hydrogen, 104, 105
  moons, 121–125, see also individual
    moons
  motions, 101–102
  orbital data, Appendix 3
  perturbation of comets orbits, 136
  physical data, Appendix 3
  radiation belts, 106
  radio map, 107
  rings, 125, 125
  satellites, 121–125, Appendix 3
  structure, 104, 104
  synchrotron radiation, 106–107
  temperature, 103
  Voyager observations, 102, 103, 107
  zones, 103

K corona, 208
K-correction, 420
k parameter (curvature of space), 483,
    485
Kamiokande II (neutron detector), 301,
    357

Keck Telescope, 178, *179*, 194
Keenan, P. C., 261
Kemp, James, 323, 322
Kepler's laws, 7–8, *7*
    first law, 7
    planets, 19
    second law (law of areas), 7, 15, *15*,
        35
        third law (harmonic law), 7–8, 235
            binary stars, 235, 239–240
            black hole, 331
            Earth-Moon system, 46, 53
            galaxies, 426, 427
            Galaxy, 278, 288, 292, 404
            gravitational collapse, 304
            mass determination, 30
            moons of Saturn, 128–129
            Newton's form, 15–16
            Pluto, 116
            pulsating stars, 340
            Roche limit, 48
            white dwarfs, 321
            X-ray binaries, 359
Kepler, Johannes, 6
    distance determination, 6–7
    laws of planetary motion, 7–8
Keplerian motions, *see* Kepler's laws,
    third law
Kinetic energy, 17
    space craft, 30
Kirkwood gaps, 27
Kirshner, R., 448
Kleinmann-Low (KL) nebula, 388
Kohlschutter, A., 261
Kramer's law, 302

Large Magellanic Cloud (LMC), 186,
    414, 429
    Supernova 1987, 355–358
Lava flows
    Io, 121
    lunar, 64
Law of Areas, 7, 15, *15*, 35
Leavitt, Henrietta, 340, 429
Lemaitre equation, 483
Lenticulars, 417
Lenz's law, 219
Leptons, 497, 497t
Leverrier, U. J.
    discovery of Neptune, 113
Librations (lunar), 54, *54*
Light
    inverse square law, 156
    velocity, 151, Appendix 7
    wave nature, 150, 155
Light curve(s)
    eclipsing binary systems, 240–244
    flare stars, 342
    novae, 349
    pulsating stars, 338
    supernovae, 352
    X-ray bursters, 363
Light scattering, 368
Light-gathering power, 177
Lighthouse model (Pulsars), 326, 331
Lightyear, 224
Limb (planets), 81
Limb brightening, 208

Limb darkening,
    eclipsing binaries, 242
    sun, 200–201, *200*
Lin, C. C., 405
Linde, A., 499
Line blanketing, 250, *251*, 255, 265
    Line locking, 470
Line of nodes, 54
Line profiles, 163–164, *164*
    equivalent width, 251–253, *251*
    Fraunhofer lines, 203–204, *203*
    21-cm line, 394, *394*
Linear momentum, 10–12
LINERS (Low Ionization Nuclear
        Emission Regions), 457
Lithosphere, 58
Local Group, 434, 439, 440, 440t, *441*
Local standard of rest (LSR), 282, 285–
    286, 288–290, 292, 293
Local Supercluster, 448, *449*, 451
Long-period variables, 339t, 341
Longitudinal compression (P) waves,
    57–58
Loop nebula, 376
Lorentz factor, 472
Lorentz force, 73, *74*
Los Alamos National Laboratory, 476
Low Ionization Nuclear Emission
        Regions (LINERS), 457
Lowell Observatory, 94
Lowell, Percival, 94
    discovery of Pluto, 115
LSR, *see* Local standard of rest
Luminosity, 170, 252, 258, 297, 308
    stellar, 230
    stellar distances, 226
Luminosity classifications, 261
Luminosity function,
    galaxies, 439, 443, *443*, *444*
    stars, 275, *275*, *276*
Lunar, *see also* moon
Lunar eclipses, 56, *56*, 57
Lunar librations, 54, *54*
Lunar occultation
    stellar diameter, 245
Lunar phases, 54, *55*
Lunar rock samples, 66–67, *66*
Lundmark, K., 431
Lyman alpha line, *see also* Hydrogen
    intergalactic, 452
    quasars, 469, 470
    solar, 207
Lyman continuum, 161
    interstellar gas, 374
Lyman line(s), 159, *160*
    ionization series limit, 161
    solar transition region, 207
Lyndon-Bell, D., 451

M101, 435
M31 (Andromeda galaxy), 272, *273*,
    400, 423, *424*, 431, 439
M33, 439, 440
M67 (NGC 2682), 313
    H-R diagram, *313*
M81, 445, *445*
M82, 417, 445, *445*, 457

M87, 461, *461*, 462, *462*, *463*
    black hole model, 474
    jets, 476
M-K luminosity classes, 261–264, 262t,
    263, Appendix 4–3
M-L relation, *see* Mass luminosity
    relationship
M/L (mass to light) ratio (galaxies),
    420, 427, 453
    missing mass, 453
Mach number, 476
Mach, Ernst, 476
Magellan spacecraft (Venus), 89, *91*
Magellanic Clouds, 279, 397
    period-luminosity relationship, 340
Magellanic stream, 397, *397*, 422
Magnetic buoyancy, 220
Magnetic field vector **B**, 151
Magnetic field(s)
    aurorae, 74–75
    Centaurus X-3, 360
    charged particle motion, 73–74, *74*
    comets, 137
    coronal loops and holes, 210
    cosmic rays, 408
    Crab nebula, 354
    earth, 71–75, 105t
    flare stars, 343
    Galaxy, 370, 407–407
    interstellar, 370–371, 408
    jets, 476
    Jupiter, 105–108, 105t
    Lorentz force, 73, *74*
    M87, 462
    magnetic variables, 265, 341, 343,
        344
    magnetosphere, 72–74, *72*
    Mars, 98
    Mercury, 84
    moon, 71
    Neptune, 114, 105t
    neutron stars, 360
    peculiar A stars, 265
    planets, 105t
    pulsars, 325, 326
    radio galaxies, 456, 461–465
    RS CVn stars, 344
    Saturn, 105t, 110–111
    solar wind, 210
    SS 433, 362
    star formation, 385
    sun, 211t, 219–221
        active regions, 214, 219–221
        bipolar magnetic regions, 215, 220
        flares, 217, 219
        sunspots, 211–212, 214, 220–221
    T Tauri stars, 341
    Uranus, 112, 105t
    Venus, 89–90
    white dwarfs, 322, 323
    X-ray binaries, 359
    Zeeman effect, 167
Magnetic flux, 322
Magnetic moment
    neutral-Hydrogen 21-cm line, 378
Magnetic monopoles, 498, 499, 500
Magnetic variables, 341, 343, 344
Magnetograph, 215
Magnetosphere, *72*
    Earth, 71–74, *72*

Magnetosphere (Cont.)
  Jupiter, 106, *106*, 107
  pulsars, 326
  Saturn, 110–111, *110*
Magnifying power, 178
Magnitudes, 225, 226
  absolute, 226
  apparent, 225
  bolometric, 230
  photovisual, 227
  systems, 226–228, 227t
  UBV system, 227, *265*
  visual, 230
Main-sequence, 258, 261, 300, 305, 313; *see also* Hertzsprung-Russell diagram
  evolution, 305, 306, 307, 309
  galactic clusters, 259
  turnoff, 313
Main-sequence fitting technique, 266, 276, 433
Mare Ibrium, 64
Mare Orientale, 64, 76
Margon, Bruce, 361
Maria (moon), 61, 64, 65, 76
Mariner Mars missions, 95
Mariner 10 (Mercury/Venus), 82, 83
Mars, 92–98, *93*
  arroyos, 95
  atmosphere, 93, 95, 98
    composition, 93
  canals, 94
  craters, 95, 97, *97*
  evolution, 98
  greenhouse effect, 93
  interior, 93, *92*, 98
  lava flows, 95
  magnetic field, 98
  Mariner missions, 95
  meteorite source, 140
  moons, 92, 119–120, *120*, Appendix 3
  motions, 92
  Olympus Mons, 96
  orbital data, Appendix 3
  physical characteristics, 92, 93, Appendix 3
  polar caps, 94, *94*
  sandstorms, 94
  soil, 95
  surface features, 93–98, *96*
  temperature, 93
  Tharsis ridge, 97
  Valles Marineris, 95, *95*
  Viking missions, 93, 94, 95
  volcanoes, 96
  water, 93, 94, 95, 98
Mascons, 64
Maser excitation, 381, 388
  energy-level diagram, *381*, *382*
Mass, 11, 13, 14–16; *see also* Stellar masses, Galaxies, etc.
  black holes, 331, 332
  Chandrasekhar limit, 319
  Earth, 53
  effect on stellar evolution, 305, 313–316, *315*, 323, 355
  neutron stars, 358–360
  planets, Appendix 3
  stars, 235–237, 240, 303

statistical masses, 240
  white dwarf, 319, 322
Mass defect, 299
Mass equation, 302
Mass exchange, *see* Mass loss
Mass fraction, 250, 296
Mass function, 239–240
  Cygnus X-1, 360
Mass loss, 309, 313, 317, 344–348, *see also* Stellar winds
  cool supergiants, 373
  emission-line stars, 265
  Galaxy (role in evolution of) 406
  giants, 345
  neutron stars, 360
  novae, 350, 351
  planetary nebulae, 347
  Population II, 311
  protostars, 385
  pulsars, 327
  shell stars, 345
  supergiants, 345
  Supernova 1987, 357
  supernovae, 352
  superwind, 309
  Wolf-Rayet stars, 347
Mass-luminosity (M/L) ratio
  clusters of galaxies, 453
  galaxies, 420, 427
Mass-luminosity relationship (M-L), 236, *236*, 237, *243*, 303, 305
  equivalent width, 253
  evolution, 305
  X-ray binaries, 359
Mass-radius relationship, 319
Massive stars, 309, 390
  evolution, 390
  formation, 388
  nucleosynthesis, 355
Matter era, 488, 491, 500, 503
Matthews, Thomas, 467
Mauna Kea, 178
Maunder minimum, 212
Maury, Antonia, 261
Maxwell Montes (Venus), 88, *88*
Maxwell's theory of radiation, 168
Maxwellian distribution, 25, *26*
Mean molecular weight, 249, 296, 302
  evolution, 305
Mean parallax, 224
Mean solar time, 36
Mercury, 78–83,
  albedo, 82
  atmosphere, 82
  basin formation, 85
  Caloris Basin, 83, *83*
  central pressure, 82
  comparison to Moon, 83, 85
  craters, 83, *82*, *84*, 85
  elongation, 4, 80
  evolution, 84–85
  hydrostatic equilibrium, 82
  interior, 81–82, *82*, 84
  lava flows, *83*, 85
  magnetic field, 84
  magnetosphere, 84, *85*
  Mariner 10 mission, 82, 83
  orbital data, Appendix 3
  physical characteristics, 81–83, Appendix 3

radar mapping, 80–81
  revolution, 80
  rotation, 80–81
  scarp, 83, *84*
  solar day, 81, *81*
  solar wind, 83, 84
  thrust faults, 84
Meridian, Appendix 10
Mesons, 498
Mesosphere, 67, *68*
Messier's catalog, 261, 270, Appendix 1
Messier, Charles, 261, 270
Metal abundances, *see also* Abundances
  Galaxy, 277, 279, 400, 406
  relation to age, 397
  stellar populations, 259, 276, 397, 406
Metal lined A stars, 343
Metallic hydrogen, 100, 110
Metastable states, 161, 209
  corona, 209
Meteors, 28–29, 137, 138, *138*, *139*
Meteor crater, 29, *29*
Meteor shower, 29
  radiant, 29
Meteorites, 28–29, 137–140, *140*
  carbonaceous chondrites, 120
  chondrites, 139
  chondrules, 139
  classification, 139
  composition, 139, 140
  iron, 139, *140*
  nebular model, 145
  orbits, 137, 138
  origin, 140, 145
  planetesimals, 145
  radiometric dating, 140
  stony, 140, *140*
  stony irons, 140
  structure, 139, 140
Meteorite impact, 64
Meteoroids, 28, 137–140, *139*
Metonic cycles, 56
Michelson, M. E.
  stellar interferometer, 245
Micrometeorites, 28, 141
Microwave background, *see* Background radiation
Microwave radiation, 152t
Milky Way Galaxy, 270, *271*, 412, 439; *see also* Galaxy
  peculiar motion, 446
  radio source, 180
  star clouds, 270
Millisecond pulsars, 324
Minkowski, R., 329
Minor planets, *see* Asteroids
Mira stars, 339t, 341
Miranda, 129, 130
Missing mass, 453; *see also* Dark matter
Mixing length, 453
Model stellar atmosphere, 249
Molecular clouds, 373, 382, 388, 380, 391
  galactic center, 402
  GMC as spiral arm tracers, 399
  solar-mass stars, 390
  stimulated emission, 381

Molecular masers, 381
Molecules
  comets, 135, 137
  Earth's atmosphere, 69–70
  electronic transitions, 163
  interstellar, 379t
  Jupiter, 103
  long period variables, 341
  Mars' atmosphere
  Neptune, 114
  rotational energy states, 163
  Saturn, 109
  spectra, 163
  Uranus, 112
  Venus' atmosphere, 86–87
  vibrational energy states, 163
Molina, Mario, 70
Momentum, 11–12, 15 *see also* Angular
    momentum,
  angular, 15
  linear, 11–12
Moon, 58–67, 75–76, 146–147
  age, 66, 75
  albedo, 61
  anorthosites, 66
  Apollo samples, 66, *66*
  atmosphere, 67
  binary accretion model, 146
  breccias, 66
  capture model, 146
  composition, 58, 64–67, 146
    surface, 61–67
  Copernicus crater, 64, *62, 63*
  core, 58–59, *59*
  cratering, 99, *99*
  craters, 62–64
  eclipses, 56
  ejecta blanket, 63
  escape velocity, 67
  evolution, 75–76, 76t, 99
  fission model, 146
  formation, 146–47
  giant impact model, 147
  gravitational perturbations, 64
  highlands, 61, 64, *65*, 76
  igneous rocks, 66
  interior, 58–59, *59*
  internal heat, 76
  iron abundance, 66
  lava, 64, 66, 67, 76
  librations, 54
  line of nodes, 54, 55
  magnetic field, 71
  mare basins, 64
  maria, 61, 64, *65*, 76
  mascons, 64
  meteoric impacts, 62–66, 75–76
  mountains, 61, 65
  orbital data, Appendix 3
  origin of meteorites, 140, 145
  phases, 54, *55*
  physical data, Appendix 3
  rays, 62–64
  regolith, 64–66
    rocks samples, 66–67, *66*
  secondary craters, 62
  seismology, 58
  soil, 66
  structure, 58–59, *59*

  surface, 64–67
    composition, 64–67
  synchronous rotation, 46, 54
  synodic orbit period, 36
  temperature, 61, 67
  tidal evolution, 46
  tidal forces, 44–46, *45*
  water, 67
Moons, (satellites) 26, 121–125,
    Appendix 3
  of asteroids, 134
  Jovian planets, 26
  Jupiter, 121–125, 121t, *121, 122, 123,
    124*
  Mars, 119–120, *120*
  Neptune, 131–132, *131, 132*
  orbital data, Appendix 3
  Pluto, 115–116, *115, 116*
  relative sizes, *53*
  Saturn, 125–127, *126, 127*
  synchronous rotation, 119, 121, 125
  terrestrial planets, 27
  Uranus, 129–131, *130*
Morgan, W. W., 261, 414, 421
Morgan-Keenan (M-K) luminosity
    classification, 261–264, 262t,
    263, Appendix 4–3
Mould, Jeremy, 436, 451
Moving cluster(s), 225, 286–287
Muons, 481, 497

Nebulae, 270; *see also* H II regions
  dark, 390
  novae, 351
Neptune, 113, *114*, 114, *see also* Jovian
    Planets
  composition, 114, 145
  internal heat, 114
  magnetic field, 114, 105t
  magnetosphere, 114
  moons, 131–132, *131, 132*, Appendix
    3
  orbit, *21*, 113, Appendix 3
  physical data, Appendix 3
  rings, 132, *133*
  Voyager 2, 114, *114*
Nereid, 131
Neutrinos, 300–301, 497, 500
  detection, 301
  mass, 498
  sun, 198, 300–301
  Supernova 1987, 357
  supernovae, 352, 355
Neutron star(s), 323, 324, 326, 352
  Chandrasekhar limit, 359
  eclipses, 360
  escape velocity, 324
  gamma-ray bursters, 363
  gravitational redshift, 324
  helium burning flashes, 363
  jets, 361
  mass, 352
  orbits, 360
  SS 433, 361
  Supernova 1987, 357
  surface gravity, 324
  X-ray binaries, 359
  X-ray bursters, 363

Neutrons, 157
  supernovae, 355
New Technology Telescope (NNT),
    180, 194
Newton (unit of force), 14, Appendix
    6
Newton's laws, 10–13
  first law (inertia), 11, 12
  second law (force), 11, 12, 482
  third law (action-reaction), 11–12, 12
  universal gravitation, 12–14, *14*, 482
Newton, Isaac, 10–13, 481
Newtonian Cosmology, 481–482
NGC 1265, *466*
NGC 1275, *457, 458*
NGC 4038, 445
NGC 4039, 445
NGC 6624, 363
Nodical month, 54, 56
Nonthermal radiation, *see also*
    Synchrotron radiation
  galactic nucleus (Sgr A), 402
  Jupiter, 105–108
  quasars, 469
  Taurus A (Crab Nebula), 354
  X-ray flares, 218
Norman, Michael, 476
Nova Aquilae, 350
Nova Cygni, *349, 350*, 351
Nova Persei, 351
Nova(e), 338, 348t, 348–351
Nuclear force(s), 299, 496, 497t, 499,
    502
  weak, 497
Nuclear reactions
  novae, 350
Nucleosynthesis, 299–301, 313–316,
    315, 315t, 496
  cosmic, 496, 503
  early Universe, 496
  r process, 355
  red giants, 306–308, 314, 355
  s process, 355
  supernovae, 355
Nucleus (atomic), 156
Nucleus (Galaxy), 272
Nutation, 47

O stars, 255, 309, 310, 345t, 346
  planetary nebulae, 348
OB associations, 345t, 390
  density-waves, 405
  galactic center, 401–402
  H II regions, 374, 382, 398
  Orion Nebula, 388–390
  spiral tracers, 398, 405
Oberon, 129, *130*, 131
Objective, 177
Objective prism, 192
Oblateness (Earth), 22, 41
Oblique rotator (magnetic variables),
    343
Occultations, 23
  lunar, 245, *245*
  Uranus, 131
Ocean of Storms (moon), 62
Oemler, A., 448

OH molecule (hydroxyl)
  comets, 134–135, 135t
  interstellar, 379–381, 379t
Olympus Mons (Mars), 88, 96, *97*
Oort comet cloud, 28, 136, 145
Oort constants, 289, 290
Oort, Jan
  comet cloud, 28, 136
  differential galactic rotation, 288
Opacity, 201, 297, 302, 304, *see also*
      Optical depth
  chromosphere, 203
  mass-luminosity relationship, 303
  photosphere, 201
  pre-main sequence star, 304
  pulsating stars, 339
  radio corona, 208
  red giant, 308
  stellar atmospheres, 252
  sun, *201*, 298
Open clusters, 259, 277, 313, 399
  distances, 286
  moving clusters, 286
  spiral structure, 398
  spiral tracers, 398
Opposition, 2, 4, *4*
  moon, 54
Optical depth, 172, 173, 201, 304, 369
  atmospheric extinction, 229
  chromosphere, 205
  photosphere, 201
  stellar atmospheres, 250
Orbits, *See also* Kepler's laws, Binary
      systems, individual planets and
      moon systems
  aphelion, 20
  artificial satellites, 31
  eccentricity, 20
  escape speed, 31
  globular clusters, 279, 406
  orbital velocity, 16
  perihelion, 20
  period, 30
  planets, 7–10, Appendix 3
  spacecraft, 19–20, *20*
Orion Nebula, 173, 388
  OB associations, 390
  dark clouds, 366, *367*
  H II regions, 173, 388, *388*
  interstellar molecules, 380, *384*
  radio spectrum, 375, *375*
  star formation, 385, 387–390, *387*,
      389
Ostriker, J. P., 416
Oxygen
  active galaxies, 456
  Earth, 67–69
Ozone (Earth's atmosphere), 69–70

P Cygni profiles, *346*, 347
  novae, 350
  quasars, 468
  supernovae, 352
  Wolf-Rayet stars, 347
P Cygni stars, 345t
P waves (seismic), 57–58
P-L relationship, 340, *341*, 429, 435
PAH (Polyclic aromatic
      hydrocarbons), 370, 372

Palomar Sky Survey, 440, 447
Panov, V. I., 482
Parabola, 9, Appendix 9
Parallactic orbit, 43
Parallax, 43, 283
  mean, 224
  statistical, 224
  trigonometric, 223
Parsec, 224
Particle physics, 496, 497
Pauli exclusion principle, 162
  white dwarf, 319
Peculiar A stars, 265, 343
Peculiar galaxies, 445
Peculiar motion, 286, 451
  earth, 446
  galaxies, 446
  superclusters, 450
Peculiar motions, 285
Peebles, P. J. E., 416
Penzias, Arno, 486
Perfect gas law, 25–26, 296
  stellar atmospheres, 249, 301–302
Perigalacticon, 293
Perigee, 20
Perihelion, 9
Period-density relationship
  pulsating stars, 340
Period-luminosity (P-L) relationship,
      340, *341*, 429, 435
  zero point, 340
Perseid meteor shower, 29
Perturbations, 49
Phase shift, 149
Phases
  moon, 54, *55*
  planets, 6
Phobos, 119–120, *120*
Phoebe, 125
Photoelectric effect, 188
Photoelectric photometry, 188, 226–
      228
Photography, 187–188, 227
  of galaxies, 413
Photoionization, 297, 298
Photometry, 226–228, *see also*
      Magnitudes
Photomultiplier tubes, 188
Photons, 155, 188
Photosphere, 199–204, 249, 262
  granulation, 199–200, *199*
  opacity, 199
  sunspots, 198, 212
Piazzi, Guiseppe, 27
Pioneer 10, 9–10,
Pioneer 11, 128
Pioneer Venus spacecraft, 86, 87
  gamma-ray burster observation, *364*
Plages, 205, 216
Planck blackbody radiation, 168–171,
      171t
  color index, 228
  flux, 165
  intensity, 170
  planets, 24
  temperature dependence, 171t, 250
Planck's constant, 155
Planck, Max, K. E. L., 168
Planetary bombardment, 142, 143–144,
      146–147, *see also* Cratering

Planetary configurations, 4–5, *4*
Planetary formation, *see* Solar system
Planetary nebulae, 309, 317, *318*, 345t,
      347–348, *347*, 376–378, *378*, *see*
      also Gaseous nebulae, H II
      regions
  jets, 476
  spectra, 377
Planetary orbits, 7–10, 19–23,
      Appendix 3; *See also* Kepler's
      laws
Planetary orbits, 7–10, 19–23,
      Appendix 3; *see also* Kepler's laws
  synchronous, 21–22
Planetary system formation, *see* Young
      stellar objects
Planetesimals, 142, 145
Planetology, 79
Planets, Appendix 3, *see also*
      individual planets by name
  accretion, 143–145
  atmospheres, 25–26, 98–99
    composition, 25
    escape speed, 26
    ionosphere, 25
    retention, 26, *27*
    speed of particles, 26
  blackbody radiation, 24, *24*
  central pressure, 58
  density, 23
  distances, 20, *21*, Appendix 3
  evolution, 94–97, 141–147
  heat loss, 76, 83, 84
  hydrostatic equilibrium, 58
  interiors, 23
  internal heat, 76
  masses, 23, Appendix 3
  moons, Appendix 3; *see* Moons
  motions, 19–23, Appendix 3
  orbits, 19–23, *21*, Appendix 3
  physical data, Appendix 3
  Planck curve, 24, *24*
  resonance, 23
  rotation, 22, Appendix 3
  satellites, Appendix 3, *see* Moons
  sizes, 23, *21*, Appendix 3
  spin-orbit coupling, 23
  surfaces, 23–25
  temperature, 24, 25,
Plasma, 210, 297
Plate scale, 177
Plate tectonics, 59, *60*
  Venus, 88, 89
Pleiades, 259, 313
  color-magnitude diagram, *260*
  H-R diagram, 312
Pluto, 114–117, *115*, *116*, *117*
  albedo, 115
  atmosphere, 115
  Charon, 116
  diameter, 115
  interior, 117, *117*
  mass, 116
  motions, 115, 116
  orbit 20, *21*, 115, Appendix 3
  physical characteristics, Appendix 3
Pluto-Charon system, *115*, *116*, 116
PMS (Pre-main-sequence stars), 303–
      305, 313, 390

Pogson, N. R.
  magnitude scale, 225
Polar icecaps, 38
Polaris, 47
Polarity reversal
  magnetic variables, 343
  solar cycle model, 220
  sunspots, 214
Polarization, 151–152, *152*
  BL Lac objects, 460, *460*
  corona, 208
  Crab nebula, 354
  galactic magnetic field, 408
  interstellar, 369
  pulsars, 325
  quasars, 470, 469, 470
  reflection nebulae, 370
  white dwarfs, 323, 322
Population I, 259–261, 265, 277, 280,
    397–399, 398t, 402, *see also*
    Stellar Populations
  Cepheids, 339, 340
  characteristics, 398t
  clusters, 313
  evolution, 305–310, *306*
  in galaxies, 414, 417, 418, 419t
  metal abundance, 406
Population II, 259–261, 265, 276–277,
    279, 310, 398t, 398–400, *see also*
    Stellar Populations
  Cepheids, 340
  characteristics, 398t
  evolution, 310–313, *311*
  globular clusters, 313
  high velocity stars, 292–293
  metal abundance, 398, 406
  planetary nebulae, 348
  RR Lyrae stars, 340
  W Virginis stars, 340
Population III, 496, 504
Positron, 300, 497
Potential energy, 17, 299
Poynting-Robertson effect, 141
PP (proton-proton) chain, 300, *300*,
    302, 307, 311
Praesepe Cluster, 287, *312*, 313
Pre-main-sequence (PMS) stars, 303–
    304, 323, 390
Precession, 37, 47, *47*, 54
Pressure broadening, 167, 262
Primeval fireball, 486–489, 500
Principle of equivalence, 482
Procyon, 321
Prominences, 216
Proper motions, 283–284, 283t, 286
  brightest stars, Appendix 4
  nearest stars, Appendix 4
  nova, 350–351
  white dwarfs, 321
Proto-Galaxy, 406
Proton decay, 498, 503
Proton-proton (PP) chain, 300, *300*,
    302, 307, 311
Protoplanets, 142, 145
Protostar(s), 303–305, 313, 383–386, *see
    also* Young stellar objects
  interstellar molecules, 380
Proxima centauri, 43
Ptolemy
  magnitude scale, 225

Pulsar(s), 324–331, 324t
  age, 324
  binary, 328
  Centaurus X-3, 360, 361
  change of period, 324
  cosmic rays, 408
  Crab nebula, 355
  distance, 325
  energy, 324, 330
  galactic magnetic fields, 408
  interpulse, 329
  magnetic fields, 325, 326
  masses, 328
  millisecond, 324, 327
  model, 326, *326*, 331
  polarized light, 325
  rotation, 326
  slowdown, 326, 330
  Vela X, 376
  X-rays, 359
Pulsating stars, 337, 338–341, *338*, 339t
  evolution, 308
  period-density relationship, 340
  period-luminosity relationship, 340–
    341
  pulsation mechanism, 339

Quadrature, 4, 54
Quanta, 155
Quantum efficiency (QE), 187, *187*,
    189, 190, 193
Quantum mechanics, 157, 158
Quarks, 498, 498t, 499
Quasars, 408, 467–476
  absorption lines, 468
  black holes, 471
  classification, 468
  double, 472, 473, *473*
  emission lines, 474
  gravitational lenses, 473
  jets, 471, 474, 476
  model, 471
  noncosmological redshifts, 475
  polarization, 469, 470
  redshift, 474, 475
  Schwarzschild radius, 471
  spectral index, 469
  spectrum, *467*, *469*
  synchrotron radiation, 456, 470
  variability, 469, 470

R process, 355
Radar astronomy, 181
  Earth-Moon distance measurement,
    54
  planetary mapping, 80–81, *80*
  planetary motion studies, 80–81
Radial velocities (speed), 192, 283,
    284, 288, 289; *see also* Doppler
    effect
Radial velocity curves, 338
Radiant (meteors), 29
Radiation era, 488, 490, 500
Radiation pressure, 296
Radiative transport, 297, 305
Radio astronomy, 180–183; *see also*
    Radio sources and telescope(s),
    21-cm line

aperture synthesis, 182
  contour map, *181*
Radio bursts
  Cygnus X-1, 359
  flare stars, 343
  Jupiter, 102–103
  solar flares, 217
Radio continuum
  galaxies, 422
  interstellar, 375
Radio galaxies, 461–467
  broad line, 474
  Centaurus A, 464
  classification, 466, *466*
  Cygnus A, 463
  double, 463
  jets, 471
  magnetic fields, 463
  synchrotron radiation, 456
Radio hydrogen recombination, 374,
    399, 402
Radio lobe, 476
Radio source(s); *see also* 21-cm line,
    Molecular clouds
  binary pulsars, 328
  Crab nebula, 354
  galactic center, 401
  Milky Way, 180
  RS CVn, 343
  SS 433, 361
  stellar winds, 318
Radio telescope(s), 181, 182
  interferometers, 182
  resolving power, 182
  Very Large Array (VLA), 182
  Very-Long-Baseline-Interferometer
    (VLBI), 183
Radio waves, 152t
Radioactive dating, 60–61
  Earth, 60–61
  meteorites, 140
  Moon, 75
Radioactive decay, 60–61, *61*
Rayet, G., 265
Rayleigh scattering law, 70
Rays (lunar), 62
Reber, Grote, 180
Recombination (H I and H II), 374
Red giants, 261–264, 308, 314
  evolution, 307–309
  mass loss, 309
  and novae, 350
  nucleosynthesis, 314, 355
Red variables, 339t, 341
Reddening, 368–369, *368*
  interstellar dust grains, 371
Redshift(s), 283, 431, 446, *see also*
    Doppler effect
  BL Lac objects, 461
  clusters of galaxies, 445
  galaxies, 432, 475
  gravitational, 322, 324
  quasars, 467, 475
  relativistic Doppler shift, 155, 468
Reduced mass, 163
Reference stars, 283
Reflection, 153
Reflection nebula, 370
Refraction, 153, *153*, 176, *190*
  Earth's atmosphere, 70
  index of, 153, 190

Regolith, 64
Regression of nodes, 56
Reifenstein, Edward C., 328
Relativistic beaming, 472
Relativity, 480
    Doppler effect, 155
    general, 480, 482, 483
    special relativity, 480, 481
Resolving power, 154, 177, 178
    radio telescopes, 181
Retention of atmospheres, 25–26, 27
Retrograde motion, 2, 3, 5
    rotation, 22
Riccioli, 61
Richter scale, 58
Right ascension, Appendix 10
Rigid body rotation, 288
Ring Nebula, 318
Rings, 125–132
    Jupiter, 125, 125
    Neptune, 132, 133
    Saturn, 127–129, 127, 128, 129
    Uranus, 131, 130
Robertson-Walker metric, 483, 484
Roche, Edouard, 48
Roche limit, 48–49
    Jupiter, 125
Roche lobe(s), 245, 348
    black hole, 334
    novae, 348, 350
    X-ray binaries, 359, 363
Rood, H., 448
Romer, Ole, 32p
ROSAT (Roentgen Satellite), 184, 358
Rotation curve (Galaxy), 291
Rowan-Robinson, M., 423
Rowland, Sherwood, 70
RR Lyrae stars, 293, 339, 339t, 340,
        400
    instability strip, 339
RS Canum Venaticorum stars (RS CVn
        stars), 341, 343, 344
    flare spectrum, 343
    magnetic activity, 343
Russell, Henry Norris, 256
RV Tauri stars, 339t
Rydberg constant, 158

S process, 355
S stars, 265
S waves (seismology), 57
Sagittarius (Sgr) A, 401–403
Saha equation, 165, 171t, 252, 262
    stellar atmospheres, 249, 255
Sandage, Allan, 414, 435, 467
Sandage-Tamman Process, 435
Sanduleak, Nicholas, 361
Saros cycle, 56
Satellites, see Moons
Saturn, 108–111, 108, 109, 455–456A,
        see also Jovian planets
    atmosphere, 108–110, 109
    blackbody radiation, 110
    composition, 109, 110
    density, 23, 108
    excess energy, 105t, 110
    formation, 145
    interior, 110, 110
    magnetic field, 110–111, 105t, 128

magnetosphere, 110–111, 110
    mass, 108
    moons, 125–127, 126, 127, 457A
    motions, 108
    orbital data, Appendix 3
    physical data, Appendix 3
    rings, 127–129, 127, 128, 129
        Cassini division, 127
        infrared observations, 129
        Roche limit, 48–49
        shepherd satellites, 128–129
    satellites, 128
    structure, 110
    Voyager observations, 109, 109, 128
Scale (telescopic), 177
Scale height, 69
    stellar atmospheres, 250
Scattering (interstellar), 368
    reflection nebulae, 370
Schechter, Paul, 443, 448
Schiaparelli, 94
Schmidt telescope, 192
Schmidt, Maarten, 467
Schwarzschild radius, 331–332
    black hole, 474
    quasars, 471
Schwarzschild, Karl, 331
Scintillation
    pulsars, 324
Seasons, 37–38, 37
Secchi, Angelo, 255
Seeing, 70–71, 79, 178, 194, 246
Seismology, 56–57, 58
Semimajor axis, 8
Semiregular red variables, 339t
Seyfert, Carl, 457
Seyfert galaxies, 457, 458, 458, 459,
        474
Sgr A (Sagittarius A), 401–403
Shane, Donald, 447
Shapley, Harlow, 340, 429, 447
Shectman, S., 448
Shell burning, 306, 308
    long period red variables, 341
Shell stars, 344, 345, 345t
Shelton, Ian, 355
Shield volcanoes, 96, 97
Shklovski, I. S., 354
Shock front, 405
Shock wave(s), 405, 476, 477
    cosmic rays, 408
    interstellar gas, 376
    molecular cloud, 390
    planetary nebula, 318
Shu, Frank, 405
Sidereal month, 54,
Sidereal period, 5, 5, 37
Sidereal rotation period, 20
Sidereal time, 34–37, 36
Sidereal year, 37
Siderius Nuncius, 6
Signal-to-noise ratio, 189
Singularity, 331, 332, 333
Sirius B, 317, 320, 320
    gravitational redshift, 322
Slipher, V. M., 431
Slowdown (pulsars), 326, 330, 355
Smarr, Larry, 476
Snell's law, 153, 190
S0 galaxies, 413, 414, 417

Soifer, B. T., 423
Solar, see also Sun
Solar activity, 211–221, 212t, see also
        Solar flares, Sunspots
    coronal holes, 210
    cosmic ray modulation, 407
    flares, 211, 216–219
    solar cycle, 212, 213
Solar antapex, 285
Solar apex, 285
Solar constant, 230
Solar cosmic rays, 216, 218
Solar cycle, 212–214, see also Solar
        activity
    flares, 218
    magnetic fields, 219–221
    model, 219–221
    polarity reversal, 214, 220
    sunspot numbers, 212–214
Solar dynamo, 219, 219
Solar flares, 216–219
    energy, 218
    magnetic fields, 210
    model, 219
    prominences disruption, 216
    radio bursts, 217
    solar activity, 210
    solar cosmic rays, 218
    solar wind, 211, 218
    X-ray bursts, 217
Solar insolation, 38, 38
Solar interior, 298
Solar luminosity, 230, 298
Solar motion, 285, 286, 431
    mean parallax, 224
Solar nebula, 142–143
    accretion, 143–145
    angular momentum, 142
    chemistry, 142, 143
    condensation 142, 143, 143
        meteorites, 145
    magnetic fields, 142, 143
    temperature, 143, 143
Solar neighborhood, see also Local
        standard of rest (LSR)
    galactic rotation, 287–288
    H-R diagram, 256–258, 258
    luminosity function, 275
Solar neutrino problem, 301
Solar prominences, 216
Solar rotation, 210, 214, 219–220
Solar system, 19–32, 20, 21, 141–145,
        see also Solar nebula
    age, 140, 432
    angular momentum, 32, 142, 144
    center of mass, 221p
    coplanar orbits, 142
    data, Appendix 3
    dynamics, 142
    formation, 141–147
Solar wind, 141, 198, 210–211
    comets, 135, 137
    coronal holes, 210
    cosmic ray modulation, 72–74, 407
    Earth's magnetosphere, 69
    flares, 211, 218
    Jupiter, 107
    magnetic field, 210
    mass loss, 309
    Mercury, 83, 84

Solar wind (Cont.)
 relation to Earth's aurorae, 72–74
 temperature, 210
 Venus, 90, *92*
Solid angle, 155
Solid body rotation, 288
Solstices, 37–38
Source function, 172
Space astronomy, 180, 184, *see also*
  Infrared, Ultraviolet, X-ray
  emission, Gamma rays, Hubble
  Space Telescope
 Apollo program, 53, 58, 64, 71
 Cosmic Background Explorer
  (COBE), 489
 Einstein X-ray Observatory (HEAO),
  184, 358, 403, 423
 Hipparcos mission, 224
 Hubble Space Telescope (HST), 184,
  185, 193, 224–225, 435, 456
 Infrared Astronomical Satellite
  (IRAS), 184, 371–372, 387, 390,
  423, 459
 International Cometary Explorer
  (ICE), 135
 International Ultraviolet Explorer
  (IUE), 184, 357, 456
 Magellan mission, 89, *91*
 Mariner missions, 82, 83, 95
 Pioneer missions, 9–10, 86, 87, 128
 Roentgen Satellite (ROSAT), 184,
  358
 Uhuru, 358, 360
 Very-Long-Baseline-Interferometry
  (VLBI), 183, 470, 471
 Voyager missions, 103, 106, 107,
  109, 112, 114, 125, 126, 128, 131
Space velocity, 284, 285
Spacetime, 331, 332, 481, 482, 484,
  485, 499
 curvature, 500
Spallation
 cosmic rays, 407
Special relativity, 480, 481
Specific heat, 298
Speckle interferometry, 246
Spectra
 galaxies, 445
Spectral classification, *see* Spectral type
Spectral index, 456
 quasars, 469
 synchrotron radiation, 456
Spectral lines
 absorption, 160, 161, *192*
 Balmer lines, 166, *192*
 broadening, 167
 emission, 161
 equivalent width, 164, *164*, 251, *251*
 hydrogen, 159, 166
 intensity, 163–164, *163*
 isoelectronic series, 162
 Lyman series, 162
 molecules, 163
 profiles, 163–164, *163*, 203, *203*, 251
 temperature dependence, 166
Spectral type, 255, 256, 256t, 258, 261,
  262, 265, Appendix 4
 distance determination, 266, *266*
 galaxies, 421, *421*, 425
 Harvard sequence, 255

luminosity classification, 261, 262,
  263, Appendix 4
 temperature, *257*
 white dwarfs, 321
Spectrograph, 190–193, *191*, 253
Spectroheliograms, *215*
Spectrometer, 253
Spectrophotometry, 192
Spectroscopic binaries, 234, 237–240,
  242
 eclipsing, 242
Spectroscopic parallaxes, 266
Spectrum, 152–153, 152t, *see also*
  Absorption lines, Blackbody
  radiation, Doppler effect,
  Electromagnetic radiation,
  Spectral type, types of objects
 blackbody radiation, 168–171
 Electromagnetic, 152t
Spectrum variables, 339
Spectrum binary, 237
Spicules, 206, *206*, 207
Spin orbit coupling, 22, 23
Spindown (pulsars), 326, 330, 355
Spiral structure, 278, 280, 393–396,
  398–399, *399*, 405 *see also* Galaxy
 galactic magnetic field, 409
 hydrogen distribution, 276, 395
 interstellar dust, 370
Spiral galaxies, 414, 416, 419t
 barred, 416
 distance scale, 435
Spitzer, Lyman, 184
SS 433, 361, *361*, 362
 jets, 476
Staelen, David H., 328
Standard deviation, 446
Standards of reference, 283
Star clouds, 270
Star counting, 274–276, 368
Star disks, 384, 385, *see also*
  Circumstellar clouds
Star formation, 277, 383, 388, 390–391,
  385, 405, *see also* Stellar
  evolution, Young stellar objects
 giant molecular cloud, *390*
 in galaxies, 425
 magnetic fields, 385
 massive stars, 390
 solar-mass stars, 390
 stages, 391
Starburst galaxies, 457
Stark effect, 167
Star(s), *see also* entries under Stellar
 atmosphere, 249–256
 binary systems, 233–247, *see also*
  Binary systems
 central pressure, 296–297
 chemical composition, 296, 302, 305;
  see also Abundances
 convection zone, 297
 core-helium-burning, 308
 data, Appendix 4
 degenerate, 308
 density, 242, 296, 302
 diameters, 241, 245, 246, *246*
 distances 223–225, *see also* Distance
  scale
 energy sources, 298–303
 energy transport, 297–298

evolution of, *see* Stellar evolution
 extended atmosphere model, 345,
  345, *346*
 flare stars, 341, 342
 flux, 230
 giants, 258, 261–264, 265
 gravitational contraction, *see*
  Gravitational contraction
 helium burning, 308
 high-velocity stars, 292
 hydrogen burning, 310
 late-type, 255
 luminosity, 242, 258, 302
 magnitude, *see* Magnitude
 mass, 296, 302, 303
 mass fractions, 296
 mass loss, 309, 317, 344
 masses
  binary systems, 233
  white dwarfs, 321
 mean molecular mass, 302
 motions, *see* Stellar motions
 nearest, Appendix 4
 pressure, 296, 302
 pulsating, 338–341, *see also* Pulsating
  stars
 radii, 241, 245, 246, *246*
 shell burning, 306, 308
 supergiants, 258
 temperature, 296–297, 302
 variable, *see* Variable stars
 X-ray emission, 266–267
Starspots
 RS CVn stars, 344
Statistical parallaxes, 224
Stefan, Josef, 170
Stefan-Boltzmann law, 171–172, 171t
 bolometric magnitude, 230
 eclipsing binaries, 241
 photospheric temperatures, 200
 planets, 24–25
 stellar atmospheres, 250
Steinhardt, P., 499
Stellar atmospheres, 172, 249–253
 absorption lines, 251
 abundances, 249
 composition, 249, 251
 density, 249
 molecules, 251
 opacity, 252
 optical depth, 250
 pressure, 249, 251
 spectral lines, 252
 surface brightness, 250
 temperature, 251, 255
Stellar classification, 255, 256, 261–263
Stellar clusters, 259, 276, 286
Stellar collapse,
 star formation, 383–385
 supernovae, 352–353
Stellar composition, *see* Abundances
Stellar diameters, 241, 246, 246t
 binary light curve, 240–241
 interferometry, 245–246
 lunar occultation, 245
Stellar evolution, 261, 277, 303–313,
  306t, 307t, 341
 evolutionary tracks, 303, 304, 308,
  310, 313
 formation of stars, 383–391

Stellar evolution (Cont.)
    main sequence, 313
    planetary nebulae, 347
    pre-main-sequence stars (PMS) 303–
        304, 323, 390
    solar mass star, 307t
    supernovae, 352, 355
    turnoff, 313
Stellar expansion, 345, 346–347
Stellar interferometry, 245–247
    Michelson, 245
Stellar luminosity, 297
Stellar magnitude, see Magnitude
Stellar masses, 235–237, 238–240, 242,
        243, 262, 303
    mass-luminosity relationship, 303
    statistical masses, 240
    supernovae, 352
    white dwarfs, 321
    X-ray binaries, 358
Stellar models, 302
Stellar motions, 282–287
Stellar parallax, 43, 223–224
Stellar populations, 259, 276, 399, 397,
        see also Population I and
        Population II
    characteristics, 398
    galaxies, 418, 425
    metal abundance, 398, 406
Stellar radii, 240–242, 245, 246, 262,
Stellar rotation, 243–244, 345
Stellar spectra, 192, 193, 254, 255,
        256t, 257, 261–264, 262t, 263,
        265
    distance determination, 266, 266
Stellar structure, 295, 297, 298, 299,
        301
Stellar winds, 318, 318, 385, see also
        Mass loss
    black hole, 360
    H-R diagram, 309, 309
    mass loss, 309
    protostars, 385
    pulsars, 327
    Superwind, 309
    T Tauri stars, 342
    Wolf-Rayet stars, 347
Stephenson, Bruce, 361
Stimulated emission, 381
Stony irons (meteorites), 140
Stratosphere, 67, 68
Stromgren sphere, 374; see also H II
        regions
Struve, F. G. W., 43, 223
Subdwarfs, 259, 261, 265
Subsolar temperature, 25
Sun, 197–221; see also entries under
        Solar
    abundances, 204
    active regions, see Solar activity
    central temperature, 297
    chromosphere, 198, 204–207, see also
        Chromosphere
        emission lines, 203
        temperature, 204, 204
    chromospheric network, 205
    composition, 204
    continuum, 201, 203
    convective zone, 197, 297, 303
    core, 197

corona, 198, 207–210, 266–267, see
        also Corona
cosmic rays, 218–219, 407
density, 197, 296
differential rotation, 214, 219
    solar cycle model, 219, 220
dynamo, 220, 219
eclipses, 55, 204
emission lines, 203, 207
filaments, 205
flares, 216–219, see also Flares
Fraunhofer absorption spectrum,
        202–203
    corona, 208
granules, 199–200, 203, 212
H-continuum, 203–204
helioseismology, 199
limb darkening, 173, 200–201
luminosity, 197, 298
luminosity class, 262
magnetic field(s)
    flux tubes, 220
    solar cycle model, 219
    solar flare, 219
    sunspots, 212
    supergranules, 203
mass, 197
neutrino flux, 198, 301
opacity, 201–202, 202
photosphere, 199–204, see also
        Photosphere
physical characteristics, Appendix 7
plages, 205
prominences, 216
rotation, 210, 214
Schwarzchild radius, 331
seismic waves, 199
solar activity, 203, 211–221; see also
        Solar activity
solar cycle model, 219
solar oscillations, 199, 203
solar wind, 210–211, 407
spectral type, 255
spectrum, 253
spicules, 198
stellar models, 302–303
structure, 197–199, 198
sunspots, 198, 211–214; see also
        Sunspots
supergranules, 203, 206
temperature structure, 200–201, 201,
        203, 203, 204, 204, 207
transition region, 203, 204, 204, 207
ultraviolet radiation, 207
X-ray emission, 266–267
Sunspot maxima/minimum, 212, 213,
        see also Solar cycle
Sunspot(s), 211–214, 219–221
    bipolar group, 212, 214
    latitude variation, 213–214
    magnetic fields, 212
    magnetic polarity, 212
    magnetic variables, 343
    Maunder minimum, 212
    penumbra, 212
    polarity, 214
    pores, 212
    solar cycle, 212–213, 219–221
    structure, 198

umbra, 212
Superclusters, 439, 446–451, 452, see
        also Clusters of galaxies
    intergalactic gas, 452
    void, 448
Supergiant(s), 262t, 309, 345–346,
        345t, 398t
    Cygnus X-1, 359
    surface gravity, 262
Supergranules, 203, 206
    solar cycle model, 220
Superior conjunction, 4
Superior planets, 4, 4, 5
Superluminal motions
    quasars, 471, 472, 472
Supernova 1987, 355–358, 356, 358
Supernova of 1054, 351, 354, see also
        Crab nebula
Supernova remnant(s), 328, 375–376,
        376, 390
    cosmic rays, 408
    Crab nebula, 353–355
    and star formation, 390
    X-ray emission, 375–376
Supernova(e), 348t, 351–353
    cosmic rays, 408
    distance indicators, 433t, 434
    in galaxies, 408, 425
    and interstellar medium, 314, 382,
        425
    neutrinos, 357
    pulsars, 328
    quasars, 475
    Type I, 425, 352, 376
    Type II, 352, 357
    Wolf-Rayet stars, 347
Supersonic flow, 476
Supersymmetry, 499
Superwind, 309
Surface gravity, 262
Symmetry, 499
Synchronous rotation, 54, 121
    binary systems, 242
    Galilean (Jupiter) moons, 121
    Martian moons, 119
    Mercury, 80
    Moon, 54
    Saturn's moons, 125
Synchrotron radiation, 354, 375, 422,
        423, 456
    cosmic rays, 408
    Crab nebula, 354
    jets, 476
    Jupiter, 102–104
    M87, 462
    Pulsars, 326
    quasars, 469, 470
    radio galaxies, 465
    RS CVn stars, 343
    self absorption, 456
    Seyfert galaxies, 458
    solar X-ray and radio bursts,
        218
    spectrum, 375
    SS 433, 362
    supernova remnants, 375, 376
Synodic month, 56
Synodic period, 4, 5

T Tauri stars, 341, 342, 390, 391, 398
Tamman, Gustav, 435
Tangential speed, 284, 289
Tarenghi, M., 448
Taurus A, 354
Taylor, Joseph, 328
Telescope(s), 176–186, 178t; *see also*
    Hubble Space Telescope; Radio
    telescopes, Space astronomy
  focal length, 176
  light-gathering power, 177
  multi-mirror, 194
  objective, 177
  optical, 176–180
  reflecting, 177
  refracting, 177
  resolving power, 177
  spherical aberration, 185
Temperature, 171, 171t, 243t
  blackbody radiation, 170
  in binary systems, 241
  kinetic, 171
  planets, 27–28, Appendix 3–3
  relation to color index, 228
  spectral type, 255–256, 257,
      Appendix 4–3
  stars, 249, 296–297, 302
  sun, 200–201, 204–205, 207, 208, 297
Terlevich, R., 451
Terrestrial planets, 78–96
  comparative evolution of, 94–96
  cratering, 99
Tharsis ridge (Mars), 97, 98
Thermal bremsstrahlung, 375
  galaxies, 423
  intergalactic gas, 452
Thermal Doppler broadening, 167,
    171t
Thermal equilibrium, 164, 298
Thermal pulses, 308
Thermal radio emission, 375
  Sagittarius A, 402
Thermodynamic equilibrium, 171, 249
Thermodynamics, second law, 297
Thermonuclear reactions, 305–301,
    315t,
  synthesis of elements, 313–316
Thermosphere, 67, 68
3C (Third Cambridge Catalog), 467
3C 48, 467
3C 206, 475
3C 273, 467, 467, 471, 472
30 Dorados, 186
Thompson, L., 448
3-kpc arm, 397
Tidal bulge, 46, 53
Tidal evolution, 46, 46
Tidal interaction(s), 44–49
  asteroids, 145
  asteroids with moons, 134
  binary systems, 242
  black holes, 332, 360
  clusters of galaxies, 444–445
  Earth-moon, 44–49, 45, 46
  galaxies, 417–418, 444–445, 459
  Galaxy, 397, 397
  Jupiter-Io, 122
  M81 & M82, 457
  planetary motion, 23
  quasars, 471

Roche limit, 48–49
RS CVn stars, 343
Seyfert galaxies, 459
spin-orbit coupling, 23
Tides, 44–46, 45
Tifft, W., 448, 475
Time, 34–37
Time zones, 36
Titan, 125–126, 126
  atmosphere, 26, 126
Titania, 129, 130, 131
Titus of Wittenberg, 20
Titus-Bode rule, 20
Tombaugh, Clyde W., 115, 447
  discovery of Pluto, 115
Transfer equation, 171, 172
  atmospheric extinction, 229
Transition region (sun), 204, 207
Transverse distortion (S) waves, 57
Trapezium cluster, 388
Trigonometric parallax, 43, 223–224
Triple-alpha process, 301, 308
Triton, 131, 132, 131
Tropic of Cancer, 37, 37
Tropic of Capricorn, 37, 37
Troposphere, 67, 68
True anomaly, 8
Tully, R. Brent, 436, 448
Tully-Fisher relationship, 436, 436
Turnoff (from main sequence), 313
Turnover frequency, 375
21-cm line, 378, 378, 379, 383, 393
  galactic center, 396
  galaxies, 422, 436
  profiles, 394, 394
Tycho (crater), 62, 64
Tycho Brahe, 6, 223
Tycho's Supernova, 376, 377

UBV magnitude system, 227, 227,
    227t, 228, 265
Uhuru X-ray instrument, 358
  Centaurus X-3, 360
Ultraviolet radiation, 152t, 180
  Fraunhofer absorption spectrum,
    203
  intergalactic. 452
  IUE, 184, 357, 456
  molecular clouds, 380
  O and B stars, 374
  observations, 180, 184
  solar transition region, 207
  stellar atmospheres, 255
  T Tauri stars, 341
Uncertainty principle, 159
Universal gravitation, 13–15, 480, 482
Universal time (UT), 36
Universe; *see also* Cosmology
  age, 432, 436
  critical density, 494
  curvature, 500
  density, 500
  early, 490, 494
    evolution, 482
    expansion, 432
    flatness, 494, 500
    history, 502, 503
    homogeneity, 494
    inflation theory, 499, 502

isotropy, 494
large-scale structure, 447
matter era, 488, 491, 500, 503
missing mass, 453
nucleosynthesis, 496
perturbations, 500, 502
radiation era, 488, 491, 500
thermal history, 496, 499, 500
world models, 485
Uranus, 111–113, 111, 113, *see also*
    Jovian planets
  atmosphere, 111, 112
  composition, 112
  density, 112
  Doppler shift measurement, 111
  interior, 112
  magnetic field, 112, 105t
  magnetosphere, 112, 112
  moons, 111, 129–131, 130, Appendix
    3
  motions, 22, 111
  orbital data, Appendix 3
  physical data, Appendix 3
  rings, 131, 130
  rotation, 22, 22, 111, 112
  satellites, 129–131, Appendix 3
  structure, 112, 112
  Voyager 2 spacecraft, 112–113, 113,
    129, 131
UV, *see* Ultraviolet radiation

Valles Marineris (Mars), 95, 98
Van Allen radiation belts, 72–74, 72,
    73
  mirror points, 73
Van Allen, James A., 72–74
Variable stars, 337–364, 339t, 348t; *see*
    also Eclipsing binaries
  cataclysmic, 348–358
  evolution, 339, 348, 352, 355, 357
  H-R diagram, 338
  nomenclature, 337
  nonpulsating, 341–344
  pulsating, 338–341
Vega (parallax), 43
Vega spacecraft, 137
Vela pulsar, 330, 331
Vela X-1, 359, 376
Velocities, 10, 12, 25–26
  escape, 26, 31
  Maxwellian distribution, 25, 26
  most probable speed, 25
  orbital, 15–16, 15, 17, 20, 20
  root mean square speed, 26
  stellar, 282–285
Velocity, 10, 12
Velocity curves
  binary systems, 237, 238, 238, 239
  pulsating stars, 338, 338
Velocity dispersion, 453
  galaxies, 445, 446, 451
  pulsar, 325
Velocity of escape, 26, 31, 482
  Galaxy, 293
  Mercury, 82
  moon, 67
  planetary atmospheres, 26, 27
  space probe, 31
Velocity of light, 34, 151, Appendix 7

Venera spacecraft, 86, 89, *90*
Venus, 85–92
  albedo, 87
  atmosphere, 86–87, *87*
    composition, 86
  cloud structure, 87, *87*
  craters, 87–89, *91*
  elongation, 4, *4*, 85
  evolution, 90–92, 99
  greenhouse effect, 86
  Hadley cells, 87
  interior, *86*, 89
  lava flows, 89
  Magellan spacecraft, 89, *91*
  magnetic field, 89–90
  mass determination, 86
  motions, 85
  mountains, 88, 89
  orbital data, Appendix 3
  phases, Galileo's observations, 6
  physical characteristics, 86,
    Appendix 3
  plate tectonics, 88, 89, 91
  radar observations, *80*, 85, 87–89, *88*
  revolution, 85
  rotation, 22, 85
  solar wind, 90, *92*
  surface, 87–89, *90*
  temperature, 86
  Venera spacecraft, 86, 89, *90*
  volcanism, 88, 89, 91, *91*
Vernal equinox, 34, 37, Appendix 10
Very high velocity clouds, 397
Very Large Array, *see* VLA
Very-Long-Baseline-Interferometer, *see*
    VLBI
Viking missions (Mars), 93, 94, 95
Virgo cluster, 435, 439, 440, 443, 448
Virial theorem, 299, 453
  galaxies, 425, 426
  Galaxy, 406
  missing mass, 453
  star formation, 383
Vis viva equation, 18
Visible light, 152, 153
Visual binaries, 234–237
VLA (Very Large Array), 182
  observations
    Cygnus A, 463
    gas flows, 385
    M87, 462
    RS CVn stars, 343
    SS 433, 362
VLBI (Very-Long-Baseline-
    Interferometer), 183
  observations
    continental drift, 60
    quasars, 470
    superluminal motion, 471
Voids, 448, 450, 451, 453, 500, 502
Volcanoes (Vulcanism)
  Io, 121, *122*
  Mars, 96
  Mercury, 85

Moon, 64
  Venus, 88, 89
von Jolly, Phillip, 14
Voyager spacecraft, 112, 114
  Jupiter, 103, 106, 107
    ring system, 125
  Neptune, 114, 131
    rings, 132
  Rings, 125, 128, 131, 132
  Saturn, 109
    rings, 128
  Titan, 126
  Uranus, 112, 113, 114, 129–131
    rings, 131

W Virginis stars, 340
W-R stars, *see* Wolf-Rayet stars
Water (H_2O)
  comets, 135t
  Earth, 59, 67t, 67, 69, 76–77
  interstellar, 380, 381
  Jupiter, 103
  Mars, 93, 94, 95, 98
  moon, 67
  Venus, 86, *87*
Wave nature of light, 150–152, *151*,
    152
WC stars, *see* Wolf-Rayet stars
Wedge diagrams, 449, *449*, *450*
Weedman, Daniel, 458
Wegner, G., 451
Weight, 13, 482
Weimann, R. J., 468
Whipple, Fred L.,
  cometary model, 135
White dwarf(s), 309, 317–323, 345
  age, 320
  binary pulsars, 328
  Chandrasekhar limit, 319, 353
  energy source, 320
  evolution, 309, 310, *315*
  gravitational redshift, 322
  H-R diagram, 321, *321*
  luminosity, 320
  magnetic fields, 322, 323
  mass, 319, 322
  mass-radius relationship, 319
  novae, 350
  planetary nebula, 317, 318
  polarization, 323
  proper motion, 321
  Sirius B, 320
  supernovae, 353
  thermal energy, 320
White hole, 333
Widmanstatten figures, 139
Wien, Wilhelm, 170
Wien's law, 24, 170, 250, 455
Wildt, Rupert, 94
Wilson, Robert, 486
Winkler, Karl-Heinz, 476
Wirtanen, Carl, 447

WN stars, *see* Wolf-Rayet stars
Wolf, C., 265
Wolf-Rayet stars, 265, 310, 345t, 346–
    347
  evolution, 310
  mass loss, 347
  planetary nebulae, 348
  supernova, 347
World models, 485, 486
Work, 17

X-ray astronomy, 180, 184
X-ray binaries, 328, 358–362
  analogy for quasar model, 417
X-ray bursters, 362–364
X-ray emission, 152t, 153, 360, 361,
    362
  AGNs (Active galaxy nuclei), 457
  binaries, 328, 358–362
  Centaurus A, 464–465
  Crab nebula, 354
  flare stars, 343
  galactic center, 403, *403*
  galaxies, 423, *424* 457, 462, 464–466
  intergalactic, 452, 453
  M82, 457
  M87, 462, 465
  novae, 351
  quasars, 469, 471
  RS CVn stars, 344
  solar flares, 216
  supernovae, 376
  stars, 266–267
  T Tauri stars, 341, 391
  X-ray bursters, 362–364
X-ray pulsar(s), 359
  Centaurus X-3, 360, 361
X-ray pulsars, 359

Year
  anomalistic, 37
  sidereal, 37
  tropical, 37
Young stellar objects (YSO), 342, 383,
    385, 390; *see also* Star formation

Z, *see* Redshift
Z components, 292
ZAMS (Zero-age main-sequence), 305,
    305, 307, 307t, 310, *310*, *311*,
    313
Zeeman Effect, 167, 408
  magnetograph, 215
  sunspots, 212
Zenith angle, 229
Zero-age main-sequence (ZAMS), *see*
    ZAMS
Zodiac, 2
Zodiacal light, 29, *30*, 141, 208
Zwicky, Fritz, 447